EMS Textbooks in Mathematics

EMS Textbooks in Mathematics is a book series aimed at students or professional mathematicians seeking an introduction into a particular field. The individual volumes are intended to provide not only relevant techniques, results and their applications, but afford insight into the motivations and ideas behind the theory. Suitably designed exercises help to master the subject and prepare the reader for the study of more advanced and specialized literature.

Jørn Justesen and Tom Høholdt, *A Course In Error-Correcting Codes*
Markus Stroppel, *Locally Compact Groups*
Peter Kunkel and Volker Mehrmann, *Differential-Algebraic Equations*
Dorothee D. Haroske and Hans Triebel, *Distributions, Sobolev Spaces, Elliptic Equations*
Thomas Timmermann, *An Invitation to Quantum Groups and Duality*
Marek Jarnicki and Peter Pflug, *First Steps in Several Complex Variables: Reinhardt Domains*
Oleg Bogopolski, *Introduction to Group Theory*

Tammo tom Dieck

Algebraic Topology

European Mathematical Society

Author:

Tammo tom Dieck
Mathematisches Institut
Georg-August-Universität Göttingen
Bunsenstrasse 3–5
37073 Göttingen
Germany

E-mail: tammo@uni-math.gwdg.de

2000 Mathematical Subject Classification: 55-01, 57-01

Key words: Covering spaces, fibrations, cofibrations, homotopy groups, cell complexes, fibre bundles, vector bundles, classifying spaces, singular and axiomatic homology and cohomology, smooth manifolds, duality, characteristic classes, bordism.

ISBN 978-3-03719-048-7

Corrected 2nd printing, 2010

The Swiss National Library lists this publication in The Swiss Book, the Swiss national bibliography, and the detailed bibliographic data are available on the Internet at http://www.helveticat.ch.

Contact address:

European Mathematical Society Publishing House
Seminar for Applied Mathematics
ETH-Zentrum SEW A21
CH-8092 Zürich
Switzerland

Email: info@ems-ph.org
Homepage: www.ems-ph.org

Typeset using the author's TEX files: I. Zimmermann, Freiburg
Printed in Germany

9 8 7 6 5 4 3

Preface

Algebraic topology is the interplay between "continuous" and "discrete" mathematics. Continuous mathematics is formulated in its general form in the language of topological spaces and continuous maps. Discrete mathematics is used to express the concepts of algebra and combinatorics. In mathematical language: we use the real numbers to conceptualize continuous forms and we model these forms with the use of the integers. For example, our intuitive idea of time supposes a continuous process without gaps, an unceasing succession of moments. But in practice we use discrete models, machines or natural processes which we define to be periodic. Likewise we conceive of a space as a continuum but we model that space as a set of discrete forms. Thus the essence of time and space is of a topological nature but algebraic topology allows their realizations to be of an algebraic nature.

Classical algebraic topology consists in the construction and use of functors from some category of topological spaces into an algebraic category, say of groups. But one can also postulate that global qualitative geometry is itself of an algebraic nature. Consequently there are two important view points from which one can study algebraic topology: homology and homotopy.

Homology, invented by Henri Poincaré, is without doubt one of the most ingenious and influential inventions in mathematics. The basic idea of homology is that we start with a geometric object (a space) which is given by combinatorial data (a simplicial complex). Then the linear algebra and boundary relations determined by these data are used to produce homology groups.

In this book, the chapters on singular homology, homology, homological algebra and cellular homology constitute an introduction to homology theory (construction, axiomatic analysis, classical applications). The chapters require a parallel reading – this indicates the complexity of the material which does not have a simple intuitive explanation. If one knows or accepts some results about manifolds, one should read the construction of bordism homology. It appears in the final chapter but offers a simple explanation of the idea of homology.

The second aspect of algebraic topology, homotopy theory, begins again with the construction of functors from topology to algebra. But this approach is important from another view point. Homotopy theory shows that the category of topological spaces has itself a kind of (hidden) algebraic structure. This becomes immediately clear in the introductory chapters on the fundamental group and covering space theory. The study of algebraic topology is often begun with these topics. The notions of fibration and cofibration, which are at first sight of a technical nature, are used to indicate that an arbitrary continuous map has something like a kernel and a cokernel – the beginning of the internal algebraic structure of topology. (The chapter on homotopy groups, which is essential to this book, should also be studied for its applications beyond our present study.) In the ensuing chapter on duality the analogy to algebra becomes clearer: For a suitable class of spaces there exists a duality theory which resembles formally the duality between a vector space and its dual space.

The first main theorem of algebraic topology is the Brouwer–Hopf degree theorem. We prove this theorem by elementary methods from homotopy theory. It is a fairly direct consequence of the Blakers–Massey excision theorem for which we present the elementary proof of Dieter Puppe. Later we indicate proofs of the degree theorem based on homology and then on differential topology. It is absolutely essential to understand this theorem from these three view points. The theorem says that the set of self-maps of a positive dimensional sphere under the homotopy relation has the structure of a (homotopically defined) ring – and this ring is the ring of integers.

The second part of the book develops further theoretical concepts (like cohomology) and presents more advanced applications to manifolds, bundles, homotopy theory, characteristic classes and bordism theory. The reader is strongly urged to read the introduction to each of the chapters in order to obtain more coherent information about the contents of the book.

Words in boldface italic are defined at the place where they appear even if there is no indication of a formal definition. In addition, there is a list of standard or global symbols. The problem sections contain exercises, examples, counter-examples and further results, and also sometimes ask the reader to extend concepts in further detail. It is not assumed that all of the problems will be completely worked out, but it is strongly recommended that they all be read. Also, the reader will find some familiarity with the full bibliography, not just the references cited in the text, to be crucial for further studies. More background material about spaces and manifolds may, at least for a while, be obtained from the author's home page.

I would like to thank Irene Zimmermann and Manfred Karbe for their help and effort in preparing the manuscript for publication.

Göttingen, September 2008 Tammo tom Dieck

Contents

Chapter 1
Topological Spaces

In this chapter we collect the basic terminology about topological spaces and some elementary results (without proofs). I assume that the reader has some experience with point-set topology including the notion of compactness. We introduce a number of examples and standard spaces that will be used throughout the book. Perhaps the reader has not met quotient spaces. Quotient spaces give precision to the intuitive concept of gluing and pasting. They comprise adjunction spaces, pushouts, attaching of spaces (in particular cells), orbit spaces of group actions. In the main text we deal with other topics: Mapping spaces and compact open topology, bundles, cell complexes, manifolds, partitions of unity, compactly generated spaces.

Transformation groups are another topic of this chapter. Whenever you study a mathematical object you should consider its symmetries. In topology one uses, of course, continuous symmetries. They are called actions of a topological group on a space or transformation groups. In this chapter we assemble notions and results about the general topology of transformation groups. We use the material later for several purposes:

- Important spaces like spheres, projective spaces and Grassmann manifolds have a high degree of symmetry which comes from linear algebra (matrix multiplication).
- The fundamental group of a space has a somewhat formal definition. In the theory of covering spaces the fundamental group is exhibited as a symmetry group. This "hidden" symmetry, which is associated to a space, will influence several other of its geometric investigations.
- The theory of fibre bundles and vector bundles makes essential use of the concept of a transformation group. Important information about a manifold is codified in its tangent bundle. We will apply the tools of algebraic topology to bundles (characteristic classes; classifying spaces).

We should point out that large parts of algebraic topology can be generalized to the setting of transformation groups (equivariant topology). At a few occasions later we point out such generalizations.

1.1 Basic Notions

A ***topology*** on a set X is a set $\mathcal{O}$ of subsets of X, called open sets, with the properties: (1) The union of an arbitrary family of open sets is open. (2) The intersection of a finite family of open sets is open. (3) The empty set $\emptyset$ and X are open. A ***topological***

space $(X, \mathcal{O})$ consists of a set X and a topology $\mathcal{O}$ on X. The sets in $\mathcal{O}$ are the ***open sets*** of the topological space $(X, \mathcal{O})$. We usually denote a topological space just by the underlying set X. A set $A \subset X$ is ***closed*** in $(X, \mathcal{O})$ if the complement $X \smallsetminus A$ is open in $(X, \mathcal{O})$. Closed sets have properties dual to (1)–(3): The intersection of an arbitrary family of closed sets is closed; the union of a finite family of closed sets is closed; the empty set $\emptyset$ and X are closed. A subset $\mathcal{B}$ of a topology $\mathcal{O}$ is a ***basis*** of $\mathcal{O}$ if each $U \in \mathcal{O}$ is a union of elements of $\mathcal{B}$. (The empty set is the union of the empty family.) A subset $\mathcal{S}$ of $\mathcal{O}$ is a ***subbasis*** of $\mathcal{O}$ if the set of finite intersections of elements in $\mathcal{S}$ is a basis of $\mathcal{O}$. (The space X is the intersection of the empty family.)

A map $f : X \to Y$ between topological spaces is ***continuous*** if the pre-image $f^{-1}(V)$ of each open set V of Y is open in X. Dually: A map is continuous if the pre-image of each closed set is closed. The identity $\mathrm{id}(X) : X \to X$ is always continuous, and the composition of continuous maps is continuous. Hence topological spaces and continuous maps form a category. We denote it by TOP. A ***homeomorphism*** $f : X \to Y$ is a continuous map with a continuous inverse $g : Y \to X$. Spaces X and Y are ***homeomorphic*** if there exists a homeomorphism between them. A map $f : X \to Y$ between topological spaces is ***open*** (***closed***) if the image of each open (closed) set is again open (closed).

In the sequel we assume that a map between topological spaces is continuous if nothing else is specified or obvious. A ***set map*** is a map which is not assumed to be continuous at the outset.

We fix a topological space X and a subset A. The intersection of the closed sets which contain A is denoted $\overline{A}$ and called ***closure*** of A in X. A set A is ***dense*** in X if $\overline{A} = X$. The ***interior*** of A is the union of the open sets contained in A. We denote the interior by A°. A point in A° is an ***interior point*** of A. A subset is ***nowhere dense*** if the interior of its closure is empty. The ***boundary*** of A in X is $\mathrm{Bd}(A) = \overline{A} \cap \overline{(X \smallsetminus A)}$.

An open subset U of X which contains A is an ***open neighbourhood*** of A in X. A set B is a ***neighbourhood*** of A if it contains an open neighbourhood. A system of neighbourhoods of the point x is a ***neighbourhood basis*** of x if each neighbourhood of x contains one of the system.

Given two topological spaces X and Y, a map $f : X \to Y$ is said to be ***continuous at*** $x \in X$ if for each neighbourhood V of $f(x)$ there exists a neighbourhood U of x such that $f(U) \subset V$; it suffices to consider a neighbourhood basis of x and $f(x)$.

Suppose $\mathcal{O}_1$ and $\mathcal{O}_2$ are topologies on X. If $\mathcal{O}_1 \subset \mathcal{O}_2$, then $\mathcal{O}_2$ is ***finer*** than $\mathcal{O}_1$ and $\mathcal{O}_1$ ***coarser*** than $\mathcal{O}_2$. The topology $\mathcal{O}_2$ is finer than $\mathcal{O}_1$ if and only if the identity $(X, \mathcal{O}_2) \to (X, \mathcal{O}_1)$ is continuous. The set of all subsets of X is the finest topology; it is the ***discrete topology*** and the resulting space a ***discrete space***. All maps $f : X \to Y$ from a discrete space X are continuous. The coarsest topology on X consists of $\emptyset$ and X alone. If $(\mathcal{O}_j \mid j \in J)$ is a family of topologies on X, then their intersection is a topology.

We list some properties which a space X may have.

(T_1) One-point subspaces are closed.

(T_2) Any two points have disjoint neighbourhoods.

(T_3) Given a point $x \in X$ and a closed subset $A \subset X$ not containing x, there exist disjoint neighbourhoods U of x and V of A.

(T_4) Any two disjoint closed subsets have disjoint neighbourhoods.

We say X satisfies the ***separation axiom*** T_j (or X is a T_j-space), if X has property T_j. The separation axioms are of a technical nature, but they serve the purpose of clarifying the concepts.

A T_2-space is called a ***Hausdorff space*** or ***separated***. A space satisfying T_1 and T_3 is said to be ***regular***. A space satisfying T_1 and T_4 is called ***normal***. In a regular space, each neighbourhood of a point contains a closed neighbourhood. A space X is called ***completely regular*** if it is separated and for each $x \in X$ and $\emptyset \neq A \subset X$ closed, $x \notin A$, there exists a continuous function $f : X \to [0,1]$ such that $f(x) = 1$ and $f(A) = \{0\}$.

A remarkable consequence of the separation property T_4 is the existence of many real-valued continuous functions. The Urysohn existence theorem (1.1.1) shows that normal spaces are completely regular.

(1.1.1) Theorem (Urysohn). *Let X be a T_4-space and suppose that A and B are disjoint closed subsets of X. Then there exists a continuous function $f : X \to [0,1]$ with $f(A) \subset \{0\}$ and $f(B) \subset \{1\}$.* □

(1.1.2) Theorem (Tietze). *Let X be a T_4-space and $A \subset X$ closed. Then each continuous map $f : A \to [0,1]$ has a continuous extension $f : X \to [0,1]$.*

A continuous map $f : A \to \mathbb{R}^n$ from a closed subset A of a T_4-space X has a continuous extension to X. □

Many examples of topological spaces arise from metric spaces. Metric spaces are important in their own right. A ***metric*** d on a set X is a map $d : X \times X \to [0, \infty[$ with the properties:

(1) $d(x, y) = 0$ if and only if $x = y$.

(2) $d(x, y) = d(y, x)$ for all $x, y \in X$.

(3) $d(x, z) \leq d(x, y) + d(y, z)$ for all $x, y, z \in X$ (***triangle inequality***).

We call $d(x, y)$ the ***distance*** between x and y with respect to the metric d. A ***metric space*** (X, d) consists of a set X and a metric d on X.

Let (X, d) be a metric space. The set $U_\varepsilon(x) = \{y \in X \mid d(x, y) < \varepsilon\}$ is the ***ε-neighbourhood*** of x. We call $U \subset X$ ***open with respect to*** d if for each $x \in X$ there exists $\varepsilon > 0$ such that $U_\varepsilon(x) \subset U$. The system $\mathcal{O}_d$ of subsets U which are open with respect to d is a topology on X, the ***underlying topology*** of the metric space, and the ε-neighbourhoods of all points are a basis for this topology. Subsets of the form $U_\varepsilon(x)$ are open with respect to d. For the proof, let $y \in U_\varepsilon(x)$

and $0 < \eta < \varepsilon - d(x,y)$. Then, by the triangle inequality, $U_\eta(y) \subset U_\varepsilon(x)$. A space $(X, \mathcal{O})$ is ***metrizable*** if there exists a metric d on X such that $\mathcal{O} = \mathcal{O}_d$. Metrizable spaces have countable neighbourhood bases of points: Take the $U_\varepsilon(x)$ with rational ε. A set U is a neighbourhood of x if and only if there exists an $\varepsilon > 0$ such that $U_\varepsilon(x) \subset U$. For metric spaces our definition of continuity is equivalent to the familiar definition of calculus: A map $f : X \to Y$ between metric spaces is continuous at $a \in X$ if for each $\varepsilon > 0$ there exists $\delta > 0$ such that $d(a,x) < \delta$ implies $d(f(a), f(x)) < \varepsilon$. Continuity only depends on the underlying topology. But a metric is a finer and more rigid structure; one can compare the size of neighbourhoods of different points and one can define uniform continuity. A map $f : (X, d_1) \to (Y, d_2)$ between metric spaces is ***uniformly continuous*** if for each $\varepsilon > 0$ there exists $\delta > 0$ such that $d_1(x,y) < \delta$ implies $d_2(f(x), f(y)) < \varepsilon$. A sequence $f_n : X \to Y$ of maps into a metric space (Y, d) ***converges uniformly*** to $f : X \to Y$ if for each $\varepsilon > 0$ there exists N such that for $n > N$ and $x \in X$ the inequality $d(f(x), f_n(x)) < \varepsilon$ holds. If the f_n are continuous functions from a topological space X which converge uniformly to f, then one shows as in calculus that f is continuous.

A set A in a metric space (X, d) is ***bounded*** if $\{d(x,y) \mid x, y \in A\}$ is bounded in $\mathbb{R}$. The supremum of the latter set is then the ***diameter*** of A. We define $d(x, A) = \inf\{d(x,a) \mid a \in A\}$ as the ***distance*** of x from $A \neq \emptyset$. The relation $|d(x,A) - d(y,A)| \le d(x,A)$ shows that the map $X \to \mathbb{R}$, $x \mapsto d(x,A)$ is uniformly continuous. The relation $d(x,A) = 0$ is equivalent to $x \in \bar{A}$.

If A and B are disjoint, non-empty, closed sets in X, then

$$f : X \to [0,1], \quad x \mapsto d(x,A)(d(x,A) + d(x,B))^{-1}$$

is a continuous function with $f(A) = \{0\}$ and $f(B) = \{1\}$. Let $0 < a < b < 1$. Then $[0, a[$ and $]b, 1]$ are open in $[0,1]$, and their pre-images under f are disjoint open neighbourhoods of A and B. Hence a metric space is normal.

A ***directed set*** $(I, \le)$ consists of a set I and a relation $\le$ on I such that: (1) $i \le i$ for all $i \in I$. (2) $i \le j$, $j \le k$ implies $i \le k$. (3) For each pair $i, j \in I$ there exists $k \in I$ such that $i \le k$, $j \le k$. We also write $j \ge i$ for $i \le j$. The set $\mathbb{N}$ with the usual order is directed. The set $\mathcal{U}(x)$ of neighbourhoods of x is directed by $U \le V \Leftrightarrow V \subset U$.

A ***net*** with directed index set I in X is a map $I \to X, i \mapsto x_i$. We write $(x_i)_{i \in I}$ or just (x_i) for such a net. A net (x_i) in a topological space X ***converges*** to x, notation $x = \lim x_i$, provided for each neighbourhood U of x there exists $i \in I$ such that $x_j \in U$ for $j \ge i$. If one chooses from each $U \in \mathcal{U}(x)$ a point x_U, then the net (x_U) with index set $\mathcal{U}(x)$ converges to x. A point x is an ***accumulation value*** of the net $(x_i)_{i \in I}$ if for each neighbourhood U of x and each $i \in I$ there exists $j \ge i$ such that $x_j \in U$. Let I and J be directed sets. A map $h : I \to J$ is ***final*** if for each $j \in J$ there exists $i \in I$ such that $h(i) \ge j$. A ***subnet*** of a net $(x_j)_{j \in J}$ is a net of the form $i \mapsto x_{h(i)}$ with a final map $h : I \to J$. If a subnet of

$(x_j)_{j\in J}$ converges to z, then z is an accumulation value. Each accumulation value is a convergence point of a subnet.

Let $(x_j)_{j\in J}$ be a net in X. For $j \in J$ let $F(j)$ be the closure of $\{x_k \mid k \geq j\}$. Then $F = \bigcap_{j\in J} F(j)$ is the set of accumulation values of the net.

1.2 Subspaces. Quotient Spaces

It is a classical idea and method to define geometric objects (spaces) as subsets of Euclidean spaces, e.g., as solution sets of a system of equations. But it is important to observe that such objects have "absolute" properties which are independent of their position in the ambient space. In the topological context this absolute property is the subspace topology.

Let $(X, \mathcal{O})$ be a topological space and $A \subset X$ a subset. Then

$$\mathcal{O}|A = \{U \subset A \mid \text{there exists } V \in \mathcal{O} \text{ with } U = A \cap V\}$$

is a topology on A. It is called the ***induced topology***, the ***subspace topology***, or the ***relative topology***. The space $(A, \mathcal{O}|A)$ is called a ***subspace*** of $(X, \mathcal{O})$; we usually say: A is a subspace of X. A continuous map $f\colon (Y, \mathcal{S}) \to (X, \mathcal{O})$ is an ***embedding*** if it is injective and $(Y, \mathcal{S}) \to (f(Y), \mathcal{O}|f(Y))$, $y \mapsto f(y)$ a homeomorphism. From the definition one verifies:

Let A be a subspace of X. Then the inclusion $i\colon A \to X, a \mapsto a$ is continuous. Let Y be a space and $f\colon Y \to X$ a set map with $f(Y) \subset A$. Then f is continuous if and only if $\varphi\colon Y \to A$, $y \mapsto f(y)$ is continuous.

(1.2.1) Proposition. *Let $i\colon Y \to X$ be an injective set map between spaces. The following are equivalent:*

(1) *i is an embedding.*

(2) *A set map $g\colon Z \to Y$ from any topological space Z is continuous if and only if $ig\colon Z \to X$ is continuous.* □

Property (2) characterizes embeddings i in categorical terms. We call this property the ***universal property of an embedding***.

Suppose $A \subset B \subset X$ are subspaces. If A is closed in B and B closed in X, then A is closed in X. Similarly for open subspaces. But in general, an open (closed) subset of B must not be open (closed) in X. The next proposition will be used many times without further reference.

(1.2.2) Proposition. *Let $f\colon X \to Y$ be a set map between topological spaces and let X be the union of the subsets $(X_j \mid j \in J)$. If the X_j are open and the maps $f_j = f|X_j$ continuous, then f is continuous. A similar assertion holds if the X_j are closed and J is finite.* □

A subset A of a space X is a ***retract*** of X if there exists a ***retraction*** $r: X \to A$, i.e., a continuous map $r: X \to A$ such that $r|A = \mathrm{id}(A)$. A continuous map $s: B \to E$ is a ***section*** of the continuous map $p: E \to B$ if $ps = \mathrm{id}(B)$. In that case s is an embedding onto its image.

In geometric and algebraic topology many of the important spaces are constructed as quotient spaces. They are obtained from a given space by an equivalence relation. Although the quotient topology is easily defined, formally, it takes some time to work with it. In several branches of mathematics quotient objects are more difficult to handle than subobjects. Even if one starts with a nice and well-known space, its quotient spaces may have strange properties; usually one has to add a number of hypotheses in order to exclude unwanted phenomena. Quotient spaces do not, in general, inherit desirable properties from the original space.

Let X be a topological space and $f: X \to Y$ a surjective map onto a set Y. Then $\mathcal{S} = \{U \subset Y \mid f^{-1}(U) \text{ open in } X\}$ is a topology on Y. This is the finest topology on Y such that f is continuous. We call $\mathcal{S}$ the ***quotient topology*** on Y with respect to f. A surjective map $f: X \to Y$ between topological spaces is called an ***identification*** or ***quotient map*** if it has the following property: $U \subset Y$ open $\Leftrightarrow f^{-1}(U) \subset X$ open. If $f: X \to Y$ is a quotient map, then Y is called a ***quotient space*** of X.

We recall that a surjective map $f: X \to Y$ is essentially the same thing as an equivalence relation on X. If R is an equivalence relation on X, then X/R denotes the set of equivalence classes. The ***canonical map*** $p: X \to X/R$ assigns to $x \in X$ its equivalence class. If $f: X \to Y$ is surjective, then $x \sim y \Leftrightarrow f(x) = f(y)$ is an equivalence relation R_f on X. There is a canonical bijection $\varphi: X/R_f \to Y$ such that $\varphi p = f$. The ***quotient space*** X/R is defined to be the set X/R together with the quotient topology of the canonical map $p: X \to X/R$.

If $A \subset X$, we denote[1] by X/A the space obtained from X by identifying A to a point. In the case that $A = \emptyset$, we understand by this symbol A together with a disjoint point (topological sum (1.3.4)).

(1.2.3) Proposition. *Let $f: X \to Y$ be a surjective map between spaces. The following are equivalent:*

(1) *f is a quotient map.*

(2) *A set map $g: Y \to Z$ into any topological space Z is continuous if and only if $gf: X \to Z$ is continuous.* □

Property (2) characterizes quotient maps f in categorical terms. We call this property the ***universal property of a quotient map***.

A subset of X is ***saturated*** with respect to an equivalence relation if it is a union of equivalence classes.

Let $j: A \subset X$ be an inclusion and $f: A \to Y$ a continuous map. We identify in the topological sum $X + Y$ for each $a \in A$ the point $a \in X$ with the point $f(a) \in Y$,

[1] A similar notation is used for factor groups and orbit spaces.

i.e., we consider the equivalence relation on the topological sum (1.3.4) $X+Y$ with equivalence classes $\{z\}$ for $z \notin A + f(A)$ and $f^{-1}(z) + \{z\}$ for $z \in f(A)$. The quotient space Z is sometimes denoted by $Y \cup_f X$ and called the ***adjunction space*** obtained by ***attaching*** X via f to Y. The canonical inclusions $X \to X+Y$ and $Y \to X+Y$ induce maps $F\colon X \to Y \cup_f X$ and $J\colon Y \to Y \cup_f X$. The diagram

$$\begin{array}{ccc} A & \xrightarrow{f} & Y \\ \downarrow{\scriptstyle j} & & \downarrow{\scriptstyle J} \\ X & \xrightarrow{F} & Z \end{array} \quad = Y \cup_f X$$

is a pushout in TOP, and J is an embedding.

(1.2.4) Proposition. *Let j be a closed embedding. Then the data of the pushout have the following properties:*

(1) *J is a closed embedding.*

(2) *F restricted to $X \smallsetminus A$ is an open embedding.*

(3) *If X, Y are T_1-spaces (T_4-spaces), then $Y \cup_f X$ is a T_1-space (T_4-space).*

(4) *If f is a quotient map, then F is a quotient map.* □

Because of (1) and (2) we identify $X \smallsetminus A$ with the open subspace $F(X \smallsetminus A)$ and Y with the closed subspace $J(Y)$. In this sense, $Y \cup_f X$ is the union of the disjoint subsets $X \smallsetminus A$ and Y.

(1.2.5) Proposition. *The space $Y \cup_f X$ is a Hausdorff space, provided the following holds: Y is a Hausdorff space, X is regular, and A is a retract of an open neighbourhood in X.* □

Problems

1. Let $f\colon X \to Y$ and $g\colon Y \to Z$ be continuous maps. If f and g are embeddings, then gf is an embedding. If gf and g are embeddings, then f is an embedding. If $gf = \mathrm{id}$, then f is an embedding. An embedding is open (closed) if and only if its image is open (closed). If $f\colon X \to Y$ is a homeomorphism and $A \subset X$, then the map $A \to f(A)$, induced by f, is a homeomorphism.

2. Let $f\colon X \to Y$ be a quotient map. Let B be open or closed in Y and set $A = f^{-1}(B)$. Then the restriction $g\colon A \to B$ of f is a quotient map.

3. Let $f\colon X \to Y$ be surjective, continuous and open (or closed). Then f is a quotient map. The restriction $f_B\colon f^{-1}(B) \to B$ is open (or closed) for each $B \subset Y$, hence a quotient map.

4. The exponential map $\exp\colon \mathbb{C} \to \mathbb{C}^* = \mathbb{C} \smallsetminus \{0\}$ is open. Similarly $p\colon \mathbb{R} \to S^1$, $t \mapsto \exp(2\pi i t)$ is open. The kernel of p is $\mathbb{Z}$. Let $q\colon \mathbb{R} \to \mathbb{R}/\mathbb{Z}$ be the quotient map onto the factor group. There is a bijective map $\alpha\colon \mathbb{R}/\mathbb{Z} \to S^1$ which satisfies $\alpha \circ q = p$.

Since p and q are quotient maps, α is a homeomorphism. The continuous periodic functions $f: \mathbb{R} \to \mathbb{R}$, $f(x+1) = f(x)$ therefore correspond to continuous maps $\mathbb{R}/\mathbb{Z} \to \mathbb{R}$ and to continuous maps $S^1 \to \mathbb{R}$ via composition with q or p. In a similar manner one obtains a homeomorphism $\mathbb{C}/\mathbb{Z} \cong \mathbb{C}^*$.

5. Let $f: A \to B$ and $g: B \to C$ be continuous. If f and g are quotient maps, then gf is a quotient map. If gf is a quotient map, then g is a quotient map. If $gf = \mathrm{id}$, then g is a quotient map.

1.3 Products and Sums

Let $((X_j, \mathcal{O}_j) \mid j \in J)$ be a family of topological spaces. The product set $X = \prod_{j \in J} X_j$ is the set of all families $(x_j \mid j \in J)$ with $x_j \in X_j$. We have the projection $\mathrm{pr}_i : X \to X_i$, $(x_j) \mapsto x_i$ into the i-th factor. Let X_j, Y_j be topological spaces and $f_j : X_j \to Y_j$ maps. The product map $\prod f_j : \prod X_j \to \prod Y_j$ is defined as $(x_j \mid j \in J) \mapsto (f_j(x_j) \mid j \in J)$. Given maps $f_j : Y \to X_j$ we denote by $(f_j) = (f_j \mid j \in J) : Y \to \prod_j X_j$ the map with components $\mathrm{pr}_i \circ (f_j) = f_i$.

The family of all pre-images $\mathrm{pr}_j^{-1}(U_j)$, $U_j \subset X_j$ open in X_j (for varying j), is the subbasis for the ***product topology*** $\mathcal{O}$ on X. We call $(X, \mathcal{O})$ the ***topological product*** of the spaces $(X_j, \mathcal{O}_j)$. The next proposition shows that $X = \prod X_j$ together with the projections pr_j is a categorical product of the family (X_j) in the category TOP. Note that for infinite J, open sets in the product are quite large; a product $\prod U_j$, $U_j \subset X_j$ open, is then in general not an open subset of $\prod X_j$.

(1.3.1) Proposition. *The product topology is the coarsest topology for which all projections* pr_j *are continuous. A set map* $f: Y \to X$ *from a space* Y *into* X *is continuous if and only if all maps* $\mathrm{pr}_j \circ f$ *are continuous. The product* $f = \prod_j f_j$ *of continuous maps* $f_j : X_j \to Y_j$ *is continuous.* □

The product of X_1, X_2 is denoted $X_1 \times X_2$, and we use $f_1 \times f_2$ for the product of maps. The "identity" $\mathrm{id}: X_1 \times (X_2 \times X_3) \to (X_1 \times X_2) \times X_3$ is a homeomorphism. In general, the topological product is associative, i.e., compatible with arbitrary bracketing. The canonical identification $\mathbb{R}^k \times \mathbb{R}^l = \mathbb{R}^{k+l}$ is a homeomorphism.

1.3.2 Pullback. Let $f: X \to B$ and $g: Y \to B$ be continuous maps. Let $Z = \{(x, y) \in X \times Y \mid f(x) = g(y)\}$ with the subspace topology of $X \times Y$. We have the projections onto the factors $F: Z \to Y$ and $G: Z \to X$. The commutative diagram

$$\begin{array}{ccc} Z & \xrightarrow{F} & Y \\ \downarrow{\scriptstyle G} & & \downarrow{\scriptstyle g} \\ X & \xrightarrow{f} & B \end{array}$$

is a pullback in TOP. The space Z is sometimes written $Z = X \times_B Y$ and called the product of X and Y over B (the product in the category TOP_B of spaces over B).

Pullbacks allow one to convert liftings into sections. Let $i: A \subset X$ and $a: A \to Y$ such that $ga = f|A$ is given. The assignment $(\sigma: X \to Z) \mapsto (F \circ \sigma: X \to Y)$ sets up a bijection between sections of G with $F\sigma|A = a$ and maps $\varphi: X \to Y$ such that $\varphi|A = a$ and $g\varphi = f$. ◇

(1.3.3) Proposition. *Let $f: X \to Y$ be surjective, continuous, and open. Then Y is separated if and only if $R = \{(x_1, x_2) \mid f(x_1) = f(x_2)\}$ is closed in $X \times X$.* □

Let $(X_j \mid j \in J)$ be a family of non-empty pairwise disjoint spaces. The set $\mathcal{O} = \{U \subset \amalg X_j \mid U \cap X_j \subset X_j \text{ open for all } j\}$ is a topology on the disjoint union $\amalg X_j$. We call $(\amalg X_j, \mathcal{O})$ the ***topological sum*** of the X_j. A sum of two spaces is denoted $X_1 + X_2$. The following assertions are easily verified from the definitions. They show that the topological sum together with the canonical inclusions $X_j \to \amalg X_j$ is a categorical sum in TOP. Given maps $f_j: X_j \to Z$ we denote by $\langle f_j \rangle: \amalg X_j \to Z$ the map with restriction f_j to X_j.

(1.3.4) Proposition. *A topological sum has the following properties: The subspace topology of X_j in $\amalg X_j$ is the original topology. Let the space X be the union of the family $(X_j \mid j \in J)$ of pairwise disjoint subsets. Then X is the topological sum of the subspaces X_j if and only if the X_j are open. $f: \amalg X_j \to Y$ is continuous if each $f|X_j: X_j \to Y$ is continuous.* □

1.3.5 Pushout. Let $j: A \to X$ and $f: A \to B$ be continuous maps and form a pushout diagram

$$\begin{array}{ccc} A & \xrightarrow{f} & B \\ \downarrow{\scriptstyle j} & & \downarrow{\scriptstyle J} \\ X & \xrightarrow{F} & Y \end{array}$$

in the category SET of sets. Then Y is obtainable as a quotient of $X + B$. We give Y the quotient topology via $\langle F, J \rangle: X + B \to Y$. Then the resulting diagram is a pushout in TOP. The space Y is sometimes written $X +^A B$ and called the sum of X and B under A (the sum in the category TOP^A of spaces under A). ◇

1.3.6 Clutching. An important method for the construction of spaces is to "paste" open subsets; see the example (1.3.8) for the simplest case. Let $(U_j \mid j \in J)$ be a family of sets. Assume that for each pair $(i, j) \in J \times J$ a subset $U_i^j \subset U_i$ is given as well as a bijection $g_i^j: U_i^j \to U_j^i$. We call the families (U_j, U_j^k, g_j^k) a ***clutching datum*** if:

(1) $U_j = U_j^j$ and $g_j^j = \text{id}$.

(2) For each triple $(i, j, k) \in J \times J \times J$ the map g_i^j induces a bijection

$$g_i^j: U_i^j \cap U_i^k \to U_j^i \cap U_j^k$$

and $g_j^k \circ g_i^j = g_i^k$ holds, considered as maps from $U_i^j \cap U_i^k$ to $U_k^j \cap U_k^i$.

Given a clutching datum, we have an equivalence relation on the disjoint sum $\amalg_{j \in J} U_j$:

$$x \in U_i \sim y \in U_j \quad \Leftrightarrow \quad x \in U_i^j \text{ and } g_i^j(x) = y.$$

Let X denote the set of equivalence classes and let $h_i : U_i \to X$ be the map which sends $x \in U_i$ to its class. Then h_i is injective. Set $U(i) = \text{image } h_i$, then $U(i) \cap U(j) = h_i(U_i^j)$.

Conversely, assume that X is a quotient of $\amalg_{j \in J} U_j$ such that each $h_i : U_i \to X$ is injective with image $U(i)$. Let $U_i^j = h_i^{-1}(U(i) \cap U(j))$ and $g_i^j = h_j^{-1} \circ h_i : U_i^j \to U_j^i$. Then the (U_i, U_i^j, g_i^j) are a clutching datum. If we apply the construction above to this datum, we get back X and the h_i. ◇

(1.3.7) Proposition. *Let (U_i, U_i^j, g_i^j) be a clutching datum. Assume that the U_i are topological spaces, the $U_i^j \subset U_i$ open subsets, and the $g_i^j : U_i^j \to U_j^i$ homeomorphisms. Let X carry the quotient topology with respect to the quotient map $p\colon \amalg_{j \in J} U_j \to X$. Then the following holds:*

(1) *The map h_i is a homeomorphism onto an open subset of X and p is open.*

(2) *Suppose the U_i are Hausdorff spaces. Then X is a Hausdorff space if and only if for each pair (i, j) the map $\gamma_i^j : U_i^j \to U_i \times U_j$, $x \mapsto (x, g_i^j(x))$ is a closed embedding.* □

1.3.8 Euclidean space with two origins. The simplest case is obtained from open subsets $V_j \subset U_j$, $j = 1, 2$, and a homeomorphism $\varphi\colon V_1 \to V_2$. Then $X = U_1 \cup_\varphi U_2$ is obtained from the topological sum $U_1 + U_2$ by identifying $v \in V_1$ with $\varphi(v) \in V_2$. Let $U_1 = U_2 = \mathbb{R}^n$ and $V_1 = V_2 = \mathbb{R}^n \smallsetminus 0$. Let $\varphi = \mathrm{id}$. Then the graph of φ in $\mathbb{R}^n \times \mathbb{R}^n$ is not closed. The resulting locally Euclidean space is not Hausdorff. If we use $\varphi(x) = x \cdot \|x\|^{-2}$, then the result is homeomorphic to S^n (see (2.3.2)). ◇

Suppose a space X is the union of subspaces $(X_j \mid j \in J)$. We say X carries the ***colimit topology*** with respect to this family if one of the equivalent statements hold:

(1) The canonical map $\coprod_{j \in J} X_j \to X$ (the inclusion on each summand) is a quotient map.

(2) C is closed in X if and only if $X_j \cap C$ is closed in X_j for each j.

(3) A set map $f : X \to Z$ into a space Z is continuous if and only if the restrictions $f|X_j : X_j \to Z$ are continuous.

(1.3.9) Example. Let X be a set which is covered by a family $(X_j \mid j \in J)$ of subsets. Suppose each X_j carries a topology such that the subspace topologies of

$X_i \cap X_j$ in X_i and X_j coincide and these subspaces are closed in X_i and X_j. Give X the quotient topology with respect to the canonical map $\coprod X_j \to X$. Then the subspace topology of $X_j \subset X$ coincides with the given topology and X_j is closed in X. Similarly if "closed" is replaced by "open".

Problems

1. Let $(X_j \mid j \in J)$ be spaces and $A_j \subset X_j$ non-empty subspaces. Then $\prod_{j\in J} \bar{A}_j = \overline{\prod_{j\in J} A_j}$. The product $\prod_{j\in J} A_j$ is closed if and only if the A_j are closed.
2. The projections $\mathrm{pr}_k \colon \prod_j X_j \to X_k$ are open maps, and in particular quotient maps. (The X_j are non-empty.)
3. A space X is separated if and only if the diagonal $D = \{(x,x) \mid x \in X\}$ is closed in $X \times X$. Let $f, g \colon X \to Y$ be continuous maps into a Hausdorff space. Then the ***coincidence set*** $A = \{x \mid f(x) = g(x)\}$ is closed in X.
4. A discrete space is the topological sum of its points. There is always a canonical homeomorphism $X \times \amalg_j Y_j \cong \amalg_j (X \times Y_j)$. For each $y \in Y$ the map $X \to X \times Y$, $x \mapsto (x, y)$ is an embedding. If $f \colon X \to Y$ is continuous, then $\gamma \colon X \to X \times Y, x \mapsto (x, f(x))$ is an embedding. If Y is a Hausdorff space, then γ is closed.

1.4 Compact Spaces

A family $A = (A_j \mid j \in J)$ of subsets of X is a ***covering*** of X if X is the union of the A_j. A covering $B = (B_k \mid k \in K)$ of X is a ***refinement*** of A if for each $k \in K$ there exists $j \in J$ such that $B_k \subset A_j$. If X is a topological space, a covering $A = (A_j \mid j \in J)$ is called ***open*** (***closed***) if each A_j is open (closed). A covering $B = (B_k \mid k \in K)$ is a ***subcovering*** of A if $K \subset J$ and $B_k = A_k$ for $k \in K$. We say B is ***finite*** or ***countable*** if K is finite or countable. A covering A is ***locally finite*** if each $x \in X$ has a neighbourhood U such that $U \cap A_j \neq \emptyset$ only for a finite number of $j \in J$. It is called ***point-finite*** if each $x \in X$ is contained only in a finite number of A_j.

A space X is ***compact*** if each open covering has a finite subcovering. (In some texts this property is called ***quasi-compact***.) By passage to complements we see: If X is compact, then any family of closed sets with empty intersection contains a finite family with empty intersection. A set A in a space X is ***relatively compact*** if its closure is compact. We recall from calculus the fundamental Heine–Borel theorem: *The unit interval* $I = [0, 1]$ *is compact.*

A space X is compact if and only if each net in X has a convergent subnet (an accumulation value). A discrete closed set in a compact space is finite. Let X be compact, $A \subset X$ closed and $f \colon X \to Y$ continuous; then A and $f(X)$ are compact.

(1.4.1) Proposition. *Let B, C be compact subsets of spaces X, Y, respectively. Let $\mathcal{U}$ be a family of open subsets of $X \times Y$ which cover $B \times C$. Then there exist open neighbourhoods U of B and V of C such that $U \times V$ is covered by a finite subfamily of $\mathcal{U}$. In particular the product of two compact spaces is compact.* □

One can show that an arbitrary product of compact spaces is compact (***theorem of Tychonoff***).

(1.4.2) Proposition. *Let B and C be disjoint compact subsets of a Hausdorff space X. Then B and C have disjoint open neighbourhoods. A compact Hausdorff space is normal. A compact subset C of a Hausdorff space X is closed.* □

(1.4.3) Proposition. *A continuous map $f: X \to Y$ from a compact space into a Hausdorff space is closed. If, moreover, f is injective (bijective), then f is an embedding (homeomorphism). If f is surjective, then it is a quotient map.* □

(1.4.4) Proposition. *Let X be a compact Hausdorff space and $f: X \to Y$ a quotient map. The following assertions are equivalent:*

(1) *Y is a Hausdorff space.*

(2) *f is closed.*

(3) *$R = \{(x_1, x_2) \mid f(x_1) = f(x_2)\}$ is closed in $X \times X$.* □

Let X be a union of subspaces $X_1 \subset X_2 \subset \cdots$. Recall that X carries the ***colimit-topology with respect to the filtration*** (X_i) if $A \subset X$ is open (closed) if and only if each intersection $A \cap X_n$ is open (closed) in X_n. We then call X the ***colimit*** of the ascending sequence (X_i). (This is a colimit in the categorical sense.)

(1.4.5) Proposition. *Suppose X is the colimit of the sequence $X_1 \subset X_2 \subset \cdots$. Suppose points in X_i are closed. Then each compact subset K of X is contained in some X_k.* □

A space is ***locally compact*** if each neighbourhood of a point x contains a compact neighbourhood. An open subset of a locally compact space is again locally compact.

Let X be a Hausdorff space and assume that each point has a compact neighbourhood. Let U be a neighbourhood of x and K a compact neighbourhood. Since K is normal, $K \cap U$ contains a closed neighbourhood L of x in K. Then L is compact and a neighbourhood of x in X. Therefore X is locally compact. In particular, a compact Hausdorff space is locally compact. If X and Y are locally compact, then $X \times Y$ is locally compact.

Let X be a topological space. An embedding $f: X \to Y$ is a ***compactification*** of X if Y is compact and $f(X)$ dense in Y.

The following theorem yields a compactification by a single point. It is called the ***Alexandroff compactification*** or the ***one-point compactification***. The additional point is the ***point at infinity***. In a general compactification $f: X \to Y$, one calls the points in $Y \smallsetminus f(X)$ the points at infinity.

(1.4.6) Theorem. *Let X be a locally compact (non-compact) Hausdorff space. Up to homeomorphism, there exists a unique compactification $f: X \to Y$ by a compact Hausdorff space such that $Y \smallsetminus f(X)$ consists of a single point.* □

(1.4.7) Proposition. *Let the locally compact space be a union of compact subsets $(K_i \mid i \in \mathbb{N})$. Then there exists a sequence $(U_i \mid i \in \mathbb{N})$ of open subsets with union X such that each $\overline{U_i}$ is compact and contained in U_{i+1}.* □

(1.4.8) Theorem. *Let the locally compact Hausdorff space $M \neq \emptyset$ be a union of closed subsets M_n, $n \in \mathbb{N}$. Then at least one of the M_n contains an interior point.* □

A subset H of a space G is called ***locally closed***, if each $x \in H$ has a neighbourhood V_x in G such that $H \cap V_x$ is closed in V_x.

(1.4.9) Proposition. (1) *A is locally closed in X if and only if $A = U \cap C$ with U open and C closed in X; one can take $C = \overline{A}$.*

(2) *A locally compact set A in a Hausdorff space X is locally closed.*

(3) *A locally closed set A in a locally compact space is locally compact.* □

Problems

1. D^n/S^{n-1} is homeomorphic to S^n. For the proof verify that

$$D^n \to S^n, \quad x \mapsto \left(2\sqrt{1-\|x\|^2}x, 2\|x\|^2 - 1\right)$$

induces a bijection $D^n/S^{n-1} \to S^n$.

2. Let $f: X \times C \to \mathbb{R}$ be continuous. Assume that C is compact and set $g(x) = \sup\{f(x,c) \mid c \in C\}$. Then $g: X \to \mathbb{R}$ is continuous.

3. Let X be the colimit of an ascending sequence of spaces $X_1 \subset X_2 \subset \cdots$. Then the X_i are subspaces of X. If $X_i \subset X_{i+1}$ is always closed, then the X_j are closed in X.

4. Let $\mathbb{R}^\infty$ be the vector space of all sequences $(x_1, x_2, \dots)$ of real numbers which are eventually zero. Let $\mathbb{R}^n$ be the subspace of sequences with $x_j = 0$ for $j > n$. Give $\mathbb{R}^\infty$ the colimit topology with respect to the subspaces $\mathbb{R}^n$. Then addition of vectors is a continuous map $\mathbb{R}^\infty \times \mathbb{R}^\infty \to \mathbb{R}^\infty$. Scalar multiplication is a continuous map $\mathbb{R} \times \mathbb{R}^\infty \to \mathbb{R}^\infty$. (Thus $\mathbb{R}^\infty$ is a topological vector space.) A neighbourhood basis of zero consists of the intersection of $\mathbb{R}^\infty$ with products of the form $\prod_{i \geq 1}]-\varepsilon_i, \varepsilon_i[$. The space $\mathbb{R}^\infty$ with this topology is not metrizable. The space has also the colimit topology with respect to the set of finite-dimensional linear subspaces. One can also consider the metric topology with respect to the metric $d((x_i),(y_i)) = \left(\sum_i (x_i - y_i)^2\right)^{1/2}$; denote it by $\mathbb{R}^\infty_d$. The identity $\mathbb{R}^\infty \to \mathbb{R}^\infty_d$ is continuous. The space $\mathbb{R}^\infty$ is separated.

1.5 Proper Maps

A continuous map $f: X \to Y$ is called ***proper*** if it is closed and the pre-images $f^{-1}(y)$, $y \in Y$ are compact.

(1.5.1) Proposition. *Let K be compact. Then* $\mathrm{pr}: X \times K \to X$ *is proper. If $f: X \to Y$ is proper and $K \subset Y$ compact, then $f^{-1}(K)$ is compact. Let f and g be proper; then $f \times g$ is proper.* □

As a generalization of the theorem of Tychonoff one can show that an arbitrary product of proper maps is proper.

(1.5.2) Proposition. *Let $f: X \to X'$ and $g: X' \to X''$ be continuous.*

(1) *If f and g are proper, then $g \circ f$ is proper.*
(2) *If $g \circ f$ is proper and f surjective, then g is proper.*
(3) *If $g \circ f$ is proper and g injective, then f is proper.* □

(1.5.3) Proposition. *Let $f: X \to Y$ be injective. Then the following are equivalent:*

(1) *f is proper.*
(2) *f is closed.*
(3) *f is a homeomorphism onto a closed subspace.* □

(1.5.4) Proposition. *Let $f: X \to Y$ be continuous.*

(1) *If f is proper, then for each $B \subset Y$ the restriction $f_B: f^{-1}(B) \to B$ of f is proper.*
(2) *Let $(U_j \mid j \in J)$ be a covering of Y such that the canonical map $p: \coprod_{j \in J} U_j \to Y$ is a quotient map. If each restriction $f_j: f^{-1}(U_j) \to U_j$ is proper, then f is proper.* □

(1.5.5) Proposition. *Let f be a continuous map of a Hausdorff space X into a locally compact Hausdorff space Y. Then f is proper if and only if each compact set $K \subset Y$ has a compact pre-image. If f is proper, then X is locally compact.* □

(1.5.6) Proposition. *Let $f: X \to X'$ and $g: X' \to X''$ be continuous and assume that gf is proper. If X' is a Hausdorff space, then f is proper.* □

(1.5.7) Theorem. *A continuous map $f: X \to Y$ is proper if and only if for each space T the product $f \times \mathrm{id}: X \times T \to Y \times T$ is closed.* □

Problems

1. A map $f: X \to Y$ is proper if and only if the following holds: For each net (x_j) in X and each accumulation value y of $(f(x_j))$ there exists an accumulation value x of (x_j) such that $f(x) = y$.
2. Let X and Y be locally compact Hausdorff spaces, let $f: X \to Y$ be continuous and $f^+: X^+ \to Y^+$ the extension to the one-point compactification. Then f^+ is continuous, if f is proper.
3. The restriction of a proper map to a closed subset is proper.
4. Let $f: X \to Y$ be proper and X a Hausdorff space. Then the subspace $f(X)$ of Y is a Hausdorff space.
5. Let $f: X \to Y$ be continuous. Let R be the equivalence relation on X induced by f, and denote by $p: X \to X/R$ the quotient map, by $h: X/R \to f(X)$ the canonical bijection, and let $i: f(X) \subset Y$. Then $f = i \circ h \circ p$ is the canonical decomposition of f. The map f is proper if and only if p is proper, h a homeomorphism, and $f(X) \subset Y$ closed.

1.6 Paracompact Spaces

Let $\mathcal{A} = (U_j \mid j \in J)$ be an open covering of the space X. An open covering $\mathcal{B} = (B_j \mid j \in J)$ is called a ***shrinking*** of $\mathcal{A}$ if for each $j \in J$ we have the inclusion $\overline{B_j} \subset U_j$.

A point-finite open covering of a normal space has a shrinking.

A space X is called ***paracompact*** if it is a Hausdorff space and if each open covering has an open, locally finite refinement. A closed subset of a paracompact space is paracompact. A compact Hausdorff space is paracompact.

A paracompact space is normal. Suppose the locally compact Hausdorff space X is a countable union of compact sets. Then X is paracompact. Let X be paracompact and K be compact Hausdorff. Then $X \times K$ is paracompact. A metric space is paracompact.

1.7 Topological Groups

A ***topological group*** $(G, m, \mathcal{O})$ consists of a group (G, m) with multiplication $m: G \times G \to G$, $(g, h) \mapsto m(g, h) = gh$ and a topology $\mathcal{O}$ on G such that the multiplication m and the inverse $\iota: G \to G, g \mapsto g^{-1}$ are continuous. We denote a topological group $(G, m, \mathcal{O})$ usually just by the letter G. The neutral element will be denoted by e (also 1 is in use and 0 for abelian groups). The ***left translation*** $l_g: G \to G, x \mapsto gx$ by $g \in G$ in a topological group is continuous, and the rules $l_g l_h = l_{gh}$ and $l_e = \mathrm{id}$ show it to be a homeomorphism. For subsets A and B of a group G we use notations like $aB = \{ab \mid b \in B\}$, $AB = \{ab \mid a \in A, b \in B\}$, $A^2 = AA$, $A^{-1} = \{a^{-1} \mid a \in A\}$, and similar ones.

A group G together with the discrete topology on the set G is a topological group, called a ***discrete*** (topological) group.

The additive groups of the real numbers $\mathbb{R}$, complex numbers $\mathbb{C}$, and quaternions $\mathbb{H}$ with their ordinary topology are topological groups, similarly the multiplicative groups $\mathbb{R}^*$, $\mathbb{C}^*$ and $\mathbb{H}^*$ of the non-zero elements. The multiplicative group $\mathbb{R}^*_+$ of the positive real numbers is an open subgroup of $\mathbb{R}^*$ and a topological group. The complex numbers of norm 1 are a compact topological group S^1 with respect to multiplication. The exponential function $\exp\colon \mathbb{R} \to \mathbb{R}^*_+$ is a continuous homomorphism with the logarithm function as a continuous inverse. The complex exponential function $\exp\colon \mathbb{C} \to \mathbb{C}^*$ is a surjective homomorphism with kernel $\{2\pi i n \mid n \in \mathbb{Z}\}$, a discrete subgroup of $\mathbb{C}$.

The main examples of topological groups are matrix groups. In the vector space $M_n(\mathbb{R})$ of real (n,n)-matrices let $\mathrm{GL}_n(\mathbb{R})$ be the subspace of the invertible matrices. Since the determinant is a continuous map, this is an open subspace. Matrix multiplication and passage to the inverse are continuous, since they are given by rational functions in the matrix entries. This makes the ***general linear group*** $\mathrm{GL}_n(\mathbb{R})$ into a topological group. Similarly for $\mathrm{GL}_n(\mathbb{C})$. The determinant is a continuous homomorphism $\det\colon \mathrm{GL}_n(\mathbb{R}) \to \mathbb{R}^*$ with kernel the ***special linear group*** $\mathrm{SL}_n(\mathbb{R})$; similarly in the complex case.

Let $\mathrm{O}(n) = \{A \in M_n(\mathbb{R}) \mid A^t \cdot A = E\}$ be the group of orthogonal (n,n)-matrices (A^t transpose of A; E unit matrix). The set $\mathrm{O}(n)$ is a compact subset in $M_n(\mathbb{R})$. Hence $\mathrm{O}(n)$ is a compact topological group (the ***orthogonal group***). The open and closed subspace $\mathrm{SO}(n) = \{A \in \mathrm{O}(n) \mid \det(A) = 1\}$ of $\mathrm{O}(n)$ is the ***special orthogonal group***. Similarly the subgroup $\mathrm{U}(n) = \{A \in M_n(\mathbb{C}) \mid A^t \cdot \bar{A} = E\}$ of unitary (n,n)-matrices is a compact topological group (***unitary group***). The topological groups $\mathrm{SO}(2)$, $\mathrm{U}(1)$, and S^1 are isomorphic. The ***special unitary group*** $\mathrm{SU}(n)$ is the compact subgroup of $\mathrm{U}(n)$ of matrices with determinant 1. The multiplicative group of quaternions of norm 1 provides S^3 with the structure of a topological group. This group is isomorphic to $\mathrm{SU}(2)$. From linear algebra one knows about a surjective homomorphism $\mathrm{SU}(2) \to \mathrm{SO}(3)$ with kernel $\pm E$ (a twofold covering); for this and other related facts see the nice discussion in [27, Kapitel IX]. For more information about matrix groups, also from the viewpoint of manifolds and Lie groups, see [29]; there you can find, among others, the symplectic groups $\mathrm{Sp}(n)$ and the Spinor groups $\mathrm{Spin}(n)$. The isomorphisms $\mathrm{SU}(2) \cong \mathrm{Spin}(3) \cong \mathrm{Sp}(1)$ hold, and these spaces are homeomorphic to S^3.

If G and H are topological groups, then the direct product $G \times H$ with the product topology is a topological group. The n-fold product $S^1 \times \cdots \times S^1$ is called an ***n-dimensional torus***.

The trivial group is often denoted by 1 (in a multiplicative notation) or by 0 (in an additive notation). The neutral element will also be denoted 1 or 0. The symbol $H \lhd G$ is used for a normal subgroup H of G. The notation $H \sim K$ or $H \sim_G K$ means that H and K are conjugate subgroups of G.

A homomorphism $f: G \to H$ between topological groups is continuous if it is continuous at the neutral element e.

If G is a topological group and $H \subset G$ a subgroup, then H, with the subspace topology, is a topological group (called a ***topological subgroup***). If $H \subset G$ is a subgroup, then the closure of H is also. If H is a normal subgroup, then $\bar{H}$ is also.

1.8 Transformation Groups

A ***left action*** of a topological group G on a topological space X is a continuous map $\rho: G \times X \to X$, $(g, x) \mapsto gx$ such that $g(hx) = (gh)x$ and $ex = x$ for $g, h \in G$, $e \in G$ the unit, and $x \in X$. A (left) ***G-space*** (X, ρ) consists of a space X and a left action ρ of G on X. The homeomorphism $l_g: X \to X$, $x \mapsto gx$ is called ***left translation*** by g. We also use ***right actions*** $X \times G \to X$, $(x, g) \mapsto xg$; they satisfy $(xh)g = x(hg)$ and $xe = x$. For $A \subset X$ and $K \subset G$ we let $KA = \{ka \mid k \in K, a \in A\}$. An action is ***effective*** if $gx = x$ for all $x \in X$ implies $g = e$. The ***trivial action*** has $gx = x$ for $g \in G$ and $x \in X$.

The set $R = \{(x, gx) \mid x \in X,\ g \in G\}$ is an equivalence relation on X. The set of equivalence classes X mod R is denoted by X/G. The quotient map $q: X \to X/G$ is used to provide X/G with the quotient topology. The resulting space X/G is called the ***orbit space*** of the G-space X. A more systematic notation for the orbit space of a left action would be $G\backslash X$. The equivalence class of $x \in X$ is the ***orbit*** Gx through x. An action is ***transitive*** if it consists of a single orbit. The set $G_x = \{g \in G \mid gx = x\}$ is a subgroup of G, the ***isotropy group*** or the ***stabilizer*** of the G-space X at x. An action is ***free*** if all isotropy groups are trivial. We have $G_{gx} = gG_xg^{-1}$. Therefore the set $\mathrm{Iso}(X)$ of isotropy groups of X consists of complete conjugacy classes of subgroups. If it contains a finite number of conjugacy classes, we say X has ***finite orbit type***.

A subset A of a G-space is called ***G-stable*** or ***G-invariant*** if $g \in G$ and $a \in A$ implies $ga \in A$. A G-stable subset A is also called a ***G-subspace***. For each subgroup H of G there is an ***H-fixed point set*** of X,

$$X^H = \{x \in X \mid hx = x \text{ for all } h \in H\}.$$

Suppose X and Y are G-spaces. A map $f: X \to Y$ is called a G-map or a ***G-equivariant*** map if for $g \in G$ and $x \in X$ the relation $f(gx) = gf(x)$ holds. In general, the term "equivariant" refers to something related to a group action. Left G-spaces and G-equivariant maps form the category G-TOP. This category has products: If $(X_j \mid j \in J)$ is a family of G-spaces, then the topological product $\prod_j X_j$ together with the ***diagonal action*** $(g, (x_j)) \mapsto (gx_j)$ is a product in this category. A G-map $f: X \to Y$ induces by passage to the orbit spaces a map $f/G: X/G \to Y/G$. We have the notion of an ***equivariant homotopy*** or ***G-homotopy*** H_t: this is a homotopy such that each H_t is a G-map.

(1.8.1) Proposition. *Let X be a G-space, $A \subset G$ and $B \subset X$. If B is open then AB is open. The orbit map $p: X \to X/G$ is open.*

Proof. $l_a(B)$ is open, since l_a is a homeomorphism. Hence $\bigcup_{a \in A} l_a(B) = AB$ is a union of open sets. Let U be open. Then $p^{-1}p(U) = \bigcup_{g \in G} l_g(U)$ is open, hence $p(U)$ is open, by definition of the quotient topology. □

(1.8.2) Proposition. (1) *Let H be a subgroup of the topological group G. Let the set G/H of cosets gH carry the quotient topology with respect to $p: G \to G/H$, $g \mapsto gH$. Then $l: G \times G/H \to G/H$, $(x, gH) \mapsto xgH$ is a continuous action.*

(2) *G/H is separated if and only if H is closed in G. In particular, G is separated if $\{e\}$ is closed.*

(3) *Let H be normal in G. Then the factor group G/H with quotient topology is a topological group.* □

A space G/H with the G-action by left multiplication is called a ***homogeneous space***. The space of left cosets Hg is $H\backslash G$; it carries a right action.

(1.8.3) Example. Homogeneous spaces are important spaces in geometry. The orthogonal group $\mathrm{O}(n+1)$ acts on the sphere S^n by matrix multiplication $(A, v) \mapsto Av$. The action is transitive. The isotropy group of $e_1 = (1, 0, \dots, 0)$ is $\mathrm{O}(n)$, here considered as the block matrices $\left(\begin{smallmatrix} 1 & 0 \\ 0 & B \end{smallmatrix}\right)$ with $B \in \mathrm{O}(n)$. We obtain a homeomorphism of $\mathrm{O}(n+1)$-spaces $\mathrm{O}(n+1)/\mathrm{O}(n) \cong S^n$. In the complex case we obtain a homeomorphism $\mathrm{U}(n+1)/\mathrm{U}(n) \cong S^{2n+1}$, in the quaternionic case a homeomorphism $\mathrm{Sp}(n+1)/\mathrm{Sp}(n) \cong S^{4n+3}$. Other important homogeneous spaces are the projective spaces, the Grassmann manifolds, and the Stiefel manifolds to be discussed later. ◇

(1.8.4) Proposition. (1) *If $x \in X$ is closed, then G_x is closed in G.*

(2) *If X is a Hausdorff space, then X^H is closed.*

(3) *Let A be a G-stable subset of the G-space X. Then A/G carries the subspace topology of X/G. In particular $X^G \to X \to X/G$ is an embedding.*

(4) *Let $B \subset X$ be closed and $A \subset X$. Then $\{g \in G \mid gA \subset B\}$ is closed in G.*

(5) *Let $B \subset X$ be closed. Then $\{g \in G \mid gB = B\}$ is closed.*

Proof. (1) The isotropy group G_x is the pre-image of x under the continuous map $G \to X$, $g \mapsto gx$. (2) The set $X^g = \{x \in X \mid gx = x\}$ is the pre-image of the diagonal under $X \to X \times X$, $x \mapsto (x, gx)$, and $X^H = \bigcap_{g \in H} X^g$. (3) Let $C \subset A/G$ be open with respect to the quotient map $A \to A/G$. Then $p^{-1}(C) \subset A$ is open, and we can write $p^{-1}(C) = A \cap U$ with an open subset $U \subset X$. We have $A \cap U = A \cap GU$, since A is G-stable. We conclude $C = p(p^{-1}C) = A/G \cap p(GU)$. Since GU is open, $p(GU)$ is open, hence C is open in the subspace topology. By continuity of $A/G \to X/G$, an open subset in the subspace topology is open in A/G. (4) $r_a: G \to X, g \mapsto ga$ is continuous, hence $r_a^{-1}(B) =$

$\{g \in G \mid ga \in B\}$ closed and therefore $\bigcap_{a\in A} r_a^{-1}(B) = \{g \in G \mid gA \subset B\}$ closed. (5) The set $\{g \mid gB = B\} = \{g \mid gB \subset B\} \cap \{g \mid g^{-1}B \subset B\}$ is closed, by (4). □

(1.8.5) Proposition. *Let $r: G \times X \to X$ be a G-action, $A \subset G$ and $B \subset X$.*

(1) *If A and B are compact, then AB is compact.*

(2) *If A is compact, then the restriction $A \times X \to X$ of r is proper. If, moreover, B is closed, then AB is closed.*

(3) *If G is compact, then the orbit map p is proper. Thus X is compact if and only if X/G is compact.*

(4) *If G is compact and X separated, then X/G is separated.*

(5) *Let G be compact, A a G-stable closed subset and U a neighbourhood of A in X. Then U contains a G-stable neighbourhood of A.*

Proof. (1) $A \times B \subset G \times X$ is compact as a product of compact spaces. Hence the continuous image AB of $A \times B$ under $r: G \times X \to X$ is compact. (2) The homeomorphism $A\times X \to A\times X$, $(s,x) \mapsto (s,sx)$ transforms r into the projection $\mathrm{pr}: A\times X \to X$. The projection is proper, since A is compact (see (1.5.1)). Hence the image AB of the closed set $A \times B$ is closed. (3) Let $A \subset X$ be closed. Then $p^{-1}p(A) = GA$ is closed, by (2). Hence $p(A)$ is closed, by definition of the quotient topology. The pre-images of points are orbits; they are compact as continuous images of G. (4) Since p is proper, so is $p \times p$. Hence the image of the diagonal under $p \times p$ is closed. (5) Let U be open. Then $p(X \smallsetminus U)$ is disjoint to $p(A)$. By (4), $X \smallsetminus p^{-1}p(X \smallsetminus U)$ is open and a G-stable neighbourhood of A contained in U. □

The ***orbit category*** $\mathrm{Or}(G)$ is the category of homogeneous G-spaces G/H, H closed in G, and G-maps. There exists a G-map $G/H \to G/K$ if and only if H is conjugate to a subgroup of K. If $a^{-1}Ha \le K$, then $R_a: G/H \to G/K$, $gH \mapsto gaK$ is a G-map and each G-map $G/H \to G/K$ has this form; moreover $R_a = R_b$ if and only if $a^{-1}b \in K$.

An action $G \times V \to V$ on a real (or complex) vector space V is called a real (or complex) ***representation*** of G if the left translations are linear maps. After choice of a basis, a representation amounts to a continuous homomorphism from G to $\mathrm{GL}_n(\mathbb{R})$ or $\mathrm{GL}_n(\mathbb{C})$. A homomorphism $G \to \mathrm{O}(n)$ or $G \to \mathrm{U}(n)$ is called an ***orthogonal*** or ***unitary representation***. Geometrically, an orthogonal representation is given by an action $G \times V \to V$ with an ***invariant scalar product*** $\langle -,- \rangle$. The latter means $\langle gv, gw \rangle = \langle v, w \rangle$ for $g \in G$ and $v, w \in V$. In an orthogonal representation, the unit sphere $S(V) = \{v \in V \mid \langle v, v \rangle = 1\}$ is G-stable.

Let E be a right G-space and F a left G-space. We denote by $E \times_G F$ the orbit space of the G-action $G \times (E \times F) \to E \times F$, $(g,(x,y)) \mapsto (xg^{-1}, gy)$. A

G-map $f\colon F_1 \to F_2$ induces a continuous map

$$\mathrm{id} \times_G f\colon E \times_G F_1 \to E \times_G F_2, \quad (x, y) \mapsto (x, f(x)).$$

If E carries a left K-action which commutes with the right G-action (i.e., $k(xg) = (kx)g$), then $E \times_G F$ carries an induced K-action $(k, (x, y)) \mapsto (kx, y)$. This construction can in particular be applied in the case that $E = K$, G a subgroup of K and the G- and K-actions on K are given by right and left multiplication. The assignments $F \mapsto K \times_G F$ and $f \mapsto \mathrm{id} \times_G f$ yield the induction functor $\mathrm{ind}_G^K\colon G\text{-}\mathrm{TOP} \to K\text{-}\mathrm{TOP}$. This functor is left adjoint to the restriction functor $\mathrm{res}_G^K\colon K\text{-}\mathrm{TOP} \to G\text{-}\mathrm{TOP}$ which is obtained by regarding a K-space as a G-space. The natural adjunction

$$\mathrm{TOP}_K(\mathrm{ind}_G^K X, Y) \cong \mathrm{TOP}_G(X, \mathrm{res}_G^K Y)$$

sends a G-map $f\colon X \to Y$ to the K-map $(k, x) \mapsto kf(x)$; in the other direction one restricts a map to $X \cong G \times_G X \subset K \times_G X$. (Here TOP_K denotes the set of K-equivariant maps.)

(1.8.6) Theorem. *Suppose the Hausdorff group G is locally compact with countable basis. Let X be a locally compact Hausdorff space and $G \times X \to X$ a transitive action. Then for each $x \in X$ the map $b\colon G \to X$, $g \mapsto gx$ is open and the induced map $\bar{b}\colon G/G_x \to X$ a homeomorphism.*

Proof. If b is open, then $\bar{b}$ is a homeomorphism. Let W be a neighbourhood of e, $(B_i \mid i \in \mathbb{N})$ a countable basis, and $g_i^{-1} \in B_i$. For each $g \in G$ there exists a j such that $B_j \subset Wg^{-1}$, $g \in g_j W$. Therefore the $g_j W$ cover the group.

Let $V \subset G$ be open and $g \in V$. There exists a compact neighbourhood W of e such that $W = W^{-1}$ and $gW^2 \subset V$. Since G is the union of the $g_j W$ and the action is transitive, $X = \bigcup g_j Wx$. Since W is compact and b continuous, $g_j Wx$ is compact and hence closed in X. By (1.4.8), there exist j such that $g_j Wx$ contains an interior point, and therefore Wx contains an interior point wx. Then x is an interior point of $w^{-1}Wx \subset W^2 x$ and hence $gx = p(g)$ an interior point of $gW^2 x \subset Vx = p(V)$. This shows that p is open. □

(1.8.7) Corollary. *Let the locally compact Hausdorff group G with countable basis act on a locally compact Hausdorff space X. An orbit is locally compact if and only if it is locally closed. An orbit is a homogeneous space with respect to the isotropy group of each of its points if and only if it is locally closed.* □

Problems

1. Let H be a normal subgroup of G and X a G-space. Restricting the group action to H, we obtain an H-space X. The orbit space $H\backslash X$ carries then an induced G/H-action.

2. Let a pushout in TOP be given with G-spaces A, B, X and G-maps j, f:

$$\begin{array}{ccc} A & \xrightarrow{f} & B \\ \downarrow{\scriptstyle j} & & \downarrow{\scriptstyle J} \\ X & \xrightarrow{F} & Y. \end{array}$$

Let G be locally compact. Then there exists a unique G action on Y such that F, J become G-maps. The diagram is then a pushout in G-TOP. Hint: (2.4.6)
3. Let Y be a K-space and G a subgroupp of K. Then $K \times_G Y \to K/G \times Y$, $(k, y) \mapsto (kG, ky)$ is a K-homeomorphism. If X is a G-space, then

$$K \times_G (X \times Y) \to (K \times_G X) \times Y, \quad (g, (x, y)) \mapsto ((g, x), gy)$$

is a K-homeomorphism.
4. Let H be a closed subgroup of G. Then G/H is a Hausdorff space and therefore $F = G/H^H$ is closed. The relation $gH \in G/H^H$ is equivalent to $g^{-1}Hg \subset H$. Hence $\{g \in G \mid g^{-1}Hg \subset H\}$ is closed in G. The normalizer $NH = \{g \in G \mid g^{-1}Hg = H\}$ of H in G is closed in G. The group H is a normal subgroup of NH and $NH/H = W_G H = WH$ is the ***Weyl group*** of H in G. The group NH always acts on the fixed set X^H, by restricting the given G-action to NH. The action

$$G/H \times WH \to G/H, \quad (gH, nH) \mapsto gnH$$

is a free right action by G-automorphisms of G/H.

1.9 Projective Spaces. Grassmann Manifolds

Let $P(\mathbb{R}^{n+1}) = \mathbb{R}P^n$ be the set of one-dimensional subspaces of the vector space $\mathbb{R}^{n+1}$. A one-dimensional subspace of V is spanned by $x \in V \smallsetminus 0$. The vectors x and y span the same subspace if and only if $x = \lambda y$ for some $\lambda \in \mathbb{R}^* = \mathbb{R} \smallsetminus 0$. We therefore consider $P(\mathbb{R}^{n+1})$ as the orbit space of the action

$$\mathbb{R}^* \times (\mathbb{R}^{n+1} \smallsetminus 0) \to \mathbb{R}^{n+1} \smallsetminus 0, \quad (\lambda, x) \mapsto \lambda x.$$

The quotient map $p\colon \mathbb{R}^{n+1} \smallsetminus 0 \to P(\mathbb{R}^{n+1})$ provides $P(\mathbb{R}^{n+1})$ with the quotient topology. The space $\mathbb{R}P^n$ is the n-dimensional ***real projective space***. We set $p(x_0, \dots, x_n) = [x_0, \dots, x_n]$ and call $x_0, \dots, x_n$ the ***homogeneous coordinates*** of the point $[x_0, \dots, x_n]$.

In a similar manner we consider the set $P(\mathbb{C}^{n+1}) = \mathbb{C}P^n$ of one-dimensional subspaces of $\mathbb{C}^{n+1}$ as the orbit space of the action

$$\mathbb{C}^* \times (\mathbb{C}^{n+1} \smallsetminus 0) \to \mathbb{C}^{n+1} \smallsetminus 0, \quad (\lambda, z) \mapsto \lambda z.$$

We have again a quotient map $p\colon \mathbb{C}^{n+1} \smallsetminus 0 \to P(\mathbb{C}^{n+1})$. The space $\mathbb{C}P^n$ is called the n-dimensional ***complex projective space***. (It is $2n$-dimensional as a manifold.)

We describe the projective spaces in a different manner as orbit spaces. The subgroup $G = \{\pm 1\} \subset \mathbb{R}^*$ acts on $S^n \subset \mathbb{R}^{n+1}$ by $(\lambda, x) \mapsto \lambda x$, called the ***antipodal involution***. The inclusion $i: S^n \to \mathbb{R}^{n+1}$ induces a continuous bijective map $\iota: S^n/G \to (\mathbb{R}^{n+1} \setminus 0)/\mathbb{R}^*$. The map $j: \mathbb{R}^{n+1} \setminus 0 \to S^n$, $x \mapsto \|x\|^{-1}x$ induces an inverse. The quotient S^n/G is compact, since S^n is compact. By (1.4.4), the quotient is a Hausdorff space. In a similar manner one treats $\mathbb{C}P^n$, but now with respect to the action $S^1 \times S^{2n+1} \to S^{2n+1}$, $(\lambda, z) \mapsto \lambda z$ of S^1 on the unit sphere $S^{2n+1} \subset \mathbb{C}^{n+1} \setminus 0$.

Projective spaces are homogeneous spaces. Consider the action of $\mathrm{O}(n+1)$ on $\mathbb{R}^{n+1}$ by matrix multiplication. If $V \in P(\mathbb{R}^{n+1})$ is a one-dimensional space and $A \in \mathrm{O}(n+1)$, then $AV \in P(\mathbb{R}^{n+1})$. We obtain an induced action

$$\mathrm{O}(n+1) \times P(\mathbb{R}^{n+1}) \to P(\mathbb{R}^{n+1}).$$

This action is transitive. The isotropy group of $[1, 0, \dots, 0]$ consists of the matrices

$$\begin{pmatrix} \lambda & 0 \\ 0 & B \end{pmatrix}, \quad \lambda \in \mathrm{O}(1),\ B \in \mathrm{O}(n).$$

We consider these matrices as the subgroup $\mathrm{O}(1) \times \mathrm{O}(n)$ of $\mathrm{O}(n+1)$. The assignment $A \mapsto Ae_1$ induces an $\mathrm{O}(n+1)$-equivariant homeomorphism

$$b: \mathrm{O}(n+1)/(\mathrm{O}(1) \times \mathrm{O}(n)) \cong P(\mathbb{R}^{n+1}).$$

The action of $\mathrm{O}(n+1)$ on $P(\mathbb{R}^{n+1})$ is continuous; this follows easily from the continuity of the action $\mathrm{O}(n+1) \times (\mathbb{R}^{n+1} \setminus 0) \to \mathbb{R}^{n+1} \setminus 0$ and the definition of the quotient topology. Therefore b is a bijective continuous map of a compact space into a Hausdorff space. In a similar manner we obtain a $\mathrm{U}(n+1)$-equivariant homeomorphism $\mathrm{U}(n+1)/(\mathrm{U}(1) \times \mathrm{U}(n)) \cong P(\mathbb{C}^{n+1})$.

Finally, one can define the ***quaternionic projective space*** $\mathbb{H}P^n$ in a similar manner as a quotient of $\mathbb{H}^{n+1} \setminus 0$ or as a quotient of S^{4n+3}.

We generalize projective spaces. Let W be an n-dimensional real vector space. We denote by $G_k(W)$ the set of k-dimensional subspaces of W. We define a topology on $G_k(W)$. Suppose W carries an inner product. Let $V_k(W)$ denote the set of orthonormal sequences $(w_1, \dots, w_k)$ in W considered as a subspace of W^k. We call $V_k(W)$ the ***Stiefel manifold*** of orthonormal k-frames in W. We have a projection $p: V_k(W) \to G_k(W)$ which sends $(w_1, \dots, w_k)$ to the subspace $[w_1, \dots, w_k]$ spanned by this sequence. We give $G_k(W)$ the quotient topology determined by p. The space $G_k(W)$ can be obtained as a homogeneous space. Let $W = \mathbb{R}^n$ with standard inner product and standard basis $e_1, \dots, e_n$. We have a continuous action of $\mathrm{O}(n)$ on $V_k(\mathbb{R}^n)$ and $G_k(\mathbb{R}^n)$ defined by $(A, (v_1, \dots, v_k)) \mapsto (Av_1, \dots, Av_k)$ and such that p becomes $\mathrm{O}(n)$-equivariant. The isotropy groups of $(e_1, \dots, e_k)$ and $[e_1, \dots, e_k]$ consist of the matrices

$$\begin{pmatrix} E_k & 0 \\ 0 & B \end{pmatrix}, \quad \begin{pmatrix} A & 0 \\ 0 & B \end{pmatrix}, \quad A \in \mathrm{O}(k),\ B \in \mathrm{O}(n-k)$$

respectively. The map $A \mapsto (Ae_1, \dots, Ae_k)$ induces equivariant homeomorphisms in the diagram

$$\begin{array}{ccc} O(n)/O(n-k) & \xrightarrow{\cong} & V_k(\mathbb{R}^n) \\ \downarrow & & \downarrow \\ O(n)/(O(k) \times O(n-k)) & \xrightarrow{\cong} & G_k(\mathbb{R}^n). \end{array}$$

This shows that $G_k(\mathbb{R}^n)$ is a compact Hausdorff space. It is called the ***Grassmann manifold*** of k-dimensional subspaces of $\mathbb{R}^n$.

In a similar manner we can work with the k-dimensional complex subspaces of $\mathbb{C}^n$ and obtain an analogous diagram of $U(n)$-spaces (complex Stiefel and Grassmann manifolds):

$$\begin{array}{ccc} U(n)/U(n-k) & \xrightarrow{\cong} & V_k(\mathbb{C}^n) \\ \downarrow & & \downarrow \\ U(n)/(U(k) \times U(n-k)) & \xrightarrow{\cong} & G_k(\mathbb{C}^n). \end{array}$$

Chapter 2

The Fundamental Group

In this chapter we introduce the homotopy notion and the first of a series of algebraic invariants associated to a topological space: the fundamental group.

Almost every topic of algebraic topology uses the homotopy notion. Therefore it is necessary to begin with this notion. A homotopy is a continuous family $h_t : X \to Y$ of continuous maps which depends on a real parameter $t \in [0, 1]$. (One may interpret this as a "time-dependent" process.) The maps f_0 and f_1 are then called homotopic, and being homotopic is an equivalence relation on the set of continuous maps $X \to Y$. This equivalence relation leads to a quotient category of the category TOP of topological spaces and continuous maps, the homotopy category h-TOP. The importance of this notion is seen from several facts.

(1) The classical tools of algebraic topology are functors from a category of spaces to an algebraic category, say of abelian groups. These functors turn out to be homotopy invariant, i.e., homotopic maps have the same value under the functor.

(2) One can change maps by homotopies and spaces by homotopy equivalences. This fact allows for a great flexibility. But still global geometric information is retained. Basic principles of topology are deformation and approximation. One idea of deformation is made precise by the notion of homotopy. Continuity is an ungeometric notion. So often one has to deform a continuous map into a map with better properties.

(3) The homotopy notion leads in an almost tautological way to algebraic structures and categorical structures. In this chapter we learn about the simplest example, the fundamental group and the fundamental groupoid.

The passage to the homotopy category is not always a suitable view-point. In general it is better to stay in the category TOP of topological spaces and continuous maps ("space level" as opposed to "homotopy level"). We thus consider homotopy as an additional structure. Then classical concepts can be generalized by using the homotopy notion. For instance one considers diagrams which are only commutative up to homotopy and the homotopies involved will be treated as additional information. One can also define generalized group objects where multiplication is only associative up to homotopy. And so on.

The passage from TOP to h-TOP may be interpreted as a passage from "continuous mathematics" to "discrete mathematics".

The homotopy notion allows us to apply algebraic concepts to continuous maps. It is not very sensible to talk about the kernel or cokernel of a continuous map.

But we will see later that there exist notions of "homotopy-kernels" (then called homotopy fibres) and "homotopy-cokernels" (then called homotopy cofibres). This is the more modern view-point of a large variety of homotopy constructions. In general terms: The idea is to replace the categorical notions limit and colimit by appropriate homotopy notions.

The prototype of a functor from spaces to groups is the fundamental group functor. Historically it is the first of such functors. It was introduced by Poincaré, in different context and terminology. In general it is difficult to determine the fundamental group of a space. Usually one builds up a space from simpler pieces and then one studies the interrelation between the groups of the pieces. This uses the functorial aspect and asks for formal properties of the functor. We prove the basic theorem of Seifert and van Kampen which roughly says that the functor transforms suitable pushouts of spaces into pushouts of groups. This may not be the type of algebra the reader is used to, and it can in fact be quite complicated. We describe some related algebra (presentation of groups by generators and relations) and discuss a number of geometric results which seem plausible from our intuition but which cannot be proved (in a systematic way) without algebraic topology. The results are of the type that two given spaces are not homeomorphic – and this follows, if their fundamental groups are different. Finally we show that each group can be realized as a fundamental group (this is the origin of the idea to apply topology to group theory).

The study of the fundamental group can be continued with the covering space theory where the fundamental group is exhibited as a symmetry group. This symmetry influences almost every other tool of algebraic topology (although we do not always carry out this influence in this text).

The chapter contains two sections on point-set topology. We discuss standard spaces like spheres, disks, cells, simplices; they will be used in many different contexts. We present the compact-open topology on spaces of continuous maps; they will be used for the dual definition of homotopy as a continuous family of paths, and this duality will henceforth be applied to many homotopy constructions and notions.

2.1 The Notion of Homotopy

A ***path*** in a topological space X from x to y is a continuous map $u\colon [a,b] \to X$ such that $u(a) = x$ and $u(b) = y$. We say that the path connects the points $u(a)$ and $u(b)$. We can reparametrize and use the unit interval as a source $[0,1] \to X$, $t \mapsto u((1-t)a + tb)$. In the general theory we mostly use the unit interval. If $u\colon [0,1] \to X$ is a path from x to y, then the ***inverse path*** $u^-\colon t \mapsto w(1-t)$ is a path from y to x. If $v\colon [0,1] \to X$ is another path from y to z, then the ***product path*** $u * v$, defined by $t \mapsto u(2t)$ for $t \le 1/2$ and $v(2t-1)$ for $t \ge 1/2$, is a path from x to z. We also have the ***constant path*** k_x with value x.

From these remarks we see that being connectible by paths is an equivalence relation on X. An equivalence class is called a ***path component*** of X. We denote by $\pi_0(X)$ the set of path components and by $[x]$ the path component of the point x. A space X is said to be ***path connected*** or ***0-connected*** if it has one of the following equivalent properties:

(1) $\pi_0(X)$ consists of a single element.
(2) Any two points can be joined by a path.
(3) Any continuous map $f\colon \partial I = \{0,1\} \to X$ has a continuous extension $F\colon I \to X$.

(Later we study the higher dimensional analogous problem of extending maps from the boundary ∂I^n of the n-dimensional cube to I^n.)

A map $f\colon X \to Y$ induces $\pi_0(f)\colon \pi_0(X) \to \pi_0(Y)$, $[x] \mapsto [f(x)]$. In this way π_0 becomes a functor from the category TOP of topological spaces to the category SET of sets[1]. We will see that this functor is the beginning of algebraic topology, although there is no algebra yet.

Thinking in terms of categories and functors is a basic method in (algebraic) topology. The size of $\pi_0(X)$ is a topological property of the space X. A functor transports isomorphisms to isomorphisms. Thus a homeomorphism f induces a bijection $\pi_0(f)$. Suppose $f\colon X \to Y$ is a homeomorphism; then f induces a homeomorphism $X \smallsetminus A \to Y \smallsetminus f(A)$ for each subset $A \subset X$. Suppose $f\colon \mathbb{R} \to \mathbb{R}^n$ is a homeomorphism; the space $\mathbb{R} \smallsetminus x$ has two path components (intermediate value theorem of calculus), and $\mathbb{R}^n \smallsetminus y$ is path connected for $n > 1$; we apply the functor π_0 and conclude that $\mathbb{R}$ is not homeomorphic to $\mathbb{R}^n$ for $n > 1$. This example seems almost trivial, but the reasoning is typical. Here is another simple example of this type: The subspace $X = \mathbb{R} \times 0 \cup 0 \times \mathbb{R}$ of $\mathbb{R}^2$ is not homeomorphic to $\mathbb{R}$ since X contains a point $x = (0,0)$ such that $\pi_0(X \smallsetminus x)$ has four elements whereas $\pi_0(\mathbb{R} \smallsetminus y)$ has always two elements.

2.1.1 Path categories. Forming the product path is not an associative composition. We can remedy this defect by using parameter intervals of different length. So let us consider paths of the form $u\colon [0,a] \to X$, $v\colon [0,b] \to X$ with $u(a) = v(0)$ and $a, b \geq 0$. Their composition $v \circ u = w$ is the path $[0, a+b] \to X$ with $w(t) = u(t)$ for $0 \leq t \leq a$ and $w(t) = v(a-t)$ for $a \leq t \leq a+b$. In this manner we obtain a category $W(X)$: Objects are the points of X; a morphism from x to y is a path $u\colon [0,a] \to X$ with $u(0) = x, u(a) = y$ for some $a \geq 0$; and composition of morphisms is as defined before; the path $[0,0] \to X$ with value x is the identity of the object x. A continuous map $f\colon X \to Y$ induces a functor $W(f)\colon W(X) \to W(Y)$, $x \mapsto f(x)$, $u \mapsto fu$. ◇

[1] Our general conventions: space = topological space, map = continuous map. A set map between spaces is a map which is not assumed to be continuous at the outset.

A space is ***connected*** if it is not the topological sum of two non-empty subspaces. Thus X is disconnected if and only if X contains a subset X which is open, closed, and different from $\emptyset$ and X. A ***decomposition*** of X is a pair U, V of open, non-empty, disjoint subsets with union X. A space X is disconnected if and only if there exists a continuous surjective map $f: X \to \{0, 1\}$; a decomposition is given by $U = f^{-1}(0)$, $V = f^{-1}(1)$. The continuous image of a connected space is connected. Recall from calculus: $A \subset \mathbb{R}$ is connected if and only if A is an interval. (An interval is a subset which contains with x, y also $[x, y]$.)

(2.1.2) Proposition. *Let $(A_j \mid j \in J)$ be a family of connected subsets of X such that $A_i \cap A_j \neq \emptyset$ for all i, j. Then $\bigcup_j A_j = Y$ is connected. Let A be connected and $A \subset B \subset \bar{A}$. Then B is connected.* □

The union of the connected sets in X which contain x is thus a closed connected subset. We call it the ***component*** $X(x)$ of x in X. If $y \in X(x)$, then $X(y) = X(x)$. A component of X is a maximal connected subset. Any space is the disjoint union of its components. A space is ***totally disconnected*** if its components consist of single points. Since intervals are connected a path connected space is connected.

A product $\prod_j X_j$ is connected if each X_j is connected. The component of $(x_j) \in \prod_j X_j$ is the product of the components of the x_j.

(2.1.3) Example. The space

$$X = [-1, 0] \times 0 \cup 0 \times [-1, 1] \cup \{(x, \sin(\pi x^{-1}) \mid 0 < x \leq 1\}$$

is connected but not path connected. The union S of X with $\{\pm 1\} \times [-2, 0] \cup [-2, 2] \times \{-2\}$ is called the ***pseudo-circle***. The complement $\mathbb{R}^2 \smallsetminus S$ has two path components.

A pseudo-circle S has a sequence $K_1 \supset K_2 \supset \cdots$ of compact neighbourhoods with $\bigcap_i K_i = S$ and K_i homeomorphic to $S^1 \times [0, 1]$. ◇

Let X and Y be topological spaces and $f, g: X \to Y$ continuous maps. A ***homotopy*** from f to g is a continuous map

$$H: X \times [0, 1] \to Y, \quad (x, t) \mapsto H(x, t) = H_t(x)$$

such that $f(x) = H(x, 0)$ and $g(x) = H(x, 1)$ for $x \in X$, i.e., $f = H_0$ and $g = H_1$. We denote this situation by $H: f \simeq g$. One can consider a homotopy as a dynamical process, the parameter t is the time and H_t is a time-dependent family of maps. One also says that f is ***deformed*** continuously into g. Another (dual) view-point of a homotopy is: a parametrized family of paths. We use the letter I for the unit interval $[0, 1]$. If we write a homotopy in the form H_t, we understand that $H: X \times I \to Y$, $(x, t) \mapsto H_t(x)$ is continuous in both variables simultaneously. We call f and g ***homotopic*** if there exists a homotopy from f to g. (One can, of course, define homotopies with $[0,1] \times X$. While this does not affect the theory, it does make a difference when orientations play a role.)

The homotopy relation $\simeq$ is an equivalence relation on the set of continuous maps $X \to Y$. Given $H: f \simeq g$, the ***inverse homotopy*** $H^-: (x,t) \mapsto H(x,1-t)$ shows $g \simeq f$. Let $K: f \simeq g$ and $L: g \simeq h$ be given. The ***product homotopy*** $K * L$ is defined by

$$(K * L)(x,t) = \begin{cases} K(x,2t), & 0 \le t \le \frac{1}{2}, \\ L(x,2t-1), & \frac{1}{2} \le t \le 1, \end{cases}$$

and shows $f \simeq h$. The ***constant homotopy*** $H(x,t) = f(x)$ shows $f \simeq f$.

The equivalence class of f is denoted $[f]$ and called the ***homotopy class*** of f. We denote by $[X,Y]$ the set of homotopy classes $[f]$ of maps $f: X \to Y$. A homotopy $H_t: X \to Y$ is said to be ***relative*** to $A \subset X$ if the restriction $H_t|A$ does not depend on t (is constant on A). We use the notation $H: f \simeq g$ (rel A) in this case.

The homotopy relation is compatible with the composition of maps: Let $H: f \simeq g: X \to Y$ and $G: k \simeq l: Y \to Z$ be given; then

$$(x,t) \mapsto G(H(x,t),t) = G_t H_t(x)$$

is a homotopy from kf to lg. We see that topological spaces and homotopy classes of maps form a quotient category of TOP, the ***homotopy category*** h-TOP, when composition of homotopy classes is induced by composition of representing maps. If $f: X \to Y$ represents an isomorphism in h-TOP, then f is called a ***homotopy equivalence*** or h-equivalence. In explicit terms this means: $f: X \to Y$ is a homotopy equivalence if there exists $g: Y \to X$, a ***homotopy inverse*** of f, such that gf and fg are both homotopic to the identity. Spaces X and Y are ***homotopy equivalent*** or of the same ***homotopy type*** if there exists a homotopy equivalence $X \to Y$. A space is ***contractible*** if it is homotopy equivalent to a point. A map $f: X \to Y$ is ***null homotopic*** if it is homotopic to a constant map; a ***null homotopy*** of f is a homotopy between f and a constant map. A null homotopy of the identity $\mathrm{id}(X)$ is a ***contraction*** of the space X.

2.1.4 Categories of homotopies. We generalize (2.1.1) and define a category $W(X,Y)$. The objects are the continuous maps $f: X \to Y$. A morphism from f to g is a homotopy $H: X \times [0,a] \to Y$ with $H_0 = f$ and $H_a = g$. Composition is defined as in (2.1.1). ◇

As in any category we also have the Hom-functors in h-TOP. Given $f: X \to Y$, we use the notation

$$f_*: [Z,X] \to [Z,Y],\ g \mapsto fg, \quad f^*: [Y,Z] \to [X,Z],\ h \mapsto hf$$

for the induced maps[2]. The reader should recall a little reasoning with Hom-functors, as follows. The map f is an h-equivalence, i.e., an isomorphism in h-TOP if and only if f_* is always bijective; similarly for f^*. If $f: X \to Y$ has a right homotopy inverse $h: Y \to X$, i.e., $fh \simeq \mathrm{id}$, and a left homotopy inverse $g: Y \to X$, i.e., $gf \simeq \mathrm{id}$, then f is an h-equivalence. If two of the maps $f: X \to Y$, $g: Y \to Z$, and gf are h-equivalences, then so is the third.

Homotopy is compatible with sums and products. Let $p_i: \prod_{j \in J} X_j \to X_i$ be the projection onto the i-th factor. Then

$$[Y, \textstyle\prod_{j \in J} X_j] \to \prod_{j \in J} [Y, X_j], \quad [f] \mapsto ([p_i \circ f])$$

is a well-defined bijection. Let $i_k: X_k \to \coprod_{j \in J} X_j$ be the canonical inclusion of the k-th summand. Then

$$[\textstyle\coprod_{j \in J} X_j, Y] \to \prod_{j \in J} [X_j, Y], \quad [f] \mapsto ([f \circ i_k])$$

is a well-defined bijection. In other words: sum and product in TOP also represent sum and product in h-TOP. (Problems arise when it comes to pullbacks and pushouts.)

Let P be a point. A map $P \to Y$ can be identified with its image and a homotopy $P \times I \to Y$ can be identified with a path. The Hom-functor $[P, -]$ is therefore essentially the same thing as the functor π_0.

2.1.5 Linear homotopy. Given maps $f, g: X \to A$, $A \subset \mathbb{R}^n$. Suppose that the line-segment from $f(x)$ to $g(x)$ is always contained in A. Then $H(x,t) = (1-t)f(x) + tg(x)$ is a homotopy from f to g (linear homotopy). It will turn out that many homotopies are constructed from linear homotopies.

A set $A \subset \mathbb{R}^n$ is ***star-shaped*** with respect to $a_0 \in A$ if for each $a \in A$ the line-segment from a_0 to a is contained in A. If A is star-shaped, then $H(a,t) = (1-t)a + ta_0$ is a null homotopy of the identity. Hence star-shaped sets are contractible. A set $C \subset \mathbb{R}^n$ is ***convex*** if and only if it is star-shaped with respect to each of its points.

Note: If $A = \mathbb{R}^n$ and $a_0 = 0$, then each H_t, $t < 1$, is a homeomorphism, and only in the very last moment is H_1 constant! This is less mysterious, if we look at the paths $t \mapsto H(x,t)$. ◇

The reader should now recall the notion of a quotient map (identification), its universal property, and the fact that the product of a quotient map by the identity of a locally compact space is again a quotient map (see (2.4.6)).

(2.1.6) Proposition. *Let $p: X \to Y$ be a quotient map. Suppose $H_t: Y \to Z$ is a family of set maps such that $H_t \circ p$ is a homotopy. Then H_t is a homotopy.*

[2]As a general principle we use a lower index for covariant functors and an upper index for contravariant functors. If we apply a (covariant) functor to a morphism f we often call the result the ***induced morphism*** and denote it simply by f_* if the functor is clear from the context.

Proof. The product $p \times \mathrm{id}\colon X \times I \to Y \times I$ is an identification, since I is compact. The composition $H \circ (p \times \mathrm{id})$ is continuous and therefore H is continuous. □

(2.1.7) Proposition. *Let $H_t\colon X \to X$ be a homotopy of the identity $H_0 = \mathrm{id}(X)$ such that the subspace $\emptyset \neq A \subset X$ is always mapped into itself, $H_t(A) \subset A$. Suppose H_1 is constant on A. Then the projection $p\colon X \to X/A$ (A identified to a point) is an h-equivalence.*

Proof. Since $H_1(A)$ is a point, there exists a map $q\colon X/A \to X$ such that $qp = H_1$. By assumption, this composition is homotopic to the identity. The map $p \circ H_t\colon X \to X/A$ factorizes over p and yields $K_t\colon X/A \to X/A$ such that $K_t p = pH_t$. By (2.1.6), K_t is a homotopy, $K_0 = \mathrm{id}$ and $K_1 = pq$. □

Problems

1. Suppose there exists a homeomorphism $\mathbb{R} \to X \times Y$. Then X or Y is a point.
2. Let $f\colon X \to Y$ be surjective. If X is (path) connected, then Y is (path) connected.
3. Let C be a countable subset of $\mathbb{R}^n$, $n \geq 2$. Show that $\mathbb{R}^n \smallsetminus C$ is path connected.
4. The unitary group $\mathrm{U}(n)$ and the general linear group $\mathrm{GL}_n(\mathbb{C})$ are path connected. The orthogonal group $\mathrm{O}(n)$ and the general linear group $\mathrm{GL}_n(\mathbb{R})$ have two path components; one of them consists of matrices with positive determinant.
5. Let $U \subset \mathbb{R}^n$ be open. The path components of U are open and coincide with the components. The set of path components is finite or countably infinite. An open subset of $\mathbb{R}$ is a disjoint union of open intervals.
6. List theorems of point-set topology which show that the product homotopy and the inverse homotopy are continuous. Do the same for the linear homotopy in 2.1.5.
7. A space X is contractible if and only if the identity $\mathrm{id}(X)$ is null homotopic.
8. gf is null homotopic, if f or g is null homotopic.
9. Let A be contractible. Then any two maps $X \to A$ are homotopic.
10. The inclusions $\mathrm{O}(n) \subset \mathrm{GL}_n(\mathbb{R})$ and $\mathrm{U}(n) \subset \mathrm{GL}_n(\mathbb{C})$ are homotopy equivalences. Let $P(n)$ denote the space of positive definite real (n,n)-matrices. Then $\mathrm{O}(n) \times P(n) \to \mathrm{GL}_n(\mathbb{R})$, $(X, P) \mapsto XP$ is a homeomorphism; $P(n)$ is star-like with respect to the unit matrix.
11. There exist contractible and non-contractible spaces consisting of two points.

2.2 Further Homotopy Notions

The homotopy notion can be adapted to a variety of other contexts and categories: Consider homotopies which preserve some additional structure of a category. We describe some examples from which the general idea emerges. This section only contains terminology.

The construction of group structures on homotopy sets uses the category of pointed spaces, as we will see shortly. We call a pair (X, x_0) consisting of a space

X and a ***base point*** $x_0 \in X$ a ***pointed space***. A ***pointed map*** $f: (X, x_0) \to (Y, y_0)$ is a continuous map $f: X \to Y$ which sends the base point to the base point. A homotopy $H: X \times I \to Y$ is ***pointed*** if H_t is pointed for each $t \in I$. We denote by $[X, Y]^0$ the set of pointed homotopy classes (fixed base points assumed) or by $[(X, x_0), (Y, y_0)]$. We obtain related notions: ***pointed homotopy equivalence, pointed contractible, pointed null homotopy***. We denote the category of pointed spaces and pointed maps by TOP^0, and by h-TOP^0 the associated homotopy category. Often a base point will just be denoted by $*$. Also a set with a single element will be denoted by its element. The choice of a base point is an additional structure. There is a functor α from TOP to TOP^0 which sends a space X to $X^+ = X + \{*\}$, i.e., to X with an additional base point added (topological sum), with the obvious extension to pointed maps. This functor is compatible with homotopies. We also have the forgetful functor β from TOP^0 to TOP. They are adjoint $\mathrm{TOP}^0(\alpha(X), Y) \cong \mathrm{TOP}(X, \beta Y)$, and similarly for the homotopy categories.

The category TOP^0 has sums and products. Suppose (X_j, x_j) is a family of pointed spaces. The family (x_j) of base points is taken as base point in the product $\prod_j X_j$; this yields the ***pointed product***. Let $\bigvee_{j \in J} X_j$ be the quotient of $\coprod_{j \in J} X_j$ where all base points are identified to a single new base point. We have canonical pointed maps $i_k: X_k \to \bigvee_j X_j$ which arise from the canonical inclusions $X_k \to \coprod_j X_j$. The ***wedge***, also called the ***bouquet***, $\bigvee_j X_j$ of the pointed spaces X_j together with the i_k is the ***pointed sum*** in TOP^0.

The sum and the product in TOP^0 also represent the sum and the product in h-TOP^0 (use (2.1.6)).

Let (A, a) and (B, b) be pointed spaces. Their ***smash product*** is

$$A \wedge B = A \times B / A \times b \cup a \times B = A \times B / A \vee B.$$

(This is not a categorical product. It is rather analogous to the tensor product.) The smash product is a functor in two variables and also compatible with homotopies: Given $f: A \to C, g: B \to D$ we have the induced map

$$f \wedge g: A \wedge B \to C \wedge D, \quad (a, b) \mapsto (f(a), g(b)),$$

and homotopies f_t, g_t induce a homotopy $f_t \wedge g_t$. Unfortunately, there are point-set topological problems with the associativity of the smash product (see Problem 14).

A ***pair (X, A) of topological spaces*** consists of a space X and a subspace A. A morphism $f: (X, A) \to (Y, B)$ between pairs is a map $f: X \to Y$ such that $f(A) \subset B$. In this way we obtain the category of pairs TOP(2). A homotopy H in this category is assumed to have each H_t a morphism of pairs. We write $[(X, A), (Y, B)]$ for the associated homotopy sets and h-TOP(2) for the homotopy category.

If (X, A) is a pair, we usually consider the quotient space X/A as a pointed space (A identified to a point) with base point $\{A\}$. If $A = \emptyset$, then $X/A = X^+$ is X with a separate base point.

(2.2.1) Note. *A continuous map* $f: (X, A) \to (Y, *)$ *into a pointed space induces a pointed map* $\bar{f}: X/A \to Y$. *The assignment* $f \mapsto \bar{f}$ *induces a bijection* $[(X, A), (Y, *)] \cong [X/A, Y]^0$. *A verification uses* (2.1.6). □

We use the notation

$$(X, A) \times (Y, B) = (X \times Y, X \times B \cup A \times Y),$$

although this is not a categorical product. With this notation $(I^m, \partial I^m) \times (I^n, \partial I^n) = (I^{m+n}, \partial I^{m+n})$. In a similar manner we treat other configurations, e.g., ***triples*** (X, A, B) of spaces $A \subset B \subset X$ and the category TOP(3) of triples.

Let K and B be fixed spaces. The category TOP^K of ***spaces under*** K has as objects the maps $i: K \to X$. A morphism from $i: K \to X$ to $j: K \to Y$ is a map $f: X \to Y$ such that $fi = j$. The category TOP_B of ***spaces over*** B has as objects the maps $p: X \to B$. A morphism from $p: X \to B$ to $q: Y \to B$ is a map $f: X \to Y$ such that $qf = p$. If B is a point, then TOP_B can be identified with TOP, since each space has a unique map to a point. If $K = \{*\}$ is a point, then TOP^K is the same as TOP^0. If $p: X \to B$ is given, then $p^{-1}(b)$ is called the ***fibre*** of p over b; in this context, B is the ***base space*** and X the ***total space*** of p. A map in TOP_B will also be called ***fibrewise*** or ***fibre preserving***.

Categories like TOP^K or TOP_B have an associated notion of homotopy. A homotopy H_t is in TOP^K if each H_t is a morphism in this category. A similar definition is used for TOP_B. A homotopy in TOP_B will also be called fibrewise or fibre preserving. Again, being homotopic is an equivalence relation in these categories. We denote by $[X, Y]^K$ the set of homotopy classes in TOP^K, and by $[X, Y]_B$ the set of homotopy classes in TOP_B. The homotopy categories are h-TOP^K and h-TOP_B. Note that a homotopy equivalence in TOP_B, i.e., a fibrewise homotopy equivalence, from $p: X \to B$ to $q: Y \to B$ induces for each $b \in B$ a homotopy equivalence $p^{-1}(b) \to q^{-1}(b)$ between the fibres over b, so this is a continuous family of ordinary homotopy equivalences, parametrized by B ([96], [97], [38], [128]).

A morphism r from $i: K \to X$ to $\mathrm{id}: K \to K$ in TOP^K is a map $r: X \to K$ such that $ri = \mathrm{id}(X)$. It is called a ***retraction*** of i. If it exists, then i is an embedding. If $i: K \subset X$ we then call K a ***retract*** of X. The retraction r of $i: K \subset X$ is a homotopy equivalence in TOP^K if and only if there exists a homotopy $h_t: X \to X$ relative to K such that $h_0 = \mathrm{id}$ and $h_1 = ir$. In this case we call K a ***deformation retract*** of X. The inclusion $S^n \subset \mathbb{R}^{n+1} \smallsetminus 0$ is a deformation retract.

A morphism s from $\mathrm{id}: B \to B$ to $p: E \to B$ in TOP_B is a map $s: B \to E$ such that $ps = \mathrm{id}(B)$. It is called a ***section*** of p. If $p: E \to B$ is homotopy equivalent in TOP_B to $\mathrm{id}(B)$ we call p ***shrinkable***. All fibres of a shrinkable map are contractible.

Problems

1. Let $((X_j, x_j) \mid j \in J)$ be a family of pointed spaces. Let $\bigvee'_j X_j$ be the subset of those points $(a_j) \in \prod_j X_j$ where all but one a_j are equal to the base point. There is a canonical bijective continuous map $\bigvee_j X_j \to \bigvee'_j X_j$. If J is finite, then this map is a homeomorphism. If J is infinite and $(X_j, x_j) = (I, 0)$, then it is not a homeomorphism.
2. The canonical maps $i_k\colon X_k \to \bigvee_j X_j$ are embeddings.
3. The ***comb space*** X is defined as $B \times [0,1] \cup [0,1] \times \{1\}$ with $B = \{n^{-1} \mid n \in \mathbb{N}\} \cup \{0\}$. Then X is contractible but not pointed contractible with respect to $(0,0)$. Let $Y = -X$ be another comb space. Then $X \cup Y \cong X \vee Y$ is not contractible. Since $(X \cup Y)/Y$ is homeomorphic to X, we see that it does not suffice in (2.1.7) to assume that A is contractible.

These counterexamples indicate the need for base points with additional (local) properties.
4. There exists a contractible subspace $X \subset \mathbb{R}^2$ which is not pointed contractible to any of its points.
5. Let the homotopy H_t in (2.1.7) be pointed with respect to some base point $a \in A$. Show that $p\colon X \to X/A$ is a pointed h-equivalence. Is (X, A) h-equivalent to $(X, \{a\})$?
6. Show that (2.1.7) yields a homotopy equivalence of pairs $(X, A) \to (X/A, *)$.
7. The inclusion $(I, \partial I) \to (I, I \smallsetminus \{1/2\})$ is not an h-equivalence in TOP(2) although the component maps $I \to I$ and $\partial I \to I \smallsetminus \{1/2\}$ are h-equivalences.
8. Let $E \subset \mathbb{R}^2$ consist of k points. Show, heuristically, that the complement $\mathbb{R}^2 \smallsetminus E$ is h-equivalent to the k-fold sum $\bigvee_1^k S^1$.
9. Remove a point from the torus $S^1 \times S^1$ and show that the result is h-equivalent to $S^1 \vee S^1$. Is there an analogous result when you remove a point from $S^m \times S^n$?
10. Construct an inclusion $A \subset X$ which is a retract and a homotopy equivalence but not a deformation retract.
11. Construct a map $p\colon E \to B$ such that all fibres $p^{-1}(b)$ are contractible but which does not have a section. Construct an h-equivalence $p\colon E \to B$ which has a section but which is not shrinkable.
12. What is the sum of two objects in TOP^K? What is the product of two objects in TOP_B?
13. A pullback of a shrinkable map is shrinkable. A pushout of a deformation retract is a deformation retract.
14. Let Y, Z be compact or X, Z be locally compact. Then the canonical bijection (the identity) $(X \wedge Y) \wedge Z \to X \wedge (Y \wedge Z)$ is a homeomorphism. (In the category of compactly generated spaces (with its associated product and smash product!) the map is always a homeomorphism. See also [155, Satz 18].)
15. Let $\bigvee_j (A_j \wedge B) \to (\bigvee_j A_j) \wedge B$ be the canonical map which is on each summand $A_k \wedge B$ induced by the inclusion $A_k \to \bigvee_j A_j$. Show that this map is a homeomorphism if the index set is finite. Show that in this case both spaces are quotients of $(\amalg_j A_j) \times B \cong \amalg_j (A_j \times B)$.
16. Let A be a compact subset of X and $p\colon X \to X/A$ be the quotient map. Then for each space Y the product $p \times \mathrm{id}(Y)$ is a quotient map. If X is a Hausdorff space, then p is proper and $p \times \mathrm{id}$ closed.
17. The canonical map $X \times I \to X \times I/\partial I \to X \wedge I/\partial I$ is a quotient map which induces a homeomorphism $\Sigma X = X \times I/(X \times \partial I \cup \{*\} \times I) \cong X \wedge I/\partial I$.
18. There is a canonical bijective continuous map $(X \times Y)/(X \times B \cup A \times Y) \to X/A \wedge Y/B$

(the identity on representatives). It is a homeomorphism if $X \times Y \to X/A \wedge Y/B$ is a quotient map, e.g., if X and Y/B are locally compact (or in the category of compactly generated spaces).

2.3 Standard Spaces

Standard spaces are Euclidean spaces, disks, cells, spheres, cubes and simplices. We collect notation and elementary results about such spaces. The material will be used almost everywhere in this book. We begin with a list of spaces. The Euclidean norm is $\|x\|$.

$\mathbb{R}^n$	Euclidean space
$D^n = \{x \in \mathbb{R}^n \mid \|x\| \le 1\}$	n-dimensional disk
$S^{n-1} = \{x \in D^n \mid \|x\| = 1\} = \partial D^n$	$(n-1)$-dimensional sphere
$E^n = D^n \smallsetminus S^{n-1}$	n-dimensional cell
$I^n = \{x \in \mathbb{R}^n \mid 0 \le x_i \le 1\}$	n-dimensional cube
$\partial I^n = \{x \in I^n \mid x_i = 0, 1 \text{ for some } i\}$	boundary of I^n
$\Delta^n = \Delta[n] = \{x \in \mathbb{R}^{n+1} \mid x_i \ge 0, \sum_i x_i = 1\}$	n-dimensional simplex
$\partial \Delta^n = \{(x_i) \in \Delta^n \mid \text{some } x_i = 0\}$	boundary of Δ^n

The spaces D^n, I^n, E^n and Δ^n are convex and hence contractible. We think of $\mathbb{R}^0 = \{0\}$. The spaces D^0, I^0, and Δ^0 are singletons, and $S^{-1} = \partial D^0, \partial \Delta^0$ are empty. In the case of Δ^n we use the indexing $t = (t_0, \dots, t_n) \in \Delta^n$; the subset $\partial_i \Delta^n = \{t \in \Delta^n \mid t_i = 0\}$ is the i-th face of Δ^n; hence $\partial \Delta^n = \bigcup_{i=0}^n \partial_i \Delta^n$.

It is useful to observe that certain standard spaces are homeomorphic. A general result of this type is:

(2.3.1) Proposition. *Let $K \subset \mathbb{R}^n$ be a compact convex subset with non-empty interior K°. Then there exists a homeomorphism of pairs $(D^n, S^{n-1}) \to (K, \partial K)$ which sends $0 \in D^n$ to a pre-assigned $x \in K^\circ$.*

Proof. Let $K \subset \mathbb{R}^n$ be closed and compact and $0 \in K^\circ$. Verify that a ray from 0 intersects the boundary ∂K of K in $\mathbb{R}^n$ in exactly one point. The map $f \colon \partial K \to S^{n-1}$, $x \mapsto x/\|x\|$ is a homeomorphism. The continuous map $\varphi \colon S^{n-1} \times [0,1] \to K$, $(x,t) \mapsto t f^{-1}(x)$ factors over $q \colon S^{n-1} \times [0,1] \to D^n$, $(x,t) \mapsto tx$ and yields a bijective map $k \colon D^n \to K$, hence a homeomorphism (use (1.4.3)). □

This proposition can be used to deduce a homeomorphism $(D^n, S^{n-1}) \cong (I^n, \partial I^n)$. The simplex Δ^n is a compact convex subset with interior points in the hyperplane $\{x \in \mathbb{R}^{n+1} \mid \sum_i x_i = 1\}$. From this fact we deduce a homeomorphism $(D^n, \partial D^n) \cong (\Delta^n, \partial \Delta^n)$.

The sphere S^n, as a homeomorphism type, will appear in many different shapes.

(2.3.2) Example. Let $N = e_{n+1} = (0, \dots, 0, 1) \in \mathbb{R}^{n+1}$. We define the ***stereographic projection*** $\varphi_N : S^n \smallsetminus \{e_{n+1}\} \to \mathbb{R}^n$; the point $\varphi_N(x)$ is the intersection of the line through e_{n+1} and x with the hyperplane $\mathbb{R}^n \times 0 = \mathbb{R}^n$. One computes $\varphi_N(x_1, \dots, x_{n+1}) = (1 - x_{n+1})^{-1}(x_1, \dots, x_n)$. The inverse map is $\pi_N : x \mapsto ((1 + \|x\|)^2)^{-1}(2x, \|x\|^2 - 1)$. We also have the stereographic projection $\varphi_S : S^n \smallsetminus \{-e_{n+1}\} \to \mathbb{R}^n$ and the transition map is $\varphi_S \circ \varphi_N^{-1}(y) = \|y\|^{-2}y$. From the stereographic projection we obtain S^n as a specific model of the one-point compactification $\mathbb{R}^n \cup \{\infty\}$ by extending $\pi_N(\infty) = e_{n+1}$. We also write $S^V = V \cup \{\infty\}$ for the one-point compactification of a finite-dimensional real vector space V. ◇

2.3.3 Spheres. Let $y \in S^n$. From (2.3.2) we see that $S^n \smallsetminus y$ is homeomorphic to $\mathbb{R}^n$ and hence contractible. Thus, if $X \to S^n$ is not surjective, it is null homotopic.

The inclusion $i : S^n \to \mathbb{R}^{n+1} \smallsetminus \{0\}$ is an h-equivalence with homotopy inverse $N : \mathbb{R}^{n+1} \smallsetminus \{0\} \to S^n$, $x \mapsto \|x\|^{-1}x$. A homotopy (rel S^n) from $i \circ N$ to the identity is the linear homotopy $(x, t) \mapsto tx + (1 - t)iN(x)$. Moreover $N \circ i = \mathrm{id}$. We see that i is a deformation retract.

Under suitable circumstances each map in a small neighbourhood of f is already homotopic to f. For a general theorem to this effect see (15.8.3). Here we only give a simple, but typical, example. Let $f, g : X \to S^n$ be maps such that $\|f(x) - g(x)\| < 2$. Then they are homotopic by a linear homotopy when viewed as maps into $\mathbb{R}^{n+1} \smallsetminus \{0\}$. We compose with N and see that $f \simeq g$.

If $f : S^m \to S^n$ is a continuous map, then there exists (by the theorem of Stone–Weierstrass, say) a C^∞-map $g : S^m \to S^n$ such that $\|f(x) - g(x)\| < 2$. This indicates another use of homotopies: Improve maps up to homotopy. If one uses some analysis, namely (the easy part of) the theorem of Sard about the density of regular values, one sees that for $m < n$ a C^∞-map $S^m \to S^n$ is not surjective and hence null homotopic. (Later we prove this fact by other methods.) There exist surjective continuous maps $S^1 \to S^2$ (Peano curves); this ungeometric behaviour of continuous maps is the source for many of the technical difficulties in topology. ◇

(2.3.4) Proposition. *The map $p : S^{n-1} \times I \to D^n$, $(x, t) \mapsto (1-t)x$ is a quotient map. Given $F : D^n \to X$, the composition $Fp : S^{n-1} \times I \to X$ is a null homotopy of $f = F|S^{n-1}$. Each null homotopy of a map $f : S^{n-1} \to X$ arises from a unique F.*

Proof. Since a null homotopy H of $f : S^{n-1} \to X$ sends $S^{n-1} \times 1$ to a point, it factors through the quotient map $q : S^{n-1} \times I \to S^{n-1} \times I/S^{n-1} \times 1$. Thus null homotopies H correspond via $\overline{H} \mapsto \overline{H}q$ to maps $\overline{H} : S^{n-1} \times I/S^{n-1} \times 1 \to X$. The map p induces a homeomorphism $\overline{p} : S^{n-1} \times I/S^{n-1} \times 1 \to D^n$ (use (1.4.3)). Hence there exists a unique F such that $F\overline{p} = \overline{H}$. □

Let us use the notation

$$S(n) = I^n/\partial I^n, \quad S^{(n)} = \mathbb{R}^n \cup \{\infty\},$$

since these spaces are homeomorphic to S^n. The canonical map $I^{n+m}/\partial I^{n+m} \to I^m \partial I^m \wedge I^n/\partial I^n$ which is the identity on representatives is a pointed homeomorphism. If V and W are finite-dimensional real vector spaces, we have a canonical pointed homeomorphism $S^V \wedge S^W \cong S^{V \oplus W}$ which is the identity away from the base point. The homeomorphism $]0,1[\to \mathbb{R}$, $s \mapsto \frac{2s-1}{s(1-s)}$ induces a homeomorphism $\gamma \colon S(1) \to S^{(1)}$ which transports $t \mapsto 1-t$ into the antipodal map $x \mapsto -x$ on $\mathbb{R}$. We obtain an induced homeomorphism

$$\gamma_n \colon S(n) = S(1) \wedge \cdots \wedge S(1) \to S^{(1)} \wedge \cdots \wedge S^{(1)} = S^{(n)}$$

of the n-fold smash products.

(2.3.5) Example. A retraction $r \colon D^n \times I \to S^{n-1} \times I \cup D^n \times 0$ is $r(x,t) = (2\alpha(x,t)^{-1} \cdot x, \alpha(x,t) - 2 + t)$ with $\alpha(x,t) = \max(2\|x\|, 2-t)$. (See Figure 2.1, a central projection from the point $(0,2)$.) Given a map $f \colon I^n \to X$ and a homotopy $h \colon \partial I^n \times I \to X$ with $h_0 = f|\partial I^n$ combine to a map $g \colon I^n \times 0 \cup \partial I^n \times I \to X$. We compose with a retraction and obtain a homotopy $H \colon I^n \to X$ which extends h and begins at $H_0 = f$. This homotopy extension property is later studied more generally under the name of cofibration. ◇

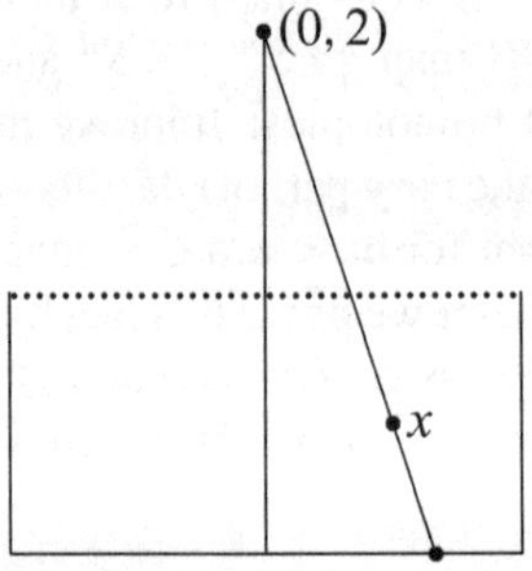

Figure 2.1. A retraction.

(2.3.6) Example. The assignment $H \colon (x,t) \mapsto (\alpha(x,t)^{-1}(1+t) \cdot x, 2 - \alpha(x,t))$ with the function $\alpha(x,t) = \max(2\|x\|, 2-t)$ yields a homeomorphism of pairs $(D^n, S^{n-1}) \times (I,0) \cong D^n \times (I,0)$, see Figure 2.2.

Similarly for $(I^n, \partial I^n)$ in place of (D^n, S^{n-1}), since these two pairs are homeomorphic. ◇

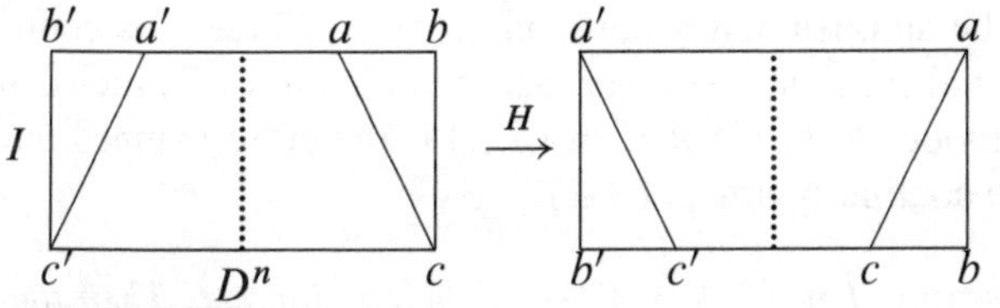

Figure 2.2. A relative homeomorphism.

Problems

1. Construct a homeomorphism $(D^m, S^{m-1}) \times (D^n, S^{n-1}) \cong (D^{m+n}, S^{m+n-1})$.
2. $\mathbb{R}^{n+1} \smallsetminus \{x\} \to S^n$, $z \mapsto (z-x)/\|z-x\|$ is an h-equivalence.
3. Let $D^n_+ = \{(x_0, \dots, x_n) \in S^n \mid x_n \geq 0\}$. Show that the quotient map $S^n \to S^n/D^n_+$ is an h-equivalence.
4. Let $f_1, \dots, f_k : \mathbb{C}^n \to \mathbb{C}$ be linearly independent linear forms ($k \leq n$). Then the complement $\mathbb{C}^n \smallsetminus \bigcup_j f_j^{-1}(0)$ is homotopy equivalent to the product of k factors S^1.
5. $S^n \to \{(x, y) \in S^n \times S^n \mid x \neq y\}$, $x \mapsto (x, -x)$ is an h-equivalence.
6. Let $f, g : X \to S^n$ be maps such that always $f(x) \neq -g(x)$. Then $f \simeq g$.
7. Let $A \subset E^n$ be star-shaped with respect to 0. Show that $S^{n-1} \subset \mathbb{R}^n \smallsetminus A$ is a deformation retract.
8. The projection $p : TS^n = \{(x, v) \in S^n \times \mathbb{R}^{n+1} \mid x \perp v\} \to S^n$, $(x, v) \mapsto x$ is called the tangent bundle of S^n. Show that p admits a fibrewise homeomorphism with $\mathrm{pr} : S^n \times S^n \smallsetminus D \to S^n$, $(x, y) \mapsto x$ (with D the diagonal).

2.4 Mapping Spaces and Homotopy

It is customary to endow sets of continuous maps with a topology. In this section we review from point-set topology the compact-open topology. It enables us to consider a homotopy $H : X \times I \to Y$ as a family of paths in Y, parametrized by X. This dual aspect of the homotopy notion will be used quite often. It can be formalized; but we use it more like a heuristic principle to dualize various constructions and notions in homotopy theory (Eckmann–Hilton duality).

We denote by Y^X or $F(X, Y)$ the set of continuous maps $X \to Y$. For $K \subset X$ and $U \subset Y$ we set $W(K, U) = \{f \in Y^X \mid f(K) \subset U\}$. The ***compact-open topology*** (CO-topology) on Y^X is the topology which has as a subbasis the sets of the form $W(K, U)$ for compact $K \subset X$ and open $U \subset Y$. In the sequel the set Y^X always carries the CO-topology. A continuous map $f : X \to Y$ induces continuous maps $f^Z : X^Z \to Y^Z$, $g \mapsto fg$ and $Z^f : Z^Y \to Z^X$, $g \mapsto gf$.

(2.4.1) Proposition. *Let X be locally compact. Then the **evaluation** $e_{X,Y} = e : Y^X \times X \to Y$, $(f, x) \mapsto f(x)$ is continuous.*

Proof. Let U be an open neighbourhood of $f(x)$. Since f is continuous and X locally compact, there exists a compact neighbourhood K of x such that $f(K) \subset U$. The neighbourhood $W(K,U) \times K$ of (f,x) is therefore mapped under e into U. This shows the continuity of e at (f,x). □

(2.4.2) Proposition. *Let $f : X \times Y \to Z$ be continuous. Then the* **adjoint map** *$f^\wedge : X \to Z^Y$, $f^\wedge(x)(y) = f(x,y)$ is continuous.*

Proof. Let $K \subset Y$ be compact and $U \subset Z$ open. It suffices to show that $W(K,U)$ has an open pre-image under $f^\wedge$. Let $f^\wedge(x) \in W(K,U)$ and hence $f(\{x\} \times K) \subset U$. Since K is compact, there exists by (1.4.1) a neighbourhood V of x in X such that $V \times K \subset f^{-1}(U)$ and hence $f^\wedge(V) \subset W(K,U)$. □

From (2.4.2) we obtain a set map $\alpha : Z^{X \times Y} \to (Z^Y)^X$, $f \mapsto f^\wedge$. Let $e_{Y,Z}$ be continuous. A continuous map $\varphi : X \to Z^Y$ induces a continuous map $\varphi^\vee = e_{Y,Z} \circ (\varphi \times \mathrm{id}_Y) : X \times Y \to Z^Y \times Y \to Z$. Hence we obtain a set map $\beta : (Z^Y)^X \to Z^{X \times Y}$, $\varphi \mapsto \varphi^\vee$.

(2.4.3) Proposition. *Let $e_{Y,Z}$ be continuous. Then α and β are inverse bijections. Thus $\varphi : X \to Y^Z$ is continuous if $\varphi^\vee : X \times Y \to Z$ is continuous, and $f : X \times Y \to Z$ is continuous if $f^\wedge : X \to Z^Y$ is continuous.* □

(2.4.4) Corollary. *If $h : X \times Y \times I \to Z$ is a homotopy, then $h^\wedge : X \times I \to Z^Y$ is a homotopy (see (2.4.2)). Hence $[X \times Y, Z] \to [X, Z^Y]$, $[f] \mapsto [f^\wedge]$ is well-defined. If, moreover, $e_{Y,Z}$ is continuous, e.g., Y locally compact, then this map is bijective (see (2.4.3)).* □

2.4.5 Dual notion of homotopy. We have the continuous evaluation $e_t : Y^I \to Y$, $w \mapsto w(t)$. A homotopy from $f_0 : X \to Y$ to $f_1 : X \to Y$ is a continuous map $h : X \to Y^I$ such that $e_\varepsilon \circ h = f_\varepsilon$ for $\varepsilon = 0, 1$. The equivalence with our original definition follows from (2.4.3): Since I is locally compact, continuous maps $X \times I \to Y$ correspond bijectively to continuous maps $X \to Y^I$. ◇

(2.4.6) Theorem. *Let Z be locally compact. Suppose $p : X \to Y$ is a quotient map. Then $p \times \mathrm{id}(Z) : X \times Z \to Y \times Z$ is a quotient map.*

Proof. We verify for $p \times \mathrm{id}$ the universal property of a quotient map: If $h : Y \times Z \to C$ is a set map and $h \circ (p \times \mathrm{id})$ is continuous, then h is continuous. The adjoint of $h \circ (p \times \mathrm{id})$ is $h^\wedge \circ p$. By (2.4.2), it is continuous. Since p is a quotient map, $h^\wedge$ is continuous. Since Z is locally compact, h is continuous, by (2.4.3). □

(2.4.7) Theorem (Exponential law). *Let X and Y be locally compact. Then the adjunction map $\alpha : Z^{X \times Y} \to (Z^Y)^X$ is a homeomorphism.*

Proof. By (2.4.3), α is continuous if $\alpha_1 = e_{X,Z^Y} \circ (\alpha \times \mathrm{id})$ is continuous. And this map is continuous if $\alpha_2 = e_{Y,Z} \circ (\alpha_1 \times \mathrm{id})$ is continuous. One verifies that $\alpha_2 = e_{X \times Y,Z}$. The evaluations which appear are continuous by (2.4.1).

The inverse α^{-1} is continuous if $e_{X \times Y,Z} \circ (\alpha^{-1} \times \mathrm{id})$ is continuous, and this map equals $e_{Y,Z} \circ (e_{X,Z^Y} \times \mathrm{id})$. □

Let (X, x) and (Y, y) be pointed spaces. We denote by $F^0(X, Y)$ the space of pointed maps with CO-topology as a subspace of $F(X, Y)$. In $F^0(X, Y)$ we use the constant map as a base point. The adjoint $f^\wedge\colon X \to F(Y, Z)$ of $f\colon X \times Y \to Z$ is a pointed map into $F^0(Y, Z)$ if and only if f sends $X \times y \cup x \times Y$ to the base point of Z. Let $p\colon X \times Y \to X \wedge Y = X \times Y/(X \times y \cup x \times Y)$ be the quotient map.

If $g\colon X \wedge Y \to Z$ is given, we denote the adjoint of $g \circ p\colon X \times Y \to X \wedge Y \to Z$ by $\alpha^0(g)$ and consider it as an element of $F^0(X, F^0(Y, Z))$. In this manner we obtain a set map $\alpha^0\colon F^0(X \wedge Y, Z) \to F^0(X, F^0(Y, Z))$.

The evaluation $F^0(X, Y) \times X \to Y$, $(f, x) \mapsto f(x)$ factors over the quotient space $F^0(X, Y) \wedge X$ and induces $e^0 = e^0_{X,Y}\colon F^0(X, Y) \wedge X \to Y$. From (2.4.1) we conclude:

(2.4.8) Proposition. *Let X be locally compact. Then $e^0_{X,Y}$ is continuous.* □

Let $e^0_{Y,Z}$ be continuous. From a pointed map $\varphi\colon X \to F^0(Y, Z)$ we obtain $\varphi^\vee = \beta^0(\varphi) = e^0_{Y,Z} \circ (\varphi \wedge \mathrm{id})\colon X \wedge Y \to Z$, and hence a set map $\beta^0\colon F^0(X, F^0(Y, Z)) \to F^0(X \wedge Y, Z)$.

(2.4.9) Proposition. *Let $e^0_{Y,Z}$ be continuous. Then α^0 and β^0 are inverse bijections.* □

(2.4.10) Corollary. *Let $h\colon (X \wedge Y) \times I \to Z$ be a pointed homotopy. Then $\alpha^0(h_t)\colon X \to F^0(Y, Z)$ is a pointed homotopy and therefore*

$$[X \wedge Y, Z]^0 \to [X, F^0(Y, Z)]^0, \quad [f] \mapsto [\alpha^0(f)]$$

is well defined. If, moreover, $e^0_{Y,Z}$ is continuous, then this map is bijective. □

By a proof formally similar to the proof of (2.4.7), we obtain the pointed version of the exponential law.

(2.4.11) Theorem (Exponential law). *Let X and Y be locally compact. Then the pointed adjunction map $\alpha^0\colon F^0(X \wedge Y, Z) \to F^0(X, F^0(Y, Z))$ is a homeomorphism.* □

(2.4.12) Lemma. *Let $k_a\colon Z \to A$ denote the constant map with value a. Then $\psi\colon X^Z \times A \to (X \times A)^Z$, $(\varphi, a) \mapsto (\varphi, k_a)$ is continuous.*

Proof. Let $\psi(f,a) \in W(K,U)$. This means: For $x \in K$ we have $(f(x),a) \in U$. There exist open neighbourhoods V_1 of $f(K)$ in X and V_2 of a in A such that $V_1 \times V_2 \subset U$. The inclusion $\psi(W(K,V_1) \times V_2) \subset W(K,U)$ shows the continuity of ψ at (f,a). □

(2.4.13) Proposition. *A homotopy $H_t \colon X \to Y$ induces homotopies H_t^Z and Z^{H_t}.*

Proof. In the first case we obtain, with a map ψ from (2.4.12), a continuous map

$$H^Z \circ \psi \colon X^Z \times I \to (X \times I)^Z \to Y^Z.$$

In the second case we use the composition

$$e \circ (\alpha \times \mathrm{id}) \circ (Z^H \times \mathrm{id}) \colon Z^Y \times I \to Z^{X \times I} \times I \to (Z^X)^I \times I \to Z^X$$

which is continuous. □

(2.4.14) Corollary. *Let f be a homotopy equivalence. Then the induced maps $F(Z,X) \to F(Z,Y)$ and $F(Y,Z) \to F(X,Z)$ are h-equivalences. If f is a pointed h-equivalence, the induced maps $F^0(Z,X) \to F^0(Z,Y)$ and $F^0(Y,Z) \to F^0(X,Z)$ are pointed h-equivalences.* □

Problems

1. Verify that f^Z and Z^f are continuous.
2. An inclusion $i \colon Z \subset Y$ induces an embedding $i^X \colon Z^X \to Y^X$.
3. The canonical map $F(\coprod_j X_j, Y) \to \prod_j F(X_j, Y)$ is always a homeomorphism.
4. The canonical map $F(X, \prod_j Y_j) \to \prod_j F(X, Y_j)$, $f \mapsto (\mathrm{pr}_j f)$ is always bijective and continuous. If X is locally compact, it is a homeomorphism.
5. Let $p \colon X \to Y$ be a surjective continuous map. Suppose the pre-image of a compact set is compact. Then $Z^p \colon Z^Y \to Z^X$ is an embedding.
6. We have a canonical bijective map $F^0(\bigvee_{j \in J} X_j, Y) \to \prod_{j \in J} F^0(X_j, Y)$, since $\bigvee_j X_j$ is the sum in TOP^0. If J is finite, it is a homeomorphism.
7. Let X, Y, U, and V be spaces. The Cartesian product of maps gives a map

$$\pi \colon U^X \times V^Y \to (U \times V)^{X \times Y}, \quad (f,g) \mapsto f \times g.$$

Let X and Y be Hausdorff spaces. Then the map π is continuous.
8. By definition of a product, a map $X \to Y \times Z$ is essentially the same thing as a pair of maps $X \to Y$, $X \to Z$. In this sense, we obtain a tautological bijection $\tau \colon (Y \times Z)^X \to Y^X \times Z^X$. Let X be a Hausdorff space. Then the tautological map τ is a homeomorphism.
9. Let X and Y be locally compact. Then composition of maps $Z^Y \times Y^X \to Z^X$, $(g,f) \mapsto g \circ f$ is continuous.
10. Let $(Y,*)$ be a pointed space, (X,A) a pair of spaces and $p \colon X \to X/A$ the quotient map. The space X/A is pointed with base point $\{A\}$. Let $F((X,A),(Y,*))$ be the subspace of $F(X,Y)$ of the maps which send A to the base point. Composition with p induces a bijective continuous map $\gamma \colon F^0(X/A, Y) \to F((X,A),(Y,*))$; and a bijection of homotopy sets

$[X/A, Y]^0 \to [(X, A), (Y, *)]$. If p has compact pre-images of compact sets, then γ is a homeomorphism.

11. Consider diagrams where the right-hand one is obtained by multiplying the left-hand one with X:

$$\begin{array}{ccc} A & \longrightarrow & B \\ \downarrow & & \downarrow \\ C & \longrightarrow & D, \end{array} \qquad \begin{array}{ccc} A \times X & \longrightarrow & B \times X \\ \downarrow & & \downarrow \\ C \times X & \longrightarrow & D \times X. \end{array}$$

If the left-hand diagram is a pushout in TOP and X locally compact, then the right-hand diagram is a pushout in TOP. In TOP^0 the smash product with a locally compact space yields again a pushout.

12. The CO-topology on the set of linear maps $\mathbb{R}^n \to \mathbb{R}$ is the standard topology.

13. Let X be a compact space and Y a metric space. Then the CO-topology on Y^X is induced by the supremum-metric.

2.5 The Fundamental Groupoid

A path in the plane can be quite ungeometric: nowhere differentiable, infinite length, surjective onto $I \times I$ (a so-called Peano curve). We introduce an equivalence relation on paths, and the equivalence classes still capture qualitative geometric properties of the path. In particular a reparametrization of a path (different "velocity") does not change basic topological properties.

We consider homotopy classes relative to ∂I of paths. A homotopy of paths is always assumed to be relative to ∂I. A ***homotopy of paths*** between paths u and v with the same end points $x_0 = u(0) = v(0)$, $x_1 = u(1) = v(1)$ is a map $H: I \times I \to X$ such that

$$\begin{aligned} H(s,0) &= u(s), \\ H(s,1) &= v(s), \\ H(0,t) &= u(0) = v(0), \\ H(1,t) &= u(1) = v(1). \end{aligned}$$

Thus for each $t \in I$ we have a path $H_t : s \mapsto H(s,t)$ and all these paths have the same end points. We write $H : u \simeq v$ for this homotopy.

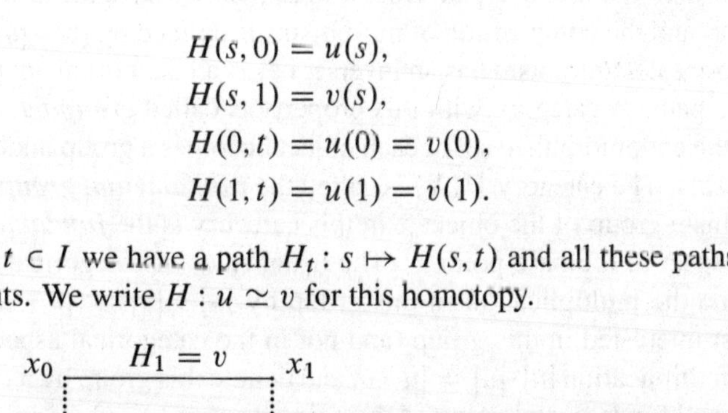

Being homotopic in this sense is an equivalence relation on the set of all paths from x to y. The product operation is compatible with this relation as the next proposition shows.

(2.5.1) Proposition. *The product of paths has the following properties:*

(1) *Let* $\alpha\colon I \to I$ *be continuous and* $\alpha(0) = 0, \alpha(1) = 1$. *Then* $u \simeq u\alpha$.

(2) $u_1 * (u_2 * u_2) \simeq (u_1 * u_2) * u_3$ *(if the products are defined).*

(3) $u_1 \simeq u_1'$ *and* $u_2 \simeq u_2'$ *implies* $u_1 * u_2 \simeq u_1' * u_2'$.

(4) $u * u^-$ *is always defined and homotopic to the constant path.*

(5) $k_{u(0)} * u \simeq u \simeq u * k_{u(1)}$.

Proof. (1) $H\colon (s,t) \mapsto u(s(1-t) + t\alpha(s))$ is a homotopy from u to $u\alpha$.

(2) The relation $(u_1 * (u_2 * u_3))\alpha = (u_1 * u_2) * u_3$ holds for α defined as $\alpha(t) = 2t$ for $t \le \frac{1}{4}$, $\alpha(t) = t + \frac{1}{4}$ for $\frac{1}{4} \le t \le \frac{1}{2}$, $\alpha(t) = \frac{t}{2} + \frac{1}{2}$ for $\frac{1}{2} \le t \le 1$.

(3) Given $F_i\colon u_i \simeq u_i'$ then $G\colon u_1 * u_2 \simeq u_1' * u_2'$ for G defined as $G(s,t) = F_1(2s,t)$ for $0 \le t \le \frac{1}{2}$ and $G(s,t) = F_2(2s-1,t)$ for $\frac{1}{2} \le t \le 1$.

(4) The map $F\colon I \times I \to X$ defined as $F(s,t) = u(2s(1-t))$ for $0 \le s \le \frac{1}{2}$ and $F(s,t) = u(2(1-s)(1-t)$ for $\frac{1}{2} \le t \le 1$ is a homotopy from $u * u^-$ to the constant path. (At time t we only use the path from 0 to $(1-t)$ and compose it with its inverse.)

(5) is proved again with the parameter invariance (1). □

From homotopy classes of paths in X we obtain again a category, denoted $\Pi(X)$. The objects are the points of X. A morphism from x to y is a homotopy class relative to ∂I of paths from x to y. A constant path represents an identity. If u is a path from a to b and w a path from b to c, then we have the product path $u * v$ from a to c, and the composition of morphisms is defined by $[v] \circ [u] = [u * v]$. In this category each morphism has an inverse, i.e. is an isomorphism, represented by the inverse path. A category with this property is called ***groupoid***. Note that in a groupoid the endomorphism set of each object becomes a group under composition of morphisms. The category $\Pi(X)$ is called the ***fundamental groupoid*** of X. The automorphism group of the object x in this category is the ***fundamental group*** of X with respect to the base point x. The usual rules of categorical notation force us to define the multiplication in this group by $[u] \circ [v] = [v * u]$. As long as we are just interested in this group (and not in the categorical aspect), we use the opposite multiplication $[u] \cdot [v] = [u * v]$ and denote this group by $\pi_1(X, x)$. This is the traditional fundamental group of the pointed space (X, x) (Poincaré 1895 [151, §12]). An element in $\pi_1(X, x)$ is represented by a closed path w based at x (i.e., $w(0) = w(1) = x$), also called a ***loop*** based at x.

(2.5.2) Remark. We can obtain the fundamental groupoid $\Pi(X)$ as a quotient category of the path category $W(X)$. In that case we call paths $u\colon [0,a]\colon I \to X$ and $v\colon [0,b] \to X$ from x to y homotopic, if there exist constant paths with image y

such that the compositions with u and v, respectively, have the same domain of definition $[0, c]$ and the resulting composed paths are homotopic rel $\{0, c\}$. ◇

(2.5.3) Remark. The set $\pi_1(X, x)$ has different interpretations. A loop based at x is a map $w\colon (I, \partial I) \to (X, x)$. It induces a pointed map $\overline{w}\colon I/\partial I \to X$. The exponential function $p_0\colon I \to S^1, t \mapsto \exp(2\pi i t)$ induces a pointed homeomorphism $q\colon I/\partial I \to S^1$ which sends the base point $\{\partial I\}$ to the base point 1. There exists a unique $u\colon S^1 \to X$ such that $uq = \overline{w}$. Altogether we obtain bijections

$$\pi_1(X, x) = [(I, \partial I), (X, x)] \cong [I/\partial I, X]^0 \cong [S^1, X]^0,$$

induced by $[w] \leftrightarrow [\overline{w}] = [uq] \leftrightarrow [u]$. ◇

It is a general fact for groupoids Π that the automorphism groups $\mathrm{Aut}(x) = \Pi(x, x)$ and $\mathrm{Aut}(y) = \Pi(y, y)$ of objects x, y in Π are isomorphic, provided there exists a morphism from x to y. If $\alpha \in \Pi(x, y)$, then

$$\Pi(x, x) \to \Pi(y, y), \quad \beta \mapsto \alpha\beta\alpha^{-1}$$

is an isomorphism. It depends on the choice of α; there is, in general, no canonical isomorphism between these groups. Thus fundamental groups associated to base points in the same path component are isomorphic, but not canonically.

A space is ***simply connected*** or ***1-connected*** if it is path connected and its fundamental group is trivial (consists of the neutral element alone).

A continuous map $f\colon X \to Y$ induces a homomorphism

$$\pi_1(f)\colon \pi_1(X, x) \to \pi_1(Y, f(x)), \quad [u] \mapsto [fu]$$

and, more generally, a functor

$$\Pi(f)\colon \Pi(X) \to \Pi(Y), \quad x \mapsto f(x),\ [u] \mapsto [fu].$$

In this way, π_1 becomes a functor from TOP^0 to the category of groups, and Π a functor from TOP to the category of small categories (small category: its objects form a set). Homotopies correspond to natural transformations:

(2.5.4) Proposition. *Let $H\colon X \times I \to Y$ be a homotopy from f to g. Each $x \in X$ yields the path $H^x\colon t \mapsto H(x, t)$ and the morphism $[H^x]$ in $\Pi(Y)$ from $f(x)$ to $g(x)$. The $[H^x]$ constitute a natural transformation $\Pi(H)$ from $\Pi(f)$ to $\Pi(g)$.*

Proof. The claim says that for each path $u\colon I \to X$ the relation $fu * H^{u(1)} \simeq H^{u(0)} * gu$ holds. We use $I \times I \to Y, (s, t) \mapsto H(u(s), t)$. We obtain $fu * H^{u(1)}$ as composition with $a * b$ and $H^{u(0)} * gu$ as composition with $c * d$, where a, b, c, and d are the sides of the square: $a(t) = (t, 0)$, $b(t) = (1, t)$, $c(t) = (0, t)$, $d(t) = (t, 1)$. But $a * b$ and $c * d$ are homotopic by a linear homotopy. □

We express the commutativity of (2.5.4) in a different way. It says that

$$t_H \circ g_* = f_*\colon \Pi X(x,y) \to \Pi Y(fx,fy),$$

where $t_H\colon \Pi(Y)(gx,gy) \to \Pi(Y)(fx,fy)$ is the bijection $a \mapsto [H^y]^{-1}a[H^x]$.

The rule $\Pi(K * L) = \Pi(L)\Pi(K)$ is obvious. Hence if f is an h-equivalence with h-inverse $g\colon Y \to X$, then $\Pi(f)\Pi(g)$ and $\Pi(g)\Pi(f)$ are naturally isomorphic to the identity functor, i.e., $\Pi(f)$ is an equivalence of categories. The natural transformation $\Pi(H)$ only depends on the homotopy class relative to $X \times \partial I$ of H.

It is a general categorical fact that a natural equivalence of categories induces a bijection of morphism sets. We prove this in the notation of our special case at hand.

(2.5.5) Proposition. *Let $f\colon X \to Y$ be a homotopy equivalence. Then the functor $\Pi(f)\colon \Pi(X) \to \Pi(Y)$ is an equivalence of categories. The induced maps between morphism sets $f_*\colon \Pi X(x,y) \to \Pi Y(fx,fy)$ are bijections. In particular,*

$$\pi_1(f)\colon \pi_1(X,x) \to \pi_1(Y,f(x)), \quad [w] \mapsto [fw]$$

is an isomorphism for each $x \in X$. A contractible space is simply connected.

Proof. Let $g\colon Y \to X$ be h-inverse to f. Consider

$$\Pi X(x,y) \xrightarrow{f_*} \Pi Y(fx,fy) \xrightarrow{g_*} \Pi X(gfx,gfy) \xrightarrow{f_*} \Pi Y(fgfx,fgfy).$$

Choose $H\colon gf \simeq \mathrm{id}(X)$. Then $g_* f_* = (gf)_* = t_H \circ (\mathrm{id})_* = t_H$ is a bijection, hence g_* is surjective. In a similar manner one proves that $f_* g_*$ is a bijection, hence g_* is also injective. Since $g_* f_*$ and g_* are bijective we see that f_* is bijective. □

The fundamental group forces us to work with pointed spaces. Usually the base points serve some technical purpose and one has to study what happens when the base point is changed. For pointed h-equivalences f it would be immediately clear that $\pi_1(f)$ is an isomorphism. For the more general case (2.5.5) one needs some argument like the one above.

(2.5.6) Proposition. *Let (X,x_0) and (Y,y_0) be pointed spaces and $i^X\colon X \to X \times Y$, $x \mapsto (x,y_0)$ and $i^Y\colon Y \to X \times Y$, $y \mapsto (x_0,y)$. Then*

$$\pi_1(X,x_0) \times \pi_1(Y,y_0) \to \pi_1(X \times Y,(x_0,y_0)), \quad (u,v) \mapsto i^X_* u \cdot i^Y_* v$$

is a well-defined isomorphism with inverse $z \mapsto (\mathrm{pr}^X_* z, \mathrm{pr}^Y_* z)$.

Proof. Since homotopy is compatible with products we know already that the second map is an isomorphism. In order to show that the first map is a homomorphism we have to verify that $i^X_* u$ commutes with $i^Y_* v$. Let now u and v be actual paths and write $w = (u \times v)\delta$ with the diagonal δ. With a notation introduced in the proof of (2.5.4) we have $(u \times v)(a * b) = i^X u * i^Y v$ and $(u \times v)(c * d) = i^Y v * i^X u$. We now use that δ, $a * b$, and $c * d$ are homotopic (linear homotopy). It should be clear that the two maps of the proposition are inverse to each other. □

2.6 The Theorem of Seifert and van Kampen

Let a space X be the union of subsets X_0, X_1. A general problem is to derive properties of X from those of X_0, X_1, and $X_{01} = X_0 \cap X_1$. (Similar problem for more general unions.) Usually the covering has to satisfy certain reasonable conditions. In this section we consider the fundamental groupoid and the fundamental group under this aspect. The basic result is the theorem (2.6.2) of Seifert [166] and van Kampen [100].

We first prove an analogous and slightly more general result for groupoids [34]. The result is more formal but the proof is (notationally) simpler because we need not take care of base points. Note that the hypothesis of the next theorem implies that X is the pushout in TOP of the inclusions $X_0 \supset X_{01} \subset X_1$. Thus (2.6.1) says that the functor Π preserves pushouts.

(2.6.1) Theorem (R. Brown). *Let X_0 and X_1 be subspaces of X such that the interiors cover X. Let $i_\nu \colon X_{01} \to X_\nu$ and $j_\nu \colon X_\nu \to X$ be the inclusions. Then*

$$\begin{array}{ccc} \Pi(X_{01}) & \xrightarrow{\Pi(i_0)} & \Pi(X_0) \\ {\scriptstyle \Pi(i_1)}\downarrow & & \downarrow{\scriptstyle \Pi(j_0)} \\ \Pi(X_1) & \xrightarrow{\Pi(j_1)} & \Pi(X) \end{array}$$

is a pushout in the category of groupoids.

Proof. Let $h_\nu \colon \Pi(X_\nu) \to \Lambda$ be functors into a groupoid such that $h_1\Pi(i_1) = h_0\Pi(i_0)$. We have to show: There exists a unique functor $\lambda \colon \Pi(X) \to \Lambda$ such that $h_1 = \lambda\Pi(j_1)$ and $h_0 = \lambda\Pi(j_0)$. We begin with a couple of remarks.

A path $w \colon [a, b] \to U$ represents a morphism $[w]$ in $\Pi(U)$ from $w(a)$ to $w(b)$ if we compose it with an increasing homeomorphism $\alpha \colon [0, 1] \to [a, b]$.

If $a = t_0 < t_1 < \cdots < t_m = b$, then w represents the composition of the morphisms $[w|[t_i, t_{i+1}]]$.

Suppose that $w \colon I \to X$ is a path. Then there exists a decomposition $0 = t_0 < t_1 < \cdots < t_{m+1} = 1$ such that $w([t_i, t_{i+1}])$ is contained in a set X_ν°. Choose $\gamma \colon \{0, \dots, m\} \to \{0, 1\}$ such that $w([t_i, t_{i+1}]) \subset X^\circ_{\gamma(i)}$. Consider $w|[t_i, t_{i+1}]$ as path w_i in $X_{\gamma(i)}$. Then

$$[w] = \Pi(j_{\gamma(m)})[w_m] \circ \cdots \circ \Pi(j_{\gamma(0)})[w_0].$$

If λ exists, then

$$\lambda[w] = h_{\gamma(m)}[w_m] \circ \cdots \circ h_{\gamma(0)}[w_0], \tag{i}$$

i.e., λ is uniquely determined.

In order to show the existence of λ, we have to define $\lambda[w]$ by (i). We have to verify that this is well-defined. The commutativity $h_0\Pi(i_0) = h_1\Pi(h_1)$ shows that a different choice of γ yields the same result.

Since h_0 and h_1 are functors, we obtain the same result if we refine the decomposition of the interval.

It remains to be shown that (i) only depends on the homotopy class of the path. Let $H: I \times I \to X$ be a homotopy of paths from x to y. There exists $n \in \mathbb{N}$ such that H sends each sub-square $[i/n, (i+1)/n] \times [j/n, (j+1)/n]$ into one of the sets X_ν° (see (2.6.4)). We consider edge-paths in the subdivided square $I \times I$ which differ by a sub-square, as indicated in the following figure ($n = 5$).

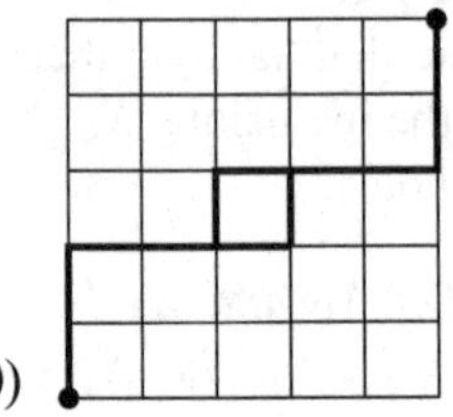

We apply H and obtain two paths in X. They yield the same result (i), since they differ by a homotopy on some subinterval which stays inside one of the sets X_ν°. Changes of this type allow us to pass inductively from the H on the lower to H on the upper boundary path from $(0,0)$ to $(1,1)$. These paths differ from H_0 and H_1 by composition with a constant path.

Finally, from the construction we conclude that λ is a functor. □

(2.6.2) Theorem (Seifert–van Kampen). *Let X_0 and X_1 be subspaces of X such that the interiors cover X. Let $i_\nu: X_{01} = X_0 \cap X_1 \to X_\nu$ and $j_\nu: X_\nu \to X$ be the inclusions. Suppose that X_0, X_1, X_{01} are path connected with base point $* \in X_{01}$. Then*

$$\begin{array}{ccc} \pi_1(X_{01}, *) & \xrightarrow{i_{1*}} & \pi_1(X_1, *) \\ \downarrow{\scriptstyle i_{0*}} & & \downarrow{\scriptstyle j_{1*}} \\ \pi_1(X_0, *) & \xrightarrow{j_{0*}} & \pi_1(X, *) \end{array}$$

is a pushout in the category of groups.

Proof. The theorem is a formal consequence of (2.6.1). In general, if Z is path connected and $z \in Z$ we have a retraction functor $r: \Pi(Z) \to \pi_1(Z, z)$ onto the full subcategory with object z. For each $z \in Z$ we choose a morphism $u_x \in \Pi(Z)$

from x to z such that $u_z = \mathrm{id}$. Then r assigns $u_y \alpha u_x^{-1}$ to a morphism $\alpha\colon x \to y$. We apply this to $Z = X_{01}, X_0, X_1, X$ and $z = *$ and choose a morphism $u_x \in \Pi(Z)$ if x is contained in Z. We obtain a commutative diagram of functors

$$\begin{array}{ccccc} \Pi(X_0) & \longleftarrow & \Pi(X_{01}) & \longrightarrow & \Pi(X_1) \\ \downarrow{\scriptstyle r_1} & & \downarrow{\scriptstyle r_{01}} & & \downarrow{\scriptstyle r_0} \\ \pi_1(X_0, *) & \longleftarrow & \pi_1(X_{01}, *) & \longrightarrow & \pi_1(X_1, *). \end{array}$$

Given homomorphisms $\varphi_\nu\colon \pi_1(X_\nu, *) \to G$ into a group G ($=$ a groupoid with a single object) which agree on $\pi_1(X_{01}, *)$, we compose with r_ν and apply (2.6.1) to obtain a functor $\Pi(X) \to G$. Its restriction to $\pi_1(X, *)$ is the unique solution of the pushout problem in (2.6.2). □

(2.6.3) Remark. From the proof of (2.6.1) we see that each morphism in $\Pi(X)$ is a composition of morphisms in $\Pi(X_0)$ and $\Pi(X_1)$. Similarly, in (2.6.2) the group $\pi_1(X, *)$ is generated by the images of j_{0*} and j_{1*}. This algebraic fact is not immediately clear from the definition of a pushout. ◇

We have used above the next fundamental result. It is impossible to prove a geometric results about continuous maps without subdivision and approximation procedures. In most of these procedures (2.6.4) will be used.

(2.6.4) Proposition (Lebesgue). *Let X be a compact metric space. Let $\mathcal{A}$ be an open covering of X. Then there exists $\varepsilon > 0$ such that for each $x \in X$ the ε-neighbourhood $U_\varepsilon(x)$ is contained in some member of $\mathcal{A}$. (An ε with this property is called a **Lebesgue number** of the covering.)* □

2.7 The Fundamental Group of the Circle

The space $\mathbb{R}$ is simply connected; $\Pi(\mathbb{R})$ has a single morphism between any two objects. We consider $\Pi(\mathbb{R})$ as a topological groupoid: The object space is $\mathbb{R}$, the morphism space is $\mathbb{R} \times \mathbb{R}$, the source is $(a, b) \mapsto a$, the range $(a, b) \mapsto b$, the identity $a \mapsto (a, a)$, and $(b, c) \circ (a, b) = (a, c)$ the composition.

The continuous map $p\colon \mathbb{R} \to S^1$ induces a functor $\Pi(p)\colon \Pi(\mathbb{R}) \to \Pi(S^1)$. It turns out that this functor is surjective on morphisms and provides us with an algebraic description of $\Pi(S^1)$. So let us define a topological groupoid G. The object space is S^1, the morphism space is $S^1 \times \mathbb{R}$, the source $(a, t) \mapsto a$, the range $(a, t) \mapsto a\exp(2\pi i t)$, the identity $a \mapsto (a, 0)$, and the composition $(b, t) \circ (a, s) = (a, s + t)$.

The assignments $a \mapsto \exp(2\pi i a)$ and $(a, b) \mapsto (\exp(2\pi i a), b - a)$ yield a continuous functor $\Pi(\mathbb{R}) \to G$. We will show that G (forgetting the topology) is $\Pi(S^1)$.

We have the open covering of $S^1 \subset \mathbb{C}$ by $X_0 = S^1 \smallsetminus \{1\}$ and $X_1 = S^1 \smallsetminus \{-1\}$ with inclusions $i_k\colon X_{01} \to X_k$ and $j_k\colon X_k \to S^1$. The sets X_k are contractible, hence simply connected. Therefore there exists a single morphism $(a,b)_k\colon a \to b$ between two objects a, b of $\Pi(X_k)$.

We have bijective maps $f_0\colon]0,1[\to X_0$ and $f_1\colon]-1/2, 1/2[\to X_1$ given by $t \mapsto \exp(2\pi i t)$. We define functors $\gamma_k\colon \Pi(X_k) \to G$ by the identity on objects and by $\gamma_k(a,b)_k = (a, f_k^{-1}(b) - f_k^{-1}(a))$. Moreover we have a functor $\zeta\colon G \to \Pi(S^1)$ which is the identity on objects and which sends the morphism (a,t) of G to the class of the path $I \to S^1$, $s \mapsto a\exp(2\pi i t s)$ from a to $a\exp(2\pi i t)$. (The idea behind the definition of G is the fact that each path in S^1 is homotopic to one of this normal form, see (2.7.9).) The following diagram is commutative.

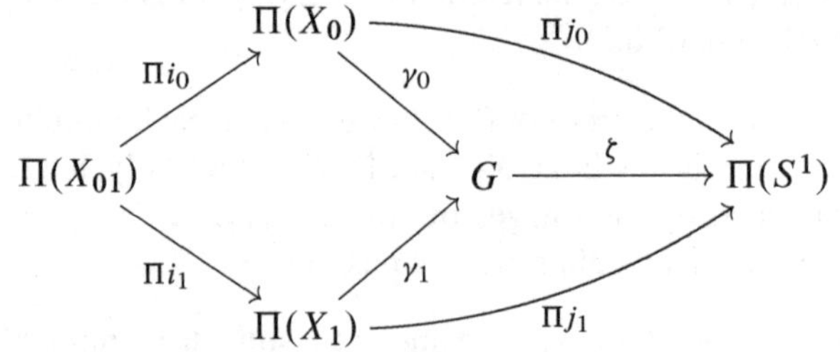

(2.7.1) Proposition. *The functor ζ is an isomorphism.*

Proof. We apply (2.6.1) to the pair (γ_0, γ_1) and obtain a functor $\gamma\colon \Pi(S^1) \to G$. The uniqueness property of a pushout solution shows $\zeta\gamma = \mathrm{id}$. In order to show $\gamma\zeta = \mathrm{id}$ we note that the morphisms of G are generated by the images of γ_0 and γ_1. Given $(a,t) \in G(a,b)$, choose a decomposition $t = t_1 + \dots + t_m$ such that $|t_r| < 1/2$ for each r. Set $a_0 = a$ and $a_r = a\exp(2\pi i(t_1 + \dots + t_r))$. Then $(a,t) = (a_{m-1}, t_m) \circ \dots \circ (a_1, t_2) \circ (a_0, t_1)$ in the groupoid G. Since $|t_r| < 1/2$ there exists for each r a $k(r) \in \{0,1\}$ such that $a_{r-1}\exp(2\pi i t_r s) \in X_{k(r)}$ for $s \in I$. Then $(a_{r-1}, t_r) = \gamma_{k(r)}(a_{r-1}, a_r)_{k(r)}$. Thus $G(a,b)$ is generated by morphisms in the images of the γ_k. □

The unit circle S^1 in the complex plane is the prototype of a loop. Typical elements in the fundamental group are obtained by running n times around the circle. Up to homotopy, there are no other possibilities. With (2.7.1) we have determined the fundamental group $\pi_1(S^1, 1)$, namely as the automorphism group in $\Pi(S^1)$ of the object 1. The automorphisms of the object 1 in G are the $(1,n)$, $n \in \mathbb{Z}$ and $\zeta(1,n)$ is the loop $t \mapsto \exp(2\pi i n t)$.

(2.7.2) Theorem. *Let $s_n\colon I \to S^1$ be the loop $t \mapsto \exp(2\pi i n t)$. The assignment $\delta\colon \mathbb{Z} \to \pi_1(S^1, 1)$, $n \mapsto [s_n]$ is an isomorphism.* □

The circle S^1 is a group with respect to multiplication of complex numbers. We show that the composition law in $\pi_1(S^1, 1)$ can also be defined using this multiplication.

More generally, assume that X is a space with a continuous multiplication

$$m\colon X \times X \to X, \quad (x, y) \mapsto m(x, y) = xy$$

and neutral element up to homotopy $e \in X$ (the base point), i.e., the maps $x \mapsto m(e, x)$ and $x \mapsto m(x, e)$ are both pointed homotopic to the identity. We call such an object a ***monoid*** in h-TOP. (We do not assume that m is associative or commutative.) We define a composition law on the pointed homotopy set $[Y, X]^0$, called the m-product, by $[f], [g] \mapsto [f] \cdot_m [g] = [f \cdot g]$; here $f \cdot g\colon y \mapsto m(f(y), g(y))$ is the ordinary pointwise multiplication. The constant map represents a two-sided unit for the m-product. In a similar manner we define by pointwise multiplication of loops the m-product on $\pi_1(X, e)$. The set $\pi_1(X, e) \cong [S^1, X]^0$ now carries two composition laws: the m-product and the $*$-product of the fundamental group.

(2.7.3) Proposition. *Let (X, m) be a monoid in* h-TOP. *Then the $*$-product and the m-product on $\pi_1(X, e)$ coincide and the product is commutative.*

Proof. Let k be the constant loop. Then for any two loops u and v the relations

$$\begin{aligned} u * v &\simeq (u \cdot k) * (k \cdot v) = (u * k) \cdot (k * v) \simeq u \cdot v, \\ u * v &\simeq u \cdot v \simeq (k * u) \cdot (v * k) = (k \cdot v) * (u \cdot k) \simeq v * u \end{aligned}$$

hold. In order to see the equalities, write down the definition of the maps. □

(2.7.4) Lemma. *The map $v\colon [S^1, S^1]^0 \to [S^1, S^1]$ which forgets the base point is a bijection.*

Proof. Given $f\colon S^1 \to S^1$ we choose a path $w\colon I \to S^1$ from 1 to $f(1)^{-1}$. Then $(x, t) \mapsto f(x)w(t)$ is a homotopy from f to a pointed map, hence v is surjective. Let $H\colon S^1 \times I \to S^1$ be a homotopy between pointed maps; then $(x, t) \mapsto H(x, t) \cdot H(1, t)^{-1}$ is a pointed homotopy between the same maps, i.e., v is injective. □

If $f, g\colon X \to S^1$ are continuous maps, then $f \cdot g\colon x \mapsto f(x)g(x)$ is again continuous. This product of functions is compatible with homotopies and induces the structure of an abelian group on $[X, S^1]$.

(2.7.5) Theorem. *From* (2.7.2), (2.7.4) *and* (2.5.3) *we obtain an isomorphism d,*

$$d\colon [S^1, S^1] \cong [S^1, S^1]^0 \cong \pi_1(S^1, 1) \cong \mathbb{Z}.$$

We call the integer $d(f) = d([f])$ the **degree** *of $f\colon S^1 \to S^1$. A standard map of degree n is $\sigma_n\colon z \mapsto z^n$. A null homotopic map has degree zero.* □

(2.7.6) Example. A polynomial function

$$g\colon \mathbb{C} \to \mathbb{C}, \quad g(z) = z^n + a_1 z^{n-1} + \cdots + a_n$$

has a root ($n \geq 1$).

Proof. Suppose $g(z) \neq 0$ for $|z| = 1$. Then $f: S^1 \to S^1$, $z \mapsto g(z)/|g(z)|$ is defined. Suppose g is non-zero for $|z| \leq 1$. Then $h(z,t) = g(tz)/|g(tz)|$ is a null homotopy of f. For $t > 0$ we have

$$k(z,t) = z^n + t(a_1 z^{n-1} + a_2 t z^{n-2} + \cdots + a_n t^{n-1}) = t^n g(z/t).$$

If g is non-zero for $|z| \geq 1$, then $H(z,t) = k(z,t)/|k(z,t)|$ is a homotopy from f to σ_n. Thus if g has no root, then σ_n is null homotopic; this contradicts (2.7.5). □

The classical approach to $\pi_1(S^1)$ uses topological properties of the exponential function $p: \mathbb{R} \to S^1$, $t \mapsto \exp(2\pi i t)$. A ***lifting*** of $w: [a,b] \to S^1$ along p is a map $W: [a,b] \to \mathbb{R}$ with $pW = w$; the value $W(a)$ is the ***initial condition*** of the lifting. Liftings always exist and depend continuously on the path and the initial condition (see (2.7.8)).

(2.7.7) Proposition. *Let $f: S^1 \to S^1$ be given. Let $F: I \to \mathbb{R}$ be a lifting of fp_0 along p. Then $F(1) - F(0)$ is the degree of f.*

Proof. Let $g = f(1)^{-1} f$. Then $\delta(d(f)) = [gp_0]$, by the definition of d in (2.7.5). There exists $a \in \mathbb{R}$ such that $f(1) = \exp(2\pi i a)$ and $F(0) = a$. Then $\Phi = F - a$ is a lifting of gp_0 with initial condition 0. Hence $F(1) - F(0) = \Phi(1) - \Phi(0) = \Phi(1) = n \in \mathbb{Z}$. The homotopy $(x,t) \mapsto (1-t)\Phi(x) + tx\Phi(1)$ is a homotopy of paths. Hence the loop gp_0 is homotopic to s_n. This shows $\delta(n) = [gp_0]$. □

The next proposition will be proved in the chapter on covering spaces. It expresses the fact that $p: \mathbb{R} \to S^1$ is fibration.

(2.7.8) Proposition. *Given a homotopy $h: X \times I \to S^1$ and an initial condition $a: X \to \mathbb{R}$ such that $pa(x) = h(x,0)$. Then there exists a unique homotopy $H: X \times I \to \mathbb{R}$ such that $H(x,0) = a(x)$ and $pH = h$.* □

(2.7.9) Example. Let $w: [0,1] \to S^1$ be a path with $w(0) = z = \exp(2\pi i a)$. Let $W: [0,1] \to \mathbb{R}$ be a lifting of w with $W(0) = a$. Suppose $W(1) = b$. Then W is, by a linear homotopy, homotopic to $t \mapsto a + t(b-a)$ and hence w homotopic to the path in normal form $t \mapsto z\exp(2\pi i(b-a)t)$. ◇

2.7.10 The winding number. Let $x \in \mathbb{C} = \mathbb{R}^2$. The map

$$r_x: \mathbb{C} \setminus \{x\} \to S^1, \quad z \mapsto (z-x)/|z-x|$$

is an h-equivalence and therefore $[S^1, \mathbb{C} \setminus \{x\}] \to [S^1, S^1]$, $[f] \mapsto [r_x f]$ a bijection. The degree of $r_x f$ is the ***winding number*** of f with respect to x. We denote it by $W(f,x)$. Maps $f_0, f_1: S^1 \to \mathbb{C} \setminus \{x\}$ are homotopic if and only if they have the same winding number. If $f: S^1 \to \mathbb{C}$ is given and $w: I \to \mathbb{C}$ a path with

$f(S^1) \cap w(I) = \emptyset$, then $(x,t) \mapsto r_{w(t)} f$ is a homotopy. Therefore the winding numbers of f with respect to $w(0)$ and $w(1)$ are equal. The complement $\mathbb{C} \setminus f(S^1)$ decomposes into open path components, and the winding number with respect to x is constant as long as x stays within a component.

Let $u: I \to \mathbb{C} \setminus \{x\}$ be a loop. Then there exists a unique continuous map $f: S^1 \to \mathbb{C} \setminus \{x\}$ such that $f \circ p_0 = u$. The winding number of f is then also called the winding number of u, and we denote it by $W(u, x)$. ◇

The notion of the degree can be extended to other situations. Let $h: S \to S^1$ be a homeomorphism and $f: S \to S$ any map; the degree of hfh^{-1} is independent of the choice of a homeomorphism h and also called the degree $d(f)$ of f.

Problems

1. Let p be a polynomial function on $\mathbb{C}$ which has no root on S^1. Then the number of roots z with $|z| < 1$ (counted with multiplicities) is equal to the winding number $W(p|S^1, 0)$. What is the winding number of the function $1/p$ with respect to 0?

2. (Properties of the degree.) $d(f \circ g) = d(f)d(g)$. A homeomorphism $S^1 \to S^1$ has degree ± 1. If $f: S^1 \to S^1$ has degree $d(f) \neq 1$, then there exists $x \in S^1$ such that $f(x) = x$. The map $z \mapsto \bar{z}$ has degree -1.

Let $u = \exp(2\pi i/n)$ be an n-th root of unity. Suppose $h: S^1 \to S^1$ satisfies $h(uz) = h(z)$. Then $d(h) \equiv 0 \bmod n$.

Let $k, j \in \mathbb{Z}$ and assume that k is coprime to n. Let $f: S^1 \to S^1$ satisfy the functional equation $f(u^k z) = u^j f(z)$. Then $k\, d(f) \equiv j \bmod n$. If, conversely, this congruence is satisfied with some integer $d(f)$, then there exists a map f of degree $d(f)$ which satisfies the functional equation. In particular an odd map f, i.e., $f(-z) = -f(z)$, has odd degree.

Suppose $f(-z) \neq f(z)$ for all z; then the degree of f is odd. Suppose $f(z) \neq g(z)$ for all z; then $d(f) = d(g)$. Suppose $d(g) \equiv 0 \bmod n$ for some $n > 0$; then there exists $h: S^1 \to S^1$ such that $g = h^n$.

3. Let $U: I \to \mathbb{C}$ be a lifting of $u: I \to \mathbb{C}^* = \mathbb{C} \setminus 0$ along the covering $P: \mathbb{C} \to \mathbb{C}^*$, $z \mapsto \exp(2\pi i z)$. Then $W(u, 0) = U(1) - U(0)$.

4. Let $\gamma: [0,1] \to \mathbb{C}^*$ be a continuously differentiable path with initial point 1. Then $\Gamma: [0,1] \to \mathbb{C}$, $t \mapsto \frac{1}{2\pi i} \int_{\gamma|[0,t]} \frac{dz}{z}$ is a continuously differentiable lifting of γ along P with initial point 0.

5. If $u: I \to \mathbb{C} \setminus \{x\}$ is a continuously differentiable loop, then $W(u, x) = \frac{1}{2\pi i} \int_u \frac{dz}{z - x}$.

6. Let $A \in \mathrm{GL}_2(\mathbb{R})$. Then the winding number of $l_A: S^1 \mapsto \mathbb{R}^2 \setminus 0$, $x \mapsto Ax$ with respect to the origin is the sign of the determinant $\det(A)$.

7. Let $v: [S^1, X]^0 \to [S^1, X]$ be the map which forgets about the base point (pointed homotopies versus free homotopies). Conjugate elements in the group $[S^1, X]^0$ have the same image under v. Hence v induces a well-defined map $\overline{v}: [S^1, X]^0/(\sim) \to [S^1, X]$ from the set of conjugacy classes. This map is injective, and surjective if X is path connected. Thus v is bijective if X is path connected and the fundamental group abelian.

2.8 Examples

The formal nature of the theorem of Seifert and van Kampen is simple, but the corresponding algebra can be complicated. The setup usually leads to groups presented by generators and relations. It may be difficult to understand a group presented in this manner. For an introduction to this type of group theory see [122]; also [171] and [39] are informative in this context. We report about some relevant algebra and describe a number of examples and different applications of the fundamental group.

2.8.1 Spheres. If a space X is covered by two open simply connected subsets with path connected intersection, then X is simply connected, since the pushout of two trivial groups is trivial. Coverings of this type exist for the spheres S^n for $n \geq 2$. Hence these spheres are simply connected. ◇

2.8.2 Removing a point. The inclusion $D^n \smallsetminus 0 \subset D^n$ induces for $n \geq 3$ an isomorphism of fundamental groups; actually both groups are zero, since D^n is contractible and $D^n \smallsetminus 0 \simeq S^{n-1}$.

Let M be a manifold of dimension $n \geq 3$ and $U \subset M$ homeomorphic to D^n under a homeomorphism that sends x to 0. Then M is the pushout of $M \smallsetminus \{x\}$ and U. Theorem (2.6.2) implies that $M \smallsetminus \{x\} \subset M$ induces an isomorphism of fundamental groups.

Often we view the space S^n as the one-point compactification $\mathbb{R}^n \cup \{\infty\}$ of the Euclidean space, see (2.3.2). Let K be a compact subset of $\mathbb{R}^n$ for $n \geq 3$. Then the inclusion $\mathbb{R}^n \smallsetminus K \subset S^n \smallsetminus K$ induces an isomorphism of fundamental groups. ◇

2.8.3 Complements of spheres. Let $S_0^m = S^{m+n+1} \cap (\mathbb{R}^{m+1} \times 0)$ and $S_1^n = S^{m+n+1} \cap (0 \times \mathbb{R}^{n+1})$. Then $X = S^{m+n+1} \smallsetminus S_1^n$ is homeomorphic to $S^m \times E^n$. A homeomorphism $S^m \times E^n \to X$ is $(x, y) \mapsto \big(\sqrt{1 - \|y\|^2}x, y\big)$. The space $Y = S^{m+n+1} \smallsetminus (S_0^m \cup S_1^n)$ is homeomorphic to $S^m \times S^n \times]0, 1[$ via $(x, y, t) \mapsto (\sqrt{1-t}x, \sqrt{t}y)$. Therefore the complement X is h-equivalent to S^m and the complement Y is h-equivalent to $S^m \times S^n$.

The fundamental group of $S^3 \smallsetminus (\{0\} \times S^1)$ is isomorphic to $\mathbb{Z}$. If we view $S^3 = \mathbb{R}^3 \cup \{\infty\}$, then we are considering the complement of the axis $Z = \{(0, 0, z) \mid z \in \mathbb{R}\} \cup \{\infty\}$. The generator of the fundamental group is a loop that runs once about the axis Z, represented by the standard sphere $W = S^1 \times \{0\}$.

It is impossible to span a 2-disk with boundary W in the complement of Z, because W represents a non-zero element in the fundamental group of the complement, see Figure 2.3. This is expressed by saying that W and Z are linked in S^3. Apply the stereographic projection (2.3.2) to $S^1 \times 0 \cup 0 \times S^1 \subset S^3$. The image yields $W \cup Z$. The fundamental group of the complement of $W \cup Z$ in $\mathbb{R}^3$ is therefore isomorphic to the fundamental group $\mathbb{Z} \times \mathbb{Z}$ of the torus $S^1 \times S^1$. The reader should draw generators of $\pi_1(\mathbb{R}^3 \smallsetminus W \cup Z)$. ◇

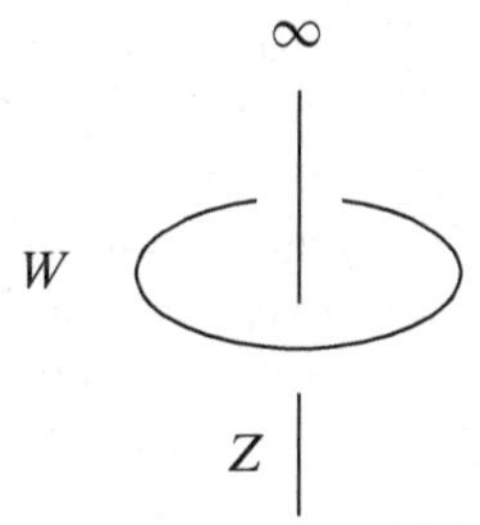

Figure 2.3. A standard circle in S^3.

2.8.4 Presentation of groups by generators and relations. Let S be a set. A ***free group*** with basis S consists of a group $F(S)$ and a set map $i: S \to F(S)$ which has the following universal property: For each set map $\alpha: S \to G$ into a group G there exists a unique homomorphism $A: F(S) \to G$ such that $A \circ i = \alpha$. It turns out that i is injective. Let us consider i as an inclusion and set $S^{-1} = \{s^{-1} \mid s \in S\}$. A word in the alphabet $X = S \amalg S^{-1}$ is a sequence $(x_1, \dots, x_m)$ of elements $x_i \in X$. The elements in $F(S)$ are the products $x_1 \dots x_m$ corresponding to the words; the neutral element belongs to the empty word; a word (x, x^{-1}) also represents the neutral element.

Let R be a set of words and $\bar{R}$ the image in $F(S)$. Let $N(R)$ be the normal subgroup generated by $\bar{R}$. The factor group $G = F(S)/N(R)$ is the group presented by the generators S and the relations R. We denote this group by $\langle S|R\rangle$. It has the following universal property: Let $\alpha: S \to H$ be a set map into a group H. Assume that for each $(x_1, \dots, x_m) \in R$ the relation $\alpha(x_1)\dots\alpha(x_m) = 1$ holds in H. Then there exists a unique homomorphism $A: G \to H$ such that $A(x) = \alpha(x)$ for each $x \in S$.

Each group can be presented in the form $\langle S \mid R\rangle$ – in many different ways. In practice one uses a less formal notation. Here are a few examples.

(i) The cyclic group of order n has the presentation $\langle a \mid a^n\rangle$.

(ii) Let $S = \{x, y\}$. Consider the word $(x, x, y^{-1}, y^{-1}, y^{-1})$ and R consisting of this word. Then we can write $G = \langle S|R\rangle$ also in the form $\langle x, y \mid x^2y^{-3}\rangle$ or $\langle x, y \mid x^2 = y^3\rangle$. The universal property says in this case that homomorphisms $G \to H$ correspond bijectively to set maps $\alpha: \{x, y\} \to H$ such that $\alpha(x)^2 = \alpha(y)^3$.

(iii) $\langle a, b \mid ab = ba\rangle$ is a presentation for the free abelian group with basis a, b. ◇

2.8.5 Free product and pushout of groups. The sum (= coproduct) in the category of groups is also called a free product. Let $(G_j \mid j \in J)$ be a family of groups. The ***free product*** of this family consists of a group $\ast_{k\in J} G_k$ together with a family of homomorphisms $\iota_j : G_j \to \ast_{k\in J} G_k$ which have the universal property of a sum in the category of groups. (The notation $G_1 * G_2$ is used for the free product of two groups.) Each family has a sum. Let $G_j = \langle S_j | R_j \rangle$ and assume that the S_j are disjoint. Let $S = \amalg_j S_j$. The R_j are then words in the alphabet $S \amalg S^{-1}$. Let $R = \bigcup_j R_j$. We have homomorphisms $\iota_j : \langle S_j | R_j \rangle \to \langle S | R \rangle$ which are induced by $S_j \subset S$. These homomorphisms are a sum in the category of groups.

Let G and H be groups and $i_1 : G \to G * H$, $j_1 : H \to G * H$ be the canonical maps which belong to the sum. Let $J : P \to G$, $I : P \to H$ be homomorphisms from a further group P. Let N be the normal subgroup of $G * H$ generated by the elements $\{i_1 J(x) \cdot j_1 I(x^{-1}) \mid x \in P\}$. Let $Q = (G * H)/N$ and denote by $i : G \to Q$, $j : H \to Q$ the composition of i_1, j_1 with the quotient map. Then (i, j) is a pushout of (J, I) in the category of groups. In the case that I and J are inclusions (but sometimes also in the general case) one writes $Q = G *_P H$.

Let S be a set and $\mathbb{Z} = \mathbb{Z}_s$ a copy of the additive group $\mathbb{Z}$ for each $s \in S$. Then the groups $F(S)$ with basis S is also the free product $\ast_{s\in S} \mathbb{Z}$. ◇

2.8.6 Free products of fundamental groups. The free product $\pi_1(X_0) * \pi_1(X_1)$ arises geometrically if X_{01} is simply connected.

Let $X = S^1 \vee S^1$ with $X_0 = Y \vee S^1$, $X_1 = S^1 \vee Y$, where $Y = S^1 \smallsetminus \{-1\}$, $* = 1$. Then $(Y, 1)$ is pointed contractible; hence the inclusion of the summands $S^1 \to X_0, X_1$ are pointed h-equivalences. One can apply (2.6.2) to the covering of X by X_0, X_1. Since $X_0 \cap X_1$ is pointed contractible, we see that $\pi_1(X)$ is the free product $\pi_1(X_0) * \pi_1(X_1)$. Hence the inclusions of the summands $S^1 \to S^1 \vee S^1$ yield a presentation of $\pi_1(S^1 \vee S^1)$ as a free product $\pi_1(S^1) * \pi_1(S^1) \cong \mathbb{Z} * \mathbb{Z}$. By induction one shows that $\pi_1(\bigvee_1^k S^1)$ is the free group of rank k. ◇

2.8.7 Plane without two points. The space $\mathbb{R}^2 \smallsetminus \{\pm 1\}$ has as a deformation retract the union X of the circles about ± 1 with radius $1/2$ and the segment from $-1/2$ to $1/2$, see Figure 2.4. (The reader should try to get an intuitive understanding of a

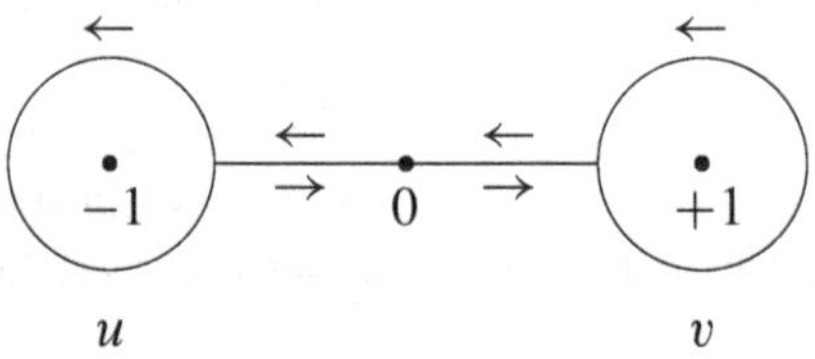

Figure 2.4. Generators u, v of $\pi_1(\mathbb{R}^2 \smallsetminus \pm 1)$.

retraction. In order to give a formal proof, without writing down explicit formulas, it is advisable to wait for the method of cofibrations.) The fundamental group $\pi_1(\mathbb{R}^2 \smallsetminus \{\pm 1\}, 0)$ is the free group $\mathbb{Z} * \mathbb{Z}$ and generators are represented by two small circles about ± 1 (of radius $1/2$, say) connected linearly to the base point. One can apply (2.6.2) to the covering of $\mathbb{R}^2 \smallsetminus \{\pm 1\}$ by the punctured half-spaces $\{(x,y) \mid x < 1/3,\ x \neq -1\}$ and $\{(x,y) \mid -1/3 < x,\ x \neq 1\}$. ◇

In the example in 2.8.6 one cannot apply (2.6.2) directly to the covering of $S^1 \vee S^1$ by the two summands, since the interiors do not cover the space. The general method in cases like this is to first "thicken" the subspaces up to h-equivalence. In the next theorem we add a hypothesis which allows for a thickening.

(2.8.8) Theorem. *Let $((X_j, x_j) \mid j \in J)$ be a family of pointed spaces with the property: The base point x_j has an open neighbourhood $U_j \subset X_j$ which is pointed contractible to the base point. The inclusions of the summands induce homomorphisms $i_j \colon \pi_1(X_j, x_j) \to \pi_1(\bigvee_{k \in J} X_k, x)$. This family is a free product of the groups $\pi_1(X_j, x_j)$.*

Proof. Let $J = \{1,2\}$. We apply (2.6.2) to the covering $X_1 \vee U_2, U_1 \vee X_2$ of $X_1 \vee X_2$. The argument is as for 2.8.4. For finite J we use induction on $|J|$. Note that $\bigvee U_j \subset \bigvee X_j$ is an open, pointed contractible neighbourhood of the base point.

Let now J be arbitrary. A path $w \colon I \to \bigvee X_j$ has, by compactness of I, an image in $(\bigvee_{e \in E} X_e) \vee (\bigvee_{j \in J \smallsetminus E} U_j)$ for a finite subset $E \subset J$. This fact and the result for E show that the canonical map $\alpha_J \colon \ast_{j \in J} \pi_1(X_j) \to \pi_1(\bigvee_{j \in J} X_j)$ is surjective. Each element $x \in \ast_{j \in J} \pi_1(X_j)$ is contained in some $\ast_{e \in E} \pi_1(X_e)$ for a finite E. Suppose x is contained in the kernel of α_J. Then a loop $w \colon I \to \bigvee_{e \in E} X_e$ representing $\alpha_E x$ is null homotopic in $\bigvee_{j \in J} X_j$ and, again by compactness, null homotopic in some larger finite wedge. The result for a finite index set now yields $x = 0$. □

2.8.9 Quotient groups. Let $i \colon K \to G$ be a homomorphism of groups and denote by $N \lhd G$ the normal subgroup generated by the image of i. Then

$$\begin{array}{ccc} K & \longrightarrow & 1 \\ \downarrow{\scriptstyle i} & & \downarrow \\ G & \xrightarrow{p} & G/N \end{array}$$

is a pushout in the category of groups, with p the quotient map.

This situation arises geometrically in (2.6.2) if one of the spaces X_ν is simply connected. ◇

2.8.10 Attaching of a 2-cell. We start with a pushout diagram of spaces

$$\begin{array}{ccc} S^1 & \xrightarrow{\varphi} & B \\ \downarrow{\scriptstyle j} & & \downarrow{\scriptstyle J} \\ D^2 & \xrightarrow{\Phi} & X. \end{array}$$

Then X is said to be obtained from B by attaching a 2-cell via the attaching map φ. (This construction will be studied in detail in the chapter on cell complexes.) Then a suitable thickening shows that we can apply (2.6.2). Since D^2 is contractible, we are in the situation of 2.8.9. Thus J_* induces an isomorphism $\pi_1(B)/\langle\varphi\rangle \cong \pi_1(X)$ where $\langle\varphi\rangle$ denotes the normal subgroup generated by $[\varphi] \in [S^1, B]^0 = \pi_1(B)$. ◇

2.8.11 Attaching of a cone. Given a map $\varphi\colon A \to B$ from a path connected space A. The cone on A is the space $A \times I/A \times 0$. Let $j : A \to CA$, $a \mapsto (a, 1)$ denote the inclusion of A into the cone. The cone is contractible, a contracting homotopy is induced by $h_t(x, s) = (x, s(1-t))$. Form a pushout

$$\begin{array}{ccc} A & \xrightarrow{\varphi} & B \\ \downarrow{\scriptstyle j} & & \downarrow{\scriptstyle J} \\ CA & \xrightarrow{\Phi} & X. \end{array}$$

Since CA is contractible, J_* induces an isomorphism $\pi_1(B)/N \cong \pi_1(X)$, where N is the normal subgroup generated by the image of φ_*. ◇

2.8.12 Realization of groups. We demonstrate that arbitrary groups can be realized as fundamental groups. Let

$$f : A = \bigvee_{k\in K} S^1 \to \bigvee_{l\in L} S^1 = B$$

be a pointed map. The inclusions of the summands yield a basis $a_k \in \pi_1(A)$ and $b_l \in \pi_1(B)$ for the free groups and $f_*(a_k) = r_k$ is a word in the b_l^t, $t \in \mathbb{Z}$. Let $N \subset \ast_{l\in L}\mathbb{Z} = G$ be the normal subgroup generated by the r_k. Then G/N is the group presented by generators and relations $\langle b_l, l \in L \mid r_k, k \in K\rangle$. Let CA be the cone on A and define X by a pushout diagram

$$\begin{array}{ccc} A & \xrightarrow{f} & B \\ \downarrow{\scriptstyle \cap} & & \downarrow \\ CA & \longrightarrow & X. \end{array}$$

Then 2.8.11 shows $\pi_1(X) \cong G/N$. Each group can be presented in the form G/N. Note that X is a 2-dimensional cell complex. ◇

2.8.13 Surfaces. The classification theory of compact connected surfaces presents a surface as a quotient space of a regular $2n$-gon, see e.g., [44, p. 75], [167], [123]. The edges are identified in pairs by a homeomorphism. The surface F is obtained as a pushout of the type

$$\begin{array}{ccc} S^1 & \xrightarrow{\varphi} & \bigvee_1^n S^1 \\ \downarrow & & \downarrow \\ D^2 & \xrightarrow{\Phi} & F. \end{array}$$

The attaching map φ is given in terms of the standard generators of $\pi_1(\bigvee_1^n S^1)$ by the so-called surface-word.

In order to save space we refer to [44, p. 83–87] for the discussion of the fundamental group of surfaces in general. We mention at least some results. They will not be used in this text.

(2.8.14) Theorem. (1) *The fundamental group of a closed connected orientable surface F_g of genus $g \geq 1$ has the presentation*

$$\pi_1(F_g) = \langle a_1, b_1, \dots, a_g, b_g \mid a_1 b_1 a_1^{-1} b_1^{-1} \dots a_g b_g a_g^{-1} b_g^{-1} \rangle.$$

(2) *The fundamental group of a closed connected non-orientable surface N_g of genus g has the presentation*

$$\pi_1(N_g) = \langle a_1, \dots, a_g \mid a_1^2 a_2^2 \dots a_g^2 \rangle.$$

(3) *The fundamental group of a compact connected surface with non-empty boundary is a free group. The number of generators is the finite number $1 - \chi(F_g)$ where $\chi(F_g)$ is the so-called Euler characteristic.*

(4) *A simply connected surface is homeomorphic to $\mathbb{R}^2$ or S^2.* □

There are many different definitions of the genus. We mention a geometric property: The genus of a closed connected orientable surface is the maximal number g of disjointly embedded circles such that their complement is connected. The genus of a closed connected non-orientable surface is the maximal number g of disjointly embedded Möbius bands such that their complement is connected. The sphere has genus zero by the Jordan separation theorem. ◇

Problems

1. Let $S^1 \subset \mathbb{R}^2 \times 0 \subset \mathbb{R}^3$ be the standard circle. Let $D = \{(0,0,t) \mid -2 \leq t \leq 2\}$ and $S^2(2) = \{x \in \mathbb{R}^3 \mid \|x\| = 2\}$. Then $S^2(2) \cup D$ is a deformation retract of $X = \mathbb{R}^3 \setminus S^1$. The space X is h-equivalent to $S^2 \vee S^1$.

2. Consider the loop based at $(0,0)$ in the plane as shown in Figure 2.5. Determine which

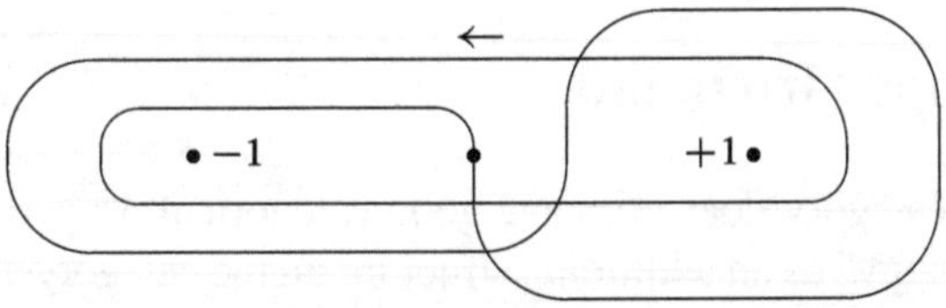

Figure 2.5.

element in $\pi_1(\mathbb{R}^2 \smallsetminus \pm 1)$ this loop represents in terms of the generators u, v in 2.8.7. Determine the winding number about points in each of the six complementary regions.

3. Let $(X_j \mid j \in J)$ be a family of subspaces of X such that the interiors X_j° cover X. Then each morphism in $\Pi(X)$ is a composition of morphisms in the X_j. If the intersections $X_i \cap X_j$ are path connected and $* \in X_i \cap X_j$, then $\pi_1(X, *)$ is generated by loops in the X_j.

4. Let i_{0*} in (2.6.2) be an isomorphism. Then j_{1*} is an isomorphism. This statement is a general formal property of pushouts. If i_{0*} is surjective, then j_{1*} is surjective.

5. Projective plane. The real projective plane P^2 is defined as the quotient of S^2 by the relation $x \sim -x$. Let $[x_0, x_1, x_2]$ denote the equivalence class of $x = (x_0, x_1, x_2)$. We can also obtain P^2 from S^1 by attaching a 2-cell

$$\begin{array}{ccc} S^1 & \xrightarrow{\varphi} & P^1 \\ \downarrow{\scriptstyle j} & & \downarrow{\scriptstyle J} \\ D^2 & \xrightarrow{\Phi} & P^2. \end{array}$$

Here $P^1 = \{[x_0, x_1, 0]\} \subset P^2$ and $\varphi(x_0, x_1) = [x_0, x_1, 0]$. The space P^1 is homeomorphic to S^1 via $[x_0, x_1, 0] \mapsto z^2, z = x_0 + ix_1$; and φ corresponds to the standard map of degree 2. The map Φ is $x = (x_0, x_1) \mapsto [x_0, x_1, \sqrt{1 - \|x\|^2}]$. As an application of 2.8.10 we obtain $\pi_1(P^2) \cong \mathbb{Z}/2$.

Another interpretation of the pushout: P^2 is obtained from D^2 by identifying opposite points of the boundary S^1. The subspace $\{(x_0, x_1) \mid \|x\| \geq 1/2\}$ becomes in P^2 a Möbius band M. Thus P^2 is obtainable from a Möbius band M and a 2-disk D by identification of the boundary circles by a homeomorphism. The projective plane cannot be embedded into $\mathbb{R}^3$, as we will prove in (18.3.7). There exist models in $\mathbb{R}^3$ with self-intersections (technically, the image of a smooth immersion.) The projective plane is a non-orientable surface.

6. Klein bottle. The Klein bottle K can be obtained from two Möbius bands M by an identification of their boundary curves with a homeomorphism, $K = M \cup_{\partial M} M$.

Apply the theorem of Seifert and van Kampen and obtain the presentation $\pi_1(K) = \langle a, b \mid a^2 = b^2 \rangle$. The elements a^2, ab generate a free abelian subgroup of rank 2 and of index 2 in the fundamental group. The element a^2 generates the center of this group, it is represented by the central loop ∂M. The quotient by the center is isomorphic to $\mathbb{Z}/2 * \mathbb{Z}/2$.

The space $M/\partial M$ is homeomorphic to the projective plane P^2. If we identify the central ∂M to a point, we obtain a map $q \colon K = M \cup_{\partial M} M \to P^2 \vee P^2$. The induced map on the fundamental group is the homomorphism onto $\mathbb{Z}/2 * \mathbb{Z}/2$.

2.9 Homotopy Groupoids

The homotopy category does not have good categorical properties. Therefore we consider "homotopy" as an additional structure on the category TOP of topological spaces. The category TOP will be enriched: The set of morphisms $\Pi(X, Y)$ between two objects carries the additional structure of a groupoid. The fundamental groupoid is the special case in which X is a point.

Recall that a category has the data: objects, morphisms, identities, and composition of morphisms. The data satisfy the axioms: composition of morphisms is associative; identities are right and left neutral with respect to composition.

Let X and Y be topological spaces. We define a category $\Pi(X, Y)$ and begin with the data. The *objects* are the continuous maps $X \to Y$. A morphism from $f: X \to Y$ to $g: X \to Y$ is represented by a homotopy $K: f \simeq g$. Two such homotopies K and L define the same morphism if they are homotopic relative to $X \times \partial I$ with $\partial I = \{0, 1\}$ the boundary of I. Let us use a second symbol $J = [0, 1]$ for the unit interval. This means: A map $\Phi: (X \times I) \times J \to Y$ is a homotopy relative to $X \times \partial I$, if $\Phi(x, 0, t)$ is independent of t and $\Phi(x, 1, t)$ is also independent of t. Therefore $\Phi_t: X \times I \to Y, (x, s) \mapsto \Phi(x, s, t)$ is for each $t \in J$ a homotopy from f to g. For this sort of relative homotopy one has, as before, the notion of a product and an inverse, now with respect to the J-variable. Hence we obtain an equivalence relation on the set of homotopies from f to g. We now define: A *morphism* $\sigma: f \to g$ in $\Pi(X, Y)$ is an equivalence class of homotopies relative to $X \times \partial I$ from f to g. *Composition of morphisms*, denoted $\circledast$, is defined by the product of homotopies

$$K: f \simeq g, L: g \simeq h, \quad [L] \circledast [K] = [K * L]: f \to h.$$

This is easily seen to be well-defined (use $\Phi_t *_I \Psi_t$). The *identity* of f in $\Pi(X, Y)$ is represented by the constant homotopy $k_f: f \simeq f$.

The verification of the category axioms is based on the fact that a reparametrization of a homotopy does not change its class.

(2.9.1) Lemma. *Let $\alpha: I \to I$ be a continuous map with $\alpha(0) = 0$ and $\alpha(1) = 1$. Then K and $K \circ (\mathrm{id} \times \alpha)$ are homotopic relative to $X \times \partial I$.*

Proof. $\Phi(x, s, t) = K(x, (1-t)s + t\alpha(s))$ is a suitable homotopy. □

(2.9.2) Proposition. *The data for $\Pi(X, Y)$ satisfy the axioms of a category. The category is a groupoid.*

Proof. The associativity of the composition follows, because

$$(K * L) * M = K * (L * M) \circ (\mathrm{id} \times \alpha),$$

with α defined by $\alpha(t) = 2t$ for $t \le \frac{1}{4}$, $\alpha(t) = t + \frac{1}{4}$ for $\frac{1}{4} \le t \le \frac{1}{2}$, $\alpha(t) = \frac{t}{2} + \frac{1}{2}$ for $\frac{1}{2} \le t \le 1$.

Similarly, for each $K: f \simeq g$ the homotopies $k_f * K$, K, and $K * k_g$ differ by a parameter change. Therefore the constant homotopies represent the identities in the category.

The inverse homotopy K^- represents an inverse of the morphism defined by K. Hence each morphism is an isomorphism. Proof: The assignments $(x, s, t) \mapsto K(x, 2s(1-t))$ for $0 \le s \le \frac{1}{2}$ and $(x, s, t) \mapsto K(x, 2(1-s)(1-t))$ for $\frac{1}{2} \le s \le 1$ yield a homotopy relative to $X \times \partial I$ from $K * K^-$ to the constant homotopy. □

The endomorphism set of an object in a groupoid is a group with respect to composition as group law. We thus see, from this view point, that the notion of homotopy directly leads to algebraic objects. This fact is a general and systematic approach to algebraic topology.

The homotopy categories of Section 2.2 have a similar enriched structure. If we work, e.g., with pointed spaces and pointed homotopies, then we obtain for pointed spaces X and Y a category $\Pi^0(X,Y)$. The objects are pointed maps. Morphisms are represented by pointed homotopies, and the equivalence is defined by homotopies Φ rel $X \times \partial I$ such that each Φ_t is a pointed homotopy.

The remainder of this section can be skipped on a first reading. We study the dependence of the groupoids $\Pi(X,Y)$ on X and Y. The formal structure of this dependence can be codified in the notion of a ***2-category***. Suppose given $\alpha\colon U \to X$ and $\beta\colon Y \to V$. Composition with α and β yield a functor

$$\beta_\# = \Pi_\#(\beta)\colon \Pi(X,Y) \to \Pi(X,V),$$

which sends f to βf and $[K]$ to $[\beta K]$ and a functor

$$\alpha^\# = \Pi^\#(\alpha)\colon \Pi(X,Y) \to \Pi(U,Y),$$

which sends f to $f\alpha$ and $[K]$ to $[K(\alpha \times \mathrm{id})]$. They satisfy $(\beta^1\beta^2)_\# = \beta^1_\#\beta^2_\#$ and $(\alpha_1\alpha_2)^\# = \alpha_2^\#\alpha_1^\#$. These functors are compatible in the following sense:

(2.9.3) Proposition. *Suppose $K\colon f \simeq g\colon X \to Y$ and $L\colon u \simeq v\colon Y \to Z$ are given. Then $[L \diamond K] = v_\#[K] \circledast f^\#[L] = g^\#[L] \circledast u_\#[K]$. Here $L \diamond K\colon uf \simeq vg\colon X \times I \to Z, (x,t) \mapsto L(K(x,t),t)$.*

Proof. We use the bi-homotopy $L \circ (K \times \mathrm{id})\colon X \times I \times I \to Z$. Restriction to the diagonal of $I \times I$ defines $L \diamond K$. Along the boundary of the square we have the following situation.

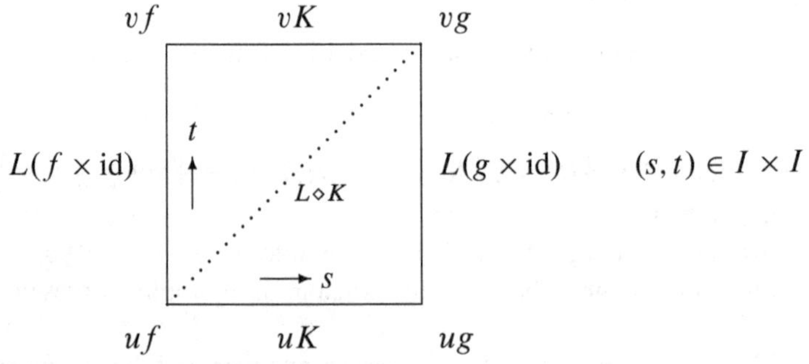

$g^\#[L] \circledast u_\#[K]$ is represented by $uK * L(g \times \mathrm{id})$. If we compose the bi-homotopy with $\mathrm{id}(X) \times \gamma$, where $\gamma(t) = (2t, 0)$ for $t \le \frac{1}{2}$ and $\gamma(t) = (1, 2t-1)$ for $t \ge \frac{1}{2}$, we obtain $uK * L(g \times \mathrm{id})$. In the same manner we obtain $L(f \times \mathrm{id}) * vK$ if we compose the bi-homotopy with $\mathrm{id}(x) \times \delta$, where $\delta(t) = (0, 2t)$ for $t \le \frac{1}{2}$ and $\delta(t) = (2t-1, 1)$ for $t \ge \frac{1}{2}$. The maps γ and δ are homotopic relative to ∂I by a linear homotopy in the square. They are also homotopic to the diagonal $t \mapsto (t,t)$ of the square. □

(2.9.4) Corollary. *The homotopy L induces a natural transformation*

$$L_\#\colon u_\# \to v_\#\colon \Pi(X,Y) \to \Pi(X,Z).$$

The value of $L_\#$ at f is $f^\#[L]$. The homotopy K induces a natural transformation

$$K^\#\colon f^\# \to g^\#\colon \Pi(Y,Z) \to \Pi(X,Z).$$

The value of $K^\#$ at u is $u_\#[K]$. □

(2.9.5) Corollary. *If $u\colon Y \to Z$ is an h-equivalence, then $u_\#$ is an equivalence of categories. Similarly in the contravariant case.* □

The data and assertions that we have obtained so far define on TOP the structure of a 2-category. In this context, the ordinary morphisms $f\colon X \to Y$ are called ***1-morphisms*** and the morphisms $[K]\colon f \simeq g$ are called ***2-morphisms***. The composition $\circledast$ of 2-morphisms is called ***vertical composition***. We also have a ***horizontal composition*** of 2-morphisms defined as $[L] \diamond [K] = [L \diamond K]$. Because of (2.9.3) we need not define $\diamond$ via the diagonal homotopy; we can use instead (2.9.3) as a definition $[L] \diamond [K] = v_\#[K] \circledast f^\#[L] = g^\#[L] \circledast u_\#[K]$.

(2.9.6) Note. *From this definition one verifies the* ***commutation rule*** *of a 2-category* $(\delta \circledast \gamma) \diamond (\beta \circledast \alpha) = (\delta \diamond \beta) \circledast (\gamma \diamond \alpha)$. □

The following figure organizes the data (horizontal – vertical).

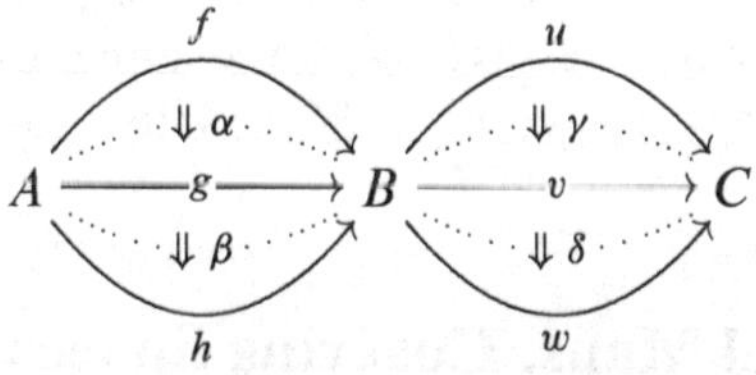

Conversely, one can derive (2.9.3) from the commutation rule (2.9.6). With the constant homotopy k_u of u we have

$$k_u \diamond \alpha = u_\#\alpha, \quad \gamma \diamond k_f = f^\#\gamma, \quad k_v \diamond \alpha = v_\#\alpha, \quad \gamma \diamond k_g = g^\#\gamma$$

and this yields

$$\gamma \diamond \alpha = (\gamma \circledast k_u) \diamond (k_g \circledast \alpha) = (\gamma \diamond k_g) \circledast (k_u \diamond \alpha) = g^\# \circledast u_\#\alpha.$$

In a similar manner one obtains $\gamma \diamond \alpha = v_\#\alpha \circledast f^\#\gamma$.

Chapter 3

Covering Spaces

A covering space is a locally trivial map with discrete fibres. Objects of this type can be classified by algebraic data related to the fundamental group. The reduction of geometric properties to algebraic data is one of the aims of algebraic topology. The main result of this chapter has some formal similarity with Galois theory.

A concise formulation of the classification states the equivalence of two categories. We denote by COV_B the category of covering spaces of B; it is the full subcategory of TOP_B of spaces over B with objects the coverings of B. Under some restrictions on the topology of B this category is equivalent to the category $\text{TRA}_B = [\Pi(B), \text{SET}]$ of functors $\Pi(B) \to \text{SET}$ and natural transformations between them. We call it the transport category. It is a natural idea that, when you move from one place to another, you carry something along with you. This transport of "information" is codified in moving along the fibres of a map (here: of a covering). We will show that the transport category is equivalent to something more familiar: group actions on sets.

The second important aspect of covering space theory is the existence of a universal covering of a space. The automorphism group of the universal covering is the fundamental group of the space – and in this manner the fundamental group appears as a symmetry group. Moreover, the whole category of covering spaces is obtainable by a simple construction (associated covering of bundle theory) from the universal covering.

In this chapter we study coverings from the view-point of the fundamental group. Another aspect belongs to bundle theory. In the chapter devoted to bundles we show for instance that isomorphism classes of n-fold coverings over a paracompact space B correspond to homotopy classes $B \to BS(n)$ into a so-called classifying space $BS(n)$.

3.1 Locally Trivial Maps. Covering Spaces

Let $p\colon E \to B$ be continuous and $U \subset B$ open. We assume that p is surjective to avoid empty fibres. A ***trivialization*** of p ***over*** U is a homeomorphism $\varphi\colon p^{-1}(U) \to U \times F$ over U, i.e., a homeomorphism which satisfies $\text{pr}_1 \circ \varphi = p$. This condition determines the space F up to homeomorphism, since φ induces a homeomorphism of $p^{-1}(u)$ with $\{u\} \times F$. The map p is ***locally trivial*** if there exists an open covering $\mathcal{U}$ of B such that p has a trivialization over each $U \in \mathcal{U}$. A locally trivial map is also called a ***bundle*** or ***fibre bundle***, and a local trivialization a ***bundle chart***. We say, p is ***trivial over*** U, if there exists a bundle chart over U. If

p is locally trivial, then the set of those $b \in B$ for which $p^{-1}(b)$ is homeomorphic to a fixed space F is open and closed in B. Therefore it suffices for most purposes to fix the homeomorphism type of the fibres. If the fibres are homeomorphic to F, we call F the ***typical fibre***. A locally trivial map is open, hence a quotient map.

A ***covering space*** or a ***covering***[1] of B is a locally trivial map $p\colon E \to B$ with discrete fibres. If F is discrete (= all subsets are open and closed), then $U \times F$ is homeomorphic to the topological sum $\amalg_{x\in F} U \times \{x\}$. The summands $U \times \{x\}$ are canonically homeomorphic to U. If $\varphi\colon p^{-1}(U) \to U \times F$ is a trivialization, then p yields via restriction a homeomorphism of $\varphi^{-1}(U \times \{x\})$ with U. A covering is therefore a local homeomorphism. The summands $\varphi^{-1}(U \times \{x\}) = U_x$ are the ***sheets*** of the covering over U; the pre-image $p^{-1}(U)$ is therefore the topological sum of the sheets U_x; the sheets are open in E and mapped homeomorphically onto U under p. If $|F| = n \in \mathbb{N}$, we talk about an n-fold covering. The ***trivial covering*** with typical fibre F is the projection $\mathrm{pr}\colon B \times F \to B$. We say, U is ***admissible*** or ***evenly covered*** if there exists a trivialization over U.

(3.1.1) Example. The exponential function $p\colon \mathbb{R} \to S^1, t \mapsto \exp(2\pi i t)$ is a covering with typical fibre $\mathbb{Z}$. For each $t \in \mathbb{R}$ and $p(t) = z$ we have a homeomorphism

$$p^{-1}(S^1 \smallsetminus z) = \coprod_{n\in\mathbb{Z}}]t+n, t+n+1[\;\cong\;]t, t+1[\times \mathbb{Z},$$

and p maps each summand homeomorphically. ◇

(3.1.2) Proposition. *Let $p\colon E \to B$ be a covering. Then the diagonal D of $E \times E$ is open and closed in $Z = \{(x, y) \in E \times E \mid p(x) = p(y)\}$.*

Proof. Let U_x be an open neighbourhood of x which is mapped homeomorphically under p. Then $Z \cap (U_x \times U_x) = W_x$ is contained in D, and W_x is an open neighbourhood of (x, x) in Z. This shows that D is open.

Let $x \neq y$ and $p(x) = p(y)$. Let $x \in U_x$ and $y \in U_y$ be the sheets of p over the open set $U \subset B$. Since $x \neq y$, the intersection $U_x \cap U_y$ is empty. Hence $Z \cap (U_x \times U_y)$ is an open neighbourhood of (x, y) in Z and disjoint to D. This shows that also the complement $Z \smallsetminus D$ is open. □

Let $p\colon E \to B$ and $f\colon X \to B$ be maps; then $F\colon X \to E$ is a ***lifting*** of f ***along*** p, if $pF = f$.

(3.1.3) Proposition (Uniqueness of liftings). *Let $p\colon E \to B$ be a covering. Let $F_0, F_1\colon X \to E$ be liftings of $f\colon X \to B$. Suppose F_0 and F_1 agree somewhere. If X is connected, then $F_0 = F_1$.*

Proof. (F_0, F_1) yield a map $F\colon X \to Z$. By assumption, $F^{-1}(D)$ is not empty, and hence, by (3.1.2), open and closed. If X is connected, then $F^{-1}(D) = X$, i.e., $F_0 = F_1$. □

[1] Observe that the term "covering" has two quite different meanings in topology.

(3.1.4) Proposition. *Let $q: E \to B \times [0,1]$ be locally trivial with typical fibre F. Then B has an open cover $\mathcal{U}$ such that q is trivial over each set $U \times [0,1]$, $U \in \mathcal{U}$.*

Proof. If q is trivial over $U \times [a,b]$ and over $U \times [b,c]$, then q is trivial over $U \times [a,c]$. Two trivializations over $U \times \{b\}$ differ by an automorphism, and this automorphism can be extended over $U \times [b,c]$. Use this extended automorphism to change the trivialization over $U \times [b,c]$, and then glue the trivializations. By compactness of I there exist $0 = t_0 < t_1 < \cdots < t_n = 1$ and an open set U such that q is trivial over $U \times [t_i, t_{i+1}]$. □

For the classification of covering spaces we need spaces with suitable local properties. A space X is called ***locally connected*** (***locally path connected***) if for each $x \in X$ and each neighbourhood U of x there exists a connected (path connected) neighbourhood V of x which is contained in U. Both properties are inherited by open subspaces.

(3.1.5) Proposition. *The components of a locally connected space are open. The path components of a locally path connected space Y are open and coincide with the components.*

Proof. Let K be the component of x. Let V be a connected neighbourhood of x. Then $K \cup V$ is connected and therefore contained in K. This shows that K is open.

Let U be a component of Y and K a path component of U. Then $U \smallsetminus K$ is a union of path components, hence open. In the case that $U \neq K$ we would obtain a decomposition of U. □

We see that each point in a locally path connected space has a neighbourhood basis of open path connected sets.

(3.1.6) Remark. Let B be path connected and locally path connected. Since a covering is a local homeomorphism, the total space E of a covering of B is locally path connected. Let E' be a component of E and $p': E' \to B$ the restriction of p. Then p' is also a covering: The sets U, over which p is trivial, can be taken as path connected, and then a sheet over U is either contained in E' or disjoint to E'. Since B is path connected, we see by path lifting (3.2.9) that p' is surjective. By (3.1.5), E is the topological sum of its components. ◇

A left action $G \times E \to E$, $(g,x) \mapsto gx$ of a discrete group G on E is called ***properly discontinuous*** if each $x \in E$ has an open neighbourhood U such that $U \cap gU = \emptyset$ for $g \neq e$. A properly discontinuous action is free. For more details about this notion see the chapter on bundle theory, in particular (14.1.12).

A ***left G-principal covering*** consists of a covering $p: E \to B$ and a properly discontinuous action of G on E such that $p(gx) = p(x)$ for $(g,x) \in G \times E$ and such that the induced action on each fibre is transitive.

(3.1.7) Example. A left G-principal covering $p\colon E \to B$ induces a homeomorphism of the orbit space E/G with B. The orbit map $E \to E/G$ of a properly discontinuous action is a G-principal covering. ◇

A covering $p\colon E \to B$ has an automorphism group $\mathrm{Aut}(p)$. An automorphism is a homeomorphism $\alpha\colon E \to E$ such that $p \circ \alpha = p$. Maps of this type are also called ***deck transformations*** of p. If p is a left G-principal covering, then each left translation $l_g\colon E \to E, x \mapsto gx$ is an automorphism of p. We thus obtain a homomorphism $l\colon G \to \mathrm{Aut}(p)$. Let E be connected. Then an automorphism α is determined by its value at a single point $x \in E$, and $\alpha(x)$ is a point in the fibre $p^{-1}(p(x))$. Since G acts transitively on each fibre, the map l is an isomorphism. Thus the connected principal coverings are the connected coverings with the largest possible automorphism group. Conversely, we can try to find principal coverings by studying the action of the automorphism group.

(3.1.8) Proposition. *Let $p\colon E \to B$ be a covering.*

(1) *If E is connected, then the action of* $\mathrm{Aut}(p)$ *(and of each subgroup of* $\mathrm{Aut}(p)$*) on E is properly discontinuous.*

(2) *Let B be locally path connected and let H be a subgroup of* $\mathrm{Aut}(p)$. *Then the map $q\colon E/H \to B$ induced by p is a covering.*

Proof. (1) Let $x \in E$ and $g \in \mathrm{Aut}(p)$. Let U be a neighbourhood of $p(x)$ which is evenly covered, and let U_x be a sheet over U containing x. For $y \in U_x \cap gU_x$ we have $p(y) = p(g^{-1}y)$, since g^{-1} is an automorphism. Hence $y = g^{-1}y$, since both elements are contained in U_x. This shows $g^{-1} = \mathrm{id}$, hence $U_x \cap gU_x = \emptyset$ for $g \neq e$, and we see that the action is properly discontinuous.

(2) Let $U \subset B$ be open, path connected, and evenly covered. Let $p^{-1}(U) = \bigcup_{j \in J} U_j$ be the decomposition into the sheets over U. An element $h \in H$ permutes the sheets, since they are the path components of $p^{-1}(U)$. The equivalence classes with respect to H are therefore open in the quotient topology of E/H and are mapped bijectively and continuously under q. Since p is open, so is q. Hence q is trivial over U. Since B is locally path connected, it has an open covering by such sets U. □

A right G-principal covering $p\colon E \to B$ gives rise to ***associated coverings***. Let F be a set with left G-action. Denote by $E \times_G F$ the quotient space of $E \times F$ under the equivalence relation $(x, f) \sim (xg^{-1}, gf)$ for $x \in E, f \in F, g \in G$. The continuous map $p_F\colon E \times_G F \to B, (x, f) \mapsto p(x)$ is a covering with typical fibre F.

A G-map $\psi\colon F_1 \to F_2$ induces a morphism of coverings

$$\mathrm{id} \times_G \psi\colon E \times_G F_1 \to E \times_G F_2, \quad (x, f) \mapsto (x, \psi(f)).$$

We thus have obtained a functor "associated coverings"

$$A(p)\colon G\text{-SET} \to \mathrm{COV}_B$$

from the category G-SET of left G-sets and G-equivariant maps. We call a G-principal covering $p\colon E \to B$ over the path connected space B ***universal*** if the functor $A(p)$ is an equivalence of categories.

3.2 Fibre Transport. Exact Sequence

The relation of a covering space to the fundamental groupoid is obtained via path lifting. For this purpose we now introduce the notion of a fibration which will be studied later in detail. A map $p\colon E \to B$ has the ***homotopy lifting property*** (HLP) for the space X if the following holds: For each homotopy $h\colon X \times I \to B$ and each map $a\colon X \to E$ such that $pa(x) = hi(x)$, $i(x) = (x, 0)$ there exists a homotopy $H\colon X \times I \to E$ with $pH = h$ and $Hi = a$. We call H a ***lifting*** of h with ***initial condition*** a. The map p is called a ***fibration*** if it has the HLP for all spaces.

(3.2.1) Example. A projection $p\colon B \times F \to B$ is a fibration. Let $a(x) = (a_1(x), a_2(x))$. The condition $pa = hi$ says $a_1(x) = h(x, 0)$. If we set $H(x,t) = (h(x,t), a_2(x))$, then H is a lifting of h with initial condition a. ◇

(3.2.2) Theorem. *A covering $p\colon E \to B$ is a fibration.*

Proof. Let the homotopy $h\colon X \times I \to B$ and the initial condition a be given. Since I is connected, a lifting with given initial condition is uniquely determined (see (3.1.3)). Therefore it suffices to find for each $x \in X$ an open neighbourhood V_x such that $h|V_x \times I$ admits a lifting with initial condition $a|V_x$. By uniqueness (3.1.3), these partial liftings combine to a well-defined continuous map.

By (3.2.3) there exists for each $x \in B$ an open neighbourhood V_x and an $n \in \mathbb{N}$ such that h maps $V_x \times [i/n, (i+1)/n]$ into a set U over which p is trivial. Since $p\colon p^{-1}(U) \to U$ is, by (3.2.1), a fibration, $h|V_x \times [i/n, (i+1)/n]$ has a lifting for each initial condition. Therefore we find by induction over i a lifting of $h|V_x \times [0, i/n]$ with initial condition $a|V_x$. □

(3.2.3) Lemma. *Let $\mathcal{U}$ be an open covering of $B \times [0,1]$. For each $b \in B$ there exists an open neighbourhood $V(b)$ of b in B and $n = n(b) \in \mathbb{N}$ such that for $0 \le i < n$ the set $V(b) \times [i/n, i/(n+1)]$ is contained in some member of $\mathcal{U}$.* □

The fact that the lifted homotopy is uniquely determined implies that for a covering $p\colon E \to B$ the diagram

$$\begin{array}{ccc} E^I & \xrightarrow{p^I} & B^I \\ \downarrow{\scriptstyle e_E^0} & & \downarrow{\scriptstyle e_B^0} \\ E & \xrightarrow{p} & B \end{array}$$

is a pullback of topological spaces. Here E^I is the space of paths in E with compact-open topology and $e_E^0(w) = w(0)$.

The homeomorphism (2.3.6) $k\colon (I^n, \partial I^n) \times (I, 0) \to I^n \times (I, 0)$ of pairs is used to solve the next homotopy lifting problem with a modified initial condition. It reduces the problem to the HLP for I^n.

(3.2.4) Proposition. *Let $p\colon E \to B$ have the* HLP *for the cube I^n. For each commutative diagram*

$$\begin{array}{ccc} I^n \times 0 \cup \partial I^n \times I & \xrightarrow{a} & E \\ \cap \downarrow i & \nearrow & \downarrow p \\ I^n \times I & \xrightarrow{h} & B \end{array}$$

there exists $H\colon I^n \times I \to E$ with $Hi = a$ and $pH = h$. □

Let $p\colon E \to B$ be a map which has the HLP for a point and for I. Write $F_b = p^{-1}(b)$. We associate to each path $v\colon I \to B$ from b to c a map $v_\#\colon \pi_0(F_b) \to \pi_0(F_c)$ which only depends on $[v] \in \Pi(B)$. Let $x \subset F_b$. Choose a lifting $V\colon I \to E$ of v with $V(0) = x$. We set $v_\#[x] = [V(1)]$. We have to show that this assignment is well-defined. For this purpose assume given:

(1) $u\colon I \to F_b$;

(2) $h\colon I \times I \to B$ a homotopy of paths from b to c;

(3) $V_0, V_1\colon I \to E$ liftings of h_0, h_1 with initial points $u(0), u(1)$.

These data yield a map $a\colon I \times \partial I \cup 0 \times I \to E$, defined by $a(s, \varepsilon) = V_\varepsilon(s)$ and $a(0, t) = u(t)$. The lifting H of h with initial condition a, according to (3.2.4), yields a path $t \mapsto H(1, t)$ in F_c from $V_0(1)$ to $V_1(1)$. This shows that the map $v_\#$ is well-defined and depends only on the morphism $[v]$ in the fundamental groupoid. The rule $w_\# v_\# = (v * w)_\#$ is easily verified from the definitions. Thus we have shown:

(3.2.5) Proposition. *The assignments $b \mapsto \pi_0(F_b)$ and $[v] \mapsto v_\#$ yield a functor $T_p\colon \Pi(B) \to \mathrm{SET}$.* □

We call $T_p = T(p)$ the ***transport functor*** associated to p.

The functor T_p provides us with

$$\pi_0(F_b) \times \pi_1(B, b) \to \pi_0(F_b), \quad (x, [v]) \mapsto v_\#(x) = x \cdot [v],$$

a right action of the fundamental group on the set $\pi_0(F_b)$. We write $\pi_0(F, x)$ if $[x]$ is chosen as base point of the set $\pi_0(F)$. We use the action to define

$$\partial_x\colon \pi_1(B, b) \to \pi_0(F_b, x), \quad [v] \mapsto x \cdot [v].$$

The map ∂_x is $\pi_1(B, b)$-equivariant, i.e., $\partial_x[v * w] = (\partial_x[v]) \cdot [w]$.

Recall that a sequence $A \xrightarrow{\alpha} B \xrightarrow{\beta} C$ of pointed maps is ***exact*** at B if the image of α equals the kernel $\beta^{-1}(*)$ of β. Similarly for longer sequences. In this context a group is pointed by its neutral element.

(3.2.6) Theorem. *Let $p(x) = b$ and $i: F_b \subset E$. The sequence*

$$\pi_1(F_b, x) \xrightarrow{i_*} \pi_1(E, x) \xrightarrow{p_*} \pi_1(B, b) \xrightarrow{\partial_x} \pi_0(F_b, x) \xrightarrow{i_*} \pi_0(E, x) \xrightarrow{p_*} \pi_0(B, b)$$

is exact.

Proof. It is easily verified from the definitions that the composition of two maps is the constant map. We consider the remaining four cases: kernel $\subset$ image.

Let $[u] \in \pi_1(E, x)$ and $h: I \times I \to B$ a null homotopy of pu. Consider the lifting problem for h with initial condition $a: I \times 0 \cup \partial I \times I \to E$ with $a(s, 0) = u(s)$ and $a(\varepsilon, t) = x$. The lifting H of h is then a homotopy of loops from u to a loop in the image of i.

Let $\partial_x[v] = [x]$. This means: There exists a lifting V of v from x to $V(1) \in [x]$. Choose a path $U: I \to F_b$ from $V(1)$ to x. Then $V * U$ is a loop in E, and its class maps under $\pi_1(p)$ to $[v]$, since pU is constant.

Let $\pi_0(i)[y] = [x]$. There exists a path $w: I \to E$ from x to y. The projection $v = pw$ is a loop and $\partial_x[v] = [y]$, by definition of ∂_x.

Let $\pi_0(p)[y] = [b]$. Thus there exists a path $v: I \to B$ from $p(y)$ to x. Let $V: I \to E$ be a lifting of v with initial point y. Then $V(1) \in F_b$, and V shows $\pi_0(i)([V(1)] = [y]$. □

There is more algebraic structure in the sequence.

(3.2.7) Proposition. *The pre-images of elements under ∂_x are the left cosets of $\pi_1(B, b)$ with respect to $p_*\pi_1(E, x)$. The pre-images of $\pi_0(i)$ are the orbits of the $\pi_1(B, b)$-action on $\pi_0(F_b, x)$.*

Proof. Let $\partial_x[u] = \partial_x[v]$. Choose liftings U, V of u, v which start in x, and let $W: I \to F_b$ be a path from $U(1)$ to $V(1)$. Then $U * W * V^-$ is a loop in E, and $p(U * W * V^-) * v \simeq u$, i.e., the elements $[u]$ and $[v]$ are contained in the same left coset. Conversely, elements in the same coset have the same image under ∂_x. (A similar assertion holds for right cosets.)

Suppose $\pi_0(i)[a] = \pi_0(i)[b]$. Then there exists a path $w: I \to E$ from a to b. Set $v = pw$. Then $[a] \cdot [v] = [b]$. Conversely, elements in the same orbit have equal image under $\pi_0(i)$. □

We can apply (3.2.6) to a covering $p: E \to B$. The fibres are discrete. Therefore $\pi_1(F_b, x)$ is the trivial group 1. Hence $p_*: \pi_1(E, x) \to \pi_1(B, b)$ is injective. We state (3.2.6) for a covering:

(3.2.8) Proposition. *Let $p\colon E \to B$ be a covering over a path connected space B. Then the sequence*

$$1 \to \pi_1(E,x) \xrightarrow{p_*} \pi_1(B,b) \xrightarrow{\partial_x} \pi_0(F_b,x) \xrightarrow{i_*} \pi_0(E,x)$$

is exact and i_* is surjective. (The sets $F_b = \pi_0(F_b)$ and $\pi_0(E)$ have x as base point, and $i\colon F_b \subset E$.) Thus E is path connected if and only if $\pi_1(B,b)$ acts transitively on F_b. The isotropy group of $x \in F_b$ is the image of $p_*\colon \pi_1(E,x) \to \pi_1(B,b)$. □

(3.2.9) Proposition (Path lifting). *Let $p\colon E \to B$ be a covering. Let $w\colon I \to B$ be a path which begins at $p(e) = w(0)$. Then there exists a unique lifting of w which begins in e. Two paths in E which start in the same point are homotopic if and only if their images in B are homotopic.*

Proof. The existence of the lifting follows from (3.2.2), applied to a point X, and the uniqueness holds by (3.1.3).

Let $h\colon I \times I \to B$ be a homotopy of paths and $H\colon I \times I \to E$ a lifting of h. Since $t \mapsto H(\varepsilon,t)$ are continuous maps into a discrete fibre, they are constant ($\varepsilon = 0,1$). Hence H is a homotopy of paths.

Let $u_0, u_1\colon I \to E$ be paths which start in x, and suppose that pu_0 and pu_1 are homotopic. If we lift a homotopy between them with constant initial condition, then the result is a homotopy between u_0 and u_1. □

Let $p\colon E \to B$ be a right G-principal covering. Each fibre F_b carries a free right transitive G-action. From the construction of the transport functor it is immediate that the fibre transport $T_p[w]\colon F_b \to F_c$ is G-equivariant. The left action $(a,x) \mapsto a \cdot x = a_\#(x)$ of $\pi_b = \Pi(B)(b,b)$ on F_b commutes with the right G-action; we say in this case that F_b is a (π_b, G)-set. Fix $x \in F_b$. For each $a \in \pi_b$ there exists a unique $\gamma_x(a) \in G$ such that $a \cdot x = x \cdot \gamma_x(a)$, since the action of G is free and transitive. The assignment $a \mapsto \gamma_x(a)$ is a homomorphism $\gamma_x\colon \pi_b \to G$. Since $\pi_1(B,b)$ is the opposite group to π_b, we set $\delta_x(a) = \gamma_x(a)^{-1}$. Then $\delta_x\colon \pi_1(B,b) \to G$ is a homomorphism. Recall the map $\partial_x\colon \pi_1(B,p(x)) \to F_b$. If we compose it with the bijection $\rho_x\colon G \to F_b$, $g \mapsto xg$, we obtain $\rho_x \delta_x = \partial_x$. Then (3.2.8) yields:

(3.2.10) Proposition. *Let $p\colon E \to B$ be a right G-principal covering with path connected total space. Then the sequence of groups and homomorphisms*

$$1 \to \pi_1(E,x) \xrightarrow{p_*} \pi_1(B,p(x)) \xrightarrow{\delta_x} G \to 1$$

is exact (for each $x \in E$). The image of p_* is a normal subgroup. The space E is simply connected if and only if δ_x is an isomorphism. Thus, if E is simply connected, then G is isomorphic to the fundamental group of B. □

If we apply this to the exponential covering $\mathbb{R} \to S^1$, a $\mathbb{Z}$-principal covering, we again obtain $\pi_1(S^1) \cong \mathbb{Z}$.

The transport functor T_p has an additional property, it is locally trivial in the following sense. Let p be trivial over U, and let $b, c \in U$. Then $T_p[w] \colon F_b \to F_c$ is independent of the path $w \colon I \to U$ from b to c. This is due to the fact that lifts of paths inside U stay within a sheet over U.

3.3 Classification of Coverings

Let $\mathrm{TRA}_B = [\Pi(B), \mathrm{SET}]$ denote the category of functors $\Pi(B) \to \mathrm{SET}$ (objects) and natural transformations between them (morphisms). We call this category the ***transport category***.

Let $p \colon E \to B$ be a covering. We have constructed the associated transport functor $T_p = T(p) \colon \Pi(B) \to \mathrm{SET}$. For a morphism $\alpha \colon p \to q$ between coverings the restrictions $\alpha_b \colon p^{-1}(b) \to q^{-1}(b)$ of α to the fibres yield a natural transformation $T(\alpha) \colon T(p) \to T(q)$ between the corresponding transport functors. So we have obtained a functor

$$T \colon \mathrm{COV}_B \to \mathrm{TRA}_B.$$

A path connected space B is called a ***transport space*** if T is an equivalence of categories.

The main theorem of this section gives conditions under which the transport functor T is an equivalence.

(3.3.1) Note. *Let $p \colon E \to B$ be a covering with simply connected E. Then B is path connected. If p is trivial over U, then two paths in U between the same points are homotopic in B.*

Proof. The space B is path connected, since p is assumed to be surjective and E is path connected. Let u_0, u_1 be paths in U between the same points. By (3.2.9) they have liftings v_0, v_1 which connect the same points. Since E is simply connected, v_0, v_1 are homotopic in E and hence u_0, u_1 are homotopic in B. □

A set $U \subset B$ is ***transport-simple*** if two paths in U between the same points are homotopic within B. A space B is ***semi-locally simply connected*** if it has an open covering by transport-simple sets. We have just seen that this condition is implied by the existence of a simply connected covering. We call B ***transport-local***, if B is path connected, locally path connected and semi-locally simply connected.

(3.3.2) Theorem (Classification I). *Let B be path connected, locally path connected and semi-locally simply connected. Then B is a transport space, i.e., T is an equivalence of categories.*

Proof. We begin by constructing a functor

$$X : \mathrm{TRA}_B \to \mathrm{COV}_B$$

in the opposite direction. Let $\Phi\colon \Pi(B) \to \mathrm{SET}$ be a functor. We construct an associated covering $p = p(\Phi)\colon X(\Phi) \to B$. As a set, $X(\Phi) = \coprod_{b\in B} \Phi(b)$, and $p(\Phi)$ sends $\Phi(b)$ to b. Let $\mathcal{U}$ be the set of open, path connected and transport-simple subsets of B. We define bundle charts over sets $U \in \mathcal{U}$. For $b \in U$ we define

$$\varphi_{U,b}\colon U \times \Phi(b) \to p^{-1}(U), \quad (u, z) \mapsto \Phi(w)z$$

with some path w in U from b to u. By our assumption on U, the map $\varphi_{U,b}$ is well-defined, i.e., the choice of w does not matter. By construction, $\varphi_{U,b}$ is bijective. We claim: There exists a unique topology on $X(\Phi)$ such that the $\varphi_{U,b}$ of this type are homeomorphisms onto open subsets. By general principles of gluing, we have to verify that the transition maps

$$\varphi_{V,c}^{-1} \circ \varphi_{U,b}\colon (U \cap V) \times \Phi(b) \to (U \cap V) \times \Phi(c)$$

are homeomorphisms. Let $x \in U \cap V$ and let $W \subset U \cap V$ be an open, path connected neighbourhood of x. Let u_x be a path from b to x inside U, and v_x a path from c to x inside V. Then for all $y \in W$

$$\varphi_{V,c}^{-1} \circ \varphi_{U,b}(y, z) = \varphi_{V,c}^{-1} \circ \varphi_{U,b}(x, z),$$

because in order to define $\varphi_{U,b}(y, z)$ we can take the product of u_x with a path w_y in W from x to y, and similarly for $\varphi_{V,c}$, so that the contribution of the piece w_y cancels. This shows that the second component of $\varphi_{V,c}^{-1} \circ \varphi_{U,b}$ is on $W \times \Phi(b)$ independent of $x \in W$. The continuity of the transition map is a consequence.

If $\alpha : \Phi_1 \to \Phi_2$ is a natural transformation, then the morphism

$$X(\alpha)\colon X(\Phi_1) \to X(\Phi_2), \quad x \in \Phi_1(b) \mapsto \alpha(x) \in \Phi_2(b),$$

induced by α, is continuous with respect to the topologies just constructed and hence a morphism of coverings. The continuity of $X(\alpha)$ follows from the fact, that bundle charts $\Phi_{U,b}$ for $X(\Phi_1)$ and $X(\Phi_2)$ transform $X(\alpha)$ into

$$\mathrm{id} \times \alpha(b)\colon U \times \Phi_1(b) \to U \times \Phi_2(b).$$

This finishes the construction of the functor $X : \mathrm{TRA}_B \to \mathrm{COV}_B$.

We now show that the functors T and X are mutually inverse equivalences of categories, i.e., that XT and TX are naturally isomorphic to the identity functor. From our constructions we see immediately a canonical homeomorphism $\rho\colon X(T(p)) \cong E(p)$ over B for each covering $p\colon E(p) \to B$, namely set-theoretically the identity. We have to show that ρ is continuous. Let $v \in F_b \subset X(T(p))$. Let W be a neighbourhood of v in $E(p)$. Then there exists an open neighbourhood $V \subset W$ of v in $E(p)$ such that V is path connected and $U = p(V) \in \mathcal{U}$. Let $b = p(v)$. Then we have the bundle charts $\varphi_{U,b}$, and $W = \varphi_{U,b}(U \times v)$ is a neighbourhood of v in $X(T(p))$. From the construction of $\varphi_{U,v}$ we see that $\rho(W) \subset V$. This shows the continuity at v. It is easily verified that the homeomorphisms ρ constitute a natural transformation. Conversely, we have to verify that the transport functor of $p(\Phi)$ is Φ. Using the bundle charts $\varphi_{U,b}$ this is first verified for paths in U. But each morphism in $\Pi(B)$ is a composition of morphisms represented by paths in such sets U. □

Fix $b \in B$. A canonical functor is the Hom-functor $\Pi(b,-)$ of the category $\Pi(B)$. Let $p^b\colon E^b \to B$ denote the associated covering. We still assume that B is transport-local. The automorphism group $\pi_b = \Pi(b,b)$ of b in $\Pi(B)$, the opposite fundamental group $\pi_1(B,b)$, acts on E^b fibrewise from the right by composition of morphisms. The action is free and transitive on each fibre. Via our bundle charts it is easily verified that the action on E^b is continuous. Thus p^b is a right π_b-principal covering. From (3.2.8) we see that E^b is simply connected. Thus we have shown:

(3.3.3) Theorem. *The canonical covering $p^b\colon E^b \to B$ associated to the Hom-functor $\Pi(b,-)$ has a simply connected total space. The right action of $\Pi(b,b)$ on the fibres by composition of morphisms is the structure of a right principal covering on p^b.* □

Problems

1. Let S be the pseudo-circle. The space S is simply connected. But S has non-trivial connected principal coverings. They can be obtained by a pullback along suitable maps $S \to S^1$ from $\mathbb{Z}/n$-principal or $\mathbb{Z}$-principal coverings of S^1. In this sense S behaves like S^1. We see that certain local properties of B are necessary in order that T is an equivalence.
2. Let $f\colon B \to C$ be a continuous map. The pullback $p\colon X \to B$ of a covering $q\colon Y \to C$ along f is a covering. Pulling back morphisms yields a functor $f^* = \mathrm{COV}(f)\colon \mathrm{COV}_C \to \mathrm{COV}_B$. The map f induces a functor $\Pi(f)\colon \Pi(B) \to \Pi(C)$, and composition with functors $\Pi(C) \to \mathrm{SET}$ yields a functor $\mathrm{TRA}(f)\colon \mathrm{TRA}_C \to \mathrm{TRA}_B$. These functors are compatible $T_B \circ \mathrm{COV}(f) = \mathrm{TRA}(f) \circ T_C\colon \mathrm{COV}_C \to \mathrm{TRA}_B$.

3.4 Connected Groupoids

In this section the space B is assumed to be path connected.

A functor $\Pi(B) \to \mathrm{SET}$ is an algebraic object. The category of these functors has an equivalent description in terms of more familiar algebraic objects, namely group actions. We explain this equivalence.

Let Π be a connected groupoid (i.e. there exists at least one morphism between any two objects) with object set B, e.g., $\Pi = \Pi(B)$ for a path connected space B.

Let $\Pi(x, y)$ denote the set of morphisms $x \to y$ and $\pi = \pi_b = \Pi(b, b)$, the automorphism group of b with respect to composition of morphisms. A functor $F\colon \Pi \to \mathrm{SET}$ has an associated set $F(b)$ with left π_b-action

$$\pi_b \times F(b) \to F(b), \quad (\alpha, x) \mapsto F(\alpha)(x).$$

A natural transformation $\alpha\colon F \to G$ yields a map $\alpha(b)\colon F(b) \to G(b)$ which is π_b-equivariant. In this manner we obtain a functor

$$\varepsilon_b\colon [\Pi, \mathrm{SET}] \to \pi_b\text{-SET}$$

from the functor category of functors $\Pi \to \mathrm{SET}$ into the category of left π_b-sets and equivariant maps.

We construct a functor η_b in the opposite direction. So let A be a π_b-set. The Hom-functor $\Pi(b, ?)$ is a functor into the right π_b-sets, namely π_b acts on $\Pi(b, x)$ by composition of morphisms. These data yield the functor $\Phi(A) = \Pi(b, ?) \times_\pi A$. (Here again $A \times_\pi B$ denotes the quotient of $A \times B$ by the equivalence relation $(ag, b) \sim (a, gb)$, $(a, g, b) \in A \times \pi \times B$ for left π-sets A and right π-sets B.) A π_b-map $f\colon A \to B$ induces a natural transformation $\Phi(f)\colon \Phi(A) \to \Phi(B)$. This finishes the definition of η_b.

(3.4.1) Proposition. *The functors ε_b and η_b are mutually inverse equivalences of categories.*

Proof. The composition $\varepsilon_b \eta_b$ associates to a π_b-set A the π_b-set $\Pi(b, b) \times_\pi A$, with π_b-action $g \cdot (f, z)$. The isomorphisms

$$\iota_A\colon \Pi(b, b) \times_\pi A \to A, \quad (f, z) \mapsto f \cdot z$$

form a natural equivalence $\iota\colon \varepsilon_b \eta_b \simeq \mathrm{Id}$.

The composition $\eta_b \varepsilon_b$ associates to a functor $F\colon \Pi \to \mathrm{SET}$ the functor $\Pi(b, -) \times_\pi F(b)$. The maps

$$\beta_F(x)\colon \Pi(b, x) \times F(b) \to F(x), \quad (f, z) \mapsto F(f)z$$

form a natural transformation, i.e., a morphism $\beta_F\colon \eta_b \varepsilon_b(F) \to F$ in $[\Pi, \mathrm{SET}]$. Since Π is a connected groupoid, the $\beta_F(x)$ are bijective, and therefore constitute an isomorphism in the functor category. The β_F are a natural equivalence $\beta\colon \eta_b \varepsilon_b \simeq \mathrm{Id}$. □

In our previous notation $\mathrm{TRA}_B = [\Pi(B), \mathrm{SET}]$. From (3.3.2) and (3.4.1) we obtain for each transport-local space an equivalence of categories $\mathrm{COV}_B \to \pi_b$-SET, the composition of the transport functor T with ε_b. It associates to a covering $p\colon E \to B$ the π_b-set F_b. The inverse equivalence associates to a π_b-set A the covering

$$X(\eta_b A) = \textstyle\coprod_{x \in B} \Pi(B)(b, x) \times_\pi A \to B.$$

It is the covering $E^b \times_\pi A \to B$ associated to the π_b-principal covering (3.3.3).

Let $p: E \to B$ be a right G-principal covering with path connected B. Then we have the functors

$$G\text{-}\mathrm{SET} \xrightarrow{A(p)} \mathrm{COV}_B \xrightarrow{T} \mathrm{TRA}_B \xrightarrow[\simeq]{\varepsilon_b} \pi_b\text{-}\mathrm{SET}.$$

The composition associates to a G-set F the π_b-set $F_b \times_G F$, where the π_b-action is induced from the left π_b-action on F_b.

Now suppose in addition that E is simply connected. Then we have a bijection $\varphi_x^F : F \to F_b \times_G F$, $z \mapsto [x, z]$ for a fixed $x \in F_b$ as well as the isomorphism $\gamma_x : \pi_b \to G$, see (3.2.10). The relation $\gamma_x(a) \cdot z = a \cdot [x, z]$ holds. So if we view G-sets via γ_x as π_b-sets, then the above composition of functors is the identity. Thus we have shown:

(3.4.2) Proposition. *Let $p: E \to B$ be a simply connected G-principal covering. Then $A(p)$ is an equivalence of categories if and only if T is an equivalence of categories.* □

(3.4.3) Theorem. *The following properties of B are equivalent:*

(1) *B is a transport space, i.e., T is an equivalence of categories.*

(2) *B has a universal right G-principal covering $p: E \to B$ with simply connected total space E.*

Proof. (1) $\Rightarrow$ (2). Since $\varepsilon_b \circ T$ is an equivalence of categories, each object of π_b-SET is isomorphic to an object in the image of $\varepsilon_b \circ T$. Thus there exists a covering $p: E \to B$ such that its π_b-set F_b is isomorphic to the π_b-set π_b. By another property of an equivalence of categories, the morphisms $p \to p$ correspond under $\varepsilon_b \circ T$ bijectively to the π_b-maps $F_b \to F_b$. The π_b-morphisms $\pi_b \to \pi_b$ are the right translations by elements of π_b. Thus π_b acts simply and transitively on E. From (3.2.8) we see that E is simply connected.

The left action of the automorphism group $\mathrm{Aut}(p)$ on E is properly discontinuous and the induced action on each fibre is transitive. If we rewrite this as a right action of the opposite group G, we obtain a right G-principal covering. Proposition (3.4.2) now says that p is universal.

(2) $\Rightarrow$ (1) is a consequence of (3.4.2). □

From a geometric view point the interesting coverings are those with connected total space.

Let $p: E \to B$ be a universal right G-principal covering. A left G-set A is the disjoint sum of its orbits. We have a corresponding sum decomposition of the total space $E \times_G A$ into the sum of $E \times_G C$, where C runs through the orbits of A. An orbit is a transitive G-set and isomorphic to a homogeneous set G/H for some subgroup H of G. The homeomorphism $E \times_G G/H \cong E/H$ shows that the summands $E \times_G C$ are path connected. The action of H on E is properly discontinuous and therefore $E \to E/H$ an H-principal covering. Also the induced map $p_H\colon E/H \to B$ is a covering.

The category of homogeneous G-sets and G-maps is the ***orbit category*** $\mathrm{Or}(G)$ of G. The sets G/K and G/L are isomorphic if and only if the subgroups K and L are conjugate in G. The isotropy groups of G/H are conjugate to H. The inclusion of the subcategory $\mathrm{Or}(G)$ into the category of transitive G-sets is an equivalence.

Let $p\colon E \to B$ be a universal right G-principal covering with simply connected E. Then the functor $A(p)$ induces an equivalence of $\mathrm{Or}(G)$ with the category of connected coverings of B. Each covering is thus isomorphic to a covering of the form $p_H\colon E/H \to B$ for a subgroup H of G.

We fix $z \in p^{-1}(b) \subset E$ and obtain an isomorphism $\delta_z\colon G \to \pi_1(B,b)$. It sends $g \in G$ to the loop $[p \circ w_g]$ where $w_g\colon I \to E$ is a path from z to zg^{-1}.

Let $q\colon X \to B$ be a connected covering. We know that the induced homomorphism $p_*\colon \pi_1(X,x) \to \pi_1(B,b)$ is injective. The image is called the ***characteristic subgroup*** $C(p,x)$ of p with respect to x. Let $u\colon I \to X$ be a path from x to $y \in p^{-1}(b)$. Then $w = pu$ is a loop and $C(p,y) = [w]C(p,x)[w]^{-1}$, thus different base points in $p^{-1}(b)$ yield conjugate characteristic subgroups. Conversely, each subgroup conjugate to $C(p,x)$ arises this way.

We apply this to the covering p_H with $\bar{z} = zH \in E/H$. Then

$$(p_H)_*(\pi_1(E/H,\bar{z}) = \delta_z(H) = C(p_H,\bar{z}).$$

We collect the results in the next theorem.

(3.4.4) Theorem (Classification II). *Let B be a transport space. The category of connected coverings of B is equivalent to the orbit category $\mathrm{Or}(\pi_1(B,b))$. The isomorphism class of a connected covering $q\colon X \to B$ corresponds under this equivalence to the isomorphism class of $\pi_1(B,b)/C(q,x)$ for any $x \in p^{-1}(b)$. The isomorphism class of a connected covering is determined by the conjugacy class of its characteristic subgroup.* □

Problems

1. The automorphism group of $p_H\colon E/H \to B$ is NH/H, where NH denotes the normalizer of H in $\pi_1(B,b)$. The covering is a principal covering (also called ***regular covering***), if and only if H is a normal subgroup of $\pi_1(B,b)$.

2. The connected coverings of S^1 are, up to isomorphism, the maps $p_n\colon z \mapsto z^n$ for $n \in \mathbb{N}$ and $p\colon \mathbb{R} \to S^1$, $t \mapsto \exp(2\pi i t)$. These coverings are principal coverings.

3. Let B be a contractible space. Is the identity $\mathrm{id}\colon B \to B$ a universal G-principal covering for the trivial group G?

3.5 Existence of Liftings

The following theorem (3.5.2) is interesting and important, because it asserts the existence of liftings under only the necessary algebraic conditions on the fundamental groups.

(3.5.1) Lemma. *Let w_0 and w_1 be paths in E beginning in x. Let $u_i = pw_i$. Then $w_0(1) = w_1(1)$ if and only if $u_0(1) = u_1(1)$ and $[u_0 * u_1^-]$ is contained in $p_*\pi_1(E, x)$.*

Proof. If $w_0(1) = w_1(1)$, then $p_*[w_0 * w_1^-] = [u_0 * u_1^-]$. Conversely: We lift $u_0 * u_1^-$ with initial point x. Since $[u_0 * u_1^-] \in p_*\pi_1(E, x)$ there exists a loop which is homotopic to $u_0 * u_1^-$, and which has a lifting with initial point x. By (3.2.9), $u_0 * u_1^-$ itself has a lifting as a loop. Therefore u_0 and u_1 have liftings with initial point x and the same end point. These liftings are then necessarily w_0 and w_1. □

(3.5.2) Theorem. *Let $p\colon E \to B$ be a covering. Suppose Z is path connected and locally path connected. Let $f : Z \to B$ be a map with $f(z) = p(x)$. Then there exists a lifting $\Phi\colon Z \to E$ of f with $\Phi(z) = x$ if and only if $f_*\pi_1(Z, z)$ is contained in $p_*\pi_1(E, x)$.*

Proof. If a lifting exists, then the inclusion of groups holds by functoriality of π_1.

Suppose $f_*\pi_1(Z, z) \subset p_*\pi_1(E, x)$. We begin by constructing Φ as a set map. Then we show its continuity.

Let $z_0 \in Z$. There exists a path w from z to z_0. Let $v\colon I \to E$ be a lifting of fw starting in x. We want to define Φ by $\Phi(z) = v(1)$. Let w_1 be another path from z to z_0 and v_1 a lifting of fw_1 starting in x. Then

$$pv(1) = fw(1) = f(z_0) = fw_1(1) = pv_1(1);$$

moreover

$$[pv * pv_1^-] = f_*[w * w_1^-] \in p_*\pi_1(E, x).$$

By (3.5.1) we have $v(1) = v_1(1)$; this shows that Φ is well-defined if we set $\Phi(z) = v(1)$.

Continuity of Φ. Let U be an open neighbourhood of $\Phi(z_0)$, such that p is trivial over $p(U) = V$, and let $p\colon U \to V$ have the inverse homeomorphism $q\colon V \to U$. Let W be a path connected neighbourhood of z_0 such that $f(W) \subset V$. We claim $\Phi(W) \subset U$. Let $z_1 \in W$ and let w_1 be a path in W from z_0 to z_1. Then $w * w_1$ is a path from z to z_1, and $v_1 = v * qfw_1$ a path with $pv_1 = f \circ (w * w_1)$ and $v_1(0) = x$. Thus $v_1(1) \in U$. □

(3.5.3) Theorem. *Let X be a topological group with neutral element x and let $p\colon E \to X$ be a covering with path connected and locally path connected E. For each $e \in p^{-1}(x)$ there exists a unique group structure on E which makes E into a topological group with neutral element e and such that p is a homomorphism.*

Proof. Construction of a group structure on E. Let $m\colon X \times X \to X$ be the group multiplication. We try to find $M\colon E \times E \to E$ as a lift $m(p \times p)$ along p with $M(e,e) = e$. This can be done, by (3.5.2), if $m_*(p \times p)_*\pi_1(E \times E) \subset p_*\pi_1(E)$. This inclusion holds, since (using (2.7.3))

$$m_*(p\times p)_*[(w_1, w_2)] = [pw_1 \cdot pw_2] = [pw_1 * pw_2] = [p(w_1 * w_2)] = p_*[w_1 * w_2].$$

From the uniqueness of liftings one shows that M is associative. In a similar manner we see that (passage to) the inverse in X has a lifting to E, and uniqueness of liftings shows that the result is an inverse for the structure M. □

A well-known result of Hermann Weyl is that a compact, connected, semi-simple Lie group has a finite simply connected covering. See [29, V.7] about fundamental groups of compact Lie groups. The group $O(n)$ has two different two-fold coverings which are non-trivial over $SO(n)$. They are distinguished by the property that the elements over the reflections at hyperplanes have order 2 or 4 ($n \geq 1$). We will see that $\pi_1(SO(n)) \cong \mathbb{Z}/2$ for $n \geq 3$. The corresponding simply connected covering groups are the spinor groups $\mathrm{Spin}(n)$; see e.g., [29, I.6].

We repeat an earlier result in a different context. We do not assume that B has a universal covering.

(3.5.4) Proposition. *Let B be path connected and locally path connected. Coverings $p_i\colon (X_i, x_i) \to (B, b)$ with path connected total space are isomorphic if and only if their characteristic subgroups are conjugate in $\pi_1(B, b)$.*

Proof. Since $C(p_1, x_1)$ and $C(p_2, x_2)$ are conjugate we can change the base point x_2 such that the groups are equal. By (3.5.2), there exist morphisms $f_1\colon (X_1, x_1) \to (X_2, x_2)$ and $f_2\colon (X_2, x_2) \to (X_1, x_1)$, and since $f_2 f_1(x_1) = x_1$, $f_1 f_2(x_2) = x_2$ both compositions are the identity.

By functoriality of π_1 we see that isomorphic coverings have conjugate characteristic subgroups. □

Problems

1. We have given a direct proof of (3.5.2), although it can also be derived from our previous classification results. The existence of a lift Φ is equivalent to the existence of a section in the covering which is obtained by pullback along f. The $\pi_1(Z, z)$-action on the fibre E_b of the pullback is obtained from the $\pi_1(B, b)$-action via $f_*\colon \pi_1(Z, z) \to \pi_1(B, b)$. The existence of a section is equivalent to E_b having a fixed point under the $\pi_1(Z, z)$-action. If the inclusion of groups holds as in the statement of the theorem, then a fixed point exists because the image of p_* is the isotropy group of the $\pi_1(B, b)$-action.

2. Let $p\colon E \to B$ be a covering with path connected and locally path connected total space. The following are equivalent: (1) $\mathrm{Aut}(p)$ acts transitively on each fibre of p. (2) $\mathrm{Aut}(p)$ acts transitively on some fibre of p. (3) The characteristic subgroup is normal in $\pi_1(B, b)$. (4) p is an $\mathrm{Aut}(p)$-principal covering.

3.6 The Universal Covering

We collect some of our results for the standard situation that B is path connected, locally path connected and semi-locally simply connected space. Let us now call a covering $p\colon E \to B$ a ***universal covering*** if E is simply connected.

(3.6.1) Theorem (Universal covering). *Let B be as above.*

(1) *There exists up to isomorphism a unique universal covering $p\colon E \to B$.*

(2) *The action of the automorphism group $\mathrm{Aut}(p)$ on E furnishes p with the structure of a left $\mathrm{Aut}(p)$-principal covering.*

(3) *The group $\mathrm{Aut}(p)$ is isomorphic to $\pi_1(B,b)$. Given $x \in p^{-1}(b)$, an isomorphism $\iota_x\colon \mathrm{Aut}(p) \to \pi_1(B,b)$ is obtained, if we assign to $\alpha \in \mathrm{Aut}(p)$ the class of the loop pw for a path w from x to $\alpha(x)$.*

(4) *The space E^b is simply connected.*

Proof. (1) Existence is shown in (3.3.3). Since B is locally path connected, the total space of each covering has the same property. Let $p_i\colon E_i \to B$ be simply connected coverings with base points $x_i \in p_i^{-1}(b)$. By (3.5.2), there exist morphisms $\alpha\colon p_1 \to p_2$ and $\beta\colon p_2 \to p_1$ such that $\alpha(x_1) = x_2$ and $\beta(x_2) = x_1$. By uniqueness of liftings, $\alpha\beta$ and $\beta\alpha$ are the identity, i.e., α and β are isomorphisms. This shows uniqueness.

(2) By (3.1.8), the action of $\mathrm{Aut}(p)$ on E is properly discontinuous. As in (1) one shows that $\mathrm{Aut}(p)$ acts transitively on each fibre of p. The map $\mathrm{Aut}(p)\backslash E \to B$, induced by p, is therefore a homeomorphism. Since $E \to \mathrm{Aut}(p)\backslash E$ is a principal covering, so is p.

(3) Since E is simply connected, there exists a unique homotopy class of paths w from x to $\alpha(x)$. Since x and $\alpha(x)$ are contained in the same fibre, pw is a loop. Therefore ι_x is well-defined. If we lift a loop u based at b to a path w beginning in x, then there exists $\alpha \in \mathrm{Aut}(p)$ such that $\alpha(x) = w(1)$. Hence ι_x is surjective. Two paths starting in x have the same end point if and only if their images in B are homotopic. Hence ι_x is injective. If v is a path from x to $\alpha(x)$ and w a path from x to $\beta(x)$, then $v * \beta w$ is a path from x to $\alpha\beta(x)$. Hence ι_x is a homomorphism.

(4) is shown in (3.3.3). □

(3.6.2) Theorem (Classification III). *Suppose that B has a universal covering $p\colon E \to B$. Then p is a $\pi_1(B,b)$-principal covering. Each connected covering of B is isomorphic to a covering of the form $E/H \to B$, $H \subset \pi_1(B,b)$ a subgroup. This covering has H as a characteristic subgroup. Two such coverings are isomorphic if and only if the corresponding subgroups of $\pi_1(B,b)$ are conjugate.* □

Problems

1. The product $\prod_1^\infty S^1$ is not semi-locally simply connected.
2. Is the product of a countably infinite number of the universal covering of S^1 a covering?
3. Identify in S^1 the open upper and the open lower hemi-sphere to a point. The resulting space X has four points. Show $\pi_1(X) \cong \mathbb{Z}$. Does X have a universal covering?
4. The quotient map $p\colon \mathbb{R}^n \to \mathbb{R}^n/\mathbb{Z}^n$ is a universal covering. The map $q\colon \mathbb{R}^n \to T^n$, $(x_j) \mapsto (\exp 2\pi i x_j)$ is a universal covering of the n-dimensional torus $T^n = S^1 \times \cdots \times S^1$. Let $f\colon T^n \to T^n$ be a continuous automorphism, and let $F\colon \mathbb{R}^n \to \mathbb{R}^n$ be a lifting of fq along q with $F(0) = 0$. The assignments $x \mapsto F(x) + F(y)$ and $x \mapsto F(x+y)$ are liftings of the same map with the same value for $x = 0$. Hence $F(x+y) = F(x) + F(y)$. From this relation one deduces that F is a linear map. Since $F(\mathbb{Z}^n) \subset \mathbb{Z}^n$, the map F is given by a matrix $A \in \mathrm{GL}_n(\mathbb{Z})$. Conversely, each matrix in $\mathrm{GL}_n(\mathbb{Z})$ gives us an automorphism of T^n. The group of continuous automorphisms of T^n is therefore isomorphic to $\mathrm{GL}_n(\mathbb{Z})$.
5. Classify the 2-fold coverings of $S^1 \vee S^1$ and of $S^1 \vee S^1 \vee S^1$. (Note that a subgroup of index 2 is normal.)
6. The k-fold ($k \in \mathbb{N}$) coverings of $S^1 \vee S^1$ correspond to isomorphism classes of $\pi = \pi_1(S^1 \vee S^1) = \langle u \rangle * \langle v \rangle$-sets of cardinality k. An action of π on $\{1,\dots,k\}$ is determined by the action of u and v, and these actions can be arbitrary permutations of $\{1,\dots,k\}$.

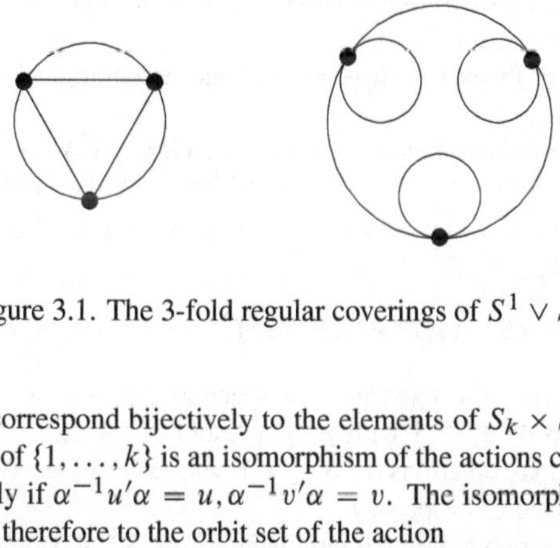

Figure 3.1. The 3-fold regular coverings of $S^1 \vee S^1$.

Hence these actions correspond bijectively to the elements of $S_k \times S_k$ (S_k the symmetric group). A bijection α of $\{1,\dots,k\}$ is an isomorphism of the actions corresponding to (u,v) and (u',v') if and only if $\alpha^{-1}u'\alpha = u, \alpha^{-1}v'\alpha = v$. The isomorphism classes of k-fold coverings correspond therefore to the orbit set of the action

$$S_k \times (S_k \times S_k) \to S_k \times S_k, \quad (\alpha, u, v) \mapsto (\alpha u \alpha^{-1}, \alpha v \alpha^{-1}).$$

Consider the case $k = 3$ and $S_3 = \langle A, B \mid A^3 = 1, B^2 = 1, BAB^{-1} = A^{-1}\rangle$. The three conjugacy classes are represented by $1, A, B$. We can normalize the first component of each orbit correspondingly. If we fix u, then the centralizer $Z(u)$ of u acts on the second component. We have $Z(1) = S_3$, $Z(B) = \{1, B\}$, and $Z(A) = \{1, A, A^2\}$. This yields the following representing pairs for the orbits:

$$(1,1), \quad (1,A)cn, \quad (1,B),$$
$$(A,1)cn, \quad (A,A)cn, \quad (A,A^2)cn, \quad (A,B)c,$$
$$(B,1), \quad (B,A)c, \quad (B,B), \quad (B,AB)c.$$

The transitive actions (which yield connected coverings) have the addition c, the normal subgroups (which yield regular coverings) have the addition n.

Draw figures for the connected coverings. For this purpose study the restrictions of the coverings to the two summands S^1; note that under restriction a connected covering may become disconnected. Over each summand one has a 3-fold covering of S^1; there are three of them.

7. Classify the regular 4-fold coverings of $S^1 \vee S^1$.

8. The Klein bottle has three 2-fold connected coverings. One of them is a torus, the other two are Klein bottles.

9. Let X be path connected and set $Y = X \times I / X \times \partial I$. Show $\pi_1(Y) \cong \mathbb{Z}$. Show that Y has a simply connected universal $\mathbb{Z}$-principal covering. Is Y always locally path connected?

10. Construct a transport space which is not locally path connected.

11. The space $\mathbb{R}^n$ with two origins is obtained from $\mathbb{R}^n + \mathbb{R}^n$ by identifying $x \neq 0$ in the first summand with the same element in the second summand. Let M be the line with two origins. Construct a universal covering of M and determine $\pi_1(M)$. What can you say about π_1 of $\mathbb{R}^n$ with two origins for $n > 1$?

12. Make the fundamental groupoid $\Pi(B)$ into a topological groupoid with object space B. (Hypothesis (3.6.1). Use (14.1.17).)

13. Let X be a compact Hausdorff space and $H(X)$ the group of homeomorphism. Then $H(X)$ together with the CO-topology is a topological group and $H(X) \times X \to X$, $(f, x) \mapsto f(x)$ a continuous group action.

14. The space $C(S^1, S^1)$ with CO-topology becomes a topological group under pointwise multiplication of maps.

15. There are two continuous homomorphisms $e\colon C(S^1, S^1) \to S^1$, $f \mapsto f(1)$ and $d\colon C(S^1, S^1) \to \mathbb{Z}$, $f \mapsto \mathrm{degree}(f)$. Let $M^0(S^1)$ be the kernel of (e, d). Let further $f_n\colon S^1 \to S^1, z \mapsto z^n$. The homomorphism $s\colon S^1 \times \mathbb{Z} \to C(S^1, S^1)$, $(\alpha, n) \mapsto \alpha f_n$ is continuous. The map

$$M^0(S^1) \times (S^1 \times \mathbb{Z}) \to C(S^1, S^1), \quad (f, (\alpha, n)) \mapsto f \cdot s(\alpha, n)$$

is an isomorphism of topological groups. The space $M^0(S^1)$ is isomorphic to the space V of continuous functions $\phi\colon \mathbb{R} \to \mathbb{R}$ with $\varphi(0) = 0$ and $\varphi(x + 2\pi) = \varphi(x)$ or, equivalently, to the space of continuous functions $\alpha\colon S^1 \to \mathbb{R}$ with $\alpha(1) = 0$. The space V carries the sup-norm and the induced CO-topology.

16. Let $M(S^1)$ be the group of homeomorphisms $S^1 \to S^1$ of degree 1 with CO-topology. Each $\lambda \in S^1$ yields a homeomorphism $f_\lambda\colon z \mapsto \lambda x$. In this way S^1 becomes a subgroup of $M(S^1)$. Let $M_1(S^1)$ be the subgroup of homeomorphisms f with $f(1) = 1$. Then

$$S^1 \times M_1(S^1) \to M(S^1), \quad (\lambda, h) \mapsto f_\lambda \circ h$$

is a homomorphism. The space $M_1(S^1)$ is homeomorphic to the space H of homeomorphism $f\colon [0,1] \to [0,1]$ with $f(0) = 0$. The space H is contractible; a contraction is $f_t(x) = (1-t)f(x) + tx$. Therefore the inclusion $S^1 \to M(S^1)$ is an h-equivalence. The space $H(S^1)$ of homeomorphisms of S^1 is h-equivalent to O(2).

Chapter 4
Elementary Homotopy Theory

Further analysis and applications of the homotopy notion require a certain amount of formal consideration. We deal with several related topics.

(1) The construction of auxiliary spaces from the basic “homotopy cylinder” $X \times I$: mapping cylinders, mapping cones, suspensions; and dual constructions based on the “path space” X^I. These elementary constructions are related to the general problem of defining homotopy limits and homotopy colimits.

(2) Natural group structures on Hom-functors in TOP^0. By category theory they arise from group and cogroup objects in this category. But we mainly work with the explicit constructions: suspension and loop space.

(3) Exact sequences involving homotopy functors based on “exact sequences” among pointed spaces (“space level”). These so-called cofibre and fibre sequences are a fundamental contribution of D. Puppe to homotopy theory [155]. The exact sequences have a three-periodic structure, and it has by now become clear that data of this type are an important structure in categories with (formal) homotopy (triangulated categories).

As an application, we use a theorem about homotopy equivalences of mapping cylinders to prove a gluing theorem for homotopy equivalences. The reader may have seen partitions of unity. In homotopy theory they are used to reduce homotopy colimits to ordinary colimits. Here we treat the simplest case: pushouts.

4.1 The Mapping Cylinder

Let $f: X \to Y$ be a map. We construct the ***mapping cylinder*** $Z = Z(f)$ of f via the pushout

$$\begin{array}{ccc} X + X & \xrightarrow{\mathrm{id}+f} & X + Y \\ \Big\downarrow{\scriptstyle \langle i_0, i_1 \rangle} & & \Big\downarrow{\scriptstyle \langle j, J \rangle} \\ X \times I & \xrightarrow[a]{} & Z(f) \end{array} \qquad \begin{array}{c} Z(f) = X \times I + Y/f(x) \sim (x,1), \\ J(y) = y, \ j(x) = (x,0). \end{array}$$

Here $i_t(x) = (x,t)$. Since $\langle i_0, i_1 \rangle$ is a closed embedding, the maps $\langle j, J \rangle$, j and J are closed embeddings. We also have the projection $q: Z(f) \to Y$, $(x,t) \mapsto f(x)$, $y \mapsto y$. The relations $qj = f$ and $qJ = \mathrm{id}$ hold. We denote elements in $Z(f)$ by their representatives in $X \times I + Y$.

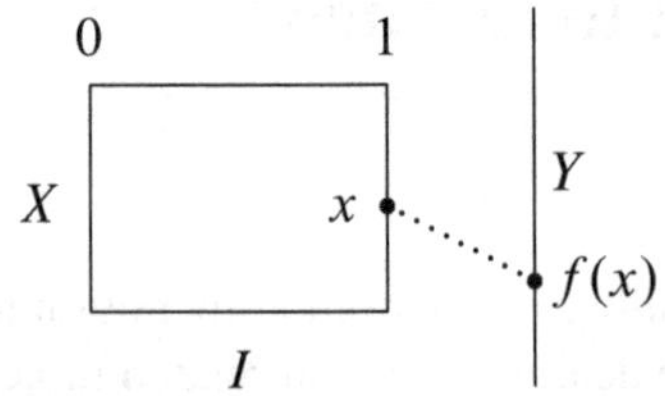

The map Jq is homotopic to the identity relative to Y. The homotopy is the identity on Y and contracts I relative 1 to 1.

We thus have a decomposition of f into a closed embedding J and a homotopy equivalence q. From the pushout property we see:

Continuous maps $\beta\colon Z(f) \to B$ correspond bijectively to pairs $h\colon X \times I \to B$ and $\alpha\colon Y \to B$ such that $h(x,1) = \alpha f(x)$.

In the following we consider $Z(f)$ as a space under $X + Y$ via the embedding (inclusion) $\langle j, J\rangle$. We now study homotopy commutative diagrams

$$\begin{array}{ccc} X & \xrightarrow{f} & Y \\ \downarrow{\scriptstyle\alpha} & & \downarrow{\scriptstyle\beta} \\ X' & \xrightarrow{f'} & Y' \end{array}$$

together with homotopies $\Phi\colon f'\alpha \simeq \beta f$. In the case that the diagram is commutative, the pair (α, β) is a morphism from f to f' in the category of arrows in TOP. We consider the data (α, β, Φ) as a generalized morphism. These data induce a map $\chi = Z(\alpha, \beta, \Phi)\colon Z(f) \to Z(f')$ defined by

$$\chi(y) = \beta(y),\ y \in Y, \quad \chi(x,s) = \begin{cases} (\alpha(x), 2s), & x \in X, s \le 1/2, \\ \Phi_{2s-1}(x), & x \in X, s \ge 1/2. \end{cases}$$

The diagram

$$\begin{array}{ccc} X + Y & \longrightarrow & Z(f) \\ \downarrow{\scriptstyle\alpha+\beta} & & \downarrow{\scriptstyle Z(\alpha,\beta,\Phi)} \\ X' + Y' & \longrightarrow & Z(f') \end{array}$$

is commutative. The composition of two such morphisms between mapping cylinders is homotopic to a morphism of the same type. Suppose we are given $f''\colon X'' \to Y''$, $\alpha'\colon X' \to X''$, $\beta'\colon Y' \to Y''$, and a homotopy $\Phi'\colon f''\alpha' \simeq \beta' f'$. These data yield a composed homotopy $\Phi' \diamond \Phi\colon f''\alpha'\alpha \simeq \beta'\beta f$ defined by

$$(\Phi' \diamond \Phi)_t = \begin{cases} \Phi'_{2t} \circ \alpha, & t \le 1/2, \\ \beta' \circ \Phi_{2t-1}, & t \ge 1/2. \end{cases}$$

(This is the product of the homotopies $\Phi'_t\alpha$ and $\beta'\Phi_t$.)

(4.1.1) Lemma. *There exists a homotopy*

$$Z(\alpha', \beta', \Phi') \circ Z(\alpha, \beta, \Phi) \simeq Z(\alpha'\alpha, \beta'\beta, \Phi' \diamond \Phi)$$

which is constant on $X + Y$.

Proof. Both maps coincide on Y and differ on $X \times I$ by a parameter transformation of I. □

We also change α and β by a homotopy. Suppose given homotopies $A^t : X \to X'$ and $B^t : Y \to Y'$ and a homotopy $\Gamma^t : f'A^t \simeq B^t f$. We assume, of course, that $\Gamma^t = (\Gamma^t_s)$ is continuous on $X \times I \times I$. We get a homotopy $Z(A^t, B^t, \Gamma^t) \colon Z(f) \to Z(f')$ which equals $A^t + B^t : X + Y \to X' + Y'$ on these subspaces.

We use the fact that A^t, B^t, Φ induce a homotopy Γ^t.

(4.1.2) Lemma. *Suppose* A^t, B^t *with* $A^0 = \alpha$, $B^0 = \beta$ *and* $\Phi \colon f'A^0 \simeq B^0 f$ *are given.Then there exists* Γ^t *with* $\Gamma^0 = \Phi$.

Proof. One applies a retraction $X \times I \times I \to X \times (\partial I \times I \cup I \times 0)$ to the map $\gamma \colon X \times (\partial I \times I \cup I \times 0) \to Y'$ defined by $\gamma(x, s, 0) = \Phi(x, s)$, $\gamma(x, 0, t) = f'A^t(x)$ and $\gamma(x, 1, t) = B^t f(x)$. □

Suppose now that $X'' = X$, $Y'' = Y$, $f'' = f$ and $\alpha'\alpha \simeq \mathrm{id}$, $\beta'\beta \simeq \mathrm{id}$, $f\alpha' \simeq \beta' f'$. We choose homotopies

$$A^t : \alpha'\alpha \simeq \mathrm{id}, \quad B^t : \beta'\beta \simeq \mathrm{id}, \quad \Phi' \colon f\alpha' \simeq \beta' f'.$$

As before, we have the composition $\Psi = \Phi' \diamond \Phi$. We use (4.1.2) to find a homotopy Γ^t with $\Gamma^0 = \Psi$ and $\Gamma^t : fA^t \simeq B^t f$. Let Γ^1_- be the inverse homotopy of Γ^1. Let $\Omega = Z(1_X, 1_Y, \Gamma^1_-) \circ Z(\alpha', \beta', \Phi') \colon Z(f') \to Z(f)$; this morphism restricts to $\alpha' + \beta' \colon X' + Y' \to X + Y$.

(4.1.3) Proposition. *There exists a homotopy from* $\Omega \circ Z(\alpha, \beta, \Phi)$ *to the identity which equals* $((k * A^t) * k + (k * B^t) * k)$ *on* $X + Y$*; here* k *denotes a constant homotopy.*

Proof. By (4.1.1) there exists a homotopy relative to $X + Y$ of the composition in question to $Z(1_X, 1_Y, \Gamma^1_-) \circ Z(\alpha'\alpha, \beta'\beta, \Psi)$. By (4.1.2) we have a further homotopy to $Z(1_X, 1_Y, \Gamma^1_-) \circ Z(1_X, 1_Y, \Gamma^1)$, which equals $A^t + B^t$ on $X + Y$, and then by (4.1.1) a homotopy to $Z(1_X, 1_Y, \Gamma^1_- \diamond \Gamma^1)$, which is constant on $X + Y$. The homotopy $\Gamma^1_- \diamond \Gamma^1 : f \simeq f$ is homotopic relative to $X \times \partial I$ to the constant homotopy k_f of f. We thus have an induced homotopy relative $X + Y$ to $Z(1_X, 1_Y, k_f)$ and finally a homotopy to the identity (Problems 1 and 2). □

(4.1.4) Theorem. *Suppose* α *and* β *are homotopy equivalences. Then the map* $Z(\alpha, \beta, \Phi)$ *is a homotopy equivalence.*

Proof. The morphism Ω in (4.1.3) has a right homotopy inverse. We can apply (4.1.1) and (4.1.3) to Ω and see that Ω also has a left homotopy inverse. Hence Ω is a homotopy equivalence. From $\Omega \circ Z(\alpha, \beta, \Phi) \simeq \mathrm{id}$ we now conclude that $Z(\alpha, \beta, \Phi)$ is a homotopy equivalence. □

Problems

1. Suppose Φ and Ψ are homotopic relative to $X \times \partial I$. Then $Z(\alpha, \beta, \Phi)$ and $Z(\alpha, \beta, \Psi)$ are homotopic relative to $X + Y$.
2. In the case that $f'\alpha = \beta f$ we have the map $Z(\alpha, \beta)\colon Z(f) \to Z(f')$ induced by $\alpha \times \mathrm{id} + \beta$. Let k be the constant homotopy. Then $Z(\alpha, \beta, k) \simeq Z(\alpha, \beta)$ relative to $X + Y$.
3. Let $[\Phi]$ denote the morphism in $\Pi(X, Y')$ represented by Φ. We think of $(\alpha, \beta, [\Phi])$ as a morphism from α to β. The composition is defined by $(\alpha', \beta', [\Phi']) \circ (\alpha, \beta, [\Phi]) = (\alpha'\alpha, \beta'\beta, [\Phi' \diamond \Phi])$. Show that we obtain in this manner a well-defined category. (This definition works in any 2-category.)

4.2 The Double Mapping Cylinder

Given a pair of maps $f\colon A \to B$ and $g\colon A \to C$. The ***double mapping cylinder*** $Z(f, g) = Z(B \xleftarrow{f} A \xrightarrow{g} C)$ is the quotient of $B + A \times I + C$ with respect to the relations $f(a) \sim (a, 0)$ and $(a, 1) \sim g(a)$ for each $a \in A$. We can also define it via a pushout

$$\begin{array}{ccc} A + A & \xrightarrow{f+g} & B + C \\ \big\downarrow{\scriptstyle \langle i_0, i_1 \rangle} & & \big\downarrow{\scriptstyle \langle j_0, j_1 \rangle} \\ A \times I & \longrightarrow & Z(f, g). \end{array}$$

The map $\langle j_0, j_1 \rangle$ is a closed embedding. In the case that $f = \mathrm{id}(A)$, we can identify $Z(\mathrm{id}(A), g) = Z(g)$. We can also glue $Z(f)$ and $Z(g)$ along the common subspace A and obtain essentially $Z(f, g)$ (up to $I \cup_{\{0\}} I \cong I$). A commutative diagram

$$\begin{array}{ccccc} B & \xleftarrow{f} & A & \xrightarrow{g} & C \\ \big\downarrow{\scriptstyle \beta} & & \big\downarrow{\scriptstyle \alpha} & & \big\downarrow{\scriptstyle \gamma} \\ B' & \xleftarrow{f'} & A' & \xrightarrow{g'} & C' \end{array}$$

induces $Z(\beta, \alpha, \gamma)\colon Z(f, g) \to Z(f', g')$, the quotient of $\beta + \alpha \times \mathrm{id} + \gamma$. We can also generalize to an h-commutative diagram as in the previous section.

(4.2.1) Theorem. *Suppose β, α, γ are h-equivalences. Then $Z(\beta, \alpha, \gamma)$ is an h-equivalence.*

In order to use the results about the mapping cylinder, we present $Z(f, g)$, up to canonical homeomorphism, also as the pushout of $j^A\colon A \to Z(f)$ and

$j^B\colon A \to Z(g)$. Here the subspace $Z(f)$ corresponds to the image of $B + A \times [0, 1/2]$ in $Z(f, g)$ and $Z(g)$ to the image of $A \times [1/2, 1] + C$. We view $Z(f, g)$ as a space under $B + A + C$. If we are given homotopies $\Phi^B\colon f'\alpha \simeq \beta f$, $\Phi^C\colon g'\alpha \simeq \gamma g$, we obtain an induced map

$$\Psi = Z(\alpha, \beta, \Phi^B) \cup_A Z(\alpha, \gamma, \Phi^C)\colon Z(f, g) \to Z(f', g')$$

which extends $\beta + \alpha + \gamma$.

(4.2.2) Theorem. *Let α be an h-equivalence with h-inverse α' and suppose β and γ have left h-inverses β', γ'. Choose homotopies $A^t\colon \alpha'\alpha \simeq \mathrm{id}$, $B^t\colon \beta'\beta \simeq \mathrm{id}$, $C^t\colon \gamma'\gamma \simeq \mathrm{id}$. Then there exists $\Omega\colon Z(f', g') \to Z(f, g)$ and a homotopy from $\Omega \circ \Psi$ to the identity which extends $((k * B^t) * k + (k * A^t) * k + (k * C^t) * k)$.*

Proof. The hypotheses imply $\beta' f' \simeq f\alpha$ and $\gamma' g' \simeq g\alpha$. We can apply (4.1.3) and find left homotopy inverses Ω^B of $Z(\alpha, \beta, \Phi^B)$ and Ω^C of $Z(\alpha, \gamma, \Phi^C)$. Then $\Omega = \Omega^B \cup_A \Omega^C$ has the desired properties. □

Theorem (4.2.1) is now a consequence of (4.2.2). The reasoning is as for (4.1.4).

In general, the ordinary pushout of a pair of maps f, g does not have good homotopy properties. One cannot expect to have a pushout in the homotopy category. A pushout is a colimit, in the terminology of category theory. In homotopy theory one replaces colimits by so-called homotopy colimits. We discuss this in the simplest case of pushouts.

Given a diagram

$$\begin{array}{ccc} X_0 & \xrightarrow{f_+} & X_+ \\ \downarrow{\scriptstyle f_-} & & \downarrow{\scriptstyle j_+} \\ X_- & \xrightarrow{j_-} & X \end{array} \tag{1}$$

and a homotopy $h\colon j_- f_- \simeq j_+ f_+$. We obtain an induced map $\varphi\colon Z(f_-, f_+) \to X$ which is the quotient of $\langle j_-, h, j_+ \rangle\colon X_- + X_0 \times I + X_+ \to X$. We define: The diagram (1) together with the homotopy h is called a ***homotopy pushout*** or ***homotopy cocartesian*** if the map φ is a homotopy equivalence. This definition is in particular important if the diagram is commutative and h the constant homotopy.

Suppose we have inclusions $f_\pm\colon X_0 \subset X_\pm$ and $j_\pm\colon X_\pm \subset X$ such that $X = X_- \cup X_+$. In the case that the interiors $X_\pm^\circ$ cover X, the space X is a pushout in the category TOP. In many cases it is also the homotopy pushout; the next proposition is implied by (4.2.4) and (4.2.5).

(4.2.3) Proposition. *Suppose the covering $X_\pm$ of X is numerable (defined below). Then X is the homotopy pushout of $f_\pm\colon X_0 \subset X_\pm$.*

For the proof we first compare $Z(f_-, f_+)$ with the subspace $N(X_-, X_+) = X_- \times 0 \cup X_0 \times I \cup X_+ \times 1$ of $X \times I$. We have a canonical bijective map

$\alpha\colon Z(f_-, f_+) \to N(X_-, X_+)$. Both spaces have a canonical projection to X (denoted p_Z, p_N), and α is a map over X with respect to these projections.

(4.2.4) Lemma. *The map α is an h-equivalence over X and under $X_\pm$.*

Proof. Let $\gamma\colon I \to I$ be defined by $\gamma(t) = 0$ for $t \le 1/3$, $\gamma(t) = 1$ for $t \ge 2/3$ and $\gamma(t) = 3t - 1$ for $1/3 \le t \le 2/3$. We define $\beta\colon N(X_-X_+) \to Z(f_-, f_+)$ as $\mathrm{id}(X_0) \times \gamma$ on $X_0 \times I$ and the identity otherwise. Homotopies $\alpha\beta \simeq \mathrm{id}$ and $\beta\alpha \simeq \mathrm{id}$ are induced by a linear homotopy in the I-coordinate. The reader should verify that β and the homotopies are continuous. □

The covering $X_\pm$ of X is ***numerable*** if the projection p_N has a section. A section σ is determined by its second component $s\colon X \to [0,1]$, and a function of this type defines a section if and only if $X \smallsetminus X_- \subset s^{-1}(0)$, $X \smallsetminus X_+ \subset s^{-1}(1)$.

(4.2.5) Lemma. *Suppose p_N has a section σ. Then p_N is shrinkable.*

Proof. A homotopy $\sigma \circ p_N \simeq \mathrm{id}$ over X is given by a linear homotopy in the I-coordinate. □

(4.2.6) Corollary. *Suppose the covering $X_\pm$ is numerable. Then p_Z is shrinkable.* □

(4.2.7) Theorem. *Let $(X, X_\pm)$ and $(Y, Y_\pm)$ be numerable coverings. Suppose that $F\colon X \to Y$ is a map with $F(X_\pm) \subset Y_\pm$. Assume that the induced partial maps $F_\pm\colon X_\pm \to Y_\pm$ and $F_0\colon X_0 \to Y_0$ are h-equivalences. Then F is an h-equivalence.*

Proof. This is a consequence of (4.2.1) and (4.2.6). □

The double mapping cylinder of the projections $X \leftarrow X \times Y \to Y$ is called the ***join*** $X \star Y$ of X and Y. It is the quotient space of $X \times I \times Y$ under the relations $(x, 0, y) \sim (x, 0, y')$ and $(x, 1, y) \sim (x', 1, y)$. Intuitively it says that each point of X is connected with each point of Y by a unit interval. The reader should verify $S^m \star S^n \cong S^{m+n+1}$. One can also think of the join as $CX \times Y \cup_{X\times Y} X \times CY$ where CX denotes the cone on X.

4.3 Suspension. Homotopy Groups

We work with pointed spaces. Each object in the homotopy groupoid $\Pi^0(X, Y)$ for TOP^0 has an automorphism group. We describe the automorphism group of the constant map in a different manner.

A map $K\colon X \times I \to Y$ is a pointed homotopy from the constant map to itself if and only if it sends the subspace $X \times \partial I \cup \{x\} \times I$ to the base point y of Y. The quotient space

$$\Sigma X = X \times I/(X \times \partial I \cup \{x\} \times I)$$

is called the ***suspension*** of the pointed space (X, x). The base point of the suspension is the set which we identified to a point.

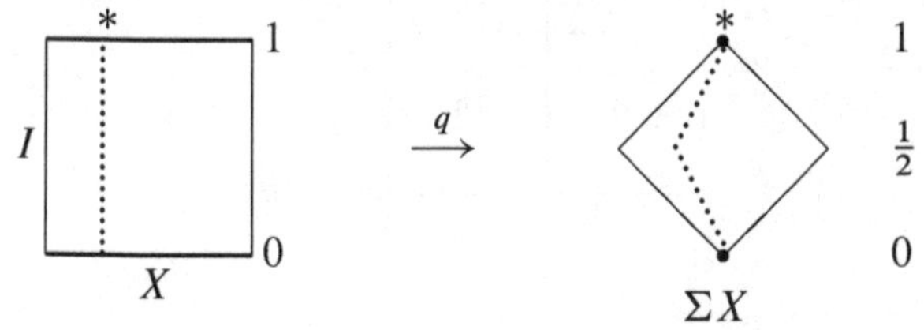

A homotopy $K\colon X \times I \to Y$ from the constant map to itself thus corresponds to a pointed map $\bar{K}\colon \Sigma X \to Y$, and homotopies relative to $X \times \partial I$ correspond to pointed homotopies $\Sigma X \to Y$. This leads us to the homotopy set $[\Sigma X, Y]^0$. This set carries a group structure (written additively) which is defined for representing maps by

$$f + g\colon (x,t) \mapsto \begin{cases} f(x, 2t), & t \le \frac{1}{2} \\ g(x, 2t-1), & \frac{1}{2} \le t. \end{cases}$$

(Again we consider the group opposite to the categorically defined group.) The inverse of $[f]$ is represented by $(x,t) \mapsto f(x, 1-t)$. For this definition we do not need the categorical considerations, but we have verified the group axioms.

If $f\colon X \to Y$ is a pointed map, then $f \times \mathrm{id}(I)$ is compatible with passing to the suspensions and induces $\Sigma f\colon \Sigma X \to \Sigma Y$, $(x,t) \mapsto (f(x), t)$. In this manner the suspension becomes a functor $\Sigma\colon \mathrm{TOP}^0 \to \mathrm{TOP}^0$. This functor is compatible with homotopies: a pointed homotopy H_t induces a pointed homotopy $\Sigma(H_t)$.

There exists a canonical homeomorphism $I^{k+l}/\partial I^{k+l} = I^k/\partial I^k \wedge I^l/\partial I^l$ which is the identity on representing elements in $I^{k+l} = I^k \times I^l$. We have for each pointed space X canonical homeomorphisms

$$(X \wedge I^k/\partial I^k) \wedge I^l/\partial I^l \cong X \wedge I^{k+l}/\partial I^{k+l}, \quad \Sigma^l(\Sigma^k X) \cong \Sigma^{k+l} X.$$

We define the k-fold suspension by $\Sigma^k X = X \wedge (I^k/\partial I^k)$. Note that $\Sigma^n X$ is canonically homeomorphic to $X \times I^n / X \times \partial I^n \cup \{x\} \times \partial I^n$. In the homotopy set $[\Sigma^k X, Y]^0$ we have k composition laws, depending on which of the I-coordinates we use:

$$(f +_i g)(x,t) = \begin{cases} f(x, t_1, \dots, t_{i-1}, 2t_i, t_{i+1}, \dots), & t_i \le \frac{1}{2}, \\ g(x, t_1, \dots, t_{i-1}, 2t_i - 1, t_{i+1}, \dots), & \frac{1}{2} \le t_i. \end{cases}$$

We show in a moment that all these group structures coincide and that they are abelian ($n \ge 2$). For the purpose of the proof one verifies directly the ***commutation rule*** (unravel the definitions)

$$(a +_1 b) +_2 (c +_1 d) = (a +_2 c) +_1 (b +_2 d).$$

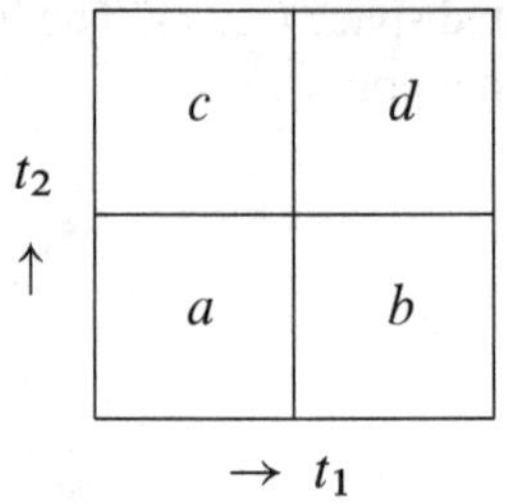

(4.3.1) Proposition. *Suppose the set M carries two composition laws $+_1$ and $+_2$ with neutral elements e_i. Suppose further that the commutation rule holds. Then $+_1 = +_2 = +$, $e_1 = e_2 = e$, and the composition $+$ is associative and commutative.*

Proof. The chain of equalities

$$\begin{aligned} a = a +_2 e_2 &= (a +_1 e_1) +_2 (e_1 +_1 e_2) = (a +_2 e_1) +_1 (e_1 +_2 e_2) \\ &= (a +_2 e_1) +_1 e_1 = a +_2 e_1 \end{aligned}$$

shows that e_1 is a right unit for $+_2$. In a similar manner one shows that e_1 is a left unit and that e_2 is a left and right unit for $+_1$. Therefore $e_1 = e_1 +_2 e_2 = e_2$. The equalities $a +_2 b = (a +_1 e) +_2 (e +_1 b) = (a +_2 e) +_1 (e +_2 b) = a +_1 b$ show $+_1 = +_2 = +$. From $b + c = (e + b) + (c + e) = (e + c) + (b + e) = c + b$ we obtain the commutativity. Finally $a + (b + c) = (a + e) + (b + c) = (a + b) + (e + c) = (a + b) + c$ shows associativity. □

The suspension induces a map $\Sigma_*\colon [A, Y]^0 \to [\Sigma A, \Sigma Y]^0$, $[f] \mapsto [\Sigma f]$, also called suspension. If $A = \Sigma X$, then Σ_* is a homomorphism, because the addition in $[\Sigma X, Y]^0$ is transformed by Σ_* into $+_1$.

Suppose $X = S^0 = \{\pm e_1\}$ with base point e_1. We have a canonical homeomorphism

$$I^n/\partial I^n \cong \Sigma^n S^0 \cong S^0 \times I^n / S^0 \times \partial I^n \cup e_1 \times I^n$$

which sends $x \in I^n$ to $(-e_1, x)$.

The classical homotopy groups of a pointed space are important algebraic invariants. The n-th homotopy group is

$$\pi_n(X) = \pi_n(X, x) = [I^n/\partial I^n, X]^0 = [(I^n, \partial I^n), (X, x)], \quad n \geq 1.$$

These groups are abelian for $n \geq 2$. We can use each of the n coordinates to define the group structure.

4.4 Loop Space

We now dualize the concepts of the previous section. Let (Y, y) be a pointed space. The ***loop space*** ΩY of Y is the subspace of the path space Y^I (with compact-open topology) consisting of the loops in Y with base point y, i.e.,

$$\Omega Y = \{w \in Y^I \mid w(0) = w(1) = y\}.$$

The constant loop k is the base point. A pointed map $f : Y \to Z$ induces a pointed map $\Omega f : \Omega Y \to \Omega Z$, $w \mapsto f \circ w$. This yields the functor $\Omega : \mathrm{TOP}^0 \to \mathrm{TOP}^0$. It is compatible with homotopies: A pointed homotopy H_t yields a pointed homotopy ΩH_t. We can also define the loop space as the space of pointed maps $F^0(I/\partial I, Y)$. The quotient map $p : I \to I/\partial I$ induces $Y^p : Y^{I/\partial I} \to Y^I$ and a homeomorphism $F^0(I/\partial I, Y) \to \Omega Y$ of the corresponding subspaces.

(4.4.1) Proposition. *The product of loops defines a multiplication*

$$m : \Omega Y \times \Omega Y \to \Omega Y, \quad (u, v) \mapsto u * v.$$

It has the following properties:

(1) *m is continuous.*

(2) *The maps $u \mapsto k * u$ and $u \mapsto u * k$ are pointed homotopic to the identity.*

(3) *$m(m \times \mathrm{id})$ and $m(\mathrm{id} \times m)$ are pointed homotopic.*

(4) *The maps $u \mapsto u * u^-$ and $u \mapsto u^- * u$ are pointed homotopic to the constant map.*

Proof. (1) By (2.4.3) it suffices to prove continuity of the adjoint map

$$\Omega Y \times \Omega Y \times I \to Y, \quad (u, v, t) \mapsto (u * v)(t).$$

This map equals on the closed subspace $\Omega Y \times \Omega Y \times [0, \frac{1}{2}]$ the evaluation $(u, v, t) \mapsto u(2t)$ and is therefore continuous.

(2) Let $h_t : I \to I$, $s \mapsto (1 - t)\min(2s, 1) + ts$. Then $\Omega Y \times I \to \Omega Y$, $(u, t) \mapsto uh_t$ is a homotopy from $u \mapsto u * k$ to the identity. Continuity is again proved by passing to the adjoint.

(3) and (4) are proved in a similar manner; universal formulae for associativity and the inverse do the job. □

The loop product induces the m-sum on $[X, \Omega Y]^0$

$$[f] +_m [g] = [m \circ (f \times g) \circ d]$$

with the diagonal $d = (\mathrm{id}, \mathrm{id}) : X \to X \times X$. The functors Σ and Ω are adjoint, see (2.4.10). We compose a map $\Sigma X \to Y$ with the quotient map $X \times I \to \Sigma X$. The adjoint $X \to Y^I$ has an image contained in ΩY. In this manner we obtain a bijection between morphisms $\Sigma X \to Y$ and $X \to \Omega Y$ in TOP^0. Moreover, this adjunction induces a bijection $[\Sigma X, Y]^0 \cong [X, \Omega Y]^0$, see (2.4.11). It transforms the μ-sum (see the next section) into the m-sum, and it is natural in the variables X and Y.

In $[\Sigma X, \Omega Y]^0$ we have two composition laws $+_\mu$ and $+_m$. They coincide and are commutative. We prove this in a formal context in the next section.

4.5 Groups and Cogroups

We work in the category TOP^0. A ***monoid*** in h-TOP^0 is a pointed space M together with a pointed map (multiplication) $m\colon M \times M \to M$ such that $x \mapsto m(*, x)$ and $x \mapsto m(x, *)$ are pointed homotopic to the identity. Spaces with this structure are called ***Hopf spaces*** or ***H-spaces***, in honour of H. Hopf [91]. In $[X, M]^0$ we have the composition law $+_m$, defined as above for $M = \Omega Y$; the constant map represents the neutral element; and $[-, M]^0$ is a contravariant functor into the category of monoids. A ***monoid*** is a set together with a composition law with neutral element. An H-space is associative if $m(m \times \mathrm{id}) \simeq m(\mathrm{id} \times m)$ and commutative if $m \simeq m\tau$ with the interchange $\tau(x, y) = (y, x)$. An inverse for an H-space is a map $\iota\colon M \to M$ such that $m(\iota \times \mathrm{id})d$ and $m(\mathrm{id} \times \iota)d$ are homotopic to the constant map (d the diagonal). An associative H-space with inverse is a ***group object*** in h-TOP^0. By a general principle we have spelled out the definition in TOP^0. The axioms of a group are satisfied up to homotopy. A homomorphism (up to homotopy) between H-spaces (M, m) and (N, n) is a map $\lambda\colon M \to N$ such that $n(\lambda \times \lambda) \simeq \lambda m$. A subtle point in this context is the problem of "coherence", e.g., can a homotopy-associative H-space be h-equivalent to a strictly associative one (by a homomorphism up to homotopy)?

The loop space $(\Omega(X), m)$ is a group object in h-TOP^0.

One can try other algebraic notions "up to homotopy". Let (M, m) be an associative H-space and X a space. A left action of M on X in h-TOP^0 is a map $r\colon M \times X \to X$ such that $r(m \times \mathrm{id}) \simeq r(\mathrm{id} \times r)$ and $x \mapsto r(*, x)$ is homotopic to the identity.

A ***comonoid*** in h-TOP^0 is a pointed space C together with a pointed map (comultiplication) $\mu\colon C \to C \vee C$ such that $\mathrm{pr}_1\, \mu$ and $\mathrm{pr}_2\, \mu$ are pointed homotopic to the identity. In $[C, Y]^0$ we have the composition law $+_\mu$ defined as

$$[f] +_\mu [g] = [\delta \circ (f \vee g) \circ \mu]$$

with the ***codiagonal*** (also called the folding map) $\delta = \langle \mathrm{id}, \mathrm{id} \rangle\colon Y \vee Y \to Y$. The functor $[C, -]^0$ is a covariant functor into the category of monoids. The comultiplication is coassociative up to homotopy if $(\mathrm{id}(C) \vee \mu) \circ \mu$ and $(\mu \vee \mathrm{id}(C)) \circ \mu$ are pointed homotopic; it is cocommutative up to homotopy if μ and $\tau\mu$ are pointed homotopic, with $\tau\colon C \vee C \to C \vee C$ the interchange map. Let (C, μ) and (D, ν) be monoids in h-TOP^0; a cohomomorphism up to homotopy $\varphi\colon C \to D$ is a pointed map such that $(\varphi \vee \varphi)\mu$ and $\nu\varphi$ are pointed homotopic. A coinverse $\iota\colon C \to C$ for the comonoid C is a map such that $\delta(\mathrm{id} \vee \iota)\mu$ and $\delta(\iota \vee \mathrm{id})\mu$ are both pointed homotopic to the constant map. A coassociative comonoid with coinverse in h-TOP^0 is called a ***cogroup*** in h-TOP^0. Let (C,μ) be a coassociative comonoid and Y a space. A left coaction of C on Y (up to homotopy) is a map $\rho\colon Y \to C \vee Y$ such that $(\mathrm{id} \vee \rho)\rho \simeq (\mu \vee \mathrm{id})\rho$ and $\mathrm{pr}_Y\, \rho \simeq \mathrm{id}$.

The suspension ΣX is such a cogroup. We define the ***comultiplication***

$$\mu\colon \Sigma X \to \Sigma X \vee \Sigma X, \quad \mu = i_1 + i_2$$

as the sum of the two injections $i_1, i_2\colon \Sigma X \to \Sigma X \vee \Sigma X$. Explicitly, $\mu(x,t) = (x,2t)$ in the first summand for $t \le \frac{1}{2}$, and $\mu(x,t) = (x, 2t-1)$ in the second summand for $\frac{1}{2} \le t$.

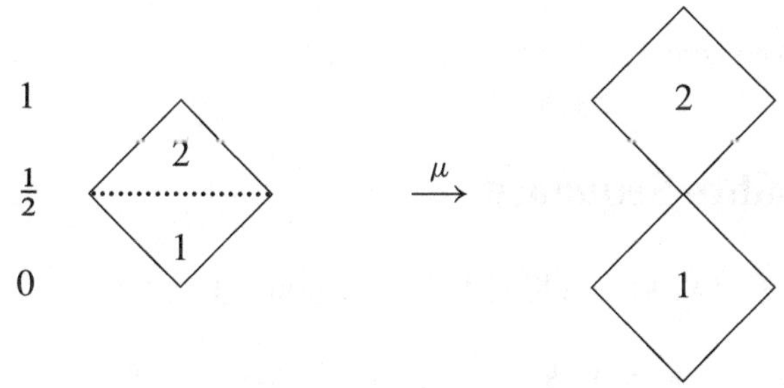

The previously defined group structure on $[\Sigma X, Y]^0$ is the μ-sum.

(4.5.1) Proposition. *Let (C,μ) be a comonoid and (M,m) a monoid in* h-TOP^0. *The composition laws $+_\mu$ and $+_m$ in $[C,M]^0$ coincide and are associative and commutative.*

Proof. We work in h-TOP^0, as we should; thus morphisms are pointed homotopy classes. We have the projections $p_k\colon M \times M \to M$ and the injections $i_l\colon C \to C \vee C$. Given $f\colon C \vee C \to M \times M$ we set $f_{kl} = p_k f i_l$. Then $mf = p_1 f +_m p_2 f$ and $f\mu = f i_1 +_\mu f i_2$. From these relations we derive the commutation rule

$$(mf)\mu = (f_{11} +_\mu f_{12}) +_m (f_{21} +_\mu f_{22}),$$

$$m(f\mu) = (f_{11} +_m f_{21}) +_\mu (f_{12} +_m f_{22}).$$

Now apply (4.3.1). □

Problems

1. Let X be a pointed space and suppose that the Hom-functor $[-,X]^0$ takes values in the category of monoids. Then X carries, up to homotopy, a unique H-space structure m which

induces the monoid structures on $[A,X]^0$ as $+_m$. There is a similar result for Hom-functors $[C,-]^0$ and comonoid structures on C.
2. Let $S(k) = I^k/\partial I^k$. We have canonical homeomorphisms

$$\Omega^k(Y) = F^0(S(k),Y) \cong F((I^k,\partial I^k),(Y,*)) \quad \text{and} \quad \Omega^k\Omega^l(Y) \cong \Omega^{k+l}(Y).$$

3. The space $F((I,0),(Y,*)) \subset Y^I$ is pointed contractible.
4. Let $F_2(I,X) = \{(u,v) \in X^I \times X^I \mid u(1) = v(0)\}$. The map $\mu_2\colon F_2(I,X) \to F(I,X)$, $(u,v) \mapsto u * v$ is continuous.
5. Let $F_3(I,X) = \{(u,v,w) \mid u(1) = v(0),\, v(1) = w(0)\}$. The two maps

$$\mu_3, \mu_3'\colon F_3(I,X) \to F(I,X), \quad (u,v,w) \mapsto (u*v)*w, u*(v*w)$$

are homotopic over $X \times X$ where the projection onto $X \times X$ is given by $(u,v,w) \mapsto (u(0), w(1))$.
6. Verify the homeomorphism $F^0(I/\partial I, Y) \cong \Omega Y$.

4.6 The Cofibre Sequence

A pointed map $f\colon (X,*) \to (Y,*)$ induces a pointed set map

$$f^*\colon [Y,B]^0 \to [X,B]^0, \quad [\alpha] \mapsto [\alpha f].$$

The kernel of f^* consists of the classes $[\alpha]$ such that αf is pointed null homotopic. We work in the category TOP^0 and often omit "pointed" in the sequel. A homotopy set $[Y,B]^0$ is pointed by the constant map. A base point is often denoted by $*$. A sequence $A \xrightarrow{\alpha} B \xrightarrow{\beta} C$ of pointed set maps is exact if $\alpha(A) = \beta^{-1}(*)$. A sequence $U \xrightarrow{f} V \xrightarrow{g} W$ in TOP^0 is called ***h-coexact*** if for each B the sequence

$$[U,B]^0 \xleftarrow{f^*} [V,B]^0 \xleftarrow{g^*} [W,B]^0$$

is exact. If we apply this to $\mathrm{id}(W)$, we see that gf is null homotopic.

A pointed homotopy $X \times I \to B$ sends $* \times I$ to the base point. Therefore we use the ***cylinder*** $XI = X \times I / * \times I$ in TOP^0 together with the embeddings $i_t\colon X \to XI$, $x \mapsto (x,t)$ and the projection $p\colon XI \to X$, $(x,t) \mapsto x$, and we consider morphisms $XI \to Y$ in TOP^0 as homotopies in TOP^0.

The (pointed) ***cone*** CX over X is now defined as $CX = X \times I / X \times 0 \cup * \times I$ with base point the identified set. The inclusion $i_1^X = i_1\colon X \to CX$, $x \mapsto (x,1)$ is an embedding. The maps $h\colon CX \to B$ correspond to the homotopies of the constant map to hi_1 (by composition with the projection $XI \to CX$).

The ***mapping cone*** of f is defined as $C(f) = CX \vee Y/(x,1) \sim f(x)$, or, more formally, via a pushout

$$\begin{array}{ccc} X & \xrightarrow{f} & Y \\ \downarrow{\scriptstyle i_1} & & \downarrow{\scriptstyle f_1} \\ CX & \xrightarrow{j} & C(f). \end{array}$$

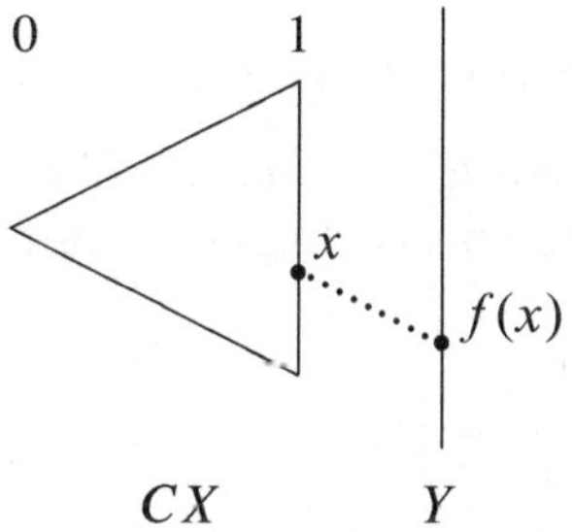

We denote the points of $C(f)$ by their representing elements in $X \times I + Y$. The inclusion $Y \subset CX+Y$ induces an embedding $f_1 \colon Y \to C(f)$, and $CX \subset CX+Y$ induces $j \colon CX \to C(f)$. The pushout property says: The pairs $\alpha \colon Y \to B$, $h \colon CX \to B$ with $\alpha f = h i_1$, i.e., the pairs of α and null homotopies of αf, correspond to maps $\beta \colon C(f) \to B$ with $\beta j = h$. If $[\alpha]$ is contained in the kernel of f^*, then there exists $\beta \colon C(f) \to B$ with $\beta f_1 = \alpha$, i.e., $[\alpha]$ is contained in the image of f_1^*. Moreover, $f_1 f \colon X \to C(f)$ is null homotopic with null homotopy j. This shows that the sequence $X \xrightarrow{f} Y \xrightarrow{f_1} C(f)$ is h-coexact.

We iterate the passage from f to f_1 and obtain the h-coexact sequence

$$X \xrightarrow{f} Y \xrightarrow{f_1} C(f) \xrightarrow{f_2} C(f_1) \xrightarrow{f_3} C(f_2) \xrightarrow{f_4} \cdots .$$

The further investigations replace the iterated mapping cones with homotopy equivalent spaces which are more appealing. This uses the suspension. It will be important that the suspension of a space arises in several ways as a quotient space; certain canonical bijections have to be proved to be homeomorphisms.

In the next diagram the left squares are pushout squares and $p, p(f), q(f)$ are quotient maps. The right vertical maps are homeomorphisms, see Problem 1. Now $\Sigma X \cong CX/i_1 X$, by the identity on representatives. Therefore we view p, $p(f)$, and $q(f)$ as morphisms to ΣX.

$$
\begin{array}{ccccccc}
X & \xrightarrow{i_1} & CX & \xrightarrow{p} & CX/i_1X & = \Sigma X \\
\downarrow f & & \downarrow j & & \downarrow & \\
Y & \xrightarrow{f_1} & C(f) & \xrightarrow{p(f)} & C(f)/f_1Y & = \Sigma X \\
\downarrow i_1 & & \downarrow f_2 & & \downarrow & \\
CY & \xrightarrow{j_1} & C(f_1) & \xrightarrow{q(f)} & C(f_1)/j_1CY & = \Sigma X
\end{array}
$$

(4.6.1) Note. *The quotient map $q(f)$ is a homotopy equivalence.*

Proof. We define a homotopy h_t of the identity of $C(f_1)$ which contracts CY along the cone lines to the cone point and drags CX correspondingly

$$
h_t(x,s) = \begin{cases} (x,(1+t)s), & (1+t)s \le 1, \\ (f(x), 2-(1+t)s), & (1+t)s \ge 1, \end{cases} \qquad h_t(y,s) = (y,(1-t)s).
$$

In order to verify continuity, one checks that the definition is compatible with the equivalence relation needed to define $C(f_1)$. The end h_1 of the homotopy has the form $s(f) \circ q(f)$ with $s(f)\colon \Sigma X \to C(f_1)$, $(x,s) \mapsto h_1(x,s)$. The composition $q(f) \circ s(f)\colon (x,s) \mapsto (x, \min(2s,1))$ is also homotopic to the identity, as we know from the discussion of the suspension. This shows that $s(f)$ is h-inverse to $q(f)$. □

We treat the next step in the same manner:

$$
\begin{array}{ccccc}
C(f) & \xrightarrow{f_2} & C(f_1) & \xrightarrow{f_3} & C(f_2) \\
& \searrow^{p(f)} & \downarrow q(f) & \searrow^{p(f_1)} & \downarrow q(f_1) \\
& & \Sigma X & \dashrightarrow_{\Sigma(f)\circ\iota} & \Sigma Y
\end{array}
$$

with an h-equivalence $q(f_1)$. Let $\iota\colon \Sigma X \to \Sigma X$, $(x,t) \mapsto (x, 1-t)$ be the inverse of the cogroup ΣX. The last diagram is not commutative if we add the morphism Σf to it. Rather the following holds:

(4.6.2) Note. $\Sigma(f) \circ \iota \circ q(f) \simeq p(f_1)$.

Proof. By (4.6.1) it suffices to study the composition with $s(f)$. We know already that $p(f_1)s(f)\colon (x,s) \mapsto (f(x), \min(1, 2(1-s))$ is homotopic to $\Sigma f \circ \iota\colon (x,s) \mapsto (f(x), 1-s)$. □

An h-coexact sequence remains h-coexact if we replace some of its spaces by h-equivalent ones. Since the homeomorphism ι does not destroy exactness of a sequence, we obtain from the preceding discussion that the sequence

$$X \xrightarrow{f} Y \xrightarrow{f_1} C(f) \xrightarrow{p(f)} \Sigma X \xrightarrow{\Sigma f} \Sigma Y$$

is h-coexact.

We can continue this coexact sequence if we apply the procedure above to Σf instead of f. The next step is then $(\Sigma f)_1 \colon \Sigma Y \to C(\Sigma f)$. But it turns out that we can also use the suspension of the original map $\Sigma(f_1) \colon \Sigma Y \to \Sigma C(f)$ in order to continue with an h-coexact sequence. This is due to the next lemma.

(4.6.3) Lemma. *There exists a homeomorphism $\tau_1 \colon C(\Sigma f) \to \Sigma C(f)$ which satisfies $\tau_1 \circ (\Sigma f)_1 = \Sigma(f_1)$.*

Proof. $C\Sigma X$ and $\Sigma C X$ are both quotients of $X \times I \times I$. Interchange of I-coordinates induces a homeomorphism $\tau \colon C\Sigma X \to \Sigma C X$ which satisfies $\tau \circ i_1^{\Sigma X} = \Sigma(i_1^X)$. We insert this into the pushout diagrams for $C(\Sigma f)$ and $\Sigma C(f)$ and obtain an induced τ_1. We use that a pushout in TOP^0 becomes a pushout again if we apply Σ (use Ω-Σ-adjunction). □

We now continue in this manner and obtain altogether an infinite h-coexact sequence.

(4.6.4) Theorem. *The sequence*

$$X \xrightarrow{f} Y \xrightarrow{f_1} C(f) \xrightarrow{p(f)} \Sigma X \xrightarrow{\Sigma f} \Sigma Y \xrightarrow{\Sigma f_1} \Sigma C(f) \xrightarrow{\Sigma p(f)} \Sigma^2 X \xrightarrow{\Sigma^2 f} \cdots$$

is h-coexact. We call it the ***Puppe-sequence*** *or the* ***cofibre sequence*** *of f* ([155]). *The functor $[-, B]^0$ applied to the Puppe-sequence yields an exact sequence of pointed sets; it consists from the fourth place onwards of groups and homomorphisms and from the seventh place onwards of abelian groups. See* [49] *for an introduction to some other aspects of the cofibre sequence.* □

The derivation of the cofibre sequence uses only formal properties of the homotopy notion. There exist several generalizations in an axiomatic context; for an introduction see [69], [101], [18].

Let $f \colon X \to Y$ be a pointed map. Define $\mu \colon C(f) \to \Sigma X \vee C(f)$ by $\mu(x,t) = ((x,2t),*)$ for $2t \le 1$, $\mu(x,t) = (*,(x,2t-1))$ for $2t \ge 1$, and $\mu(y) = y$. This map is called the ***h-coaction*** of the h-cogroup ΣX on $C(f)$. This terminology is justified by the next proposition.

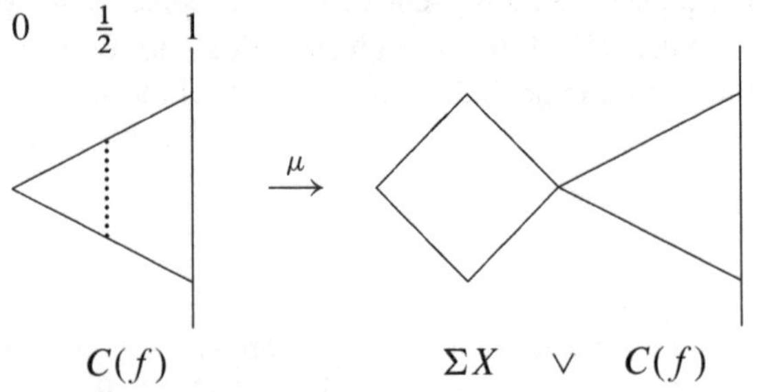

(4.6.5) Proposition. *The map μ induces a left action*

$$[\Sigma X, B]^0 \times [C(f), B]^0 \cong [\Sigma X \vee C(f), B]^0 \xrightarrow{\mu^*} [C(f), B]^0, \quad (\alpha, \beta) \mapsto \alpha \odot \beta.$$

This action satisfies $\alpha_1 \odot p(f)^*\alpha_2 = p(f)^*(\alpha_1\alpha_2)$. Moreover, $f_1^*(\beta_1) = f_1^*(\beta_2)$ if and only if there exists α such that $\alpha \odot \beta_1 = \beta_2$. Thus f_1^* induces an injective map of the orbits of the action.

Proof. That μ^* satisfies the axioms of a group action is proved as for the group axioms involving ΣX. Also the property involving $p(f)$ is proved in the same manner. It remains to verify the last statement.

Assume that $f, g\colon C(f) \to B$ are maps which become homotopic when restricted to Y. Consider the subspaces $C_0 = \{(x,t) \mid 2t \le 1\} \subset C(f)$ and $C_1 = \{(x,t) \mid 2t \ge 1\} \cup Y \subset C(f)$. These inclusions are cofibrations (see the next chapter). Therefore we can change f and g within their homotopy classes such that $f|C_0$ is constant and $g|C_1 = f|C_1$. Then g is constant on $C_0 \cap C_1$. Therefore there exists $h\colon \Sigma X \vee C(f) \to Y$ such that $h\mu = g$ and $h|C(f) = f$. Let $k = h|\Sigma X$. Then $[k] \odot [f] = [g]$. Conversely, $\mu \circ \langle a, b\rangle$ and b have the same restriction to Y. □

Problems

1. Let the left square in the next diagram be a pushout with an embedding j and hence an embedding J. Then F induces a homeomorphism $\overline{F}$ of the quotient spaces.

$$\begin{array}{ccccc} A & \xrightarrow{j} & X & \xrightarrow{p} & X/A \\ \downarrow{\scriptstyle f} & & \downarrow{\scriptstyle F} & & \downarrow{\scriptstyle \overline{F}} \\ B & \xrightarrow{J} & Y & \xrightarrow{q} & Y/B \end{array}$$

2. The map $p(f)^*\colon [\Sigma X, B]^0 \to [C(f), B]^0$ induces an injective map of the left (or right) cosets of $[\Sigma X, B]^0$ modulo the subgroup $\operatorname{Im} \Sigma(f)^*$.

4.7 The Fibre Sequence

The investigations in this section are dual to those of the preceding section. For the purpose of this section we describe homotopies $B \times I \to Y$ in a dual (adjoint) form $B \to Y^I$ as maps into function spaces. A pointed map $f : (X, *) \to (Y, *)$ induces a pointed set map

$$f_* : [B, X]^0 \to [B, Y]^0, \quad [\alpha] \mapsto [f\alpha].$$

A sequence $U \xrightarrow{f} V \xrightarrow{g} W$ in TOP^0 is called ***h-exact***, if for each B the sequence

$$[B, U]^0 \xrightarrow{f_*} [B, V]^0 \xrightarrow{g_*} [B, W]^0$$

is exact. If we apply this to $\mathrm{id}(U)$, we see that gf is null homotopic.

We need the dual form of the cone. Let $F(Y) = \{w \in Y^I \mid w(0) = *\}$ be the space of paths which start in the base point of Y, with the constant path $k_*(t) = *$ as base point, and evaluation $e^1 : FY \to Y$, $w \mapsto w(1)$. Via adjunction we have $F^0(B, FY) \cong F^0(CB, Y)$. The pointed maps $h : B \to FY$ correspond to pointed homotopies from the constant map to $e^1 h$, if we pass from h to the adjoint map $B \times I \to Y$. We define $F(f)$ via a pullback

$$\begin{array}{ccc} F(f) & \xrightarrow{q} & FY \\ \downarrow{\scriptstyle f^1} & & \downarrow{\scriptstyle e^1} \\ X & \xrightarrow{f} & Y \end{array} \qquad \begin{array}{c} F(f) = \{(x, w) \in X \times FY \mid f(x) = w(1)\} \\ f^1(x, w) = x, \quad q(x, w) = w. \end{array}$$

The maps $\beta : B \to F(f)$ correspond to pairs $\alpha = f^1\beta : B \to X$ together with the null homotopies $q\beta : B \to FY$ of $f\alpha$. This shows that $F(f) \xrightarrow{f^1} X \xrightarrow{f} Y$ is h-exact.

We now iterate the passage from f to f^1

$$\cdots \xrightarrow{f^4} F(f^2) \xrightarrow{f^3} F(f^1) \xrightarrow{f^2} F(f) \xrightarrow{f^1} X \xrightarrow{f} Y$$

and show that $F(f^1)$ and $F(f^2)$ can be replaced, up to h-equivalence, by ΩY and ΩX. We begin with the remark that

$$(f^1)^{-1}(*) = \{(x, w) \mid w(0) = *,\ w(1) = f(x),\ x = *\}$$

can be identified with ΩY, via $w \mapsto (*, w)$. Let $i(f) : \Omega Y \to F(f)$ be the associated inclusion of this fibre of f^1. The space (by its pullback definition)

$$F(f^1) = \{(x, w, v) \mid w(0) = *,\ w(1) = f(x),\ x = v(1),\ v(0) = *\}$$

can be replaced by the homeomorphic space

$$F(f^1) = \{(w, v) \in FY \times FX \mid w(1) = fv(1)\}.$$

Then f^2 becomes $f^2\colon F(f^1) \to F(f)$, $(w, v) \mapsto (v(1), w)$. The map

$$j(f)\colon \Omega Y \to F(f^1), \quad w \mapsto (w, k_*)$$

satisfies $f^2 \circ j(f) = i(f)$.

(4.7.1) Note. *The injection $j(f)$ is a homotopy equivalence.*

Proof. We construct a homotopy h_t of the identity of $F(f^1)$ which shrinks the path v to its beginning point and drags behind the path w correspondingly. We write $h_t(w, v) = (h_t^1(w, v), h_t^2(w, v)) \in FY \times FX$ and define

$$h_t^1(s) = \begin{cases} w(s(1+t)), & s(1+t) \le 1, \\ fv(2-(1+t)s), & s(1+t) \ge 1, \end{cases} \qquad h_t^2(s) = v(s(1-t)).$$

The end h_1 of the homotopy has the form $j(f) \circ r(f)$ with

$$r(f)\colon F(f^1) \to \Omega Y, \quad (w, v) \mapsto w * (fv)^-.$$

The relation $(r(f) \circ j(f))(w) = w * k_*^-$ shows that also $r(f) \circ j(f)$ is homotopic to the identity. The continuity of h_t is proved by passing to the adjoint maps. □

We treat the next step in a similar manner.

$$\begin{array}{ccc} F(f^2) & \xrightarrow{f^3} & F(f^1) \\ {\scriptstyle j(f^1)}\uparrow & \nearrow{\scriptstyle i(f^1)} & \uparrow{\scriptstyle j(f)} \\ \Omega X & \xrightarrow[\Omega f]{} & \Omega Y \end{array} \qquad i(f^1)(v) = (k_*, v).$$

The upper triangle is commutative, and (4.7.2) applies to the lower one. The map $i(f^1)$ is the embedding of the fibre over the base point. Let $\iota\colon \Omega Y \to \Omega Y$, $w \mapsto w^-$ be the inverse.

(4.7.2) Note. $j(f) \circ \iota \circ \Omega f \simeq i(f^1)$.

Proof. We compose both sides with the h-equivalence $r(f)$ from the proof of (4.7.1). Then $r(f) \circ i(f^1)$ equals $v \mapsto k_* * (fv)^-$, and this is obviously homotopic to $\iota \circ \Omega f\colon v \mapsto (fv)^-$. □

As a consequence of the preceding discussion we see that the sequence

$$\Omega X \xrightarrow{\Omega f} \Omega Y \xrightarrow{i(f)} F(f) \xrightarrow{f^1} X \xrightarrow{f} Y$$

is h-exact.

(4.7.3) Lemma. *There exists a homeomorphism* $\tau^1\colon F(\Omega f) \to \Omega F(f)$ *such that* $(\Omega f)^1 = \Omega(f^1) \circ \tau^1$.

Proof. From the definitions and standard properties of mapping spaces we have $\Omega F(f) \subset \Omega X \times \Omega FY$ and $F\Omega(f) \subset \Omega X \times F\Omega Y$. We use the exponential law for mapping spaces and consider ΩFY and $F\Omega Y$ as subspaces of $Y^{I\times I}$. In the first case we have to use all maps which send $\partial I \times I \cup I \times 0$ to the base point, in the second case all maps which send $I \times \partial I \cup 0 \times I$ to the base point. Interchanging the I-coordinates yields a homeomorphism and it induces τ^1. □

We now continue as in the previous section.

(4.7.4) Theorem. *The sequence*

$$\cdots \xrightarrow{\Omega^2 f} \Omega^2 Y \xrightarrow{\Omega i(f)} \Omega F(f) \xrightarrow{\Omega f^1} \Omega X \xrightarrow{\Omega f} \Omega Y \xrightarrow{i(f)} F(f) \xrightarrow{f^1} X \xrightarrow{f} Y$$

is h-exact. We call it the ***fibre sequence*** *of* f. *When we apply the functor* $[B,-]^0$ *to the fibre sequence we obtain an exact sequence of pointed sets which consists from the fourth place onwards of groups and homomorphisms and from the seventh place onwards of abelian groups* ([147]). □

Problems

1. Work out the dual of (4.6.5).

2. Describe what happens to the fibre sequence under adjunction. A map

$$a\colon T \to F(f) = \{(x,w) \in X \times FY \mid f(x) = w(1)\}$$

has two components $b\colon T \to X$ and $\beta\colon T \to FY$. Under adjunction, β corresponds to a map $B\colon CT \to Y$ from the cone over T. The condition $f(x) = w(1)$ is equivalent to the commutativity $fb = Bi_1$. This transition is also compatible with pointed homotopies, and therefore we obtain a bijection $[a] \in [T, F(f)]^0 \cong [i_1, f]^0 \ni [B,b]$. This bijection transforms f^1_* into the restriction $[B,b] \in [i_1, f]^0 \to [T,X]^0 \ni [b]$. In the next step we have

$$\begin{array}{ccccc} [T,\Omega Y]^0 & \xrightarrow{\cong} & [\Sigma T, Y]^0 & & [\gamma] \\ \downarrow{\scriptstyle i(f)_*} & & \downarrow & & \downarrow \\ [T, F(f)]^0 & \xrightarrow{\cong} & [i_1, f]^0, & & [\gamma \circ p, c]. \end{array}$$

The image of γ is obtained in the following manner: With the quotient map $p\colon CT \to \Sigma T$ we have $B = \gamma \circ p$, and b is the constant map c.

3. There exist several relations between fibre and cofibre sequences.

The adjunction (Σ, Ω) yields in TOP^0 the maps $\eta\colon X \to \Omega\Sigma X$ (unit of the adjunction) and $\varepsilon\colon \Sigma\Omega X \to X$ (counit of the adjunction). These are natural in the variable X. For each $f\colon X \to Y$ we also have natural maps

$$\eta\colon F(f) \to \Omega C(f), \quad \varepsilon\colon \Sigma F(f) \to C(f)$$

defined by

$$\eta(x,w)(t) = \begin{cases} [x,2t], & t \le 1/2, \\ w(2-2t), & t \ge 1/2, \end{cases}$$

and ε adjoint to η. Verify the following assertions from the definitions.

(1) The next diagram is homotopy commutative

$$\begin{array}{ccccc} \Omega Y & \xrightarrow{i(f)} & F(f) & \xrightarrow{f^1} & X \\ \downarrow{\scriptstyle \iota} & & \downarrow{\scriptstyle \eta} & & \downarrow{\scriptstyle \eta} \\ \Omega Y & \xrightarrow{\Omega f_1} & \Omega C(f) & \xrightarrow{\Omega p(f)} & \Omega \Sigma X. \end{array}$$

(2) Let $i\colon X \to Z(f)$ be the inclusion and $r\colon Z(f) \to Y$ the retraction. A path in $Z(f)$ that starts in $*$ and ends in $X \times 0$ yields under the projection to $C(f)$ a loop. This gives a map $\pi\colon F(i) \to \Omega C(f)$. The commutativity $\eta \circ F(r) \simeq \iota \circ \pi$ holds.

(3) The next diagram is homotopy commutative

$$\begin{array}{ccc} F(f_1) & \xrightarrow{i(f_1)} & Y \\ {\scriptstyle \Omega q(f_1)\circ\eta}\downarrow & & \downarrow{\scriptstyle \iota\circ\eta} \\ \Omega\Sigma X & \xrightarrow{\Omega\Sigma f} & \Omega\Sigma Y. \end{array}$$

Chapter 5
Cofibrations and Fibrations

This chapter is also devoted to mostly formal homotopy theory. In it we study the homotopy extension and lifting property.

An ***extension*** of $f: A \to Y$ ***along*** $i: A \to X$ is a map $F: X \to Y$ such that $Fi = f$. If $i: A \subset X$ is an inclusion, then this is an extension in the ordinary sense. Many topological problems can be given the form of an extension problem. It is important to find conditions on i under which the extendibility of f only depends on the homotopy class of f. If this is the case, then f is called a cofibration.

The dual of the extension problem is the lifting problem. Suppose given maps $p: E \to B$ and $f: X \to B$. A ***lifting*** of f ***along*** p is a map $F: X \to E$ such that $pF = f$. We ask for conditions on p such that the existence of a lifting only depends on the homotopy class of f. If this is the case, then f is called a fibration.

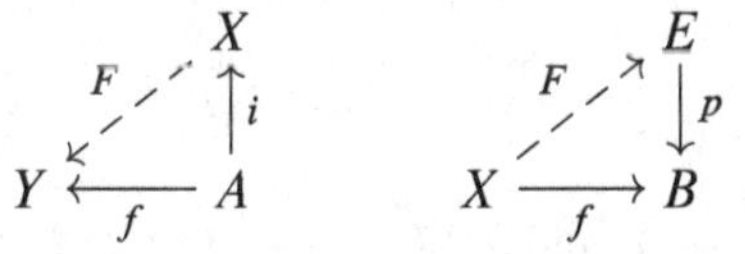

Each map is the composition of a cofibration and a homotopy equivalence and (dually) the composition of a homotopy equivalence and a fibration. The notions are then used to define homotopy fibres ("homotopy kernels") and homotopy cofibres ("homotopy cokernels"). Axiomatizations of certain parts of homotopy theory ("model categories") are based on these notions. The notions also have many practical applications, e.g., to showing that maps are homotopy equivalences with additional properties like fibrewise homotopy equivalences.

Another simple typical example: A base point $x \in X$ is only good for homotopy theory if the inclusion $\{x\} \subset X$ is a cofibration (or the homotopy invariant weakening, a so-called h-cofibration). This is then used to study the interrelation between pointed and unpointed homotopy constructions, like pointed and unpointed suspensions.

5.1 The Homotopy Extension Property

A map $i: A \to X$ has the ***homotopy extension property*** (HEP) for the space Y if for each homotopy $h: A \times I \to Y$ and each map $f: X \to Y$ with $fi(a) = h(a, 0)$ there exists a homotopy $H: X \times I \to Y$ with $H(x, 0) = f(x)$ and $H(i(a), t) = h(a, t)$. We call H an ***extension*** of h with ***initial condition*** f. The map $i: A \to X$

is a ***cofibration*** if it has the HEP for all spaces. The data of the HEP are displayed in the next diagram. We set $i_t^X : X \to X \times I, x \mapsto (x,t)$ and $e^0(w) = w(0)$.

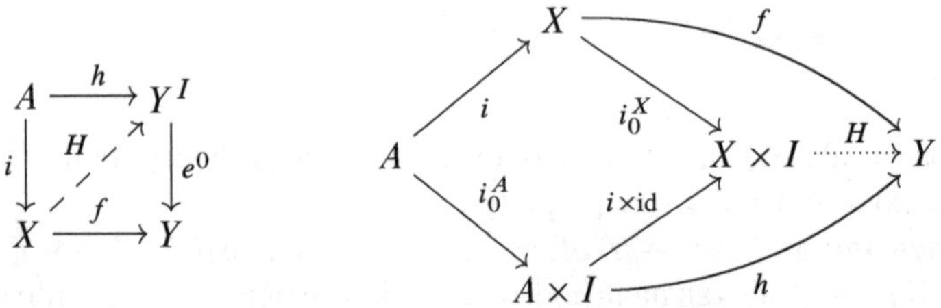

For a cofibration $i : A \to X$, the extendibility of f only depends on its homotopy class.

From this definition one cannot prove directly that a map is a cofibration, but it suffices to test the HEP for a universal space Y, the mapping cylinder $Z(i)$ of i. Recall that $Z(i)$ is defined by a pushout

$$\begin{array}{ccc} A & \xrightarrow{\;i\;} & X \\ \downarrow{\scriptstyle i_0^A} & & \downarrow{\scriptstyle b} \\ A \times I & \xrightarrow{\;k\;} & Z(i). \end{array}$$

Pairs of maps $f : X \to Y$ and $h : A \times I \to Y$ with $hi_0^A = fi$ then correspond to maps $\sigma : Z(i) \to Y$ with $\sigma b = f$ and $\sigma k = h$. We apply this to the pair $i_0^X : X \to X \times I$ and $i \times \mathrm{id} : A \times I \to X \times I$ and obtain $s : Z(i) \to X \times I$ such that $sb = i_0^X$ and $sk = i \times \mathrm{id}$.

Now suppose that i is a cofibration. We use the HEP for the space $Z(i)$, the initial condition b and the homotopy k. The HEP then provides us with a map $r : X \times I \to Z(i)$ such that $ri_0^X = b$ and $r(i \times \mathrm{id}) = k$. We conclude from $rsb = ri_0^X = b, rsk = r(i \times \mathrm{id}) = k$ and the pushout property that $rs = \mathrm{id}(Z(i))$, i.e., s is an embedding and r a retraction. Let r be a retraction of s. Given f and h, find σ as above and set $H = \sigma r$. Then H extends h with initial condition f. Altogether we have shown:

(5.1.1) Proposition. *The following statements about* $i : A \to X$ *are equivalent:*

(1) i *is a cofibration.*

(2) i *has the* HEP *for the mapping cylinder* $Z(i)$.

(3) $s : Z(i) \to X \times I$ *has a retraction.* □

A cofibration $i : A \to X$ is an embedding; and $i(A)$ is closed in X, if X is a Hausdorff space (Problem 1). Therefore we restrict attention to closed cofibrations whenever this simplifies the exposition. A pointed space (X, x) is called ***well-pointed*** and the base point ***nondegenerate*** if $\{x\} \subset X$ is a closed cofibration.

(5.1.2) Proposition. *If $i\colon A \subset X$ is a cofibration, then there exists a retraction $r\colon X \times I \to X \times 0 \cup A \times I$. If A is closed in X and if there exists a retraction r, then i is a cofibration.*

Proof. Let $Y = X \times 0 \cup A \times I$, $f(x) = (x, 0)$, and $h(a,t) = (a,t)$. Apply the HEP to obtain a retraction $r = H$.

If A is closed in X, then $g\colon X \times 0 \cup A \times I \to Y, g(x,0) = f(x), (a,t) = h(a,t)$ is continuous. A suitable extension H is given by gr. □

(5.1.3) Example. Let $r\colon X \times I \to X \times 0 \cup A \times I$ be a retraction. Set $r(x,t) = (r_1(x,t), r_2(x,t))$. Then

$$H\colon X \times I \times I \to X \times I, \quad (x,t,s) \mapsto (r_1(x, t(1-s)), st + (1-s)r_2(x,t))$$

is a homotopy relative to $X \times 0 \cup A \times I$ of r to the identity, i.e., a deformation retraction. □

(5.1.4) Example. The inclusions $S^{n-1} \subset D^n$ and $\partial I^n \subset I^n$ are cofibrations. A retraction $r\colon D^n \to S^{n-1} \times I \cup D^n \times 0$ was constructed in (2.3.5). ◇

It is an interesting fact that one need not assume A to be closed. Strøm [180, Theorem 2] proved that an inclusion $A \subset X$ is a cofibration if and only if the subspace $X \times 0 \cup A \times I$ is a retract of $X \times I$.

If we multiply a retraction by $\mathrm{id}(Y)$ we obtain again a retraction. Hence $A \times Y \to X \times Y$ is a (closed) cofibration for each Y, if $i\colon A \to X$ is a (closed) cofibration. Since we have proved (5.1.2) only for closed cofibrations, we mention another special case, to be used in a moment. Let Y be locally compact and $i\colon A \to X$ a cofibration. Then $i \times \mathrm{id}\colon A \times Y \to X \times Y$ is a cofibration. For a proof use the fact that via adjunction and the exponential law for mapping spaces the HEP of $i \times \mathrm{id}$ for Z corresponds to the HEP of i for Z^Y.

(5.1.5) Proposition. *Let $A \subset X$ and assume that $A \times I \subset X \times I$ has the* HEP *for Y. Given maps $\varphi\colon A \times I \times I \to Y$, $H\colon X \times I \to Y$, $f^\varepsilon\colon X \times I \to Y$ such that*

$$\varphi(a,s,0) = H(a,s), \quad f^\varepsilon(x,0) = H(x,\varepsilon), \quad f^\varepsilon(a,t) = \varphi(a,\varepsilon,t)$$

$\varepsilon \in \{0,1\}$, $a \in A$, $x \in X$, $s,t \in I$. Then there exists $\Phi\colon X \times I \times I \to Y$ such that

$$\Phi(a,s,t) = \varphi(a,s,t), \quad \Phi(x,s,0) = H(x,s), \quad \Phi(x,\varepsilon,t) = f^\varepsilon(x,t).$$

Proof. H and f^ε together yield a map $\alpha\colon X \times (I \times 0 \cup \partial I \times I) \to Y$ defined by $\alpha(x,s,0) = H(x,s)$ and $\alpha(x,\varepsilon,t) = f^\varepsilon(x,t)$. By our assumptions, α and φ coincide on $A \times (I \times 0 \cup \partial I \times I)$. Let $k\colon (I \times I, I \times 0 \cup \partial I \times I) \to (I \times I, I \times 0)$ be a homeomorphism of pairs. Since $A \times I \to X \times I$ has the HEP for Y, there exists $\Psi\colon X \times I \times I \to Y$ which extends $\varphi \circ (1 \times k^{-1})$ and $\alpha \circ (1 \times k^{-1})$. The map $\Phi = \Psi \circ (1 \times k)$ solves the extension problem. □

(5.1.6) Proposition. *Let $i\colon A \subset X$ be a cofibration. Then $X \times \partial I \cup A \times I \subset X \times I$ is a cofibration.*

Proof. Given $h\colon (A \times I \cup X \times \partial I) \times I \to Y$ and an initial condition $H\colon X \times I \to Y$, we set $\varphi = h|A \times I \times I$ and $f^{\varepsilon}(x,t) = h(x,\varepsilon,t)$. Then we apply (5.1.5). □

For $A = \emptyset$ we obtain from (5.1.5) that $X \times \partial I \subset X \times I$ is a cofibration, in particular $\partial I \subset I$ and $\{0\} \subset I$ are cofibrations. Induction over n shows again that $\partial I^n \subset I^n$ is a cofibration.

We list some special cases of (5.1.5) for a cofibration $A \subset X$.

(5.1.7) Corollary. (1) *Let $\Phi\colon X \times I \to Y$ be a homotopy. Suppose $\varphi = \Phi|A \times I$ is homotopic rel $A \times \partial I$ to ψ. Then Φ is homotopic rel $X \times \partial I$ to $\Psi\colon X \times I \to Y$ such that $\Psi|A \times I = \psi$.*

(2) *Let Φ solve the extension problem for (φ, f) and Ψ the extension problem for (ψ, g). Suppose $f \simeq g$ rel A and $\varphi \simeq \psi$ rel $A \times \partial I$. Then $\Phi_1 \simeq \Psi_1$ rel A.*

(3) *Let $\Phi, \Psi\colon X \times I \to Y$ solve the extension problem for (h, f). Then there exists a homotopy $\Gamma\colon \Phi \simeq \Psi$ rel $X \times 0 \cup A \times I$.* □

(5.1.8) Proposition. *Let a pushout diagram in* TOP *be given.*

$$\begin{array}{ccc} A & \xrightarrow{f} & B \\ \downarrow{\scriptstyle j} & & \downarrow{\scriptstyle J} \\ X & \xrightarrow{F} & Y \end{array}$$

If j has the HEP *for Z, then J has the* HEP *for Z. If j is a cofibration, then J is a cofibration.*

Proof. Suppose $h\colon B \times I \to Z$ and $\varphi\colon Y \to Z$ are given such that $h(b,0) = \varphi J(b)$ for $b \in B$. We use the fact that the product with I of a pushout is again a pushout. Since j has the HEP for Z, there exists $K_t\colon X \to Z$ such that $K_0 = \varphi f$ and $K_t j = h_t f$. By the pushout property, there exists $H_t\colon Y \to Z$ such that $H_t F = K_t$ and $H_t J = h_t$. The uniqueness property shows $H_0 = \varphi$, since both maps have the same composition with F and J. □

We call J the cofibration ***induced*** from j via ***cobase change*** along f.

Example. If $A \subset X$ is a cofibration, then $\{A\} \subset X/A$ is a cofibration. $S^{n-1} \subset D^n$ is a cofibration, hence $\{S^{n-1}\} \subset D^n/S^{n-1}$ is a cofibration. The space D^n/S^{n-1} is homeomorphic to S^n; therefore $(S^n, *)$ is well-pointed. ◇

Example. If (X_j) is a family of well-pointed spaces, then the wedge $\bigvee_j X_j$ is well-pointed. ◇

Our next result, the homotopy theorem for cofibration says, among other things, that homotopic maps induce h-equivalent cofibrations from a given cofibration under a cobase change.

Let $j: A \to X$ be a cofibration and $\varphi_t: f \simeq g: A \to B$ a homotopy. We consider two pushout diagrams.

$$\begin{array}{ccc} A & \xrightarrow{f} & B \\ \downarrow{\scriptstyle j} & & \downarrow{\scriptstyle j_f} \\ X & \xrightarrow{F} & Y_f \end{array} \qquad\qquad \begin{array}{ccc} A & \xrightarrow{g} & B \\ \downarrow{\scriptstyle j} & & \downarrow{\scriptstyle j_g} \\ X & \xrightarrow{G} & Y_g \end{array}$$

Since j is a cofibration, there exists a homotopy $\Phi_t: X \to Y_f$ with initial condition $\Phi_0 = F$ and $\Phi_t j = j_f \varphi_t$. The pushout property of the Y_g-diagram provides us with a unique map $\kappa = \kappa_\varphi$ such that $\kappa j_g = j_f$ and $\kappa G = \Phi_1$. (We use the notation κ_φ although the map depends on Φ_1.) Thus κ_φ is a morphism of cofibrations $\kappa: j_g \to j_f$ between objects in TOP^B. Moreover $\kappa G \simeq F$. We now verify that the homotopy class of κ is independent of some of the choices involved. Let ψ_t be another homotopy from f to g which is homotopic to φ_t relative to $A \times \partial I$. Let $\Psi_t: X \to Y_f$ be an extension of $j_f \psi_t$ with initial condition $\Psi_0 = F$. Let $\gamma: A \times I \times I \to B$ be a homotopy rel $A \times \partial I$ from φ to ψ. These data give us on $X \times 0 \times I \cup X \times I \times \partial I$ a map Γ into Y_f such that

$$\Gamma(x,0,t) = F(x), \quad \Gamma(x,s,0) = \Phi(x,s), \quad \Gamma(x,s,1) = \Psi(x,s).$$

By (5.1.5) there exists an extension, still denoted Γ, to $X \times I \times I$ such that $j_f \gamma = \Gamma(j \times \mathrm{id} \times \mathrm{id})$. We multiply the Y_g diagram by I and obtain again a pushout. It provides us with a unique homotopy $K: Y_g \times I \to Y_f$ such that $K \circ (G \times \mathrm{id}) = \Gamma_1$ and $K \circ (j_g \times \mathrm{id}) = j_f \circ \mathrm{pr}$ where $\Gamma_1: X \times I \to Y_f$, $(x,t) \mapsto \Gamma(x,1,t)$. By construction, K is a homotopy under B from κ_φ to a corresponding map κ_ψ obtained from ψ_t and Ψ_t. We thus have shown that the homotopy class $[\kappa]^B$ under B of κ only depends on the morphism $[\varphi]$ from f to g in the groupoid $\Pi(A,B)$. Let us write $[\kappa] = \beta[\varphi]$.

We verify that β is a functor $\beta([\psi] \circledast [\varphi]) = \beta[\varphi] \circ \beta[\psi]$. Let $\psi: g \simeq h: A \to B$. Choose a homotopy $\Psi_t: X \to Y_g$ with $\Psi_0 = G$ and $\Psi_t j = j_g \psi_t$. Then $\kappa_\psi: Y_h \to Y_g$ is determined by $\kappa_\psi j_h = j_g$ and $\kappa_\psi H = \Psi_1$. (Here (H, j_h) are the pushout data for (j,h).) Since $\kappa_\varphi \Psi_0 = \kappa_\varphi G = \Phi_1$, we can form the product homotopy $\Phi * \kappa_\varphi \Psi$. It has the initial condition F and satisfies $(\Phi * \kappa_\varphi \Psi)(j \times \mathrm{id}) = j_f \varphi * \kappa_\varphi j_g \psi = j_f(\varphi * \psi)$. Hence $\kappa_{\varphi * \psi}$, constructed with this homotopy, is determined by $\kappa_{\varphi*\psi} H = \kappa_\varphi \Psi_1 = \kappa_\varphi \kappa_\psi H$ and $\kappa_{\varphi*\psi} j_h = j_f = \kappa_\varphi j_g = \kappa_\varphi \kappa_\psi j_h$. Therefore $\kappa_\varphi \kappa_\psi$ represents $\beta([\psi] \circledast [\varphi])$.

Let h-COF^B denote the full subcategory of h-TOP^B with objects the cofibrations under B. Then we have shown above:

(5.1.9) Theorem. *Let $j: A \to X$ be a cofibration. We assign to the object $f: A \to B$ in $\Pi(A, B)$ the induced cofibration $j_f: B \to Y_f$ and to the morphism $[\varphi]: f \to g$ in $\Pi(A, B)$ the morphism $[\kappa_\varphi]: j_g \to j_f$. These assignments yield a contravariant functor $\beta_j: \Pi(A, B) \to$ h-COF^B.* □

Since $\Pi(A, B)$ is a groupoid, $[\kappa_\varphi]$ is always an isomorphism in h-TOP^B. We refer to this fact as the ***homotopy theorem*** for cofibrations.

(5.1.10) Proposition. *In the pushout* (5.1.8) *let j be a cofibration and f a homotopy equivalence. Then F is a homotopy equivalence.*

Proof. With an h-inverse $g: B \to A$ of f we form a pushout

$$\begin{array}{ccc} B & \xrightarrow{g} & A \\ \downarrow J & & \downarrow i \\ Y & \xrightarrow{G} & Z. \end{array}$$

Since $gf \simeq \mathrm{id}$, there exists, by (5.1.9), an h-equivalence $\kappa: Z \to X$ under A such that $\kappa GF \simeq \mathrm{id}$. Hence F has a left h-inverse and G a right h-inverse. Now interchange the roles of F and G. □

Problems

1. A cofibration is an embedding. For the proof use that $i_1: A \to Z(i)$, $a \mapsto (a, 1)$ is an embedding. From $i_1 = rsi_1 = ri_1^X i$ then conclude that i is an embedding.

Consider a cofibration as an inclusion $i: A \subset X$. The image of $s: Z(i) \to X \times I$ is the subset $X \times 0 \cup A \times I$. Since s is an embedding, this subset equals the mapping cylinder, i.e., one can define a continuous map $X \times 0 \cup A \times I$ by specifying its restrictions to $X \times 0$ and $A \times I$. (This is always so if A is closed in X, and is a special property of $i: A \subset X$ if i is a cofibration.)

Let X be a Hausdorff space. Then a cofibration $i: A \to X$ is a closed embedding. Let $r: X \times I \to X \times 0 \cup A \times I$ be a retraction. Then $x \in A$ is equivalent to $r(x, 1) = (x, 1)$. Hence A is the coincidence set of the maps $X \to X \times I$, $x \mapsto (x, 1)$, $x \mapsto r(x, 1)$ into a Hausdorff space and therefore closed.

2. If $i: K \to L$, $j: L \to M$ have the HEP for Y, then ji has the HEP for Y. A homeomorphism is a cofibration. $\emptyset \subset X$ is a cofibration. The sum $\amalg i_j: \amalg A_j \to \amalg X_j$ of cofibrations $i_j: A_j \to X_j$ is a cofibration.

3. Let $p: P \to Q$ be an h-equivalence and $i: A \subset B$ a cofibration. Then $f: A \to P$ has an extension to B if and only if pf has an extension to B. Suppose $f_0, f_1: B \to P$ agree on A. If pf_0 and pf_1 are homotopic rel A so are f_0, f_1.

4. Compression. Let $A \subset X$ be a cofibration and $f: (X, A) \to (Y, B)$ a map which is homotopic as a map of pairs to $k: (X, A) \to (B, B)$. Then f is homotopic relative to A to a map g such that $g(X) \subset B$.

5. Let $A \subset X$ be a cofibration and A contractible. Then the quotient map $X \to X/A$ is a homotopy equivalence.

6. The space $C'X = X \times I/X \times 1$ is called the unpointed ***cone*** on X. We have the closed inclusions $j: X \to C'X$, $x \mapsto (x,0)$ and $b: \{*\} \to C'X$, $* \mapsto \{X \times 1\}$. Both maps are cofibrations.

7. Let $f: A \subset X$ be an inclusion. We have a pushout diagram

$$\begin{array}{ccc} A & \xrightarrow{j} & C'A \\ \downarrow{\scriptstyle f} & & \downarrow{\scriptstyle F} \\ X & \xrightarrow{J} & X \cup C'A. \end{array}$$

Since j is a cofibration, so is J. If f is a cofibration, then F is a cofibration. There exists a canonical homeomorphism $X \cup C'A/C'A \cong X/A$; it is induced by J. Since $C'A$ is contractible, we obtain a homotopy equivalence $X \cup C'A \to X \cup C'A/C'A \cong X/A$.

8. The unpointed ***suspension*** $\Sigma' X$ of a space X is obtained from $X \times I$ if we identify each of the sets $X \times 0$ and $X \times 1$ to a point. If $*$ is a basepoint of X, we have the embedding $j: I \to \Sigma' X$, $t \mapsto (*, t)$. If $\{*\} \subset X$ is a closed cofibration, then j is a closed (induced) cofibration. The quotient map $\Sigma' X \to \Sigma X$ is a homotopy equivalence.

5.2 Transport

Let $i: K \to A$ be a cofibration and $\varphi: K \times I \to X$ a homotopy. We define a map

$$\varphi^\#: [(A,i),(X,\varphi_0)]^K \to [(A,i),(X,\varphi_1)]^K,$$

called ***transport along*** φ, as follows: Let $f: A \to X$ with $fi = \varphi_0$ be given. Choose a homotopy $\Phi_t: A \to X$ with $\Phi_0 = f$ and $\Phi_t i = \varphi_t$. We define $\varphi^\#[f] = [\Phi_1]$. Then (5.1.5) shows that $\varphi^\#$ is well defined and only depends on the homotopy class of φ rel $K \times \partial I$, i.e., on the morphism $[\varphi] \in \Pi(K,X)$. From the construction we see $(\varphi * \psi)^\# = \psi^\# \varphi^\#$. Altogether we obtain:

(5.2.1) Proposition. *Let $i: K \to A$ be a cofibration. The assignments $\varphi_0 \mapsto [i,\varphi_0]^K$ and $[\varphi] \mapsto \varphi^\#$ yield a **transport functor** from $\Pi(K,X)$ to sets. Since $\Pi(K,X)$ is a groupoid, $\varphi^\#$ is always bijective.* □

The transport functor measures the difference between "homotopic" in TOP^K and in TOP. The following is a direct consequence of the definitions.

(5.2.2) Proposition. *Let $i: K \to A$ be a cofibration. Let $f: (A,i) \to (X,g)$ and $f': (A,i) \to (X,g')$ be morphisms in TOP^K. Then $[f] = [f']$, if and only if there exists $[\varphi] \in \Pi(K,X)$ from (X,g) to (X,g') with $[f']^K = \varphi^\#[f]^K$.* □

(5.2.3) Proposition. *Let $i: K \to A$ be a cofibration, $g: K \to X$ a map, and $\psi: X \times I \to Y$ a homotopy. Then $(\psi \circ (g \times \mathrm{id}))^\# \circ \psi_{0*} = \psi_{1*}$, if we set $\psi_{i*}[f] = [\psi_i f]$.*

Proof. Use that $\psi_t f$ is an extension of $\psi_t g$ and apply the definition. □

(5.2.4) Proposition. *Let $f\colon X \to Y$ be an ordinary homotopy equivalence and $i\colon K \to A$ a cofibration. Then $f_*\colon [(A,i),(X,g)]^K \to [(A,i),(Y,fg)]^K$ is bijective.*

Proof. Let g be h-inverse to f and choose $\varphi\colon \mathrm{id} \simeq gf$. Consider

$$[i,v]^K \xrightarrow{f_*} [i,fv]^K \xrightarrow{g_*} [i,gfv]^K \xrightarrow{f_*} [i,fgfv]^K.$$

Since $g_*f_* = (gf)_* = [\varphi(v \times \mathrm{id})]^\# \,\mathrm{id}_*$, we conclude from (5.2.1) and (5.2.3) that g_*f_* is bijective, hence g_* is surjective. The bijectivity of f_*g_* shows that g_* is also injective. Therefore g_* is bijective and hence f_* is bijective too. □

(5.2.5) Proposition. *Let $i\colon K \to X$ and $j\colon K \to Y$ be cofibrations and $f\colon X \to Y$ an h-equivalence such that $fi = j$. Then f is an h-equivalence under K.*

Proof. By (5.2.4), we have a bijective map

$$f_*\colon [(Y,j),(X,i)]^K \to [(Y,j),(Y,j)]^K.$$

Hence there exists $[g]$ with $f_*[g]^K = [fg]^K = [\mathrm{id}]^K$. Since f is an h-equivalence, so is g. Since also g_* is bijective, g has a homotopy right inverse under K. Hence g and f are h-equivalences under K. □

(5.2.6) Proposition. *Let $i\colon A \to X$ be a cofibration and an h-equivalence. Then i is a deformation retract.*

Proof. The map i is a morphism from $\mathrm{id}(A)$ to i. By (5.2.5), i is an h-equivalence under A. This means: There exists a homotopy $X \times I \to X$ rel A from the identity to a map $r\colon X \to A$ such that $ri = \mathrm{id}(A)$, and this is what was claimed. □

(5.2.7) Proposition. *Given a commutative diagram*

$$\begin{array}{ccccc} A & \xrightarrow{u} & Y & \xleftarrow{\xi} & X \\ \downarrow{\scriptstyle i} & & \downarrow{\scriptstyle g} & & \downarrow{\scriptstyle f} \\ A' & \xrightarrow{u'} & Y' & \xleftarrow{\xi'} & X' \end{array}$$

with a cofibration i and h-equivalences ξ and ξ'. Given $v\colon A \to X$ and $\varphi\colon \xi v \simeq u$. Then there exists $v'\colon A' \to X'$ and $\varphi'\colon \xi' v' \simeq u'$ such that $v'i = fv$ and $\varphi'_t i = g\varphi_t$.

Proof. We have bijective maps (note $\xi' fv = g\xi v \simeq gu = u'i$)

$$(g\varphi)^\# \circ \xi'_*\colon [(A',i),(X',fv)]^A \to [(A',i),(Y',\xi' fv)]^A \to [(A',i),(Y',u'i)]^A.$$

Let $v'\colon A' \to X'$ be chosen such that $(g\varphi)^\# \xi'_*[v']^A = [u']^A$. This means: $v'i = fv$; and $\xi' v'$ has a transport along $g\varphi$ to a map which is homotopic under A to u'.

This yields a homotopy $\varphi''\colon \xi'v' \simeq u'$ such that $\varphi''(i(a),t) = g\varphi(a,\min(2t,1))$. The homotopy $\varphi * k\colon (a,t) \mapsto \varphi(a,\min(2t,1))$ is homotopic rel $A \times \partial I$ to φ. We now use (5.1.6) in order to change this φ'' into another homotopy φ' with the desired properties. □

If we apply (5.2.7) in the case that u and u' are the identity we obtain the next result (in different notation). It generalizes (5.2.5).

(5.2.8) Proposition. *Given a commutative diagram*

$$\begin{array}{ccc} A & \xrightarrow{f} & B \\ \downarrow{\scriptstyle i} & & \downarrow{\scriptstyle j} \\ X & \xrightarrow{F} & Y \end{array}$$

with cofibrations i, j and h-equivalences f and F. Given $g\colon B \to A$ and $\varphi\colon gf \simeq$ id, there exists $G\colon Y \to X$ and $\Phi\colon GF \simeq \mathrm{id}$ such that $Gj = ig$ and $\Phi_t i = i\varphi_t$. In particular: (F,f) is an h-equivalence of pairs, and there exists a homotopy inverse of the form $(G,g)\colon j \to i$. □

(5.2.9) Proposition. *Suppose a commutative diagram*

$$\begin{array}{ccccccc} X_0 & \xrightarrow{a_1} & X_1 & \xrightarrow{a_2} & X_2 & \longrightarrow & \cdots \\ \downarrow{\scriptstyle f_0} & & \downarrow{\scriptstyle f_1} & & \downarrow{\scriptstyle f_2} & & \\ Y_0 & \xrightarrow{b_1} & Y_2 & \xrightarrow{b_2} & Y_2 & \longrightarrow & \cdots \end{array}$$

is given with cofibration a_j, b_j and h-equivalences f_j. Let X be the colimit of the a_j and Y the colimit of the b_j. Then the map $f\colon X \to Y$ induced by the f_j is a homotopy equivalence.

Proof. We choose inductively h-equivalences $F_n\colon Y_n \to X_n$ such that $a_n F_{n-1} = F_n b_n$ and homotopies $\varphi_n\colon X_n \times I \to X_n$ from $F_n f_n$ to $\mathrm{id}(X_n)$ such that $a_n\varphi_{n-1} = \varphi_n(a_n \times \mathrm{id})$. This is possible by (5.2.7). The F_n and φ_n induce $F\colon Y \to X$ and $\varphi\colon X \times I \to X$, $Ff \simeq \mathrm{id}$. Hence F is a left homotopy inverse of f. □

Problems

1. Let $i\colon K \to A$ and $j\colon K \to B$ be cofibrations. Let $\alpha\colon (B,j) \to (A,i)$ be a morphism under K, $\xi\colon X \to Y$ a continuous map, and $\varphi\colon K \times I \to X$ a homotopy. Then

$$\begin{array}{ccc} [(A,i),(X,\varphi_0)]^K & \xrightarrow{\varphi^\#} & [(A,i),(X,\varphi_1)]^K \\ \downarrow{\scriptstyle [\alpha,\xi]^K} & & \downarrow{\scriptstyle [\alpha,\xi]^K} \\ [(B,j),(Y,\xi\varphi_0)]^K & \xrightarrow{(\xi\varphi)^\#} & [(B,j),(Y,\xi\varphi_1)]^K \end{array}$$

commutes; here $[\alpha,\xi]^K[f] = [\xi f \alpha]$.
2. Apply the transport functor to pointed homotopy sets. Assume that the inclusion $\{*\} \subset A$ is a cofibration. For each path $w\colon I \to X$ we have the transport

$$w^{\#}\colon [A,(X,w(0))]^0 \to [A,(X,w(1))]^0.$$

As a special case we obtain a right action of the fundamental group (transport along loops)

$$[A,X]^0 \times \pi_1(X,*) \to [A,X]^0, \quad (x,\alpha) \mapsto x \cdot \alpha = \alpha^{\#}(x).$$

Let $v\colon [A,X]^0 \to [A,X]$ denote the forgetful map which disregards the base point. The map v induces an injective map from the orbits of the π_1-action into $[A,X]$. This map is bijective, if X is path connected.

A space is said to be ***A*-simple** if for each path w the transport $w^{\#}$ only depends on the endpoints of w; equivalently, if for each $x \in X$ the fundamental group $\pi_1(X,x)$ acts trivially on $[A,(X,x)]^0$. If $A = S^n$, then we say ***n*-simple** instead of A-simple. We call X ***simple*** if it is A-simple for each well-pointed A.
3. The action on $[I/\partial I, X]^0 = \pi_1(X)$ is given by conjugation. Hence this action is trivial if and only if the fundamental group is abelian.
4. Let $[A,X]^0$ carry a composition law induced by a comultiplication on A. Then $w^{\#}$ is a homomorphism. In particular $\pi_1(X)$ acts by homomorphisms. (Thus, if the composition law on $[A,X]^0$ is an abelian group, then this action makes this group into a right module over the integral group ring $\mathbb{Z}\pi_1(X)$.)
5. Write $S(1) = I/\partial I$ and $\pi_1(X) = [S(1),X]^0$. Then we can identify $[A,X]^0 \times \pi_1(X,*) \cong [A \vee S(1), X]^0$. The action of the previous problem is induced by a map $\nu\colon A \to A \vee S(1)$ which can be obtained as follows. Extend the homotopy $I \to A \vee S(1)$, $t \mapsto t \in S(1)$ to a homotopy $\varphi\colon A \times I \to A \vee S(1)$ with the initial condition $A \subset A \vee S(1)$ and set $\nu = \varphi_1$. Express in terms of ν and the comultiplication of $S(1)$ the fact that the induced map is a group action (ν is a coaction up to homotopy).
6. Let (X,e) be a path connected monoid in h-TOP^0. Then the $\pi_1(X,e)$-action on $[A,X]^0$ is trivial.

5.3 Replacing a Map by a Cofibration

We recall from Section 4.1 the construction of the mapping cylinder. Let $f\colon X \to Y$ be a map. We construct the ***mapping cylinder*** $Z = Z(f)$ of f via the pushout

$$\begin{array}{ccc} X + X & \xrightarrow{f+\mathrm{id}} & Y + X \\ \downarrow{\scriptstyle \langle i_0, i_1\rangle} & & \downarrow{\scriptstyle \langle s, j\rangle} \\ X \times I & \xrightarrow[a]{} & Z(f) \end{array} \qquad \begin{array}{c} Z(f) = X \times I + Y/f(x) \sim (x,0), \\ s(y) = y, \quad j(x) = (x,1). \end{array}$$

Since $\langle i_0, i_1\rangle$ is a closed cofibration, the maps $\langle s, j\rangle$, s and j are closed cofibrations. We also have the projection $q\colon Z(f) \to Y$, $(x,t) \mapsto f(x)$, $y \mapsto y$. In the case that $f\colon X \subset Y$, let $p\colon Y \to Y/X$ be the quotient map. We also have the quotient map $P\colon Z(f) \to C(f) = Z(f)/j(X)$ onto the ***mapping cone*** $C(f)$. (Now we consider the unpointed situation. The "direction" of the unit interval is different from the one in the previous chapter.) We display the data in the next diagram. The map r is induced by q.

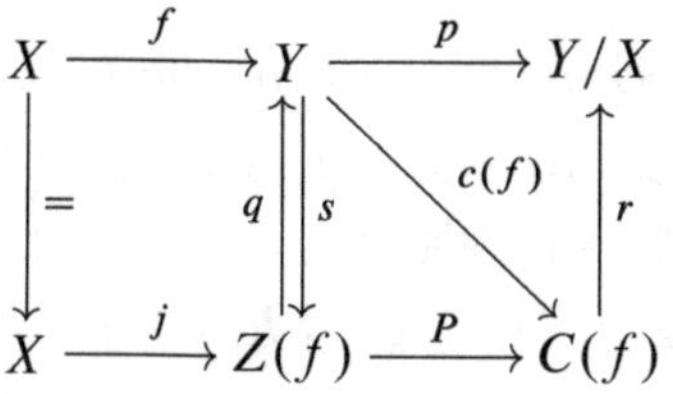

$$qj = f, \quad qs = \mathrm{id}$$
$$Ps = c(f), \quad pq = rP$$

(5.3.1) Proposition. *The following assertions hold:*

(1) *j and s are cofibrations.*

(2) *sq is homotopic to the identity relative to Y. Hence s is a deformation retraction with h-inverse q.*

(3) *If f is a cofibration, then q is a homotopy equivalence under X and r the induced homotopy equivalence.*

(4) *$c(f)\colon Y \to C(f)$ is a cofibration.*

Proof. (1) was already shown.

(2) The homotopy contracts the cylinder $X \times I$ to $X \times 0$ and leaves Y fixed, $h_t(x,c) = (x, tc + 1 - t)$, $h_t(y) = y$.

(3) is a consequence of (5.2.5).

(4) $c(f)$ is induced from the cofibration $i_0\colon X \to X \times I/X \times 1$ via cobase change along f. □

We have constructed a factorization $f = qj$ into a (closed) cofibration and a homotopy equivalence q. Factorizations of this type are unique in the following sense. Suppose $f = q'j'\colon X \to Z' \to Y$ is another such factorization. Then $iq'\colon Z' \to Z$ satisfies $iq'j' \simeq i$. Since j' is a cofibration, we can change $iq' \simeq k$ such that $kj' = j$. Since iq' is an h-equivalence, the map k is an h-equivalence under X, by (5.2.5). Also $qk \simeq q'$. This expresses a uniqueness of the factorization. If $f = qj\colon X \to Z \to Y$ is a factorization into a cofibration j and a homotopy equivalence q, then $Z/j(X)$ is called the (homotopical) ***cofibre*** of f. The uniqueness of the factorization implies uniqueness up to homotopy equivalence of the cofibre. If $f\colon X \subset Y$ is already a cofibration, then $Y \to Y/X$ is the projection onto the cofibre; in this case $q\colon Z \to Y$ is an h-equivalence under X.

The factorization of a map into a cofibration and a homotopy equivalence is a useful technical tool. The proof of the next proposition is a good example.

(5.3.2) Proposition. *Let a pushout diagram*

$$\begin{array}{ccc} A & \xrightarrow{f} & B \\ \downarrow{\scriptstyle j} & & \downarrow{\scriptstyle J} \\ X & \xrightarrow{F} & Y \end{array}$$

with a cofibration j be given. Then the diagram is a homotopy pushout.

Proof. Let $qi: A \to Z(j) \to X$ be the factorization of j. Since q is a homotopy equivalence under A, it induces a homotopy equivalence

$$q \cup_A \mathrm{id}: Z(f) \cup_A B = Z(f, j) \to X \cup_A B = Y$$

of the adjunction spaces. □

(5.3.3) Proposition. *Let a commutative diagram*

$$\begin{array}{ccccccc} A' & & & \xrightarrow{k'} & & & C' \\ & \nwarrow{\scriptstyle \alpha} & & & & \nearrow{\scriptstyle \gamma} & \\ \downarrow{\scriptstyle l'} & & A & \xrightarrow{k} & C & & \downarrow{\scriptstyle L'} \\ & & \downarrow{\scriptstyle l} & & \downarrow{\scriptstyle L} & & \\ & & B & \xrightarrow{K} & D & & \\ & \swarrow{\scriptstyle \beta} & & & & \searrow{\scriptstyle \delta} & \\ B' & & & \xrightarrow{K'} & & & D' \end{array}$$

be given. Suppose the inner and the outer square are homotopy cocartesian. If α, β, γ are homotopy equivalences, then δ is a homotopy equivalence.

Proof. From the data of the diagram we obtain a commutative diagram

$$\begin{array}{ccc} Z(k, l) & \xrightarrow{Z(\beta,\alpha,\gamma)} & Z(k', l') \\ \downarrow{\scriptstyle \varphi} & & \downarrow{\scriptstyle \varphi'} \\ D & \xrightarrow{\delta} & D' \end{array}$$

where φ and φ' are the canonical maps. By hypothesis, φ and φ' are homotopy equivalences. By (4.2.1) the map $Z(\beta, \alpha, \gamma)$ is a homotopy equivalence. □

(5.3.4) Proposition. *Given a commutative diagram as in the previous proposition. Assume that the squares are pushout diagrams. Then δ is induced by α, β, γ. Suppose that α, β, γ are homotopy equivalences and that one of the maps k, l and one of the maps k', l' is a cofibration. Then δ is a homotopy equivalence.*

Proof. From (5.3.2) we see that the squares are homotopy cocartesian. Thus we can apply (5.3.3). □

Problems

1. A map $f\colon X \to Y$ has a left homotopy inverse if and only if $j\colon X \to Z(f)$ has a retraction $r\colon Z(f) \to X$. The map f is a homotopy equivalence if and only if j is a deformation retract.
2. In the case of a pointed map $f\colon (X,*) \to (Y,*)$ one has analogous factorizations into a cofibration and a homotopy equivalence. One replaces the mapping cylinder $Z(f)$ with the pointed mapping cylinder $Z^0(f)$ defined by the pushout

$$\begin{array}{ccc} X \vee X & \xrightarrow{f \vee \mathrm{id}} & Y \vee X \\ {\scriptstyle \langle i_0, i_1 \rangle}\downarrow & & \downarrow{\scriptstyle \langle s, j \rangle} \\ XI & \longrightarrow & Z^0(f) \end{array}$$

with the pointed cylinder $XI = X \times I/\{*\} \times I$. The maps $\langle i_0, i_1\rangle$, $\langle s, j\rangle$, s and j are pointed cofibrations. We have a diagram as for (5.3.1) with pointed homotopy equivalences s, q and $C^0(f) = Z^0(f)/j(X)$ the pointed mapping cone, the pointed cofibre of f.
3. $\langle i_0, i_1\rangle\colon X \vee X \to XI$ is an embedding.
4. Let $f\colon X \to Y$ and $g\colon Y \to Z$ be pointed maps. We have canonical maps $\alpha\colon C(f) \to C(gf)$ and $\beta\colon C(gf) \to C(g)$; α is the identity on the cone and maps Y by g, and β is the identity on Z and maps the cone by $f \times \mathrm{id}$. Show that β is the pointed homotopy cofibre of α.

5.4 Characterization of Cofibrations

We look for conditions on $A \subset X$ which imply that this inclusion is a cofibration. We begin by reformulating the existence of a retraction (5.1.2).

(5.4.1) Proposition. *There exists a retraction $r\colon X \times I \to A \times I \cup X \times 0$ if and only if the following holds: There exists a map $u\colon X \to [0, \infty[$ and a homotopy $\varphi\colon X \times I \to X$ such that:*

(1) $A \subset u^{-1}(0)$

(2) $\varphi(x,0) = x$ *for* $x \in X$

(3) $\varphi(a,t) = a$ *for* $(a,t) \in A \times I$

(4) $\varphi(x,t) \subset A$ *for* $t > u(x)$.

Proof. Suppose we are given a retraction r. We set $\varphi(x,t) = \mathrm{pr}_1 \circ r(x,t)$ and $u(x) = \max\{t - \mathrm{pr}_2 \circ r(x,t) \mid t \in I\}$. For (4) note the following implications: $t > u(x)$, $\mathrm{pr}_2\, r(x,t) > 0$, $r(x,t) \in A \times I$, $\varphi(x,t) \in A$. The other properties are immediate from the definition. Conversely, given u and φ, then $r(x,t) = (\varphi(x,t), \max(t - u(x), 0))$ is a retraction. □

(5.4.2) Note. *Let $t_n > u(x)$ be a sequence which converges to $u(x)$. Then* (4) *implies $\varphi(x, u(x)) \in \bar{A}$. If $u(x) = 0$, then $x = \varphi(x,0) = \varphi(x,u(x)) \in \bar{A}$. Thus*

$\bar{A} = u^{-1}(0)$. *Therefore in a closed cofibration* $A \subset X$ *the subspace* A *has the remarkable property of being the zero-set of a continuous real-valued function.* □

(5.4.3) Lemma. *Let* $u\colon X \to I$ *and* $A = u^{-1}(0)$. *Let* $\Phi\colon f \simeq g\colon X \to Z$ *rel* A. *Then there exists* $\tilde{\Phi}\colon f \simeq g$ *rel* A *such that* $\tilde{\Phi}(x,t) = \tilde{\Phi}(x,u(x)) = \tilde{\Phi}(x,1)$ *for* $t \ge u(x)$.

Proof. We define $\tilde{\Phi}$ by $\tilde{\Phi}(x,t) = \Phi(x,1)$ for $t \ge u(x)$ and by $\Phi(x,tu(x)^{-1})$ for $t < u(x)$. For the continuity of $\tilde{\Phi}$ on $C = \{(x,t) \mid t \le u(x)\}$ see Problem 1. □

We call (X, A) a ***neighbourhood deformation retract*** (***NDR***), if there exist a homotopy $\psi\colon X \times I \to X$ and a function $v\colon X \to I$ such that:

(1) $A = v^{-1}(0)$

(2) $\psi(x,0) = x$ for $x \in X$

(3) $\psi(a,t) = a$ for $(a,t) \in A \times I$

(4) $\psi(x,1) \in A$ for $1 > v(x)$.

The pair (ψ, v) is said to be an ***NDR-presentation*** of (X, A).

(5.4.4) Proposition. (X, A) *is a closed cofibration if and only if it is an* NDR.

Proof. If $A \subset X$ is a closed cofibration, then an NDR-presentation is obtained from (5.4.1) and (5.4.2). For the converse, we modify an NDR-presentation (ψ, u) by (5.4.3) and apply (5.4.1) to the result $(\tilde{\psi}, u)$. □

(5.4.5) Theorem (Union theorem). *Let* $A \subset X$, $B \subset X$, *and* $A \cap B \subset X$ *be closed cofibrations. Then* $A \cup B \subset X$ *is a cofibration.*

Proof ([112]). Let $\varphi\colon (A \cup B) \times I \to Z$ be a homotopy and $f\colon X \to Z$ an initial condition. There exist extensions $\Phi^A\colon X \times I \to Z$ of $\varphi|A \times I$ and $\Phi^B\colon B \times I \to Z$ of $\varphi|B \times I$ with initial condition f. The homotopies Φ^A and Φ^B coincide on $(A \cap B) \times I$. Therefore there exists $\Psi\colon \Phi^A \simeq \Phi^B$ rel $(A \cap B) \times I \cup X \times 0$.

Let $p\colon X \times I \to X \times I/\sim$ be the quotient map which identifies each interval $\{c\} \times I$, $c \in A \cap B$ to a point. Let $T\colon I \times I \to I \times I$ switch the factors. Then $\Psi \circ (\mathrm{id} \times T)$ factors over $p \times \mathrm{id}$ and yields $\Omega\colon (X \times I/\sim) \times I \to Z$.

Let $u\colon X \to I$ and $v\colon X \to I$ be functions such that $A = u^{-1}(0)$ and $B = v^{-1}(0)$. Define $j\colon X \to X \times I/\sim$ by $j(x) = (x, u(x)/(u(x)+v(x)))$ for $x \notin A \cap B$ and by $j(x) = (x,0) = (x,t)$ for $x \in A \cap B$. Using the compactness of I one shows the continuity of j.

An extension of φ and f is now given by $\Omega \circ (j \times \mathrm{id})$. □

(5.4.6) Theorem (Product theorem). *Let* $A \subset X$ *and* $B \subset Y$ *be closed cofibrations. Then the inclusion* $X \times B \cup A \times Y \subset X \times Y$ *is a cofibration.*

Proof. $A \times X \subset X \times Y$, $X \times B \subset X \times Y$, and $A \times B = (A \times Y) \cap (X \times B) \subset X \times B \subset X \times Y$ are cofibrations. Now apply (5.4.5). □

Problems

1. Let $C = \{(x,t) \mid t \le u(x)\}$ and $q\colon X \times I \to C$, $(x,t) \mapsto (x, tu(x))$. Then $\tilde{\Phi} q = \Phi$. It suffices to show that q is a quotient map. The map $\Gamma\colon X \times I \to X \times I \times I$, $(x,t) \mapsto (x,t,u(x))$ is an embedding onto a closed subspace D. The map $m\colon I \times I \to I$, $(a,b) \mapsto ab$ is proper, hence $M = \mathrm{id} \times m$ is closed. The restriction of M to D is closed, hence $M\Gamma = q$ is closed and therefore a quotient map.
2. The inclusion $0 \cup \{n^{-1} \mid n \in \mathbb{N}\} \subset [0,1]$ is not a cofibration. The inclusions $A_j = \{0, j^{-1}\} \subset I$ are cofibrations. Hence (5.4.5) does not hold for an infinite number of cofibrations.
3. Set $X = \{a,b\}$ with open sets $\emptyset, \{a\}, X$ for its topology. Then $A = \{a\} \subset X$ is a non-closed cofibration. The product $X \times A \cup A \times X \subset X \times X$ is not a cofibration.
4. Let $A_j \subset X$ be closed cofibrations $(1 \le j \le n)$. For all $\sigma \subset \{1,\dots,n\}$ let $A_\sigma = \bigcap_{j \in \sigma} A_j \subset X$ be a cofibration. Then $\bigcup_1^n A_j \subset X$ is a cofibration.
5. Let A and B be well-pointed spaces. Then $A \wedge B$ is well-pointed.

5.5 The Homotopy Lifting Property

A map $p\colon E \to B$ has the ***homotopy lifting property*** (HLP) for the space X if the following holds: For each homotopy $h\colon X \times I \to B$ and each map $a\colon X \to E$ such that $pa(x) = h(x,0)$ there exists a homotopy $H\colon X \times I \to E$ with $pH = h$ and $H(x,0) = a(x)$. We call H a ***lifting*** of h with ***initial condition*** a. The map p is called a ***fibration*** (sometimes ***Hurewicz fibration***) if it has the HLP for all spaces. It is called a ***Serre fibration*** if it has the HLP for all cubes I^n, $n \in \mathbb{N}_0$. Serre fibrations suffice for the investigation of homotopy groups. In order to see the duality we can use the dual definition of homotopy and specify the data in the right diagram. It uses the evaluation $e_E^0\colon E^I \to E$, $w \mapsto w(0)$.

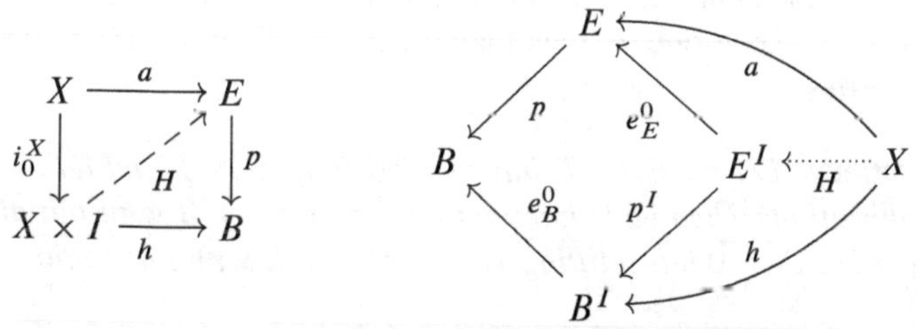

We begin by introducing the dual $W(p)$ of the mapping cylinder. It is defined by the pullback

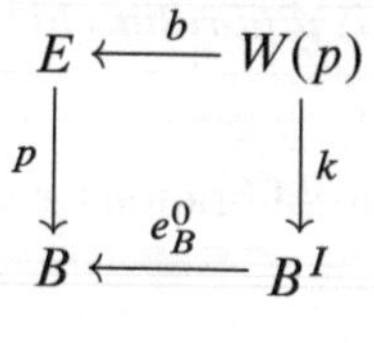

$$W(p) = \{(x,w) \in E \times B^I \mid p(x) = w(0)\},$$

$$k(x,w) = w, \quad b(x,w) = x.$$

If we apply the pullback property to e_E^0, p^I, we obtain a unique map $r\colon E^I \to W(p)$, $v \mapsto (v(0), pv)$ such that $br = e_E^0$ and $kr = p^I$. If we apply the HLP to $(W(p), b, k)$, we obtain a map $s\colon W(p) \to E^I$ such that $e_E^0 s = b$ and $p^I s = k$. The relations $brs = e_E^0 s = b$ and $krs = p^I s = k$ imply $rs = \mathrm{id}$, by uniqueness. Therefore s is a section of r. Conversely, given data (a, h) for a homotopy lifting problem. They combine to a map $\rho\colon X \to W(p)$. The composition $H = s\rho$ with a section s is a solution of the lifting problem. Therefore we have shown:

(5.5.1) Proposition. *The following statements about $p\colon E \to B$ are equivalent:*

(1) *p is a fibration.*

(2) *p has the* HLP *for $W(p)$.*

(3) *$r\colon E^I \to W(p)$ has a section.* □

(5.5.2) Proposition. *Let $p\colon E \to B$ have the* HLP *for X. Let $i\colon A \subset X$ be a closed cofibration and an h-equivalence. Let $f\colon X \to B$ be given and $a\colon A \to E$ a lifting of f over A, i.e., $pa = fi$. Then there exists a lifting F of f which extends a.*

Proof. By (5.2.6) and (5.4.2) we know: There exists $u\colon X \to I$ and $\varphi\colon X \times I \to X$ rel A such that $A = u^{-1}(0)$, $\varphi_1 = \mathrm{id}(X)$, $\varphi_0(X) \subset A$. Set $r\colon X \to A$, $x \mapsto \varphi_0(x)$. Define a new homotopy $\Phi\colon X \times I \to X$ by $\Phi(x,t) = \varphi(x, tu(x)^{-1})$ for $t < u(x)$ and $\Phi(x,t) = \varphi(x,1) = x$ for $t \geq u(x)$. We have seen in (5.4.3) that Φ is continuous. Apply the HLP to $h = f\Phi$ with initial condition $b = ar\colon X \to E$. The verification

$$h(x,0) = f\Phi(x,0) = f\varphi(x,0) = fr(x) = par(x) = pb(x)$$

shows that b is indeed an initial condition. Let $H\colon X \times I \to E$ solve the lifting problem for h, b. Then one verifies that $F\colon X \to E$, $x \mapsto H(x, u(x))$ has the desired properties. □

(5.5.3) Corollary. *Let $p\colon E \to B$ have the* HLP *for $X \times I$ and let $i\colon A \subset X$ be a closed cofibration. Then each homotopy $h\colon X \times I \to B$ with initial condition given on $A \times I \cup X \times 0$ has a lifting $H\colon X \times I \to E$ with this initial condition.*

Proof. This is a consequence of (5.1.3) and (5.5.2). □

(5.5.4) Proposition. *Let $i\colon A \subset B$ be a (closed) cofibration of locally compact spaces. The restriction from B to A yields a fibration $p\colon Z^B \to Z^A$.*

Let $p\colon X \to B$ be a fibration. Then $p^Z\colon X^Z \to B^Z$ is a fibration for locally compact Z.

Proof. Use adjunction and the fact that $X \times A \to X \times B$ is a cofibration for each X. □

(5.5.5) Proposition. *Let $p\colon E \to B$ be a fibration. Then $r\colon E^I \to W(p)$, $v \mapsto (v(0), pv)$ is a fibration.*

Proof. A homotopy lifting problem for X and r is transformed via adjunction into a lifting problem for p and $X \times I$ with initial condition given on the subspace $X \times (I \times 0 \cup 0 \times I)$. □

(5.5.6) Proposition. *Let $p\colon E \to B$ be a fibration, $B_0 \subset B$ and $E_0 = p^{-1}(B_0)$. If $B_0 \subset B$ is a closed cofibration, then $E_0 \subset E$ is a closed cofibration.*

Proof. Let $u\colon B \to I$ and $h\colon B \times I \to B$ be an NDR-presentation of $B_0 \subset B$. Let $H\colon X \times I \to X$ solve the homotopy lifting problem for $h(p \times \mathrm{id})$ with initial condition $\mathrm{id}(X)$. Define $K\colon X \to X$ by $K(x,t) = H(x, \min(t, up(x)))$. Then (K, up) is an NDR-presentation for $X_0 \subset X$. □

The proof of the next formal proposition is again left to the reader.

(5.5.7) Proposition. *Let a pullback in* TOP *be given.*

$$\begin{array}{ccc} Y & \xrightarrow{F} & X \\ {\scriptstyle q}\downarrow & & \downarrow{\scriptstyle p} \\ C & \xrightarrow[f]{} & B \end{array}$$

If q has the HLP *for Z, then so also has p. If p is a fibration, then q is a fibration.* □

We call q the fibration ***induced*** from the fibration p via ***base change*** along f. In the case that $f\colon C \subset B$ the restriction $p\colon p^{-1}(C) \to C$ can be taken as the induced fibration.

(5.5.8) Example. $X^I \to X^{\partial I} \cong X \times X\colon w \mapsto (w(0), w(1))$ is a fibration (5.5.4). The evaluation $e^1\colon FY \to Y$, $w \mapsto w(1)$ is a fibration (restriction to $* \times Y$). Hence we have the induced fibration $f^1\colon F(f) \to Y$. ◇

The homotopy theorem for fibrations says, among other things, that homotopic maps induce h-equivalent fibrations.

Let $p\colon X \to B$ be a fibration and $\varphi\colon f \simeq g\colon C \to B$ a homotopy. We consider two pullback diagrams.

$$\begin{array}{ccc} Y_f & \xrightarrow{F} & X \\ {\scriptstyle p_f}\downarrow & & \downarrow{\scriptstyle p} \\ C & \xrightarrow{f} & B \end{array} \qquad\qquad \begin{array}{ccc} Y_g & \xrightarrow{G} & X \\ {\scriptstyle p_g}\downarrow & & \downarrow{\scriptstyle p} \\ C & \xrightarrow{g} & B \end{array}$$

There exists a homotopy $\Phi_t\colon Y_f \to X$ such that $\Phi_0 = F$ and $p\Phi_t = \varphi_t p_f$. The pullback property of the right square yields a map $\kappa = \kappa_\varphi\colon Y_f \to Y_g$ such that $G\kappa = \Phi_1$ and $p_g \kappa = p_f$. Let $\psi_t\colon f \simeq g$ be homotopic to φ_t by a homotopy

$\gamma\colon C\times I\times I\to B$ relative to $C\times\partial I$. We obtain in a similar manner a map κ_ψ from a lifting Ψ_t of $\psi_t p_f$. Claim: The maps κ_φ and κ_ψ are homotopic over C. In order to verify this, we lift the homotopy $\gamma\circ(p_f\times\mathrm{id}\times\mathrm{id})\colon Y_f\times I\times I\to B$ to a homotopy Γ with initial data $\Gamma(y,s,0)=\Phi(y,s)$, $\Gamma(y,s,1)=\Psi(y,s)$, and $\Gamma(y,0,t)=F(t)$ by an application of (5.1.5). The homotopy $H\colon(y,t)\mapsto\Gamma(y,1,t)$ yields, by the pullback property of the right square, a homotopy $K\colon Y_f\times I\to Y_g$ such that $GK=H$ and $p_gK=\mathrm{pr}\circ p_f$. By construction, K is a homotopy over C from κ_f to κ_g. The reader should now verify the functoriality $[\kappa_{\varphi*\psi}]=[\kappa_\psi][\kappa_\varphi]$.

Let h-FIB$_C$ be the full subcategory of h-TOP$_C$ with objects the fibrations over C.

(5.5.9) Proposition. *Let $p\colon X\to B$ be a fibration. We assign to $f\colon C\to B$ the induced fibration $p_f\colon Y_f\to C$ and to the morphism $[\varphi]\colon f\to g$ in $\Pi(C,B)$ the morphism $[\kappa_\varphi]$. This yields a functor $\Pi(C,B)\to$ h-FIB$_C$.* □

Since $\Pi(C,B)$ is a groupoid, $[\kappa_\varphi]$ is always an isomorphism in h-TOP$_B$. This fact we call the ***homotopy theorem for fibrations***.

As a special case of (5.5.9) we obtain the ***fibre transport***. It generalizes the fibre transport in coverings. Let $p\colon E\to B$ be a fibration and $w\colon I\to B$ a path from b to c. We obtain a homotopy equivalence $T_p[w]\colon F_b\to F_c$ which only depends on the homotopy class $[w]$ of w, and $T_p[u*v]=T_p[v]T_p[u]$. This yields a functor $T_p\colon\Pi(B)\to$ h-TOP. In particular the fibres over points in the same path component of B are h-equivalent.

(5.5.10) Proposition. *In the pullback* (5.5.7) *let p be a fibration and f a homotopy equivalence. Then F is a homotopy equivalence.*

Proof. The proof is based on (5.5.9) and follows the pattern of (5.1.10). □

(5.5.11) Remark. The notion of fibration and cofibration are not homotopy invariant. The projection $I\times 0\cup 0\times I\to I$, $(x,t)\mapsto x$ is not a fibration, but the map is over I h-equivalent to id. One definition of an ***h-fibration*** $p\colon E\to B$ is that homotopies $X\times I\to B$ which are constant on $X\times[0,\varepsilon]$, $\varepsilon>0$ can be lifted with a given initial condition; a similar definition for homotopy extensions gives the notion on an ***h-cofibration***. In [46] you can find details about these notions.

Problems

1. A composition of fibrations is a fibration. A product of fibrations is a fibration. $\emptyset\to B$ is a fibration.

2. Suppose $p\colon E\to B$ has the HLP for $Y\times I^n$. Then each homotopy $h\colon Y\times I^n\times I\to B$ has a lifting to E with initial condition given on $Y\times(I^n\times 0\cup\partial I^n\times I)$.

3. Let $p\colon E\to B\times I$ be a fibration and $p_0\colon E_0\to B$ its restriction to $B\times 0=B$. Then there exists a fibrewise h-equivalence from $p_0\times\mathrm{id}(I)$ to p which is over $B\times 0$ the inclusion $E_0\to E$.

4. Go through the proof of (5.5.9) and verify a relative version. Let (C, D) be a closed cofibration. Consider only maps $C \to B$ with a fixed restriction $d: D \to B$ and homotopies relative to D. Let $p_D: Y_D \to B$ be the pullback of p along d. Then the maps p_f have the form $(p_f, p_D): (Y_f, Y_D) \to (C, D)$. By (5.5.6), (Y_f, Y_D) is a closed cofibration, and by (5.5.3) the homotopies Φ_t can be chosen constant on Y_D. The maps $\kappa_\varphi: Y_f \to Y_g$ are then the identity on Y_D. The homotopy class of κ_φ is unique as a map over C and under Y_D.
5. Let (B, C) be a closed deformation retract with retraction $r: B \to C$. Let $p: X \to B$ be a fibration and $p_C: X_C \to C$ its restriction to C. Then there exists a retraction $R: X \to X_C$ such that $p_C R = rp$.

5.6 Transport

We construct a dual transport functor. Let $p: E \to B$ be a fibration, $\varphi: Y \times I \to B$ a homotopy and $\Phi: Y \times I \to E$ a lifting along p with initial condition f. We define

$$\varphi^{\#}: [(Y, \varphi_0), (E, p)]_B \to [(Y, \varphi_1), (E, p)]_B, \quad [f] \mapsto [\Phi_1].$$

One shows that this map is well defined and depends only on the homotopy class of φ relative to $Y \times \partial I$ (see the analogous situation for cofibrations). Moreover, $(\varphi * \psi)^{\#} = \psi^{\#}\varphi^{\#}$.

(5.6.1) Proposition. *The assignments $f \mapsto [f, p]_B$ and $[\varphi] \mapsto \varphi^{\#}$ are a functor, called **transport functor**, from $\Pi(Y, B)$ into the category of sets. Since $\Pi(Y, B)$ is a groupoid, $\varphi^{\#}$ is always bijective.* □

(5.6.2) Note. *Let $p: E \to B$ be a fibration and $\psi: X \times I \to Y$ be a homotopy. Then $\psi_1^* = [g\psi_t]^{\#}\psi_0^*$; here $\psi_0^*: [g, p] \to [g\psi_0, p]$ is the composition with ψ_0.* □

(5.6.3) Theorem. *Let $f: X \to Y$ be an h-equivalence and $p: E \to B$ be a fibration. Then $f^*: [v, p]_B \to [vf, p]_B$ is bijective for each $v: Y \to B$.*

Proof. The proof is based on (5.6.1) and (5.6.2) and formally similar to the proof of (5.2.4). □

(5.6.4) Theorem. *Let $p: X \to B$ and $q: Y \to B$ be fibrations. Let $h: X \to Y$ be an h-equivalence and a map over B. Then f is an h-equivalence over B.*

Proof. The proof is based on (5.6.3) and formally similar to the proof of (5.2.5). □

(5.6.5) Corollary. *Let $q: Y \to C$ be a fibration and a homotopy equivalence. Then q is shrinkable.*

Let $p: E \to B$ be a fibration. Then the canonical map $r: E^I \to W(p)$ is shrinkable (see (5.5.5)). □

5.7 Replacing a Map by a Fibration

Let $f: X \to Y$ be a map. Consider the pullback

$$\begin{array}{ccc} W(f) & \longrightarrow & Y^I \\ {\scriptstyle (q,p)}\downarrow & & \downarrow{\scriptstyle (e^0,e^1)} \\ X \times Y & \xrightarrow[f\times \mathrm{id}]{} & Y \times Y \end{array} \qquad \begin{aligned} W(f) &= \{(x,w) \in X \times Y^I \mid f(x) = w(0)\}, \\ q(x,w) &= x, \quad p(x,w) = w(1). \end{aligned}$$

Since (e^0, e^1) is a fibration (see (5.5.8)), the maps (q, p), q and p are fibrations. Let $s: X \to W(f)$, $x \mapsto (x, k_{f(x)})$, with k_y the constant path with value y. Then $qs = \mathrm{id}$ and $ps = f$. (The "direction" of the unit interval is again different from the one in the previous chapter.) We display the data and some other to be explained below in a diagram.

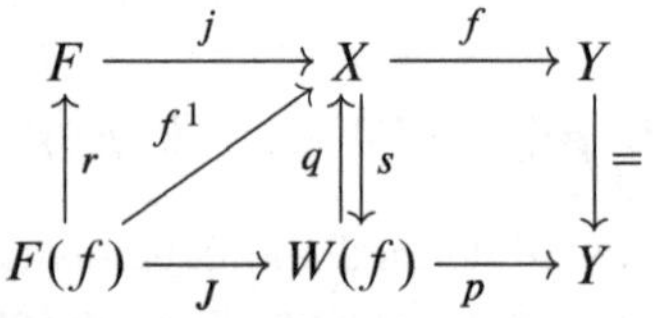

The map s is a shrinking of q; a homotopy $h_t: sq \simeq \mathrm{id}$ is given by $h_t(x, w) = (x, w^t)$, $w^t(s) = w((1-t)s)$. We therefore have a factorization $f = ps$ into a homotopy equivalence s and a fibration p. If $f = p's'$ is another factorization of this type, then there exists a fibrewise homotopy equivalence $k: W(f) \to W'$ such that $p'k = p$ and $ks \simeq s'$. This expresses the uniqueness of the factorization.

Now suppose f is a pointed map with base points $*$. Then $W(f)$ is given the base point $(*, k_*)$. The maps p, q, s become pointed maps, and the homotopy h_t is pointed too. One verifies that q and p are pointed fibrations. Let $F(f) = p^{-1}(*)$ and $F = f^{-1}(*)$ be the fibres over the base point, with j and J the inclusions. The map q induces r. We call $F(f)$ the ***homotopy fibre*** of f. We use the same notion for the fibre of any replacement of f by a fibration as above. If f is already a fibration, then q is a fibrewise homotopy equivalence (5.6.4) and r the induced homotopy equivalence; hence the actual fibre is also the homotopy fibre.

A map $f: X \to Y$ has a right homotopy inverse if and only if $p: W(f) \to X$ has a section. It is a homotopy equivalence if and only if p is shrinkable.

Chapter 6
Homotopy Groups

The first fundamental theorem of algebraic topology is the Brouwer–Hopf degree theorem. It says that the homotopy set $[S^n, S^n]$ has for $n \geq 1$ a homotopically defined ring structure. The ring is isomorphic to $\mathbb{Z}$, the identity map corresponds to $1 \in \mathbb{Z}$ and the constant map to $0 \in \mathbb{Z}$. The integer associated to a map $f: S^n \to S^n$ is called the degree of f. We have proved this already for $n = 1$. Also in the general case "degree n" roughly means that f winds S^n n-times around itself. In order to give precision to this statement, one has to count the number of pre-images of a "regular" value with signs. This is related to a geometric interpretation of the degree in terms of differential topology.

Our homotopical proof of the degree theorem is embedded into a more general investigation of homotopy groups. It will be a simple formal consequence of the so-called excision theorem of Blakers and Massey. The elegant elementary proof of this theorem is due to Dieter Puppe. It uses only elementary concepts of homotopy theory, it does not even use the group structure. (The excision isomorphism is the basic property of the homology groups introduced later where it holds without any restrictions on the dimensions.)

Another consequence of the excision theorem is the famous suspension theorem of Freudenthal. There is a simple geometric construction (the suspension) which leads from $[S^m, S^n]$ to $[S^{m+1}, S^{n+1}]$. Freudenthal's theorem says that this process after a while is "stable", i.e., induces a bijection of homotopy sets. This is the origin of the so-called stable homotopy theory – a theory which has developed into a highly technical mathematical field of independent interest and where homotopy theory has better formal and algebraic properties. (Homology theory belongs to stable homotopy.)

The degree theorem contains the weaker statement that the identity of S^n is not null homotopic. It has the following interpretation: If you extend the inclusion $S^{n-1} \subset \mathbb{R}^n$ to a continuous map $f: D^n \to \mathbb{R}^n$, then there exists a point x with $f(x) = 0$. For $n = 1$ this is the intermediate value theorem of calculus; the higher dimensional analogue has other interesting consequences which we discuss under the heading of the Brouwer fixed point theorem.

This chapter contains the fundamental non-formal results of homotopy theory. Based on these results, one can develop algebraic topology from the view-point of homotopy theory. The chapter is essentially independent of the three previous chapters. But in the last section we refer to the definition of a cofibration and a suspension.

6.1 The Exact Sequence of Homotopy Groups

Let I^n be the Cartesian product of n copies of the unit interval $I = [0,1]$, and $\partial I^n = \{(t_1,\dots,t_n) \in I^n \mid t_i \in \{0,1\} \text{ for at least one } i\}$ its combinatorial boundary ($n \geq 1$). We set $I^0 = \{z\}$, a singleton, and $\partial I^0 = \emptyset$. In $I^n/\partial I^n$ we use ∂I^n as base point. (For $n = 0$ this yields $I^0/\partial I^0 = \{z\} + \{*\}$, an additional disjoint base point $*$.) The n-th homotopy group of a pointed space $(X,*)$ is

$$\pi_n(X,*) = [(I^n, \partial I^n), (X, \{*\})] \cong [I^n/\partial I^n, X]^0$$

with the group structure defined below. For $n = 1$ it is the fundamental group. The definition of the set $\pi_n(X,*)$ also makes sense for $n = 0$, and it can be identified with the set $\pi_0(X)$ of path components of X with $[*]$ as a base point. The composition law on $\pi_n(X,*)$ for $n \geq 1$ is defined as follows. Suppose $[f]$ and $[g]$ in $\pi_n(X,*)$ are given. Then $[f] + [g]$ is represented by $f +_i g$:

$$(1) \qquad (f +_i g)(t_1,\dots,t_n) = \begin{cases} f(t_1,\dots,t_{i-1}, 2t_i, \dots, t_n) & \text{for } t_i \leq \frac{1}{2}, \\ g(t_1,\dots,t_{i-1}, 2t_i - 1, \dots, t_n) & \text{for } \frac{1}{2} \leq t_i. \end{cases}$$

As in the case of the fundamental group one shows that this composition law is a group structure. The next result is a consequence of (4.3.1); a direct verification along the same lines is easy. See also (2.7.3) and the isomorphism (2) below.

(6.1.1) Proposition. *For $n \geq 2$ the group $\pi_n(X,*)$ is abelian, and the equality $+_1 = +_i$ holds for $i \geq 2$.* □

We now define relative homotopy groups (sets) $\pi_k(X,A,*)$ for a pointed pair (X,A). For $n \geq 1$, let $J^n = \partial I^n \times I \cup I^n \times \{0\} \subset \partial I^{n+1} \subset I^n \times I$ and set $J^0 = \{0\} \subset I$. We denote by $\pi_{n+1}(X,A,*)$ the set of homotopy classes of maps of triples $f\colon (I^{n+1}, \partial I^{n+1}, J^n) \to (X,A,*)$. (Recall that this means $f(\partial I^{n+1}) \subset A$, $f(J^n) \subset \{*\}$, and for homotopies H we require H_t for each $t \in I$ to be a map of triples.) Thus, with notation introduced earlier,

$$\pi_{n+1}(X,A,*) = [(I^{n+1}, \partial I^{n+1}, J^n), (X,A,*)].$$

A group structure $+_i$, $1 \leq i \leq n$ is defined again by the formula (1) above. There is no group structure in the case $n = 0$.

We now consider π_n as a functor. Composition with $f\colon (X,A,*) \to (Y,B,*)$ induces $f_*\colon \pi_n(X,A,*) \to \pi_n(Y,B,*)$; this is a homomorphism for $n \geq 2$. Similarly, $f\colon (X,*) \to (Y,*)$ induces for $n \geq 1$ a homomorphism $f_*\colon \pi_n(X,*) \to \pi_n(Y,*)$. The functor properties $(gf)_* = g_* f_*$ and $\mathrm{id}_* = \mathrm{id}$ are clear. The morphism $j_*\colon \pi_n(X,*) \to \pi_n(X,A,*)$ is obtained by interpreting the first group as $\pi_n(X,\{*\},*)$ and then using the map induced by the inclusion $(X,\{*\},*) \subset (X,A,*)$. Maps which are pointed homotopic induce the same homomorphisms. The group $\pi_n(X,A,*)$ is commutative for $n \geq 3$.

Let $h\colon (I^{n+1}, \partial I^{n+1}, J^n) \to (X, A, *)$ be given. We restrict to $I^n = I^n \times \{1\}$ and obtain a map $\partial h\colon (I^n, \partial I^n) \to (A, *)$. Passage to homotopy classes then yields the ***boundary operator*** $\partial\colon \pi_{n+1}(X, A, *) \to \pi_n(A, *)$. The boundary operator is a homomorphism for $n \geq 1$. For $n = 0$ we have $\partial[h] = [h(1)]$.

We rewrite the homotopy groups in terms of mapping spaces. This is not strictly necessary for the following investigations but sometimes technically convenient.

Let $\Omega^k(X, *)$ be the space of maps $I^k \to X$ which send ∂I^k to the base point; the constant map is the base point. The space $\Omega^1(X) = \Omega(X)$ is the loop space of X. Given a map $(I^n, \partial I^n) \to (X, *)$ we have the induced map $\bar{f}\colon I^{n-k} \to \Omega^k(X, *)$ which sends $u \in I^{n-k}$ to $I^k \to X$, $(t_1, \dots, t_k) \mapsto f(t_1, \dots, t_k, u_1, \dots, u_{n-k})$. This adjunction is compatible with homotopies and induces a bijection

(2)
$$\pi_n(X, *) \cong \pi_{n-k}(\Omega^k(X, *), *).$$

Adjunction as above also yields a bijection

(3)
$$\pi_{n+1}(X, A, *) \cong \pi_{n+1-k}(\Omega^k(X), \Omega(A)^k, *).$$

These isomorphisms are natural in $(X, A, *)$, compatible with the boundary operators, and the group structures.

(6.1.2) Theorem (Exact homotopy sequence). *The sequence*

$$\cdots \longrightarrow \pi_n(A, *) \xrightarrow{i_*} \pi_n(X, *) \xrightarrow{j_*} \pi_n(X, A, *)$$
$$\xrightarrow{\partial} \cdots \longrightarrow \pi_1(X, A, *) \xrightarrow{\partial} \pi_0(A, *) \xrightarrow{i_*} \pi_0(X, *)$$

is exact. The maps i_* and j_* are induced by the inclusions.

Proof. We prove the exactness for the portion involving π_0 and π_1 in an elementary manner. Exactness at $\pi_0(A, *)$ and the relations $\partial j_* = 0$ and $j_* i_* = 0$ are left to the reader.

Let $w\colon I \to X$ represent an element in $\pi_1(X, A, *)$ with $\partial[w] = 0$. This means: There exists a path $u\colon I \to A$ with $u(0) = w(1)$ and $u(1) = *$. The product $w * u$ is then a loop in X. The homotopy H which is defined by

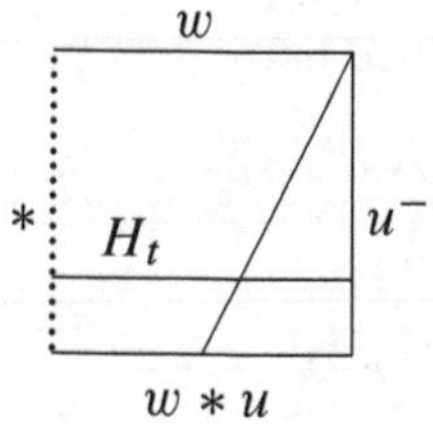

$$H_t(s) = \begin{cases} w(2s/(1+t)), & 2s \leq 1+t, \\ u(2s - t - 1), & 2s \geq 1+t, \end{cases}$$

shows $j_*[w * u] = [w]$. Thus we have shown exactness at $\pi_1(X, A, *)$.

Given a loop $w\colon I \to X$. Let $H\colon (I, \partial I, 0) \times I \to (X, A, *)$ be a homotopy from w to a constant path. Then $u\colon s \mapsto H(1,s)$ is a loop in A. We restrict H to the boundary of the square and compose it with a linear homotopy to prove $k_* * w \simeq k_* * u$.

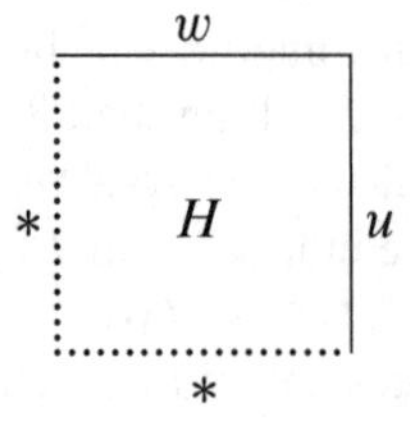

$$k_* * w \simeq k_* * u,\quad u(t) = H(1,t).$$

Hence $i_*[u] = i_*[k_* * u] = [k_* * w] = [w]$.

We now apply this part of the exact sequence to the pairs $(\Omega^n(X), \Omega^n(A))$ and obtain the other pieces of the sequence via adjunction. □

(6.1.3) Remark. We previously introduced the mapping space

$$F(\iota) = \{(a, w) \in A \times X^I \mid w(0) = *, w(1) = a\},$$

with base point $(*, k)$, $k\colon I \to \{*\}$ the constant path. This space is homeomorphic to

$$F(X, A) = \{w \in X^I \mid w(0) = *, w(1) \in A\},$$

$i(\iota)$ becomes the inclusion $\Omega(X) \subset F(X, A)$, and ι^1 the evaluation $F(X, A) \to A$, $w \mapsto w(1)$. For $n \geq 1$ we assign to $f\colon (I^{n+1}, \partial I^{n+1}, J^n) \to (X, A, *)$ the adjoint map $f^\wedge\colon I^n \to F(X, A)$, defined by $f^\wedge(t_1, \dots, t_n)(t) = f(t_1, \dots, t_n, t)$. It sends ∂I^n to the base point and induces a pointed map $\bar{f}\colon I^n/\partial I^n \to F(X, A)$. By standard properties of adjunction we see that the assignment $[f] \mapsto [\bar{f}]$ is a well-defined bijection

$$\pi_{n+1}(X, A, *) \cong \pi_n(F(X, A), *), \tag{4}$$

and in fact a homomorphism with respect to the composition laws $+_i$ for $1 \leq i \leq n$. These considerations also make sense for $n = 0$. In the case that $A = \{*\}$, the space $F(X, A)$ is the loop space $\Omega(X)$.

The exact sequence is also obtained from the fibre sequence of ι. Under the identifications (4) the boundary operator is transformed into

$$\iota^1_*\colon [I^n/\partial I^n, F(X, A)]^0 \to [I^n/\partial I^n, A]^0,$$

and $\pi_{n+1}(X, *) \to \pi_{n+1}(X, A, *)$ is transformed into

$$i(\iota)_*\colon [I^n/\partial I^n, \Omega(X)]^0 \to [I^n/\partial I^n, F(X, A, *)]^0.$$

Now apply $B = I^n/\partial I^n$ to the fibre sequence (4.7.4) of $\iota\colon A \subset X$ to see the exactness of a typical portion of the homotopy sequence. ◇

The sequence (6.1.2) is compatible with maps $f\colon (X, A, *) \to (Y, B, *)$. In particular $f_*\partial = \partial f_*$.

(6.1.4) Remark. In the sequel it will be useful to have different interpretations for elements in homotopy groups. (See also the discussion in Section 2.3.) We set $S(n) = I^n/\partial I^n$ and $D(n+1) = CS(n)$, the pointed cone on $S(n)$. We have homeomorphisms

$$S(n) \to \partial I^{n+1}/J^n, \quad D(n+1) \to I^{n+1}/J^n,$$

the first one $x \mapsto (x, 1)$, the second one the identity on representatives in $I^n \times I$; moreover we have the embedding $S(n) \to D(n+1)$, $x \mapsto (x, 1)$ which we consider as an inclusion. These homeomorphisms allow us to write

$$\pi_n(X, *) \cong [I^n/\partial I^n, X]^0 = [S(n), X]^0,$$

$$\pi_{n+1}(X, A, *) \cong [(I^{n+1}/J^n, \partial I^{n+1}/J^n), (X, A)]^0 \cong [(D(n+1), S(n)), (X, A)]^0,$$

and $\partial\colon \pi_{n+1}(X, A, *) \to \pi_n(A, *)$ is induced by the restriction from $D(n+1)$ to $S(n)$.

The pointed cone on S^n is D^{n+1}: We have a homeomorphism

$$S^n \times I/(S^n \times 0 \cup e_{n+1} \times I) \to D^{n+1}, \quad (x, t) \mapsto (1-t)e_{n+1} + tx.$$

Therefore we can also represent elements in $\pi_{n+1}(X, A, *)$ by pointed maps $(D^{n+1}, S^n) \to (X, A)$ and elements in $\pi_n(X, *)$ by pointed maps $S^n \to X$. In comparing these different models for the homotopy groups it is important to remember the homeomorphism between the standard objects (disks and spheres), since there are two homotopy classes of homeomorphisms. ◇

Problems

1. $\pi_n(A, A, a) = 0$. Given $f\colon (I^n, \partial I^n, J^{n-1}) \to (A, A, a)$. Then a null homotopy is $f_t(x_1, \dots, x_n) = f(x_1, x_2, \dots, (1-t)x_n)$.
2. Let $* \in X_0 \subset X_1 \subset X_2 \subset \cdots$ be a sequence of T_1-spaces (i.e., points are closed). Give $X = \bigcup_{n\ge 1} X_n$ the colimit topology. Then a compact subset $K \subset X$ is contained in some X_n. Use this to show that the canonical maps $\pi_n(X_i, *) \to \pi_n(X, *)$ induce an isomorphism $\operatorname{colim}_i \pi_n(X_i, *) \cong \pi_n(X, *)$.
3. Let (X, A, B, b) be a pointed triple. Define the boundary operator $\partial\colon \pi_n(X, A, b) \to \pi_{n-1}(A, b) \to \pi_{n-1}(A, B, b)$ as the composition of the previously defined operator with the map induced by the inclusion. Show that the sequence

$$\cdots \to \pi_n(A, B, b) \to \pi_n(X, B, b) \to \pi_n(X, A, b) \xrightarrow{\partial} \pi_{n-1}(A, B, b) \to \cdots$$

is exact. The sequence ends with $\pi_1(X, A, b)$.
4. The group structure in $\pi_{n+1}(X, A, *)$ is induced by an h-cogroup structure on $(D(n+1), S(n))$ in the category of pointed pairs.

5. Let $f\colon (X,x) \to (Y,y)$ be a pointed map. One can embed the induced morphism $f_*\colon \pi_n(X,x) \to \pi_n(Y,y)$ into an exact sequence which generalizes the case of an inclusion f. Let $Z(f)$ be the pointed mapping cylinder of f and $f = pi : X \to Z(f) \to Y$ the standard factorization into an inclusion and a homotopy equivalence, as explained in (5.3.1). We can now insert the isomorphism $p_*\colon \pi_n(Z(f),*) \to \pi_n(Y,*)$ into the exact sequence of the pair and obtain an exact sequence

$$\cdots \to \pi_n(X,*) \xrightarrow{f_*} \pi_n(Y,*) \to \pi_n(Z(f),X),*) \to \cdots.$$

One can define the groups $\pi_n(Z(f),X,*)$ without using the mapping cylinder. Consider commutative diagrams with pointed maps φ and Φ.

$$\begin{array}{ccc} \partial I^n/J^{n-1} & \xrightarrow{\varphi} & X \\ {\scriptstyle j}\downarrow\cap & & \downarrow{\scriptstyle f} \\ I^n/J^{n-1} & \xrightarrow{\Phi} & Y \end{array}$$

We consider $(\varphi,\Phi)\colon j \to f$ as a morphism in the category of pointed arrows. Let $\pi_n(f)$ denote the set of homotopy classes of such morphisms. For $f\colon X \subset Y$ we obtain the previously defined $\pi_n(Y,X,*)$. The projection $p\colon i \to f$ induces an isomorphism $\pi_n(Z(f),X,*) = \pi_n(i) \to \pi_n(f)$. One can also use the fibre sequence of f.

6.2 The Role of the Base Point

We have to discuss the role of the base point. This uses the transport along paths. Let a path $v\colon I \to X$ and $f\colon (I^n,\partial I^n) \to (X, v(0))$ be given. We consider v as a homotopy $\hat{v}$ of the constant map $\partial I^n \to \{v(0)\}$. We extend the homotopy v_t to a homotopy $V_t\colon I^n \to X$ with initial condition $f = V_0$. An extension exists because $\partial I^n \subset I^n$ is a cofibration. The next proposition is a special case of (5.2.1) and problems in that section. In order to be independent of that section, we also repeat a proof in the present context.

(6.2.1) Proposition. *The assignment $[V_0] \mapsto [V_1]$ is a well-defined map*

$$v_\#\colon \pi_n(X,v(0)) \to \pi_n(X,v(1))$$

which only depends on the morphism $[v]$ in the fundamental groupoid $\Pi(X)$. The relation $(v*w)_\# = w_\# \circ v_\#$ holds, and thus we obtain a transport functor from $\Pi(X)$ which assigns to $x_0 \in X$ the group $\pi_n(X,x_0)$ and to a path v the morphism $v_\#$. The map $v_\#$ is a homomorphism. □

Proof. Let $\varphi\colon f \simeq g$ be a homotopy of maps $(I^n,\partial I^n) \to (X,x_0)$ and $\psi\colon v \simeq w$ a homotopy of paths from x_0 to x_1. Let $V_t\colon I^n \to X$ be a homotopy which extends $(f,\hat{v})$ and W_t a homotopy which extends $(g,\hat{w})$. These data combine to a map on $T = I^n \times 0 \times I \cup \partial I^n \times I \times I \cup I^n \times I \times \partial I \subset I^{n+2}$ as follows: On $I^n \times 0 \times I$

we use φ, on $\partial I^n \times I \times I$ we use $\hat{\psi}$, on $I^n \times I \times 0$ we use V, and on $I^n \times I \times 1$ we use W. If we interchange the last two coordinates then T is transformed into J^{n+1}. Therefore our map has an extension to I^{n+2}, and its restriction to $I^n \times 1 \times I$ is a homotopy from V_1 to W_1. This shows the independence of the representatives f and v. The other properties are clear from the construction. □

There is a similar transport functor in the relative case. We start with a function $f\colon (I^n, \partial I^n, J^{n-1}) \to (X, A, a_0)$ and a path $v\colon I \to A$ from a_0 to a_1. We consider the path as a homotopy of the constant map $J^{n-1} \to \{a_0\}$. Then we extend this homotopy to a homotopy $V_t\colon (I^n, \partial I^n) \to (X, A)$. An extension exists because $J^{n-1} \subset \partial I^n$ and $\partial I^n \subset I^n$ are cofibrations.

(6.2.2) Proposition. *The assignment* $[V_0] \mapsto [V_1]$ *is a well-defined map*

$$v_\#\colon \pi_n(X, A, a_0) \to \pi_n(X, A, a_1)$$

which only depends on the morphism $[v]$ *in the fundamental groupoid* $\Pi(A)$. *For* $n \geq 2$ *the map* $v_\#$ *is a homomorphism. As above we have a transport functor from* $\Pi(A)$. □

Since $v_\#$ is always bijective, homotopy groups associated to base points in the same path component are isomorphic.

We list some naturality properties of the transport functors. As a special case of the functor property we obtain right actions of the fundamental groups:

$$\pi_n(X, x) \times \pi_1(X, x) \to \pi_n(X, x), \qquad (\alpha, \beta) \mapsto \alpha \cdot \beta = \beta_\#(\alpha),$$
$$\pi_n(X, A, a) \times \pi_1(A, a) \to \pi_n(X, A, a), \quad (\alpha, \beta) \mapsto \alpha \cdot \beta = \beta_\#(\alpha).$$

We also have an action of $\pi_1(A, a)$ on $\pi_n(X, a)$ via the natural homomorphism; more generally, we can make the $\pi_n(X, a)$ into a functor on $\Pi(A)$ by viewing a path in A as a path in X. From the constructions we see:

(6.2.3) Proposition. *The morphisms in the exact homotopy sequence of the pair* (X, A) *are natural transformations of transport functors on* $\Pi(A)$. *In particular, they are* $\pi_1(A, *)$*-equivariant with respect to the actions above.* □

Continuous maps $f\colon (X, A) \to (Y, B)$ are compatible with the transport functors

$$f_*(w_\#(\alpha)) = (fw)_\#(f_*(\alpha)).$$

Let $f_t\colon (X, A) \to (Y, B)$ be a homotopy and set $w\colon t \mapsto f(a, t)$. Then the diagram

$$\begin{array}{ccc} & & \pi_n(Y, B, f_0 a) \\ & \overset{f_{0*}}{\nearrow} & \\ \pi_n(X, A, a) & & \downarrow w_\# \\ & \underset{f_{1*}}{\searrow} & \\ & & \pi_n(Y, B, f_1 a) \end{array}$$

is commutative. As in the proof of (2.5.5) one uses this fact to show:

(6.2.4) Proposition. *Let $f\colon (X,A) \to (Y,B)$ be an h-equivalence. Then the induced map $f_*\colon \pi_n(X,A,a) \to \pi_n(Y,B,fa)$ is bijective.* □

Suppose that f that induces isomorphisms $\pi_j(A) \to \pi_j(B)$ and $\pi_j(X) \to \pi_j(Y)$ for $j \in \{n, n+1\}$, $n \ge 1$. Then the Five Lemma (11.1.4) implies that $f_*\colon \pi_{n+1}(X,A,*) \to \pi_{n+1}(Y,B,*)$ is an isomorphism. With some care, this also holds for $n = 0$, see Problem 3.

Let $f\colon (X,A) \to (Y,B)$ be a map of pairs such that the individual maps $X \to Y$ and $A \to B$ induce for each base point in A isomorphisms for all π_n, then $f_*\colon \pi_n(X,A,a) \to \pi_n(Y,B,f(a))$ is bijective for each $n \ge 1$ and each $a \in A$. For the case $n = 1$ see Problem 3.

The transport functors have special properties in low dimensions.

(6.2.5) Proposition. *Let $v\colon I \to X$ be given. Then $v_\#\colon \pi_1(X,v(0)) \to \pi_1(X,v(1))$ is the map $[w] \mapsto [v^-][w][v]$. In particular, the right action of $\pi_1(X,x)$ on itself is given by conjugation $\alpha \cdot \beta = \beta^{-1}\alpha\beta$.* □

(6.2.6) Proposition. *Let $x_1, x_2 \in \pi_2(X,A,*)$ be given. Let $z = \partial x_2 \in \pi_1(A,*)$. Then $x_1 \cdot z = (x_2)^{-1}x_1x_2$ (multiplicative notation for π_2).*

Proof. We first prove the claim in a universal situation and then transport it by naturality to the general case. Set $D = D(2)$, $S = S(1)$.

Let $\iota_1, \iota_2 \in \pi_2(D \vee D, S \vee S)$ be the elements represented by the inclusions of the summands $(D,S) \to (D \vee D, S \vee S)$. Set $\zeta = \partial(\iota_2) \in \pi_1(S \vee S)$. From (6.2.3) and (6.2.5) we compute

$$\partial(\iota_1 \cdot \zeta) = (\partial\iota_1) \cdot \zeta = \zeta^{-1}(\partial\iota_1)\zeta = (\partial\iota_2)^{-1}(\partial\iota_1)(\partial\iota_2) = \partial(\iota_2^{-1}\iota_1\iota_2).$$

Since $D \vee D$ is contractible, ∂ is an isomorphism, hence $\iota_1 \cdot \iota_2 = \iota_2^{-1}\iota_1\iota_2$.

Let now $h\colon (D \vee D, S \vee S) \to (X,A)$ be a map such that $h\iota_k$ represents x_k, i.e., $h_*(\iota_k) = x_k$. The computation

$$x_1 \cdot z = (h_*\iota_1) \cdot (\partial h_*\iota_2) = h_*(\iota_1 \cdot \zeta) = h_*(\iota_2^{-1}\iota_1\iota_2) = x_2^{-1}x_1x_2$$

proves the assertion in the general case. □

(6.2.7) Corollary. *The image of the natural map $\pi_2(X,*) \to \pi_2(X,A,*)$ is contained in the center.* □

The actions of the fundamental group also explain the difference between pointed and free (= unpointed) homotopy classes.

(6.2.8) Proposition. *Let $[S(n),X]^0/(\sim)$ denote the orbit set of the $\pi_1(X,*)$-action on $[S(n),X]^0$. The map $[S(n),X]^0 \to [S(n),X]$ which forgets the base point induces an injective map $v\colon [S(n),X]^0/(\sim) \to [S(n),X]$. For path connected X the map v is bijective. The forgetful map*

$$\pi_n(X,A,*) = [(D(n),S(n-1)),(X,A)]^0 \to [(D(n),S(n-1)),(X,A)]$$

induces an injective map of the orbits of the $\pi_1(A,*)$-action; this map is bijective if A is path connected ($n \geq 2$). □

Problems

1. Let A be path connected. Each element of $\pi_1(X,A,a)$ is represented by a loop in (X,a). The map $j_*\colon \pi_1(X,a) \to \pi_1(X,A,a)$ induces a bijection of $\pi_1(X,A,a)$ with the right (or left) cosets of $\pi_1(X,a)$ modulo the image of $i_*\colon \pi_1(A,a) \to \pi_1(X,a)$.
2. Let $x \in \pi_1(X,A,a)$ be represented by $v\colon I \to X$ with $v(1) \in A$ and $v(0) = a$. Let $w\colon I \to X$ be a loop in (X,a). The assignment $([w],[v]) \mapsto [w * v] = [w]\cdot[v]$ defines a left action of the group $\pi_1(X,a)$ on the set $\pi_1(X,A,a)$. The orbits of this action are the pre-images of elements under $\partial\colon \pi_1(X,A,a) \to \pi_0(A,a)$. Let $(F,f)\colon (X,A) \to (Y,B)$ be a map of pairs. Then $F_*\colon \pi_1(X,A,a) \to \pi_1(Y,B,f(a))$ is equivariant with respect to the homomorphism $F_*\colon \pi_1(X,a) \to \pi_1(Y,f(a))$. Let $[v] \in \pi_1(X,A,a)$ with $v(1) = u \in A$. The isotropy group of $[v]$ is the image of $\pi_1(A,u)$ in $\pi_1(X,a)$ with respect to $[w] \mapsto [v * w * v^-]$. Find an example $\alpha_0, \alpha_1 \in \pi_1(X,A,a)$ such that α_0 has trivial and α_1 non-trivial isotropy group. It is in general impossible to define a group structure on $\pi_1(X,A,a)$ such that $\pi_1(X,a) \to \pi_1(X,A,a)$ becomes a homomorphism.
3. Although there is only a restricted algebraic structure at the beginning of the exact sequence we still have a Five Lemma type result. Let $f\colon (X,A) \to (Y,B)$ be a map of pairs. If $f_*\colon \pi_0(A) \to \pi_0(B)$ and $f_*\colon \pi_1(X,a) \to \pi_1(Y,f(a))$ are surjective and $f_*\colon \pi_0(X) \to \pi_0(Y)$ is injective, then $f_*\colon \pi_1(X,A,a) \to \pi_1(Y,B,f(a))$ is surjective. Suppose that for each $c \in A$ the maps $f_*\colon \pi_1(X,c) \to \pi_1(Y,f(c))$ and $f_*\colon \pi_0(A) \to \pi_0(B)$ are injective and $f_*\colon \pi_1(A,c) \to \pi_1(B,f(c))$ is surjective, then $f_*\colon \pi_1(X,A,a) \to \pi_1(Y,B,f(a))$ is injective for each $a \in A$.
4. Let (X,A) be a pair such that X is contractible. Then $\partial\colon \pi_{q+1}(X,A,a) \to \pi_q(A,a)$ is for each $q \geq 0$ and each $a \in A$ a bijection.
5. Let $A \subset X$ be an h-equivalence. Then $\pi_n(X,A,a) = 0$ for $n \geq 1$ and $a \in A$.
6. Let X carry the structure of an h-monoid. Then $\pi_1(X)$ is abelian and the action of the fundamental group on $\pi_n(X,*)$ is trivial.
7. Give a proof of (6.2.8).
8. The $\pi_1(X,*)$-action on $\pi_n(X,*)$ is induced by a map $\mu_n\colon S(n) \to S(n) \vee S(1)$ by an application of the functor $[-,X]^0$. If we use the model D^n/S^{n-1} for the n-sphere, then an explicit map μ_n is $x \mapsto (2x,*)$ for $2\|x\| \leq 1$ and $x \mapsto (*, 2\|x\| - 1)$ for $2\|x\| \geq 1$.

6.3 Serre Fibrations

The notion of a Serre fibration is adapted to the investigation of homotopy groups, only the homotopy lifting property for cubes is used.

(6.3.1) Theorem. *Let $p\colon E \to B$ be a Serre fibration. For $B_0 \subset B$ set $E_0 = p^{-1}B_0$. Choose base points $* \in B_0$ and $* \in E_0$ with $p(*) = *$. Then p induces for $n \geq 1$ a bijection $p_*\colon \pi_n(E,E_0,*) \to \pi_n(B,B_0,*)$.*

Proof. p_* surjective. Let $x \in \pi_n(B, B_0, *)$ be represented by

$$h\colon (I^n, \partial I^n, J^{n-1}) \to (B, B_0, *).$$

By (3.2.4), there exists a lifting $H\colon I^n \to E$ with $H(J^{n-1}) = \{*\}$ and $pH = h$. We then have $H(\partial I^n) \subset E_0$, and therefore H represents a pre-image of x under p_*.

p_* injective. Let $x_0, x_1 \in \pi_n(E, E_0, *)$ be represented by f_0, f_1 and have the same image under p_*. Then there exists a homotopy $\phi_t\colon (I^n, \partial I^n, J^{n-1}) \to (B, B_0, *)$ such that $\phi_0(u) = pf_0(u)$, $\phi_1(u) = pf_1(u)$ for $u \in I^n$. Consider the subspace $T = I^n \times \partial I \cup J^{n-1} \times I$ and define $G\colon T \to E$ by

$$G(u,t) = \begin{cases} f_t(u), & u \in I^n,\ t \in \{0,1\}, \\ *, & u \in J^{n-1},\ t \in I. \end{cases}$$

The set $T \subset \partial(I^n \times I)$ is transformed into J^n, if one interchanges the last two coordinates. By (3.2.4) again, there exists a map $H\colon I^n \times I \to E$ such that $H|T = G$ and $pH = \phi$. We can view H as a homotopy from f_0 to f_1. □

We use the isomorphism $\pi_n(E, F, *) \cong \pi_n(B, *)$, $F = p^{-1}(*)$ in the exact sequence of the pair $(E, F, *)$ and obtain as a corollary to (6.3.1) the ***exact sequence of a Serre fibration***:

(6.3.2) Theorem. *For a Serre fibration $p\colon E \to B$ with inclusion $i\colon F = p^{-1}(b) \subset E$ and $x \in F$ the sequence*

$$\cdots \to \pi_n(F,x) \xrightarrow{i_*} \pi_n(E,x) \xrightarrow{p_*} \pi_n(B,b) \xrightarrow{\partial} \pi_{n-1}(F,x) \to \cdots$$

is exact. The sequence ends with $\pi_0(E,x) \to \pi_0(B,b)$. □

The new map ∂ has the following description: Let $f\colon (I^n, \partial I^n) \to (B, b)$ be given. View f as $I^{n-1} \times I \to B$. Lift to $\phi\colon I^n \to E$, constant on J^{n-1}. Then $\partial[f]$ is represented by $\phi|I^{n-1} \times 1$. The very end of the sequence requires a little extra argument. For additional algebraic structure at the beginning of the sequence see the discussion of the special case in (3.2.7).

(6.3.3) Theorem. *Let $p\colon E \to B$ be a continuous map and $\mathcal{U}$ a set of subsets such that the interiors cover B. Assume that for $U \in \mathcal{U}$ the map $p_U\colon p^{-1}(U) \to U$ induced by p is a Serre fibration. Then p is a Serre fibration.*

Proof. A subdivision of width $\delta = 1/N$, $N \in \mathbb{N}$ of I^n consists of the cubes $I(a_1, \dots, a_n) = \prod_{j=1}^n I(a_j)$ where $I(k) = [k/N, (k+1)/N]$ for $0 \le k < N$, $k \in \mathbb{Z}$. A k-dimensional face of $I(a_1, \dots, a_n)$ is obtained by replacing $n-k$ of the intervals $I(a_j)$ by one of its boundary points. (The a_j are integers, $0 \le a_j < n$.)

It suffices to work with an open covering $\mathcal{U}$. Choose N such that each cube $I(a_1, \dots, a_n) \times I(b)$ is mapped under h into some $U \in \mathcal{U}$. This is possible by the Lebesgue lemma (2.6.4). Let $V^k \subset I^n$ denote the union of the k-dimensional faces of the subdivision of I^n.

We have to solve a lifting problem for the space I^n with initial condition a. We begin by extending a over $I^n \times [0, \delta]$ to a lifting of h. We solve the lifting problems

$$\begin{array}{ccc} I^n \times 0 \cup V^{k-1} \times [0,\delta] & \xrightarrow{H(k-1)} & E \\ \cap\downarrow & \overset{H(k)}{\nearrow} & \downarrow p \\ I^n \times 0 \cup V^k \times [0,\delta] & \xrightarrow[h]{} & B \end{array}$$

for $k = 0, \dots, n$ with $V^{-1} = \emptyset$ and $H(-1) = a$ by induction over k. Let W be a k-dimensional cube and ∂W the union of its $(k-1)$-dimensional faces. We can solve the lifting problems

$$\begin{array}{ccc} W \times 0 \cup \partial W \times [0,\delta] & \xrightarrow{H(k-1)} & p^{-1}U \\ \cap\downarrow & \overset{H_W}{\nearrow} & \downarrow p_U \\ W \times [0,\delta] & \xrightarrow[h]{} & U \end{array}$$

by a map H_W, since p_U is a Serre fibration; here $U \in \mathcal{U}$ was chosen such that $h(W \times [0, \delta]) \subset U$.

The H_W combine to a continuous map $H(k)\colon V^k \times [0, \delta] \to E$ which lifts h and extends $H(k-1)$. We define H on the first layer $I^n \times [0, \delta]$ as $H(n)$. We now treat $I^n \times [\delta, 2\delta]$ similarly with initial condition given by $H(n)|I^n \times \{\delta\}$ and continue in this manner inductively. □

(6.3.4) Example. Since a product projection is a fibration we obtain from (6.3.3): A locally trivial map is a Serre fibration. ◇

(6.3.5) Example. Let $p\colon E \to B$ be a covering with typical fibre F. Since each map $I^n \to F$ is constant, $\pi_n(F, *)$ is for $n \geq 1$ the trivial group. The exact sequence of p then shows $p_*\colon \pi_n(E) \cong \pi_n(B)$ for $n \geq 2$. The covering $\mathbb{R} \to S^1$ then yields $\pi_n(S^1) \cong 0$ for $n \geq 2$. Moreover we have the exact sequence

$$1 \to \pi_1(E, *) \xrightarrow{p_*} \pi_1(B, *) \xrightarrow{\partial} \pi_0(F, *) \xrightarrow{i_*} \pi_0(E, *) \xrightarrow{p_*} \pi_0(B, *) \to 1$$

with the inclusion $i\colon F = p^{-1}(*) \subset E$ and $\pi_0(F, *) = F$. It yields for $p\colon \mathbb{R} \to S^1$ the bijection $\partial\colon \pi_1(S^1) \cong \mathbb{Z}$. A lifting of the loop $s_n\colon I \to S^1$, $t \mapsto \exp(2\pi i n t)$ with initial condition 0 is $t \mapsto nt$. Hence $\partial[s_n] = n$. Thus we have another method for the computation of $\pi_1(S^1)$. ◇

(6.3.6) Example. Recall the Hopf fibration $p\colon S^{2n+1} \to \mathbb{C}P^n$ (14.1.9). The exact sequence (6.3.2) and $\pi_i(S^1) = 0$ for $i > 1$ yield the isomorphisms

$$p_*\colon \pi_i(S^{2n+1}) \cong \pi_i(\mathbb{C}P^n), \quad \text{for } i \geq 3;$$

and in particular $\pi_i(S^3) \cong \pi_i(S^2)$ for $i \geq 3$, since $\mathbb{C}P^1$ is homeomorphic to S^2 (the Riemann sphere). ◇

(6.3.7) Example. From linear algebra one knows a surjective homomorphism $\mathrm{SU}(2) \to \mathrm{SO}(3)$ with kernel $\{\pm E\} \cong \mathbb{Z}/2$. The space $\mathrm{SU}(2)$ is homeomorphic to S^3. Hence $\mathrm{SO}(3)$ is homeomorphic to $\mathbb{R}P^3$ and $\pi_1(\mathrm{SO}(3)) \cong \mathbb{Z}/2$.

(6.3.8) Proposition. *Let $p\colon (E_1, E_0) \to B$ be a relative Serre fibration, i.e., $p\colon E_1 \to B$ is a Serre fibration and the restriction of p to E_0 is also a Serre fibration. Let (F_1^b, F_0^b) be the pair of fibres over $p(e) = b \in B$. Then:*

(1) *The inclusion induces bijections $\pi_n(F_1^b, F_0^b, e) \cong \pi_n(E_1, E_0, e)$.*

(2) *$\pi_0(E_0) \to \pi_0(E_1)$ is surjective if and only if $\pi_0(F_0^b) \to \pi_0(F_1^b)$ is surjective for each $b \in B$.*

Proof. (1) We first prove the claim for $n = 1$ and begin with the surjectivity. Let $f\colon (I, \partial I, 0) \to (E_1, E_0, e)$ be given. The path $(pf)^-\colon I \to B$ is lifted to $g\colon I \to E_0$ with initial point $f(1)$. Then $g(1) \in F_0$, and f and $f * g$ represent the same element in $\pi_1(E_1, E_0, e)$. The projection $p(f * g)$ is a null homotopic loop with base point b. We lift a null homotopy to E_1 with initial condition $f * g$ on $I \times 0$ and constant on $\partial I \times I$. The lifting is a homotopy $(I, \partial I, 0) \times I \to (E_1, E_0, e)$ from $f * g$ to a map into (F_1, F_0, e). This proves the surjectivity.

Suppose $f_0, f_1\colon (I, \partial I, 0) \to (F_1, F_0, e)$ are given, and let $K\colon (I, \partial I, 0) \times I \to (E_1, E_0, e)$ be a homotopy from f_0 to f_1. We lift pK^- to $L\colon I \times I \to E_0$ with initial condition $L(s, 0) = K(s, 1)$ and $L(0, t) = L(1, t) = e$. The homotopy $p(K *_2 L)$ is a homotopy of loops which is relative to ∂I^2 homotopic to the constant map. We lift a homotopy to E_1 with initial condition $K *_2 L$ on $I^2 \times 0$ and constant on $\partial I^2 \times I$. The end is a homotopy from $f_0 * k_e$ to $f_1 * k_e$. This proves the injectivity.

The higher dimensional case is obtained by an application to the relative Serre fibration $(\Omega^n F_1, \Omega^n F_0) \to (\Omega^n E_1, \Omega^n E_0) \to B$.

(2) Suppose $\pi_0(E_0) \to \pi_0(E_1)$ is surjective. The argument above for the surjectivity is used to show the surjectivity of $\pi_0(F_0^b) \to \pi_0(F_1^b)$. The other implication is easy. □

Problems

1. The 2-fold covering $S^n \to \mathbb{R}P^n$ yields for $n \geq 2$ the isomorphism $\pi_1(\mathbb{R}P^n) \cong \mathbb{Z}/2$.

2. Prove directly the exactness of the sequence (6.3.2) without using (6.1.2).

3. The map $\mathbb{C} \to \mathbb{C}$, $z \mapsto z^2$ has the HLP for I^0 but not for I^1.

4. Let $p\colon (E, e) \to (B, b)$ be a Serre fibration with fibre $F = p^{-1}(b)$. Then

$$\Omega^n(p)\colon \Omega^n(E, e) \to \Omega^n(B, b)$$

is a Serre fibration with fibre $\Omega^n(F,e)$.

6.4 The Excision Theorem

A basic result about homotopy groups is the ***excision theorem of Blakers and Massey*** [22].

(6.4.1) Theorem (Blakers–Massey). *Let Y be the union of open subspaces Y_1 and Y_2 with non-empty intersection $Y_0 = Y_1 \cap Y_2$. Suppose that*

$$\pi_i(Y_1, Y_0, *) = 0 \quad \textit{for } 0 < i < p,\ p \geq 1$$
$$\pi_i(Y_2, Y_0, *) = 0 \quad \textit{for } 0 < i < q,\ q \geq 1$$

for each base point $* \in Y_0$. Then the ***excision map****, induced by the inclusion,*

$$\iota\colon \pi_n(Y_2, Y_0, *) \to \pi_n(Y, Y_1, *)$$

is surjective for $1 \leq n \leq p+q-2$ and bijective for $1 \leq n < p+q-2$ (for each choice of the base point $* \in Y_0$). In the case that $p = 1$, there is no condition on $\pi_i(Y_1, Y_0, *)$.

We defer the proof of this theorem for a while and begin with some applications and examples. We state a special case which has a somewhat simpler proof and already interesting applications. It is also a special case of (6.7.9).

(6.4.2) Proposition. *Let Y be the union of open subspaces Y_1 and Y_2 with non-empty intersection Y_0. Suppose (Y_2, Y_0) is q-connected. Then (Y, Y_1) is q-connected.* □

We apply the excision theorem (6.4.1) to the homotopy group of spheres. We use the following subspaces of S^n, $n \geq 0$,

$$D^n_\pm = \{(x_1, \dots, x_{n+1}) \in S^n \mid \pm x_{n+1} \geq 0\} \subset H^n_\pm = \{x \in S^n \mid x \neq \mp e_{n+1}\}.$$

We use $\mathbb{R}^n \subset \mathbb{R}^{n+1}$, $(z_1, \dots, z_n) \mapsto (z_1, \dots, z_n, 0)$ and similar inclusions for subsets of $\mathbb{R}^n$. We choose $* = -e_1$ as a base point; e_i is the standard unit vector.

(6.4.3) Lemma. *We have isomorphisms $\partial\colon \pi_{i+1}(D^{n+1}_-, S^n, *) \to \pi_i(S^n, *)$ for $i \geq 0$, $n \geq 0$ and $\pi_i(S^n, *) \to \pi_i(S^n, D^n_\pm, *)$ for $i \geq 1$, $n \geq 1$.*

Proof. In the first case we use the exact sequence of the pair (D^{n+1}_-, S^n). The space D^{n+1}_- is contractible and hence $\pi_i(D^{n+1}_-, *) = 0$ for $i \geq 0$ and $n \geq 0$.

In the second case we consider similarly the exact sequence of $(S^n, D^n_\pm)$. Note that $* = -e_1 \in D^n_\pm$ for $n \geq 1$. □

For $n \geq 0$ we have a diagram with the isomorphisms (6.4.3)

$$\begin{array}{ccc} \pi_i(S^n, *) & \xrightarrow{E} & \pi_{i+1}(S^{n+1}, *) \\ \cong \uparrow \partial & & \downarrow \cong \\ \pi_{i+1}(D^{n+1}_-, S^n, *) & \xrightarrow{\iota} & \pi_{i+1}(S^{n+1}, D^{n+1}_+, *). \end{array}$$

The morphism ι is induced by the inclusion and E is defined so as to make the diagram commutative. Note that the inductive proof of (1) in the next theorem only uses (6.4.2).

(6.4.4) Theorem. (1) *$\pi_i(S^n) = 0$ for $i < n$.*

(2) *The homomorphism ι is an isomorphism for $i \leq 2n-2$ and an epimorphism for $i = 2n-1$. A similar statement holds for E.*

Proof. Let $N(n)$ be the statement (1) and $E(n)$ the statement (2). Obviously $N(1)$ holds. Assume $N(n)$ holds. We then deduce $E(n)$. We apply the excision theorem to $(Y, Y_1, Y_2, Y_0) = (S^{n+1}, D^{n+1}_+, D^{n+1}_-, S^n)$. By $N(n)$ and (6.4.3) we have $\pi_i(S^n) \cong \pi_{i+1}(D^{n+1}_\pm, S^n) = 0$ for $0 \leq i < n$. We use the excision theorem for $p = q = n+1$ and see that ι is surjective for $i+1 \leq 2n$ and bijective for $i+1 \leq 2n-1$. Finally, $E(n)$ and $N(n)$ imply $N(n+1)$.

In order to have the correct hypotheses for the excision theorem, we thicken the spaces, replace $D^n_\pm$ by $H^n_\pm$ and note that the inclusions $D^n \pm \subset H^n_\pm$ and $S^{n-1} \subset H^n_+ \cap H^n_-$ are h-equivalences. □

(6.4.5) Proposition. *The homomorphism $\pi_i(D^{n+1}_-, S^n, *) \to \pi_i(D^{n+1}_-/S^n, *)$ induced by the quotient map is an isomorphism for $i \leq 2n-1$ and an epimorphism for $i = 2n$.*

Proof. Consider the commutative diagram

$$\begin{array}{ccc} \pi_i(D^{n+1}_-, S^n, *) & \longrightarrow & \pi_i(D^{n+1}_-/S^n, *) \\ \downarrow \iota & & \downarrow (1) \\ \pi_i(S^{n+1}, D^{n+1}_+, *) & \xrightarrow{(2)} & \pi_i(S^{n+1}/D^{n+1}_+, *). \end{array}$$

The map (1) is induced by a homeomorphism and the map (2) by a homotopy equivalence, hence both are isomorphisms. Now apply (6.4.4). □

The homomorphism E is essentially the suspension homomorphism. In order to see this, let us work with (6.1.4). The suspension homomorphism Σ_* is the composition

$$\Sigma_* \colon \pi_n(X, *) \xleftarrow[\cong]{\partial} \pi_{n+1}(CX, X, *) \xrightarrow{q_*} \pi_{n+1}(CX/X, *) = \pi_{n+1}(\Sigma X, *)$$

with the quotient map $q\colon D(n+1) \to D(n+1)/S(n) = S(n+1)$.

The next result is the famous ***suspension theorem of Freudenthal*** ([66]).

(6.4.6) Theorem. *The suspension $\Sigma_*\colon \pi_i(S(n)) \to \pi_{i+1}(S(n+1))$ is an isomorphism for $i \le 2n-2$ and an epimorphism for $i = 2n-1$.*

Proof. We have to show that $q_*\colon \pi_{i+1}(CX, X) \to \pi_{i+1}(CX/X)$ is for $X = S(n)$ an isomorphism (epimorphism) in the appropriate range. This follows from (6.4.5); one has to use that S^n is homeomorphic to $S(n)$ and that D^{n+1}_- is the (pointed) cone on S^n. □

(6.4.7) Theorem. *$\pi_n(S(n)) \cong \mathbb{Z}$ and $\Sigma_*\colon \pi_n(S(n)) \to \pi_{n+1}(S(n+1))$ is an isomorphism ($n \ge 1$). The group $\pi_n(S(n))$ is generated by the identity of $S(n)$.*

Proof. From the exact sequence $\pi_2(S^3) \to \pi_2(S^2) \to \pi_1(S^1) \to \pi_1(S^3)$ of the Hopf fibration $S^1 \to S^3 \to S^2$ and $\pi_j(S^3) = 0$ for $j = 1, 2$ we obtain an isomorphism $\partial\colon \pi_2(S^2) \cong \pi_1(S^1) \cong \mathbb{Z}$. From (6.4.6) we obtain a surjection $\Sigma_*\colon \pi_1(S(1)) \to \pi_2(S(2))$; this is an isomorphism, since both groups are isomorphic to $\mathbb{Z}$. For $n \ge 2$, (6.4.6) gives directly an isomorphism Σ_*. We know that $\pi_1(S(1)) \cong \mathbb{Z}$ is generated by the identity, and Σ_* respects the identity. □

(6.4.8) Example. We continue the discussion of the Hopf fibrations (6.3.6). The Hopf fibration $S^1 \to S^{2n+1} \to \mathbb{C}P^n$ and $\pi_i(S^{2n+1}) = 0$ for $i \le 2n$ yield $\pi_2(\mathbb{C}P^n) \cong \pi_1(S^1) \cong \mathbb{Z}$ and $\pi_i(\mathbb{C}P^n) = 0$ for $0 \le i \le 2n$, $i \ne 2$. The inclusion $S^{2n+1} \to S^{2n+3}$, $z \mapsto (z, 0)$ induces an embedding $\mathbb{C}P^n \subset \mathbb{C}P^{n+1}$. We compare the corresponding Hopf fibrations and their exact sequences and conclude $\pi_2(\mathbb{C}P^n) \cong \pi_2(\mathbb{C}P^{n+1})$. Let $\mathbb{C}P^\infty = \bigcup_{n \ge 1} \mathbb{C}P^n$ be the colimit. The canonical inclusion $\mathbb{C}P^n \subset \mathbb{C}P^\infty$ induces $\pi_i(\mathbb{C}P^n) \cong \pi_i(\mathbb{C}P^\infty)$ for $i \le 2n$. A proof uses the fact that a compact subset of $\mathbb{C}P^\infty$ is contained in some finite $\mathbb{C}P^N$. Therefore $\mathbb{C}P^\infty$ is a space with a single non trivial homotopy group $\pi_2(\mathbb{C}P^\infty) \cong \mathbb{Z}$.

Note also the special case $\pi_3(S^2) \cong \pi_3(S^3) \cong \mathbb{Z}$.

We have similar results for real projective spaces. The twofold coverings $\mathbb{Z}/2 \to S^n \to \mathbb{R}P^n$ are used to show that $\pi_1(\mathbb{R}P^2) \cong \pi_1(\mathbb{R}P^3) \cong \cdots \cong \pi_1(\mathbb{R}P^\infty) \cong \mathbb{Z}/2$, induced by the inclusions, $\pi_i(\mathbb{R}P^n) \cong \pi_i(\mathbb{R}P^{n+1})$ for $i < n$ and $\pi_i(\mathbb{R}P^n) = 0$ for $0 \le i < n$, $i \ne 1$. The space $\mathbb{R}P^\infty$ has a single non-trivial homotopy group $\pi_1(\mathbb{R}P^\infty) \cong \mathbb{Z}/2$. ◇

6.5 The Degree

Let $d\colon \pi_n(S(n)) \to \mathbb{Z}$ be the isomorphism which sends [id] to 1. If $f\colon S(n) \to S(n)$ is a pointed map, then $f_*\colon \pi_n(S(n)) \to \pi_n(S(n))$ is the multiplication by the integer $d(f) = d(f_*[\mathrm{id}]) = d([f])$. Since the map $[S(n), S(n)]^0 \to [S(n), S(n)]$ which forgets about the base point is bijective (see (6.2.8)), we can transport d to a bijection $d\colon [S(n), S(n)] \to \mathbb{Z}$. The functoriality $f_* g_* = (fg)_*$ shows

$d(fg) = d(f)d(g)$; therefore $d(h) = \pm 1$ if h is a homeomorphism. The suspension sends $[f]$ to $[f \wedge \mathrm{id}]$; hence $d(f) = d(f \wedge \mathrm{id})$.

(6.5.1) Proposition. *Given pointed maps $f\colon S(m) \to S(m)$ and $g\colon S(n) \to S(n)$. Then $d(f \wedge g) = d(f)d(g)$.*

Proof. We use the factorization $f \wedge g = (f \wedge \mathrm{id})(\mathrm{id} \wedge g)$. The map $f \wedge \mathrm{id}$ is a suspension of f, and suspension does not change the degree. Let $\tau\colon S(m) \wedge S(n) \to S(n) \wedge S(m)$ interchange the factors. From $\tau(g \wedge \mathrm{id})\tau = \mathrm{id} \wedge g$ we conclude $d(\mathrm{id} \wedge g) = d(g \wedge \mathrm{id}) = d(g)$. □

Let $k_n\colon S(n) \to S^n$ be a homeomorphism. The bijection

$$[S^n, S^n] \to [S(n), S(n)], \quad [f] \mapsto [k_n f k_n^{-1}]$$

is independent of the choice of k_n. We use this bijection to transport d to a bijection $d\colon [S^n, S^n] \to \mathbb{Z}$. If $d([f]) = k$ we call k the ***degree*** $d(f)$ of f. We still have the properties $d(f)d(g) = d(fg)$, $d(\mathrm{id}) = 1$, $d(h) = \pm 1$ for a homeomorphism h. By a similar procedure we define the degree $d(f)$ for any self-map f of a space S which is homeomorphic to $S(n)$.

Matrix multiplication $l_A\colon \mathbb{R}^n \to \mathbb{R}^n$, $x \mapsto Ax$ induces for each $A \in \mathrm{GL}_n(\mathbb{R})$ a pointed map $L_A\colon S^{(n)} \to S^{(n)}$. For the notation see (6.1.4).

(6.5.2) Proposition. *The degree of L_A is the sign of the determinant* $\det(A)$.

Proof. Let $w\colon I \to \mathrm{GL}_n(\mathbb{R})$, $t \mapsto A(t)$ be a path. Then $(x,t) \mapsto L_{A(t)}x$ is a homotopy. Hence $d(L_A)$ only depends on the path component of A in $\mathrm{GL}_n(\mathbb{R})$. The group $\mathrm{GL}_n(\mathbb{R})$ has two path components, distinguished by the sign of the determinant. Thus it suffices to show that for some A with $\det(A) = -1$ we have $d(L_A) = -1$. By the preceding discussion and (6.1.4) we see that $(x_1, \dots, x_n) \mapsto (-x_1, x_2, \dots, x_n)$ has degree -1. □

The stereographic projection (2.3.2) now shows that the map $S^n \to S^n$ which changes the sign of the first coordinate has degree -1.

(6.5.3) Proposition. *Let $A \in \mathrm{O}(n+1)$. Then $\lambda_A\colon S^n \to S^n$, $x \mapsto Ax$ has degree* $\det(A)$.

Proof. Again it suffices to verify this for appropriate elements in the two path components of $\mathrm{O}(n+1)$, and this we have already achieved. □

(6.5.4) Corollary. *The map $S^n \to S^n$, $x \mapsto -x$ has degree $(-1)^{n+1}$.* □

A ***vector field*** on S^n is a continuous map $F\colon S^n \to \mathbb{R}^{n+1}$ such that for each $x \in S^n$ the vector $F(x)$ is orthogonal to x. For the maximal number of linearly independent vector fields see [3].

(6.5.5) Theorem. *There exists a vector field F on S^n such that $F(x) \neq 0$ for each $x \in S^n$ if and only if n is odd.*

Proof. Let $n = 2k - 1$. Then

$$(x_1, x_2, \dots, x_{2k-1}, x_{2k}) \mapsto (x_2, -x_1, \dots, x_{2k}, -x_{2k-1})$$

is a vector field with the desired property.

Let F be a vector field such that $F(x) \neq 0$. Set $V(x) = F(x)/\|F(x)\|$. Then $(x, t) \mapsto \cos \pi t \cdot x + \sin \pi t \cdot V(x)$ is a homotopy from the identity to the antipodal map. Hence the antipodal map has degree 1. By (6.5.4), n is odd. □

(6.5.6) Proposition. *Let $\tau \colon S(m) \wedge S(n) \to S(n) \wedge S(m)$ interchange the factors. Then $d(\tau) = (-1)^{mn}$.*

Proof. By (6.5.2) we know the analogous assertion for the models $S^{(m)}$. □

6.6 The Brouwer Fixed Point Theorem

We prove the ***fixed point theorem of Brouwer*** and a number of equivalent results. As an application we discuss the problem of topological dimension.

Theorem (6.6.1) collects nine statements, one of which is the Brouwer fixed point theorem, and they are shown to be equivalent by elementary homotopy arguments. In our setting we have proved statement (3) of (6.6.1): The identity of S^{n-1} has degree one and is therefore not null homotopic.

Let us first introduce some notation. Consider the cube

$$W = W^n = \{(x_i) \in \mathbb{R}^n \mid -1 \leq x_i \leq 1\}$$

with the faces $C_i(\pm) = \{x \in W^n \mid x_i = \pm 1\}$. We say, $B_i \subset W^n$ separates $C_i(+)$ and $C_i(-)$, if B_i is closed in W, and if

$$W \smallsetminus B_i = B_i(+) \cup B_i(-), \quad \emptyset = B_i(+) \cap B_i(-), \quad C_i(\pm) \subset B_i(\pm),$$

with open subsets $B_i(+)$ and $B_i(-)$ of $W \smallsetminus B_i$. The n dimensional standard simplex is Δ^n. Its boundary $\partial \Delta^n$ is the union of the faces $\partial_i \Delta^n = \{(t_0, \dots, t_n) \in \Delta^n \mid t_i = 0\}$.

(6.6.1) Theorem. *The following statements are equivalent:*

(1) *A continuous map $b \colon D^n \to D^n$ has a fixed point (Brouwer fixed point theorem).*

(2) *There does not exist a continuous map $r \colon D^n \to S^{n-1}$ which is the identity on S^{n-1} (retraction theorem).*

(3) *The identity of S^{n-1} is not null homotopic (homotopy theorem).*

(4) *Let* $f: D^n \to \mathbb{R}^n$ *be a continuous map such that* $f(z) = z$ *for* $z \in S^{n-1}$. *Then* D^n *is contained in the image of* f.

(5) *Let* $g: D^n \to \mathbb{R}^n$ *be continuous. Then there exists a fixed point or there exists* $z \in S^{n-1}$ *such that* $g(z) = \lambda z$ *with* $\lambda > 1$.

(6) *Let* $v_i: W^n \to \mathbb{R}$, $1 \le i \le n$ *be functions such that* $v_i(x) < 0$ *for* $x \in C_i(-)$ *and* $v_i(x) > 0$ *for* $x \in C_i(+)$. *Then there exists* $x \in W^n$ *such that* $v_i(x) = 0$ *for each* i *(intermediate value theorem).*

(7) *Suppose* B_i *separates* $C_i(-)$ *and* $C_i(+)$ *for* $1 \le i \le n$. *Then the intersection* $B_1 \cap B_2 \cap \cdots \cap B_n$ *is non-empty.*

(8) *Let* $B_0, \dots, B_n$ *be a closed covering of* Δ^n *such that* $e_i \notin B_i$ *and* $\partial_i \Delta^n \subset B_i$. *Then* $\bigcap_{i=0}^n B_i \neq \emptyset$. *The same conclusion holds if we assume that the* B_i *are open.*

(9) *Let* $B_0, \dots, B_n$ *be a closed covering of* Δ^n *such that* $e_i \in B_i$ *and* $\partial_i \Delta^n \cap B_i = \emptyset$. *Then* $\bigcap_{i=0}^n B_i \neq \emptyset$.

The fixed point theorem expresses a topological property of D^n. If $h: X \to D^n$ is a homeomorphism and $f: X \to X$ a self-map, then hfh^{-1} has a fixed point z and therefore f has the fixed point $h(z)$. We can apply (2) to the pairs $(W^n, \partial W^n)$ and $(\Delta^n, \partial\Delta^n)$, since they are homeomorphic to (D^n, S^{n-1}). Statement (3) is also equivalent to the inclusion $S^{n-1} \subset \mathbb{R}^n \smallsetminus \{0\}$ not being null homotopic (similarly for ∂W^n in place of S^{n-1}).

Proof. (1) $\Rightarrow$ (2). Suppose r is a retraction. Then $x \mapsto -r(x)$ is a map without fixed point.

(2) $\Rightarrow$ (3). The map $r: D^n \to S^{n-1}$ which corresponds by (2.3.4) to a null homotopy of the identity is a retraction.

(3) $\Rightarrow$ (1). Suppose b has no fixed point. Then

$$S^{n-1} \times I \to S^{n-1}, \quad (x,t) \mapsto \frac{x - tb(x)}{\|x - tb(x)\|} = N(x - tb(x))$$

is a homotopy from the identity to the map $f: x \mapsto N(x - b(x))$. Since b has no fixed point, the formula for f defines a map on the whole of D^n, and then $(x,t) \mapsto f(tx)$ is a homotopy from the constant map to f. Thus f is null homotopic, and therefore also $\mathrm{id}(S^{n-1})$.

(2) $\Rightarrow$ (4). If x is contained in the interior of D^n, then there exists a retraction $r: \mathbb{R}^n \smallsetminus x \to S^{n-1}$ of $S^{n-1} \subset \mathbb{R}^n \smallsetminus x$. If x is not contained in the image of f, then $r \circ f: D^n \to S^{n-1}$ contradicts the retraction theorem.

(4) $\Rightarrow$ (5). Define a map $f: D^n \to \mathbb{R}^n$ by

$$\begin{aligned} &\text{(i)} \quad f(x) = 2x - g(2x), && \|x\| \le 1/2, \\ &\text{(ii)} \quad f(x) = \|x\|^{-1}x - 2(1 - \|x\|)g(\|x\|^{-1}x), && \|x\| \ge 1/2. \end{aligned}$$

For $\|x\| = \frac{1}{2}$ we obtain in both cases $2x - g(2x)$. Thus f is a well-defined continuous map.

For $\|x\| = 1$ we have $f(x) = x$. By (4), there exists y with $f(y) = 0$. If $\|y\| \le \frac{1}{2}$, then (i) shows that $2y$ is a fixed point. If $\|y\| > \frac{1}{2}$, then $\|y\| \ne 1$, and (ii) shows the second case with $\lambda = (2 - 2\|y\|)^{-1} > 1$.

(5) $\Rightarrow$ (1). A special case.

(3) $\Rightarrow$ (6). Set $v\colon W^n \to \mathbb{R}^n, x \mapsto (v_1(x), \dots, v_n(x))$. Suppose $v(x) \ne 0$ for each $x \in W^n$. Then $v\colon W^n \to \mathbb{R}^n \smallsetminus 0$. Consider $h\colon (t,x) \mapsto (1-t)x + tv(x)$. If $x \in C_i(-)$, i.e., $x_i < 0$, then $(1-t)x_i + tv_i(x) < 0$ for each $t \in I$. Hence $h_t\colon \partial W^n \to \mathbb{R}^n \smallsetminus 0$ is a homotopy from the inclusion to v. Since v has an extension to W^n, it is null homotopic, but the inclusion is not null homotopic. A contradiction.

(6) $\Rightarrow$ (7). Let d denote the Euclidean distance. Define $v_i\colon W \to \mathbb{R}$ by

$$v_i(x) = \begin{cases} -d(x, B_i), & x \in B_i(-), \\ d(x, B_i), & x \in B_i(+) \cup B_i, \end{cases}$$

and apply (6).

(7) $\Rightarrow$ (2). Let $r\colon W^n \to \partial W^n$ be a retraction. We define $B_i(\pm) = r^{-1}(\pm x_i > 0)$ and $B_i = r^{-1}(x_i = 0)$. We apply (7) and obtain a contradiction.

(3) $\Rightarrow$ (8). We use the functions $v_i(x) = d(x, B_i)$. Our assumptions imply $v_i(e_i) > 0$, and $v_i(x) = 0$ provided $x \in \partial_i \Delta^n$. If the B_i have empty intersection, then $v(x) = (v_0(x), \dots, v_n(x)) \ne 0$ for every $x \in \Delta^n$. This gives us a map

$$\alpha\colon \Delta^n \to \partial\Delta^n, \quad x \mapsto (\textstyle\sum v_i(x))^{-1} v(x),$$

because, since the B_i cover Δ^n, for each x at least one coordinate $v_i(x)$ is zero. If $x \in \partial_i \Delta^n$, then $\alpha(x) \in \partial_i \Delta^n$, hence $(1-t)x + tv(x) \in \partial_i \Delta^n$ for each $t \in [0,1]$. The identity of $\partial\Delta^n$ is therefore homotopic to $\beta = \alpha|\partial\Delta^n$. Since β has the extension α it is null homotopic, and therefore also $\mathrm{id}(\partial\Delta^n)$ is null homotopic. This contradicts (3).

Now suppose the B_i are open. By a general result of point-set topology there exist closed sets $C_i \subset B_i$ and the C_i still form a covering. In order to make sure that the C_i satisfy the hypotheses of (8) we can replace the C_i by $C_i \cup \partial_i \Delta^n$. The first part of the proof now shows that the C_i have non-empty intersection.

(8) $\Rightarrow$ (9). Set $U_i = \Delta^n \smallsetminus B_i$. Suppose the B_i have empty intersection. Then the U_i cover Δ^n. Since the B_i are a covering, the U_i have empty intersection. By construction, $e_i \notin U_i$ and $\partial_i \Delta^n \subset U_i$. We therefore can apply (8) in the case of the open covering by the U_i and see that the U_i have non-empty intersection. Contradiction.

(9) $\Rightarrow$ (2). Let $A_j = \{(t_0, \dots, t_n) \in \partial\Delta^n \mid t_j \ge 1/n\}$. Let $r\colon \Delta^n \to \partial\Delta^n$ be a retraction and set $B_j = r^{-1}(A_j)$. Then (9) tells us that the B_j have non-empty intersection, and this is impossible. □

Some statements of (6.6.1) have interesting different proofs. For a proof which uses only basic results in differential topology see [79]. Another interesting proof is based on a combinatorial result, called Sperner's lemma [173]. The proof of (3) does not need the full strength of the bijection $[S^n, S^n] \cong \mathbb{Z}$. It suffices to have a homotopy invariant functor which distinguishes the constant map from the identity. A standard argument for this uses homology theory and the determination of the homology of spheres.

The retraction theorem does not hold for infinite-dimensional spaces. In [70, Chapter 19] you can find a proof that the unit disk of an infinite-dimensional Banach space admits a retraction onto its unit sphere.

Does there exist a sensible topological notion of dimension for suitable classes of spaces? Greatest generality is not necessary at this point. As an example we introduce the covering dimension of compact metric spaces X. (For dimension theory in general see [94].) Let $\mathcal{C}$ be a finite covering of X and $\varepsilon > 0$ a real number. We call $\mathcal{C}$ an ***ε-covering***, if each member of $\mathcal{C}$ has diameter less than ε. We say $\mathcal{C}$ has ***order*** m, if at least one point is contained in m members but no point in $m+1$. The compact metric space X has ***covering dimension*** $\dim X = k$, if there exists for each $\varepsilon > 0$ a finite closed ε-covering of X of order $k+1$ and $k \in \mathbb{N}_0$ is minimal with this property. Thus X is zero-dimensional in this sense, if there exists for each $\varepsilon > 0$ a finite partition of X into closed sets of diameter at most ε. We verify that this notion of dimension is a topological property.

(6.6.2) Proposition. *Let X and Y be homeomorphic compact metric spaces. If X is k-dimensional then also is Y.*

Proof. Let $h: X \to Y$ be a homeomorphism. Fix $\varepsilon > 0$ and let $\mathcal{U}$ be the covering of Y by the open ε-balls $U_\varepsilon(y) = \{x \mid d(x, y) < \varepsilon\}$. (We use d for the metrics.) Let δ be a Lebesgue number of the covering $(h^{-1}(U) \mid U \in \mathcal{U})$. Since $\dim X = k$, there exists a finite closed δ-covering $\mathcal{C}$ of X of order $k+1$. The finite closed covering $\mathcal{D} = (h(C) \mid C \in \mathcal{C})$ of Y has then the order $k+1$, and since each member of C is contained in a set $h^{-1}(U)$, the covering $\mathcal{D}$ is an ε-covering. Thus we have shown $\dim Y \le k$.

We now show that $\dim Y \ge k$, i.e., there exists $\delta > 0$ such that each finite closed δ-covering has order at least $k+1$. Let $\varepsilon > 0$ be a corresponding number for X. A homeomorphism $g: Y \to X$ is uniformly continuous: There exists a $\delta > 0$ such that $d(y_1, y_2) < \delta$ implies $d(g(y_1), g(y_2)) < \varepsilon$. So if $\mathcal{C}$ is a δ-covering of Y, then $\mathcal{D} = (g(C) \mid C \in \mathcal{C})$ is an ε-covering of X. Since $\mathcal{D}$ has order at least $k+1$, so has $\mathcal{C}$. □

(6.6.3) Proposition. *There exists $\varepsilon > 0$ such that each finite closed ε-covering $(B_j \mid j \in J)$ of Δ^n has order at least $n+1$.*

Proof. Let ε be a Lebesgue number of the covering $U_i = \Delta^n \smallsetminus \partial_i \Delta^n$, $i = 0, \dots, n$. Hence for each $j \in J$ there exist i such that $B_j \subset U_i$, and the latter is equivalent to $B_j \cap \partial_i \Delta^n = \emptyset$. Suppose $e_k \in B_j$. Since $e_k \in \partial_i \Delta^n$ for $i \neq k$, we cannot have $B_j \subset U_i$; thus $e_k \in B_j$ implies $B_j \subset U_k$. Since each e_k is contained in at least one of the sets B_j we conclude $|J| \geq n+1$. For each $j \in J$ we now choose $g(j) \in \{0, \dots, n\}$ such that $B_j \cap \partial_{g(j)} \Delta^n = \emptyset$ and set $A_k = \cup\{B_j \mid g(j) = k\}$; this is a closed set because J is finite. Each B_j is contained in some A_k, hence the A_k cover Δ^n. Moreover, by construction, $A_k \cap \partial_k \Delta^n = \emptyset$. We can therefore apply part (9) of (6.6.1) and find an x in the intersection of the A_k. Hence for each k there exists i_k such that $x \in B_{i_k}$. Since each B_j is contained in exactly one of the sets A_k, the element x is contained in the $n+1$ members B_{i_k}, $k = 0, \dots, n$ of the covering. □

We can now compare the covering dimension and the algebraic dimension.

(6.6.4) Theorem. *Δ^n has covering dimension n. A compact subset of $\mathbb{R}^n$ has covering dimension at most n.*

Proof. By (6.6.3), Δ^n has covering dimension at least n. It remains to construct finite closed ε-coverings of order $n+1$ for each ε. See Problem 4. □

Problems

1. Let U, V be an open covering of I^2. Then there exists either a path $u \colon I \to U$ such that $u(0) \in I \times 0, u(1) \in I \times 1$ or a path $v \colon I \to V$ such that $v(0) \in 0 \times I, v(1) \in 1 \times I$.
2. Let U, V be an open covering of Δ^2. Then there exists a path component $\tilde{U}$ of U such that $\tilde{U} \cap \partial_i \Delta^2 \neq \emptyset$ for each i or a path component of V with a similar property.
3. Generalize the preceding two exercises to n dimensions.
4. The following figure indicates the construction of closed ε-coverings of order 3 for the square.

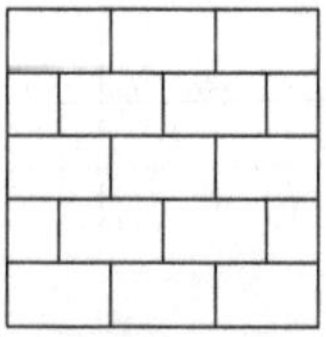

Generalize this construction to the cube I^n by a suitable induction.
5. Suppose I^n is the union of a finite number of closed sets, none of which contains points of two opposite faces. Then at least $n+1$ of these closed sets have a common point.

6.7 Higher Connectivity

For many applications it is important to know that the homotopy groups of a space vanish in a certain range. We discuss several reformulations of this fact. In the following $\pi_0(X, x) = \pi_0(X)$ with base point $[x]$. The space D^0 is a singleton and $S^{-1} = \emptyset$.

(6.7.1) Proposition. *Let $n \geq 0$. The following are equivalent:*

(1) $\pi_n(X,x) = 0$ *for each* $x \in X$.

(2) *Each map* $S^n \to X$ *has an extension to* D^{n+1}.

(3) *Each map* $\partial I^{n+1} \to X$ *has an extension to* I^{n+1}.

Proof. The case $n = 0$ is trivial. The equivalence of (2) and (3) is a consequence of the homeomorphism $(D^{n+1}, S^n) \cong (I^{n+1}, \partial I^{n+1})$. Suppose $f \colon S^n \to X$ is given. Use $e_1 = (1, 0, \dots) \in S^n$ as a base point and think of f representing an element in $\pi_n(X,x)$. If (1) holds, then f is pointed null homotopic. A null homotopy $S^n \times I \to X$ factors over the quotient map $S^n \times I \to D^{n+1}$, $(x,t) \mapsto (1-t)e_1 + tx$ and yields an extension of f. Conversely, let an element α of $\pi_n(X,x)$ be represented by a pointed map $f \colon (S^n, e_1) \to (X,x)$. If this map has an extension F to D^{n+1}, then (F,f) represents $\beta \in \pi_{n+1}(X,X,x) = 0$ with $\partial\beta = \alpha$. □

(6.7.2) Proposition. *Let $n \geq 0$. Let $f \colon (D^n, S^{n-1}) \to (X,A)$ be homotopic as a map of pairs to a map $k \colon (D^n, S^{n-1}) \to (A,A)$. Then f is relative to S^{n-1} homotopic to a map g such that $g(D^n) \subset A$.*

Proof. The case $n = 0$ is trivial. Let $G_t \colon (D^n, S^{n-1}) \to (X,A)$ be a homotopy from f to k according to the assumption. Define $\lambda \colon D^n \times I \to D^n \times I$ by $\lambda(x,t) = (2\alpha(x,t)^{-1} \cdot x, 2 - \alpha(x,t))$ with the function $\alpha(x,t) = \max(2\|x\|, 2-t)$. Then $H = G \circ \lambda$ is a homotopy with the desired property from f to $g = H_1$. □

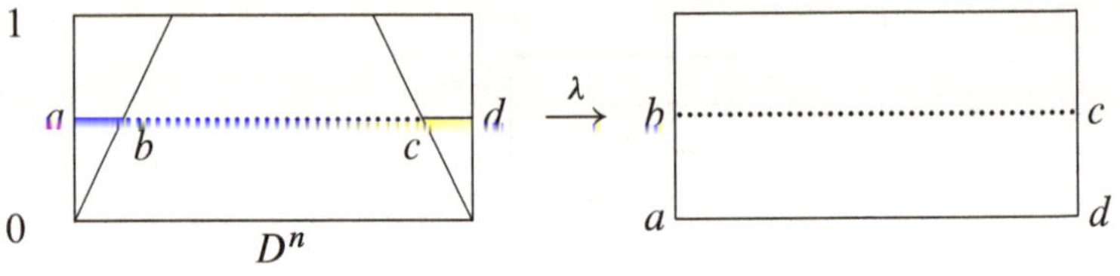

(6.7.3) Proposition. *Let $n \geq 1$. The following assertions about (X,A) are equivalent:*

(1) $\pi_n(X,A,*) = 0$ *for each choice of* $* \in A$.

(2) *Each map* $f \colon (I^n, \partial I^n) \to (X,A)$ *is as a map of pairs homotopic to a constant map.*

(3) *Each map* $f \colon (I^n, \partial I^n) \to (X,A)$ *is homotopic rel* ∂I^n *to a map into* A.

Proof. (1) ⇒ (2). Let $f \colon (I^n, \partial I^n) \to (X,A)$ be given. Since J^{n-1} is contractible, there exists a homotopy of the restriction $f \colon J^{n-1} \to A$ to a constant map. Since $J^{n-1} \subset \partial I^n$ and $\partial I^n \subset I^n$ are cofibrations, f is as a map of pairs homotopic to

$g\colon (I^n, \partial I^n) \to (X, A)$ such that $g(J^{n-1}) = \{a_0\}$. Since $\pi_n(X, A, a_0) = 0$, the map $g\colon (I^n, \partial I^n, J^{n-1}) \to (X, A, a_0)$ is null homotopic as a map of triples.

(2) $\Rightarrow$ (3). (6.7.2).

(3) $\Rightarrow$ (1). Let $f\colon (I^n, \partial I^n, J^{n-1}) \to (X, A, *)$ be given. By assumption (3) $[f]$ is contained in the image of $\pi_n(A, A, *) \to \pi(X, A, *)$. Now use $\pi_n(A, A, *) = 0$. □

We call (X, A) ***n*-compressible** if one of the assertions in (6.7.3) holds. More generally, we call a map $f\colon X \to Y$ n-compressible if the following holds: For each commutative diagram

$$\begin{array}{ccc} \partial I^n & \xrightarrow{\varphi} & X \\ \downarrow \cap & & \downarrow f \\ I^n & \xrightarrow{\Phi} & Y \end{array}$$

there exists $\Psi\colon I^n \to X$ such that $\Psi|\partial I^n = \varphi$ and $f\Psi \simeq \Phi$ relative to ∂I^n. (This amounts to part (3) in (6.7.3).) This notion is homotopy invariant in the following sense:

(6.7.4) Proposition. *Given $f\colon X \to Y$ and a homotopy equivalence $p\colon Y \to Z$. Then f is n-compressible if and only pf is n-compressible.* □

(6.7.5) Proposition. *Let $n \ge 0$. The following assertions about (X, A) are equivalent:*

(1) *Each map $f\colon (I^q, \partial I^q) \to (X, A)$, $q \in \{0, \dots, n\}$ is relative to ∂I^q homotopic to a map into A.*

(2) *The inclusion $j\colon A \to X$ induces for each base point $a \in A$ a bijection $j_*\colon \pi_q(A, a) \to \pi_q(X, a)$ for $q < n$ and a surjection for $q = n$.*

(3) *$\pi_0(A) \to \pi_0(X)$ is surjective, and $\pi_q(X, A, a) = 0$ for $q \in \{1, \dots, n\}$ and each $a \in A$.*

Proof. (1) $\Leftrightarrow$ (3). The surjectivity of $\pi_0(A) \to \pi_0(X)$ is equivalent to (1) for $q = 0$. The other cases follow from (6.7.3).

(2) $\Leftrightarrow$ (3). This follows from the exact sequence (6.1.2). □

A pair (X, A) is called ***n*-connected** if (1)–(3) in (6.7.5) hold. We call (X, A) ***∞*-connected** if the pair is n-connected for each n. A pair is ∞-connected if and only if $j_*\colon \pi_n(A, a) \to \pi_n(X, a)$ is always bijective. If $X \ne \emptyset$ but $A = \emptyset$ we say that (X, A) is (-1)-connected, and $(\emptyset, \emptyset)$ is ∞-connected.

(6.7.6) Proposition. *Let $n \ge 0$. The following assertions about X are equivalent:*

(1) *$\pi_q(X, x) = 0$ for $0 \le q \le n$ and $x \in X$.*

(2) *The pair (CX, X) is $(n + 1)$-connected.*

(3) *Each map $f\colon \partial I^q \to X$, $0 \le q \le n + 1$ has an extension to I^q.*

Proof. The cone CX is contractible. Therefore $\partial\colon \pi_{q+1}(CX, X, *) \cong \pi_q(X, *)$. This and (6.7.5) shows the equivalence of (1) and (2). The equivalence of (1) and (3) uses (6.7.1). □

A space X is ***n*-connected** if (1)–(3) in (6.7.6) hold for X. Note that this is compatible with our previous definitions for $n = 0, 1$.

Let $f\colon X \to Y$ be a map and $X \subset Z(f)$ the inclusion into the mapping cylinder. Then f is said to be n-connected if $(Z(f), X)$ is n-connected. We then also say that f is an ***n*-equivalence**. Thus f is n-connected if and only if $f_*\colon \pi_q(X, x) \to \pi_q(Y, f(x))$ is for each $x \in X$ bijective (surjective) for $q < n$ $(q = n)$. If f is an ∞-equivalence we also say that f is a ***weak (homotopy) equivalence***. Thus f is a weak equivalence if and only if $f_*\colon \pi_n(X, x) \to \pi_n(Y, f(x))$ is bijective for each $n \geq 0$ and each $x \in X$.

(6.7.7) Proposition. *Let $(p_1, p_0)\colon (E_1, E_0) \to B$ be a relative Serre fibration. Let F_j^b denote the fibre of p_j over b. Then the following are equivalent:*

(1) *(E_1, E_0) is n-connected.*

(2) *(F_1^b, F_0^b) is n-connected for each $b \in B$.*

Proof. This is a direct consequence of (6.3.8). □

The compression properties of an n-connected map can be generalized to pairs of spaces which are regular unions of cubes of dimension at most n. We use this generalization in the proof of theorem (6.7.9). Consider a subdivision of a cube I^n. Let us call B a cube-complex if B is the union of cubes of this subdivision. A subcomplex A of B is then the union of a subset of the cubes in B. We understand that B and A contain with each cube all of its faces. The k-skeleton $B(k)$ of B consists of the cubes in B of dimension $\leq k$; thus $A(k) = B(k) \cap A$.

(6.7.8) Proposition. *Let $f\colon X \to Y$ be n-connected. Suppose (C, A) is a pair of cube-complexes of dimension at most n. Then to each commutative diagram*

$$\begin{array}{ccc} A & \xrightarrow{\varphi} & X \\ \big\downarrow{\scriptstyle\cap} & & \big\downarrow{\scriptstyle f} \\ C & \xrightarrow{\Phi} & Y \end{array}$$

there exists $\Psi\colon C \to X$ such that $\Psi|A = \varphi$ and $f\Psi \simeq \Phi$ relative to A.

Proof. Induction over the number of cubes. Let $A \subset B \subset C$ such that C is obtained from B by adding a cube W of highest dimension. Then $\partial W \subset B$. By induction there exists $\Psi'\colon B \to X$ such that $\Psi'|A = \varphi$ and a homotopy $H\colon f\Psi' \simeq \Phi|B$ relative to A. Extend H to a homotopy of Φ. The end Φ_1 of this homotopy satisfies $\Phi_1|B = f\Psi'$. We now use that f is n-connected and extend Ψ' over W to $\Psi\colon C \to X$ such that $f\Psi \simeq \Phi_1$ relative to B. Altogether we have $f\Psi \simeq_B \Phi_1 \simeq_A \Phi$ and $\Phi|A = \Psi'|A = \varphi$. □

(6.7.9) Theorem. *Let $\varphi\colon (X, X_0, X_1) \to (Y, Y_0, Y_1)$ be a map such that the restrictions $\varphi_i : X_i \to Y_i$ are n-connected and $\varphi_{01} : X_0 \cap X_1 \to Y_0 \cap Y_1$ is $(n-1)$-connected. Suppose $X = X_0^\circ \cup X_1^\circ$ and $Y = Y_0^\circ \cup Y_1^\circ$. Then φ is an n-equivalence.*

Proof. We use mapping cylinders to reduce to the case of inclusions $\varphi\colon X \subset Y, \varphi_i : X_i \subset Y_i$. Let $(F, f)\colon (I^n, \partial I^n) \to (Y, X)$ be given. We have to show that this map is homotopic relative to ∂I^n to a map into X. Let

$$A_i = F^{-1}(Y \smallsetminus Y_i^\circ) \cup f^{-1}(X \smallsetminus X_i^\circ).$$

These sets are closed and disjoint. By the Lebesgue lemma we choose a cubical subdivision of I^n such that no cube W of the subdivision intersects both A_0 and A_1. Let K_j be the union of the cubes W which satisfy

$$F(W) \subset Y_i^\circ, \quad f(W \cap \partial I^n) \subset X_i^\circ.$$

Then K_i is a cubical subcomplex and

$$I^n = K_0 \cup K_1, \quad F(K_i) \subset Y_i^\circ, \quad f(K_i \cap \partial I^n) \subset X_i^\circ.$$

We denote by $K^\bullet$ the $(n-1)$-skeleton of a cubical complex; then $K \cap \partial I^n = K^\bullet \cap \partial I^n$ and $K_0 \cap K_1^\bullet = K_0^\bullet \cap K_1^\bullet$. We have a commutative square

$$\begin{array}{ccc} X_{01} & \longrightarrow & Y_{01} \\ \uparrow f_{01} & \overset{g_{01}}{\nwarrow} & \uparrow F_{01} \\ \partial I^n \cap K_{01} & \longrightarrow & K_{01}^\bullet . \end{array}$$

Since (Y_{01}, X_{01}) is $(n-1)$-connected there exists a homotopy relative to $\partial I^n \cap K_{01}$ from F_{01} to a map $g_{01} : K_{01}^\bullet \to X_{01}$. Define $g_0 : K_0 \cap (\partial I^n \cup K_\bullet) \to X_0$ by

$$g_0 | K_0 \cap \partial I^n = f_0, \quad g_0 | K_0 \cap K_1^\bullet = g_{01}.$$

(Both maps agree on the intersection.) The homotopy $F_{01} \simeq g_{01}$ and the constant homotopy of f_0 combine to a homotopy of $F_0 | K_0 \cap (\partial I^n \cup K_1^\bullet)$ to g_0 which is constant on $K_0 \cap \partial I^n$. Since the inclusion of a cube complex into another one is a cofibration, this homotopy can be extended to a homotopy $\psi : K_0 \times I \to Y_0$ from F_0 to H_0. We obtain a diagram

$$\begin{array}{ccc} X_0 & \longrightarrow & Y_0 \\ \uparrow g_0 & \overset{h_0}{\nwarrow} & \uparrow H_0 \\ K_0 \cap (\partial I^n \cup K_1^\bullet) & \longrightarrow & K_0 \end{array}$$

where H_0 is homotopic to $h_0 : K_0 \to X_1$ relative to $K_0 \cap (\partial I^n \cup K_1^\bullet)$, since (Y_0, X_0) is n-connected.

We prove the second part similarly. We obtain a map $g_1 \colon K_1 \cap (\partial I^n \cup K_0^\bullet) \to Y_1$ with $g_1 | K_1 \cap \partial I^n = f_1$ and $g_1 | K_1 \cap K_0^\bullet = g_{01}$ and then

$$\begin{array}{ccc} X_1 & \longrightarrow & Y_1 \\ \uparrow{\scriptstyle g_1} & \overset{h_1}{\nwarrow} & \uparrow{\scriptstyle H_1} \\ K_1 \cap (\partial I^n \cup K_0^\bullet) & \longrightarrow & K_1. \end{array}$$

The maps h_0 and h_1 coincide on $K_{01}^\bullet$ and yield a map $h \colon K_0^\bullet \cup K_1^\bullet \to X$ which is homotopic relative ∂I^n to $F | K_0^\bullet \cup K_1^\bullet$; moreover $h | \partial I^n = f$. Let now W be an n-dimensional cube, say with $W \subset K_0$. Then $\partial W \subset K_0^\bullet$ and $h(\partial W) = h_0(\partial W) \subset X_0$. Since (Y_0, X_0) is n-connected, we can deform the map relative to ∂W to a map into X_0. □

(6.7.10) Corollary. *Let $f \colon X \to Y$ be an n-connected map between well-pointed spaces. Then $\Sigma f \colon \Sigma X \to \Sigma Y$ is $(n+1)$-connected. If X is n-connected, then ΣX is $(n+1)$-connected. The sphere S^{k+1} is k-connected.*

Proof. Let $\Sigma' X$ denote the unpointed suspension of X. This is a quotient of $X \times I$ and covered by the open cones $C_0 = X \times [0,1[/ X \times 0$ and $C_1 = X \times]0,1] / X \times 1$ with intersection $X \times]0,1[$. We can apply (6.7.7) directly; the cones are contractible and therefore the induced maps $C_j(X) \to C_j(Y)$ ∞-connected. In the case of a well-pointed space X the quotient map $\Sigma' X \to \Sigma X$ is an h-equivalence. □

(6.7.11) Theorem. *Let $f \colon X \to Y$ be a continuous map. Let $(U_j \mid j \in J)$ and $(V_j \mid j \in J)$ open coverings of X and Y such that $f(U_j) \subset V_j$. Suppose that for each finite $E \subset J$ the induced map $f_E \colon \bigcap_{j \in E} U_j \to \bigcap_{j \in E} V_j$ is a weak equivalence. Then f is a weak equivalence*

Proof. By passage to the mapping cylinder we can assume that f is an inclusion. Let $h \colon (I^n, \partial I^n) \to (Y, X)$ be given. We have to deform h relative to ∂I^n into X. By compactness of I^n it suffices to work with finite J. A simple induction reduces the problem to $J = \{0, 1\}$. Then we apply (6.7.9). □

Problems

1. Let $Y = \{0\} \cup \{n^{-1} \mid n \in \mathbb{N}\}$ and X the same set with the discrete topology. Then the identity $X \to Y$ is a weak equivalence but there does not exist a weak equivalence $Y \to X$.
2. Identify in S^1 the open sets $\{(x,y) \mid y > 0\}$ and $\{(x,y) \mid y < 0\}$ to a point. The quotient map $S^1 \to S$ onto the quotient space S, consisting of four points, is a weak equivalence (but not a homotopy equivalence). In particular $\pi_1(S) \cong \mathbb{Z}$. Show that S has a universal covering.

6.8 Classical Groups

We use exact sequences of Serre fibrations and deduce from our knowledge of $\pi_i(S^n)$ other results about homotopy groups of classical groups and Stiefel manifolds. We use a uniform notation for the (skew) fields $\mathbb{F} = \mathbb{R}, \mathbb{C}, \mathbb{H}$ and the corresponding groups (orthogonal, unitary, symplectic)

$$\begin{aligned} \mathrm{O}(n) &= \mathrm{O}(n,\mathbb{R}), \quad \mathrm{SO}(n) = \mathrm{SO}(n,\mathbb{R}), \\ \mathrm{U}(n) &= \mathrm{O}(n,\mathbb{C}), \quad \mathrm{SU}(n) = \mathrm{SO}(n,\mathbb{C}), \\ \mathrm{Sp}(n) &= \mathrm{O}(n,\mathbb{H}). \end{aligned}$$

Let $d = \dim_{\mathbb{R}} \mathbb{F}$. The starting point are the (Serre) fibrations which arise from the action of the orthogonal groups on the unit spheres by matrix multiplication

$$\mathrm{O}(n,\mathbb{F}) \xrightarrow{j} \mathrm{O}(n+1,\mathbb{F}) \to S^{d(n+1)-1},$$

$$\mathrm{SO}(n,\mathbb{F}) \xrightarrow{j} \mathrm{SO}(n+1,\mathbb{F}) \to S^{d(n+1)-1}.$$

The inclusions j of the groups arise from $A \mapsto \left(\begin{smallmatrix} A & 0 \\ 0 & 1 \end{smallmatrix}\right)$. We also pass to the colimit and obtain $\mathrm{O}(\infty,\mathbb{F}) = \mathrm{colim}_n\, \mathrm{O}(n,\mathbb{F})$ and $\mathrm{SO}(\infty,\mathbb{F}) = \mathrm{colim}_n\, \mathrm{SO}(n,\mathbb{F})$. From $\pi_i(S^n) = 0$, $i < n$ and the exact homotopy sequences of the fibrations we deduce that the inclusions $j\colon \mathrm{O}(n,\mathbb{F}) \to \mathrm{O}(n+1,\mathbb{F})$ and $j\colon \mathrm{SO}(n,\mathbb{F}) \to \mathrm{SO}(n+1,\mathbb{F})$ are $d(n+1)-2$ connected. By induction and passage to the colimit we obtain

(6.8.1) Proposition. *For $n < m \le \infty$, the inclusions $\mathrm{O}(n,\mathbb{F}) \to \mathrm{O}(m,\mathbb{F})$ and $\mathrm{SO}(n,\mathbb{F}) \to \mathrm{SO}(m,\mathbb{F})$ are $d(n+1)-2$ connected; in particular, the homomorphisms $\pi_i(\mathrm{O}(n,\mathbb{F})) \to \pi_i(\mathrm{O}(m,\mathbb{F}))$ are isomorphisms in the range $i \le n-2$ ($\mathbb{R}$), $i \le 2n-1$ ($\mathbb{C}$), and $i \le 4n+1$ ($\mathbb{H}$).* □

We turn our attention to Stiefel manifolds of orthonormal k-frames in $\mathbb{F}^n$:

$$\begin{aligned} V_k(\mathbb{R}^n) &\cong \mathrm{O}(n)/\mathrm{O}(n-k) \cong \mathrm{SO}(n)/\mathrm{SO}(n-k), \\ V_k(\mathbb{C}^n) &\cong \mathrm{U}(n)/\mathrm{U}(n-k) \cong \mathrm{SU}(n)/\mathrm{SU}(n-k), \\ V_k(\mathbb{H}^n) &\cong \mathrm{Sp}(n)/\mathrm{Sp}(n-k). \end{aligned}$$

We have the corresponding (Serre) fibrations of the type $H \to G \to G/H$ for these homogeneous spaces. We use (6.8.1) in the exact homotopy sequences of these fibrations and obtain:

(6.8.2) Proposition. *$\pi_i(V_k(\mathbb{F}^n)) = 0$ for $i \le d(n-k+1)-2$.* □

We have the fibration

$$p\colon V_{k+1}(\mathbb{F}^{n+1}) \to V_1(\mathbb{F}^{n+1}), \quad (v_1,\dots,v_{k+1}) \mapsto v_{k+1}.$$

The fibre over e_{k+1} is homeomorphic to $V_k(\mathbb{F}^n)$; with $\iota\colon v \mapsto (v,0)$ we obtain a homeomorphism $j\colon (v_1,\dots,v_k) \mapsto (\iota v_1,\dots,\iota v_k, e_{k+1})$ onto this fibre. From the homotopy sequence of this fibration we obtain

(6.8.3) Proposition. *$j_*\colon \pi_i(V_k(\mathbb{F}^n)) \to \pi_i(V_{k+1}(\mathbb{F}^{n+1}))$ is an isomorphism for $i \le d(n+1) - 3$.* □

We use $V_1(\mathbb{F}^n) = S^{dm-1}$ and $\pi_t(S^t) \cong \mathbb{Z}$ and obtain from (6.8.3) by induction

(6.8.4) Proposition. $\pi_{2(n-k)+1}(V_k(\mathbb{C}^n)) \cong \mathbb{Z}$, $\pi_{4(n-k)+3}(V_k(\mathbb{H}^n)) \cong \mathbb{Z}$. □

The real case is more complicated. The result is

(6.8.5) Proposition.

$$\pi_{n-k}(V_k(\mathbb{R}^n)) \cong \begin{cases} \mathbb{Z}, & k = 1, \text{ or } n-k \text{ even}, \\ \mathbb{Z}/2, & k \ge 2,\ n-k \text{ odd}. \end{cases}$$

Proof. By (6.8.3) and induction it suffices to consider the case $k = 2$. Later we compute the homology groups of $V_2(\mathbb{R}^n)$, and the theorem of Hurewicz will then give us the desired result. □

Problems

1. The group O(n) has two path components. The groups SO(n), U(n), SU(n), and Sp(n) are path connected.
2. In low dimensions we have some special situations, namely

$$\begin{aligned} &\mathrm{U}(1) \cong \mathrm{SO}(2) \cong S^1, \\ &\mathrm{Spin}(3) \cong \mathrm{SU}(2) \cong \mathrm{Sp}(1) \cong S^3, \\ &\mathbb{Z}/2 \to \mathrm{SU}(2) \to \mathrm{SO}(3), \text{ a 2-fold covering}, \\ &\mathrm{SU}(n) \to \mathrm{U}(n) \to S^1, \text{ a fibration}. \end{aligned}$$

Use these data in order to verify

$$\begin{aligned} &\pi_1(\mathrm{SO}(2)) \cong \pi_1(\mathrm{O}(2)) \cong \mathbb{Z}, \\ &\pi_1(\mathrm{SO}(3)) \cong \pi_1(\mathrm{SO}(n)) \cong \mathbb{Z}/2, && n \ge 3, \\ &\pi_1(\mathrm{U}(1)) \cong \pi_1(\mathrm{U}(n)) \cong \mathbb{Z}, && n \ge 1, \\ &\pi_1(\mathrm{SU}(n)) \cong \pi_1(\mathrm{Sp}(n)) \cong 0, && n \ge 1, \\ &\pi_2(\mathrm{SU}(n)) \cong \pi_2(\mathrm{U}(n)) \cong \pi_2(\mathrm{Sp}(n)) \cong 0, && n \ge 1, \\ &\pi_2(\mathrm{SO}(n)) \cong 0, && n \ge 3, \\ &\pi_3(\mathrm{U}(2)) \cong \pi_3(\mathrm{U}(k)) \cong \mathbb{Z}, && k \ge 2, \\ &\pi_3(\mathrm{SU}(2)) \cong \pi_3(\mathrm{SU}(k)) \cong \mathbb{Z}, && k \ge 2, \\ &\pi_3(\mathrm{Sp}(1)) \cong \pi_3(\mathrm{Sp}(k)) \cong \mathbb{Z}, && k \ge 1, \\ &\pi_3(\mathrm{SO}(3)) \cong \mathbb{Z}. \end{aligned}$$

6.9 Proof of the Excision Theorem

In this section we present an elementary proof of the excision theorem (6.4.1). The proof is due to D. Puppe [46]. We derive the excision theorem from a more conceptual reformulation (6.9.3). The reformulation is more satisfactory, because it is "symmetric" in Y_1, Y_2. In (6.4.1) we have a second conclusion with the roles of Y_1 and Y_2 interchanged.

We begin with a technical lemma used in the proof.

A cube in $\mathbb{R}^n$, $n \geq 1$ will be a subset of the form

$$W = W(a, \delta, L) = \{x \in \mathbb{R}^n \mid a_i \leq x_i \leq a_i + \delta \text{ for } i \in L, a_i = x_i \text{ for } i \notin L\}$$

for $a = (a_1, \dots, a_n) \in \mathbb{R}^n$, $\delta > 0$, $L \subset \{1, \dots, n\}$. ($L$ can be empty.) We set $\dim W = |L|$. A face of W is a subset of the form

$$W' = \{x \in W \mid x_i = a_i \quad \text{for } i \in L_0,\ x_j = a_j + \delta \quad \text{for } j \in L_1\}$$

for some $L_0 \subset L$, $L_1 \subset L$. (W' can be empty.) Let ∂W denote the union of all faces of W which are different from W. We use the following subsets of $W = W(a, \delta, L)$:

$$K_p(W) = \{x \in W \mid x_i < a_i + \tfrac{\delta}{2} \text{ for at least } p \text{ values } i \in L\},$$
$$G_p(W) = \{x \in W \mid x_i > a_i + \tfrac{\delta}{2} \text{ for at least } p \text{ values } i \in L\}.$$

Here $1 \leq p \leq n$. For $p > \dim W$ we let $K_p(W)$ and $G_p(W)$ be the empty set.

(6.9.1) Lemma. *Let $f\colon W \to Y$ and $A \subset Y$ be given. Suppose that for $p \leq \dim W$ the inclusions*

$$f^{-1}(A) \cap W' \subset K_p(W') \quad \textit{for all } W' \subset \partial W$$

hold. Then there exists a map g which is homotopic to f relative to ∂W such that $g^{-1}(A) \subset K_p(W)$. (Similarly for G_p in place of K_p.)

Proof. We can assume that $W = I^n$, $n \geq 1$. We define $h\colon I^n \to I^n$ in the following manner: Let $x = (\frac{1}{4}, \dots, \frac{1}{4})$. For a ray y which begins in x we consider its intersection $P(y)$ with $\partial[0, \frac{1}{2}]^n$ and its intersection $Q(y)$ with ∂I^n. Let h map the segment from $P(y)$ to $Q(y)$ onto the single point $Q(y)$ and the segment from x to $P(y)$ affinely to the segment from x to $Q(y)$. Then h is homotopic relative to ∂I^n to the identity. We set $g = fh$. Let $z \in I^n$ and $g(z) \in A$. If $z_i < \frac{1}{2}$ for all i, then $z \in K_n(I^n) \subset K_p(I^n)$. Suppose now that for at least one i we have $z_i \geq \frac{1}{2}$, then $h(z) \in \partial I^n$ and hence $h(z) \in W'$ for some face W' with $\dim W' = n - 1$. Since also $h(z) \in f^{-1}(A)$, by assumption $h(z) \in K_p(W')$. Hence we have for at least p coordinates $\frac{1}{2} > h(z)_i$. By definition of h, we have $h(z)_i = \frac{1}{4} + t(z_i - \frac{1}{4})$ with $t \geq 1$. We conclude that for at least p coordinates $\frac{1}{2} > z_i$. □

The next theorem is the basic technical result. In it we deform a map $I^n \to Y$ into a kind of normal form. We call it the ***preparation theorem***. Let Y be the union of open subspaces Y_1, Y_2 with non-empty intersection Y_0. Let $f: I^n \to Y$ be given. By the Lebesgue lemma (2.6.4) there exists a subdivision of I^n into cubes W such that either $f(W) \subset Y_1$ or $f(W) \subset Y_2$ for each cube. In this situation we claim:

(6.9.2) Theorem. *Suppose (Y_1, Y_0) is p-connected and (Y_2, Y_0) is q-connected $(p, q \geq 0)$. Then there exists a homotopy f_t of f with the following properties:*

(1) *If $f(W) \subset Y_j$, then $f_t(W) \subset Y_j$.*

(2) *If $f(W) \subset Y_0$, then f_t is constant on W.*

(3) *If $f(W) \subset Y_1$, then $f_1^{-1}(Y_1 \smallsetminus Y_0) \cap W \subset K_{p+1}(W)$.*

(4) *If $f(W) \subset Y_2$, then $f_1^{-1}(Y_2 \smallsetminus Y_0) \cap W \subset G_{q+1}(W)$.*

Here W is any cube of the subdivision.

Proof. Let C^k be the union of the cubes W with $\dim W \leq k$. We construct the homotopy inductively over $C^k \times I$.

Let $\dim W = 0$. If $f(W) \subset Y_0$ we use condition (2). If $f(W) \subset Y_1$, $f(W) \not\subset Y_2$, there exists a path in Y_1 from $f(W)$ to a point in Y_0, since (Y_1, Y_0) is 0-connected. We use this path as our homotopy on W. Then (1) and (3) hold. Similarly if $f(W) \subset Y_2$, $f(W) \not\subset Y_1$. Thus we have found a suitable homotopy on C^0. We extend this homotopy to the higher dimensional cubes by induction over the dimension; we use that $\partial W \subset W$ is a cofibration, and we take care of (1) and (2).

Suppose we have changed f by a homotopy such that (1) and (2) hold and (3), (4) for cubes of dimension less than k. Call this map again f. Let $\dim W = k$. If $f(W) \subset Y_0$, we can use (2) for our homotopy. Let $f(W) \subset Y_1$, $f(W) \not\subset Y_2$. If $\dim W \leq p$ there exists a homotopy $f_t^W : W \to Y_1$ relative to ∂W of $f|W$ with $f_1^W(W) \subset Y_0$, since (Y_1, Y_0) is p-connected. If $\dim W > p$ we use (6.9.1) in order to find a suitable homotopy of $f|W$. We treat the case $f(W) \subset Y_2$, $f(W) \not\subset Y_1$ in a similar manner. Again we extend the homotopy to the higher dimensional cubes. This finishes the induction step. □

Let us denote by $F(Y_1, Y, Y_2)$ the path space $\{w \in Y^I \mid w(0) \in Y_1, w(1) \in Y_2\}$. We have the subspace $F(Y_1, Y_1, Y_0)$.

(6.9.3) Theorem. *Under the hypothesis of the previous theorem the inclusion $F(Y_1, Y_1, Y_0) \subset F(Y_1, Y, Y_2)$ is $(p+q-1)$-connected.*

Proof. Let a map $\varphi: (I^n, \partial I^n) \to (F(Y_1, Y, Y_2), F(Y_1, Y_1, Y_0))$ be given where $n \leq p + q - 1$. We have to deform this map of pairs into the subspace. By adjunction, a map of this type corresponds to a map $\Phi: I^n \times I \to Y$ with the following properties:

(1) $\Phi(x,0) \in Y_1$ for $x \in I^n$,
(2) $\Phi(x,1) \in Y_2$ for $x \in I^n$,
(3) $\Phi(y,t) \in Y_1$ for $y \in \partial I^n$ and $t \in I$.

Let us call maps of this type admissible. The claim of the theorem is equivalent to the statement, that Φ can be deformed as an admissible map into a map with image in Y_1. We apply the preparation theorem to Φ and obtain a certain map Ψ. The deformation in (6.9.2) stays inside admissible maps. Consider the projection $\pi\colon I^n \times I \to I^n$. We claim that the images of $\Psi^{-1}(Y \smallsetminus Y_1)$ and $\Psi^{-1}(Y \smallsetminus Y_2)$ under π are disjoint. Let $y \in \pi\Psi^{-1}(Y \smallsetminus Y_2)$, $y = \pi(z)$ and $z \in \Psi^{-1}(Y \smallsetminus Y_2) \cap W$ for a cube W. Then $z \in K_{p+1}(W)$ and hence y has at least p small coordinates. In a similar manner we conclude from $y \in \pi\Psi^{-1}(Y \smallsetminus Y_1)$ that y has at least q large coordinates. In the case that $n < p+q$ the point y cannot have p small and q large coordinates.

The set $\pi\Psi^{-1}(Y \smallsetminus Y_1)$ is disjoint to ∂I^n, since $\Psi(\partial I^n \times I) \subset Y_1$. There exists a continuous function $\tau\colon I^n \to I$ which assumes the value 0 on $\pi\Psi^{-1}(Y \smallsetminus Y_1)$ and the value 1 on $\partial I^n \cup \pi\Psi(Y \smallsetminus Y_2)$. The homotopy

$$((x,t),s) \mapsto \Psi(x,(1-s)t + st\tau(x))$$

is a homotopy of admissible maps from Ψ to a map with image in Y_1. □

(6.9.4) Theorem. *Under the hypothesis of* (6.9.2) *the inclusion induces an isomorphism* $\pi_j(Y_1,Y_0) \to \pi_j(Y,Y_2)$ *for* $j < p+q$ *and an epimorphism for* $j = p+q$.

Proof. We have the path fibration $F(Y,Y,Y_2) \to Y$, $w \mapsto w(0)$. The pullback along $Y_1 \subset Y$ yields the fibration $F(Y_1,Y,Y_2) \to Y_1$, $w \mapsto w(0)$. The fibre over $*$ is $F(*,Y,Y_2)$. We obtain a commutative diagram of fibrations:

$$\begin{array}{ccc} F(*,Y_1,Y_0) & \xrightarrow{\beta} & F(*,Y,Y_2) \\ \downarrow & & \downarrow \\ F(Y_1,Y_1,Y_0) & \xrightarrow{\alpha} & F(Y_1,Y,Y_2) \\ \downarrow & & \downarrow \\ Y_1 & \xrightarrow{=} & Y_1. \end{array}$$

The inclusion α is $(p+q-1)$-connected (see (6.9.3)). Hence β has the same connectivity (see (6.7.8)), i.e., the inclusion $(Y_1,Y_0) \subset (Y,Y_2)$,

$$\begin{array}{ccc} \pi_n(F(*,Y_1,Y_0)) & \longrightarrow & \pi_n(F(*,Y,Y_2)) \\ \downarrow\cong & & \downarrow\cong \\ \pi_{n+1}(Y_1,Y_0,*) & \longrightarrow & \pi_{n+1}(Y,Y_2,*) \end{array}$$

induces an isomorphism for $n < p+q-1$ and an epimorphism for $n = p+q-1$.
□

Problems

1. The hypothesis of (6.4.1) is a little different from the hypothesis of (6.9.4), since we did not assume in (6.4.1) that (Y_1, Y_0) and (Y_2, Y_0) are 0-connected. Let Y' be the subset of points that can be connected by a path to Y_0. Show that Y' has the open cover Y_1', Y_2' and the inclusion induces isomorphisms $\pi_*(Y_1', Y_0) \cong \pi_*(Y_1, Y_0)$ and $\pi_*(Y, Y_2) \cong \pi_*(Y', Y_2')$. This reduces (6.4.1) to (6.9.4).
2. The map $Y_0 \to F(Y_1, Y_1, Y_0)$ which sends $y \in Y_0$ to the constant path with value y is an h-equivalence.
3. The map $a_1\colon F(Y, Y, Y_1) \to Y, w \mapsto w(0)$ replaces the inclusion $Y_1 \to Y$ by a fibration. There is a pullback diagram

$$\begin{array}{ccc} F(Y_1, Y, Y_2) & \longrightarrow & F(Y, Y, Y_1) \\ \downarrow & & \downarrow a_1 \\ F(Y, Y, Y_2) & \xrightarrow{a_2} & Y. \end{array}$$

Thus (6.9.3) compares the pushout Y of $Y_1 \leftarrow Y_0 \rightarrow Y_2$ and the pullback of a_1, a_2 with Y_0. For generalizations see [73].
4. Show that the proof of (6.4.2) along the lines of this section does not need (6.9.1).

6.10 Further Applications of Excision

The excision theorem is a fundamental result in homotopy theory. For its applications it is useful to verify that it holds under different hypotheses. In the next proposition we show and use that Y is the homotopy pushout.

(6.10.1) Proposition. *Let a pushout diagram be given with a cofibration j,*

$$\begin{array}{ccc} A & \xrightarrow{f} & B \\ \downarrow j & & \downarrow J \\ X & \xrightarrow{F} & Y. \end{array}$$

Suppose $\pi_i(X, A, a) = 0$ for $0 < i < p$ and each $a \in A$, and $\pi_i(f, a) = 0$ for $0 < i < q$ and each $a \in A$. Then the map $(F, f)_*\colon \pi_n(X, A, a) \to \pi_n(Y, B, f(a))$ is surjective for $1 \le n \le p + q - 2$ and bijective for $1 \le n < p + q - 2$.

Proof. We modify the spaces up to h-equivalence such that (6.4.1) can be applied. Let $Z(f) = B \cup_f A \times [0,1] = B + A \times [0,1]/f(a) \sim (a, 0)$ be the mapping cylinder of f with inclusion $k\colon A \to Z(f)$, $a \mapsto (a, 1)$ and projection $p\colon Z(f) \to B$ a homotopy equivalence. We form the pushout diagrams

$$\begin{array}{ccccc} A & \xrightarrow{k} & Z(f) & \xrightarrow{p} & B \\ \downarrow j & & \downarrow L & & \downarrow J \\ X & \xrightarrow{K} & Z & \xrightarrow{P} & Y \end{array}$$

with $pk = f$ and $PK = F$. Then P is a homotopy equivalence by (5.1.10) and (P, p) induces an isomorphism of homotopy groups. Therefore it suffices to analyze $(K, k)_*$. The space Z can be constructed as

$$Z = B \cup_f A \times [0,1] \cup X = Z(f) + X/(a,1) \sim a.$$

The map (K, k) is the composition of

$$(X, A, a) \to (A \times]0,1] \cup X, A \times]0,1], (a,1)), \quad x \mapsto (x, 1)$$

with the inclusion ι into $(Z, Z(f), (a, 1))$. The first map induces an isomorphism of homotopy groups, by homotopy equivalence. In order to exhibit $\pi_n(\iota)$ as an isomorphism, we can pass to the base point $(a, 1/2)$, by naturality of transport. With this base point we have a commutative diagram

$$\begin{array}{ccc} \pi_n(A\times]0,1] \cup X, A\times]0,1]) & \longrightarrow & \pi_n(Z, Z(f)) \\ \uparrow & & \uparrow \\ \pi_n(A\times]0,1] \cup X, A\times]0,1[) & \longrightarrow & \pi_n(Z, B \cup A \times [0,1[). \end{array}$$

The vertical maps are isomorphism, by homotopy invariance. We apply (6.4.1) to the bottom map. Note that $\pi_i(A\times]0,1] \cup X, A\times]0,1]) \cong \pi_i(X, A)$ and

$$\pi_i(B \cup A \times [0,1[, A\times]0,1[) \cong \pi_i(Z(f), A),$$

again by homotopy invariance. □

(6.10.2) Theorem (Quotient theorem). *Let $A \subset X$ be a cofibration. Let further $p\colon (X, A) \to (X/A, *)$ be the map which collapses A to a point. Suppose that for each base point $a \in A$,*

$$\pi_i(CA, A, a) = 0 \ \textit{for}\ 0 < i < m, \quad \pi_i(X, A, a) = 0 \ \textit{for}\ 0 < i < n.$$

Then $p_*\colon \pi_i(X, A, a) \to \pi_i(X/A, *)$ is bijective for $0 < i < m + n - 2$ and surjective for $i = n + m - 2$. (See also Problem 14.)

Proof. By pushout excision, $\pi_i(X, A) \to \pi_i(X \cup CA, CA)$ is bijective (surjective) in the indicated range. Note that $\partial\colon \pi_i(CA, A, a) \cong \pi_{i-1}(A, a)$, so that the first hypothesis is a property of A. The inclusion $CA \subset X \cup CA$ is an induced cofibration. Since CA is contractible, the projection $p\colon X \cup CA \to X \cup CA/CA \cong X/A$ is a homotopy equivalence. □

(6.10.3) Corollary. *Let $A \subset X$ be a cofibration. Assume that $\pi_i(A) = 0$ for $0 \le i \le m - 1$ and $\pi_i(X) = 0$ for $0 \le i \le m \ge 2$. Then $\pi_i(X, A) \to \pi_i(X/A)$*

is an isomorphism for $0 < i \le 2m-1$. We use this isomorphism in the exact sequence of the pair (X,A) and obtain an exact sequence

$$\begin{aligned}\pi_{2m-1}(A) \to \pi_{2m-1}(X) \to \pi_{2m-1}(X/A) \to \pi_{2m-2}(A)\\ \to \cdots \to \pi_{m+1}(X) \to \pi_{m+1}(X/A) \to \pi_m(A) \to 0.\end{aligned}$$

A similar exact sequence exists for an arbitrary pointed map $f: A \to X$ between well-pointed spaces where a typical portion comes from the cofibre sequence $\pi_i(A) \xrightarrow{f_*} \pi_i(X) \xrightarrow{f_{1*}} \pi_i(C(f))$. (See also Problem 15.) □

We now generalize the suspension theorem. Let $(X,*)$ be a pointed space. Recall the suspension ΣX and the homomorphism $\Sigma_*: \pi_n(X) \to \pi_{n+1}(\Sigma X)$.

(6.10.4) Theorem. *Let X be a well-pointed space. Suppose $\pi_i(X) = 0$ for $0 \le i \le n$. Then $\Sigma_*: \pi_j(X) \to \pi_{j+1}(\Sigma X)$ is bijective for $0 \le j \le 2n$ and surjective for $j = 2n+1$.*

Proof. Let $CX = X \times I/(X \times 0 \cup \{*\} \times I)$ be the cone on X. We have an embedding $i: X \to CX$, $x \mapsto [x,0]$ which we consider as an inclusion. The quotient CX/X can be identified with ΣX. From the assumption that $\{*\} \subset X$ is a cofibration one concludes that i is a cofibration (Problem 6). Since CX is contractible, the exact sequence of the pair (CX,X) yields an isomorphism $\partial: \pi_{j+1}(CX,X) \simeq \pi_j(X)$. The inverse isomorphism sends an element represented by $f: I^n \to X$ to the element represented by $f \times \mathrm{id}(I)$. From this fact we see

$$\Sigma_* = p_* \circ \partial^{-1}: \pi_j(X) \xleftarrow{\partial} \pi_{j+1}(CX,X) \xrightarrow{p_*} \pi_{j+1}(\Sigma X),$$

with the quotient map $p: CX \to CX/X = \Sigma X$. We can therefore prove the theorem by showing that p_* is bijective or surjective in the same range. This follows from the quotient theorem (6.10.2). □

(6.10.5) Theorem. *Let X and Y be well-pointed spaces. Assume $\pi_i(X) = 0$ for $i < p$ (≥ 2) and $\pi_i(Y) = 0$ for $i < q$ (≥ 2). Then the inclusion $X \vee Y \to X \times Y$ induces an isomorphism of the π_i-groups for $i \le p+q-2$. The groups $\pi_i(X \times Y, X \vee Y)$ and $\pi_i(X \wedge Y)$ are zero for $i \le p+q-1$.*

Proof. We first observe that $j: \pi_i(X \vee Y) \to \pi_i(X \times Y)$, induced by the inclusion, is always surjective. The projections onto the factors induce isomorphisms $k: \pi_i(X \times Y) \cong \pi_i(X) \times \pi_i(Y)$. Let $j^X: X \to X \vee Y$ and $j^Y: Y \to X \vee Y$ denote the inclusions. Let

$$s: \pi_i(X) \times \pi_i(Y) \to \pi_i(X \vee Y), \quad (x,y) \mapsto j_*^X(x) + j_*^Y(y).$$

Then sk is right inverse to j. Hence the exact sequence of the pair $(X \times Y, X \vee Y)$ yields an exact sequence

$$(*) \qquad 0 \to \pi_{i+1}(X \times Y, X \vee Y) \to \pi_i(X \vee Y) \to \pi_i(X \times Y) \to 0.$$

In the case that $i \geq 2$, the sequence splits, since we are then working with abelian groups; hence

$$\pi_i(X \vee Y) \cong \pi_i(X) \oplus \pi_i(Y) \oplus \pi_{i+1}(X \times Y, X \vee Y), \quad i \geq 2.$$

Since the spaces are well-pointed, we can apply the theorem of Seifert–van Kampen to $(X \vee Y, X, Y)$ and see that $\pi_1(X \vee Y) = 0$. We now consider the diagram

$$\begin{array}{ccccc} & \overset{(1)}{\nearrow} & \pi_i(X \vee Y, Y) & & \\ & & \uparrow & & \\ \pi_i(X) & \longrightarrow & \pi_i(X \vee Y) & \longrightarrow & \pi_i(X \vee Y, X) \\ & & \uparrow & \overset{(2)}{\nearrow} & \\ & & \pi_i(Y) & & \end{array}$$

with exact row and column. The diagonal arrows are always injective and split; this is seen by composing with the projections.

Since the spaces are well-pointed, we can apply the pushout excision to the triad $(X \vee Y, X, Y, *)$. It says that (1) and (2) are surjective for $i \leq p+q-2$, and hence bijective (since we already know the injectivity).

We now apply the Sum Lemma (11.1.2) to the diagram and conclude that $\langle j_*^X, j_*^Y \rangle$ is an isomorphism, and therefore also the map of the theorem is an isomorphism. The exact sequence now yields $\pi_i(X \times Y, X \vee Y) = 0$ for $i \leq p+q-1$.

We apply (6.10.2) to $\pi_i(X \times Y, X \vee Y) \to \pi_i(X \wedge Y)$. By what we have already proved, we can apply this theorem with the data $n = p + q - 1$ and $m = \min(p-1, q-1)$. We also need that $X \vee Y \to X \times Y$ is a cofibration. This is a consequence of the product theorem for cofibrations. □

(6.10.6) Proposition. *Let $(Y_j \mid j \in J)$ be the family of path components of Y and $c^j : Y_j \to Y$ the inclusion. Then*

$$\langle c_*^j \rangle \colon \bigoplus_{j \in J} \pi_k(Y_j^+ \wedge S^n) \to \pi_k(Y^+ \wedge S^n)$$

is an isomorphism for $k \leq n \geq 2$.

Proof. Suppose Y is the topological sum of its path components. Then we have a homeomorphism $Y^+ \wedge S^n \cong \bigvee_{j \in J} Y_j^+ \wedge S^n$, and the assertion follows for finite J by induction on the cardinality of J from (6.10.5) and for general J then by a compactness argument. For general Y it suffices to find a 1-connected map $X \to Y$ such that X is the topological sum of its path components, because then $X^+ \wedge S^n \to Y^+ \wedge S^n$ is $(n+1)$-connected by (6.7.10) (and similarly for the path components). □

(6.10.7) Proposition. *Let Y be k-connected ($k \geq 0$) and Z be l-connected ($l \geq -1$) and well-pointed. Then the natural maps*

$$\pi_j(Z) \to \pi_j(Y \times Z, Y \times *) \to \pi_j(Y \times Z/Y \times *) \to \pi_j(Z)$$

are isomorphisms for $0 < j \leq k + l + 1$.

Proof. The first map is always bijective for $j \geq 1$; this is a consequence of the exact sequence of the pair $(Y \times Z, Y \times *)$ and the isomorphism $\pi_j(Y) \times \pi_j(Z) \cong \pi_j(Y \times Z)$. Since the composition of the maps is the identity, we see that the second map is always injective and the third one surjective. Thus if $p_* \colon \pi_j(Y \times Z, Y \times *) \to \pi_j(Y \times Z/Y \times *)$ is surjective, then all maps are bijective. From our assumption about Z we conclude that $\pi_j(Y \times Z, Y \times *) = 0$ for $0 < j \leq l$ (thus there is no condition for $l = 0, -1$). The quotient theorem now tells us that p_* is surjective for $0 < j \leq k + l + 1$. □

(6.10.8) Corollary. *Let Y be path connected. The natural maps*

$$\pi_k(S^n) \to \pi_k(Y \times S^n, Y \times *) \to \pi_k(Y \times S^n/Y \times *) \to \pi_k(S^n)$$

are isomorphisms for $1 \leq k \leq n$. □

(6.10.9) Proposition. *Suppose $\pi_i(X) = 0$ for $i < p$ (≥ 0) and $\pi_i(Y) = 0$ for $i < q$ (≥ 0). Then $\pi_i(X \star Y) = 0$ for $i < p + q + 1$.*

Proof. In the case that $p = 0$ there is no condition on X. From the definition of the join we see that $X \star Y$ is always path connected. For $p = 0$ we claim that $\pi_i(X \star Y) = 0$ for $i < q + 1$. Consider the diagram

$$\begin{array}{ccccc} X & \xleftarrow{\text{pr}} & X \times Y & \xrightarrow{\text{pr}} & Y \\ \downarrow & & \downarrow{\scriptstyle \text{pr}} & & \downarrow \\ X & \longleftarrow & X & \longrightarrow & \{*\} \end{array}$$

and apply (6.7.9). In the general case the excision theorem says that the map $\pi_i(CX \times Y, X \times Y) \to \pi_i(X \star Y, X \times CY)$ is an epimorphism for $i < p + q + 1$. Now use diagram chasing in the diagram

$$\begin{array}{ccccccc} \pi_i(X \times CY) & \longrightarrow & \pi_i(X \star Y) & \longrightarrow & \pi_i(X \star Y, X \times CY) & \longrightarrow & \pi_{i-1}(X \times CY) \\ \uparrow & & \uparrow & & \uparrow & & \uparrow \\ \pi_i(X \times Y) & \longrightarrow & \pi_i(CX \times Y) & \longrightarrow & \pi_i(CX \times Y, X \times Y) & \longrightarrow & \pi_{i-1}(X \times Y) \end{array}$$

(a morphism between exact homotopy sequences). □

The excision theorem in the formulation of (6.9.4) has a dual. Suppose given a pullback diagram

$$\begin{array}{ccc} E & \xrightarrow{F} & X \\ \downarrow{\scriptstyle G} & & \downarrow{\scriptstyle f} \\ Y & \xrightarrow[g]{} & B \end{array}$$

with fibrations f and g. The double mapping cylinder $Z(F,G)$ can be considered as the fibrewise join of f and g. It has a canonical map $\rho\colon Z(F,G) \to B$.

(6.10.10) Proposition. *Suppose f is p-connected and g is q-connected. Then ρ is $p+q+1$-connected.*

Proof. Use fibre sequences and (6.10.9). □

Problems

1. Let $\pi_0(X) = 0$ and $\pi_i(Y) = 0$ for $i < q(\geq 2)$. Then $\pi_i(X) \to \pi_i(X \vee Y)$ is an isomorphism for $i < q$. Show also $\pi_2(X \times Y, X \vee Y) = 0$.
2. Let X and Y be 0-connected and well-pointed. Show $\pi_1(X \wedge Y) = 0$.
3. Show that $\pi_3(D^2, S^1) \to \pi_3(D^2/S^1)$ is not surjective.
4. Show $\pi_1(S^2 \vee S^1, S^1) = 0$. Show that $\pi_2(S^2 \vee S^1, S^1) \to \pi_2(S^2 \vee S^1/S^1) \cong \pi_2(S^2)$ is surjective but not injective.
5. For $X = Y = S^1$ and $i = 1$ the sequence $(*)$ does not split. The fundamental group $\pi_1(S^1 \vee S^1) = \mathbb{Z} * \mathbb{Z}$ has no subgroup isomorphic to $\mathbb{Z} \oplus \mathbb{Z}$.
6. Show that the diagram

$$\begin{array}{ccc} X \times \partial I \cup \{*\} \times I & \xrightarrow{p} & X \\ \downarrow{\scriptstyle \cap} & & \downarrow{\scriptstyle i} \\ X \times I & \longrightarrow & CX \end{array}$$

with $p(x,0) = x$, $p(x,1) = *$, $p(*,t) = *$ is a pushout.
7. If X is well-pointed, then ΣX is well-pointed.
8. Some hypothesis like e.g. well-pointed is necessary in both (6.10.1) and (6.10.4). Let $A = \{0\} \cup \{n^{-1} \mid n \in \mathbb{N}\}$ and $A = A \times 0 \subset X = A \times I / A \times 1$ with base point $(0,0)$. Then $\pi_1(\Sigma X)$ and $\pi_1(\Sigma A)$ are uncountable; $\Sigma_*\colon \pi_0(X) \to \pi_1(\Sigma X)$ is not surjective. Note: $A \subset X$ is a cofibration and X/A is well-pointed. The inclusions $\{0\} = \{*\} \to A$ and $A \to CA$ are not cofibrations, ΣA is not well-pointed.
9. Let $e_1, \ldots, e_{n+1}$ be the standard basis of unit vectors in $\mathbb{R}^{n+1}$, and let e_1 be the base point of S^n. A pointed homeomorphism $h_n\colon \Sigma S^n \cong S^{n+1}$ is

$$h_n\colon \Sigma S^n \to S^{n+1},\ (x,t) \mapsto \tfrac{1}{2}(e_1 + x) + \tfrac{1}{2}\cos 2\pi t \cdot (e_1 - x) + \tfrac{1}{2}\,|e_1 - x| \sin 2\pi t \cdot e_{n+2}$$

where $\mathbb{R}^{n+1} = \mathbb{R}^{n+1} \times 0 \subset \mathbb{R}^{n+2}$.
10. Let $K \subset \mathbb{R}^{n+1}$ be compact. Show that each map $f\colon K \to S^n$ has an extension to the complement $\mathbb{R}^{n+1} \smallsetminus E$ of a finite set E. One can choose E such that each component of $\mathbb{R}^{n+1} \smallsetminus K$ contains at most one point of E.

11. Determine $\pi_{2n-1}(S^n \vee S^n)$ for $n \geq 2$.
12. Let f_j be a self-map of $S^{n(j)}$. Show $d(f_1 \star f_2) = d(f_1)d(f_2)$.
13. Let $H\colon \pi_3(S^2) \to \mathbb{Z}$ be the isomorphism which sends (the class of the) Hopf map $\eta\colon S^3 \to S^2$ to 1 (the Hopf invariant). Show that for $f: S^3 \to S^3$ and $g\colon S^2 \to S^2$ the relations $H(\alpha \circ f) = d(f)H(\alpha)$ and $H(g \circ \alpha) = d(g)^2 H(\alpha)$ hold.
14. Concerning the quotient theorem: If $m = 1$ there is no assumption about A. For $m \geq 2$ the assumption is equivalent to $\pi_j(A, a) = 0$ for $0 \leq j \leq m - 2$. For $m = 1$ the theorem then asserts that $\pi_i(X/A, *) = 0$ for $1 \leq i \leq n-1$. For $n = 1$ there is no assumption about (X, A). In that case $\pi_i(X, A, a) \to \pi_i(X/A, *)$ is bijective (surjective) for $1 \leq i \leq m - 2$ $(i = m - 1)$.
15. Let $f: A \to X$ be a pointed map between well-pointed spaces. Then the canonical inclusion $A \to Z(f)$ into the pointed mapping cylinder is a cofibration. For the proof consider the diagram

$$\begin{array}{ccccc} A \times 0 \cup \{*\} \times I \cup A \times 1 & \longrightarrow & A \vee A & \longrightarrow & A \vee X \\ \downarrow{\scriptstyle (1)} & & \downarrow{\scriptstyle (2)} & & \downarrow{\scriptstyle (3)} \\ A \times I & \longrightarrow & A \times I/\{*\} \times I & \longrightarrow & Z(f) \end{array}$$

in which the squares are pushout diagrams. If A is well-pointed, (1) is a cofibration and (2), (3) are induced cofibration. If X is well-pointed, the canonical inclusion $A \to A \vee X$ is a cofibration.

Chapter 7

Stable Homotopy. Duality

The suspension theorem of Freudenthal indicates that homotopy theory simplifies by use of iterated suspensions. We use this idea to construct the simplest stable homotopy category. Its construction does not need extensive technical considerations, yet it has interesting applications. The term "stable" refers to the fact that iteration of suspension induces after a while a bijection of homotopy classes.

We use the stable category to give an introduction to homotopical duality theory. In this theory the stable homotopy type of a closed subspace $X \subset \mathbb{R}^n$ and its complement $\mathbb{R}^n \smallsetminus X$ are compared. This elementary treatment of duality theory is based on ideas of Albrecht Dold and Dieter Puppe; see in particular [54]. It is related to the classical Alexander duality of homology theory and to Spanier–Whitehead duality.

We introduce a naive form of spectra and use them to define spectral homology and cohomology theories. The homotopical Euclidean complement duality is then used to give a simple proof for the Alexander duality isomorphism. In a later chapter we reconsider duality theory in the context of product structures.

7.1 A Stable Category

Pointed spaces X and Y are called ***stably homotopy equivalent***, in symbols $X \simeq_s Y$, if there exists an integer $k \geq 0$ such that the suspensions $\Sigma^k X$ and $\Sigma^k Y$ are homotopy equivalent. Pointed maps $f, g\colon X \to Y$ are called **stably homotopic**, in symbols $f \simeq_s g$, if for some integer k the suspensions $\Sigma^k f$ and $\Sigma^k g$ are homotopic. We state some of the results to be proved in this chapter which use these notions.

(7.1.1) Theorem (Stable complement theorem). *Let X and Y be homeomorphic closed subsets of the Euclidean space $\mathbb{R}^n$. Then the complements $\mathbb{R}^n \smallsetminus X$ and $\mathbb{R}^n \smallsetminus Y$ are either both empty or they have the same stable homotopy type with respect to arbitrary base points.*

In general the complements themselves can have quite different homotopy type. A typical example occurs in knot theory, the case that $X \cong Y \cong S^1$ are subsets of $\mathbb{R}^3$. On the other hand the stable homotopy type still carries some interesting geometric information: see (7.1.10).

(7.1.2) Theorem (Component theorem). *Let X and Y be closed homeomorphic subsets of $\mathbb{R}^n$. Then $\pi_0(\mathbb{R}^n \smallsetminus X)$ and $\pi_0(\mathbb{R}^n \smallsetminus Y)$ have the same cardinality.*

Later we give another proof of Theorem (7.1.2) based on homology theory, see (10.3.3). From the component theorem one can deduce classical results: The Jordan separation theorem (10.3.4) and the invariance of domain (10.3.7).

Theorem (7.1.1) is a direct consequence of (7.1.3). One can also compare complements in different Euclidean spaces. The next result gives some information about how many suspensions suffice.

(7.1.3) Theorem. *Let $X \subset \mathbb{R}^n$ and $Y \subset \mathbb{R}^m$ be closed subsets and $h\colon X \to Y$ a homeomorphism. Suppose $n \le m$. Then the following holds:*

(1) *If $\mathbb{R}^n \ne X$, then $\mathbb{R}^m \ne Y$, and h induces a canonical homotopy equivalence $\Sigma^{m+1}(\mathbb{R}^n \smallsetminus X) \simeq \Sigma^{n+1}(\mathbb{R}^m \smallsetminus Y)$ with respect to arbitrary base points.*

(2) *If $\mathbb{R}^n = X$ and $\mathbb{R}^m \ne Y$, then $n < m$ and $\Sigma^{n+1}(\mathbb{R}^m \smallsetminus Y) \simeq S^m$, i.e., $\mathbb{R}^m \smallsetminus Y$ has the stable homotopy type of S^{m-n-1}.*

(3) *If $\mathbb{R}^n = X$ and $\mathbb{R}^m = Y$, then $n = m$.*

In many cases the number of suspensions is not important. Since it also depends on the situation, it is convenient to pass from homotopy classes to stable homotopy classes. This idea leads to the simplest stable category.

The objects of our new category ST are pairs (X, n) of pointed spaces X and integers $n \in \mathbb{Z}$. The consideration of pairs is a technical device which allows for a better formulation of some results. Thus we should comment on it right now.

The pair $(X, 0)$ will be identified with X. The subcategory of the objects $(X, 0) = X$ with morphisms the so-called stable homotopy classes is the geometric input. For positive n the pair (X, n) replaces the n-fold suspension $\Sigma^n X$. But it will be convenient to have the object (X, n) also for negative n ("desuspension"). Here is an interesting example. In the situation of (7.1.3) the homotopy equivalence $\Sigma^{m+1}(\mathbb{R}^n \smallsetminus X) \to \Sigma^{n+1}(\mathbb{R}^m \smallsetminus Y)$ induced by h represents in the category ST an isomorphism $h_\bullet\colon (\mathbb{R}^n \smallsetminus X, -n) \to (\mathbb{R}^m \smallsetminus Y, -m)$. In this formulation it then makes sense to say that the assignment $h \mapsto h_\bullet$ is a functor. (Otherwise we would have to use a mess of different suspensions.) Thus if X is a space which admits an embedding $i\colon X \to \mathbb{R}^n$ as a closed subset for some n, then the isomorphism type of $(\mathbb{R}^n \smallsetminus i(X), -n)$ in ST is independent of the choice of the embedding. Hence we have associated to X a "dual object" in ST (up to canonical isomorphism).

Let $\Sigma^t X = X \wedge S^t$ be the t-fold suspension of X. As a model for the sphere S^t we use either the one-point compactification $\mathbb{R}^t \cup \{\infty\}$ or the quotient space $S(t) = I^t/\partial I^t$. In these cases we have a canonical associative homeomorphism $S^a \wedge S^b \cong S^{a+b}$ which we usually treat as identity. Suppose $n, m, k \in \mathbb{Z}$ are integers such that $n + k \ge 0, m + k \ge 0$. Then we have the suspension morphism

$$\Sigma\colon [X \wedge S^{n+k}, Y \wedge S^{m+k}]^0 \to [X \wedge S^{n+k+1}, Y \wedge S^{m+k+1}]^0, \quad f \mapsto f \wedge \mathrm{id}(S^1).$$

We form the colimit over these morphisms, $\operatorname{colim}_k [X \wedge S^{n+k}, Y \wedge S^{m+k}]^0$. For $n + k \ge 2$ the set $[X \wedge S^{n+k}, Y \wedge S^{m+k}]^0$ carries the structure of an abelian group and Σ is a homomorphism. The colimit inherits the structure of an abelian group. We define as morphism group in our category ST

$$\mathrm{ST}((X,n),(Y,m)) = \operatorname{colim}_k [X \wedge S^{n+k}, Y \wedge S^{m+k}]^0.$$

Formation of the colimit means the following: An element f of $\mathrm{ST}((X,n),(Y,m))$ is represented by pointed maps $f_k \colon X \wedge S^{n+k} \to Y \wedge S^{m+k}$, and f_k, $f_l, l \geq k$ represent the same element of the colimit if $\Sigma^{l-k} f_k \simeq f_l$. Composition of morphisms is defined by composition of representatives. Let $f_k \colon X \wedge S^{n+k} \to Y \wedge S^{m+k}$ and $g_l \colon Y \wedge S^{m+l} \to Z \wedge S^{p+l}$ be representatives of morphisms and let $r \geq k, l$. Then the following composition of maps represents the composition of the morphisms (dotted arrow):

$$\begin{array}{ccc} X \wedge S^{n+r} = X \wedge S^{n+k} \wedge S^{r-k} & \xrightarrow{\Sigma^{r-k} f_k} & Y \wedge S^{m+k} \wedge S^{r-k} \\ \vdots & & \downarrow = \\ Z \wedge S^{p+r} = Z \wedge S^{p+l} \wedge S^{r-l} & \xleftarrow{\Sigma^{r-l} g_l} & Y \wedge S^{m+l} \wedge S^{r-l}. \end{array}$$

One verifies that this definition does not depend on the choice of representatives. The group structure is compatible with the composition

$$\beta \circ (\alpha_1 + \alpha_2) = \beta \circ \alpha_1 + \beta \circ \alpha_2, \quad (\beta_1 + \beta_2) \circ \alpha = \beta_1 \circ \alpha + \beta_2 \circ \alpha.$$

The category ST has formal suspension automorphisms $\Sigma^p \colon \mathrm{ST} \to \mathrm{ST}$, $p \in \mathbb{Z}$

$$(X,n) \mapsto (X, n+p), \quad f \mapsto \Sigma^p f.$$

If $f \colon (X,n) \to (Y,m)$ is represented by $f_k \colon \Sigma^{n+k} X \to \Sigma^{m+k} Y$ (with $n+k \geq 0$, $m+k \geq 0$, $k \geq |p|$), then $\Sigma^p f$ is represented by

$$\begin{aligned} (\Sigma^p f)_k &= \Sigma^p (f_k) \colon \Sigma^{n+k+p} X \to \Sigma^{m+k+p} Y, & p \geq 0, \\ (\Sigma^p f)_{k+|p|} &= f_k \colon \Sigma^{n+k+p+|p|} X \to \Sigma^{m+k+p+|p|} Y, & p \leq 0. \end{aligned}$$

The rules $\Sigma^0 = \mathrm{id}(\mathrm{ST})$ and $\Sigma^p \circ \Sigma^q = \Sigma^{p+q}$ show that Σ^p is an automorphism. For $p > 0$ we call Σ^p the p-fold suspension and for $p < 0$ the p-fold desuspension. We have a canonical isomorphism $\kappa^p \colon (X,n) \to (\Sigma^p X, n-p)$; it is represented by the identity $X \wedge S^{n+k} \to (X \wedge S^p) \wedge S^{n+k-p}$ for $n+k \geq p \geq 0$. We write X for the object $(X,0)$. Thus for positive n the object (X,n) can be replaced by $\Sigma^n X$.

(7.1.4) Example. Pointed spaces X, Y are stably homotopy equivalent if and only if they are isomorphic in ST. The image $\mathrm{ST}(f)$ of $f \colon X \to Y$ in $\mathrm{ST}(X,Y)$ is called the stable homotopy class of f. Maps $f, g \colon X \to Y$ are stably homotopic if and only if they represent the same element in $\mathrm{ST}(X,Y)$. The groups $\mathrm{ST}(S^k, S^0) = \operatorname{colim}_n \pi_{n+k}(S^n)$ are the stable homotopy groups of the spheres. ◇

(7.1.5) Example. It is in general difficult to determine morphism groups in ST. But we know that the category in non-trivial. The suspension theorem and the degree theorem yield

$$\mathrm{ST}(S^n, S^n) = \mathrm{colim}_k [S^{n+k}, S^{n+k}]^0 \cong \mathbb{Z}.$$

The composition of morphisms corresponds to multiplication of integers. ◇

(7.1.6) Proposition. *Let Y be pathwise connected. We have the embedding $i : S^n \to Y^+ \wedge S^n$, $x \mapsto (*, x)$ and the projection $p : Y^+ \wedge S^n \to S^n$, $(y, x) \mapsto x$ with $pi = \mathrm{id}$. They induce isomorphisms of the π_k-groups for $k \le n \ge 1$.*

Proof. This is a consequence of (6.10.6). One has to use the homeomorphism $Y^+ \wedge S^n \cong Y \times S^n / Y \times \{*\}$, see Problem 2. □

(7.1.7) Proposition. *Let $(Y_j \mid j \in J)$ be the family of path components of Y and $c^j : Y_j \to Y$ the inclusion. Let $n \ge 2$. Then*

$$\langle c^j_* \rangle : \bigoplus_{j \in J} \pi_k(Y_j^+ \wedge S^n) \to \pi_k(Y^+ \wedge S^n)$$

is an isomorphism for $0 \le k \le n$. In particular $\pi_n(Y^+ \wedge S^n)$ is a free abelian group of rank $|\pi_0(Y)|$.

Proof. (7.1.6) and (6.10.6). □

(7.1.8) Proposition. *Let Y be well-pointed and $n \ge 2$. Then $\pi_n(Y \wedge S^n)$ is a free abelian group of rank $|\pi_0(Y)| - 1$ and the suspension $\pi_n(Y \wedge S^n) \to \pi_{n+1}(Y \wedge S^{n+1})$ is an isomorphism.*

From the exact homotopy sequence of the pair $(Y^+ \wedge S^n, S^n)$ we conclude that $\pi_k(Y^+ \wedge S^n, S^n) = 0$ for $0 < k < n$. The quotient theorem (6.10.2) shows that $\pi_k(Y^+ \wedge S^n, S^n) \to \pi_k(Y^+ \wedge S^n / S^n) \cong \pi_k(Y \wedge S^n)$ is bijective (surjective) for $0 < k \le 2n - 2$ ($k = 2n - 1$). From the exact sequence $0 \to \pi_n(S^n) \to \pi_n(Y^+ \wedge S^n) \to \pi_n(Y^+ \wedge S^n, S^n) \to 0$ we deduce a similar exact sequence where the relative group is replaced by $\pi_n(Y \wedge S^n)$. The inclusion of $\pi_n(S^n)$ splits. Now we can use (7.1.7).

(7.1.9) Corollary. *Let X be a well-pointed space. Then $\mathrm{ST}(S^0, X)$ is a free abelian group of rank $|\pi_0(X)| - 1$.* □

The group $\mathrm{ST}(S^0, X)$ only depends on the stable homotopy type of X. Therefore we can state:

(7.1.10) Corollary. *Let X and Y be well-pointed spaces of the same stable homotopy type. Then $|\pi_0(X)| = |\pi_0(Y)|$. Therefore (7.1.2) is a consequence of (7.1.3).* □

The category ST has a "product structure" induced by the smash product. The category ST together with this additional structure is called in category theory a symmetric tensor category (also called a symmetric monoidal category). The tensor product of objects is defined by

$$(X, m) \otimes (Y, n) = (X \wedge Y, m + n).$$

Let

$$f_k : X \wedge S^{m+k} \to X' \wedge S^{m'+k}, \quad g_l : Y \wedge S^{n+l} \to Y' \wedge S^{n'+l}$$

be representing maps for morphisms $f : (X, m) \to (X', m')$ and $g : (Y, n) \to (Y', n')$. A representing morphism $(f \otimes g)_{k+l}$ is defined to be $(-1)^{k(n+n')}$ times the composition $\tau' \circ (f_k \wedge g_l) \circ \tau$ (dotted arrow)

$$\begin{array}{ccc} X \wedge Y \wedge S^{m+k+n+l} & \xrightarrow{\tau} & X \wedge S^{m+k} \wedge Y \wedge S^{n+l} \\ \vdots & & \downarrow{\scriptstyle f_k \wedge g_l} \\ X' \wedge Y' \wedge S^{m'+k+n'+l} & \xleftarrow{\tau'} & X' \wedge S^{m'+k} \wedge Y' \wedge S^{n'+l} \end{array}$$

where τ and τ' interchange two factors in the middle. Now one has to verify: (1) The definition does not depend on the representatives; (2) the functor property $(f' \otimes g')(f \otimes g) = f'f \otimes g'g$ holds; (3) the tensor product is associative. These requirements make it necessary to introduce signs in the definition. The neutral object is $(S^0, 0)$. The symmetry $c : (X, m) \otimes (Y, n) \to (Y, n) \otimes (X, m)$ is $(-1)^{mn}$ times the morphism represented by the interchange map $X \wedge Y \to Y \wedge X$. If f_k is a representing map then $-f_k$ is the inverse in the homotopy set, i.e., not well-defined as a map. The product structure uses the associativity of the $\wedge$-product, so one should work with compactly generated spaces.

Problems

1. The spaces $S^1 \times S^1$ and $S^1 \vee S^1 \vee S^2$ are not homotopy equivalent. They have different fundamental group. Their suspensions are homotopy equivalent.
2. The inclusion $X \times Y \to X^+ \times Y$ induces for each pointed space Y a homeomorphism $(X \times Y)/(X \times \{*\}) \cong X^+ \wedge Y$.
3. Let X and Y be well-pointed spaces. Then $Y \to (X \times Y)/(X \times \{*\})$, $y \mapsto (*, y)$ is a cofibration.
4. Let P be a point. We have an embedding $P^+ \wedge Y \to X^+ \wedge Y$ and a canonical homeomorphism $X \wedge Y \to X^+ \wedge Y / P^+ \wedge Y$.
5. Let Y be pathwise connected. Then $Y^+ \wedge S^1 \cong Y \times S^1 / Y \times \{*\}$ is pathwise connected. The base point of Y^+ is non-degenerate. Hence the quotient $\Sigma'(Y^+) \to \Sigma(Y^+)$ from the unreduced suspension to the reduced suspension is an h-equivalence. The projection $Y \to P$ onto a point induces a 2-connected map between double mapping cylinders

$$\Sigma'(Y^+) = Z(* \leftarrow Y + \{*\} \to *) \to Z(* \leftarrow P + \{*\} \to *) = \Sigma'(P^+) \cong S^1.$$

From this fact one deduces the assertion (7.1.6) for $n = 1$. Deduce now the general assertion for $n \ge 2$ by the suspension theorem.

7.2 Mapping Cones

We need a few technical results about mapping cones. Let $f: X \to Y$ be a pointed map. We use as a model for the (unpointed) mapping cone $C(f)$ the double mapping cylinder $Z(Y \leftarrow X \to *)$; it is the quotient of $Y + X \times I + \{*\}$ under the relations $f(x) \sim (x,0), (x,1) \sim *$. The image of $*$ is the basepoint. For an inclusion $\iota: A \subset X$ we write $C(X,A) = C(\iota)$. For empty A we have $C(X,\emptyset) = X^+$. Since we will meet situations where products of quotient maps occur, we work in the category of compactly generated spaces where such products are again quotient maps. The mapping cone is a functor $C: \mathrm{TOP}(2) \to \mathrm{TOP}^0$; a map of pairs $(F,f): (X,A) \to (Y,B)$ induces a pointed map $C(F,f): C(X,A) \to C(Y,B)$, and a homotopy (F_t, f_t) induces a pointed homotopy $C(F_t, f_t)$. We note for further use a consequence of (4.2.1):

(7.2.1) Proposition. *If F and f are h-equivalences, then $C(F,f)$ is a pointed h-equivalence.* □

(7.2.2) Example. We write $C^n = C(\mathbb{R}^n, \mathbb{R}^n \smallsetminus 0)$. This space will be our model for the homotopy type of S^n. In order to get a homotopy equivalence $C^n \to S^n$, we observe that S^n is homeomorphic to the double mapping cylinder $Z(* \leftarrow S^{n-1} \to *)$. We have the canonical projection from $C^n = Z(\mathbb{R}^n \leftarrow \mathbb{R}^n \smallsetminus 0 \to *)$. An explicit homotopy equivalence is $(x,t) \mapsto (\sin \pi t \frac{x}{\|x\|}, \cos \pi t)$, $x \mapsto (0,\dots,0,1)$. ◇

(7.2.3) Example. Let $X \subset \mathbb{R}^n$ be a closed subspace. Then

$$C(\mathbb{R}^n, \mathbb{R}^n \smallsetminus X) = Z(\mathbb{R}^n \leftarrow \mathbb{R}^n \smallsetminus X \to *) \simeq Z(* \leftarrow \mathbb{R}^n \smallsetminus X \to *) = \Sigma'(\mathbb{R}^n \smallsetminus X),$$

the unpointed suspension. If $X = \mathbb{R}^n$, then this space is h-equivalent to S^0. If $X \neq \mathbb{R}^n$, then $\mathbb{R}^n \smallsetminus X$ is well-pointed with respect to any point and $\Sigma'(\mathbb{R}^n \smallsetminus X)$ is h-equivalent to the pointed suspension $\Sigma(\mathbb{R}^n \smallsetminus X)$. ◇

We are mainly interested in the homotopy type of $C(X,A)$ (under $\{*\} + X$). It is sometimes convenient to provide the set $C(X,A)$ with a possibly different topology which does not change the homotopy type. Set theoretically we can view $C(X,A)$ as the quotients $C_1(X,A) = (X \times 0 \cup A \times I)/A \times 1$ or $C_2(X,A) = (X \times 0 \cup A \times I \cup X \times 1)/X \times 1$. We can provide C_1 and C_2 with the quotient topology. Then we have canonical continuous maps $p: C(X,A) \to C_1(X,A)$ and $q: C_1(X,A) \to C_2(X,A)$ which are the identity on representative elements.

(7.2.4) Lemma. *The maps p and q are homotopy equivalences under $\{*\} + X$.*

Proof. Define $\bar{p}\colon C_1(X, A) \to C(X, A)$ by

$$\bar{p}(x,t) = x,\ t \le 1/2, \quad \bar{p}(a,t) = (a, \max(2t-1, 0)), \quad \bar{p}(a,1) = *.$$

One verifies that this assignment is well-defined and continuous. A homotopy $p\bar{p} \simeq \mathrm{id}$ is given by $((x,t),s) \mapsto (x, st + (1-s)\max(2t-1,0))$. A similar formula works for $\bar{p}p \simeq \mathrm{id}$. Define $\bar{q}\colon C_2(X, A) \to C_1(X, A)$ by

$$\bar{q}(x,t) = (x, \min(2t, 1)),\ t < 1, \quad \bar{q}(x,t) = * = \{A \times 1\},\ t \ge 1/2.$$

Again linear homotopies in the t-coordinate yield homotopies from $q\bar{q}$ and $\bar{q}q$ to the identity. □

(7.2.5) Proposition (Excision). *Let $U \subset A \subset X$ and suppose there exists a function $\tau\colon X \to I$ such that $U \subset \tau^{-1}(0)$ and $\overline{\tau^{-1}[0,1[} \subset A$. Then the inclusion of pairs induces a pointed h-equivalence $g\colon C(X \smallsetminus U, A \smallsetminus U) \to C(X, A)$.*

Proof. Set $\sigma(x) = \max(2\tau(x) - 1, 0)$. A homotopy inverse of g is the map $f\colon (x,t) \mapsto (x, \sigma(x)t)$. The definition of f uses the notation C_2 for the mapping cone. The homotopies from fg and gf to the identity are obtained by a linear homotopy in the t-coordinate. □

(7.2.6) Remark. Mapping cones of inclusions are used at various occasions to relate the category TOP(2) of pairs with the category TOP^0 of pointed spaces. We make some general remarks which concern the relations. They will be relevant for the investigation of homology and cohomology theories.

Let $\tilde{h}\colon \mathrm{TOP}^0 \to \mathcal{C}$ be a homotopy invariant functor. We define an associated functor $h = P\tilde{h}\colon \mathrm{TOP}(2) \to \mathcal{C}$ by composition with the mapping cone functor $(X, A) \mapsto C(X, A)$. The functor $P\tilde{h}$ is homotopy invariant in a stronger sense: If $f\colon (X, A) \to (Y, B)$ is a map of pairs such that the components $f\colon X \to Y$ and $f\colon A \to B$ are h-equivalences, then the induced map $h(X, A) \to h(Y, B)$ is an isomorphism (see (7.2.1)). Moreover h satisfies excision: Under the hypothesis of (7.2.5) the inclusion induces an isomorphism $h(X \smallsetminus U, A \smallsetminus U) \cong h(X, A)$.

Conversely, let $h\colon \mathrm{TOP}(2) \to \mathcal{C}$ be a functor. We define an associated functor $Rh = \tilde{h}$ on objects by $Rh(X) = h(X, *)$ and with the obvious induced morphisms. If h is homotopy invariant, then also Rh.

The composition PR is given by $PRh(X, A) = h(C(X, A), *)$. We have natural morphisms

$$h(C(X,A), *) \to h(C(X,A), CA) \leftarrow h(C(X,A) \smallsetminus U, CA \smallsetminus U) \leftarrow h(X, A).$$

Here CA is the cone on A and $U \subset CA$ is the subspace with t-coordinates in $[1/2, 1]$. If h is strongly homotopy invariant and satisfies excision, then these morphisms are isomorphisms, i.e., PR is naturally isomorphic to the identity.

The composition RP is given by $RP\tilde{h}(X) = \tilde{h}(C(X,*))$. There is a canonical projection $C(X,*) \to X$. It is a pointed h-equivalence, if the inclusion $\{*\} \to X$ is a cofibration. Thus if $\tilde{h}$ is homotopy invariant, the composition RP is naturally isomorphic to the identity on the subcategory of well-pointed spaces. ◇

Let (X,A) and (Y,B) be two pairs. We call $A \times Y, X \times B$ ***excisive*** in $X \times Y$ if the canonical map $p\colon Z(A \times Y \leftarrow A \times B \to X \times B) \to A \times Y \cup X \times B$ is a homotopy equivalence.

(7.2.7) Proposition (Products). *Let $(A \times Y, X \times B)$ be excisive. Then there exists a natural pointed homotopy equivalence*

$$\alpha\colon C(X,A) \wedge C(Y,B) \to C((X,A) \times (Y,B)).$$

It is defined by the assignments

$$\begin{aligned}(x,y) &\mapsto (x,y),\\ (a,s,y) &\mapsto (a,y,s),\\ (x,b,s) &\mapsto (x,b,s),\\ (a,s,b,t) &\mapsto (a,b,\max(s,t)).\end{aligned}$$

(See the proof for an explanation of notation.)

Proof. In the category of compactly generated spaces $C(X,A) \wedge C(Y,B)$ is a quotient of

$$X \times Y + A \times I \times Y + X \times B \times I + A \times I \times B \times I$$

under the following relations: $(a,0,y) \sim (a,y)$, $(x,b,0) \sim (x,b)$, $(a,0,b,t) \sim (a,b,t)$, $(a,s,b,0) \sim (a,s,b)$, and $A \times 1 \times Y + X \times B \times 1 + A \times 1 \times B \times I \cup A \times I \times B \times 1$ is identified to a base point $*$.

In a first step we show that the smash product is homeomorphic to the double mapping cylinder $Z(X \times Y \xleftarrow{\iota p} Z \to \{*\})$ where

$$Z = Z(A \times Y \leftarrow A \times B \to X \times B).$$

This space is the quotient of

$$X \times Y + (A \times Y + A \times B \times I + X \times B) \times I + A \times B \times (I \times I / I \times 0)$$

under the following relations: $(x,b,0) \sim (x,b)$, $(a,y,0) \sim (a,y)$, $(a,b,t,0) \sim (a,b)$, $(a,b,1,s) \sim (a,b,s)$, and $(A \times Y + A \times B \times I + X \times B) \times 1$ is identified to a base point $*$.

The assignment

$$I \times I \to I \times I, \quad (u,v) \mapsto \begin{cases} (2uv, v), & u \le 1/2,\\ (v, 2v(1-u)), & u \ge 1/2.\end{cases}$$

induces a homeomorphism $\gamma_0 \colon I \times I/(I \times 0) \to I \times I$. Its inverse β_0 has the form

$$I \times I \smallsetminus \{(0,0)\} \to I \times (I \smallsetminus \{0\}), \quad (s,t) \mapsto \begin{cases} (1 - t/2s, s), & s \geq t, \\ (s/2t, t), & s \leq t. \end{cases}$$

A homeomorphism $\beta \colon C(X,A) \wedge C(Y,B) \to C(\iota p)$ is now defined by $\beta(x,y) = (x,y)$, $\beta(a,s,y) = (a,y,s)$, $\beta(x,b,t) = (x,b,t)$, $\beta(a,s,b,t) = (a,b,\beta_0(s,t))$. The diagram

$$\begin{array}{ccccc} X \times Y & \xleftarrow{\iota p} & Z & \longrightarrow & \{*\} \\ \downarrow{=} & & \downarrow{p} & & \downarrow{=} \\ X \times Y & \xleftarrow{\iota} & A \times Y \cup X \times B & \longrightarrow & \{*\} \end{array}$$

induces $\pi \colon C(\iota p) \to C(\iota)$. It is a pointed h-equivalence if p is an h-equivalence. One verifies that $\alpha = \pi\beta$. □

(7.2.8) Remark. The maps α are associative: For three pairs (X,A), (Y,B), (Z,C) the relation $\alpha(\alpha \wedge \mathrm{id}) = \alpha(\mathrm{id} \wedge \alpha)$ holds. They are also compatible with the interchange map. Finally, they yield a natural transformation. ◇

Problems

1. Verify that the map f in the proof of 7.2.5 is continuous. Similar problem for the homotopies.
2. Let $(F,f) \colon (X,A) \to (Y,B)$ be a map of pairs. If F is n-connected and f $(n-1)$-connected, then $C(F,f)$ is n-connected.
3. Let $X = A \cup B$ and suppose that the interiors A°, B° still cover X. Then the inclusion induces a weak homotopy equivalence $C(B, A \cap B) \to C(X,A)$.
4. Construct explicit h-equivalences $C^n \to \mathbb{R}^n \cup \{\infty\} = S^{(n)}$ such that

$$\begin{array}{ccc} C^m \wedge C^n & \xrightarrow{\alpha} & C^{m+n} \\ \downarrow & & \downarrow \\ S^{(m)} \wedge S^{(n)} & \xrightarrow{\cong} & S^{(m+n)} \end{array}$$

is homotopy commutative.

7.3 Euclidean Complements

This section is devoted to the proof of (7.1.3). We need an interesting result from general topology.

(7.3.1) Proposition. *Let $A \subset \mathbb{R}^m$ and $B \subset \mathbb{R}^n$ be closed subsets and let $f : A \to B$ be a homeomorphism. Then there exists a homeomorphism of pairs*

$$F : (\mathbb{R}^m \times \mathbb{R}^n, A \times 0) \to (\mathbb{R}^n \times \mathbb{R}^m, B \times 0)$$

such that $F(a, 0) = (f(a), 0)$ for $a \in A$.

Proof. By the extension theorem of Tietze (1.1.2) there exists a continuous extension $\varphi : \mathbb{R}^m \to \mathbb{R}^n$ of $f : A \to B \subset \mathbb{R}^n$. The maps

$$\Phi_\pm : \mathbb{R}^m \times \mathbb{R}^n \to \mathbb{R}^m \times \mathbb{R}^n, \quad (x, y) \mapsto (x, y \pm \varphi(x))$$

are inverse homeomorphisms. Let $G(f) = \{(a, f(a)) \mid a \in A\}$ denote the graph of f. Then Φ_+ sends $A \times 0$ homeomorphically to $G(f)$ by $(a, 0) \mapsto (a, f(a))$.

Let $\psi : \mathbb{R}^n \to \mathbb{R}^m$ be a Tietze extension of the inverse g of f. Then we have similar homeomorphisms

$$\Psi_\pm : \mathbb{R}^n \times \mathbb{R}^m \to \mathbb{R}^n \times \mathbb{R}^m, \quad (y, x) \mapsto (y, x \pm \psi(y)).$$

The desired homeomorphism F is the composition $\Psi_- \circ \tau \circ \Phi_+$ where τ interchanges $\mathbb{R}^m$ and $\mathbb{R}^n$ (and sends $G(f)$ to $G(g)$). □

Let $X \subset \mathbb{R}^n$ and $Y \subset \mathbb{R}^m$ be closed subsets and $f : X \to Y$ a homeomorphism. The induced homeomorphism F from (7.3.1) can be written as a homeomorphism

$$F : (\mathbb{R}^n, \mathbb{R}^n \smallsetminus X) \times (\mathbb{R}^m, \mathbb{R}^m \smallsetminus 0) \to (\mathbb{R}^m, \mathbb{R}^m \smallsetminus Y) \times (\mathbb{R}^n, \mathbb{R}^n \smallsetminus 0).$$

We apply the mapping cone functor to F and use (7.2.2) and (7.2.7). The result is a homotopy equivalence

$$C(\mathbb{R}^n, \mathbb{R}^n \smallsetminus X) \wedge S^m \simeq C(\mathbb{R}^m, \mathbb{R}^m \smallsetminus Y) \wedge S^n.$$

If $X \neq \mathbb{R}^m$ and $Y \neq \mathbb{R}^m$ we obtain together with (7.2.3)

$$\Sigma^{m+1}(\mathbb{R}^n \smallsetminus X) \simeq \Sigma^{n+1}(\mathbb{R}^m \smallsetminus Y)$$

If $X \neq \mathbb{R}^n$ then we have $C(\mathbb{R}^n, \mathbb{R}^n \smallsetminus X) \simeq \Sigma(\mathbb{R}^n \smallsetminus X)$, and if $X = \mathbb{R}^n$ then we have $C(\mathbb{R}^n, \mathbb{R}^n \smallsetminus X) \simeq S^0$.

Suppose $X \neq \mathbb{R}^n$ but $Y = \mathbb{R}^m$. Then $\Sigma^{m+1}(\mathbb{R}^n \smallsetminus X) \simeq S^n$. Since $n \leq m$ the homotopy group $\pi_n(\Sigma^{m+1}(\mathbb{R}^n \smallsetminus X)) = 0$ and $\pi_n(S^n) \cong \mathbb{Z}$. This contradiction shows that $Y \neq \mathbb{R}^m$.

Suppose $X = \mathbb{R}^n$ and $Y \neq \mathbb{R}^m$. Then $n = m$ is excluded by the previous proof. Thus

$$S^m \simeq C(\mathbb{R}^n, \mathbb{R}^n \smallsetminus X) \wedge S^m \simeq C(\mathbb{R}^m, \mathbb{R}^m \smallsetminus Y) \wedge S^n \simeq \Sigma^{m+1}(\mathbb{R}^m \smallsetminus Y).$$

If $X = \mathbb{R}^n$ and $Y = \mathbb{R}^m$, then

$$S^n \simeq \Sigma^n C(\mathbb{R}^n, \mathbb{R}^n \smallsetminus X) \simeq \Sigma^m C(\mathbb{R}^m, \mathbb{R}^m \smallsetminus Y) \simeq S^m$$

and therefore $m = n$.

This finishes the proof of (7.1.3).

7.4 The Complement Duality Functor

The complement duality functor is concerned with the stable homotopy type of Euclidean complements $\mathbb{R}^n \smallsetminus X$ for closed subsets $X \subset \mathbb{R}^n$. We consider an associated category $\mathcal{E}$. The objects are pairs $(\mathbb{R}^n, X)$ where X is closed in $\mathbb{R}^n$. A morphism $(\mathbb{R}^n, X) \to (\mathbb{R}^m, Y)$ is a proper map $f : X \to Y$. The duality functor is a contravariant functor $D : \mathcal{E} \to \mathrm{ST}$ which assigns to $(\mathbb{R}^n, X)$ the object $\Sigma^{-n} C(\mathbb{R}^n, \mathbb{R}^n \smallsetminus X) = (C(\mathbb{R}^n, \mathbb{R}^n \smallsetminus X), -n)$. The associated morphism $D(f) : \Sigma^{-m} C(\mathbb{R}^m, \mathbb{R}^m \smallsetminus Y) \to \Sigma^{-n} C(\mathbb{R}^n, \mathbb{R}^n \smallsetminus X)$ will be constructed via a representing morphism $D(f)_{m+n}$. Its construction needs some preparation.

Given the data $X \subset \mathbb{R}^n, Y \subset \mathbb{R}^m$ and a proper map $f : X \to Y$. Henceforth we use the notation $A|B = (A, A \smallsetminus B)$ for pairs $B \subset A$. Note that in this notation $A|B \times C|D = A \times C | B \times D$. The basic step in the construction of the functor will be an associated homotopy class

$$D_{\#} f : \mathbb{R}^n | D^n \times \mathbb{R}^m | Y \to \mathbb{R}^n | X \times \mathbb{R}^m | 0.$$

Here D^n again denotes the n-dimensional standard disk. A ***scaling function*** for a proper map $f : X \to Y$ is a continuous function $\varphi : Y \to]0, \infty[$ with the property

$$\varphi(f(x)) \geq \|x\|, \quad x \in X.$$

The next lemma shows the existence of scaling functions with an additional property.

(7.4.1) Lemma. *There exists a positive continuous function $\psi : [0, \infty[\to]0, \infty[$ such that the inequality $\psi(\|fx\|) \geq \|x\|$ holds for $x \in X$. A scaling function in the sense of the definition is then $y \mapsto \psi(\|y\|)$.*

Proof. The set $f^{-1}D(t) = \max\{x \in X \mid \|fx\| \leq t\}$ is compact, since f is proper. Let $\tilde{\psi}(t)$ be its norm maximum $\max\{\|x\| \mid x \in X, \|fx\| \leq t\}$. Then $\tilde{\psi}(\|fx\|) = \max\{\|a\| \mid a \in X, \|fa\| \leq \|fx\|\} \geq \|x\|$. The function $\tilde{\psi} : [0, \infty[\to [0, \infty[$ is increasing. There exists a continuous increasing function $\psi : [0, \infty[\to]0, \infty[$ such that $\psi(t) \geq \tilde{\psi}(t)$ for each $t \geq 0$. □

The set of scaling functions is a positive convex cone. Let φ_1, φ_2 be scaling functions and $0 \leq \lambda \leq 1$; then $\lambda\varphi_1 + (1 - \lambda)\varphi_2$ is a scaling function. Let $\tilde{\varphi} > \varphi$; if φ is a scaling function then also $\tilde{\varphi}$.

Let φ be a scaling function and set $M(\varphi) = \{(x, y) \mid \varphi(y) \geq \|x\|\}$. Then we have a homeomorphism

$$\mathbb{R}^{n+m} | D^n \times Y \to \mathbb{R}^{n+m} | M(\varphi), \quad (x, y) \mapsto (\varphi(y) \cdot x, y).$$

The graph $G(f) = \{(x, fx) \mid x \in X\}$ of f is contained in $M(\varphi)$. We thus can continue with the inclusion and obtain a map $D_1(f, \varphi)$ of pairs

$$\mathbb{R}^{n+m} | D^n \times Y \to \mathbb{R}^{n+m} | G(f), \quad (x, y) \mapsto (\varphi(y) \cdot x, y).$$

The homotopy class of $D_1(f,\varphi)$ does not depend on the choice of the scaling function: If φ_1, φ_2 are scaling functions, then

$$(x,y,t) \mapsto ((t\varphi_1(y) + (1-t)\varphi_2(y)) \cdot x, y)$$

is a homotopy from $D_1(f,\varphi_2)$ to $D_1(f,\varphi_1)$. A continuous map $f: X \to Y$ has a Tietze extension $\tilde{f}: \mathbb{R}^n \to \mathbb{R}^m$. The homeomorphism $(x,y) \mapsto (x, y - \tilde{f}(x))$ of $\mathbb{R}^{n+m}$ sends $(x, f(x))$ to $(x,0)$. We obtain a homeomorphism of pairs

$$D_2(f,\tilde{f}): \mathbb{R}^{n+m}|G(f) \to \mathbb{R}^{n+m}|X \times 0.$$

The homotopy class is independent of the choice of the Tietze extension: The homotopy $(x,y,t) \mapsto (x, y - (1-t)\tilde{f}_1(x) - t\tilde{f}_2(x))$ proves this assertion. The duality functor will be based on the composition

$$D_\#(f) = D_2(f,\tilde{f}) \circ D_1(f,\varphi): \mathbb{R}^n|D^n \times \mathbb{R}^m|Y \to \mathbb{R}^n|X \times \mathbb{R}^m|0.$$

We have written $D_\#(f)$, since the homotopy class is independent of the choice of the scaling function and the Tietze extension. The morphism $Df: \Sigma^{-m}C(\mathbb{R}^m|Y) \to \Sigma^{-n}C(\mathbb{R}^n|X)$ is defined by a representative of the colimit:

$$(Df)_{n+m}: \Sigma^n C(\mathbb{R}^m|Y) \to \Sigma^m C(\mathbb{R}^n|X).$$

Consider the composition

$$\begin{array}{ccccc}
C(\mathbb{R}^m|Y) \wedge C^n & \xrightarrow{\tau} & C^n \wedge C(\mathbb{R}^m|Y) & \xleftarrow{\simeq} & C(\mathbb{R}^n|D^n) \wedge C(\mathbb{R}^m|Y) \\
\Big\downarrow {\scriptstyle (-1)^{nm}(Df)_{n+m}} & & & & \Big\downarrow {\scriptstyle \alpha} \\
C(\mathbb{R}^n|X) \wedge C^m & \xleftarrow{\alpha^{-1}} & C(\mathbb{R}^n|X \times \mathbb{R}^m|0) & \xleftarrow{CD_\# f} & C(\mathbb{R}^n|D^n \times \mathbb{R}^m|Y).
\end{array}$$

Explanation. τ interchanges the factors; the inclusion $\mathbb{R}|D^n \to \mathbb{R}^n|0$ induces a homotopy equivalence $C(\mathbb{R}^n|D^n) \to C^n$; the morphisms α comes from (7.2.7); and $CD_\# f$ is obtained by applying the mapping cone to $D_\# f$; finally, we multiply the homotopy class of the composition by $(-1)^{nm}$. We take the freedom to use $(Df)_{n+m}: C(\mathbb{R}^m|Y) \wedge C^n \to C(\mathbb{R}^n|X) \wedge C^m$ as our model for Df, i.e., we do not compose with the h-equivalences of the type $C^n \to S^n$ obtained in (7.2.2).

(7.4.2) Lemma. *Let f be an inclusion, $f: X \subset Y \subset \mathbb{R}^n$. Then Df has as a representative the map $C(\mathbb{R}^n|Y) \to C(\mathbb{R}^n|X)$ induced by the inclusion. In particular the identity of X is send to the identity.*

Proof. We take the scaling function $y \mapsto \|y\| + 1$ and extend f by the identity. Then $D_\#(f)$ is the map $(x,y) \mapsto ((\|y\|+1)\cdot x, y - (\|y\|+1)\cdot x)$. The map D_1 is $(x,y) \mapsto ((\|y\|+1)\cdot x, y)$ and the homotopy

$$(x,y,t) \mapsto ((1-t)(\|y\|+1)\cdot x + t(x+y), y)$$

is a homotopy of pairs from D_1 to $(x, y) \mapsto (x+y, y)$. Hence $D_2 \circ D_1$ is homotopic to $(x, y) \mapsto (x + y, -x)$ and the homotopy $(x, y, t) \mapsto ((1-t)x + y, -x)$ shows it to be homotopic to $(x, y) \mapsto (y, -x)$. Now interchange the factors and observe that $x \mapsto -x$ has the degree $(-1)^n = (-1)^{n \cdot n}$. □

Next we consider the case of a homeomorphism $f : X \to Y$ with inverse g. Let $\tilde{g} : \mathbb{R}^m \to \mathbb{R}^n$ be a Tietze extension of g. Then:

(7.4.3) Lemma. *The maps $(x, y) \mapsto (\varphi(\|y\|) \cdot x, y)$ and $(x, y) \mapsto (x + \tilde{g}(y), y)$ are as maps of pairs $\mathbb{R}^n | D^n \times \mathbb{R}^m | Y \to \mathbb{R}^{n+m} | G(f)$ homotopic. Here $\psi : [0, \infty[\to]0, \infty[$ is a function such that $\psi(\|fx\|) \geq \|x\|$ and $\varphi(r) = 1 + \psi(r)$.*

Proof. We use the linear homotopy $((1-t)(x + \tilde{g}(y)) + t\varphi(\|y\|) \cdot x, y)$. Suppose this element is contained in $G(f)$. Then $y \in Y$ and hence $g(y) = \tilde{g}(y)$, and the first component equals $g(y)$. We solve for x and obtain

$$x = \frac{t}{1 + t\psi(\|y\|)} g(y).$$

Then we take the norm

$$\|x\| = \frac{t}{1 + t\psi(\|y\|)} \|gy\| \leq \frac{t\psi(\|y\|)}{1 + t\psi(\|y\|)} < 1.$$

Hence $(x, y) \in D^n \times Y$. □

In the situation of the previous lemma the map $D_{\#}(f)$ is homotopic to the restriction of the homeomorphism $\mathbb{R}^n | 0 \times \mathbb{R}^m | Y \to \mathbb{R}^n | X \times \mathbb{R}^m | 0$ obtainable from (7.3.1). Another special case is obtained from a homeomorphism h of $\mathbb{R}^m$ and $X \subset \mathbb{R}^m$, $Y = h(X) \subset \mathbb{R}^m$. In this case h and h^{-1} are Tietze extensions.

For the verification of the functor property we start with the following data: $(\mathbb{R}^n, X)$, $(\mathbb{R}^m, Y)$, $(\mathbb{R}^p, Z)$ and proper maps $f : X \to Y$, $g : Y \to Z$. We have the inclusion $G(g) \subset \mathbb{R}^m \times \mathbb{R}^p$ and the proper map $h : X \to G(g)$, $x \mapsto (fx, gfx)$.

(7.4.4) Proposition. *The diagram*

$$\begin{array}{ccc}
\mathbb{R}^n|D^n \times \mathbb{R}^m|D^m \times \mathbb{R}^p|Z & \xrightarrow{\tau_{nm} \times 1} & \mathbb{R}^m|D^m \times \mathbb{R}^n|D^n \times \mathbb{R}^p|Z \\
\downarrow{\scriptstyle 1 \times D_{\#} g} & & \downarrow{\scriptstyle 1 \times D_{\#}(gf)} \\
\mathbb{R}^n|D^n \times \mathbb{R}^m|Y \times \mathbb{R}^p|0 & & \mathbb{R}^m|D^m \times \mathbb{R}^n|X \times \mathbb{R}^p|0 \\
\downarrow{\scriptstyle D_{\#} f \times 1} & & \downarrow{\scriptstyle \cap} \\
\mathbb{R}^n|X \times \mathbb{R}^m|0 \times \mathbb{R}^p|0 & \xrightarrow{\tau_{nm} \times 1} & \mathbb{R}^m|0 \times \mathbb{R}^n|X \times \mathbb{R}^p|0
\end{array}$$

is homotopy commutative. Here τ_{nm} are the appropriate interchange maps.

Proof. For the proof we use the intermediate morphism $D_\#(h)$. In the sequel we skip the notation for the scaling function and the Tietze extension. If φ_f is a scaling function for f and φ_{gf} a scaling function for gf, then

$$\psi_1\colon Y\times Z \to]0,\infty[\,,\ (y,z)\mapsto \varphi_f(y), \quad \psi_2\colon Y\times Z \to]0,\infty[\,,\ (y,z)\mapsto \varphi_{gf}(z)$$

are scaling functions for h. We have a factorization $D_2(h) = D_2^2(h)D_2^1(h)$ where $D_2^1(h)(x,y,z) = (x, y - \tilde{f}(x), z)$ and $D_2^2(h)(x,y,z) = (x,y,z - \widetilde{gf}(x))$ and we use $\tilde{h} = (\tilde{f}, \widetilde{gf})$ with $\widetilde{gf} = \tilde{g}\tilde{f}$. The diagram

$$\begin{array}{ccc} \mathbb{R}^{n+m+p}|\{x,0,gfx\} & \xrightarrow{\tau_{nm}\times 1} & \mathbb{R}^{m+n+p}|\{0,x,gfx\} \\ \downarrow{\scriptstyle D_2^2(h)} & & \downarrow{\scriptstyle 1\times D_2(gf)} \\ \mathbb{R}^{n+m+p}|\{x,0,0\} & \xrightarrow{\tau_{nm}\times 1} & \mathbb{R}^{m+n+p}|\{0,x,0\} \end{array}$$

commutes. The notation $\{(x,0,gfx)\}$ means that we take the set of all element of the given form where $x \in X$. We verify that the diagram

$$\begin{array}{ccc} \mathbb{R}^n|D^n \times \mathbb{R}^m|D^m \times \mathbb{R}^p|Z & \xrightarrow{\tau_{nm}\times 1} & \mathbb{R}^m|D^m \times \mathbb{R}^n|D^n \times \mathbb{R}^p|Z \\ \downarrow{\scriptstyle \zeta} & & \downarrow{\scriptstyle \iota\times D_1(gf)} \\ \mathbb{R}^{n+m+p}|\{x,0,gfx\} & \xrightarrow{\tau_{nm}\times 1} & \mathbb{R}^{m+n+p}|\{0,x,gfx\} \end{array}$$

with $\zeta = D_2^1 h \circ D_1 h \circ (1 \times D_1 g)$ and $\iota\colon \mathbb{R}^m|D^m \subset \mathbb{R}^m|0$ commutes up to homotopy. The map ζ is, with the choice $\varphi_h = \varphi_{gf}$, the assignment

$$(x,y,z) \mapsto (\varphi_{gf}(z)\cdot x, \varphi_g(z)\cdot y - \tilde{f}(\varphi_{gf}(z)\cdot x), z).$$

We use the linear homotopy $(\varphi_{gf}(z)\cdot x, s(\varphi_g(z)\cdot y - \tilde{f}(\varphi_{gf}(z)\cdot x) + (1-s)y, z)$. We verify that this is a homotopy of pairs, i.e., an element $\{\tilde{x}, 0, gf(\tilde{x})\}$ only occurs as the image of an element $(x,y,z) \in D^n \times D^m \times Z$. Thus assume

(i) $\tilde{x} = \varphi_{gf}(z)\cdot x \in X$;

(ii) $s\varphi_g(z)\cdot y - s\tilde{f}(\varphi_{gf}(z)\cdot x) + (1-s)y = 0$;

(iii) $z = gf(\tilde{x})$.

Since $\varphi_{gf}(z)\cdot x \in X$, we can replace in (ii) $\tilde{f}$ by f. We apply φ_{gf} to (iii) and obtain

$$\varphi_{gf}(z) = \varphi_{gf}(gf(\varphi_{gf}(z)\cdot x)) \geq \varphi_{gf}(z)\cdot \|x\|,$$

hence $\|x\| \leq 1$. The equation (ii) for $s = 0$ says $y = 0$, hence $y \in D^m$. Thus assume $s \neq 0$. Then $f(\varphi_{gf}(z)\cdot x) = (\varphi_g(z) + s^{-1} - 1)\cdot y$. We apply g to this equation and use (ii):

$$z = gf(\varphi_{gf}(z)\cdot x) = g((\varphi_g(z) + s^{-1} - 1)\cdot y).$$

Finally we apply φ_g to this equation and obtain

$$\varphi_g(z) = \varphi_g g((\varphi_g(z) + s^{-1} - 1) \cdot y) \geq (\varphi_g(z) + s^{-1} - 1)\|y\| \geq \varphi_g(z)\|y\|,$$

and therefore $\|y\| \leq 1$.

Finally we show that the diagram

$$\begin{array}{ccc} \mathbb{R}^n|D^n \times \mathbb{R}^{m+p}|G(g) & \xrightarrow{D_\#(h)} & \mathbb{R}^{n+m+p}|\{x,0,0\} \\ \downarrow{\scriptstyle 1\times D_2 g} & \nearrow{\scriptstyle D_\#(f)\times 1} & \\ \mathbb{R}^n|D^n \times \mathbb{R}^m|Y \times \mathbb{R}^p|0 & & \end{array}$$

commutes up to homotopy. In this case we use for h the scaling function ψ_1. Then $D_1 h\colon (x,y,z) \mapsto (\varphi_f(y) \cdot x, y, z)$ and

$$(D_\# f \times 1) \circ (1 \circ D_2 g)(x,y,z) = (\varphi_f(y) \cdot x, y - \tilde{f}(\varphi_f(y) \cdot x), z - \tilde{g}(y)),$$

$$D_\#(h)(x,y,z) = (\varphi_f(y) \cdot x, y - \tilde{f}(\varphi_f(y) \cdot x), z - \tilde{g}\tilde{f}(\varphi_f(y) \cdot x)).$$

Again we use a linear homotopy with $z - \tilde{g}((1-t)y + t\tilde{f}(\varphi_f(y) \cdot x))$ as the third component and have to verify that it is a homotopy of pairs. Suppose the image is contained in $\{x,0,0\}$. Then

(i) $\varphi_f(y) \cdot x \in X$;

(ii) $y = \tilde{f}(\varphi_f(y) \cdot x) \overset{\text{(i)}}{=} f(\varphi_f(y) \cdot x) \in Y$;

(iii) $z = \tilde{g}((1-t)y + tf(\varphi_f(y) \cdot x)) \overset{\text{(ii)}}{=} g(y)$.

(iii) shows that $(y,z) \in G(g)$. We apply φ_f to (ii) and see that $\|x\| \leq 1$. The three diagrams in this proof combine to the h-commutativity of the diagram in (7.4.4). □

(7.4.5) Proposition. *Suppose that $h\colon X \times I \to Y$ is a proper homotopy. Then $D(h_0) = D(h_1)$.*

Proof. Let $j_0\colon X \to X \times I$, $x \mapsto (x,0)$. The map is the composition of the homeomorphism $a\colon X \to X \times 0$ and the inclusion $b\colon X \times 0 \subset X \times I$. Thus Da is an isomorphism and Db is induced by the inclusion $\mathbb{R}^{n+1}|X \times I \subset \mathbb{R}^{n+1}|X \times 0$; hence Db is an isomorphism, since induced by a homotopy equivalence (use (7.4.2)). Thus Dj_0 is an isomorphism. The composition $\mathrm{pr} \circ j_0$ is the identity; hence $D(\mathrm{pr})$ is inverse to $D(j_0)$. A similar argument for j_1 shows that $Dj_0 = Dj_1$. We conclude that the maps $h_t = h \circ j_t$ have the same image under D. □

(7.4.6) Remark. The construction of the dual morphism is a little simpler for a map between compact subsets of Euclidean spaces. Let $X \subset \mathbb{R}^n$ be compact. Choose a disk D such that $X \subset D$. Then the dual morphism is obtained from

$$R^n|0 \times R^m|Y \supset \mathbb{R}^n|D \times \mathbb{R}^m|Y \subset \mathbb{R}^{n+m}|G(f) \to \mathbb{R}^n|X \times \mathbb{R}^m|0$$

where the last morphism is as before $(x, y) \mapsto (x, y - \tilde{f}(x))$. Also the proof of the functoriality (7.4.4) is simpler in this case.

The composition $(D_\# f \times 1)(1 \times D_\# g)$ is $(x, y, z) \mapsto (x, y - \tilde{f}(x), z - \tilde{g}(y))$. The other composition is $(x, y, z) \mapsto (x, y, z - \tilde{g}\tilde{f}(x))$. Then we use the homotopies of pairs $(x, y - \tilde{f}(x), z - \tilde{g}((1-t)y + t\tilde{f}(x)))$ and $(x, y - t\tilde{f}(x), z - \tilde{g}\tilde{f}(x))$. ◇

Problems

1. Verify in detail that the commutativity of the diagram in (7.4.4) implies that D is a functor.

2. Use the homotopy invariance of the duality functor and generalize (7.1.3) as follows. Suppose $X \subset \mathbb{R}^n$ and $Y \subset \mathbb{R}^m$ are closed subsets which are properly homotopy equivalent. Let $n \le m$.

(1) If $\mathbb{R}^n \smallsetminus X \ne \emptyset$, then $\mathbb{R}^m \smallsetminus Y \ne \emptyset$.

(2) Let $\mathbb{R}^n \ne X$. For each choice of a base point $\mathbb{R}^m \smallsetminus Y$ has the same stable homotopy type as $\Sigma^{n-m}(\mathbb{R}^n \smallsetminus X)$.

(3) If $\mathbb{R}^n \smallsetminus X$ is empty and $\mathbb{R}^m \smallsetminus Y$ is non-empty, then $\mathbb{R}^m \smallsetminus Y$ has the stable homotopy type of S^{m-n-1}.

(4) If $n = m$ then the complements of X and Y have the same number of path components.

3. Let $X \subset \mathbb{R}^n$ and $Y \subset \mathbb{R}^m$ be closed subsets and $f \colon X \to Y$ a proper map. Consider the closed subspace $W \subset \mathbb{R}^m \times \mathbb{R} \times \mathbb{R}^n$ of points

$$(y, t, x) = \begin{cases} y \in Y,\ t = 0,\ x = 0 \\ ((1-t)f(x), t, tx), & x \in X,\ t \in I. \end{cases}$$

Then W is homeomorphic to the mapping cylinder $Z(f)$ of f.

4. Let $X \subset \mathbb{R}^n$ and $f \colon X \to \mathbb{R}^n \times \mathbb{R}^m$ the standard embedding $x \mapsto (x, 0)$. Then Df is represented by the homotopy equivalence

$$C(\mathbb{R}^n | X \times \mathbb{R}^m | 0) \xleftarrow{\alpha} C(\mathbb{R}^n | X) \wedge C^m \simeq C(\mathbb{R}^n | X) \wedge S^m.$$

(Direct proof or an application of (7.4.3).)

5. Let $X \subset \mathbb{R}^n$ be compact. Suppose $\|x\| \le r > 0$ for $x \in X$. Then the constant function $\varphi(t) = r$ is a scaling function for each $f \colon X \to Y$. Show that the map

$$C(\mathbb{R}^m | Y) \wedge C^n \xrightarrow{\tau} C^n \wedge C(\mathbb{R}^m | Y) \to C(\mathbb{R}^n | X) \wedge C^m$$

which is obtained from the definition in (7.4.6) is homotopic to the map $C(D_\# f)$ of the general definition.

7.5 Duality

We have associated to a proper map between closed subsets of Euclidean spaces a dual morphism in the stable category ST. If $X \subset \mathbb{R}^n$ then the stable homotopy type of $\mathbb{R}^n \smallsetminus X$ or $C(\mathbb{R}^n|X)$ is to be considered as a dual object of X. There is a categorical notion of duality in tensor categories.

Let A and B be pointed spaces. An ***n-duality*** between (A, B) consists of an ***evaluation***

$$\varepsilon \colon B \wedge A \to S^n$$

and a ***coevaluation***

$$\eta \colon S^n \to A \wedge B$$

such that the following holds:

(1) The composition

$$(1 \wedge \varepsilon)(\eta \wedge 1) \colon S^n \wedge A \to A \wedge B \wedge A \to A \wedge S^n$$

is homotopic to the interchange map τ.

(2) The composition

$$(\varepsilon \wedge 1)(1 \wedge \eta) \colon B \wedge S^n \to B \wedge A \wedge B \to S^n \wedge B$$

is homotopic to $(-1)^n \tau$.

We now construct an n-duality for $(B, A) = (C(\mathbb{R}^n|K), K^+)$ where $K \subset \mathbb{R}^n$ is a suitable space. In the general definition of an n-duality above we now replace S^n by C^n. The evaluation is defined to be

$$\varepsilon \colon C(\mathbb{R}^n|K) \wedge C(K, \emptyset) \to C(\mathbb{R}^n|K \times K|K) \xrightarrow{C(d)} C(\mathbb{R}^n|0)$$

where d is the difference map

$$d \colon (\mathbb{R}^n \times K, (\mathbb{R}^n \smallsetminus K) \times K) \to (\mathbb{R}^n, \mathbb{R}^n \smallsetminus 0), \quad (x, k) \mapsto x - k$$

as a map of pairs. This definition works for arbitrary $K \subset \mathbb{R}^n$.

Let $K \subset \mathbb{R}^n$ be compact, $D \subset \mathbb{R}^n$ a disk which contains K, and V an open neighbourhood of K. Consider the following diagram with $\Delta \colon x \mapsto (x, x)$

$$\mathbb{R}^n|0 \xleftarrow[i]{\supset} \mathbb{R}^n|D \xrightarrow{\subset} \mathbb{R}^n|K \xleftarrow[j]{\supset} V|K \xrightarrow{\Delta} V|V \times \mathbb{R}^n|K.$$

We apply the mapping cone functor and (7.2.1) (7.2.7). The maps i and j induce h-equivalences. We obtain $\eta_V \colon C^n \to V^+ \wedge C(\mathbb{R}^n|K)$. Let now $r \colon V \to K$ be a retraction of $K \subset V$. We compose with $r^+ \colon V^+ \to K^+$ and obtain a coevaluation

$$\eta \colon C^n \to C(K, \emptyset) \wedge C(\mathbb{R}^n|K).$$

A closed subspace $K \subset \mathbb{R}^n$ is a Euclidean neighbourhood retract (= ENR) if there exists a retraction $r \colon V \to K$ from a neighbourhood V of K in $\mathbb{R}^n$. The existence of such a retraction does not depend on the particular embedding into a Euclidean space; see (18.4.1).

(7.5.1) Proposition. *The maps ε and η are an n-duality for $(K^+, C(\mathbb{R}^n|K))$.*

Proof. For the proof of axiom (1) we consider the diagram

$$\begin{array}{ccccc}
\mathbb{R}^n|D \times K|K & \longrightarrow & \mathbb{R}^n|K \times K|K & \xleftarrow{j\times 1} & V|K \times K|K \\
& \searrow^{\alpha} & \downarrow{\scriptstyle\beta} & & \downarrow{\scriptstyle\Delta\times 1} \\
& & V|V \times \mathbb{R}^n|0 & \xleftarrow{1\times d} & V|V \times \mathbb{R}^n|K \times K|K
\end{array}$$

with $\alpha(x,y) = (y,x)$, $\beta(x,y) = (y, x-y)$, and $\gamma = (1 \times d)(\Delta \times 1)\colon (x,y) \mapsto (x, x-y)$. The homotopy $(x,y,t) \mapsto (y, x-ty)$ shows that the triangle is h-commutative. The square becomes h-commutative after passage to mapping cones. For the proof of this claim consider $\lambda\colon V \times K \times I \to \mathbb{R}^n\colon (x,y,t) \mapsto tx + (1-t)y$. Since $\lambda(\Delta_K \times I) \subset K$, there exists an open neighbourhood L of Δ_K such that $\lambda(L \times I) \subset V$. Then consider, with the excision ι, the diagram

$$(L, L \smallsetminus \Delta_K) \xrightarrow{\iota} (V \times K, V \times K \smallsetminus \Delta_K) \longleftarrow (V \times K, (V \smallsetminus K) \times K).$$

The three terms are mapped by β, γ into $V|V \times \mathbb{R}^n|0$ and the left most β, γ are homotopic by the homotopy λ. The axiom (1) of an n-duality now follows if we write out the morphisms according to their definition and use the result just proved.

For the proof of the axiom (2) we start with the diagram

$$\begin{array}{ccccc}
\mathbb{R}^n|K \times \mathbb{R}^n|D & \xrightarrow{1\times j'} & \mathbb{R}^n|K \times \mathbb{R}^n|K & \xleftarrow{1\times j} & \mathbb{R}^n|K \times V|K \\
& \searrow^{\alpha} & \downarrow{\scriptstyle\beta} & \swarrow^{\gamma} & \downarrow{\scriptstyle(1\times r\times 1)(1\times\Delta)} \\
& & \mathbb{R}^n|D \times \mathbb{R}^n|K & \xleftarrow{d\times 1} & \mathbb{R}^n|K \times K|K \times \mathbb{R}^n|K
\end{array}$$

with $\alpha(x,y) = (-y,x)$, $\beta(x,y) = (x-y,x)$, and $\gamma(x,y) = (x-y,y)$. The homotopy $(x,y,t) \mapsto (x - ty - (1-t)r(y), y)$ shows that the bottom triangle is h-commutative; the homotopy $(x,y,t) \mapsto (x-y, (1-t)x + ty)$ shows $\gamma \simeq \beta(1 \times j)$; the homotopy $(x,y,t) \mapsto (tx - y, x)$ shows $\alpha \simeq \beta(1 \times j')$. Again we write out the morphisms according to their definition and use this result. □

Given a natural duality for objects via evaluations and coevaluations one can define the dual of an induced map. We verify that we recover in the case of compact ENR the morphisms constructed in the previous section. The following three proposition verify that the duality maps have the properties predicted by the categorical duality theory.

(7.5.2) Proposition. *Let $X \subset \mathbb{R}^n$ be compact and a retract of a neighbourhood V. The following diagram is homotopy commutative*

$$\begin{array}{ccc} C(\mathbb{R}^m|Y) \wedge C^n & \xrightarrow{1\wedge\eta_X} & C(\mathbb{R}^m|Y) \wedge C(X,\emptyset) \wedge C(\mathbb{R}^n|X) \\ \downarrow{\scriptstyle \tau(Df_{m+n})\tau} & & \downarrow{\scriptstyle 1\wedge C(f)\wedge 1} \\ C^m \wedge C(\mathbb{R}^n|X) & \xleftarrow{\varepsilon_Y\wedge 1} & C(\mathbb{R}^m|Y) \wedge C(Y,\emptyset) \wedge C(\mathbb{R}^n|X). \end{array}$$

Proof. We reduce the problem to maps of pairs. We use the simplified definition (7.4.6) of the duality map. First we have the basic reduction

$$\mathbb{R}^m|Y \times \mathbb{R}^n|0 \leftarrow \mathbb{R}^m|Y \times \mathbb{R}^n|D \rightarrow \mathbb{R}^m|Y \times \mathbb{R}^n|X \leftarrow \mathbb{R}^m|Y \times V|X.$$

Then the remaining composition $\mathbb{R}^m|Y \times V|X \to \mathbb{R}^m|0 \times \mathbb{R}^n|X$ which involves η, $C(f)$, ε is the assignment $(y,x) \mapsto (y - fr(x), x)$. The other map is $(y,x) \mapsto (y - \tilde{f}(x), x)$. Now we observe that we can arrange that $\tilde{f}|V = fr$ (by possibly passing to a smaller neighbourhood, see Problem 1). □

Dual maps are adjoint with respect to evaluation and coevaluation. This is the content of (7.5.3) and (7.5.4).

(7.5.3) Proposition. *The following diagram is homotopy commutative*

$$\begin{array}{ccc} C(\mathbb{R}^n|D) \wedge C(\mathbb{R}^m|Y) \wedge C(X,\emptyset) & \xrightarrow{D_\# f\wedge 1} & C(\mathbb{R}^n|X) \wedge C(\mathbb{R}^m|0) \wedge C(X,\emptyset) \\ \downarrow{\scriptstyle 1\wedge 1\wedge C(f)} & & \downarrow{\scriptstyle (\varepsilon\wedge 1)\tau} \\ C(\mathbb{R}^n|D) \wedge C(\mathbb{R}^m|Y) \wedge C(Y,\emptyset) & \xrightarrow{i\wedge\varepsilon} & C^n \wedge C^m. \end{array}$$

Proof. Consider the diagram

$$\begin{array}{ccc} \mathbb{R}^n|D^n \times \mathbb{R}^m|Y \times X|X & \xrightarrow{D_\# f\times 1} & \mathbb{R}^n|X \times \mathbb{R}^m|0 \times X|X \\ \downarrow{\scriptstyle 1\times 1\times f} & & \downarrow{\scriptstyle (d\times 1)\tau} \\ \mathbb{R}^n|D^n \times \mathbb{R}^m|Y \times Y|Y & \xrightarrow{i\times d} & \mathbb{R}^n|0 \times \mathbb{R}^m|0. \end{array}$$

The composition down-right sends the element (a,b,x) to $(a, b - f(x))$ and the composition right-down to $(\varphi(b)\cdot a - x, b - \tilde{f}(\varphi(b)\cdot a))$. We use the homotopies $(\varphi(b)\cdot a - x, b - \tilde{f}(t\varphi(b)\cdot a + (1-t)x))$ and then $((1-t)a + t(\varphi(b)\cdot a - x), b - f(x))$. □

(7.5.4) Proposition. *Let X and Y be compact and retracts of open neighbourhoods. Then the following diagram is homotopy commutative*

$$\begin{array}{ccc} C^n \wedge C^m & \xrightarrow{\eta\wedge 1} & C(X,\emptyset)\wedge C(\mathbb{R}^n|X)\wedge C^m \\ \downarrow{\scriptstyle \tau(1\wedge\eta)} & & \downarrow{\scriptstyle C(f)\wedge 1\wedge 1} \\ C(Y,\emptyset)\wedge C(\mathbb{R}^n|D)\wedge C(\mathbb{R}^m|Y) & \xrightarrow{1\wedge D_\# f} & C(Y,\emptyset)\wedge C(\mathbb{R}^n|X)\wedge C^m. \end{array}$$

Proof. We unravel the definitions and deform suitable maps between pairs. The composition $(1\wedge D_\# f)(\tau\wedge 1)(1\wedge\eta)$ is induced by maps

$$\begin{array}{ccccc} \mathbb{R}^n|0\times\mathbb{R}^m|0 & \longleftarrow & \mathbb{R}^n|D\times\mathbb{R}^m|D & \longrightarrow & \mathbb{R}^n|D\times\mathbb{R}^m|Y \\ & & & & \uparrow \\ & & Y|Y\times\mathbb{R}^n|X\times\mathbb{R}^m|0 & \xleftarrow{\alpha} & \mathbb{R}^n|D\times W|Y \end{array}$$

with $\alpha(x,y)=(r_Y(y),x,y-\tilde f(x))$. Further investigations concern α. We use the next diagram

$$\begin{array}{ccccc} \mathbb{R}^n|D\times W|Y & \longrightarrow & \mathbb{R}^n|X\times W|Y & \longleftarrow & V|X\times W|Y \\ & \searrow{\scriptstyle\alpha} & \downarrow{\scriptstyle\alpha} & \swarrow{\scriptstyle\alpha} & \downarrow \\ & & Y|Y\times\mathbb{R}^n|X\times\mathbb{R}^n|0 & \xleftarrow{\alpha} (V\times W)|G(f) & \longleftarrow U|G(f). \end{array}$$

Let

$$\lambda\colon V\times W\times I\to\mathbb{R}^m,\quad (x,y,t)\mapsto ty+(1-t)fr_X(x).$$

This homotopy is constant on $G(f)$. Hence there exists an open neighbourhood U of $G(f)$ such that $\lambda(U\times I)\subset W$. On U we consider the homotopy of α given by $(r_Y(ty+(1-t)fr_X(x)),x,y-\tilde f(x))$. For $t=0$ we obtain the morphism $(fr_X(x),x,y-\tilde f(x))$ which is defined on $V|X\times\mathbb{R}^m|Y$. Consider the composition $(C(f)\wedge 1\wedge 1)(\eta\wedge 1)$. It is induced by

$$V|X\times\mathbb{R}^m|0\to Y|Y\times\mathbb{R}^n|X\times\mathbb{R}^m|0,\quad (x,y)\mapsto(fr_X(x),x,y).$$

Now we use the homotopy $(fr_X(x),x,y-t\tilde f(x))$. For $t=1$ this homotopy is defined on $V|X\times\mathbb{R}^m|Y$. □

(7.5.5) Remark. Let $X\subset\mathbb{R}^n$ be a compact ENR and $f\colon X\to X$ a continuous map. From the associated n-duality we obtain a homotopy class

$$\lambda_f\colon S^n\xrightarrow{\eta}X^+\wedge C(\mathbb{R}^n|X)\xrightarrow{\tau}C(\mathbb{R}^n|X)\wedge X^+\xrightarrow{1\wedge f^+}C(\mathbb{R}^n|X)\wedge X^+\xrightarrow{\varepsilon}S^n.$$

The degree $d(\lambda_f)=L(f)\in\mathbb{Z}$ is an interesting invariant of the map f, the ***Lefschetz fixed point index***. If f is the identity, then $L(\mathrm{id})$ is the Euler characteristic of X ([51], [54]). ◇

Problems

1. Let A be a closed subset of a normal space X. Let $r: W \to A$ be a retraction of an open neighbourhood. Choose open sets U, V such that

$$A \subset U \subset \bar{U} \subset V \subset \bar{V} \subset W.$$

Choose a continuous function $\varphi: X \to [0,1]$ such that $\varphi(U) = \{1\}$ and $\varphi(X \smallsetminus V) = \{0\}$. Let $f: A \to [0,1]$ be continuous. Define $F: X \to [0,1]$ by $F(x) = \varphi(x) \cdot fr(x)$ for $x \in \bar{V}$ and $F(x) = 0$ for $x \in X \smallsetminus V$. Then F is a Tietze extension of f and $F|U = fr|U$.
2. Verify directly that the homotopy class of the coevaluation η does not depend on the choice of the retraction $r: V \to X$.
3. The n-dualities which we have constructed can be interpreted as representative elements for morphisms in the category ST. We obtain

$$\varepsilon: (C(\mathbb{R}^n|X), -n) \otimes (X^+, 0) \to (S^0, 0)$$
$$\eta: (S^0, 0) \to (X^+, 0) \otimes (C(\mathbb{R}^n|X), -n).$$

They satisfy the relations

$$(1 \wedge \varepsilon)(\eta \wedge 1) = \mathrm{id}, \quad (\varepsilon \wedge 1)(1 \wedge \eta) = \mathrm{id}$$

which define dualities in tensor categories.

7.6 Homology and Cohomology for Pointed Spaces

A ***homology theory for pointed spaces*** with values in the category R-MOD of left modules over the commutative ring R consists of a family $(\tilde{h}_n \mid n \in \mathbb{Z})$ of functors $\tilde{h}_n$: TOP$^0 \to R$-MOD and a family $(\sigma_{(n)} \mid n \in \mathbb{Z})$ of natural suspension isomorphisms $\sigma = \sigma_{(n)}: \tilde{h}_n \to \tilde{h}_{n+1} \circ \Sigma$. These data are required to satisfy the following axioms.

(1) **Homotopy invariance.** For each pointed homotopy f_t the equality $\tilde{h}_n(f_0) = \tilde{h}_n(f_1)$ holds.

(2) **Exactness.** For each pointed map $f: X \to Y$ the induced sequence $\tilde{h}_n(X) \xrightarrow{f_*} \tilde{h}_n(Y) \xrightarrow{f_1*} \tilde{h}_n(C(f))$ is exact.

Let $(X_j \mid j \in J)$ be a family of well-pointed spaces with inclusions $\iota_\nu: X_\nu \to \bigvee_{j \in J} X_j$ of the summands. The theory is called ***additive***, if

$$\textstyle\bigoplus_{j \in J} \tilde{h}_n(X_j) \to \tilde{h}_n(\bigvee_{j \in J} X_j), \quad (x_j) \mapsto \sum_{j \in J} (i_j)_*(x_j)$$

is always an isomorphism.

As a variant of the axioms we require the suspension isomorphisms and the exact sequences only for well-pointed spaces.

If we apply the exactness axiom to the identity of a point P we see that $\tilde{h}_n(P) = 0$. If X and Y are well-pointed, then the inclusion and projection give a cofibre sequence $X \to X \vee Y \to Y$. This is used to verify that the additivity isomorphism holds for a finite number of well-pointed spaces. The groups $\tilde{h}_n(S^0)$ are the coefficient groups of the theory.

A natural transformation of homology theories for pointed spaces consists of a family of natural transformations $\tilde{h}_n(-) \to \tilde{k}_n(-)$ which commute with the suspension isomorphisms.

A ***cohomology theory for pointed spaces*** consists of a family of contravariant functors $\tilde{h}^n\colon \mathrm{TOP}^0 \to R\text{-}\mathrm{MOD}$ and natural suspension isomorphisms $\sigma = \sigma^{(n)}\colon \tilde{h}^n \to \tilde{h}^{n+1} \circ \Sigma$ such that the analogous axioms (1) and (2) hold. The theory is called ***additive***, if

$$\tilde{h}^n(\textstyle\bigvee_{j\in J} X_j) \to \prod_{j\in J} \tilde{h}^n(X_j), \quad x \mapsto ((i_j)^*(x))$$

is always an isomorphism for well-pointed spaces X_j.

In Chapter 10 we define homology theories by the axioms of Eilenberg and Steenrod. They involve functors on TOP(2). We show in Section 10.4 that they induce a homology theory for pointed spaces as defined above.

Given a homology theory $\tilde{h}_*$ for pointed spaces we construct from it a homology theory for pairs of spaces as follows. We set $h_n(X,A) = \tilde{h}_n(C(X,A))$. It should be clear that the h_n are part of a homotopy invariant functor $\mathrm{TOP}(2) \to R\text{-}\mathrm{MOD}$. We define the boundary operator as the composition

$$\partial\colon h_n(X,A) = \tilde{h}_n(C(i^+)) \xrightarrow{p(i)_*} \tilde{h}_n(\Sigma(A^+)) \cong \tilde{h}_{n-1}(A^+) = h_{n-1}(A).$$

The isomorphism is the given suspension isomorphism of the theory $\tilde{h}_*$. The Eilenberg–Steenrod exactness axioms holds; it is a consequence of the assumption that $\tilde{h}_*$ transforms a cofibre sequence into an exact sequence and of the naturality of the suspension isomorphism. The excision isomorphism follows from (7.2.5). We need the additional hypothesis that the covering is numerable. Remark (7.2.6) is relevant for the passage from one set of axioms to the other.

7.7 Spectral Homology and Cohomology

In this section we report about the homotopical construction of homology and cohomology theories. We work in the category of compactly generated spaces. A ***pre-spectrum*** consists of a family $(Z(n) \mid n \in \mathbb{Z})$ of pointed spaces and a family $(e_n\colon \Sigma Z(n) \to Z(n+1) \mid n \in \mathbb{Z})$ of pointed maps. Since we only work with pre-spectra in this text, we henceforth just call them spectra. A spectrum is called an ***Ω-spectrum***, if the maps $\varepsilon_n\colon Z(n) \to \Omega Z(n+1)$ which are adjoint to e_n are pointed homotopy equivalences.

Let $Z = (Z(n), \varepsilon_n)$ be an Ω-spectrum. We define $Z^n(X) = [X, Z(n)]^0$ for a pointed space X. Since $Z(n)$ is up to h-equivalence a double loop space, namely

$Z(n) \simeq \Omega^2 Z(n+2)$, we see that $Z^n(X)$ is an abelian group, and we can view Z^n as a contravariant and homotopy invariant functor $\mathrm{TOP}^0 \to \mathbb{Z}\text{-}\mathrm{MOD}$. We define $\sigma\colon Z^n(X) \cong Z^{n+1}(\Sigma X)$ via the structure maps and adjointness as

$$[X, Z(n)]^0 \xrightarrow{(\varepsilon_n)_*} [X, \Omega Z(n+1)]^0 \cong [\Sigma X, Z(n+1)]^0.$$

We thus have the data for a cohomology theory on TOP^0. The axioms are satisfied (Puppe sequence). The theory is additive.

We now associate a cohomology theory to an arbitrary spectrum $Z = (Z(n), e_n)$. For $k \geq 0$ we have morphisms

$$b_n^k\colon [\Sigma^k X, Z(n)]^0 \xrightarrow{\Sigma} [\Sigma(\Sigma^k X), \Sigma Z(n)]^0 \xrightarrow{(e_n)_*} [\Sigma^{k+1} X, Z(n+1)]^0 .$$

Let $Z^{n-k}(X)$ be the colimit over this system of morphisms. The b_n^k are compatible with pointed maps $f\colon X \to Y$ and induce homomorphisms of the colimit groups. In this manner we consider Z^n as a homotopy invariant, contravariant functor $\mathrm{TOP}^0 \to \mathbb{Z}\text{-}\mathrm{MOD}$. (The b_n^k are for $k \geq 2$ homomorphisms between abelian groups.) The exactness axiom again follows directly from the cofibre sequence. The suspension isomorphism is obtained via the identity

$$[\Sigma^{k+1} X, Z(n+k+1)]^0 \cong [\Sigma^k(\Sigma X), Z(n+k+1)]^0$$

which gives in the colimit $Z^n(X) \cong Z^{n+1}(\Sigma X)$. If the spectrum is an Ω-spectrum, we get the same theory as before, since the canonical morphisms $[X, Z(n)]^0 \to Z^n(X)$ are natural isomorphisms of cohomology theories. Because of the colimit process we need the spaces $Z(k)$ only for $k \geq k_0$. We use this remark in the following examples.

7.7.1 Sphere spectrum. We define $Z(n) = S(n)$ and $e_n\colon \Sigma S(n) \cong S(n+1)$ the identity. We set $\omega^k(X) = \mathrm{colim}_n[\Sigma^n X, S^{n+k}]^0$ and call this group the k-th stable ***cohomotopy group*** of X. ◇

7.7.2 Suspension spectrum. Let Y be a pointed space. We define a spectrum with spaces $\Sigma^n Y$ and $e_n\colon \Sigma(\Sigma^n Y) \cong \Sigma^{n+1} Y$. ◇

7.7.3 Smash product. Let $Z = (Z(n), e_n)$ be a spectrum and Y a pointed space. The spectrum $Y \wedge Z$ consists of the spaces $Y \wedge Z(n)$ and the maps

$$\mathrm{id} \wedge e_n\colon \Sigma(Y \wedge Z(n)) \cong Y \wedge \Sigma Z(n) \to Y \wedge Z(n+1).$$

(Note that $\Sigma A = A \wedge I/\partial I$. Here and in other places we have to use the associativity of the $\wedge$-product. For this purpose it is convenient to work in the category of k-spaces.) We write in this case

$$Z^k(X; Y) = \mathrm{colim}_n[\Sigma^n X, Y \wedge Z(n+k)]^0.$$

The functors $Z^k(-;Y)$ depend covariantly on Y: A pointed map $f\colon Y_1 \to Y_2$ induces a natural transformation of cohomology theories $Z^k(-;Y_1) \to Z^k(-;Y_2)$. ◇

In general, the definition of the cohomology theory $Z^*(-)$ has to be improved, since this theory may not be additive.

We now construct homology theories. Let

$$E = (E(n), e_n\colon E(n) \wedge S^1 \to E(n+1) \mid n \in \mathbb{Z})$$

be a spectrum. We use spheres as pointed spaces and take as standard model the one-point compactification $S^n = \mathbb{R}^n \cup \{\infty\}$. If V is a vector space one often writes $S^V = V \cup \{\infty\}$ with base point ∞. Then we have a canonical homeomorphism $S^V \wedge S^W \cong S^{V \oplus W}$, the identity away from the base point. A linear isomorphism $f\colon V \to W$ induces a pointed map $S^f\colon S^V \to S^W$.

The homology group $E_k(X)$ of a pointed space X is defined as colimit over the maps

$$b\colon [S^{n+k}, X \wedge E(n)]^0 \to [S^{n+k} \wedge S^1, X \wedge E(n) \wedge S^1]^0 \to [S^{n+k+1}, X \wedge E(n+1)]^0.$$

The first map is $- \wedge S^1$ and the second map is induced by $\mathrm{id} \wedge e_n$. For $n+k \geq 2$ the morphism b is a homomorphism between abelian groups.

It should be clear from the definition that $E_k(-)$ is a functor on TOP^0. We need the suspension morphisms. We first define suspension morphisms

$$\sigma_l\colon E_k(Z) \to E_{k+1}(S^1 \wedge Z).$$

They arise from the suspensions $S^1 \wedge -$

$$[S^{n+k}, Z \wedge E(n)]^0 \to [S^{n+k+1}, S^1 \wedge Z \wedge E(n)]^0,$$

which are compatible with the maps b above. Then we set $\sigma = (-1)^k \tau_* \sigma_l$ where the map $\tau\colon S^1 \wedge Z \to Z \wedge S^1$ interchanges the factors.

(7.7.4) Lemma. *σ_l is an isomorphism.*

Proof. Let $x \in E_k(Z)$ be contained in the kernel of σ_l. Then there exists $[f] \in [S^{n+k}, Z \wedge E(n)]^0$ representing x such that $1 \wedge f$ is null homotopic. Consider the diagram

$$\begin{array}{ccccc} S^1 \wedge S^{n+k} & \xrightarrow{1\wedge f} & S^1 \wedge Z \wedge E(n) & & \\ \downarrow \tau_1 & & \downarrow \tau_2 & & \\ S^{n+k} \wedge S^1 & \xrightarrow{f \wedge 1} & Z \wedge E(n) \wedge S^1 & \xrightarrow{1 \wedge e} & Z \wedge E(n+1) \end{array}$$

with permutation of factors τ_1 and τ_2. Since $1 \wedge f$ is null homotopic, the representative $(1 \wedge e) \circ (f \wedge 1)$ of x is also null homotopic. This shows that σ_l is injective.

In order to prove surjectivity we consider the two-fold suspension. Let $x \in E_{k+2}(S^2 \wedge Z)$ have the representative $g\colon S^{n+k+2} \to S^2 \wedge Z \wedge E(n)$. Then

$$f\colon S^{n+k+2} \xrightarrow{g} S^2 \wedge Z \wedge E(n) \xrightarrow{\tau} Z \wedge E(n) \wedge S^2 \xrightarrow{e^2} Z \wedge E(n+2)$$

represents an element $y \in E_k(Z)$. Here e^2 is the composition of the spectral structure maps

$$E(n) \wedge S^2 = (E(n) \wedge S^1) \wedge S^1 \to E(n+1) \wedge S^1 \to E(n+2).$$

We show $\sigma_l^2(y) = x$. Once we have proved this we see that the second σ_l is surjective and injective and hence the same holds for the first σ_l.

The proof of the claim is based on the next diagram with interchange maps τ, τ', τ''.

$$\begin{array}{ccc}
S^2 \wedge S^{n+k+2} & \xrightarrow{\tau''} & S^{n+k+2} \wedge S^2 \\
\downarrow{\scriptstyle 1\wedge g} & & \downarrow{\scriptstyle g\wedge 1} \\
S^2 \wedge (S^2 \wedge Z \wedge E(n)) & \xrightarrow{\tau''_1} & S^2 \wedge Z \wedge E(n) \wedge S^2 \\
\downarrow{\scriptstyle 1\wedge\tau} & \swarrow{\scriptstyle \tau'} & \\
S^2 \wedge (Z \wedge E(n) \wedge S^2) & & \downarrow{\scriptstyle 1\wedge e^2} \\
\downarrow{\scriptstyle 1\wedge e^2} & & \\
S^2 \wedge Z \wedge E(n+2) & \xrightarrow{=} & S^2 \wedge Z \wedge E(n+2)
\end{array}$$

The maps τ' and τ'' are homotopic to the identity, since we are interchanging a sphere with an even-dimensional sphere. The composition of the left verticals represents $\sigma_l^2(y)$, and the composition of the right verticals represents x. □

(7.7.5) Proposition. *For each pointed map $f: Y \to Z$ the sequence*

$$E_k(Y) \xrightarrow{f_*} E_k(Z) \xrightarrow{f_{1*}} E_k(C(f))$$

is exact.

Proof. The exactness is again a simple consequence of the cofibre sequence. But since the cofibre sequence is inserted into the "wrong" covariant part, passage to the colimit is now essential. Suppose $z \in E_k(Z)$ is contained in the kernel of f_{1*}. Then there exists a representing map $h\colon S^{n+k} \to Z \wedge E(n)$ such that

$(f_1 \wedge 1) \circ h$ is null homotopic. The next diagram compares the cofibre sequences of $\mathrm{id}\colon S^{n+k} \to S^{n+k}$ and $f \wedge 1\colon Y \wedge E(n) \to Z \wedge E(n)$.

$$\begin{array}{ccccccc}
S^{n+k} & \xrightarrow{\mathrm{id}_1} & C(\mathrm{id}) & \xrightarrow{p(\mathrm{id})} & S^{n+k} \wedge S^1 & \xrightarrow{\mathrm{id}} & S^{n+k} \wedge S^1 \\
\downarrow{\scriptstyle h} & & \downarrow{\scriptstyle H} & & \downarrow{\scriptstyle \beta} & & \downarrow{\scriptstyle h\wedge 1} \\
Z \wedge E(n) & \xrightarrow{(f\wedge 1)_1} & C(f \wedge 1) & \xrightarrow{p(f\wedge 1)} & Y \wedge E(n) \wedge S^1 & \xrightarrow{f\wedge 1\wedge 1} & Z \wedge E(n) \wedge S^1 \\
& \searrow{\scriptstyle f_1\wedge 1} & \downarrow{\scriptstyle \varphi} & & \downarrow{\scriptstyle 1\wedge e} & & \downarrow{\scriptstyle 1\wedge e} \\
& & C(f) \wedge E(n) & & Y \wedge E(n+1) & \xrightarrow{f\wedge 1} & Z \wedge E(n+1)
\end{array}$$

The map φ is the canonical homeomorphism (in the category of k-spaces) which makes the triangle commutative. Since $(f \wedge 1)_1 \circ h$ is null homotopic, there exists H such that the first square commutes. The map β is induced from (h, H) by passing to the quotients, therefore the second square commutes. It is a simple consequence of the earlier discussion of the cofibre sequence that the third square is h-commutative (Problem 1). The composition $(1\wedge e)(h\wedge 1)$ is another representative of z, and the diagram shows that $(1 \wedge e)\beta$ represents an element $y \in E_k(Y)$ such that $f_* y = z$. □

A similar proof shows that the $Z^k(X; Y)$ form a homology theory in the variable Y.

Problems

1. Consider the cofibre sequences of two maps $f\colon A \to B$ and $f'\colon A' \to B'$. In the diagram

$$\begin{array}{ccccccccccc}
A & \xrightarrow{f} & B & \xrightarrow{f_1} & C(f) & \xrightarrow{p(f)} & \Sigma A & \xrightarrow{\Sigma f} & \Sigma B & \longrightarrow & \cdots \\
& & \downarrow{\scriptstyle h} & & \downarrow{\scriptstyle H} & & \downarrow{\scriptstyle \beta} & & \downarrow{\scriptstyle \Sigma h} & & \\
A' & \xrightarrow{f'} & B' & \xrightarrow{f'_1} & C(f') & \xrightarrow{p(f')} & \Sigma A' & \xrightarrow{\Sigma f'} & \Sigma B' & \longrightarrow & \cdots
\end{array}$$

assume given h and H such that the first square commutes. The map β is induced from (h, H) by passing to the quotients. Show that the third square commutes up to homotopy (use (4.6.2)).

2. Show that the homology theory defined by a spectrum is additive (for families of well-pointed spaces).

3. Show that a weak pointed h-equivalence between well-pointed spaces induces an isomorphism in spectral homology. The use of k-spaces is therefore not essential.

4. Let Z be the sphere spectrum (7.7.1). Then, in the notation of (7.7.3),

$$\mathrm{ST}((X,n),(Y,m)) = Z^{m-n}(X;Y),$$

the morphism set of the category ST of Section 7.1.

7.8 Alexander Duality

Let $E = (E_n, e(n)\colon E_n \wedge S^1 \to E_{n+1})$ be a spectrum. Let $\eta\colon S^n \to B \wedge A$, $\varepsilon\colon A \wedge B \to S^n$ be an n-duality. The compositions

$$[A \wedge S^t, E_{k+t}]^0 \xrightarrow{B\wedge -} [B \wedge A \wedge S^t, B \wedge E_{k+t}]^0 \xrightarrow{\eta^*} [S^n \wedge S^t, B \wedge E_{k+t}]^0$$

are compatible with the passage to the colimit and induce a homomorphism

$$D^\bullet\colon E^k(A) \to E_{n-k}(B).$$

The compositions

$$[S^{t+n-k}, B \wedge E_t]^0 \xrightarrow{A\wedge} [A \wedge S^{t+n-k}, A \wedge B \wedge E_t]^0 \xrightarrow{\varepsilon_*} [A \wedge S^{n+k-t}, S^n \wedge E_t]^0$$

$$\xrightarrow{\tau_*} [A \wedge S^{n+k-t}, E_t \wedge S^n]^0 \xrightarrow{e_*} [A \wedge S^{t+n-k}, E_{t+n}]^0$$

with the interchange map τ are compatible with the passage to the colimit if we multiply them by $(-1)^{nt}$. They induce a homomorphism

$$D_\bullet\colon E_{n-k}(B) \to E^k(A).$$

(7.8.1) Theorem (Alexander duality). *The morphisms $D^\bullet$ and $D_\bullet$ are isomorphisms. They satisfy $D_\bullet D^\bullet = (-1)^{nk}\,\mathrm{id}$ and $D^\bullet D_\bullet = (-1)^{nk}\,\mathrm{id}$.*

Proof. The relations of the theorem are a direct consequence of the defining properties of an n-duality. The composition $\varepsilon_* \circ (A \wedge -) \circ \eta^* \circ (B \wedge -)$ equals

$$\tau_* \circ \Sigma^n\colon [A \wedge S^t, E_{k+t}]^0 \to [S^n \wedge A \wedge S^t, S^n \wedge E_{k+t}]^0 \to [A \wedge S^n \wedge S^t, S^n \wedge E_{k+t}]^0.$$

The definition of $D_\bullet$ then involves the sign $(-1)^{n(k+t)}$. This morphism differs from a map in the direct system for $E^k(A)$ by the interchange $S^n \wedge S^t \to S^t \wedge S^n$, hence by a sign $(-1)^{nt}$. Hence the sign $(-1)^{nk}$ remains. The second relation is verified similarly. □

Let $PE_*(-)$ and $Ph^*(-)$ be the homology and cohomology theories on TOP(2) constructed from the theories $E_*(-)$ and $E^*(-)$. If we use the n-duality between X^+ and $C(\mathbb{R}^n|X)$ for a compact ENR in $\mathbb{R}^n$ we obtain isomorphisms

$$PE_{n-k}(\mathbb{R}^n, \mathbb{R}^n \smallsetminus X) \cong PE^k(X), \quad PE_{n-k}(X) \cong PE^k(\mathbb{R}^n, \mathbb{R}^n \smallsetminus X).$$

This is the usual appearance of Alexander duality.

In this setting one can also work with the bi-variant theory $Z^k(X;Y) = Z_{-k}(X;Y)$. Then one obtains from an n-duality an adjointness isomorphism $Z^k(A \wedge X; Y) \cong Z_{n-k}(X; B \wedge Y)$.

A homology theory $h_*(-)$ is defined on the category ST. Here one defines $h_l((X,n)) = h_{l-n}(X)$. Let a morphism $f \in \mathrm{ST}((X,n),(Y,m))$ be represented by $f_k\colon X \wedge S^{n+k} \to Y \wedge S^{m+k}$. The induced morphism is defined by commutativity of the next diagram

$$\begin{array}{ccccc} h_l((X,n)) & \xrightarrow{=} & h_{l-n}(X) & \xrightarrow{\Sigma^{n+k}} & h_{l+k}(X \wedge S^{n+k}) \\ \downarrow{\scriptstyle h_l(f)} & & & & \downarrow{\scriptstyle (f_k)_*} \\ h_l((Y,m)) & \xrightarrow{=} & h_{l-m}(Y) & \xrightarrow{\Sigma^{m+k}} & h_{l+k}(Y \wedge S^{m+k}). \end{array}$$

Given a homology theory $h_*(-)$ one can define via the complement duality functor a sort of cohomology for spaces which admit an embedding as a closed subset of a Euclidean space. Let X be such a space. Choose an embedding $i\colon X \to \mathbb{R}^n$ and define $h^k(iX) = h_{n-k}(C(\mathbb{R}^n|iX)) = h_{-k}(C(\mathbb{R}^n|iX),-n)$. If $j\colon X \to \mathbb{R}^m$ is another embedding, we have the homeomorphism $ji^{-1}\colon iX \to jY$ and we have the duality map $D(ji^{-1})$. The set of embeddings together with the morphisms $D(ji^{-1})$ from i to j form a contractible groupoid; it is a complicated replacement for the space X. We obtain the induced contractible groupoid of the $h^k(iX)$. It is equivalent to a group which we denote $h^k(X)$. From the complement duality functor we obtain a well-defined homomorphism $h^k(f)\colon h^k(Y) \to h^k(X)$ for a proper map $f\colon X \to Y$; in this way $h^k(-)$ becomes a contravariant functor. We do not discuss in what sense the $h^k(X)$ can be made into a cohomology theory. This cohomology theory is the "correct" one for duality theory in the sense that the Alexander duality isomorphism $h^k(X) \cong h_{-k}(C(\mathbb{R}^n|X),-n)$ holds for all spaces in question (and not only for compact ENR). A similar device can be applied to a given cohomology theory. One obtains a homology theory which is again the "correct" one for duality theory.

7.9 Compactly Generated Spaces

Several constructions in homotopy theory lead to problems in general topology. A typical problem arises from the fact that a product of quotient maps is in general no longer a quotient map. In this auxiliary section we report about some devices to deal with such problems. The idea is to construct a category with better formal properties. One has to pay a price and change some of the standard notions, e.g., redefine topological products.

A compact Hausdorff space will be called a ch-space. For the purpose of the following investigations we also call a ch-space a ***test space*** and a continuous map $f\colon C \to X$ of a test space C a ***test map***. A space X is called ***weakly hausdorff*** or ***wh-space***, if the image of each test map is closed.

(7.9.1) Proposition. *A Hausdorff space is a wh-space. A wh-space is a T_1-space. A space X is a wh-space if and only if each test map $f: K \to X$ is proper. If X is a wh-space, then the image of each test map is a Hausdorff space. A subspace of a wh-space is a wh-space. Products of wh-spaces are wh-spaces.* □

A subset A of a topological space $(X, \mathcal{T})$ is said to be ***k-closed*** (***k-open***), if for each test map $f: K \to X$ the pre-image $f^{-1}(A)$ is closed (open) in K. The k-open sets in $(X, \mathcal{T})$ form a topology $k\mathcal{T}$ on X. A closed (open) subset is also k-closed (k-open). Therefore $k\mathcal{T}$ is finer than $\mathcal{T}$ and the identity $\iota = \iota_X : kX \to X$ is continuous. We set $kX = k(X) = (X, k\mathcal{T})$. Let $f: K \to X$ be a test map. The same set map $f: K \to kX$ is then also continuous. For if $U \subset kX$ is open, then $U \subset X$ is k-open, hence $f^{-1}(U) \subset K$ is open. Therefore ι_X induces for each ch-space K a bijection.

$$\mathrm{TOP}(K, kX) \xrightarrow{\cong} \mathrm{TOP}(K, X), \quad f \mapsto \iota_X \circ f.$$

Hence X and kX have the same k-open sets, i.e., $k(kX) = kX$. A topological space X is called ***k-space***, if the k-closed sets are closed, i.e., if $X = kX$. Because of $k(kX) = kX$ the space kX is always a k-space. A k-space is also called ***compactly generated***. We let k-TOP be the full subcategory of TOP with objects the k-spaces. A whk-space is a space which is a wh-space and a k-space.

The next proposition explains the definition of a k-space. We call a topology $\mathcal{S}$ on X ch-definable, if there exists a family $(f_j : K_j \to X \mid j \in J)$ of test maps such that: $A \subset X$ is $\mathcal{S}$-closed $\Leftrightarrow$ for each $j \in J$ the pre-image $f_j^{-1}(A)$ is closed in K_j. We can rephrase this condition: The canonical map $\langle f_j \rangle : \coprod_j K_j \to (X, \mathcal{S})$ is a quotient map. A ch-definable topology is finer than $\mathcal{T}$. We define a partial ordering on the set of ch-definable topologies by $\mathcal{S}_1 \le \mathcal{S}_2 \Leftrightarrow \mathcal{S}_1 \supset \mathcal{S}_2$.

(7.9.2) Proposition. *The topology $k\mathcal{T}$ is the maximal ch-definable topology with respect to the partial ordering.*

Proof. By Zorn's lemma there exists a maximal ch-definable topology $\mathcal{S}$. If this topology is different from $k\mathcal{T}$, then there exists an $\mathcal{S}$-open set U, which is not k-open. Hence there exists a test map $t: K \to X$ such that $t^{-1}(U)$ is not open. If we adjoin this test map to the defining family of $\mathcal{S}$, we see that $\mathcal{S}$ is not maximal. □

(7.9.3) Corollary. *The k-spaces are the spaces which are quotients of a topological sum of ch-spaces.* □

(7.9.4) Proposition. *The following are equivalent:*

(1) *X is a k-space.*

(2) *A set map $f: X \to Y$ is continuous if and only if for each test map $t: K \to X$ the composition ft is continuous.*

Proof. (1) ⇒ (2). Let $U \subset Y$ be open. In order to see that $f^{-1}(U)$ is open it suffices to show that this set is k-open, since X is a k-space. Let $t: K \to X$ be a test map and ft continuous. Then $k^{-1}(f^{-1}(U))$ is open, and this shows what we want.

(2) ⇒ (1). We show that the identity $X \to kX$ is continuous. This holds by (2) and because X and kX have the same test maps. □

(7.9.5) Proposition. *Let $f: X \to Y$ be continuous. Then the same set map $kf: kX \to kY$ is continuous.* □

The assignments $X \mapsto kX$, $f \mapsto kf$ yield a functor k; moreover, we have the inclusion functor i,

$$k\colon \mathrm{TOP} \to \text{k-}\mathrm{TOP}, \quad i\colon \text{k-}\mathrm{TOP} \to \mathrm{TOP}.$$

(7.9.6) Proposition. *The functor k is right adjoint to the functor i.*

Proof. A natural bijection is $\text{k-TOP}(Y, kX) \cong \mathrm{TOP}(iY, X)$, $f \mapsto \iota \circ f$. This map is certainly injective. If Y is a k-space and $f: Y \to X$ continuous, then $kf: Y = kY \to kX$ is continuous; this is used to show surjectivity. □

(7.9.7) Proposition. *Let X be a wh-space. Then $A \subset X$ is k-closed if and only if for each ch-space $K \subset X$ the set $A \cap K$ is closed in K. In particular a wh-space X is a k-space if and only if: $A \subset X$ is closed ⇔ for each ch-space $K \subset X$ the intersection $A \cap K$ is closed in K.*

Proof. Let A be k-closed. The inclusion $K \subset X$ of a ch-space is a test map. Hence $A \cap K$ is closed in K.

Conversely, suppose that A satisfies the stated condition and let $f: L \to X$ be a test map. Since X is a wh-space, $f(L)$ is a ch-space and therefore $f(L) \cap A$ is closed in $f(L)$. Then $f^{-1}(A) = f^{-1}(f(L) \cap A)$ is closed in $L = f^{-1}f(L)$. This shows: A is k-closed. □

Thus we see that wh-spaces have an internal characterization of their k-closed sets. For wh-spaces therefore $k(X)$ can be defined from internal properties of X. If X is a wh-space, so is kX.

(7.9.8) Theorem. *X is a k-space under one of the following conditions:*

(1) *X is metrizable.*

(2) *Each point of X has a countable neighbourhood basis.*

(3) *Each point of X has a neighbourhood which is a ch-space.*

(4) *For $Q \subset X$ and $x \in \overline{Q}$ there exists a ch-subspace $K \subset X$ such x is contained in the closure of $Q \cap K$ in K.*

(5) *For each $Q \subset X$ the following holds: $Q \cap K$ open (closed) in K for each test space $K \subset X$ implies Q open (closed) in X.*

Proof. (1) is a special case of (2).

(2) Let $Q \subset X$ and suppose that $f^{-1}(Q)$ is closed for each test map $f: C \to X$. We have to show that Q is closed. Thus let $a \in \bar{Q}$ and let $(U_n \mid n \in \mathbb{N})$ be a neighbourhood basis of a. For each n choose $a_n \in Q \cap U_1 \cap \cdots \cap U_n$. Then the sequence (a_n) converges to a. The subspace $K = \{0, 1, 2^{-1}, 3^{-1}, \dots\}$ of $\mathbb{R}$ is compact. The map $f: K \to X$, $f(0) = a$, $f(n^{-1}) = a_n$ is continuous, and $n^{-1} \in f^{-1}(Q)$. By assumption, $f^{-1}(Q)$ is closed in K, hence $0 \in f^{-1}(Q)$, and therefore $a = f(0) \in Q$.

(3) $\Rightarrow$ (4). Let $Q \subset X$ and suppose $a \in \bar{Q}$. We choose a ch-neighbourhood K of a and show that a is contained in the closure of $Q \cap K$ in K. Thus let U be a neighbourhood of a in K. Then there exists a neighbourhood U' of a in X such that $U' \cap K \subset U$. Since $U' \cap K$ is a neighbourhood of a in X and $a \in \bar{Q}$, we conclude

$$U \cap (Q \cap K) \supset (U' \cap K) \cap (Q \cap K) = (U' \cap K) \cap Q \neq \emptyset.$$

Hence a is contained in the closure of $Q \cap K$ in K.

(4) $\Rightarrow$ (5). Suppose $Q \cap K$ is closed in K for every test subspace $K \subset X$. Let $a \in \bar{Q}$. By (4), there exists a test subspace K_0 of X, such that a is contained in the closure of $Q \cap K_0$ in K_0. By the assumption (5), $Q \cap K_0$ is closed in K_0; and hence $a \in Q \cap K_0 \subset Q$.

(5) Let $f^{-1}(Q)$ be closed in K for each test map $f: K \to X$. Then, in particular, for each test subspace $L \subset X$ the set $Q \cap L$ is closed in L. The assumption (5) then says that Q is closed in X. This shows that X is a k-space. □

(7.9.9) Theorem. *Let $p: Y \to X$ be a quotient map and Y a k-space. Then X is a k-space.*

Proof. Let $B \subset X$ be k-closed. We have to show that B is closed, hence, since p is a quotient map, that $p^{-1}(B)$ is closed in Y. Let $g: D \to Y$ be a test map. Then $g^{-1}(p^{-1}(B)) = (pg)^{-1}(B)$ is closed in D, because B is k-closed. Since Y is a k-space, $p^{-1}(B)$ is closed in Y. □

(7.9.10) Proposition. *A closed (open) subspace of a k-space is a k-space. The same holds for whk-spaces.*

Proof. Let A be closed and $B \subset A$ a subset such that $f^{-1}(B)$ is closed in C for test maps $f: C \to A$. We have to show: B is closed in A or, equivalently, in X.

If $g: D \to X$ is a test map, then $g^{-1}(A)$ is closed in D and hence compact, since D is compact. The restriction of g yields a continuous map $h: g^{-1}(A) \to A$. The set $h^{-1}(B) = g^{-1}(B)$ is closed in $g^{-1}(A)$ and therefore in D, and this shows that B is closed in X.

Let U be open in the k-space X. We write X as quotient $q\colon Z \to X$ of a sum Z of ch-spaces (see 7.9.3). Then $q\colon q^{-1}(U) \to U$ is a quotient map and $q^{-1}(U)$ as the topological sum of locally compact Hausdorff spaces is a k-space. Therefore the quotient U is a k-space.

The second assertion follows, if we take (7.9.1) into account. □

In general, a subspace of a k-space is not a k-space (see (7.9.23)). Let X be a k-space and $i: A \subset X$ the inclusion. Then the map $k(i)\colon k(A) \to X = k(X)$ is continuous. The next proposition shows that $k(i)$ has in the category k-TOP the formal property of a subspace.

(7.9.11) Proposition. *A map $h\colon Z \to k(A)$ from a k-space Z into $k(A)$ is continuous if and only if $k(i) \circ h$ is continuous.*

Proof. If h is continuous then also is $k(i) \circ h$. Conversely, let $k(i) \circ h$ be continuous. We have $k(i) = i \circ \iota_A$. Since i is the inclusion of a subspace, $\iota_A \circ h$ is continuous; (7.9.6) now shows that h is continuous. □

(7.9.12) Theorem. *The product in* TOP *of a k-space X with a locally compact Hausdorff space Y is a k-space.*

Proof. By (7.9.8), a locally compact Hausdorff space is a k-space. We write X as quotient of Z, where Z is a sum of ch-spaces, see (7.9.3). Since the product of a quotient map with a locally compact space is again a quotient map, we see that $X \times Y$ is a quotient of the locally compact Hausdorff space, hence k-space, $Z \times Y$, and therefore a k-space by (7.9.9). □

A product of k-spaces is not always a k-space (see (7.9.23)). Therefore one is looking for a categorical product in the category k-TOP. Let $(X_j \mid j \in J)$ be a family of k-spaces and $\prod_j X_j$ its product in the category TOP, i.e., the ordinary topological product. We have a continuous map

$$p_j = k(\mathrm{pr}_j)\colon k(\textstyle\prod_j X_j) \to k(X_j) = X_j.$$

The following theorem is a special case of the fact that a right adjoint functor respects limits.

(7.9.13) Theorem. *$(p_j : k(\prod_j X_j) \to X_j \mid j \in J)$ is a product of $(X_j \mid j \in J)$ in the category* k-TOP.

Proof. We use (7.9.6) and the universal property of the topological product and obtain, in short-hand notation, for a k-space B the canonical bijection

$$\text{k-TOP}\,(B, k(\textstyle\prod X_j)) = \text{TOP}\,(B, \prod X_j) \cong \prod \text{TOP}(B, X_j) = \prod \text{k-TOP}(B, X_j),$$

and this is the claim. □

In the case of two factors, we use the notation $X \times_k Y$ for the product in k-TOP just defined. The next result shows that the wh-spaces are the formally hausdorff spaces in the category k-TOP.

(7.9.14) Proposition. *A k-space X is a wh-space if and only if the diagonal D_X of the product $X \times_k X$ is closed.*

Proof. Let X be a wh-space. In order to verify that D_X is closed, we have to show that for each test map $f: K \to X \times_k X$ the pre-image $f^{-1}(D_X)$ is closed. Let $f_j: K \to X$ be the j-th component of f. Then $L_j = f_j(K)$ is a ch-space, since X is a wh-space. Hence $L = L_1 \cup L_2 \subset X$ is a ch-space. The relation $f^{-1}D_X = f^{-1}((L \times L) \cap D_X)$ shows that this set is closed.

Let D_X be closed in $X \times_k X$ and $f: K \to X$ a test map. We have to show that $f(K) \subset X$ is closed. Let $g: L \to X$ be another test map. Since X is a k-space, we have to show that $g^{-1}f(K) \subset L$ is closed. We use the relation

$$g^{-1}f(K) = \mathrm{pr}_2((f \times g)^{-1}D_X).$$

Since D_X is closed, the pre-image under $f \times g$ is closed and therefore also its image under pr_2 as a compact set in a Hausdorff space. □

Recall the mapping space $F(X, Y)$ with compact-open topology.

(7.9.15) Theorem. *Let X and Y be k-spaces, and let $f: X \times_k Y \to Z$ be continuous. The adjoint map $f^\wedge: X \to kF(Y, Z)$, which exists as a set map, is continuous.*

Proof. The map $f^\wedge: X \to kF(Y, Z)$ is continuous, if for each test map $t: C \to X$ the composition $f^\wedge \circ t$ is continuous. We use $f^\wedge \circ t = (f \circ (t \times \mathrm{id}_Y))^\wedge$. Therefore it suffices to assume that X is a ch-space. But then, by (7.9.12), $X \times_k Y = X \times Y$ and therefore $f^\wedge: X \to F(Y, Z)$ is continuous and hence also $f^\wedge: X \to kF(Y, Z)$, by (7.9.4). □

(7.9.16) Theorem. *Let Y be a k-space. Then the evaluation*

$$e_{Y,Z}: kF(Y, Z) \times_k Y \to Z, \quad (f, y) \mapsto f(y)$$

is continuous.

Proof. Let $t: C \to kF(Y, Z) \times_k Y$ be a test map. We have to show the continuity of $e_{Y,Z} \circ t$. Let $t_1^\wedge: C \to F(Y, Z)$ and $t_2: C \to Y$ be the continuous components of t. We show first: The adjoint $t_1: C \times Y \to Z$ of $t_1^\wedge$ is continuous. By (2.4.3), this continuity is equivalent to the continuity of the second adjoint map $t_1^\vee: Y \to F(C, Z)$. In order to show its continuity, we compose with a test map $s: D \to Y$. But $t_1^\vee \circ s = F(s, Z) \circ t_1^\wedge$ is continuous. Moreover we have $e_{Y,Z} \circ t = t_1 \circ (\mathrm{id}, t_2)$, and the right-hand side is continuous. □

A combination of (7.9.15) and (7.9.16) now yields the ***universal property of the evaluation*** $e_{Y,Z}$ for k-spaces:

(7.9.17) Proposition. *Let X and Y be k-spaces. The assignments $f \mapsto f^\wedge$ and $g \mapsto e_{Y,Z} \circ (g \times_k \mathrm{id}_Y) = g^\sim$ are inverse bijections*

$$\mathrm{TOP}(X \times_k Y, Z) \cong \mathrm{TOP}(X, kF(Y,Z))$$

between these sets. □

(7.9.18) Theorem. *Let X, Y and Z be k-spaces. Since $e_{Y,Z}$ is continuous, we have an induced set map*

$$\lambda\colon kF(X, kF(Y,Z)) \to kF(X \times_k Y, Z), \quad f \mapsto e_{Y,Z} \circ (f \times_k \mathrm{id}_Y) = f^\sim.$$

The map λ is a homeomorphism.

Proof. We use the commutative diagram

$$\begin{array}{ccc} kF(X,kF(Y,Z)) \times_k X \times_k Y & \xrightarrow{e_1 \times \mathrm{id}} & kF(Y,Z) \times_k Y \\ \downarrow{\scriptstyle \lambda \times \mathrm{id} \times \mathrm{id}} & & \downarrow{\scriptstyle e_2} \\ kF(X \times_k Y, Z) \times_k X \times_k Y & \xrightarrow{\quad e_3 \quad} & Z \end{array}$$

with $e_1 = e_{X,kF(Y,Z)}$, $e_2 = e_{Y,Z}$, and $e_3 = e_{X \times_k Y, Z}$. Since $e_1 \times \mathrm{id}$ and e_2 are continuous, the universal property of e_3 shows that λ is continuous; namely, using the notation from (7.9.17), we have $e_2 \circ (e_1 \times \mathrm{id}) = \lambda^\sim$. The universal property of e_1 provides us with a unique continuous map

$$\mu\colon kF(X \times_k Y, Z) \to kF(X, kF(Y,Z)), \quad f \mapsto f^\wedge,$$

such that $e_1 \circ (\mu \times \mathrm{id}(X)) = e_3^\wedge$, where $e_3^\wedge\colon kF(X \times_k Y, Z) \times_k X \to kF(Y,Z)$ is the adjoint of e_3 with respect to the variable Y. One checks that λ and μ are inverse to each other, hence homeomorphisms. □

(7.9.19) Theorem. *Let X and Y be k-spaces, and $f\colon X \to X'$ and $g\colon Y \to Y'$ be quotient maps. Then $f \times g\colon X \times_k Y \to X' \times_k Y'$ is a quotient map.*

Proof. It suffices to treat the case $g = \mathrm{id}$, since a composition of quotient maps is a quotient map. Using (7.9.18), the proof is now analogous to (2.4.6). □

(7.9.20) Proposition. *Let $f\colon X \to Y$ be a quotient map and X a whk-space. Then Y is a whk-space if and only if $R = \{(x_1, x_) \mid f(x_1) = f(x_2)\}$ is closed in $X \times_k X$.*

Proof. The set R is the pre-image of D_Y under $f \times f$. Since $f \times_k f$ is a quotient map (7.9.19), D_Y is closed if and only if R is closed. Now apply (7.9.9) and (7.9.14). □

(7.9.21) Proposition. *Let Y and Z be k-spaces and assume that Z is a wh-space. Then the mapping space $kF(Y,Z)$ is a wh-space. In particular, if Y and Z are whk-spaces, then $kF(Y,Z)$ is a whk-space.*

Proof. Let $f^\wedge\colon K \to kF(Y,Z)$ be a test map. We have to show that it has a closed image hence is k-closed. For this purpose let $g^\wedge\colon L \to kF(Y,Z)$ be another test map. It remains to show that the pre-image M of $f^\wedge(K)$ under $g^\wedge$ is closed. We use the adjoint maps $f\colon K \times Y \to Z$ and $g\colon L \times Y \to Z$. For $y \in Y$ let $i_y\colon K \times L \to (K \times Y) \times_k (L \times Y)$, $(k,l) \mapsto (k,y,l,y)$. Then $M = \mathrm{pr}_2\big(\bigcap_{y \in Y}((f \times g)i_y)^{-1}D_Z\big)$. Since Z is a wh-space and therefore the diagonal D_Z is closed by (7.9.14), we see that M is closed. □

We now consider pointed spaces. Let $(X_j \mid j \in J)$ be a family of pointed k-spaces. Let $\prod_j^k X_j$ be its product in k-TOP. Let $W_J X_j$ be the subset of the product of those points for which at least one component equals the base point. The ***smash product*** $\bigwedge_j^k X_j$ is the quotient space $\big(\prod_j^k X_j\big)/W_J X_j$. In the case that $J = \{1,\dots,n\}$ we denote this space by $X_1 \wedge_k \cdots \wedge_k X_n$. A family of pointed maps $f_j\colon X_j \to Y_j$ induces a pointed map $\bigwedge^k f_j\colon \bigwedge_j^k X_j \to \bigwedge_j^k Y_j$.

Let X and Y be pointed k-spaces. Let $F^0(X,Y) \subset F(X,Y)$ be the subspace of pointed maps. We compose a pointed map $f\colon X \wedge_k Y \to Z$ with the projections $p\colon X \times_k Y \to X \wedge_k Y$. The adjoint $(fp)^\wedge\colon X \to kF(Y,Z)$ is continuous and has an image contained in $kF^0(Y,Z)$. We obtain a continuous map $X \to kF^0(Y,Z)$ which will be denoted by $f^\wedge$.

The evaluation $e_{Y,Z}$ induces $e^0_{Y,Z}$ which makes the following diagram commutative:

$$\begin{array}{ccc} kF^0(Y,Z) \times_k X & \xrightarrow{k(i)\times \mathrm{id}} & kF(Y,Z) \times_k X \\ \downarrow{\scriptstyle p} & & \downarrow{\scriptstyle e_{Y,Z}} \\ kF^0(Y,Z) \wedge_k X & \xrightarrow{e^0_{Y,Z}} & Y. \end{array}$$

i is the inclusion and p the quotient map. The continuity of $k(i)$ and $e_{Y,Z}$ implies the continuity of the pointed evaluation $e^0_{Y,Z}$. In analogy to (7.9.18) one proves:

(7.9.22) Theorem. *Let X, Y and Z be pointed k-spaces. The assignment*

$$\mu^0\colon kF^0(X \wedge_k Y, Z) \to kF^0(X, kF^0(Y,Z)), \quad f \mapsto f^\wedge$$

is a homeomorphism. □

(7.9.23) Example. Let $\mathbb{R}/\mathbb{Z}$ be obtained from $\mathbb{R}$ by identifying the subset $\mathbb{Z}$ to a point (so this is not the factor group!). We denote by $p\colon \mathbb{R} \to \mathbb{R}/\mathbb{Z}$ the quotient map.

(1) The product $p \times \mathrm{id}\colon \mathbb{R} \times \mathbb{Q} \to \mathbb{R}/\mathbb{Z} \times \mathbb{Q}$ of quotient maps is not a quotient map.

(2) The product $\mathbb{R}/\mathbb{Z} \times \mathbb{Q}$ is not a k-space, but the factors are k-spaces (see (7.9.4)).

(3) The product $\mathbb{R}/\mathbb{Z} \times \mathbb{R}$ is a k-space (see (7.9.9) and (7.9.12)), but the subspace $\mathbb{R}/\mathbb{Z} \times \mathbb{Q}$ is not a k-space by (2).

If $K \subset \mathbb{R}/\mathbb{Z}$ is compact, then there exists $l \in \mathbb{N}$ such that $K \subset p[-l, l]$.

Let $(r_n \mid n \in \mathbb{N})$ be a strictly decreasing sequence of rational numbers with limit $\sqrt{2}$. The set $F = \{(m + \frac{1}{2n}, \frac{r_n}{m}) \mid n, m \in \mathbb{N}\} \subset \mathbb{R} \times \mathbb{Q}$ is saturated with respect to $p \times \mathrm{id}$ and closed in $\mathbb{R} \times \mathbb{Q}$.

The set $G = (p \times \mathrm{id})(F)$ is not closed in $\mathbb{R}/\mathbb{Z} \times \mathbb{Q}$. Note that $z = (p(0), 0) \notin G$; but we show that $z \in \bar{G}$. Let U be a neighbourhood of z. Then there exists a neighbourhood V of $p(0)$ in $\mathbb{R}/\mathbb{Z}$ and $\varepsilon > 0$ such that $V \times (]-\varepsilon, \varepsilon[\cap \mathbb{Q}) \subset U$. Choose $m \in \mathbb{N}$ such that $m^{-1}\sqrt{2} < 2^{-1}\varepsilon$. The set $p^{-1}(V)$ is then a neighbourhood of m in $\mathbb{R}$, since $m \in p^{-1}p(0) \subset p^{-1}(V)$. Hence there exists $\delta > 0$ such that $]m - \delta, m + \delta[\subset p^{-1}(V)$. Now choose $n \in \mathbb{N}$ such that $\frac{1}{2n} < \delta$ and $r_n - \sqrt{2} < m\frac{\varepsilon}{2}$. Then $(p \times \mathrm{id})(m + \frac{1}{2n}, \frac{r_n}{m}) \in V \times (]-\varepsilon, \varepsilon[\cap \mathbb{Q}) \subset U$ holds, because $m + \frac{1}{2n} \in]m-\delta, m+\delta[\subset p^{-1}(V)$ and $0 < \frac{r_n}{m} = \frac{\sqrt{2}}{m} + \frac{r_n - \sqrt{2}}{m} < \frac{\varepsilon}{2} + \frac{\varepsilon}{2} = \varepsilon$. We see that $U \cap G \neq \emptyset$. This finishes the proof that $z \in \bar{G}$.

We now see that $p \times \mathrm{id}$ is not a quotient map, since there exists a saturated closed set F with non-closed image G.

The space $\mathbb{R}/\mathbb{Z} \times \mathbb{Q}$ is not a k-space. Let $s \colon K \to \mathbb{R}/\mathbb{Z} \times \mathbb{Q}$ be an arbitrary test map. We show that $s^{-1}(G)$ is closed in K although G is not closed (this could not occur in a k-space). The two projections $\mathrm{pr}_i\, s(K)$ are compact and Hausdorff. Hence there exists $l \in \mathbb{N}$ such that $\mathrm{pr}_1\, s(K) \subset p[-l, l]$. The inclusion

$$s(K) \subset \mathrm{pr}_1\, s(K) \times \mathrm{pr}_2\, s(K) \subset p[-l - l] \times \mathrm{pr}_2\, s(K)$$

then shows that we have $s^{-1}(G) = s^{-1}(G \cap p[-l, l] \times \mathrm{pr}_2\, s(K))$. But the set $G \cap p[-l, l] \times \mathrm{pr}_2\, s(K)$ is finite: By construction, F is a closed discrete subspace of $\mathbb{R} \times \mathbb{Q}$; moreover, $F \cap [-l, l] \times \mathrm{pr}_2\, s(K)$ is finite as a closed discrete subspace of the compact space $[-l, l] \times \mathrm{pr}_2\, s(K)$; therefore also

$$(p \times \mathrm{id})(F \cap [-l, l] \times \mathrm{pr}_2\, s(K)) = G \cap p[-l, l] \times \mathrm{pr}_2\, s(K)$$

is finite. A finite set in a Hausdorff space is closed, and therefore $s^{-1}(G)$ as preimage of a closed set is closed itself. ◇

(7.9.24) Example. It is stated already in [155, p. 336] that $(\mathbb{Q} \wedge \mathbb{Q}) \wedge \mathbb{N}_0$ and $\mathbb{Q} \wedge (\mathbb{Q} \wedge \mathbb{N}_0)$ are not homeomorphic. In [128, p. 26] it is proved that the canonical continuous bijection from the first to the second is not a homeomorphism. ◇

Problems

1. A space is a k-space if and only if it is a quotient of a locally compact Hausdorff space.
2. Let $X_1 \subset X_2 \subset \cdots$, let X_j be a whk-space and let $X_j \subset X_{j+1}$ be closed. Then $X = \bigcup_j X_j$, with colimit topology, is a whk-space. If the X_i are k-spaces, then X is a k-space, being a quotient of the k-space $\coprod_i X_i$. If the X_i are wh-spaces, hence T_1-spaces, then each test map $f\colon K \to X$ has an image which is contained in some X_i and therefore closed. If each inclusion is $X_i \subset X_{i+1}$ closed, the image is also closed in X and therefore X is a wh-space.
3. Let X and Y be k-spaces. Passage to adjoint maps induces bijections of homotopy sets $[X \times_k Y, Z] \cong [X, kF(Y,Z)]$ and $[X \wedge Y, Z]^0 \cong [X, kF^0(Y,Z)]^0$.
4. Let $(X_j \mid j \in J)$ be a family of k-spaces. Then the topological sum $\coprod_{j \in J} X_j$ is a k-space. The product in k-TOP is compatible with sums.
5. Let a pushout of topological spaces with closed $j\colon A \subset X$ be given.

$$\begin{array}{ccc} A & \xrightarrow{f} & B \\ \downarrow{\scriptstyle j} & & \downarrow{\scriptstyle J} \\ X & \xrightarrow{F} & Y \end{array}$$

Let X and B be whk-spaces. Then Y is a whk-space.

Chapter 8
Cell Complexes

The success of algebraic topology is largely due to the fact that one can describe spaces of interest by discrete (or even finite) combinatorial data. Purely combinatorial objects are simplicial complexes. Given such a complex, one defines from its data by simple linear algebra the homology groups. It is then a remarkable fact that these groups are independent of the combinatorial description and even homotopy invariant. Simplicial complexes are a very rigid structure. A weakening of this structure is given by the cell complexes (CW-complexes in the sense of J. H. C. Whitehead). They are more flexible and better adapted to homotopy theory.

An n-cell in a space is a subset which is homeomorphic to the standard n-cell $E^n = \{x \in \mathbb{R}^n \mid \|x\| < 1\}$. A cell complex is a decomposition of a space into cells. In order that one obtains something interesting, one has to add conditions about the closure of the cells, and one has to relate the topology of the space to the topology of the closed cells.

A finite cell complex is easily defined: a Hausdorff space X which is the union of a finite number of cells. It e is an n-cell of this decomposition, then it is required that there exists a continuous map $\varphi\colon D^n \to X$ which induces a homeomorphism $E^n \to e$ and sends S^{n-1} into the union of the cells up to dimension $n-1$.

From these data one obtains already an interesting invariant of X, the so-called Euler characteristic. Let $n(i)$ denote the number of i-cells. Define the combinatorial Euler characteristic to be the alternating sum $\chi(X) = \sum_{i \ge 0}(-1)^i n(i)$. It is a non-trivial fact that h-equivalent finite complexes have the same Euler characteristic. The origin of the notion is the famous result of Leonhard Euler ($\sim$ 1752) that for a sphere S^2 each polyhedral decomposition yields the value $n(0) - n(1) + n(2) = 2$.

In this chapter we present some point-set topology and elementary homotopy theory of cell complexes. Then we demonstrate the use of cell complexes in the construction of spaces with specific properties. In particular we construct so-called Eilenberg–Mac Lane spaces $K(\pi, n)$. They have a single non-vanishing homotopy group $\pi_n(K(\pi, n)) \cong \pi$ (here π can be an arbitrary abelian group). Eilenberg–Mac Lane spaces can be used as building blocks for general homotopy types (Postnikov systems). They also yield a homotopical definition of cohomology (and homology) groups: The homotopy set $[X, K(\pi, n)]$ carries a natural structure of an abelian group and is known to be a version of a cohomology group denoted $H^n(X; \pi)$.

8.1 Simplicial Complexes

Simplicial complexes are the objects of combinatorial topology.

A ***simplicial complex*** $K = (E, S)$ consists of a set E of ***vertices*** and a set S of finite non-empty subsets of E. A set $s \in S$ with $q + 1$ elements is called a q-***simplex*** of K. We require the following axioms:

(1) $\{e\} \in S$ for each $e \in E$.

(2) If $t \in S$ and $s \subset t$ is non-empty, then $s \in S$.

If $s \in S$ is a q-simplex, then q is called the ***dimension*** of s. If $t \subset s$, then t is a ***face*** of s. A 1-simplex of K is also called an ***edge*** of K. The 0-simplices of K correspond to the elements of E; a 0-simplex is called a ***vertex***. A simplex is determined by its 0-faces.

A simplicial complex is n-***dimensional***, if it contains at least one n-simplex but no $(n+1)$-simplices. A ***subcomplex*** L of K consists of a set of simplices of K which contains with s also the faces of s. A 1-dimensional complex is a ***graph***. A complex $K = (E, S)$ is ***finite*** if E is finite and ***locally finite*** if each vertex is contained in a finite number of simplices. The n-skeleton $K^n = (E, S^n)$ of $K = (E, S)$ is the subcomplex with $S^n = \{s \in S \mid \dim s \le n\}$.

(8.1.1) Example. Let $\mathcal{U} = (U_j \mid j \in J)$ be a covering of a set X by non-empty sets U_j. For a finite non-empty set $E \subset J$ let $U_E = \bigcap_{j \in E} U_j$ and let $E(J) = \{E \subset J \mid U_E \neq \emptyset\}$. Then $(J, E(J))$ is a simplicial complex, called the ***nerve*** $N(\mathcal{U})$ of the covering $\mathcal{U}$. ◇

(8.1.2) Example. Let P be a set with a partial ordering $\le$. The simplicial complex (P, S_P) associated to a partially ordered set has as simplices the totally ordered finite subsets of P. ◇

(8.1.3) Example. Let $K = (E, S)$ be a simplicial complex. Define a partial order on S by $s \le t \Leftrightarrow s \subset t$. The simplicial complex K' associated to this ordered set is called the ***barycentric subdivision*** of K. ◇

Let $K = (E, S)$ be a simplicial complex. We denote by $|K|$ the set of functions $\alpha \colon E \to I$ such that

(1) $\{e \in E \mid \alpha(e) > 0\}$ is a simplex of K.

(2) $\sum_{e \in E} \alpha(e) = 1$.

We regard $|K|$ as a subset of the product I^E. Let $|K|_p$ denote this set with the subspace topology of the product topology. We have a metric d on $|K|$ defined by

$$d(\alpha, \beta) = \left(\textstyle\sum_{e \in E} (\alpha(e) - \beta(e))^2\right)^{\frac{1}{2}}.$$

We denote $|K|$ with this metric topology by $|K|_m$. Each vertex $e \in E$ gives us a continuous map $e \colon |K|_m \to I$, $\alpha \mapsto \alpha(e)$. Therefore the identity $|K|_m \to |K|_p$ is continuous. We leave it as an exercise to show that this map is actually a homeomorphism. The numbers $(\alpha(e) \mid e \in E)$ are the ***barycentric coordinates*** of α.

We define a further topology on $|K|$. For $s \in S$ let $\Delta(s)$ be the standard simplex $\{(t_e) \in |K| \mid t_e = 0 \text{ for } e \notin s\}$. Then $|K|$ is the union of the $\Delta(s)$, and we write $|K|_c$ for $|K|$ with the quotient topology defined by the canonical map $\coprod_{s\in S} \Delta(s) \to |K|$. The identity $|K|_c \to |K|_p$ is continuous but not, in general, a homeomorphism. The next proposition will be proved in the more general context of simplicial diagrams.

(8.1.4) Proposition. *$|K|_c \to |K|_p$ is a homotopy equivalence.* □

In the sequel we write $|K| = |K|_c$ and call this space the ***geometric realization*** of K. We define $|s| \subset |K|$ as $|s| = \{\alpha \in |K| \mid \alpha(e) \neq 0 \Rightarrow e \in s\}$ and call this set a ***closed simplex*** of $|K|$. For each simplex s of K the ***open simplex*** $\langle s\rangle \subset |K|$ is the subspace $\langle s\rangle = \{\alpha \in |K| \mid \alpha(e) \neq 0 \Leftrightarrow e \in s\}$. The complement $|s| \setminus \langle s\rangle = \partial|s|$ is the ***combinatorial boundary*** of $|s|$; it is the geometric realization of the subcomplex which consists of the proper faces of s. The set $|K|$ is the disjoint union of the $\langle s\rangle, s \in S$.

Let $|K|^n$ be the union of the $\Delta(s)$ with $\dim s \le n$.

(8.1.5) Proposition. *The space $|K|$ is the colimit of the $|K|^n$. The equality $|K^n| = |K|^n$ holds. The canonical diagram*

$$\begin{array}{ccc} \coprod_{s,\dim s=n} \partial\Delta(s) & \longrightarrow & |K|^{n-1} \\ \downarrow & & \downarrow \\ \coprod_{s,\dim s=n} \Delta(s) & \longrightarrow & |K|^n \end{array}$$

is a pushout. □

A homeomorphism $t\colon |K| \to X$ is called a ***triangulation*** of X. The triangulation of surfaces was proved by Radó [161], the triangulation of 3-dimensional manifolds by Moise (see [141] for references and proofs; [197, 7.5.1]). Differentiable manifolds can be triangulated, and the triangulation can be chosen in such a way that it is on each simplex a smooth embedding ([193]; [143]).

Since $|K|_m$ is separated and $\mathrm{id}\colon |K| \to |K|_m$ continuous, $|K|$ is separated. For finite K the identity $|K| \to |K|_m$ is a homeomorphism.

For each vertex $e \in E$ the set $\mathrm{St}(e) = \{\alpha \in |K| \mid \alpha(e) \neq 0\}$ is called the ***star*** of e. Since $\alpha \mapsto \alpha(e)$ is continuous, the set $\mathrm{St}(e)$ is open in $|K|_d$ and therefore also in $|K|$. If we identify e with the function $\alpha(e) = 1, \alpha(e') = 0$ for $e \neq e'$, then $\mathrm{St}(e)$ is an open neighbourhood of e.

Points $e_0, \dots, e_k$ of $\mathbb{R}^n$ are ***affinely independent***, if the relations $\Sigma\lambda_i e_i = 0$ and $\Sigma\lambda_i = 0$ imply that each $\lambda_i = 0$. If $e_0, \dots, e_k$ are affinely independent, then the simplex $\left\{\sum_{i=0}^k \lambda_i e_i \mid \lambda_i \ge 0,\ \Sigma\lambda_i = 1\right\}$ spanned by $e_0, \dots, e_k$ is the convex hull of this set and homeomorphic to the k-dimensional standard simplex.

Let $K = (E, S)$ be a simplicial complex and $(x_e \mid e \in E)$ a family of points in $\mathbb{R}^n$. Consider the continuous map

$$f : |K| \to \mathbb{R}^n, \quad \alpha \mapsto \sum_{e \in E} \alpha(e) x_e.$$

If f is an embedding, we call the image of f a ***simplicial polyhedron*** in $\mathbb{R}^n$ of type K, and $f(|K|)$ is a realization of K as a polyhedron in $\mathbb{R}^n$.

Standard tools for the application of simplicial complexes in algebraic topology are subdivision and simplicial approximation [67, p. 124].

Problems

1. $\mathrm{id} : |K|_m \to |K|_p$ is a homeomorphism.
2. Let $K = (\mathbb{N}_0, S)$ be the simplicial complex where S consists of all finite subsets of $\mathbb{N}_0$. The canonical map $|K|_c \to |K|_p$ is not a homeomorphism.
3. Let L be a subcomplex of K. We can identify $|L|$ with a subset of $|K|$, and $|L|$ carries then the subspace topology of $|K|$. If $(L_j \mid j \in J)$ is a family of subcomplex of K, then $\bigcup L_j$ and $\bigcap L_j$ are subcomplexes and the relations $\bigcup |L_j| = |\bigcup L_j|$ and $\bigcap |L_j| = |\bigcap L_j|$ hold.
4. Let K be a simplicial complex. Then the following assertions are equivalent: (1) K is locally finite. (2) $|K|$ is locally compact. (3) The identity $|K| \to |K|_d$ is a homeomorphism. (4) $|K|$ is metrizable. (5) Each point of $|K|$ has a countable neighbourhood basis. (See [44, p. 65].)
5. Let K be a countable, locally finite simplicial complex of dimension at most n. Then K has a realization as a polyhedron in $\mathbb{R}^{2n+1}$. (See [44, p. 66].)

8.2 Whitehead Complexes

We use the standard subsets of Euclidean spaces $S^{n-1}, D^n, E^n = D^n \setminus S^{n-1}$, $(n \geq 1)$. We set $S^{-1} = \emptyset$ and let D^0 be a point, hence $E^0 = D^0$. A ***k-dimensional cell*** (a k-cell) in a space X is a subset e which is, in its subspace topology, homeomorphic to E^k. A point is always a 0-cell.

A ***Whitehead complex*** is a space X together with a decomposition into cells $(e_\lambda \mid \lambda \in \Lambda)$ such that:

(W1) X is a Hausdorff space.

(W2) For each n-cell e_λ there exists a ***characteristic map*** $\Phi_\lambda : D^n = D^n_\lambda \to X$ which induces a homeomorphism $E^n \to e_\lambda$ and sends S^{n-1} into the union X^{n-1} of the cells up to dimension $n-1$.

(W3) The closure $\overline{e}_\lambda$ of each cell e_λ intersects only a finite number of cells.

(W4) X carries the colimit topology with respect to the family $(\overline{e}_\lambda \mid \lambda \in \Lambda)$.

A subset A of a Whitehead complex is a ***subcomplex*** if it is a union of cells and the closure of each cell in A is contained in A. We will see that a subcomplex together with its cells is itself a Whitehead complex. From the definition of a subcomplex we see that intersections and unions of subcomplexes are again subcomplexes. Therefore there exists a smallest subcomplex $X(L)$ which contains a given set L.

The decomposition of a Hausdorff space into its points always satisfies (W1)–(W3). We see that (W4) is an important condition. Condition (W3) is also called (C), for closure finite. Condition (W4) is called (W), for weak topology. This is the origin for the name CW-complex. In the next section we consider these complexes from a different view-point and introduce the notion of a CW-complex.

(8.2.1) Lemma. *Let $\Phi\colon D^n \to X$ be a continuous map into a Hausdorff space. Let $e = \Phi(E^n)$. Then $\Phi(D^n) = \overline{e}$. In particular $\overline{e}$ is compact.*

Proof. $\Phi(D^n)$ is a compact subset of a Hausdorff space and therefore closed. This yields $\overline{e} = \overline{\Phi(E^n)} \subset \overline{\Phi(D^n)} = \Phi(D^n) = \Phi(\overline{E^n}) \subset \overline{\Phi(E^n)} = \overline{e}$. □

(8.2.2) Example. Suppose X has a cell decomposition into a finite number of cells such that properties (W1) and (W2) hold. Then X is a finite union of closures $\overline{e}$ of cells and therefore compact by (8.2.1). Properties (W3) and (W4) are satisfied and X is a Whitehead complex. ◇

(8.2.3) Examples. The sphere S^n has the structure of a Whitehead complex with a single 0-cell and a single n-cell. The map

$$\Phi\colon D^n \to S^n, \quad x \mapsto (2\sqrt{1-\|x\|^2}\cdot x, 2\|x\|^2 - 1)$$

sends S^{n-1} to the 0-cell $e_{n+1} = (0,\dots,0,1)$ and induces a homeomorphism of E^n with $S^n \smallsetminus \{e_{n+1}\}$, hence is a characteristic map for the n-cell.

From this cell decomposition we obtain a cell decomposition of D^{n+1} by adding another $(n+1)$-cell E^{n+1} with characteristic map the identity.

Another cell-composition of S^n has two j-cells for each $j \in \{0,\dots,n\}$ and is obtained inductively from $D^n_\pm = \{(x_i) \in S^n \mid \pm x_{n+1} \ge 0\}$ with intersection $S^{n-1} = S^{n-1} \times 0$. A characteristic map is $D^n \to D^n_\pm$, $x \mapsto (x, \pm\sqrt{1-\|x\|^2})$.◇

(8.2.4) Proposition. *Let X be a Whitehead complex.*

(1) *A compact set K in X meets only a finite number of cells.*

(2) *A subcomplex which consists of a finite number of cells is compact and closed in X.*

(3) *$X(e) = X(\overline{e})$ is for each cell e a finite subcomplex.*

(4) *A compact subset of a Whitehead complex is contained in a finite subcomplex.*

(5) *X carries the colimit topology with respect to the finite subcomplexes.*

(6) *A subcomplex A is closed in X.*

Proof. (1) Let E be the set of cells which meet K. For each $e \in E$ we choose a point $x_e \in K \cap e$ and set $Z = \{x_e \mid e \in E\}$. Let $Y \subset Z$ be any subset. For each cell f of X the closure $\overline{f}$ is contained in the union of a finite number of cells. Thus

$Y \cap \overline{f}$ is a finite set, hence closed in $\overline{f}$ since $\overline{f}$ is a Hausdorff space. The condition (W4) now says that Y is closed in X and hence in Z. This tells us that Z carries the discrete topology and is closed in X. A discrete closed set in a compact space is finite.

(2) Let $A = e_1 \cup \cdots \cup e_r$ be a finite union of cells e_j. Then $\overline{A} = \overline{e}_1 \cup \cdots \cup \overline{e}_r \subset A$, by definition of a subcomplex. By (8.2.1), $A = \overline{A}$ is compact and closed.

(3) Induction over $\dim(e)$. If e is a 0-cell, then e is a point and closed, hence a subcomplex and $X(e) = X(\overline{e}) = e$. Suppose $X(f)$ is finite for each cell f with $\dim(f) < n$. Let e be an n-cell with characteristic map Φ.

The set $\Phi(S^{n-1})$ is contained in the union of cells of dimension at most $n-1$, hence is contained in $\overline{e} \smallsetminus e$.

Then $\overline{e} \smallsetminus e = \Phi(S^{n-1})$ is compact, hence contained in a finite number of cells $e_1, \dots, e_k$, by (1), which are contained in X^{n-1}, by (W2). By induction hypothesis, the set $C = e \cup X(e_1) \cup \cdots \cup X(e_k)$ is a finite subcomplex which contains e and hence $X(e)$. Therefore $X(e)$ is finite. Since $X(e)$ is closed, by (2), we have $\overline{e} \subset X(e)$ and $X(\overline{e}) \subset X(e)$.

(4) This is a consequence of (1) and (3).

(5) We show: $A \subset X$ is closed if and only if for each finite subcomplex Y the intersection $A \cap Y$ is closed in Y.

Suppose the condition is satisfied, and let f be an arbitrary cell. Then $A \cap X(f)$ is closed in $X(f)$, hence, by (2) and (3), closed in X; therefore $A \cap \overline{f} = A \cap X(f) \cap \overline{f}$ is closed in X and in $\overline{f}$, hence closed in X by condition (W4).

(6) If Y is a finite subcomplex, then $A \cap Y$ is a finite subcomplex, hence closed. By (5), A is closed. □

(8.2.5) Proposition. *A subcomplex Y of a Whitehead complex X is a Whitehead complex.*

Proof. Let e be a cell in Y and $\Phi\colon D^n \to X$ a characteristic map. Then $\Phi(D^n) = \overline{e} \subset Y$, since Y is closed. Hence Φ can be taken as a characteristic map for Y.

It remains to verify condition (W4). Let $L \subset Y$ and suppose $L \cap \overline{e}$ is closed in $\overline{e}$ for each cell e in Y. We have to show that L is closed in Y. We show that L is closed in X. Let f be a cell of X. By (W3), $\overline{f}$ is contained in a finite union $e_1 \cup \cdots \cup e_k$ of cells. Let $e_1, \dots, e_j$ be those which are contained in Y. Then

$$\overline{f} \cap Y \subset e_1 \cup \cdots \cup e_j \subset \overline{e}_1 \cup \cdots \cup \overline{e}_j \subset Y$$

since Y is a subcomplex. Hence

$$\overline{f} \cap Y = (\overline{f} \cap \overline{e}_1) \cup \cdots \cup (\overline{f} \cap \overline{e}_j), \quad \overline{f} \cap L = \overline{f} \cap L \cap Y = \textstyle\bigcup_{k=1}^{j} (\overline{f} \cap \overline{e}_k \cap L).$$

By assumption, $\overline{e}_k \cap L$ is closed in $\overline{e}_k$; hence $\overline{f} \cap \overline{e}_k \cap L$ is closed in $\overline{f}$; therefore $\overline{f} \cap L$ is a finite union of sets which are closed in X. □

(8.2.6) Proposition. *Let X be a Whitehead complex. Then:*

(1) *X carries the colimit topology with respect to the family $(X^n \mid n \in \mathbb{N}_0)$.*

(2) *Let $(e_\lambda \mid \lambda \in \Lambda(n))$ be the family of n-cells of X with characteristic maps $\Phi_\lambda \colon D^n_\lambda \to X^n$ and restrictions $\varphi_\lambda \colon S^{n-1}_\lambda \to X^{n-1}$. Then*

$$\begin{array}{ccc} \coprod_\lambda S^{n-1}_\lambda & \xrightarrow{\varphi=\langle\varphi_\lambda\rangle} & X^{n-1} \\ \downarrow{\scriptstyle i} & & \downarrow{\scriptstyle \cap} \\ \coprod_\lambda D^n_\lambda & \xrightarrow{\Phi=\langle\Phi_\lambda\rangle} & X^n \end{array}$$

is a pushout in TOP. $(X^{-1} = \emptyset.)$

Proof. (1) Suppose $A \cap X^n$ is closed in X^n for each n. Then for each n-cell e of X the set $A \cap \bar{e} = A \cap \bar{e} \cap X^n$ is closed in $\bar{e}$. By (W4), A is closed in X.

(2) The diagram is a pushout of sets. Give X^n the pushout topology and denote this space by Z. By construction, the identity $\kappa \colon Z \to X^n$ is continuous. We show that κ is also closed. Let $V \subset Z$ be closed. By definition of the pushout topology this means:

(i) $V \cap X^{n-1}$ is closed in X^{n-1}.

(ii) $\Phi^{-1}(V) \cap D^n_\lambda$ is closed in D^n_λ, hence also compact.

We conclude that $\Phi(\Phi^{-1}(V) \cap D^n_\lambda) = V \cap \Phi(D^n_\lambda) = V \cap \bar{e}_\lambda$ is closed in $\bar{e}_\lambda$, being a continuous image of a compact space in a Hausdorff space. From (i) and (ii) we therefore conclude that for each cell e of X^n the set $V \cap \bar{e}$ is closed in $\bar{e}$. Since X^n is a Whitehead complex, V is closed in X^n. □

(8.2.7) Proposition. *Let X be a Whitehead complex, pointed by a 0-cell. The inclusions of the finite pointed subcomplexes $F \subset X$ induce a canonical map* $\operatorname{colim}_F \pi_k(F,*) \to \pi_k(X,*)$. *This map is an isomorphism.* □

Recall from Section 7.9 the notion of a k-space and the k-space $k(X)$ obtained from a space X.

(8.2.8) Proposition. *Let X have a cell decomposition such that* (W1)–(W3) *hold and such that each compact set is contained in a finite number of cells. Then $k(X)$ is a Whitehead complex with respect to the given cell decomposition and the same characteristic maps. Moreover, X is a Whitehead complex if and only if $k(X) = X$.*

Proof. Let $\Phi \colon D^n \to X$ be a characteristic map for the cell e. Since $\bar{e}$ is compact it has the same topology in $k(X)$. Hence $\Phi \colon D^n \to k(X)$ is continuous. Since Φ is a quotient map and $\Phi^{-1}(e) = E^n$, we see that e has the same topology in $k(X)$ and X. Thus e is a cell in $k(X)$ with characteristic map Φ.

Let $A \cap \bar{e}$ be closed in $\bar{e}$ for each cell e. Let $K \subset k(X)$ be compact. By hypothesis, K is contained in a finite number of cells, say $K \subset e_1 \cup \dots \cup e_k$. Then $A \cap K = ((A \cap \bar{e}_1) \cup \dots \cup (A \cap \bar{e}_k)) \cap K$ is closed in K. Hence A is k-closed. □

Let X and Y be Whitehead complexes and $e \subset X$, $f \subset Y$ cells. Then $e \times f \subset X \times Y$ is a cell. From characteristic maps $\Phi\colon D^m \to X$, $\Psi\colon D^n \to Y$ for e, f we obtain $\Phi \times \Psi\colon D^m \times D^n \to X \times Y$, and this can be considered as a characteristic map for $e \times f$. For this purpose use a homeomorphism

$$(D^{m+n}, S^{m+n-1}) \to (D^m \times D^n, D^m \times S^{n-1} \cup S^{m-1} \times D^n).$$

With this cell structure, $X \times Y$ satisfies conditions (W1)–(W3) in the definition of a Whitehead complex. In general, property (W4) may not hold. In this case one re-topologizes $X \times Y$ such that the compact subsets do not change. The space $X \times_k Y = k(X \times Y)$ is then a Whitehead complex (see (8.2.8)).

Problems

1. $\mathbb{R}$ carries the structure of a Whitehead complex with 0-cells $\{n\}$, $n \in \mathbb{Z}$ and 1-cells $]n, n+1[$, $n \in \mathbb{Z}$. There is an analogous Whitehead complex structure $W(\delta)$ on $\mathbb{R}^n$ with 0-cells the set of points $\delta(k_1, \dots, k_n)$, $k_j \in \mathbb{Z}$, $\delta > 0$ fixed and the associated δ-cubes. Thus, given a compact set $K \subset \mathbb{R}^n$ and a neighbourhood U of K, there exists another neighbourhood L of K contained in U such that L is a subcomplex of the complex $W(\delta)$. In this sense, compact subsets can be approximated by finite complexes.
2. The geometric realization of a simplicial complex is a Whitehead complex.

8.3 CW-Complexes

We now use (8.2.6) as a starting point for another definition of a cell complex. Let (X, A) be a pair of spaces. We say, X is obtained from A by ***attaching an n-cell***, if there exists a pushout

$$\begin{array}{ccc} S^{n-1} & \xrightarrow{\varphi} & A \\ \downarrow \cap & & \downarrow \cap \\ D^n & \xrightarrow{\Phi} & X. \end{array}$$

Then A is closed in X and $X \setminus A$ is homeomorphic to E^n via Φ. We call $X \setminus A$ an ***n-cell*** in X, φ its ***attaching map*** and Φ its ***characteristic map***.

(8.3.1) Proposition. *Let a commutative diagram with closed embeddings j, J be given:*

$$\begin{array}{ccc} A & \xrightarrow{f} & Y \\ \downarrow j & & \downarrow J \\ X & \xrightarrow{F} & Z. \end{array}$$

Suppose F induces a bijection $X \smallsetminus A \to Z \smallsetminus Y$. Then the diagram is a pushout, provided that (1) *$F(X) \subset Z$ is closed;* (2) *$F\colon X \to F(X)$ is a quotient map. Condition* (2) *holds if X is compact and Z Hausdorff.*

Proof. Let $g \colon X \to U$ and $h \colon Y \to U$ be given such that $gj = hf$. The diagram is a set-theoretical pushout. Therefore there exists a unique set map $\varphi \colon Z \to U$ with $\varphi F = g$, $\varphi J = h$. Since J is a closed embedding, $\varphi | J(Y)$ is continuous. Since F is a quotient map, $\varphi | F(X)$ is continuous. Thus φ is continuous, since $F(X)$ and $J(Y)$ are closed sets which cover Z. □

(8.3.2) Note. *Let X be a Hausdorff space and A a closed subset. Suppose there exists a continuous map $\Phi \colon D^n \to X$ which induces a homeomorphism $\Phi \colon E^n \to X \setminus A$. Then X is obtained from A by attaching an n-cell.*

Proof. We show $\Phi(S^{n-1}) \subset A$. Suppose there exists $s \in S^{n-1}$ with $\Phi(s) \in X \setminus A$. Then there exists a unique $t \in E^n$ with $\Phi(s) = \Phi(t)$. Let $V \subset E^n$, $W \subset D^n$ be disjoint open neighbourhoods of t, s. Then $\Phi(V) \subset X \setminus A$ is open in X, since $\Phi \colon E^n \to X \setminus A$ is a homeomorphism and A is closed in X. Since Φ is continuous, there exists an open neighbourhood $W_1 \subset W$ of s with $\Phi(W_1) \subset \Phi(V)$. This contradicts the injectivity of $\Phi | E^n$.

Thus Φ provides us with a map $\varphi \colon S^{n-1} \to A$. We now use (8.3.1). □

(8.3.3) Example. The projective space $\mathbb{R}P^n$ is obtained from $\mathbb{R}P^{n-1}$ by attaching an n-cell. The projective space $\mathbb{C}P^n$ is obtained from $\mathbb{C}P^{n-1}$ by attaching a $2n$-cell.

We recall that $\mathbb{C}P^{n-1}$ is obtained from S^{2n-1} by the equivalence relation $(z_1, \dots, z_n) \sim (\lambda z_1, \dots, \lambda z_n)$, $\lambda \in S^1$, or from $\mathbb{C}^n \setminus 0$ by $z \sim \lambda z$, $\lambda \in \mathbb{C}^*$. The class of z is denoted $[z_1, \dots, z_n]$. A characteristic map $\Phi \colon D^{2n} \to \mathbb{C}P^n$ is $x \mapsto [x, \sqrt{1 - \|x\|^2}]$.

The space $\mathbb{R}P^{n-1}$ is obtained from S^{n-1} by the relation $z \sim -z$, or from $\mathbb{R}^n \setminus 0$ by $z \sim \lambda z$, $\lambda \in \mathbb{R}^*$. A characteristic map $\Phi \colon D^n \to \mathbb{R}P^n$ is given by the same formula as in the complex case. ◇

We can also attach several n-cells simultaneously. We say X is obtained from A by ***attaching n-cells*** if there exists a pushout

$$\begin{array}{ccc} \coprod_{j \in J} S_j^{n-1} & \xrightarrow{\varphi} & A \\ \big\downarrow \cap & & \big\downarrow \cap \\ \coprod_{j \in J} D_j^n & \xrightarrow{\Phi} & X. \end{array}$$

The index j just enumerates different copies of the same space. Again, A is then closed in X and Φ induces a homeomorphism of $\coprod_j E_j^n$ with $X \setminus A$. Therefore $X \setminus A$ is a union of components and each component is an n-cell. (By invariance of dimension, the integer n is determined by $X \setminus A$.) We allow $J = \emptyset$; in that case $A = X$. We write $\Phi_j = \Phi | D_j^n$ and $\varphi_j = \varphi | S^{n-1}$ and call Φ_j the ***characteristic map*** of the n-cell $\Phi(E_j^n)$ and φ_j its ***attaching map***.

Let us give another interpretation: $X = X(\varphi)$ is the double mapping cylinder of $J \xleftarrow{\text{pr}} S^n \times J \xrightarrow{\varphi} A$ where J is a discrete set. From this setting we see: If φ is replaced by a homotopic map ψ, then $X(\varphi)$ and $X(\psi)$ are h-equivalent under A.

Let $f : A \to Y$ be a given map. Assume that X is obtained from A by attaching n-cells via attaching maps $\langle \varphi_j \rangle : \coprod_j S_j^{n-1} \to A$. From the pushout definition of the attaching process we obtain:

(8.3.4) Note. *There exists an extension $F : X \to Y$ of f if and only if the maps $f\varphi_j$ are null homotopic. We view a null homotopy of $f\varphi_j$ as an extension to D_j^n. Then the extensions F correspond to the set of null homotopies of the $f\varphi_j$.* □

In view of this note we call the homotopy classes $[f\varphi_j]$ the ***obstructions*** to extending f.

Let A be a subspace of X. A ***CW-decomposition*** of (X, A) consists of a sequence of subspaces $A = X^{-1} \subset X^0 \subset X^1 \subset \cdots \subset X$ such that:

(1) $X = \bigcup_{n \geq 0} X^n$.

(2) For each $n \geq 0$, the space X^n is obtained from X^{n-1} by attaching n-cells.

(3) X carries the colimit topology with respect to the family (X^n).

X_k is a subspace of the colimit X of a sequence $X_j \subset X_{j+1} \subset \cdots$. If the inclusions are closed, then X_k is closed in X. This is an immediate consequence of the definition of the colimit topology.

A pair (X, A) together with a CW-decomposition $(X^n \mid n \geq -1)$ is called a ***relative CW-complex***. In the case $A = \emptyset$ we call X a ***CW-complex***. The space X^n is the ***n-skeleton*** of (X, A) and $(X^n \mid n \geq -1)$ is the ***skeleton filtration***. The cells of $X^n \setminus X^{n-1}$ are the n-cells of (X, A). We say, (X, A) is ***finite*** (***countable*** etc.) if $X \setminus A$ consists of a finite (countable etc.) number of cells. If $X = X^n$, $X \neq X^{n-1}$ we denote by $n = \dim(X, A)$ the ***cellular dimension*** of (X, A). If $A = \emptyset$, then A is suppressed in the notation. We call X a ***CW-space*** if there exists some cellular decomposition $X^0 \subset X^1 \subset \cdots$ of X.

Let X be a Whitehead complex. From (8.2.6) we obtain a CW-decomposition of X. The converse also holds: From a CW-decomposition we obtain a decomposition into cells and characteristic maps; it remains to verify that X is a Hausdorff space and carries the colimit topology with respect to the closures of cells (see (8.3.8)).

In the context of CW-complexes (X, A), the symbol X^n usually denotes the n-skeleton and not the n-fold Cartesian product.

(8.3.5) Note. *If (X, A) is a relative CW-complex, then also (X, X^n) and (X^n, A) are relative CW-complexes, with the obvious skeleton-filtration inherited from $(X^n \mid n \geq -1)$.* □

(8.3.6) Example. From (8.3.3) we obtain cellular decompositions of $\mathbb{C}P^n$ and $\mathbb{R}P^n$. The union of the sequence $\mathbb{R}P^n \subset \mathbb{R}^{n+1} \subset \cdots$ defines the infinite projective space $\mathbb{R}P^\infty$ as a CW-complex. It has a single n-cell for each $n \geq 0$. Similarly, we obtain $\mathbb{C}P^\infty$ with a single cell in each even dimension. ◇

(8.3.7) Example. The sphere S^n has a CW-decomposition with a single 0-cell and a single n-cell, and another CW-composition with two j-cells for each $j \in \{0, \dots, n\}$, see (8.2.3). The quotient map $S^n \to \mathbb{R}P^n$ sends each cell of the latter homeomorphically onto a cell of $\mathbb{R}P^n$ in the decomposition (8.3.6). We can also form the colimit S^∞ of $S^n \subset S^{n+1} \subset \cdots$, a CW-complex with two cells in each dimension. ◇

The general topology of adjunction spaces and colimit topologies gives us the next results.

(8.3.8) Proposition. *Let (X, A) be a relative CW-complex. If A is a T_1-space, then X is a T_1-space and a compact subset of X meets only a finite number of cells. If A is a Hausdorff space, then X is a Hausdorff space. If A is normal, then X is normal. If A is a Hausdorff space, then X carries the colimit topology with respect to the family which consists of A and the closures of cells.*

Proof. We only verify the last statement. Let C be a subset of X and suppose $A \cap C$ in closed in A and $A \cap \overline{e}$ closed in $\overline{e}$ for each cell e. We show inductively, that $C \cap X^n$ is closed in X^n. This holds for $n = -1$ by assumption. The space X^n is a quotient of $Z^n = X^{n-1} + \coprod D^n_j$. Each characteristic map $\Phi_j : D^n_j \to \overline{e}_j$ is a quotient map, since X is Hausdorff. From the assumptions we see that $X^n \cap C$ has a closed pre-image in Z^n. □

The considerations so far show that a CW-complex is a Whitehead complex.

(8.3.9) Proposition. *Let (X, A) be a relative CW-complex. Then $A \subset X$ is a cofibration.*

Proof. We know that $\coprod S^{n-1}_j \to \coprod D^n_j$ is a cofibration. Hence $X^{n-1} \subset X^n$ is an induced cofibration. Therefore the compositions $X^n \subset X^{n+k}$ are cofibrations. Given $f : X \to Z$ and a homotopy $h^{-1} : X^{-1} \times I \to Z$ of $f|X^{-1}$, we can extend this inductively to homotopies $h^n : X^n \times I \to Z$ such that $h^{n+1}|X^n \times I = h^n$. Since $X \times I$ is the colimit of the $X^n \times I$, the h^n combine to a homotopy $h : X \times I \to Z$. □

Problems

1. The attaching map for the n-cells yields a homeomorphism $\bigvee_j (D^n/S^{n-1})_j \cong X/A$.
2. Let (X, A) and (Y, B) be relative CW-complexes. Consider $X \times Y$ with the closed subspaces

$$(X \times Y)^n = \textstyle\bigcup_{i=-1}^{n+1} X^i \times Y^{n-i}, \quad n \geq -1.$$

In favorable cases, the filtration $((X \times Y)^n \mid n \geq -1)$ is a CW-decomposition of the pair $(X \times Y,\ A \times B)$.

Let Y be locally compact. Then $(X \times Y)^n$ is obtained from $(X \times Y)^{n-1}$ by attaching n-cells.

3. Let (X, A) be a relative CW-complex and let $C \subset A$. Then $(X/C, A/C)$ is a relative CW-complex with CW-decomposition (X^n/C). Moreover, X/A is a CW-complex.

4. Let $A \subset X$ be a subcomplex. Then X/A is a CW-complex.

5. Let A and B be subcomplexes of X. Then $A/(A \cap B)$ is a subcomplex of X/B.

6. Let A be a subcomplex of B and Y another CW-complex. Then $A \wedge_k Y$ is a subcomplex of $B \wedge_k Y$.

7. Let A be a CW-complex. Suppose X is obtained from A by attaching n-cells via attaching maps $\varphi\colon \coprod_j S_j^{n-1} \to A^{n-1}$. Then X is a CW-complex with CW-decomposition $X^j = A^j$ for $j < n$ and $X^j = A^j \cup (X \smallsetminus A)$ for $j \geq n$, and A is a subcomplex of X.

8. Let $\varphi_0, \varphi_1 : \coprod_j S_j^{n-1} \to A$ be homotopic attaching maps. The spaces $X(0), X(1)$ which are obtained by attaching n-cells with φ_0, φ_1 are h-equivalent under A. (Homotopy theorem for cofibrations.)

9. Let X be a pointed CW-complex with base point $*$ a 0-cell. Then the cone CX and the suspension ΣX are CW-complexes. (In statements of this type the reader is asked to find a canonical cell decomposition induced from the initial data.)

10. Let $(X_j \mid j \in J)$ be a family of pointed CW-complexes with base point a 0-cell. Then $\bigvee_{j \in J} X_j$ has the structure of a CW-complex such that the summands are subcomplexes.

11. Let $p\colon E \to B$ be a Serre fibration and (X, A) a CW-pair. Then each homotopy $h\colon X \times I \to B$ has a lifting along p with given initial condition on $X \times 0 \cup A \times I$.

12. Suppose X is obtained from A by attaching n-cells. Let $p\colon E \to X$ be a covering and $E' = p^{-1}(A)$. Then E is obtained from E' by attaching n-cells.

13. Let X be a CW-complex with n-skeleton X^n and $p\colon E \to X$ a covering. Then E is a CW-complex with n-skeleton $E^n = p^{-1}(B^n)$ such that p maps the cells of E homeomorphically to cells of X. An automorphism of p maps cells of E homeomorphically to cells.

14. Each neighbourhood U of a point x of a CW-complex contains a neighbourhood V which is pointed contractible to x. A connected CW-complex has a universal covering. The universal covering has a cell decomposition such that its automorphism group permutes the cells freely.

15. Let X and Y be countable CW-complexes. Then $X \times Y$ is a CW complex in the product topology.

8.4 Weak Homotopy Equivalences

We now study the notion of an n-connected map and of a weak homotopy equivalence in the context of CW-complexes.

(8.4.1) Proposition. *Let (Y, B) be n-connected. Then a map $f : (X, A) \to (Y, B)$ from a relative CW-complex (X, A) of dimension* $\dim(X, A) \leq n$ *is homotopic* *relative to A to a map into B. In the case that* $\dim(X, A) < n$ *the homotopy class of $X \to B$ is unique relative to A.*

Proof. Induction over the skeleton filtration. Suppose X is obtained from A by attaching q-cells via $\varphi\colon \coprod_k S_k^{q-1} \to A$, $q \le n$. Consider a commutative diagram

$$\begin{array}{ccccc} \coprod S_k^{q-1} & \xrightarrow{\varphi} & A & \xrightarrow{f} & B \\ \downarrow & & \cap\downarrow i & & \cap\downarrow j \\ \coprod D_k^q & \xrightarrow{\Phi} & X & \xrightarrow{F} & Y. \end{array}$$

Since (Y, B) is n-connected, $F\Phi$ is homotopic relative to $\coprod S_k^{q-1}$ to a map into B. Since the left square is a pushout, we obtain a homotopy of F from a pair of homotopies of $F\Phi$ and Fi which coincide on $\coprod S_k^{q-1}$. Since we have homotopies of $F\Phi$ relative to $\coprod S_k^{q-1}$, we can use on A the constant homotopy. Altogether we obtain a homotopy of F relative to A to a map into B.

For an arbitrary (X, A) with $\dim(X, A) \le n$ we apply this argument inductively. Suppose we have a homotopy of f relative to A to a map g which sends X^k into B. By the argument just given we obtain a homotopy of $g|X^{k+1}$ relative to X^k which sends X^{k+1} into B. Since $X^{k+1} \subset X$ is a cofibration, we extend this homotopy to X. In the case that $n = \infty$, we have to concatenate an infinite number of homotopies. We use the first homotopy on $[0, 1/2]$ the second on $[1/2, 3/4]$ and so on. (Compare the proof of (8.5.4).) Suppose $\dim(X, A) < n$. Let $F_0, F_1\colon X \to B$ be homotopic relative to A to f. We obtain from such homotopies a map $(X \times I, X \times \partial I \cup A \times I) \to (Y, B)$ which is the constant homotopy on A. We apply the previous argument to the pair $(X \times I, X \times \partial I \cup A \times I)$ of dimension $\le n$ and see that the homotopy class of the deformation $X \to B$ of f is unique relative to A. □

(8.4.2) Theorem. *Let $h\colon B \to Y$ be n-connected, $n \ge 0$. Then $h_*\colon [X, B] \to [X, Y]$ is bijective (surjective) if X is a CW-complex with* $\dim X < n$ ($\dim X \le n$)*. If $h\colon B \to Y$ is pointed, then $h_*\colon [X, B]^0 \to [X, Y]^0$ is injective (surjective) in the same range.*

Proof. By use of mapping cylinders we can assume that h is an inclusion. The surjectivity follows if we apply (8.4.1) to the pair $(X, \emptyset)$. The injectivity follows, if we apply it to the pair $(X \times I, X \times \partial I)$. In the pointed case we deform $(X, *) \to (Y, B)$ rel $\{*\}$ to obtain surjectivity, and for the proof of injectivity we apply (8.4.1) to the pair $(X \times I, X \times \partial I \cup * \times I)$. □

(8.4.3) Theorem. *Let $f\colon Y \to Z$ be a map between CW-complexes.*

(1) *f is a homotopy equivalence, if and only if for each $b \in Y$ and each $q \ge 0$ the induced map $f_*\colon \pi_q(Y, b) \to \pi_q(Z, f(b))$ is bijective.*

(2) *Suppose* $\dim Y \le k$, $\dim Z \le k$. *Then* f *is a homotopy equivalence if* f_* *is bijective for* $q \le k$.

Proof. (1) is a special case of (2).

(2) $f_*\colon [Z,Y] \to [Z,Z]$ is surjective, since f is k-connected (see (8.4.2)). Hence there exists $g\colon Z \to Y$ such that $fg \simeq \mathrm{id}(Z)$. Then $g_*\colon \pi_q(Z) \to \pi_q(Y)$ is bijective for $q \le k$, since $f_*g_* = \mathrm{id}$ and f_* is bijective. Hence there exists $h\colon Y \to Z$ with $gh \simeq \mathrm{id}(Y)$. Thus g has a left and a right h-inverse and is therefore an h-equivalence. From $fg \simeq \mathrm{id}$ we then conclude that f is an h-equivalence. □

The importance of the last theorem lies in the fact that "homotopy equivalence" can be tested algebraically. Note that the theorem does not say: If $\pi_q(Y) \cong \pi_q(Z)$ for each q, then Y and Z are homotopy equivalent; it is important to have a map which induces an isomorphism of homotopy groups. Mapping a space to a point gives:

(8.4.4) Corollary. *A CW-complex* X *is contractible if and only if* $\pi_q(X) = 0$ *for* $q \ge 0$. □

(8.4.5) Example. From $\pi_j(S^n) = 0$ for $j < n$ and $\pi_j(S^\infty) = \mathrm{colim}_n\, \pi_j(S^n)$ we conclude that the homotopy groups of S^∞ are trivial. Hence S^∞ is contractible.◇

(8.4.6) Example. A simply connected 1-dimensional complex is contractible. A contractible 1-dimensional CW-complex is called a ***tree***. ◇

(8.4.7) Theorem. *A connected CW-complex* X *contains a maximal (with respect to inclusion) tree as subcomplex. A tree in* X *is maximal if and only if it contains each* 0*-cell.*

Proof. Let $\mathcal{B}$ denote the set of all trees in X, partially ordered by inclusion. Let $\mathcal{T} \subset B$ be a totally ordered subset. Then $C = \bigcup_{T\in\mathcal{T}} T$ is contractible: $\pi_1(C) = 0$, since a compact subset of C is contained in a finite subcomplex and therefore in some $T \in \mathcal{T}$. Thus, by Zorn's lemma, there exist maximal trees.

Let B be a maximal tree. Consider the 1-cells which have at least one end point in B. If the second end point is not contained in B, then B is obviously not maximal. Therefore the union V of these 1-cells together with B form a subcomplex of X^1, and the remaining 1-cells together with their end points form a subcomplex $X^1 \smallsetminus V$. Since X is connected so is X^1, hence $V = X^1$, and $B^0 = V^0 = X^0$.

Let B be a tree which contains X^0. Let $B' \supset B$ be a strictly larger tree. Since B is contractible, B' and B'/B are h-equivalent. Hence B'/B is contractible. Since $X^0 \subset B$, the space B'/B has the form $\bigvee S^1$ and is not simply connected. Contradiction. □

We now generalize the suspension theorem (6.10.4). Let X and Y be pointed spaces. We have the suspension map $\Sigma_*\colon [X,Y]^0 \to [\Sigma X, \Sigma Y]^0$. We use the adjunction $[\Sigma X, \Sigma Y]^0 \cong [X, \Omega\Sigma Y]^0$. The resulting map $[X, Y]^0 \to [X, \Omega\Sigma Y]^0$ is then induced by the pointed map $\sigma\colon Y \to \Omega\Sigma Y$ which assigns to $y \in Y$ the loop $t \mapsto [y, t]$ in ΣY.

(8.4.8) Theorem. *Suppose $\pi_i(Y) = 0$ for $0 \le i \le n$. Then the suspension $\Sigma_*\colon [X, Y]^0 \to [\Sigma X, \Sigma Y]^0$ is bijective (surjective) if X is a CW-complex of dimension* $\dim X \le 2n$ ($\dim X \le 2n + 1$).

Proof. By the suspension theorem (6.10.4), the map σ is $(2n+1)$-connected. Now use the pointed version of (8.4.2). □

(8.4.9) Theorem. *Let X be a finite pointed CW-complex. Then*

$$\Sigma_*\colon [\Sigma^k X, \Sigma^k Y]^0 \to [\Sigma^{k+1} X, \Sigma^{k+1} Y]^0$$

is bijective for $\dim(X) \le k - 1$.

Proof. We have $\dim \Sigma^k X = k + \dim X$. The space ΣY is path connected. By the theorem of Seifert and van Kampen, $\Sigma^2 Y$ is simply connected. From the suspension theorem we conclude that $\pi_j(\Sigma^k Y) = 0$ for $0 \le j \le k - 1$. By the previous theorem, Σ_* is a bijection for $k + \dim X \le 2(k-1)$. □

8.5 Cellular Approximation

(8.5.1) Proposition. *Suppose X is obtained from A by attaching $(n+1)$-cells. Then (X, A) is n-connected.*

Proof. We know that (D^{n+1}, S^n) is n-connected. The assertion is therefore a consequence of (6.10.1). After a little re-writing one can also use (6.4.2). □

(8.5.2) Proposition. *Let X be obtained from A by attaching n-cells ($n \ge 1$). Suppose A is simply connected. Then the quotient map induces an isomorphism $\pi_n(X, A) \to \pi_n(X/A)$.*

Proof. (8.5.1) and (6.10.2). □

(8.5.3) Proposition. *For each relative CW-complex (X, A) the pair (X, X^n) is n-connected.*

Proof. From (8.5.1) we obtain by induction on k that (X^{n+k}, X^n) is n-connected. The compactness argument (8.3.8) finally shows (X, X^n) to be n-connected. □

Let X and Y be CW-complexes. A map $f\colon X \to Y$ is ***cellular***, if $f(X^n) \subset Y^n$ for each $n \in \mathbb{N}_0$. The ***cellular approximation theorem*** (8.5.4) is an application of (8.4.1).

(8.5.4) Theorem. *A map $f: X \to Y$ is homotopic to a cellular map $g: X \to Y$. If $B \subset X$ is a subcomplex and $f|B$ cellular, then the homotopy $f \simeq g$ can be chosen relative to B.*

Proof. We show inductively that there exist homotopies $H^n: X \times I \to Y$ such that

(1) $H_0^0 = f$, $H_1^{n-1} = H_0^n$ for $n \geq 1$;

(2) $H_1^n(X^i) \subset Y^i$ for $i \leq n$;

(3) H^n is constant on $X^{n-1} \cup B$.

For the induction step we assume $f(X^i) \subset Y^i$ for $i < n$. Let $\Phi: (D^n, S^{n-1}) \to (X^n, X^{n-1})$ be a characteristic map of an n-cell not contained in B. The map $f \circ \Phi$ is homotopic relative to S^{n-1} to a map into Y^n, since (Y, Y^n) is n-connected. A corresponding homotopy is used to define a homotopy of f on the associated closed n-cells. This process defines the homotopy on $B \cup X^n$; and we extend it to X, using the fact that $B \cup X^n \subset X$ is a subcomplex and hence a cofibration. We now concatenate the homotopies H^n:

$$H(x,t) = \begin{cases} H^i(x, 2^{i+1}(t - 1 + 2^{-i})), & 1 - 2^{-i} \leq t \leq 1 - 2^{-i-1}, \\ H^i(x, 1), & x \in X^i,\ t = 1. \end{cases}$$

This map is continuous on $X^i \times I$ and hence on $X \times I$, since this space is the colimit of the $X^i \times I$. □

(8.5.5) Corollary. *Let $f_0, f_1: X \to Y$ be cellular maps which are homotopic. Then there exists a homotopy f between them such that $f(X^n \times I) \subset Y^{n+1}$. If f_0, f_1 are homotopic rel B, then f can be chosen rel B.*

Proof. Choose a homotopy $f: f_0 \simeq f_1$ rel B. Then f maps $(X \times \partial I \cup B \times I)^n = X^n \times \partial I \cup B^{n-1} \times I$ into Y^n; hence f is cellular on $X \times \partial I \cup B \times I$. □

Problems

1. Let $A \subset X$ be a subcomplex and $f: A \to Y$ a cellular map. Then $Y = X \cup_f Y$ is a CW-complex.

2. A CW-complex is path connected if and only if the 1-skeleton is path connected. The components are equal to the path components, and the path components are open.

8.6 CW-Approximation

We show in this section, among other things, that each space is weakly homotopy equivalent to a CW-complex. Our first aim is to raise the connectivity of a map.

(8.6.1) Theorem. *Let $f: A \to Y$ be a k-connected map, $k \geq -1$. Then there exists for each $n > k$ a relative CW-complex (X, A) with cells only in dimensions*

$j \in \{k+1,\dots,n\}$, $n \le \infty$, and an n-connected extension $F\colon X \to Y$ of f. If A is CW-complex, then A can be chosen as a subcomplex of X.

Proof. (Induction over n.) Recall that the map f is k-connected if the induced map $f_*\colon \pi_j(A,*) \to \pi_j(Y, f(*))$ is bijective for $j < k$ and surjective for $j = k$ (no condition for $k = -1$). If we attach cells of dimension greater than k and extend, then the extension remains k-connected. This fact allows for an inductive construction.

Let $n = 0$, $k = -1$. Suppose $f_*\colon \pi_0(A) \to \pi_0(Y)$ is not surjective. Let $C = \{c_j \mid j \in J\}$ be a family of points in Y which contains one element from each path component $\pi_0(Y) \setminus f_*\pi_0(A)$. Set $X = A + \coprod D^0_j$ and define $F\colon X \to Y$ by $F|A = f$ and $F(D^0_j) = \{c_j\}$. Then X is obtained from A by attaching 0-cells and F is a 0-connected extension of f.

$n = 1$. Suppose $f\colon A \to Y$ is 0-connected. Then $f_*\colon \pi_0(A) \to \pi_0(Y)$ is surjective. Let c_-, c_+ be points in different path components of A which have the same image under f_*. Then $\varphi\colon S^0 \to A$, $\varphi(\pm 1) = c_\pm$, is an attaching map for a 1-cell. We can extend f over $A \cup_\varphi D^1$ by a path from $f(c_-)$ to $f(c_+)$. Treating other pairs of path components similarly, we obtain an extension $F'\colon X' \to Y$ of f over a relative 1-complex (X', A) such that $F'_*\colon \pi_0(X') \to \pi_0(Y)$ is bijective. The bijectivity of F'_* follows from these facts: We have $F'_* j_* = f_*$ with the inclusion $j\colon A \to X'$; the map j is 0-connected; path components with the same image under f_* have, by construction, the same image under j_*.

We still have to extend $F'\colon X' \to Y$ to a relative 1-complex $X \supset X'$ such that $F_*\colon \pi_1(X,x) \to \pi_1(Y, f(x))$ is surjective for each $x \in X$. Let $F_j\colon (D^1, S^0) \to (Y, y)$ be a family of maps such that the $[F_j] \in \pi_1(Y,y)$ together with $F'_*(\pi_1(X',x))$ generate $\pi_1(Y,y)$, $y = F'(x)$. Let $X \supset X'$ be obtained from X' by attaching 1-cells with characteristic map $(\Phi_j, \varphi_j)\colon (D^1, S^0) \to (X, X')$, $\varphi_j(S^0) = \{x\}$. We extend F' to F such that $F \circ \Phi_j = F_j$. Then $F_*\colon \pi_1(X,x) \to \pi_1(Y,y)$ is surjective.

$n \ge 2$. Suppose $f\colon A \to Y$ is $(n-1)$-connected. By the use of mapping cylinders, we can assume that f is an inclusion. Let $(\Phi_j, \varphi_j)\colon (D^n, S^{n-1}, e_0) \to (Y, A, a)$ be a set of maps such that the $y_j = [\Phi_j, \varphi_j] \in \pi_n(Y, A, a)$ generate the $\pi_1(A,a)$-module $\pi_n(Y, A, a)$. We attach n-cells to A by attaching maps φ_j to obtain X and extend f to F by the null homotopies Φ_j of $f\varphi_j$. The characteristic map of the n-cell with attaching map φ_j represents $x_j \in \pi_n(X, A, a)$ and $F_* x_j = y_j$. The map F induces a morphism of the exact homotopy sequence of (X, A, a) into the sequence of (Y, A, a), and $F_*\colon \pi_n(X, A, a) \to \pi_n(Y, A, a)$ is surjective by construction. Consider the diagram

$$\begin{array}{ccccccccccc}
\pi_n(A) & \longrightarrow & \pi_n(X) & \longrightarrow & \pi_n(X,A) & \longrightarrow & \pi_{n-1}(A) & \longrightarrow & \pi_{n-1}(X) & \longrightarrow & 0 \\
\downarrow{=} & & (1)\downarrow{F_*} & & (2)\downarrow{F_*} & & \downarrow{=} & & (3)\downarrow{F_*} & & \\
\pi_n(A) & \longrightarrow & \pi_n(Y) & \longrightarrow & \pi_n(Y,A) & \longrightarrow & \pi_{n-1}(A) & \longrightarrow & \pi_{n-1}(Y) & \longrightarrow & 0.
\end{array}$$

The sequences end with 0, since $\pi_{n-1}(X, A) = 0$ and $\pi_{n-1}(A) \to \pi_{n-1}(Y)$ is surjective by assumption. (2) is surjective. The Five Lemma shows us that (1) is surjective and (3) injective. By induction hypothesis, (3) is already surjective. Hence F is n-connected.

In order to obtain A as a subcomplex of X, one works with cellular attaching maps. □

(8.6.2) Theorem. *Let Y be a CW-complex such that $\pi_i(Y) = 0$ for $0 \le i \le k$. Then Y is homotopy equivalent to a CW-complex X with $X^k = \{*\}$.*

Proof. Start with the k-connected map $f : A = \{*\} \to Y$ and extend it to a weak equivalence $F : X \to Y$ by attaching cells of dimension greater than k. □

(8.6.3) Proposition. *Let A and B be pointed CW-complexes. Assume that A is $(m-1)$-connected and B is $(n-1)$-connected. Then $A \wedge_k B$ is $(m+n-1)$-connected.*

Proof. We can assume that A has no cells in dimensions less than m and n no cells in dimensions less than n (except the base point). Then $A \wedge_k B$ has no cells in dimensions less than $m+n$. □

(8.6.4) Theorem. *Let $(X_j \mid j \in J)$ be a family of $(n-1)$-connected CW-complexes. Let $\iota_k : X_k \to \bigvee_{j \in J} X_j$ be the inclusion of the k-th summand. Then*

$$\alpha_J = \langle \iota_{j*} : \rangle \bigoplus_{j \in J} \pi_n(X_j) \to \pi_n(\bigvee_{j \in J} X_j)$$

is an isomorphism ($n \ge 2$).

Proof. Let J be finite. Up to h-equivalence we can assume that X_j has no cells in dimensions less than n, except the base point. Then $\prod_j X_j$ is obtained from $\bigvee_j X_j$ by attaching cells of dimension $\ge 2n$. Hence $\pi_m(\bigvee X_j) \to \pi_m(\prod X_j)$ is an isomorphism for $m \le 2n-2$. From the diagram

$$\begin{array}{ccc} \pi_n(\bigvee X_j) & \xrightarrow{\cong} & \pi_n(\prod X_j) \\ {\scriptstyle \alpha_J}\uparrow & & \downarrow{\scriptstyle \cong\ (p_{j*})} \\ \bigoplus \pi_n(X_j) & \underset{\cong}{\xrightarrow{(1)}} & \prod \pi_n(X_j) \end{array}$$

we conclude that α_J is an isomorphism.

Let now J be arbitrary. For each $x \in \pi_n(\bigvee_{j \in J} X_j)$ there exists a finite $E \subset J$ such that x is contained in the image of $\pi_n(\bigvee_E X_j) \to \pi_n(\bigvee_J X_j)$, since a compact subset is contained in a finite wedge. The result for E now shows that α_J is surjective. If x_1 and x_2 have the same image under α_J, then these elements are contained in some finite sum $\bigoplus_E$ and, again by a compactness argument, they have the same image under some α_E, if E is chosen large enough. This shows the injectivity of α_J. □

(8.6.5) Proposition. *Suppose $\pi_j(Y) = 0$ for $j > n$. Let X be obtained from A by attaching cells of dimension $\geq n+2$. Then $A \subset X$ induces a bijection $[X,Y] \to [A,Y]$.*

Proof. Surjective. Let $f\colon A \to Y$ be given. Attach $(n+2)$-cells via maps $\varphi\colon S^{n+1} \to A$. Since $f\varphi\colon S^{n+1} \to Y$ is null homotopic, we can extend f over the $(n+2)$-cells. Continue in this manner.

Injective. Use the same argument for $(X \times I, X \times \partial I \cup A \times I)$. The cells of this relative complex have a dimension $> n+2$. □

(8.6.6) Theorem. *Let A be an arbitrary space and $k \in \mathbb{N}_0$. There exists a relative CW-complex (X, A) with cells only in dimensions $j \geq k+2$, such that $\pi_n(X, x) = 0$ for $n > k$ and $x \in X$, and the induced map $\pi_n(A, a) \to \pi_n(X, a)$ is an isomorphism for $n \leq k$ and $a \in A$.*

Proof. We construct inductively for $t \geq 2$ a sequence $A = X^{k+1} \subset X^{k+2} \subset \cdots \subset X^{k+t}$ such that $\pi_n(A,a) \cong \pi_n(X^{k+t}, a)$ for $n \leq k$, $\pi_n(X^{k+t}, a) = 0$ for $k < n \leq k+t-1$, and X^{m+1} is obtained from X^m by attaching $(m+1)$-cells.

The induction step: If we attach $(m+1)$-cells to X^m by the attaching maps $\varphi_j\colon (S^m_j, e_0) \to (X^m, a)$ to obtain X^{m+1}, then $\pi_n(X^m, a) \cong \pi_n(X^{m+1}, a)$ for $n \leq m-1$. The exact sequence

$$\pi_{m+1}(X^{m+1}, X^m, a) \xrightarrow{\partial} \pi_m(X^m, a) \to \pi_m(X^{m+1}, a) \to 0$$

shows that the $[\varphi_j]$ are in the image of ∂. Thus, if the $[\varphi_j]$ generate $\pi_m(X^m, a)$, then $\pi_m(X^{m+1}, a) = 0$. □

(8.6.7) Example. We can attach cells of dimension $\geq n+2$ to S^n to obtain a space $K(\mathbb{Z}, n)$ which has a single non-trivial homotopy group $\pi_n(K(\mathbb{Z}, n)) \cong \mathbb{Z}$. See the section on Eilenberg–Mac Lane spaces for a generalization. ◇

Let $i_n^X\colon X \to X[n]$ be an inclusion of the type constructed in (8.6.6), namely $X[n]$ is obtained by attaching cells of dimension greater than $n+1$ such that $\pi_k(X[n]) = 0$ for $k > n$ and i_n^X induces an isomorphism $\pi_k(i_n^X)$ for $k \leq n$. Given a map $f\colon X \to Y$ and $i_m^Y\colon Y \to Y[m]$ for $m \leq n$, there exists a unique homotopy class $f_{n,m}\colon X[n] \to Y[m]$ such that $i_m^Y \circ f = f_{n,m} \circ i_n^X$; this is a consequence of (8.6.5). We let $j_n^X\colon X\langle n\rangle \to X$ be the homotopy fibre of i_n^X. We call j_n^X the ***n-connective covering*** of X. The induced map $\pi_i(j_n^X)\colon \pi_i(X\langle n\rangle) \to \pi_i(X)$ is an isomorphism for $i > n$ and $\pi_i(X\langle n\rangle) = 0$ for $i \leq n$. The universal covering has such properties in the case that $n = 1$. So we have a generalization, in the realm of fibrations. Objects of this type occur in the theory of Postnikov decompositions of a space, see e.g., [192].

As a consequence of (8.6.1) for $A = \emptyset$ we see that for each space Y there exists a CW-complex X and a weak equivalence $f\colon X \to Y$. We call such a weak equivalence a ***CW-approximation*** of Y. Note that a weak equivalence between CW-complexes is a homotopy equivalence (8.4.3). We show that CW-approximations are unique up to homotopy and functorial in the homotopy category.

(8.6.8) Theorem. *Let $f: Y_1 \to Y_2$ be a continuous map and let $\alpha_j: X_j \to Y_j$ be CW-approximations. Then there exists a map $\varphi: X_1 \to X_2$ such that $f\alpha_1 \simeq \alpha_2\varphi$, and the homotopy class of φ is uniquely determined by this property.*

Proof. Since α_2 is a weak equivalence, $(\alpha_2)_*: [X_1, X_2] \to [X_1, Y_2]$ is bijective. Hence there exists a unique homotopy class φ such that $f\alpha_1 \simeq \alpha_2\varphi$. □

A ***domination*** of X by K consists of maps $i: X \to K, p: K \to X$ and a homotopy $pi \simeq \mathrm{id}(X)$.

(8.6.9) Proposition. *Suppose M is dominated by a CW-complex X. Then M has the homotopy type of a CW-complex.*

Proof. Suppose $i: M \to X$ and $r: X \to M$ are given such that ri is homotopic to the identity. There exists a CW-complex $\iota: X \subset Y$ and an extension $R: Y \to M$ of r such that R induces an isomorphism of homotopy groups. Let $j = \iota i: M \to X \to Y$. Since $Rj = ri \simeq \mathrm{id}$, the composition Rj induces isomorphisms of homotopy groups, hence so does j. From $jRj \simeq j$ we conclude that jR induces the identity on homotopy groups and is therefore a homotopy equivalence. Let k be h-inverse to jR, then $j(Rk) \simeq \mathrm{id}$. Hence j has the left inverse R and the right inverse Rk and is therefore a homotopy equivalence. □

A (half-exact) ***homotopy functor*** on the category C^0 of pointed connected CW-spaces is a contravariant functor $h: C^0 \to \mathrm{SET}^0$ into the category of pointed sets with the properties:

(1) (Homotopy invariance) Pointed homotopic maps induce the same morphism.

(2) (Mayer–Vietoris property) Suppose X is the union of subcomplexes A and B. If $a \in h(A)$ and $b \in h(B)$ are elements with the same restriction in $h(A \cap B)$, then there exists an element $x \in h(X)$ with restrictions a and b.

(3) (Additivity) Let $X = \bigvee_j X_j$ with inclusions $i_j: X_j \to X$. Then

$$h(X) \to \textstyle\prod_j h(X_j), \quad x \mapsto (h(i_j)x)$$

is bijective.

(8.6.10) Theorem (E. H. Brown). *For each homotopy functor $h: C^0 \to \mathrm{SET}^0$ there exist $K \in C^0$ and $u \in h(K)$ such that*

$$[X, K]^0 \to h(X), \quad [f] \mapsto f^*(u)$$

is bijective for each $X \in C^0$. □

In category theory one says that K is a representing object for the functor h. The theorem is called the ***representability theorem of E. H. Brown***. For a proof see [31], [4].

(8.6.11) Example. Let $h(X) = [X, Z]^0$ for a connected pointed space Z. Then h is a homotopy functor. From (8.6.10) we obtain $K \in C^0$ and $f : K \to Z$ such that $f_* : [X, K]^0 \to [X, Z]^0$ is always bijective, i.e., f is a weak h-equivalence. Thus we have obtained a CW-approximation X of Z. ◇

Problems

1. As a consequence of (8.6.8) one can extend homotopy functors from CW-complexes to arbitrary spaces. Let F be a functor from the category of CW-complexes such that homotopic maps $f \simeq g$ induce the same morphism $F(f) = F(g)$. Then there is, up to natural isomorphism, a unique extension of F to a homotopy invariant functor on TOP which maps weak equivalences to isomorphisms.
2. A point is a CW-approximation of the pseudo-circle.
3. Determine the CW-approximation of $\{0\} \cup \{n^{-1} \mid n \in \mathbb{N}\}$.
4. Let X and Y be CW-complexes. Show that the identity $X \times_k Y \to X \times Y$ is a CW-approximation.
5. Let $(Y_j \mid j \in J)$ be a family of well-pointed spaces and $\alpha_j : X_j \to Y_j$ a family of pointed CW-approximations. Then $\bigvee_j \alpha_j$ is a CW-approximation. Give a counterexample (with two spaces) in the case that the spaces are not well-pointed.
6. Let $f : A \to B$ and $g : C \to D$ be pointed weak homotopy equivalences between well-pointed spaces. Then $f \wedge g$ is a weak homotopy equivalence.
7. Verify from the axioms of a homotopy functor that $h(P)$ for a point P contains a single element.
8. Verify from the axioms of a homotopy functor that for each inclusion $A \subset X$ in C^0 the canonical sequence $h(X/A) \to h(X) \to h(A)$ is an exact sequence of pointed sets.

8.7 Homotopy Classification

In favorable cases the homotopy class of a map is determined by its effect on homotopy groups.

(8.7.1) Theorem. *Let X be an $(n-1)$-connected pointed CW-complex. Let Y be a pointed space such that $\pi_i(Y) = 0$ for $i > n \geq 2$. Then*

$$h_X : [X, Y]^0 \to \mathrm{Hom}(\pi_n(X), \pi_n(Y)), \quad [f] \mapsto f_*$$

is bijective.

Proof. The assertion only depends on the pointed homotopy type of X. We use (8.6.2) and assume $X^{n-1} = \{*\}$. The h_X constitute a natural transformation in the variable X. Since (X, X^{n+1}) is $(n+1)$-connected, the inclusion $X^{n+1} \subset X$ induces an isomorphism on π_n. By (8.6.5), the restriction $r\colon [X,Y]^0 \to [X^{n+1},Y]^0$ is a bijection. Therefore it suffices to consider the case, that X has, apart from the base point, only cells of dimension n and $n+1$. Moreover, by the homotopy theorem for cofibrations, we can assume that the attaching maps for the $(n+1)$-cells are pointed. In this case X is the mapping cone of a pointed map $f\colon A = \bigvee S^n_k \to \bigvee S^n_j = B$. We therefore have the exact cofibre sequence

$$[A,Y]^0 \xleftarrow{f^*} [B,Y]^0 \leftarrow [X,Y]^0 \leftarrow [\Sigma A, Y]^0.$$

Our assumption about Y yields $[\Sigma A, Y]^0 = [\bigvee \Sigma S^n_k, Y]^0 \cong \prod_k \pi_{n+1}(Y) = 0$. We apply the natural transformation h and obtain a commutative diagram

$$\begin{array}{ccccccc}
[A,Y]^0 & \xleftarrow{f^*} & [B,Y]^0 & \xleftarrow{f_1^*} & [X,Y]^0 & \longleftarrow & 0 \\
\downarrow{\scriptstyle h_A} & & \downarrow{\scriptstyle h_B} & & \downarrow{\scriptstyle h_X} & & \\
\operatorname{Hom}(\pi_n A, \pi_n Y) & \longleftarrow & \operatorname{Hom}(\pi_n B, \pi_n Y) & \longleftarrow & \operatorname{Hom}(\pi_n X, \pi_n Y) & \longleftarrow & 0.
\end{array}$$

As one of the consequences of the excision theorem we showed that the sequence $\pi_n(A) \to \pi_n(B) \to \pi_n(X) \to 0$ is exact, and therefore the bottom sequence of the diagram is exact. We show that h_A and h_B are isomorphisms. If $A = S^n$, then

$$h_A\colon \pi_n(Y) = [S^n, Y]^0 \to \operatorname{Hom}(\pi_n(S^n), \pi_n(Y))$$

is an isomorphism, since $\pi_n(S^n)$ is generated by the identity. In the case that $A = \bigvee S^n_k$, we have a commutative diagram

$$\begin{array}{ccc}
[\bigvee S^n_k, Y]^0 & \xrightarrow{\cong} & \prod [S^n_k, Y] \\
\downarrow{\scriptstyle h_{\bigvee S^n_k}} & & \cong\downarrow{\scriptstyle \prod h_{S^n_k}} \\
\operatorname{Hom}(\pi_n(\bigvee S^n_k), \pi_n(Y)) & \xrightarrow{(1)} & \prod \operatorname{Hom}(\pi_n(S^n_k), \pi_n(Y)).
\end{array}$$

The map (1) is induced by the isomorphism $\bigoplus \pi_n(S^n_k) \cong \pi_n(\bigvee S^n_k)$ and therefore an isomorphism. We now want to conclude from the diagram by a Five Lemma type argument that h_X is bijective. The proof of surjectivity does not use the group structure. Injectivity follows, if f_1^* is injective. In order to see this, one can use the general fact that $[\Sigma A, Y]^0$ acts on $[X,Y]^0$ and the orbits are mapped injectively, or one uses that f is, up to homotopy, a suspension, because $\Sigma_*\colon [\bigvee S^{n-1}_k, \bigvee S^{n-1}_j]^0 \to [\bigvee S^n_k, \bigvee S^n_j]^0$ is surjective. □

8.8 Eilenberg–Mac Lane Spaces

Let π be an abelian group. An ***Eilenberg–Mac Lane space of type*** $\boldsymbol{K(\pi, n)}$ is a CW-complex $K(\pi,n)$ such that $\pi_n(K(\pi,n)) \cong \pi$ and $\pi_j(K(\pi,n)) \cong 0$ for $j \neq n$.

In the cases $n = 0, 1$, the group π can be non-abelian. In the case $n = 0$, we think of $K(\pi, 0) = \pi$ with the discrete topology.

(8.8.1) Theorem. *Eilenberg–Mac Lane spaces $K(\pi, n)$ exist.*

Proof. Let $n \geq 2$. There exists an exact sequence

$$0 \longrightarrow F_1 \xrightarrow{\alpha} F_0 \xrightarrow{\beta} \pi \longrightarrow 0$$

with free abelian groups F_0 and F_1. We fix a basis $(a_k \mid k \in K)$ of F_1 and $(b_j \mid j \in J)$ of F_0. Then α is determined by the matrix $\alpha(a_k) = \sum_j n(j,k)b_j$. We now construct a geometric realization of this algebraic situation. The group $\pi_n(\bigvee_k S_k^n) \cong \bigoplus_k \pi_n(S_k^n)$ is free abelian (8.6.4). A basis is given by the canonical inclusions $S_l^n \to \bigvee S_k^n$. There exists a unique homotopy class $f : A = \bigvee_k S_k^n \to \bigvee S_j^n = B$ which realizes the matrix $(n(j,k))$ with respect to these bases. Let $X = C(f)$ be the mapping cone of f. Then the sequence

$$\pi_n(A) \xrightarrow{f_*} \pi_n(B) \xrightarrow{f_{1*}} \pi_n(X) \longrightarrow 0$$

is exact. Hence $\pi_n(X) \cong \pi$. Also $\pi_i(X) = 0$ for $i < n$. We can now attach cells of dimensions $\geq n + 2$ to X in order to obtain a $K(\pi, n)$, see (8.6.6). □

(8.8.2) Examples. The space S^1 is a $K(\mathbb{Z}, 1)$. We know $\pi_1(S^1) \cong \mathbb{Z}$, and from the exact sequence of the universal covering $\mathbb{R} \to S^1$ we know that $\pi_n(S^1) = 0$ for $n \geq 2$. The space $\mathbb{C}P^\infty$ is a model for $K(\mathbb{Z}, 2)$. The space $\mathbb{R}P^\infty$ is a $K(\mathbb{Z}/2, 1)$. ◇

The adjunction $[\Sigma X, Y]^0 \cong [X, \Omega Y]^0$ shows that $\Omega K(\pi, n+1)$ has the homotopy groups of a $K(\pi, n)$. By a theorem of Milnor [132], [67], ΩY has the homotopy type of a CW-complex if Y is a CW-complex. If one does not want to use this result one has the weaker result that there exists a weak homotopy equivalence $K(\pi, n) \to \Omega K(\pi, n+1)$.

We now establish further properties of Eilenberg–Mac Lane spaces. We begin by showing that Eilenberg–Mac Lane spaces are H-spaces. Then we construct product pairings $K(\pi, m) \wedge K(\rho, n) \to K(\pi \otimes \rho, m+n)$. In this context $\pi \otimes \rho$ denotes the tensor product of the abelian groups π and ρ (alias $\mathbb{Z}$-modules) over $\mathbb{Z}$.

We call the space $K(\pi, n)$ ***polarized***, if we have chosen a fixed isomorphism $\alpha : \pi_n(K(\pi, n)) \to \pi$. If $(K(\pi, n), \alpha)$ and $(K(\rho, n), \beta)$ are polarized complexes, the product $K(\pi, n) \times K(\rho, n)$ will be polarized by

$$\pi_n(K(\pi, n) \times K(\rho, n)) \cong \pi_n(K(\pi, n)) \times \pi_n(K(\rho, n)) \xrightarrow{\alpha \times \beta} \pi_1 \times \pi_2.$$

(8.8.3) Proposition. *Having chosen polarizations, we obtain from* (8.7.1) *an isomorphism* $[K(\pi, n), K(\rho, n)]^0 \cong \mathrm{Hom}(\pi, \rho)$. □

(8.8.4) Theorem. *Let π be an abelian group. Then an Eilenberg–Mac Lane complex $K(\pi, n)$ is a commutative group object in* h-TOP.

Proof. Let $K = (K(\pi, n), \alpha)$ be a polarized complex with base point a 0-cell e. For an abelian group π, the multiplication $\mu\colon \pi \times \pi \to \pi$, $(g, h) \mapsto gh$ is a homomorphism. Therefore there exists a map $m\colon K \times K \to K$, unique up to homotopy, which corresponds under (8.8.3) to μ. Similarly, $\iota\colon \pi \to \pi, g \mapsto g^{-1}$ is a homomorphism and yields a map $i : K \to K$. Claim: (K, m, i) is an associative and commutative H-space. The maps $m \circ (m \times \mathrm{id})$ and $m \circ (\mathrm{id} \times m)$ induce the same homomorphism when π_n is applied; hence these maps are homotopic. In a similar manner one shows that $x \mapsto m(x, e)$ is homotopic to the identity. Since $K \vee K \subset K \times K$ is a cofibration, we can change m by a homotopy such that $m(x, e) = m(e, x) = x$. We write $x \mapsto m(x, i(x))$ as composition

$$K \xrightarrow{d} K \times K \xrightarrow{\mathrm{id} \times i} K \times K \xrightarrow{m} K$$

and apply π_n; the result is the constant homomorphism. Hence this map is null homotopic. Commutativity is verified in a similar manner by applying π_n. See also Problem 1. □

For the construction of the product pairing we need a general result about products for homotopy groups. We take the smash product of representatives $f\colon I^m/\partial I^m \to X$, $g\colon I^n/\partial I^n \to Y$ and obtain a well-defined map

$$\pi_m(X) \times \pi_n(Y) \to \pi_{m+n}(X \wedge_k Y), \quad ([f], [g]) \mapsto [f \wedge g] = [f] \wedge [g].$$

We call this map the $\wedge$-product for homotopy groups. It is natural in the variables X and Y.

(8.8.5) Proposition. *The $\wedge$-product is bi-additive.*

Proof. The additivity in the first variable follows directly from the definition of the addition, if we use $+_1$. We see the additivity in the second variable, if we use the composition laws $+_1$ and $+_{m+1}$ in the homotopy groups. □

(8.8.6) Proposition. *Let A be an $(m-1)$-connected and B an $(n-1)$-connected CW-complex. Then $A \wedge B$ is $(m+n-1)$-connected and the $\wedge$-product*

$$\pi_m(A) \otimes \pi_n(B) \to \pi_{m+n}(A \wedge_k B)$$

is an isomorphism $(m, n \geq 2)$. *If m or n equals* 1*, then one has to use the abelianized groups.*

Proof. The assertion about the connectivity was shown in (8.6.3). For $A = S^m$ the assertion holds by the suspension theorem. For $A = \bigvee_j S_j^m$ we use the commutative diagram

$$\begin{array}{ccc}
\pi_m(\bigvee S_j^m) \otimes \pi_n(B) & \xrightarrow{\wedge} & \pi_{m+n}((\bigvee S_j^m) \wedge_k B) \\
\uparrow {\scriptstyle (1)} & & \uparrow {\scriptstyle (2)} \\
(\bigoplus \pi_m(S_j^m)) \otimes \pi_n(B) & & \pi_{m+n}(\bigvee (S_j^m \wedge B)) \\
\uparrow {\scriptstyle (3)} & & \uparrow {\scriptstyle (4)} \\
\bigoplus (\pi_m(S_j^m) \otimes \pi_n(B)) & \xrightarrow{(5)} & \bigoplus \pi_{m+n}(S_j^m \wedge B).
\end{array}$$

(1) is an isomorphism by (8.6.4). (2) is induced by a homeomorphism. (3) is an isomorphism by algebra. (4) is an isomorphism by (8.6.4). (5) is an isomorphism by the suspension theorem. This settles the case of a wedge of m-spheres. Next we let A be the mapping cone of a map $f : C \to D$ where C and D are wedges of m-spheres. Then we have a commutative diagram with exact rows:

$$\begin{array}{ccccccc}
\pi_m(C) \otimes \pi_m(Y) & \longrightarrow & \pi_m(D) \otimes \pi_n(B) & \longrightarrow & \pi_m(A) \otimes \pi_n(B) & \longrightarrow & 0 \\
\downarrow \cong & & \downarrow \cong & & \downarrow & & \\
\pi_{m+n}(C \wedge_k B) & \longrightarrow & \pi_{m+n}(D \wedge_k B) & \longrightarrow & \pi_{m+n}(A \wedge_k B) & \longrightarrow & 0.
\end{array}$$

The general case now follows from the observation that the inclusion $A^{m+1} \to A$ induces an isomorphism on π_m and $A^{m+1} \wedge_k B \to A \wedge_k B$ induces an isomorphism on π_{m+n}. □

Let $(K(G,m),\alpha)$, $(K(H,n),\beta)$ and $(K(G \otimes H, m+n),\gamma)$ be polarized Eilenberg–Mac Lane complexes for abelian groups G and H. A ***product*** is a map

$$\gamma_{m,n} \colon K(G,m) \wedge_k K(H,n) \to K(G \otimes H, m+n)$$

such that the diagram

$$\begin{array}{ccc}
\pi_m(K(G,m)) \otimes \pi_n(K(H,n)) & \xrightarrow{\wedge} & \pi_{m+n}(K(G,m) \wedge_k K(H,n)) \\
\uparrow {\scriptstyle \alpha \otimes \beta} & & \downarrow {\scriptstyle (\gamma_{m,n})_*} \\
G \otimes H & \xrightarrow{\gamma} & \pi_{m+n}(K(G \otimes H, m+n))
\end{array}$$

is commutative. Here we have use the $\wedge$-product (8.8.5).

(8.8.7) Theorem. *There exists a product. It is unique up to homotopy.*

Proof. Let G and H be abelian groups. The first non-trivial homotopy group of $K(G,m) \wedge_k K(H,n)$ is π_{m+n} and it is isomorphic to $G \otimes H$, see (8.8.6). By (8.6.6) there exists an inclusion

$$\gamma_{m,n}\colon K(G,m) \wedge_k K(H,n) \to K(G \otimes H, m+n).$$

We can choose the polarization γ so that the diagram above becomes commutative. Uniqueness follows from (8.7.1). □

The products (8.8.7) are associative, i.e.,

$$\gamma_{m+n,p} \circ (\gamma_{m,n} \times \mathrm{id}) \simeq \gamma_{m,n+p} \circ (\mathrm{id} \times \gamma_{n,p}).$$

The products are graded commutative in the following sense:

$$K(\tau, m+n) \circ \gamma_{m,n} \simeq (-1)^{mn} \tau' \circ \gamma_{n,m}$$

with the interchange maps $\tau'\colon K(m,G) \wedge K(n,H) \to K(n,H) \wedge K(m,G)$ and $\tau\colon G \otimes H \to H \otimes G$.

Let R be a commutative ring with 1. We think of the multiplication as being a homomorphism $\mu\colon R \otimes R \to R$ between abelian groups. From this homomorphism we obtain a unique homotopy class $K(\mu)\colon K(R \otimes R, m) \to K(R,m)$.

We compose $K(\mu)$ with $\gamma_{k,l}$ and obtain a product map

$$m_{k,l}\colon K(R,k) \wedge_k K(R,l) \to K(R,k+l).$$

Also these products are associative and graded commutative.

In the associated homotopy groups $H^k(X;R) = [X^+, K(R,k)]^0$ we obtain via $(f,g) \mapsto m_{k,l}(f \wedge g)$ products

$$H^k(X;R) \otimes H^l(Y;R) \to H^{k+l}(X \times Y; R),$$

which are also associative and graded commutative. (See also Problem 3.) In a similar manner we can start from an R-module structure $R \otimes M \to M$ on M. Later, when we study singular cohomology, we show that for a CW-complex X the group $[X^+, K(R,k)]^0$ is naturally isomorphic to the singular cohomology group $H^k(X;R)$ with coefficients in the ring R. This opens the way to a homotopical study of cohomology. The product (8.8.7) can then be used to construct the so-called cup product in cohomology.

Once singular cohomology theory is constructed one obtains from the representability theorem of Brown Eilenberg–Mac Lane spaces as representing objects.

8.8.8 Eilenberg–Mac Lane spectra. Let A be an abelian group. The Eilenberg–Mac Lane spectrum HA consists of the family $(K(A,n) \mid n \in \mathbb{N}_0)$ of Eilenberg–Mac Lane CW-spaces and maps $e_n\colon \Sigma K(A,n) \to K(A,n+1)$ (an inclusion of subcomplexes; attach cells to $\Sigma K(A,n)$ to obtain a $K(A,n+1)$). This spectrum is an Ω-spectrum. We have proved in any case that $\varepsilon_n\colon K(A,n) \to \Omega K(A,n+1)$ is a weak homotopy equivalence. This suffices if one wants to define the cohomology theory only for pointed CW-spaces. ◇

Problems

1. From the natural isomorphism $[X, K(\pi,n)]^0 \cong [X, \Omega^2 K(\pi, n+2)]^0$ we see that $X \mapsto [X, K(\pi,n)]^0$ is a contravariant functor into the category of abelian groups. Therefore, by category theory, there exists a unique (up to homotopy) structure of a commutative h-group on $K(\pi,n)$ inducing the group structures of this functor.
2. Let $\alpha \in \pi_m(X)$, $\beta \in \pi_n(Y)$, and $\tau\colon X \wedge_K Y \to Y \wedge_K X$ the interchange map. Then $\alpha \wedge \beta = (-1)^{mn} \beta \wedge \alpha$.
3. Let M be an R-module. A left translation $l_r\colon M \to M$, $x \mapsto rx$ is a homomorphism of the abelian group M and induces therefore a map $L_r\colon K(M,k) \to K(M,k)$. Use these maps to define a natural structure of an R-module on $[X, K(M,k)]^0$.
4. The simply connected surfaces are S^2 and $\mathbb{R}^2$ [44, p. 87]. If a surface is different from S^2 and $\mathbb{R}P^2$, then it is a $K(\pi,1)$.
5. Let $E(\pi) \to B(\pi)$ be a π-principal covering with contractible $E(\pi)$. Then $B(\pi)$ is a $K(\pi,1)$. Spaces of the type $B(\pi)$ will occur later as classifying spaces. There is a bijection $[K(\pi,1), K(\rho,1)] \cong \mathrm{Hom}(\pi,\rho)/\sim$ between homotopy classes and group homomorphisms up to inner automorphisms.
6. Let S^∞ be the colimit of the unit spheres $S(\mathbb{C}^n) \subset S(\mathbb{C}^{n+1}) \subset \cdots$. This space carries a free action of the cyclic group $\mathbb{Z}/m \subset S^1$ by scalar multiplication. Show that S^∞ with this action is a $\mathbb{Z}/m$-principal covering. The quotient space is a CW-space $B(\mathbb{Z}/m)$ and hence a $K(\mathbb{Z}/m, 1)$.
7. A connected 1-dimensional CW-complex X is a $K(\pi,1)$. Determine π from the topology of X.
8. A connected non-closed surface (with or without boundary) is a $K(\pi,1)$.

Chapter 9
Singular Homology

Homology is the most ingenious invention in algebraic topology. Classically, the definition of homology groups was based on the combinatorial data of simplicial complexes. This definition did not yield directly a topological invariant. The definition of homology groups and (dually) cohomology groups has gone through various stages and generalizations.

The construction of the so-called singular homology groups by Eilenberg [56] was one of the definitive settings. This theory is very elegant and almost entirely algebraic. Very little topology is used as an input. And yet the homology groups are defined for arbitrary spaces in an invariant manner. But one has to pay a price: The definition is in no way intuitively plausible. If one does not mind jumping into cold water, then one may well start algebraic topology with singular homology. Also interesting geometric applications are easily obtainable.

In learning about homology, one has to follow three lines of thinking at the same time: (1) The construction. (2) Homological algebra. (3) Axiomatic treatment.

(1) The construction of singular homology groups and the verification of its main properties, now called the axioms of Eilenberg and Steenrod.

(2) A certain amount of algebra, designed for use in homology theory (but also of independent algebraic interest). It deals with diagrams, exact sequences, and chain complexes. Later more advanced topics are needed: Tensor products, linear algebra of chain complexes, derived functors and all that.

(3) The object that one constructs with singular homology is now called a homology theory, defined by the axioms of Eilenberg and Steenrod. Almost all applications of homology are derived from these axioms. The axiomatic treatment has other advantages. Various other homology and cohomology theories are known, either constructed by special input (bordism theories, K-theories, de Rham cohomology) or in a systematic manner via stable homotopy and spectra.

The axioms of a homology or cohomology theory are easily motivated from the view-point of homotopy theory. But we should point out that many results of algebraic topology need the idea of homology: The reduction to combinatorial data via cell complexes, chain complexes, spectral sequences, homological algebra, etc.

Reading this chapter requires a parallel reading of the chapter on homological algebra. Already in the first section we use the terminology of chain complexes and their homology groups and results about exact sequences of homology groups.

9.1 Singular Homology Groups

The ***n-dimensional standard simplex*** is

$$\Delta^n = \Delta[n] = \{(t_0, \dots, t_n) \in \mathbb{R}^{n+1} \mid \textstyle\sum_{i=0}^n t_i = 1, t_i \geq 0\} \subset \mathbb{R}^{n+1}.$$

We set $[n] = \{0, \dots, n\}$. A weakly increasing map $\alpha \colon [m] \to [n]$ induces an affine map

$$\Delta(\alpha)\colon \Delta[m] \to \Delta[n], \quad \textstyle\sum_{i=0}^m t_i e_i \mapsto \sum_{i=0}^m t_i e_{\alpha(i)}.$$

Here e_i is the standard unit vector, thus $\sum_{i=0}^m t_i e_i = (t_0, \dots, t_m)$. These maps satisfy the rules of a functor $\Delta(\alpha \circ \beta) = \Delta(\alpha) \circ \Delta(\beta)$ and $\Delta(\mathrm{id}) = \mathrm{id}$. Let $\delta_i^n \colon [n-1] \to [n]$ be the injective map which misses the value i.

(9.1.1) Note. $\delta_j^{n+1}\delta_i^n = \delta_i^{n+1}\delta_{j-1}^n$, $i < j$. *(The composition misses i and j.) We write $d_i^n = \Delta(\delta_i^n)$. By functoriality, the d_i^n satisfy the analogous commutation rules.* □

A continuous map $\sigma \colon \Delta^n \to X$ is called a ***singular n-simplex*** in X. The ***i-th face*** of σ is $\sigma \circ d_i^n$. We denote by $S_n(X)$ the free abelian group with basis the set of singular n-simplices in X. (We also set, for formal reasons, $S_n(X) = 0$ in the case that $n < 0$ but disregard mostly this trivial case. If $X = \emptyset$, we let $S_n(X) = 0$.) An element $x \in S_n(X)$ is called a ***singular n-chain***. We think of x as a formal finite linear combination $x = \sum_\sigma n_\sigma \sigma$, $n_\sigma \in \mathbb{Z}$. In practice, we skip a summand with $n_\sigma = 0$; also we write $1 \cdot \sigma = \sigma$. We use without further notice the algebraic fact that a homomorphism from $S_n(X)$ is determined by its values on the basis elements $\sigma \colon \Delta^n \to X$, and these values can be prescribed arbitrarily. The ***boundary operator*** ∂_q is defined for $q \geq 1$ by

$$\partial_q \colon S_q(X) \to S_{q-1}(X), \quad \sigma \mapsto \textstyle\sum_{i=0}^q (-1)^i \sigma d_i^q,$$

and for $q \leq 0$ as the zero map. Basic for everything that follows is the

9.1.2 Boundary relation. $\partial_{q-1}\partial_q = 0$.

Proof. We decompose the sum $\partial\partial\sigma = \sum_{j=0}^q \sum_{i=0}^{q-1} (-1)^{i+j} \sigma d_j^q d_i^{q-1}$ into the parts $\sum_{i<j}$ and $\sum_{i \geq j}$. When we rewrite the first sum using (9.1.1), the result is the negative of the second sum. □

The singular chain groups $S_q(X)$ and the boundary operators ∂_q form a chain complex, called the ***singular chain complex*** $S_\bullet(X)$ of X. Its n-th homology group is denoted $H_n(X) = H_n(X; \mathbb{Z})$ and called ***singular homology group*** of X (with coefficients in $\mathbb{Z}$). A continuous map $f \colon X \to Y$ induces a homomorphism

$$f_\# = S_q(f) \colon S_q(X) \to S_q(Y), \quad \sigma \mapsto f\sigma.$$

The family of the $S_q(f)$ is a chain map $S_\bullet(f): S_\bullet(X) \to S_\bullet(Y)$. Thus we have induced homomorphisms $f_* = H_q(f): H_q(X) \to H_q(Y)$. In this manner, the H_q become functors from TOP into the category ABEL of abelian groups.

Let now $i: A \subset X$ be an inclusion. We define $S_n(X, A)$ as the cokernel of $S_n(i): S_n(A) \to S_n(X)$. Less formally: The group $S_n(X, A)$ is free abelian and has as a basis the singular simplices $\sigma: \Delta^n \to X$ with image not contained in A. Since $S_n(\emptyset) = 0$, we identify canonically $S_n(X) = S_n(X, \emptyset)$. The boundary operator of $S_\bullet(X)$ induces a boundary operator $\partial_n: S_n(X, A) \to S_{n-1}(X, A)$ such that the family of quotient homomorphisms $S_n(X) \to S_n(X, A)$ is a chain map. The homology groups $H_n(X, A) = H_n(X, A; \mathbb{Z})$ of $S_\bullet(X, A)$ are the ***relative singular homology groups*** of the pair (X, A) (with coefficients in $\mathbb{Z}$). A continuous map $f: (X, A) \to (Y, B)$ induces a chain map $f_\bullet: S_\bullet(X, A) \to S_\bullet(Y, B)$ and homomorphisms $f_* = H_q(f): H_q(X, A) \to H_q(Y, B)$. In this way, H_q becomes a functor from TOP(2) to ABEL.

We apply (11.3.2) to the exact sequence of singular chain complexes

$$0 \to S_\bullet(A) \to S_\bullet(X) \to S_\bullet(X, A) \to 0$$

and obtain the associated ***exact homology sequence***:

(9.1.3) Theorem. *For each pair (X, A) the sequence*

$$\cdots \xrightarrow{\partial} H_n(A) \to H_n(X) \to H_n(X, A) \xrightarrow{\partial} H_{n-1}(A) \to \cdots$$

is exact. The sequence terminates with $H_0(X) \to H_0(X, A) \to 0$. The undecorated arrows are induced by the inclusions $(A, \emptyset) \subset (X, \emptyset)$ and $(X, \emptyset) \subset (X, A)$. □

Let (X, A, B) be a triple, i.e., $B \subset A \subset X$. The inclusion $S_\bullet(A) \to S_\bullet(X)$ induces by passage to factor groups an inclusion $S_\bullet(A, B) \to S_\bullet(X, B)$, and its cokernel can be identified with $S_\bullet(X, A)$. We apply (11.3.2) to the exact sequence of chain complexes

$$0 \to S_\bullet(A, B) \to S_\bullet(X, B) \to S_\bullet(X, A) \to 0$$

and obtain the ***exact sequence of a triple***

$$\cdots \xrightarrow{\partial} H_n(A, B) \to H_n(X, B) \to H_n(X, A) \xrightarrow{\partial} H_{n-1}(A, B) \to \cdots.$$

The boundary operator $\partial: H_n(X, B) \to H_{n-1}(A, B)$ in the exact sequence of a triple is the composition of the boundary operator for (X, A) followed by the map $H_{n-1}(A) \to H_{n-1}(A, B)$ induced by the inclusion.

It remains to verify that the connecting morphisms ∂ constitute a natural transformation, i.e., that for each map between triples $f: (X, A, B) \to (X', A', B')$ the diagram

$$\begin{array}{ccc} H_k(X,A) & \xrightarrow{\partial} & H_{k-1}(A,B) \\ \downarrow f_* & & \downarrow f_* \\ H_k(X',A') & \xrightarrow{\partial} & H_{k-1}(A',B') \end{array}$$

is commutative. This is a special case of an analogous fact for morphisms between short exact sequences of chain complexes and their associated connecting morphisms. We leave this as an exercise.

One cannot determine the groups $H_q(X,A)$ just from its definition (except in a few trivial cases). Note that for open sets in Euclidean spaces the chain groups have an uncountable basis. So it is clear that the setup only serves theoretical purposes. Before we prove the basic properties of the homology functors (the axioms of Eilenberg and Steenrod) we collect a few results which follow directly from the definitions.

9.1.4 Point. Let $X = P$ be a point. There is a unique singular n-simplex, hence $S_n(P) \cong \mathbb{Z}, n \geq 0$. The boundary operators ∂_0 and ∂_{2i+1} are zero and $\partial_2, \partial_4, \dots$ are isomorphisms. Hence $H_i(P) = 0$ for $i \neq 0$; and $H_0(P) \cong \mathbb{Z}$, via the homomorphism which sends the unique 0-simplex to $1 \in \mathbb{Z}$. ◇

9.1.5 Additivity. Let $(X_j \mid j \in J)$ be the path components of X, and let $\iota^j \colon (X_j, X_j \cap A) \to (X, A)$ be the inclusion. Then

$$\bigoplus_{j\in J} S_n(X_j, X_j \cap A) \to S_n(X,A), \quad (x_j) \mapsto \sum_j \iota^j_\#(x_j)$$

is an isomorphism. Similarly for H_n instead of S_n. The reason is that Δ^n is path connected, and therefore $\sigma \colon \Delta^n \to X$ has an image in one of the X_j, so we can sort the basis elements of $S_n(X)$ according to the components X_j. ◇

9.1.6 The groups H_0. The group $H_0(X)$ is canonically isomorphic to the free abelian group $\mathbb{Z}\pi_0(X)$ over the set $\pi_0(X)$ of path components. We identify a singular 0-simplex $\sigma \colon \Delta^0 \to X$ with the point $\sigma(\Delta^0)$. Then $S_0(X)$ is the free abelian group on the points of X. A singular 1-simplex $\sigma \colon \Delta^1 \to X$ is essentially the same thing as a path, only the domain of definition has been changed from I to Δ^1. We associate to σ the path $w_\sigma \colon I \to X, t \mapsto \sigma(1-t,t)$. Then $\partial_0\sigma = w_\sigma(1)$ and $\partial_1\sigma = w_\sigma(0)$, hence $\partial\sigma = \partial_0\sigma - \partial_1\sigma$ corresponds to the orientation convention $\partial w = w(1) - w(0)$. If two points $a, b \in X$ are in the same path component, then the zero-simplices a and b are homologous. Hence we obtain a homomorphism from $\mathbb{Z}\pi_0(X)$ into $H_0(X)$, if we assign to the path component of a its homology class. We also have a homomorphism $S_0(X) \to \mathbb{Z}\pi_0(X)$ which sends the singular simplex of $a \in X$ to the path component of a. This homomorphism sends the image of $\partial \colon S_1(X) \to S_0(X)$ to zero. Hence we obtain an inverse homomorphism $H_0(X) \to \mathbb{Z}\pi_0(X)$. ◇

9.2 The Fundamental Group

The signs which appear in the definition of the boundary operator have an interpretation in low dimensions. They are a consequence of orientation conventions. A singular 1-simplex $\sigma\colon \Delta^1 \to X$ is essentially the same thing as a path, only the domain of definition has changed from $[0,1]$ to Δ^1. We associate to σ the path $I \to X$, $t \mapsto \sigma(1-t,t)$. The inverse path is then $\sigma^-(t_0,t_1) = \sigma(t_1,t_0)$. The product of paths has now the form

$$(\sigma * \tau)(t_0,t_1) = \begin{cases} \sigma(2t_0-1, 2t_1), & t_1 \le 1/2, \\ \tau(2t_0, 2t_1-1), & t_1 \ge 1/2. \end{cases}$$

If we define $\omega\colon \Delta^2 \to X$, $(t_0,t_1,t_2) \mapsto (\sigma * \tau)(t_0 + t_1/2, t_1/2 + t_2)$, then one verifies $\partial\omega = \sigma - \sigma * \tau + \tau$. A loop $\sigma\colon \Delta^1 \to X$ is a 1-cycle; let $[\sigma]$ be its homology class. Thus for loops σ, τ we have

$$(1) \qquad [\sigma * \tau] = [\sigma] + [\tau].$$

(Here $[z]$ denotes the homology class of the cycle z.) Let $k\colon \Delta^1 \times I \to X$ be a homotopy of paths $\Delta^1 \to X$. The map k factors over the quotient map $q\colon \Delta^1 \times I \to \Delta^2$, $(t_0,t_1,t) \mapsto (t_0, t_1(1-t), t_1 t)$ and yields $\sigma\colon \Delta^2 \to X$. We compute $\partial\sigma = c - k_1 + k_0$, with a constant c. A constant 1-simplex is a boundary. Hence $k_0 - k_1$ is a boundary

$$[k_0] = [k_1] \in C_1(X)/B_1(X).$$

In particular, homotopic loops yield the same element in $H_1(X)$. Thus we obtain a well-defined map $h'\colon \pi_1(X,x_0) \to H_1(X)$; by (1), it is a homomorphism. The fundamental group is in general non-abelian. Therefore we modify h' algebraically to take this fact into account. Each group G has the associated ***abelianized*** factor group $G^{ab} = G/[G,G]$; the ***commutator group*** $[G,G]$ is the normal subgroup generated by all commutators $xyx^{-1}y^{-1}$. A homomorphism $G \to A$ to an abelian group A factorizes uniquely over G^{ab}. We apply this definition to h' and obtain a homomorphism

$$h\colon \pi_1(X,x_0)^{ab} \to H_1(X).$$

(9.2.1) Theorem. *Let X be path connected. Then h is an isomorphism.*

Proof. We construct a homomorphism in the other direction. For $x \in X$ we choose a path $u(x)$ from x_0 to x. We assign to a 1-simplex $\sigma\colon \Delta^1 \to X$ from $\sigma_0 = \sigma(1,0)$ to $\sigma_1 = \sigma(0,1)$ the class of the loop $(u(\sigma_0) * \sigma) * u(\sigma_1)^-$. We extend this assignment linearly to a homomorphism $l'\colon C_1(X) \to \pi_1(X,x_0)^{ab}$. Let $\tau\colon \Delta^2 \to X$ be a 2-simplex with faces $\tau_j = \tau d_j$. Since Δ^2 is contractible, $\tau_2 * \tau_0 \simeq \tau_1$. This implies

$$l'([\tau_2]) + l'([\tau_0]) = l'([\tau_2] + [\tau_0]) = l'([\tau_2 * \tau_0]) = l'([\tau_1]).$$

Hence l' factors over $C_1(X)/B_1(X)$ and induces $l\colon H_1(X) \to \pi_1(X,x_0)^{ab}$. By construction, $lh = \mathrm{id}$. We show that h is surjective. Let $\sum a_\sigma\sigma \in C_1(X)$ be a cycle. Then

$$\sum a_\sigma[\sigma] = \sum a_\sigma([u(\sigma_0)] + [\sigma] - [u(\sigma_1)]) = \sum a_\sigma[(u(\sigma_0) * \sigma) * u(\sigma_1)^-],$$

and the last element is contained in the image of h. □

One of the first applications of the homology axioms is the computation $H_1(S^1) \cong \mathbb{Z}$. Granted the formal result that $\pi_1(S^1)$ is abelian, we obtain yet another proof for $\pi_1(S^1) \cong \mathbb{Z}$.

9.3 Homotopy

We prove in this section the homotopy invariance of the singular homology groups. We begin with a special case.

9.3.1 Cone construction. Let X be a contractible space. Define a chain map $\varepsilon = (\varepsilon_n)\colon S_\bullet(X) \to S_\bullet(X)$ by $\varepsilon_n = 0$ for $n \neq 0$ and by $\varepsilon_0(\sum n_\sigma\sigma) = (\sum n_\sigma)\sigma_0$ where $\sigma_0\colon \Delta^0 \to \{x_0\}$. We associate to each homotopy $h\colon X \times I \to X$ from the identity to the constant map with value x_0 a chain homotopy $s = (s_n)$ from ε to the identity. The homomorphisms $s\colon S_{n-1}(X) \to S_n(X)$ are obtained from a cone construction. Let

$$q\colon \Delta^{n-1} \times I \to \Delta^n, \quad ((\mu_0,\dots,\mu_{n-1}),t) \mapsto (t,(1-t)\mu_0,\dots,(1-t)\mu_{n-1}).$$

Given $\sigma\colon \Delta^{n-1} \to X$, there exists a unique simplex $s(\sigma) = s\sigma\colon \Delta^n \to X$ such that $h \circ (\sigma \times \mathrm{id}) = s(\sigma) \circ q$, since q is a quotient map. For the faces of $s\sigma$ we verify $(s\sigma)d_i = s(\sigma d_{i-1})$, for $i > 0$, and $(s\sigma)d_0 = \sigma$. From these data we compute for $n > 1$,

$$\begin{aligned}\partial(s\sigma) &= (s\sigma)d_0 - \textstyle\sum_1^n(-1)^{i-1}(s\sigma)d_i = \sigma - \sum_0^{n-1}(-1)^j s(\sigma d_{j-1})\\ &= \sigma - s(\partial\sigma)\end{aligned}$$

and $\partial(s\sigma) = \sigma - \sigma_0$ for a 0-simplex σ. These relations imply $\partial s + s\partial = \mathrm{id} - \varepsilon$. Note that ε induces the zero map in dimensions $n \neq 0$. ◇

(9.3.2) Proposition. *Let X be contractible. Then $H_n(X) = 0$ for $n \neq 0$.* □

The inclusions $\eta^t(X) = \eta^t\colon X \to X \times I$, $x \mapsto (x,t)$ induce chain maps $(\eta^t_n) = \eta^t_\bullet\colon S_\bullet(X) \to S_\bullet(X \times I)$. We consider these chain maps as natural transformations between functors; the naturality says that for each continuous map $f\colon X \to Y$ the commutation relation $(f \times \mathrm{id})_\bullet\eta^t(X)_\bullet = \eta^t(Y)_\bullet f_\bullet$ holds.

(9.3.3) Theorem. *There exists a natural chain homotopy $s_\bullet$ from $\eta^0_\bullet$ to $\eta^1_\bullet$.*

Proof. We apply (11.5.1) to $\mathcal{C} = \mathrm{TOP}$, $F_*(X) = S_\bullet(X)$, $G_*(X) = S_\bullet(X \times I)$ and the natural transformation $\eta^t_\bullet$. The model set for F_n consists of Δ^n, and the corresponding b-element is the identity of Δ^n considered as a singular simplex. From (9.3.2) we see that G_* is acyclic. It should be clear that $\eta^0_\bullet$ and $\eta^1_\bullet$ induce the same transformations in H_0. □

For the convenience of the reader we also rewrite the foregoing abstract proof in explicit terms. See also Problem 2 for an explicit chain homotopy and its geometric meaning.

Proof. We have to show: There exist morphisms $s^X_n : S_n(X) \to S_{n+1}(X \times I)$ such that

$$(K_n) \qquad \partial s^X_n + s^X_{n-1}\partial = \eta^1_n(X) - \eta^0_n(X)$$

(chain homotopy), and such that for continuous $X \to Y$ the relations

$$(N_n) \qquad (f \times \mathrm{id})_\# \circ s^X_n = s^Y_n \circ f_\#$$

hold (naturality). We construct the s_n inductively.

$n = 0$. In this case, s_0 sends the 0-simplex $\sigma : \Delta^0 \to \{x\} \subset X$ to the 1-simplex $s_0\sigma : \Delta^1 \to X \times I$, $(t_0, t_1) \mapsto (x, t_1)$. Then the computation

$$\partial(s_0\sigma) = (s_0\sigma)d_0 - (s_0\sigma)d_1 = \eta^1_0\sigma - \eta^0_0\sigma$$

shows that (K_0) holds, and also (N_0) is a direct consequence of the definitions.

Now suppose that the s_k for $k < n$ are given, and that they satisfy (K_k) and (N_k). The identity of Δ^n is a singular n-simplex; let $\iota_n \in S_n(\Delta^n)$ be the corresponding element. The chain to be constructed $s_n\iota_n$ should satisfy

$$\partial(s_n\iota_n) = \eta^1_n(\iota_n) - \eta^0_n(\iota_n) - s_{n-1}\partial(\iota_n).$$

The right-hand side is a cycle in $S_n(\Delta^n \times I)$, as the next computation shows.

$$\begin{aligned}
&\partial(\eta^1_n(\iota_n) - \eta^0_n(\iota_n) - s_{n-1}\partial(\iota_n))\\
&\quad = \eta^1_{n-1}(\partial\iota_n) - \eta^0_{n-1}(\partial\iota_n) - \partial s_{n-1}(\partial\iota_n)\\
&\quad = \eta^1_{n-1}(\partial\iota_n) - \eta^0_{n-1}(\partial\iota_n) - (\eta^1_{n-1}(\partial\iota_n) - \eta^0_{n-1}(\partial\iota_n) - s_{n-2}\partial\partial\iota_n) = 0.
\end{aligned}$$

We have used the relation (K_{n-1}) for $\partial s_{n-1}\partial(\iota_n)$ and that the $\eta^t_\bullet$ are chain maps. Since $\Delta^n \times I$ is contractible, there exists, by (9.3.2), an $a \in S_{n+1}(\Delta^n \times I)$ with the property $\partial a = \eta^1_n(\iota_n) - \eta^0_n(\iota_n) - s_{n-1}\partial(\iota_n)$. We choose an a with this property and define $s_n(\iota_n) = a$ and in general $s_n(\sigma) = (\sigma \times \mathrm{id})_\# a$ for $\sigma : \Delta^n \to X$; the required naturality (N_n) forces us to do so. We now verify (K_n) and (N_n). We compute

$$\begin{aligned}\partial s_n(\sigma) &= \partial(\sigma\times \mathrm{id})_\# a = (\sigma\times\mathrm{id})_\#\partial a\\ &= (\sigma\times\mathrm{id})_\#(\eta^1_n\iota_n - \eta^0_n\iota_n - s_{n-1}\partial\iota_n)\\ &= \eta^1_n\sigma_\#\iota_n - \eta^0_n\sigma_\#\iota_n - s_{n-1}\sigma_\#\partial\iota_n\\ &= \eta^1_n\sigma - \eta^0_n\sigma - s_{n-1}\partial\sigma.\end{aligned}$$

We have used: $(\sigma\times\mathrm{id})_\#$ is a chain map; choice of a; naturality of $\eta^1_\bullet$, $\eta^0_\bullet$, and (N_{n-1}); $\sigma_\#\iota_n = \sigma$; $\sigma_\#$ is a chain map. Thus we have shown (K_n). The equalities

$$(f\times\mathrm{id})_\# s_n(\sigma) = (f\times\mathrm{id})_\#(\sigma\times\mathrm{id})_\# a = (f\sigma\times\mathrm{id})_\# a = s_n(f\sigma) = s_n f_\#\sigma$$

finally show the naturality (N_n). □

With (9.3.3) we control the universal situation. Let $f\colon (X,A)\times I\to (Y,B)$ be a homotopy in TOP(2) from f^0 to f^1. The s_n in (9.3.3) induce by naturality also a chain homotopy $s_n\colon S_n(X,A)\to S_{n+1}(X\times I, A\times I)$. The computation

$$\partial(f_\#\circ s_n) + (f_\#\circ s_{n-1})\partial = f_\#\partial s_n + f_\# s_{n-1}\partial = f_\#(\eta^1-\eta^0) = f^1_\# - f^0_\#$$

proves the $f_\# s_n$ to be a chain homotopy from $f^0_\#$ to $f^1_\#$. Altogether we see:

(9.3.4) Theorem. *Homotopic maps induce homotopic chain maps and hence the same homomorphisms between the homology groups.* □

(9.3.5) Example. Let $a^0(X), a^1(X)\colon S_\bullet(X)\to S_\bullet(X)$ be chain maps, natural in X, which coincide on $S_0(X)$. Then there exists a natural chain homotopy from a^0 to a^1. This is a consequence of (11.5.1) for $F_* = G_*$ and the models Δ^n as in the proof of (9.3.3). ◇

Problems

1. Let $\tau_n\colon \Delta^n\to\Delta^n$, $(\lambda_0,\dots,\lambda_n)\mapsto(\lambda_n,\dots,\lambda_0)$. Verify that $S_n(X)\to S_n(X)$, $\sigma\mapsto(-1)^{(n+1)n/2}\sigma\tau_n$ is a natural chain map. By (9.3.5), it is naturally homotopic to the identity.

2. One can prove the homotopy invariance by constructing an explicit chain homotopy. A natural construction would associate to a singular n-simplex $\sigma\colon\Delta^n\to X$ the singular prism $\sigma\times\mathrm{id}\colon \Delta^n\times I\to X\times I$. The combinatorial (set-theoretic) boundary of $\Delta^n\times I$ is $\Delta^n\times 1\cup\Delta^n\times 0\cup(\partial\Delta^n)\times I$, and this corresponds exactly to the definition of a chain homotopy, if one takes orientations into account. This idea works; one has to decompose $\Delta^n\times I$ into simplices, and it suffices to do this algebraically.

In the prism $\Delta^n\times I$ let $0,1,\dots,n$ denote the vertices of the base and $0',1',\dots,n'$ those of the top. In the notation for affine singular simplices introduced later, show that an explicit formula for $a = s_n\iota_n$ is

$$s_n\iota_n = \textstyle\sum_{i=0}^n(-1)^i[0,1,\dots,i,i',(i+1)',\dots,n'].$$

(This is a special case of the Eilenberg–Mac Lane shuffle morphism to be discussed in the section on homology products.)

9.4 Barycentric Subdivision. Excision

The basic property of homology is the excision theorem (9.4.7). It is this theorem which allows for effective computations. Its proof is based on subdivision of standard simplices. We have to work out the algebraic form of this subdivision first.

Let $D \subset \mathbb{R}^n$ be convex, and let $v_0, \dots, v_p$ be elements in D. The affine singular simplex $\sigma\colon \Delta^p \to D, \sum_i \lambda_i e_i \mapsto \sum_i \lambda_i v_i$ will be denoted $\sigma = [v_0, \dots, v_p]$. With this notation

$$\partial[v_0, \dots, v_p] = \textstyle\sum_{i=0}^p (-1)^i [v_0, \dots, \widehat{v_i}, \dots, v_p],$$

where $\widehat{v_i}$ means that v_i has to be omitted from the string of vertices. For each $v \in D$ we have the contracting homotopy $D \times I \to D$, $(x,t) \mapsto (1-t)x + tv$. If we apply the cone construction 9.3.1 to $[v_0, \dots, v_p]$ we obtain $[v, v_0, \dots, v_p]$. We denote the chain homotopy associated to the contraction by $S_p(D) \to S_{p+1}(D)$, $c \mapsto v \cdot c$. We have for $c \in S_p(D)$:

$$(1) \qquad \partial(v \cdot c) = \begin{cases} c - v \cdot \partial c, & p > 0, \\ c - \varepsilon(c)v, & p = 0, \end{cases}$$

with $\varepsilon\colon S_0(D) \to \mathbb{Z}, \sum n_\sigma \sigma \mapsto \sum n_\sigma$.

The ***barycenter*** of $\sigma = [v_0, \dots, v_p]$ is $\sigma^\beta = \frac{1}{p+1}\sum_{i=0}^p v_i$. We define inductively

$$\mathcal{B}_p(X) = \mathcal{B}_p\colon S_p(X) \to S_p(X)$$

to be the homomorphism which sends $\sigma\colon \Delta^p \to X$ to $\mathcal{B}_p(\sigma) = \sigma_\# \mathcal{B}_p(\iota_p)$, where $\mathcal{B}_p(\iota_p)$ is defined inductively as

$$(2) \qquad \mathcal{B}_p(\iota_p) = \begin{cases} \iota_0, & p = 0, \\ \iota_p^\beta \cdot \mathcal{B}_{p-1}(\partial \iota_p), & p > 0. \end{cases}$$

(9.4.1) Proposition. *The $\mathcal{B}_p$ constitute a natural chain map which is naturally homotopic to the identity.*

Proof. The equalities

$$f_\# \mathcal{B}\sigma = f_\# \sigma_\# \mathcal{B}(\iota_p) = (f\sigma)_\# \mathcal{B}(\iota_p) = \mathcal{B}(f\sigma) = \mathcal{B} f_\# \sigma$$

prove the naturality. We verify by induction over p that we have a chain map. Let $p = 1$. Then $\partial \mathcal{B}(\iota_1) = \partial(\iota_1^\beta \cdot \mathcal{B}(\partial \iota_1)) = \partial \iota_1 = \mathcal{B}\partial(\iota_1)$. For $p > 1$ we compute

$$\partial \mathcal{B}\iota_p = \partial(\iota_p^\beta \cdot \mathcal{B}(\partial \iota_p)) = \mathcal{B}\partial\iota_p - \iota_p^\beta \cdot \partial\mathcal{B}\partial\iota_p = \mathcal{B}\partial\iota_p - \iota_p^\beta \cdot \mathcal{B}\partial\partial\iota_p = \mathcal{B}\partial\iota_p.$$

We have used: Definition; (1); inductive assumption; $\partial\partial = 0$. We now use this special case and the naturality

$$\mathcal{B}\partial\sigma = \mathcal{B}\partial\sigma_\#\iota_p = \mathcal{B}\sigma_\#\partial\iota_p = \sigma_\#\mathcal{B}\partial\iota_p = \sigma_\#\partial\mathcal{B}\iota_p = \partial\sigma_\#\mathcal{B}\iota_p = \partial\mathcal{B}\sigma,$$

and this computation covers the general case.

The chain map $\mathcal{B}$ is naturally homotopic to the identity (see (9.3.5)). □

Let $\mathcal{U}$ be a family of subsets of X such that their interiors cover X. We call a singular simplex $\mathcal{U}$-small, if its image is contained in some member of $\mathcal{U}$. The subgroup spanned by the $\mathcal{U}$-small simplices is a subcomplex $S_\bullet^{\mathcal{U}}(X)$ of $S_\bullet(X)$ with homology groups denoted by $H_n^{\mathcal{U}}(X)$.

(9.4.2) Lemma. *The diameter $d(v_0, \dots, v_p)$ of the affine simplex $[v_0, \dots, v_p]$ with respect to the Euclidean norm is the maximum of the $\|v_i - v_j\|$.*

Proof. Let $x, y \in [v_0, \dots, v_p]$ and $x = \sum \lambda_j v_j$. Then, because of $\sum \lambda_j = 1$,

$$\|x - y\| = \|\sum \lambda_j (v_j - y)\| \le \sum \lambda_j \|v_j - y\| \le \max_j \|v_j - y\|.$$

This shows in particular $\|y - v_i\| \le \max_j \|v_j - v_i\|$; we insert this in the above and obtain $\|x - y\| \le \max_{i,j} \|v_i - v_j\|$; hence the diameter is at most as stated. On the other hand, this value is clearly attained as the distance between two points. □

(9.4.3) Lemma. *Let $v_0, \dots, v_p \in \mathbb{R}^n$. Then $\mathcal{B}_p[v_0, \dots, v_p]$ is a linear combination of affine simplices with diameter at most $\frac{p}{p+1} d(v_0, \dots, v_p)$.*

Proof. From the inductive definition (2) and the naturality of $\mathcal{B}$ we conclude

(3) $$\mathcal{B}[v_0, \dots, v_p] = \textstyle\sum_{j=0}^{p} (-1)^j \sigma^\beta \cdot \mathcal{B}[v_0, \dots, \widehat{v_j}, \dots, v_p]$$

where $\sigma = [v_0, \dots, v_p]$.

We prove the claim by induction over p. The assertion is obvious for $p = 0$, a point has diameter zero. By induction hypothesis, the simplices in the chain $\mathcal{B}[v_0, \dots, \widehat{v_j}, \dots, v_p]$ are affine of diameter at most $\frac{p-1}{p} d(v_0, \dots, \widehat{v_j}, \dots, v_p) \le \frac{p-1}{p} d(v_0, \dots, v_p)$. The simplices in $\mathcal{B}[v_0, \dots, v_p]$ have vertices σ^β and vertices from simplices in $\mathcal{B}[v_0, \dots, \widehat{v_j}, \dots, v_p]$. It suffices to evaluate the distance of σ^β from such vertices. It is less than or equal to $\sup(\|\sigma^\beta - x\| \mid x \in [v_0, \dots, v_p])$. Let $x = \sum \lambda_j v_j$. Then $\|\sigma^\beta - x\| \le \max \|\sigma^\beta - v_j\|$, as in the proof of (9.4.2). Moreover we have

$$\begin{aligned} \|\sigma^\beta - v_j\| &= \|\tfrac{1}{p+1}(\textstyle\sum_i v_i) - v_j\| \le \tfrac{1}{p+1} \textstyle\sum_i \|v_i - v_j\| \\ &\le \tfrac{p}{p+1} \max_{i,j} \|v_i - v_j\| = \tfrac{p}{p+1} d(v_0, \dots, v_p). \end{aligned}$$

Since $(p-1)/p < p/(p+1)$ we have verified, altogether, the claim. □

(9.4.4) Lemma. *Let $\sigma\colon \Delta^p \to X$ be a singular simplex. Then there exists a $k \in \mathbb{N}$ such that each simplex in the chain $\mathcal{B}^k \sigma$ has an image contained in a member of $\mathcal{U}$. (Here $\mathcal{B}^k$ is the k-fold iteration of $\mathcal{B}$.)*

Proof. We consider the open covering $(\sigma^{-1}(U^\circ)), U \in \mathcal{U}$ of Δ^p. Let $\varepsilon > 0$ be a Lebesgue number of this covering. The simplices of $\mathcal{B}^k\sigma$ arise by an application of σ to the simplices in $\mathcal{B}^k\iota_p$. From (9.4.3) we see that the diameter of these simplices is at most $(\frac{p}{p+1})^k d(e_0,\dots,e_p)$. If k is large enough, this number is smaller than ε. □

(9.4.5) Theorem. *The inclusion of chain complexes $S^{\mathcal{U}}_\bullet(X) \subset S_\bullet(X)$ induces an isomorphism $H^{\mathcal{U}}_*(X) \to H_*(X)$.*

Proof. Let $a \in S^{\mathcal{U}}_n(X)$ be a cycle which represents a homology class in the kernel. Thus $a = \partial b$ with some $b \in S_{n+1}(X)$. By (9.4.4), there exists k such that $\mathcal{B}^k(b) \in S^{\mathcal{U}}_{n+1}(X)$ (apply (9.4.4) to the finite number of simplices in the linear combination of b). By (9.4.1), there exists a natural chain homotopy T_k between $\mathcal{B}^k$ and the identity. Therefore

$$\mathcal{B}^k(b) - b = T_k(\partial b) + \partial T_k(b) = T_k(a) + \partial T_k(b),$$

and we conclude

$$\partial\mathcal{B}^k(b) - \partial b = \partial T_k(a), \quad a = \partial b = \partial(\mathcal{B}^k(b) - T_k(a)).$$

From the naturality of T_k and the inclusion $a \in S^{\mathcal{U}}_n(X)$ we see $T_k(a) \in S^{\mathcal{U}}_{n+1}(X)$. Therefore a is a boundary in $S^{\mathcal{U}}_\bullet(X)$. This shows the injectivity of the map in question.

Let $a \in S_n(X)$ be a cycle. By (9.4.4), there exists k such that $\mathcal{B}^k a \in S^{\mathcal{U}}_n(X)$. We know that

$$\mathcal{B}^k a - a = T_k(\partial a) + \partial T_k(a) = \partial T_k(a).$$

Since $\mathcal{B}^k$ is a chain map, $\mathcal{B}^k a$ is a cycle. From the last equality we see that a is homologous to a cycle in $S^{\mathcal{U}}_n(X)$. This shows the surjectivity of the map in question. □

Let now (X, A) be a pair of spaces. We write $\mathcal{U} \cap A = (U \cap A \mid U \in \mathcal{U})$ and define the chain complex $S^{\mathcal{U}}_\bullet(X, A) = S^{\mathcal{U}}_\bullet(X)/S^{\mathcal{U}\cap A}_\bullet(A)$ with homology groups $H^{\mathcal{U}}_*(X, A)$. We obtain a commutative diagram of chain complexes with exact rows:

$$\begin{array}{ccccccccc} 0 & \longrightarrow & S^{\mathcal{U}\cap A}_\bullet(A) & \longrightarrow & S^{\mathcal{U}}_\bullet & \longrightarrow & S^{\mathcal{U}}_\bullet(X, A) & \longrightarrow & 0 \\ & & \downarrow & & \downarrow & & \downarrow & & \\ 0 & \longrightarrow & S_\bullet(A) & \longrightarrow & S_\bullet(X) & \longrightarrow & S_\bullet(X, A) & \longrightarrow & 0. \end{array}$$

Each row has its long exact homology sequence. We apply (9.4.5) to $(X, \mathcal{U})$ and $(A, \mathcal{U} \cap A)$, use the Five Lemma (11.2.7), and obtain:

(9.4.6) Theorem. *The inclusion of chain complexes $\iota\colon S^{\mathcal{U}}_\bullet(X, A) \to S(X, A)$ induces an isomorphism $H^{\mathcal{U}}_*(X, A) \cong H_*(X, A)$. By an application of* (11.6.3) *we see that the inclusion ι is actually a chain equivalence.* □

(9.4.7) Theorem (Excision theorem). *Let* $Y = Y_1^\circ \cup Y_2^\circ$. *Then the inclusion induces an isomorphism* $H_*(Y_2, Y_1 \cap Y_2) \cong H_*(Y, Y_1)$. *Let* $B \subset A \subset X$ *and suppose that* $\overline{B} \subset A^\circ$. *Then the inclusion* $(X \smallsetminus B, A \smallsetminus B) \to (X, A)$ *induces an isomorphism* $H_*(X \smallsetminus B, A \smallsetminus B) \cong H_*(X, A)$. *Again we can invoke* (11.6.3) *and conclude that the inclusion actually induces chain equivalences between the chain complexes under consideration.*

Proof. The covering $\mathcal{U} = (Y_1, Y_2)$ satisfies the hypothesis of (9.4.5). By definition, we have $S_n^{\mathcal{U}}(X) = S_n(Y_1) + S_n(Y_2)$ and also $S_n(Y_1 \cap Y_2) = S_n(Y_1) \cap S_n(Y_2)$. The inclusion $S_\bullet(Y_2) \to S_\bullet(Y)$ induces therefore, by an isomorphism theorem of elementary algebra,

$$\frac{S_n(Y_2)}{S_n(Y_1 \cap Y_2)} = \frac{S_n(Y_2)}{S_n(Y_1) \cap S_n(Y_2)} \cong \frac{S_n(Y_1) + S_n(Y_2)}{S_n(Y_1)} = \frac{S_n^{\mathcal{U}}(Y)}{S_n(Y_1)}.$$

By (9.4.5) and (11.2.7) we see, firstly, that $S_\bullet^{\mathcal{U}}(Y)/S_\bullet(Y_1) \to S_\bullet(Y)/S_\bullet(Y_1)$ and, altogether, that $S_\bullet(Y_2)/S_\bullet(Y_1 \cap Y_2) \to S_\bullet(Y)/S_\bullet(Y_1)$ induces an isomorphism in homology. The second statement is equivalent to the first; we use $X = Y$, $A = Y_1$, $X \smallsetminus B = Y_2$. □

Problems

1. Let $D \subset \mathbb{R}^m$ and $E \subset \mathbb{R}^n$ be convex and let $f \colon D \to E$ be the restriction of a linear map. Then $f_\#(v \cdot c) = f(v) \cdot f_\#(c)$.

2. Although not necessary for further investigations, it might be interesting to describe the chain $\mathcal{B}[v_0, \dots, v_p]$ in detail. We use (3) in the proof of (9.4.3). By (2), formula (3) also holds for $[v_0, \dots, v_p]$. This yields $\mathcal{B}[v_0, v_1] = [v_{01}, v_1] - [v_{01}, v_0]$ with barycenter v_{01}, and for $\mathcal{B}[v_0, v_1, v_2]$ we obtain in short-hand notation what is illustrated by the next figure.

$$[012, 12, 2] - [012, 12, 1] - [012, 02, 2] + [012, 02, 0] + [012, 01, 1] - [012, 01, 0].$$

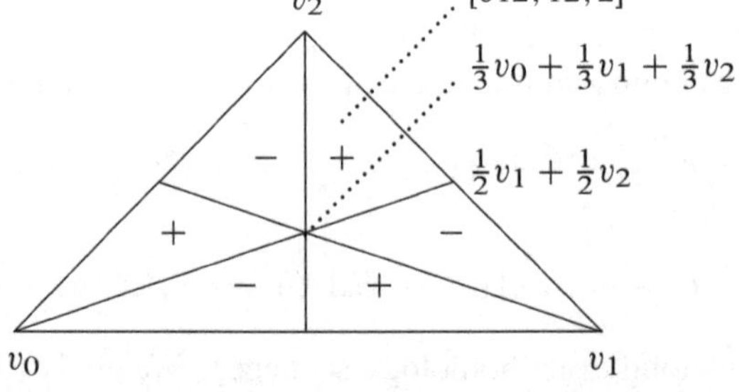

One continues inductively in this manner. Let $S(p+1)$ denote the permutation group of $\{0, \dots, p\}$. We associate to $\sigma = [v_0, \dots, v_p]$ and $\pi \in S(p+1)$ the simplex $\sigma^\pi = [v_0^\pi, \dots, v_p^\pi]$, where $v_r^\pi = [v_{\pi(r)}, \dots, v_{\pi(p)}]^\beta$. With this notation the following holds: $\mathcal{B}\sigma = \sum_{\pi \in S(p+1)} \operatorname{sign}(\pi)\sigma^\pi$.

9.5 Weak Equivalences and Homology

Although singular homology groups are defined for arbitrary topological spaces, they only capture combinatorial information. The theory is determined by its values on cell complexes. Technically this uses two facts: (1) a weak homotopy equivalence induces isomorphisms of homology groups; (2) every topological space is weakly equivalent to a CW-complex. One can use cell complexes to give proofs by induction over the skeleta. Usually the situation for a single cell is quite transparent, and this fact makes the inductive proofs easy to follow and to remember. Once a theorem is known for cell complexes, it can formally be extended to general topological spaces. We now prove this invariance property of singular homology [56], [21].

Let $(X, A, *)$ be a pointed pair. Let $\Delta[k]^n$ be the n-skeleton of the standard simplicial complex $\Delta[k]$ (this is the reason for switching the notation for the standard k-simplex). Let $S_k^{(n,A)}(X)$ for $n \geq 0$ denote the subgroup of $S_k(X)$ spanned by the singular simplices $\sigma\colon \Delta[k] \to X$ with the property

$$(\#) \qquad \sigma(\Delta[k]^n) \subset A.$$

The groups $(S_k^{(n,A)}(X) \mid k \geq 0)$ form the ***Eilenberg subcomplex*** $S_\bullet^{(n,A)}(X)$ of $S_\bullet(X)$.

(9.5.1) Theorem. *Let (X, A) be n-connected. Then the inclusion of the Eilenberg subcomplex $\alpha\colon S_\bullet^{(n,A)}(X) \to S_\bullet(X)$ is a chain equivalence.*

Proof. We assign to a simplex $\sigma\colon \Delta[k] \to X$ a homotopy $P(\sigma)\colon \Delta[k] \times I \to X$ such that

(1) $P(\sigma)_0 = \sigma$,

(2) $P(\sigma)_1$ satisfies (#),

(3) $P(\sigma)_t = \sigma$, provided σ satisfies already (#),

(4) $P(\sigma) \circ (d_i^k \times \mathrm{id}) = P(\sigma \circ d_i^k)$.

According to (3), the assignment P is defined for simplices which satisfy (#). For the remaining simplices we use an inductive construction.

Suppose $k = 0$. Then $\sigma(\Delta[0]) \in X$ is a point. Since (X, A) is 0-connected, there exists a path from this point to a point in A. We choose a path of this type as $P(\sigma)$.

Suppose P is given for j-simplices, $j < k$. Then for each k-simplex σ the homotopy $P(\sigma \circ d_i^k)$ is already defined, and the $P(\sigma \circ d_i^k)$ combine to a homotopy $\partial\Delta[k] \times I \to X$. Moreover $P(\sigma)_0$ is given. Altogether we obtain

$$\tilde{P}(\sigma)\colon (\Delta[k] \times 0 \cup \partial\Delta[k] \times I, \partial\Delta[k] \times 1) \to (X, A).$$

Let $k \leq n$. Then $\Delta[k]^n = \Delta[k]$, and similarly for the faces. By the inductive assumption, $\tilde{P}(\sigma)$ sends $\partial\Delta[k] \times 1$ into A.

There exists a homeomorphism $\kappa\colon \Delta[k]\times I \to \Delta[k]\times I$ which induces homeomorphisms (see (2.3.6))

$$\begin{aligned}\Delta[k]\times 0 &\cong \Delta[k]\times 0 \cup \partial\Delta[k]\times I,\\ \partial\Delta[k]\times 0 &\cong \partial\Delta[k]\times 1,\\ \partial\Delta[k]\times I\cup\Delta[k]\times 1 &\cong \Delta[k]\times 1.\end{aligned}$$

Since (X,A) is k-connected, the map

$$\tilde{P}(\sigma)\circ\kappa\colon (\Delta[k]\times 0, \partial\Delta[k]\times 0)\to (X,A)$$

can be extended to a homotopy $Q\colon \Delta[k]\times I \to X$ which is constant on $\partial\Delta[k]\times I$ and sends $\Delta[k]\times 1$ into A. We now set $P(\sigma) = Q\circ\kappa^{-1}$. Then $P(\sigma)$ extends $\tilde{P}(\sigma)$, hence (1) and (4) are satisfied, and (2) also holds by construction.

Let $k > n$. We use the cofibration $(\Delta[k],\partial\Delta[k])$ in order to extend $\tilde{P}(\sigma)$ to $P(\sigma)$. Since $\Delta[k]^n \subset \partial\Delta[k]$, we see that $P(\sigma)_1$ satisfies (#).

We now define $\rho\colon S_k(X)\to S_k^{(n,A)}(X)$ by $\sigma\mapsto P(\sigma)_1$. Property (4) shows that ρ is a chain map, and $\rho\circ\alpha = \mathrm{id}$ holds by construction. We define $s\colon S_k(X)\to S_{k+1}(X)$ by $s(\sigma) = P(\sigma)_\# h(\iota_k)$

$$\iota_k \in S_k(\Delta[k]) \xrightarrow{\;h\;} S_{k+1}(\Delta[k]\times I) \xrightarrow{P(\sigma)_\#} S_{k+1}(X) \ni s(\sigma)$$

where h is the natural chain homotopy between $i^0_\#$ and $i^1_\#$, see (9.3.3). The computations

$$\begin{aligned}\partial s(\sigma) &= \partial(P(\sigma)_\# h(\iota_k)) = P(\sigma)_\#\partial h(\iota_k)\\ &= P(\sigma)_{1\#}(\iota_k) - P(\sigma)_{0\#}(\iota_k) - P(\sigma)_\# h(\partial\iota_k)\\ &= \rho(\sigma) - \sigma - P(\sigma)_\# h(\partial\iota_k),\\ s\partial(\sigma) &= s\left(\sum(-1)^i\sigma\circ d_i^k\right) = \sum(-1)^i P(\sigma\circ d_i^k)_\# h(\iota_{k-1})\\ &= P(\sigma)_\#\left(\sum(-1)^i d^k_{i\#}h(\iota_{k-1})\right) = P(\sigma)_\# h(\partial\iota_k)\end{aligned}$$

show that s is a chain homotopy between $\alpha\circ\rho$ and id. □

For $k\le n$ we have $\Delta[k]^n = \Delta[k]$ and therefore $S_k^{(n,A)}(X) = S_k(A)$. The chain equivalence (9.5.1) and the exact homology sequence of (X,A) now yield:

(9.5.2) Theorem. *Let (X,A) be n-connected. Then $H_k(A)\xrightarrow{\cong} H_k(X)$ and $H_k(X,A)=0$ for $k\le n$.* □

Let $f\colon X\to Y$ be a weak homotopy equivalence. We can assume that f is an inclusion (mapping cylinder and homotopy invariance).

(9.5.3) Theorem. *A weak homotopy equivalence induces isomorphisms of the singular homology groups.* □

(9.5.4) Remark. Suppose that $(X, A, *)$ is a pointed pair and A is pathwise connected. Then we can define a subcomplex $S_\bullet^{(X,A,*)}(X)$ of $S_\bullet(X)$ where we require in addition to (#) that $\sigma(\Delta[k]^0) = \{*\}$. Again the inclusion is a chain equivalence. ◇

9.6 Homology with Coefficients

Let $C_\bullet = (C_n, c_n)$ be a chain complex of abelian groups and let G be a further abelian group. Then the groups $C_n \otimes G$ and the boundary operators $c_n \otimes \mathrm{id}$ form again a chain complex (the tensor product is taken over $\mathbb{Z}$). We denote it by $C_\bullet \otimes G$. We apply this process to the singular complex $S_\bullet(X, A)$ and obtain the complex $S_\bullet(X, A) \otimes G$ of singular ***chains with coefficients in G***. Its homology group in dimension n is denoted $H_n(X, A; G)$. The cases $G = \mathbb{Z}, \mathbb{Q}, \mathbb{Z}/p$ are often referred to as ***integral, rational, mod(p) homology***. Chains in $S_n(X, A) \otimes G$ can be written as finite formal linear combinations $\sum_\sigma a_\sigma \sigma$, $a_\sigma \in G$ of singular n-simplices σ; this accounts for the name "chain with coefficients". The sequence $0 \to S_\bullet(A) \to S_\bullet(X) \to S_\bullet(X, A) \to 0$ remains exact when tensored with G, i.e., $S_n(X, A) \otimes G \cong S_n(X) \otimes G / S_n(A) \otimes G$. Therefore we still have the exact homology sequence (11.3.2)

$$\cdots \to H_n(A; G) \to H_n(X; G) \to H_n(X, A; G) \xrightarrow{\partial} H_{n-1}(A; G) \to \cdots$$

and the analogous sequence for triples. The boundary operators ∂ are again natural transformations. If $0 \to G' \to G \to G'' \to G$ is an exact sequence of abelian groups, then the tensor product with $S_\bullet(X, A)$ yields again an exact sequence of chain complexes and we obtain from (11.3.2) an exact sequence of the form

$$\cdots \to H_n(X, A; G') \to H_n(X, A; G) \to H_n(X, A; G'') \to H_{n-1}(X, A; G') \to \cdots .$$

The passage from $C_\bullet$ to $C_\bullet \otimes G$ is compatible with chain maps and chain homotopies. A chain equivalence induces a chain equivalence. This fact yields the homotopy invariance of the homology groups $H_n(X, A; G)$. The excision theorem still holds. This is a consequence of (9.4.7): Under the hypothesis of the excision theorem, the chain equivalence $S_\bullet(Y_1, Y_1 \cap Y_2) \to S_\bullet(Y, Y_2)$ induces a chain equivalence when tensored with G. Hence the functors $H_n(X, A; G)$ satisfy the axioms of Eilenberg and Steenrod for a homology theory. The dimension axiom holds: We have a canonical isomorphism $\varepsilon_P : H_0(P) \cong G$ for a point P, which maps the homology class of the chain $a\sigma$ to a, where σ is the unique 0-simplex.

The application of (11.9.1) to topology uses the fact that the singular chain complex consists of free abelian groups. Therefore we obtain:

(9.6.1) Theorem (Universal coefficients). *Let R be a principal ideal domain and G an R-module. There exists an exact sequence*

$$0 \to H_n(X,A;R) \otimes_R G \xrightarrow{\alpha} H_n(X,A;G) \to \operatorname{Tor}(H_{n-1}(X,A;R),G) \to 0.$$

The sequence is natural in (X,A) and G. The sequence splits, the splitting is natural in G, but not in (X,A). □

The splitting statement means that $H_n(X,A;G)$ can be determined as an abelian group from homology with coefficients in $\mathbb{Z}$, but the functor $H_n(-;G)$ is not the direct sum of the functors $H_n(-;\mathbb{Z}) \otimes G$ and $\operatorname{Tor}(H_{n-1}(-;\mathbb{Z}),G)$. Here is a consequence of (9.6.1): If $f\colon (X,A) \to (Y,B)$ induces an isomorphism $f_*\colon H_*(X,A) \cong H_*(Y,B)$ for $* = n-1, n$, then it induces also an isomorphism $H_n(X,A;G) \cong H_n(Y,B;G)$.

9.7 The Theorem of Eilenberg and Zilber

We study the homology of products. For this purpose we compare the chain complexes $S_\bullet(X) \otimes S_\bullet(Y)$ and $S_\bullet(X \times Y)$. Both are values at (X,Y) of a functor $\mathrm{TOP} \times \mathrm{TOP} \to \mathrm{CH}_+$ into the category of chain complexes which are zero in negative degrees. In dimension zero they essentially coincide. For $x \in X$ let $x \in S_0(X)$ also denote the basis element given by the singular simplex $\Delta^0 \to \{x\} \subset X$. Then $S_\bullet(X \times Y)$ has the basis (x,y) and $S_\bullet(X) \otimes S_\bullet(Y)$ the basis $x \otimes y$ for $(x,y) \in X \times Y$. Natural transformations $P\colon S_\bullet(-) \otimes S_\bullet(-) \to S_\bullet(- \times -)$ and $Q\colon S_\bullet(- \times -) \to S_\bullet(-) \otimes S_\bullet(-)$ are called an ***Eilenberg–Zilber morphisms*** if in dimension zero always $P(x \otimes y) = (x,y)$ and $Q(x,y) = x \otimes y$. Both functors are free and acyclic in the sense of (11.5.1). For $(S_\bullet(-) \otimes S_\bullet(-))_n$ we use the models (Δ^k, Δ^{n-k}) and the elements $\mathrm{id} \otimes \mathrm{id}$; for $S_n(- \times -)$ we use the models (Δ^n, Δ^n) and the diagonal maps $\Delta^n \to \Delta^n \times \Delta^n$. They account for the freeness. The homology of the chain complexes $S_\bullet(\Delta^p \times \Delta^q)$ is zero in positive dimensions, since $\Delta^p \times \Delta^q$ is contractible; the homology of $S_\bullet(\Delta^p) \otimes S_\bullet(\Delta^q)$ is zero in positive dimensions, since the tensor product of chain complexes is compatible with chain homotopies, and the chain complexes $S_\bullet(\Delta^p)$ are homotopy equivalent to the trivial complex. Similar statements hold for the analogous functors in three (or more) variables like $S_\bullet(X \times Y \times Z)$ or the corresponding three-fold tensor products. As an application of (11.5.1) we obtain:

(9.7.1) Theorem. (1) *Eilenberg–Zilber morphisms P and Q exist. For each pair (P,Q) of Eilenberg–Zilber morphisms the compositions $P \circ Q$ and $Q \circ P$ are naturally homotopic to the identity. Hence the $P_{X,Y}$ and $Q_{X,Y}$ are chain equivalences and any two Eilenberg–Zilber morphisms P, P' are naturally homotopic (similarly for Q, Q').*

(2) *An Eilenberg–Zilber morphism P is associative and commutative up to natural homotopy, i.e., the natural transformations $P_{X \times Y, Z} \circ (P_{X,Y} \otimes 1)$ and*

$P_{X,Y\times Z} \circ (1 \otimes P_{Y,Z})$ from $S_\bullet(X) \otimes S_\bullet(Y) \otimes S_\bullet(Z)$ to $S_\bullet(X \times Y \times X)$ are naturally homotopic and the transformations $(t_{X,Y})_\# \circ P_{X,Y}$ and $P_{Y,X} \circ \tau_{X,Y}$ are naturally homotopic. Here $t_{X,Y} : X \times Y \to Y \times X$ interchanges the factors and $\tau_{X,Y}(x \otimes y) = (-1)^{|x||y|} y \otimes x$.

(3) *An Eilenberg–Zilber morphism Q is coassociative and cocommutative up to natural homotopy, i.e., the natural transformations $(Q_{X,Y} \otimes 1) \circ Q_{X\times Y,Z}$ and $(1 \otimes Q_{Y,Z}) \circ Q_{X,Y\times Z}$ are naturally homotopic, and the transformations $\tau_{X,Y} \circ Q_{X,Y}$ and $Q_{Y,X} \circ (t_{X,Y})_\#$ are naturally homotopic.* □

As a consequence one can determine the homology of $X \times Y$ from the chain complex $S_\bullet(X) \otimes S_\bullet(Y)$. We now turn to relative chain complexes and abbreviate $S = S_\bullet$.

(9.7.2) Proposition. *For Eilenberg–Zilber transformations P, Q and pairs of spaces (X, A), (Y, B) we have a commutative diagram with short exact rows*

$$\begin{array}{ccccc}
S(A) \otimes S(Y) + S(X) \otimes S(B) & \longrightarrow & S(X) \otimes S(Y) & \longrightarrow & S(X,A) \otimes S(Y,B) \\
Q' \uparrow\downarrow P' & & Q \uparrow\downarrow P & & Q'' \uparrow\downarrow P'' \\
S(A \times Y) + S(X \times B) & \longrightarrow & S(X \times Y) & \longrightarrow & \dfrac{S(X \times Y)}{S(A \times Y) + S(X \times B)}.
\end{array}$$

The vertical maps are induced by P and Q. The compositions $P'Q'$, $Q'P'$, $P''Q''$, $Q''P''$ are naturally homotopic to the identity.

Proof. The naturality of P shows $P(S(A) \otimes S(Y)) \subset S(A \times Y)$ and similarly for Q. This shows that P, Q induce by restriction P', Q', and P'', Q'' are the homomorphisms induced on the quotients. Since the homotopy $PQ \simeq \mathrm{id}$ is natural, it maps $S(A \times Y) + S(X \times B)$ into itself and shows $P'Q' \simeq \mathrm{id}$. □

9.7.3 We can compose

$$P : S(X, A) \otimes S(Y, B) \to S(X \times Y)/(S(A \times Y) + S(X \times B))$$

with the map induced by the inclusion $S(A \times Y) + S(X \times B) \subset S(A \times Y \cup X \times B)$ and obtain altogether natural chain maps

$$P : S(X, A) \otimes S(Y, B) \to S((X, A) \times (Y, B)).$$

We call the pair $(A \times Y, X \times B)$ ***excisive***, if this chain map is a chain equivalence. ◇

9.7.4 The natural chain map P induces natural chain maps for singular chain complexes with coefficients. Let R be a commutative ring and M, N R-modules.

$$\begin{aligned} S(X,A;M) \otimes S(Y,B;N) &= (S(X,A) \otimes M) \otimes_R (S(Y,B) \otimes N) \\ &\to (S(X,A) \otimes S(Y,B)) \otimes (M \otimes_R N) \\ &\to (S((X,A) \times (Y,B))) \otimes (M \otimes_R N). \end{aligned}$$

In many cases this chain map is followed by a homomorphism induced by a linear map $M \otimes_R N \to L$. Examples are $R \otimes_R R \to R$, $x \otimes y \mapsto xy$ in the case of a ring R and $R \otimes_R N \to N$, $x \otimes n \mapsto xn$ in the case of an R-module N. ◇

Problems

1. There exist explicit Eilenberg–Zilber morphisms which have further properties. Let $p, q \in \mathbb{N}$. We use the notation $[n] = \{0, 1, \dots, n\}$. A ***(p, q)-shuffle*** is a map $\lambda\colon [p+q] \to [p] \times [q]$ with $\lambda(0) = (0,0)$ and $\lambda(p+q) = (p,q)$ such that both components of $\lambda = (\lambda_1, \lambda_2)$ are (weakly) increasing. Given λ, there exists a permutation $(\mu, \nu) = (\mu_1, \dots, \mu_p, \nu_1, \dots, \nu_q)$ of $1, 2, \dots, p+q$ such that

$$1 \le \mu_1 < \dots < \mu_p \le p+q, \quad 1 \le \nu_1, \dots, \nu_q \le p+q$$

and $\lambda_1(\mu_j) > \lambda_1(\mu_j - 1)$ and $\lambda_2(\nu_k) > \lambda_2(\nu_k - 1)$. We denote the signum of the permutation (μ, ν) by $\varepsilon(\lambda)$. If we interpret the points $\lambda(0), \dots, \lambda(p+q)$ in the integral lattice $[p] \times [q]$ as the vertices of an edge-path from $(0,0)$ to (p,q), then the step $\lambda(j) \to \lambda(j+1)$ is horizontal or vertical of length 1. In the convex set $\Delta^p \times \Delta^q$ we have the affine $(p+q)$-simplex $[\lambda(0), \dots, \lambda(p+q)]$, also denoted λ, and the set of these simplices form a triangulation of the product when λ runs through the (p,q)-shuffles $\Sigma(p,q)$. (We do not need this geometric fact, but it explains the idea of the construction.) Define

$$P^s_{p,q}\colon S_p(X) \otimes S_q(Y) \to S_{p+q}(X,Y), \quad \sigma \otimes \tau \mapsto \textstyle\sum_{\lambda \in \Sigma(p,q)} \varepsilon(\lambda)((\sigma \times \tau) \circ \lambda)$$

on a pair σ, τ of singular simplices.

The $P^s_{p,q}$ are a strictly associative Eilenberg–Zilber morphism. We call it the ***shuffle morphism*** or the ***Eilenberg–Mac Lane morphism***.

2. An ***approximation of the diagonal*** is a natural chain map $D\colon S_\bullet(X) \to S_\bullet(X) \otimes S_\bullet(X)$ which coincides in dimension zero with $x \mapsto x \otimes x$. (The name refers to the fact that the diagonal of a cellular complex is not a cellular map, and so one looks for a homotopic cellular approximation.) By an application of (11.5.1) one shows that any two approximations of the diagonal are naturally chain homotopic.

3. The classical approximation of the diagonal is the ***Alexander–Whitney map***.

Let $\sigma\colon \Delta_n \to X$ be an n-simplex, $n = p+q$, $0 \le p, q \le n$. We have the affine maps $a_p\colon \Delta_p \to \Delta_n$, $e_i \mapsto e_i$ and $b_q\colon \Delta_q \to \Delta_n$, $e_i \mapsto e_{n-q+i}$. They are used to define $\sigma^1_p = \sigma \circ a_p$ and $\sigma^2_q = \sigma \circ b_q$.

The ***Alexander–Whitney approximation of the diagonal*** is defined by

$$D\sigma_n = \textstyle\sum_{p+q=n} \sigma^1_p \otimes \sigma^2_q, \quad \sigma_n\colon \Delta_n \to X$$

and linear extension.

4. Given an approximation D of the diagonal one constructs from it an Eilenberg–Zilber morphism Q as the composition

$$S_\bullet(X\times Y) \xrightarrow{D_{X\times Y}} S_\bullet(X\times Y)\otimes S_\bullet(X\times Y) \xrightarrow{\mathrm{pr}^X_\# \otimes \mathrm{pr}^Y_\#} S_\bullet(X)\otimes S_\bullet(Y).$$

Let Q^{AW} be the Eilenberg–Zilber morphism obtained from the Alexander–Whitney approximation of the diagonal and call it the ***Alexander–Whitney morphism***. The Alexander–Whitney morphism is strictly coassociative.

5. The Eilenberg–Mac Lane morphism EM and the Alexander–Whitney morphism AW are also compatible in a certain sense:

$$\begin{array}{ccc}
S_\bullet(W\times X)\otimes S_\bullet(Y\times Z) & \xrightarrow{EM} & S_\bullet(W\times X\times Y\times Z) \\
\Big\downarrow{\scriptstyle AW\otimes AW} & & \Big\downarrow{\scriptstyle (1\times t_{X,Y}\times 1)_\#} \\
S_\bullet(W)\otimes S_\bullet(X)\otimes S_\bullet(Y)\otimes S_\bullet(Z) & & S_\bullet(W\times Y\times X\times Z) \\
\Big\downarrow{\scriptstyle 1\otimes\tau_{X,Y}\otimes 1} & & \Big\downarrow{\scriptstyle AW} \\
S_\bullet(W)\otimes S_\bullet(Y)\otimes S_\bullet(X)\otimes S_\bullet(Z) & \xrightarrow{EM\otimes EM} & S_\bullet(W\times Y)\otimes S_\bullet(X\times Z)
\end{array}$$

commutes.

9.8 The Homology Product

We pass to homology from the chain map P in (9.7.4)

$$\begin{aligned}
H_*(X,A;M)\otimes H_*(Y,B;N) &= H_*(S(X,A;M))\otimes H_*(S(Y,B;N)) \\
&\to H_*(S(X,A;M)\otimes S(Y,B;N)) \\
&\to H_*(S((X,A)\times(Y,B));M\otimes N)
\end{aligned}$$

These maps are natural transformations, and we call them the homology product. We use the notation $x\otimes y\mapsto x\times y$ for the homology product. In the case of $M=N=R$ we combine with the map induced by the canonical isomorphism $R\otimes_R R\to R$ and obtain a homology product

$$H_i(X,A;R)\otimes H_j(Y,B;R)\to H_{i+j}((X,A)\times(Y,B);R).$$

In general we can compose with a bilinear map $M\otimes N\to P$; for instance we can use an R-module structure $R\otimes M\to M$ on M. We list some formal properties of the homology product, for simplicity of notation only for homology with coefficients in R. We use the following notation: $f\colon (X,A)\to(X',A')$ and $g\colon (Y,B)\to(Y',B')$ are continuous maps. Let $(X\times B, A\times Y)$ be excisive in $X\times Y$. Then we have two boundary operators

$$\partial'\colon H_m((X,A)\times(Y,B))\to H_{m-1}(X\times B\cup A\times Y, X\times B)\leftarrow H_{m-1}(A\times(Y,B)),$$

$$\partial'' \colon H_m((X,A)\times(Y,B)) \to H_{m-1}(X\times B\cup A\times Y, A\times Y) \leftarrow H_{m-1}((X,A)\times B).$$

t is the interchange map $t(u,v) = (v,u)$. Let $C = \{c\}$ be a point and $1 \in H_0(C;R)$ be represented by $c \otimes 1$.

9.8.1 Properties of the homology product.

$$\begin{aligned} (f\times g)_*(x\times y) &= f_*x \times g_*(y), \\ \partial x \times y &= \partial'(x\times y), \\ x\times \partial y &= (-1)^{|x|}\partial''(x\times y), \\ (x\times y)\times z &= x\times(y\times z), \\ x\times y &= (-1)^{|x||y|}t_*(y\times x), \\ 1\times y &= y. \end{aligned}$$

The algebraic Künneth formula (11.10.1) yields

(9.8.2) Theorem (Künneth formula). *Let R be an integral domain. Further, let $(A\times Y, X\times B)$ be excisive in $X\times Y$ for singular homology. Then we have a natural exact sequence*

$$\begin{aligned} 0 \to \bigoplus_{i+j=n} H_i(X,A;R)\otimes H_j(Y,B;R) &\to H_n((X,A)\times(Y,B);R) \\ \to \bigoplus_{i+j=n-1} H_i(X,A;R) * H_j(Y,B;R) &\to 0. \end{aligned}$$

The sequence splits, but the splitting is not natural in the variable (X,A). For homology with coefficients in a field k we obtain an isomorphism

$$H_*(X,A;k)\otimes_k H_*(Y,B;k) \cong H_*((X,A)\times(Y,B);k)$$

as a special case. This isomorphism holds for an arbitrary commutative ring R if the homology groups $H_*(X,A;R)$ are free R-modules. □

(9.8.3) Example. The homology class $e \in H_1(\mathbb{R}, \mathbb{R}\smallsetminus 0; R) \cong R$ represented by $\sigma\otimes 1$ with the singular simplex $\sigma\colon \Delta^1 \to \mathbb{R}$, $(t_0,t_1)\mapsto 1-2t_0$ is a generator. The product $y \mapsto e\times y$ with e is an isomorphism

$$H_n(Y,B;N) \cong H_{n+1}((\mathbb{R},\mathbb{R}\smallsetminus 0)\times(Y,B);N)$$

for each R-module N. This isomorphism can also be deduced from the axiomatic properties 9.8.1 (see a similar deduction (17.3.1) in the case of cohomology theories). The n-fold product $e_n = e\times\cdots\times e \in H_n(\mathbb{R}^n, \mathbb{R}^n\smallsetminus 0; R)$ serves as a canonical generator (a homological R-orientation of $\mathbb{R}^n$). ◇

Problems

1. Let $v_0, \dots v_n$ be affinely independent points in $\mathbb{R}^n$ and suppose that the origin $0 \in \mathbb{R}^n$ is contained in the interior of the affine simplex $v = [v_0, \dots, v_n]$. We then have the singular simplex $\sigma\colon \Delta^n \to \mathbb{R}^n$ determined by $\sigma(e_i) = v_i$. Show that σ represents a generator x_v of $H_n(\mathbb{R}^n, \mathbb{R}^n \smallsetminus 0; \mathbb{Z})$. We therefore have a relation $x_v = \pm e_n$. Determine the sign, depending on v. You might first consider the case $n = 2$ and make an intuitive guess. (This problem indicates that keeping track of signs can be a nuisance.)
2. Verify the properties 9.8.1 of the homology product.
3. Study the axioms for a multiplicative cohomology theory and use 9.8.1 to define multiplicative homology theories axiomatically.

Chapter 10
Homology

In this chapter we define homology theories via the axioms of Eilenberg and Steenrod. From these axioms we derive some classical results: the Jordan separation theorem; invariance of domain and dimension; degree and its determination by local data; the theorem of Borsuk–Ulam. The theorem of Borsuk–Ulam is used for a problem in combinatorics: the determination of the chromatic number of Kneser graphs.

A second topic of the chapter is the derivation of some results of a general nature: reduced homology; additivity; suspension isomorphisms; Mayer–Vietoris sequences; compatibility of homology with colimits.

10.1 The Axioms of Eilenberg and Steenrod

Recall the category TOP(2) of pairs of topological spaces. We use the functor $\kappa\colon \mathrm{TOP}(2) \to \mathrm{TOP}(2)$ which sends (X, A) to $(A, \emptyset)$ and $f\colon (X, A) \to (Y, B)$ to the restriction $f\colon (A, \emptyset) \to (B, \emptyset)$.

A ***homology theory*** for pairs of spaces consists of a family $(h_n \mid n \in \mathbb{Z})$ of covariant functors $h_n\colon \mathrm{TOP}(2) \to R\text{-}\mathrm{MOD}$ and a family $(\partial_n \mid n \in \mathbb{Z})$ of natural transformations $\partial_n\colon h_n \to h_{n-1} \circ \kappa$. These data are required to satisfy the following axioms of Eilenberg and Steenrod [58], [59]:

(1) **Homotopy invariance.** For each homotopy f_t in TOP(2) the equality $h_n(f_0) = h_n(f_1)$ holds.

(2) **Exact sequence.** For each pair (X, A) the sequence

$$\cdots \to h_{n+1}(X, A) \xrightarrow{\partial} h_n(A, \emptyset) \to h_n(X, \emptyset) \to h_n(X, A) \xrightarrow{\partial} \cdots$$

is exact. The undecorated homomorphisms are induced by the inclusions.

(3) **Excision.** Let (X, A) be a pair and $U \subset A$ such that $\bar{U} \subset A^\circ$. Then the inclusion $(X \setminus U, A \setminus U) \to (X, A)$ induces an excision isomorphism $h_n(X \setminus U, A \setminus U) \cong h_n(X, A)$.

The excision axiom can be expressed in a different form. Suppose Y_1, Y_2 are subspaces of Y such that $Y = Y_1^\circ \cup Y_2^\circ$. Then the inclusion induces an isomorphism $h_n(Y_1, Y_1 \cap Y_2) \cong h_n(Y, Y_2)$.

The module $h_n(X, A)$ is the n-th homology group (module) of (X, A) in the homology theory (we also say in degree or in dimension n). We set $h_n(X, \emptyset) = h_n(X)$. The groups $h_n(X)$ are the absolute groups and the groups $h_n(X, A)$ are the relative groups. The homomorphisms $\partial = \partial_n$ are the ***boundary operators***. We often write $h_n(f) = f_*$ and call (as already above) f_* the induced morphism. The homology groups $h_n(P)$ for a point P are the ***coefficient groups*** of the theory. (More precisely, a group h_n together with a compatible family of isomorphisms $\varepsilon_P\colon h_n \to h_n(P)$ for each point P is given.) In the case that $h_n = 0$ for $n \neq 0$, we say that the homology theory satisfies the ***dimension axiom*** and call the homology theory an ordinary or classical one.

The exact sequence of a pair of spaces can be extended slightly to an exact sequence for triples $B \subset A \subset X$ (see [59, p. 24], [189]). The boundary operator for a triple is defined by

$$\partial\colon h_n(X,A) \to h_{n-1}(A) \to h_{n-1}(A,B);$$

the first map is the previous boundary operator and the second map is induced by the inclusion.

(10.1.1) Proposition. *For each triple (X,A,B) the sequence*

$$\cdots \to h_{n+1}(X,A) \xrightarrow{\partial} h_n(A,B) \to h_n(X,B) \to h_n(X,A) \xrightarrow{\partial} \cdots$$

is exact. The undecorated homomorphisms are induced by the inclusions. □

We do not derive this proposition from the axioms right now (see 10.4.2 for a proof which uses the homotopy invariance). In most constructions of homology theories one verifies this more general exact sequence directly from the definitions; so we can treat it as an extended axiom.

A homology theory is called additive, if the homology groups are compatible with topological sums. We formulate this as another axiom.

(4) **Additivity.** Let (X_j, A_j), $j \in J$ be a family of pairs of spaces. Denote by $i_j\colon (X_j, A_j) \to (\amalg_j X_j, \amalg_j A_j)$ the canonical inclusions into the topological sum. Then the additivity axiom says that

$$\textstyle\bigoplus_{j\in J} h_n(X_j, A_j) \to h_n(\amalg_j X_j, \amalg_j A_j), \quad (x_j) \mapsto \sum_{j\in J} h_n(i_j)(x_j)$$

is always an isomorphism. For finite families the additivity follows from the axioms (see (10.2.1)).

Singular homology theory has further properties which may also be required in an axiomatic treatment.

(5) **Weak equivalence.** A weak equivalence $f\colon X \to Y$ induces isomorphisms $f_*\colon h_*(X) \cong h_*(Y)$ of the homology groups.

(6) **Compact support.** For each $x \in h_n(X,A)$ there exists a map $f\colon (K,L) \to (X,A)$ from a pair (K,L) of compact Hausdorff spaces and $z \in h_n(K,L)$ with $f_* z = x$.

If Axiom (5) holds, then the theory is determined by its restriction to CW-complexes (actually to simplicial complexes).

Sometimes we have to compare different homology theories. Let $h_* = (h_n, \partial_n)$ and $k_* = (k_n, \partial'_n)$ be homology theories. A natural transformation $\varphi_*\colon h_* \to k_*$ of homology theories consists of a family $\varphi_n\colon h_n \to k_n$ of natural transformations which are compatible with the boundary operators $\partial'_n \circ \varphi = \varphi_{n-1} \circ \partial_n$.

Problems

1. Let (h_n, ∂_n) be a homology theory with values in R-MOD. Let $(\varepsilon_n \in R)$ be a family of units of the ring R. Then $(h_n, \varepsilon_n \partial_n)$ is again a homology theory. It is naturally isomorphic to the original theory.

2. Given a homology theory (h_n, ∂_n) we can define a new theory by shifting the indices $k_n = h_{n+t}$.

3. Let h_* be a homology theory. For a fixed space Y we define a new homology theory whose ingredients are the groups $k_n(X, A) = h_n(X \times Y, A \times Y)$. The boundary operators for the new theory are the boundary operators of the pair $(X \times Y, A \times Y)$. The projections $\mathrm{pr}\colon X \times Y \to X$ induce a natural transformation $k_* \to h_*$ of homology theories. If h_* is additive then k_* is additive.

4. Let h_* be a homology theory with values in R-MOD. Let M be a flat R-module, i.e., the tensor product $\otimes_R M$ preserves exact sequences. Then the $h_n(-) \otimes_R M$ and $\partial \otimes_R M$ are the data of a new homology theory. Since the tensor product preserves direct sums, the new theory is additive if h_* was additive. In the case that $R = \mathbb{Z}$ one can take for M a subring T of the rational numbers $\mathbb{Q}$, in particular $\mathbb{Q}$ itself. It turns out that the rationalized theories $h_*(-) \otimes \mathbb{Q}$ are in many respects simpler than the original ones but still contain interesting information.

5. If ${}_jh_*$ is a family of homology theories $(j \in J)$, then their direct sum $\bigoplus_j {}_jh_*$ is again a homology theory. One can combine this device with the shift of indices.

10.2 Elementary Consequences of the Axioms

We assume given a homology theory h_* and derive some consequences of the axioms of Eilenberg and Steenrod.

Suppose the inclusion $A \subset X$ induces for $j = n, n+1$ an isomorphism $h_j(A) \cong h_j(X)$. Then $h_n(X, A) = 0$. In particular $h_n(X, X) = 0$ always, and $h_n(\emptyset) = h_n(\emptyset, \emptyset) = 0$. This is an immediate consequence of the exact sequence.

Let $f\colon X \to Y$ be an h-equivalence. Then $f_*\colon h_n(X) \to h_n(Y)$ is an isomorphism, by functoriality and homotopy invariance. If f is, in addition, an inclusion, then $h_n(X, Y) = 0$.

Let $f\colon (X, A) \to (Y, B)$ be a map such that the components $f\colon X \to Y$ and $f\colon A \to B$ induce isomorphisms of homology groups, e.g., the components are h-equivalences. Then the induced maps $f_*\colon h_n(X,A) \to h_n(Y,B)$ are isomorphisms. The map f induces a morphism from the homology sequence of (X,A) into the homology sequence of (Y,B), by functoriality and naturality of ∂. The claim is then a consequence of the Five Lemma (11.1.4).

If X and A are contractible, then $h_n(X,A) = 0$. If (X,A,B) is a triple and X, B are contractible, then the exact sequence of a triple shows that $\partial\colon h_n(X,A) \to h_{n-1}(A,B)$ is an isomorphism.

Let the homology theory satisfy the dimension axiom. If X is contractible, then $h_k(X) = 0$ for $k \neq 0$. A null homotopic map $Y \to Y$ therefore induces the zero morphism on $h_k(Y)$ for $k \neq 0$.

Let $i\colon A \subset X$ and suppose there exists $r\colon X \to A$ such that $ri \simeq \mathrm{id}$. From $\mathrm{id} = \mathrm{id}_* = (ri)_* = r_*i_*$ we see that r_* is a retraction of i_*, hence i_* is always injective. Therefore the exact homology sequence decomposes into split short exact sequences $0 \to h_n(A) \to h_n(X) \to h_n(X,A) \to 0$.

(10.2.1) Proposition (Finite additivity). *Let (X_j, A_j), $j \in J$ be a finite family of pairs of spaces. Denote by $i_j\colon (X_j, A_j) \to (\amalg_j X_j, \amalg_j A_j)$ the canonical inclusions into the topological sum. Then*

$$\bigoplus_{j\in J} h_n(X_j, A_j) \to h_n(\amalg_j X_j, \amalg_j A_j), \quad (x_j) \mapsto \sum_{j\in J} h_n(i_j)(x_j)$$

is an isomorphism.

Proof. As a consequence of the excision axiom we see that the inclusion always induces an isomorphism $h_n(A,B) \cong h_n(A+C, B+C)$. It suffices to consider the case $J = \{1,2\}$ and, by the Five Lemma, to deal with the absolute groups. One applies the Sum Lemma (11.1.2) with $A_k = h_n(X_k)$, $B_k = h_n(X, X_k)$, $C = h_n(X_1 + X_2)$. □

(10.2.2) Proposition. *The identity ι_n of Δ^n is a cycle modulo $\partial\Delta^n$ in singular homology theory. The group $H_n(\Delta^n, \partial\Delta^n)$ is isomorphic to $\mathbb{Z}$ and $[\iota_n]$ is a generator.*

Proof. The proof is by induction on n. Denote by $s(i) = d_i^n\Delta^{n-1}$ the i-th face of Δ^n. Consider

$$h_{k-1}(\Delta^{n-1}, \partial\Delta^{n-1}) \xrightarrow[\cong]{(d_i^n)_*} h_{k-1}(\partial\Delta^n, \partial\Delta^n \smallsetminus s(i)^\circ) \xleftarrow[\cong]{\partial} h_k(\Delta^n, \partial\Delta^n).$$

The space $\partial\Delta^n \smallsetminus s(i)^\circ$ is contractible, a linear homotopy contracts it to the point e_i. Since Δ^n is also contractible, ∂ is an isomorphism. The inclusion d_i^n maps $(\Delta^{n-1}, \partial\Delta^{n-1})$ into the complement of e_i, and as such it is a deformation retraction of pairs. The excision of e_i induces an isomorphism. Therefore $(d_i^n)_*$ is the composition of two isomorphisms. If we always work with the first face map d_0^n, we obtain by iteration an isomorphism $h_{k-n}(\Delta^0) \cong h_k(\Delta^n, \partial\Delta^n)$. (So far we can work with any homology theory.)

Now consider the special case $h_n = H_n$ of singular homology with coefficients in $\mathbb{Z}$. By definition of the boundary operator, $\partial[\iota_n] = (-1)^i [d_i^n]$, since the d_j^n for $j \neq i$ are zero in the relative group. If we again work with d_0^n, we see that the isomorphism above sends the generator $[\iota_{n-1}]$ to $[\iota_n]$.

If we work with H_k for $k \neq n$, then the isomorphism above and the dimension axiom tell us that $H_k(\Delta^n, \partial\Delta^n) = 0$. The pair $(\Delta^n, \partial\Delta^n)$ is homeomorphic to (D^n, S^{n-1}). So we also see that $H_k(D^n, S^{n-1})$ is zero for $k \neq n$ and isomorphic to $\mathbb{Z}$ for $k = n$. □

In an additive homology theory we also have an additivity isomorphism for pointed spaces if the base points are well-behaved. For a finite number of summands we do not need the additivity axiom in (10.2.3).

(10.2.3) Proposition. *Let (X_j, P_j) be a family of pointed spaces and $\bigvee_j X_j = (X, P)$ the pointed sum with embedding $i^j \colon X_j \to X$ of a summand. Assume that the closure of P in X has a neighbourhood U such that $h_*(U, P) \to h_*(X, P)$ is the zero map. Then $\langle i_*^j \rangle \colon \bigoplus_j h_*(X_j, P_j) \to h_*(X, P)$ is an isomorphism.*

Proof. Set $U_j = U \cap X_j$. Consider the diagram

$$\begin{array}{ccccccc}
0 \longrightarrow & h_*(X,P) & \xrightarrow{(1)} & h_*(X,U) & \xrightarrow{\partial} & h_*(U,P) \\
& \uparrow{\scriptstyle (2)} & & \uparrow{\scriptstyle (3)} & & \uparrow{\scriptstyle (4)} \\
& h_*(\amalg X_j, \amalg P_j) & \longrightarrow & h_*(\amalg X_j, \amalg U_j) & \xrightarrow{\partial} & h_*(\amalg U_j, \amalg P_j).
\end{array}$$

The vertical morphisms are induced by the quotient maps. The horizontal morphisms are part of the exact sequence of triples. The hypothesis implies that (1) is injective. We use the additivity in order to conclude that (2) and (4) are injective. This is due to the fact that we have the projections $p^j \colon X \to X_j$, and $p_*^k i_*^j = \delta^{kj}$. We show that (3) is an isomorphism. Consider the diagram

$$\begin{array}{ccc}
h_*(X \smallsetminus P, U \smallsetminus P) & \xleftarrow{=} & h_*(\amalg(X_j \smallsetminus P_j), \amalg(U_j \smallsetminus P_j)) \\
\downarrow{\scriptstyle \cong} & & \downarrow{\scriptstyle \cong} \\
h_*(X,U) & \xleftarrow{(3)} & h_*(\amalg X_j, \amalg U_j).
\end{array}$$

The isomorphisms hold by excision; here we use the assumption $\overline{P} \subset U^\circ$. Diagram chasing (Five Lemma) now shows that (2) is also surjective. □

(10.2.4) Remark. The hypothesis of (10.2.3) is satisfied if P_j is closed in X_j and has a neighbourhood U_j such that $U_j \subset X_j$ is pointed homotopic to the constant map. These conditions hold if the spaces (X_j, P_j) are well-pointed; see (5.4.4). ◇

10.2.5 Suspension. We define the ***homological suspension*** isomorphism $\sigma = \sigma^{(X,A)}$ by the commutative diagram

$$\begin{array}{ccc} h_n(X,A) & \xleftarrow{\cong} & h_n(1\times X, 1\times A) \\ \downarrow \sigma & & \cong \downarrow (1) \\ h_{n+1}((I,\partial I)\times(X,A)) & \xrightarrow[\cong]{\partial} & h_n(\partial I\times X\cup I\times A, 0\times X\cup I\times A). \end{array}$$

(1) is an isomorphism: excise $0\times X$ and then use an h-equivalence. The boundary operator ∂ for the triple sequence of $(I\times X, \partial I\times X\cup I\times A, 0\times X\cup I\times A)$ is an isomorphism, since $0\times X\cup I\times A\subset I\times X$ is an h-equivalence. We can interchange the roles of 0 and 1; let σ^- denote this suspension isomorphism. By applying the Hexagon Lemma (11.1.3) with center group $h_n(I\times X, I\times A\cup\partial I\times X)$ we obtain $\sigma = -\sigma^-$ (draw the appropriate diagram). For some purposes of homotopy theory one has to use a similar suspension isomorphism defined with $X\times I$.

The n-fold iteration of σ provides us in particular with an isomorphism $\sigma^n\colon h_k\cong h_k(I^0)\cong h_{k+1}(I,\partial I)\cong\cdots\cong h_{k+n}(I^n,\partial I^n)$. ◇

10.3 Jordan Curves. Invariance of Domain

As a first application of homology theory we prove a general duality theorem. We then apply this general result to prove classical results: A generalized Jordan separation theorem and the invariance of domain and dimension. For this section see [53].

The propositions (10.3.1) and (10.3.2) can be proved in a similar manner for an arbitrary homology (or, mutatis mutandis, cohomology) theory. For the applications one needs homology groups which determine the cardinality of $\pi_0(X)$ for open subsets X of Euclidean spaces, and this holds, e.g., for singular homology $H_*(-)$. (If one knows a little analysis, one can use de Rham cohomology for open subsets of Euclidean spaces.)

(10.3.1) Proposition. *Let $A\subset\mathbb{R}^n$ be a closed subset. Then $H_k(\mathbb{R}^n,\mathbb{R}^n\smallsetminus A)$ and $H_{k+1}(\mathbb{R}^n\times\mathbb{R},\mathbb{R}^n\times\mathbb{R}\smallsetminus A\times 0)$ are isomorphic.*

Proof. We use the open subsets of $\mathbb{R}^{n+1}$

$$\begin{aligned} H_+ &= (\mathbb{R}^n\smallsetminus A)\times\,]-1,\infty[\,\cup A\times\,]0,\infty[\\ H_- &= (\mathbb{R}^n\smallsetminus A)\times\,]-\infty,1[\,\cup A\times\,]-\infty,0[\end{aligned}$$

$$\begin{aligned} H_+\cup H_- &= \mathbb{R}^{n+1}\smallsetminus A\times 0 \\ H_+\cap H_- &= (\mathbb{R}^n\smallsetminus A)\times\,]-1,1[. \end{aligned}$$

These data occur in the diagram

$$\begin{array}{ccccc} H_k(\mathbb{R}^n, \mathbb{R}^n \smallsetminus A) & & \dashrightarrow & & H_{k+1}(\mathbb{R}^{n+1}, \mathbb{R}^{n+1} \smallsetminus A \times 0) \\ \downarrow{\scriptstyle (3)} & & & & \downarrow{\scriptstyle \partial} \\ H_k(\mathbb{R}^{n+1}, H_+ \cap H_-) & \xleftarrow{(2)} & H_k(H_+, H_+ \cap H_-) & \xrightarrow{(1)} & H_k(H_+ \cup H_-, H_-). \end{array}$$

The map ∂ is the boundary operator of the triple $(\mathbb{R}^{n+1}, H_+ \cup H_-, H_-)$. The maps (1), (2), and (3) are induced by the inclusions. We show that the morphisms in the diagram are isomorphisms. The proof uses the fact that H_+ and H_- are contractible; the homotopy $h_t \colon H_+ \to H_+$, $(x,s) \mapsto (x, s+t)$ starts at the identity and has an image in the contractible space $\mathbb{R}^n \times]0, \infty[$. The map ∂ is an isomorphism, because $H_k(\mathbb{R}^{n+1}, H_-) = 0$, by contractibility of H_-. The map (1) is an excision. The maps (2) and (3) are isomorphisms by homotopy invariance ($\mathbb{R}^n = \mathbb{R}^n \times 0$). □

(10.3.2) Theorem (Duality theorem). *Let A and B be closed homeomorphic subsets of $\mathbb{R}^n$. Then the groups $H_k(\mathbb{R}^n, \mathbb{R}^n \smallsetminus A)$ and $H_k(\mathbb{R}^n, \mathbb{R}^n \smallsetminus B)$ are isomorphic ($k \in \mathbb{Z}$).*

Proof. A homeomorphism $f \colon A \to B$ yields a homeomorphism $\alpha \colon \mathbb{R}^n \times \mathbb{R}^n \to \mathbb{R}^n \times \mathbb{R}^n$, which sends $A \times 0$ via $f \times 0$ to $B \times 0$ (see (7.3.1)). We obtain an induced isomorphism

$$H_{k+n}(\mathbb{R}^n \times \mathbb{R}^n, \mathbb{R}^n \times \mathbb{R}^n \smallsetminus A \times 0) \cong H_{k+n}(\mathbb{R}^n \times \mathbb{R}^n, \mathbb{R}^n \times \mathbb{R}^n \smallsetminus B \times 0).$$

Now apply (10.3.1) n times. □

(10.3.3) Theorem (Component theorem). *Let A and B be closed homeomorphic subsets of $\mathbb{R}^n$. Then $\pi_0(\mathbb{R}^n \smallsetminus A)$ and $\pi_0(\mathbb{R}^n \smallsetminus B)$ have the same cardinality.*

Proof. We use the fact that $H_0(\mathbb{R}^n \smallsetminus A)$ is the free abelian group on $\pi_0(\mathbb{R}^n \smallsetminus A)$ and the algebraic fact a free abelian group determines the cardinality of a basis (= the rank). Suppose that $A \neq \mathbb{R}^n$. Then we have an exact sequence

$$0 \to H_1(\mathbb{R}^n, \mathbb{R}^n \smallsetminus A) \to H_0(\mathbb{R}^n \smallsetminus A) \to H_0(\mathbb{R}^n) \to 0.$$

This shows the rank of $H_1(\mathbb{R}^n, \mathbb{R}^n \smallsetminus A)$ is one less than the rank of $H_0(\mathbb{R}^n \smallsetminus A)$. Hence if A and B are different from $\mathbb{R}^n$, then the result follows from (10.3.2). We see in the next section that $A = \mathbb{R}^n$ implies that B is open in $\mathbb{R}^n$ and therefore, since $\mathbb{R}^n$ is connected, also equal to $\mathbb{R}^n$. □

An injective continuous map $f \colon S^1 \to \mathbb{R}^2$ is an embedding, and its image is called a ***Jordan curve***.

(10.3.4) Theorem (Jordan separation theorem). *Let $S \subset \mathbb{R}^n$ be homeomorphic to S^{n-1} $(n \geq 2)$. Then $\mathbb{R}^n \smallsetminus S$ has two path components, the bounded interior J and the unbounded exterior A. Moreover, S is the set of boundary points of J and of A.*

Proof. The assertion holds in the elementary case $S = S^{n-1}$. Hence, by (10.3.3), the complement of S has two components. It remains to study the boundary points.

Let $x \in S$ and let V be an open neighbourhood of x in $\mathbb{R}^n$. Then $C = S \smallsetminus (S \cap V)$ is closed in S and homeomorphic to a proper closed subset D of S^{n-1}. Therefore $\mathbb{R}^n \smallsetminus D$ is path connected and hence, by (10.3.3), also $\mathbb{R}^n \smallsetminus C$. Let $p \in J$ and $q \in A$ and $w\colon [0,1] \to \mathbb{R}^n \smallsetminus C$ a path from p to q. Then $w^{-1}(S) \neq \emptyset$. Let t_1 be the minimum and t_2 the maximum of $w^{-1}(S)$. Then $w(t_1)$ and $w(t_2)$ are contained in $S \cap V$. Therefore $w(t_1)$ is limit point of $w([0, t_1[) \subset J$ and $w(t_2)$ limit point of $w(]t_2, 1]) \subset A$. Hence there exist $t_3 \in [0, t_1[$ with $w(t_3) \in J \cap V$ and $t_4 \in]t_2, 1]$ with $w(t_4) \in A \cap V$. This shows that x is contained in the boundary of J and of A. □

(10.3.5) Remarks. For $n = 2$ one can improve the separation theorem. There holds the ***theorem of Schoenflies*** (for a topological proof see [141]):

Let $J \subset \mathbb{R}^2$ be a Jordan curve. Then there exists a homeomorphism $f\colon \mathbb{R}^2 \to \mathbb{R}^2$ which maps J onto the standard circle.

There exists a stronger result. By the Riemann mapping theorem there exists a holomorphic isomorphism from the interior of J onto the interior of S^1; and one can show that this isomorphism can be extended to a homeomorphism of the closures. See e.g., [149].

For $n \geq 3$ it is in general not true that the interior of an embedding $S^{n-1} \to \mathbb{R}^n$ is homeomorphic to an n-cell. One can construct an embedding $D^3 \to \mathbb{R}^3$ such that the complement is not simply connected. Under some regularity conditions on the embedding the situation still resembles the standard embedding. There holds the ***theorem of M. Brown*** [32] (see also [25, p. 236]):

Let $f\colon S^{n-1} \times [-1,1] \to S^n$ be an embedding $(n \geq 1)$. Then the closure of each component of $S^n \smallsetminus f(S^{n-1} \times 0)$ is homeomorphic to D^n.

From the duality theorem (18.3.3) one can conclude that both components of $S^n \smallsetminus S$ have for an arbitrary embedding the integral singular homology groups of a point. ◇

(10.3.6) Theorem. *Let $A \subset \mathbb{R}^n$ be homeomorphic to D^k, $k \leq n$. Then $\mathbb{R}^n \smallsetminus A$ is path connected $(n > 1)$.*

Proof. D^k is compact, hence A is compact too. Therefore A is closed in $\mathbb{R}^n$ and the assertion follows from (10.3.3), since D^k obviously has a path connected complement. □

(10.3.7) Theorem (Invariance of domain). *Let $U \subset \mathbb{R}^n$ be open and $f\colon U \to \mathbb{R}^n$ an injective continuous map. Then $f(U)$ is open in $\mathbb{R}^n$, and f maps U homeomorphically onto $f(U)$.*

Let $V \subset \mathbb{R}^n$ be homeomorphic to an open subset of $U \subset \mathbb{R}^n$. Then V is open in $\mathbb{R}^n$.

Proof. It suffices to show that $f(U)$ is open. It then follows that f is open.

Let $D = \{x \in \mathbb{R}^n \mid \|x - a\| \le \delta\} \subset U$, and let S be the boundary of D. It suffices to show that $f(D^\circ)$ is open. We consider the case $n \ge 2$ and leave the case $n = 1$ as exercise. The set S as well as $T = f(S)$ are homeomorphic to S^{n-1}. Suppose U_1, U_2 are the (open) components of $\mathbb{R}^n \smallsetminus T$. Let U_2 be unbounded. By (10.3.6), $\mathbb{R}^n \smallsetminus f(D)$ is path connected, and therefore contained in U_1 or U_2. Since $f(D)$ is compact, the complement is unbounded. This implies $T \cup U_1 = \mathbb{R}^n \smallsetminus U_2 \subset f(D)$ and then $U_1 \subset f(D^\circ)$. Since D° is connected, so is $f(D^\circ)$. The inclusion $f(D^\circ) \subset U_1 \cup U_2$ shows that $f(D^\circ) \subset U_1$. Therefore $f(D^\circ) = U_1$, and this set is open.

The second statement follows from the first, since by hypothesis there exists an injective continuous map $f : U \to \mathbb{R}^n$ with image V. □

(10.3.8) Theorem (Invariance of dimension). *Let $U \subset \mathbb{R}^m$ and $V \subset \mathbb{R}^n$ be non-empty homeomorphic open subsets. Then $m = n$.*

Proof. Let $m < n$. Then, by (10.3.7), $U \subset \mathbb{R}^m \subset \mathbb{R}^n$ is open in $\mathbb{R}^n$, which is impossible. □

10.4 Reduced Homology Groups

The coefficient groups of a homology theory are important data of the theory but they contain no information about spaces. We therefore split off these groups from the homology groups $h_n(X)$.

Let X be a non-empty space and $r : X \to P$ the unique map to a point. We set

$$\tilde{h}_n(X) = \text{kernel}\,(r_* : h_n(X) \to h_n(P))$$

and call this group a ***reduced homology group***. The homomorphism $h_n(f)$ for a continuous map $f : X \to Y$ restricts to $\tilde{h}_n(f) : \tilde{h}_n(X) \to \tilde{h}_n(Y)$. In this way $\tilde{h}_n$ becomes a homotopy invariant functor TOP $\to$ R-MOD. Let $(X, *)$ be a pointed space and $i : P = \{*\} \subset X$ be the inclusion of the base point. Then we have a short exact sequence

$$0 \to h_n(P) \xrightarrow{i_*} h_n(X) \xrightarrow{j_*} h_n(X, P) \to 0$$

with a splitting r_* of i_*. Thus j_* induces isomorphisms $\tilde{h}_n(X) \cong h_n(X, P) \cong \text{Coker}(i_*)$. By(11.1.1) we also have an isomorphism

$$h_n(X) = \tilde{h}_n(X) \oplus i_*(h_n(P)) \cong \tilde{h}_n(X) \oplus h_n$$

which is natural for pointed maps (but not, in general, for arbitrary maps), and the canonical diagram

$$\begin{array}{ccccc} \tilde{h}_n(X) & \longrightarrow & h_n(X) & \longrightarrow & \operatorname{Coker} i_* \\ & \underset{\cong}{\searrow} & \downarrow{\scriptstyle j_*} & \underset{\cong}{\swarrow} & \\ & & h_n(X,P) & & \end{array}$$

is commutative.

(10.4.1) Proposition. *Let A be non-empty. The image of the boundary operator $\partial\colon h_n(X,A) \to h_{n-1}(A)$ is contained in $\tilde{h}_{n-1}(A)$. The images of $h_n(X)$ and $\tilde{h}_n(X)$ in $h_n(X,A)$ coincide. The homology sequence for the reduced groups*

$$\cdots \to \tilde{h}_n(A) \to \tilde{h}_n(X) \to h_n(X,A) \to \tilde{h}_{n-1}(A) \to \cdots$$

is exact.

Proof. Map the exact sequence of (X, A) into the exact sequence of (P, P) and perform diagram chasing. The exactness is also a special case of (11.3.2). □

From (10.4.1) we see: If A is contractible, then $\tilde{h}_n(A) = 0$ and $\tilde{h}_n(X) \to h_n(X,A)$ is an isomorphism. If X is contractible, then $\partial\colon h_n(X,A) \to \tilde{h}_{n-1}(A)$ is an isomorphism.

10.4.2 Triple sequence. Let $(X, A, *)$ be a pointed pair. The reduced homology sequence of (X, A) is canonically isomorphic to the homology sequence of the triple $(X, A, *)$. Hence the latter is exact.

Let CB denote the cone over B. The homology sequence of a triple (X, A, B) is, via excision, isomorphic to the sequence of $(X \cup CB, A \cup CB, CB)$, and the latter, via h-equivalence isomorphic to the sequence of $(X \cup CB, A \cup CB, *)$. This proves the exactness of the triple sequence. ◇

Under the hypothesis of (10.2.3) or (10.2.4) we have for an additive homology theory an isomorphism $\bigoplus_j \tilde{h}_*(X_j) \cong \tilde{h}_*(\bigvee_j X_j)$. For a finite number of summands we do not need the additivity axiom. We call $\bigvee_j X_j$ ***decomposable*** with respect to $\tilde{h}_n$ if the canonical map $\bigoplus_j \tilde{h}_n(X_j) \to \tilde{h}_n(\bigvee_j X_j)$ is an isomorphism.

(10.4.3) Proposition. *Suppose $X \vee Y$ is decomposable with respect to h_n, h_{n+1}. Then the homology sequence of $(X \times Y, X \vee Y)$ yields a split short exact sequence*

$$0 \to \tilde{h}_n(X \vee Y) \to \tilde{h}_n(X \times Y) \to h_n(X \times Y, X \vee Y) \to 0.$$

Proof. The projections onto the factors induce $\tilde{h}_n(X \times Y) \to \tilde{h}_n(X) \oplus \tilde{h}_n(Y)$, and together with the assumed decomposition isomorphism we obtain a left inverse to the morphism $\tilde{h}_n(X \vee Y) \to \tilde{h}_n(X \times Y)$. □

Let (C, μ) be a comonoid in h-TOP^0, i.e., $\mu\colon C \to C \vee C$ is a pointed map such that the composition with projection onto the summands is pointed homotopic to the identity. Then we have the μ-sum in each homotopy set $[C, Y]^0$, defined by $[f] + [g] = [\delta(f \vee g)\mu]$ with the folding map $\delta = \langle \mathrm{id}, \mathrm{id} \rangle\colon Y \vee Y \to Y$. Let us write $h = \tilde{h}_n$.

(10.4.4) Proposition. *Assume that $C \vee C$ is decomposable with respect to h_n. Then the morphism*

$$\omega\colon [C, Y]^0 \to \mathrm{Hom}(h(C), h(Y)), \quad [f] \mapsto f_*$$

is a homomorphism.

Proof. Consider the commutative diagram

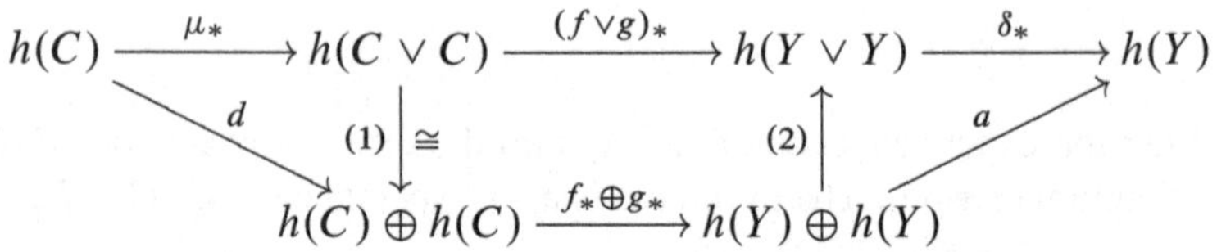

with the diagonal d and the addition $a(y, z) = y + z$. By our assumption about C, the isomorphism (1) is induced by the projections onto the summands, and by our assumption about μ, the left triangle commutes. The morphism (2) and the inverse of (1) are induced by the injection of the summands; this shows that the rectangle and the right triangle commute. Now observe that $a(f_* \oplus g_*)d = f_* + g_*$. □

The hypothesis of (10.4.4) holds for the suspension $C = \Sigma X$ of a well-pointed space X. We thus obtain in particular a homomorphism

$$\omega\colon \pi_n(X) \to \mathrm{Hom}(\tilde{h}_n(S^n), \tilde{h}_n(X))$$

for each pointed space X.

(10.4.5) Proposition. *Let $i\colon A \subset X$ be a cofibration and let $p\colon (X, A) \to (X/A, *)$ be the map which identifies A to a base point $*$. Then $p_*\colon h_n(X, A) \to h_n(X/A, *)$ is an isomorphism. We can write this isomorphism also in the form $q\colon h_n(X, A) \cong \tilde{h}_n(X/A)$.*

Proof. Let $X \cup CA = (CA + X)/(a, 1) \sim i(a)$ be the mapping cone of i. The inclusion $j\colon (X, A) \to (X \cup CA, CA)$ induces an isomorphism in homology: Excise the cone point $A \times 0$ and apply a homotopy equivalence. For a cofibration we have an h-equivalence $p\colon X \cup CA \to X \cup CA/CA \cong X/A$. Hence $q_* = p_* j_*$ is the composition of two isomorphisms. □

The isomorphism q also holds for $A = \emptyset$. In this case $X/A = X + \{*\}$ and p is the inclusion $(X, \emptyset) \to (X, \{*\})$. We use q to modify the exact sequence (10.4.1) in the case of a cofibration $A \subset X$

$$\cdots \to \tilde{h}_n(A) \to \tilde{h}_n(X) \to \tilde{h}_n(X/A) \to \tilde{h}_{n-1}(A) \to \cdots .$$

(10.4.6) Proposition. *Let $j : A \subset X$ be a cofibration and attach X to B via $f : A \to B$. Then the map $h_n(X, A) \to h_n(X \cup_A B, B)$ induced by the inclusion is an isomorphism.*

Proof. We apply (10.4.5) to the homeomorphism $X/A \to X \cup_A B/B$. □

(10.4.7) Proposition. *Let $f : X \to Y$ be a pointed map between well-pointed spaces and $X \xrightarrow{f} Y \xrightarrow{f_1} C(f)$ the beginning of the cofibre sequence. Then the sequence $\tilde{h}_n(X) \xrightarrow{f_*} \tilde{h}_n(Y) \xrightarrow{f_{1*}} \tilde{h}_n(C(f))$ is exact.*

Proof. Let $Z(f) = (X \times I + Y)/(x, 1) \sim f(x)$ be the (unpointed) mapping cylinder of f and $X \subset Z(f)$, $x \mapsto (x, 0)$ the canonical inclusion, a cofibration. Consider the commuting diagram

$$\begin{array}{ccccc}
\tilde{h}_n(X) & \longrightarrow & \tilde{h}_n(Z(f)) & \longrightarrow & h_n(Z(f), X) \\
\downarrow = & & \downarrow \cong & & \downarrow \cong \\
\tilde{h}_n(X) & \xrightarrow{f_*} & \tilde{h}_n(Y) & \xrightarrow{c(f)_*} & \tilde{h}_n(Z(f)/X)
\end{array}$$

with the canonical inclusion $c(f) : Y \subset Z(f)/X$. The top row is the exact sequence of the pair. The isomorphisms hold by homotopy invariance and (10.4.5). Now we use that for a well-pointed pair the quotient map $Z(f)/X \to C(f)$ is a homotopy equivalence, since a unit interval which is embedded as a cofibration is identified to a point. □

For a well-pointed space X the inclusion $X \times \partial I \cup \{*\} \times I \subset X \times I$ is a cofibration. The quotient is the suspension ΣX. From (10.4.5) we obtain an isomorphism

$$q : h_n((X, *) \times (I, \partial I)) \to h_n(\Sigma X, *) \cong \tilde{h}_n(\Sigma X)$$

and a suspension isomorphism $\tilde{\sigma} : \tilde{h}_n(X) \cong \tilde{h}_{n+1}(\Sigma X)$ which makes the diagram

$$\begin{array}{ccc}
h_n(X, *) & \xrightarrow{\sigma} & h_{n+1}((X, *) \times (I, \partial I)) \\
\downarrow q & & \downarrow q \\
\tilde{h}_n(X) & \xrightarrow{\tilde{\sigma}} & \tilde{h}_{n+1}(\Sigma X)
\end{array}$$

commutative. In particular, we obtain $h_m \cong \tilde{h}_m(I^0/\partial I^0) \cong h_{m+n}(I^n/\partial I^n)$.

Problems

1. Let $f_q\colon S^n \to S^n$ be a map of degree q, and denote by $M(q,n)$ its mapping cone. Determine the groups and homomorphisms in the sequence (9.6.1) for the space $M(q,n)$.
2. Let $M(q,1) = M(q)$. Use the cofibre sequence of $\mathrm{id}_X \wedge f_q$ in order to derive an exact sequence

$$0 \to \tilde{h}_n(X) \otimes \mathbb{Z}/q \to \tilde{h}_{n+1}(X \wedge M(q)) \to \mathrm{Tor}(\tilde{h}_{n-1}(X), \mathbb{Z}/q) \to 0.$$

This suggests defining for any homology theory $\tilde{h}_*$ a theory with coefficients $\mathbb{Z}/q$ by $\tilde{h}_*(- \wedge M(q))$. Unfortunately the homotopy situation is more complicated than one would expect (or wish), see [8]. Spaces of the type $M(q,n)$ are sometimes called ***Moore spaces***.

10.5 The Degree

In 10.2.5 we determined the homology groups of spheres from the axioms of a homology theory h_*. We describe yet another variant.

We use the standard subspaces $D^n, S^{n-1}, E^n = D^n \smallsetminus S^{n-1}$ of $\mathbb{R}^n$ and $D^n_\pm = \{(x_1,\dots,x_{n+1}) \in S^n \mid \pm x_{n+1} \ge 0\}$. The space $D^0 = \{D\}$ is a singleton and $S^{-1} = \emptyset$.

We define a suspension isomorphism σ_+ as the composition

$$\sigma_+\colon \tilde{h}_{k-1}(S^{n-1}) \xleftarrow[\cong]{\partial} h_k(D^n_-, S^{n-1}) \xrightarrow[\cong]{s_+} h_k(S^n, D^n_+) \xleftarrow[\cong]{j} \tilde{h}_k(S^n).$$

The maps j and ∂ are isomorphisms, since $D^n_\pm$ is contractible, and s_+ is an isomorphism (compare (6.4.4)). Iteration of σ_+ yields (suspension) isomorphisms

$$\sigma^{(n)}\colon h_{k-n} \cong \tilde{h}_{k-n}(S^0) \cong \tilde{h}_k(S^n), \quad h_k(S^n) \cong h_{k-n} \oplus h_n.$$

The first isomorphism is determined by

$$h_m \to \tilde{h}_m(S^0) \subset h_m(S^0) \cong h_m(+1) \oplus h_m(-1) \cong h_m \oplus h_m, \quad x \mapsto (x,-x).$$

We also have an isomorphism $\partial\colon h_k(D^n,S^{n-1}) \cong h_{k-1}(S^{n-1},e) \cong \tilde{h}_{k-1}(S^{n-1})$. In the case of ordinary homology $H_*(-;G)$ with coefficients in G we obtain:

$$H_k(S^n;G) \cong \begin{cases} G, & k = n > 0,\ n > k = 0,\\ G \oplus G, & k = n = 0,\\ 0, & \text{otherwise.}\end{cases}$$

Moreover $H_n(D^n,S^{n-1};G) \cong H_n(\mathbb{R}^n, \mathbb{R}^n \smallsetminus 0; G) \cong G$.

A generator of $H_n(D^n,S^{n-1};\mathbb{Z}) \cong \mathbb{Z}$ is called a homological orientation of D^n; and a generator of $H_n(S^n)$, $n \ge 1$ a homological orientation of S^n. Given an orientation $z \in H_n(S^n)$ and a continuous map $f\colon S^n \to S^n$, we define the

(homological) degree $d(f) \in \mathbb{Z}$ of f by $f_*(z) = d(f)z$. A different choice of a generator does not effect the degree. We also define the degree for $n = 0$: The identity has degree $+1$, the antipodal map the degree -1 and a constant map the degree 0. From the definition we see directly some properties of the degree: $d(f \circ g) = d(f)d(g)$, $d(\mathrm{id}) = 1$; a homotopy equivalence has degree ± 1; a null homotopic map has degree zero.

(10.5.1) Proposition. *Let h_* be a homology theory such that $h_0(\mathrm{Point}) = \mathbb{Z}$. Then $\omega\colon [S^n, S^n] \to \mathrm{Hom}(\tilde{h}_n(S^n), \tilde{h}_n(S^n))$, $f \mapsto f_*$ is an isomorphism ($n \geq 1$).*

Proof. The suspension isomorphism and the hypothesis yield $\tilde{h}_n(S^n) \cong \mathbb{Z}$. Thus the Hom-group is canonically isomorphic to $\mathbb{Z}$. We now use that $\pi_n(S^n) \cong [S^n, S^n] \cong \mathbb{Z}$. The identity of S^n is mapped to 1. By (10.4.4), ω is a homomorphism, hence necessarily an isomorphism. □

We have already defined a (homotopical) degree. From (10.5.1) we see that the homotopical and the homological degree coincide. If one starts algebraic topology with (singular) homology, then one has in any case the important homotopy invariant "degree". Proposition (10.5.4) is not immediate from the homological definition.

(10.5.2) Proposition. *Define an isomorphism $\sigma_-\colon \tilde{h}_{k-1}(S^{n-1}) \to \tilde{h}_k(S^n)$ as in the case of σ_+, but with the roles of $D^n_\pm$ interchanged. Then $\sigma_+ = -\sigma_-$.*

Proof. Consider the commutative diagram

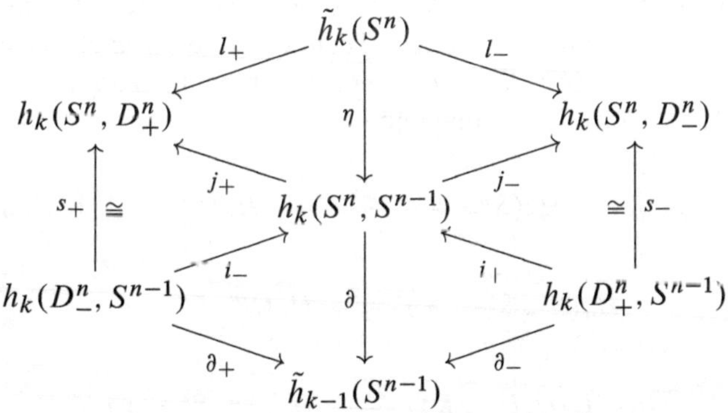

and apply the Hexagon Lemma (11.1.3). □

(10.5.3) Proposition. *Let $A \in \mathrm{O}(n+1)$ and $l^A\colon S^n \to S^n$, $x \mapsto Ax$. Then l^A_* is on $\tilde{h}_k(S^n)$ the multiplication by $\det(A)$. The antipodal map $x \mapsto -x$ on S^n has the degree $(-1)^{n+1}$.*

Proof. Let $t\colon S^n \to S^n$ change the sign of the first coordinate. Then $t_*\sigma_+ = \sigma_-$. Hence $t_* = -\mathrm{id}$, by (10.5.2). The group $\mathrm{O}(n+1)$ has two path components. If A and B are contained in the same component, then l^A and l^B are homotopic. Representatives of the components are the unit matrix E and the diagonal matrix $T = \mathrm{Diag}(-1, 1, \dots, 1)$. The relations $l^T = t$ and $l^E = \mathrm{id}$ now finish the proof. □

(10.5.4) Proposition. *Let $f\colon S^n \to S^n$ be a map of degree d. Then f induces on $\tilde{h}_k(S^n)$ the multiplication by d.*

Proof. The cases $d = 1, 0$ are clear and $d = -1$ follows from (10.5.3). The general case is then a consequence of (10.4.4) and our knowledge of $\pi_n(S^n)$. □

(10.5.5) Proposition. *Let $A \in \mathrm{GL}_n(\mathbb{R})$ and $l^A\colon \mathbb{R}^n \to \mathbb{R}^n$, $x \mapsto Ax$. Then l_*^A is on $h_k(\mathbb{R}^n, \mathbb{R}^n \smallsetminus 0)$ the multiplication by the sign $\varepsilon(A) = \det(A)/|\det(A)|$ of the determinant.*

Proof. We have isomorphisms $h_k(\mathbb{R}^n, \mathbb{R}^n \smallsetminus 0) \xrightarrow{\partial} \tilde{h}_{k-1}(\mathbb{R}^n \smallsetminus 0) \cong \tilde{h}_{k-1}(S^{n-1})$ which are compatible with the action of l_*^A if $A \in \mathrm{O}(n)$. In this case the claim follows from (10.5.3). In the general case we use that $\mathrm{GL}_n(\mathbb{R})$ has two path components which are characterized by the sign of the determinant. □

Let $S_r^n(a) = \{x \in \mathbb{R}^{n+1} \mid \|x-a\| = r\}$. We have a canonical homeomorphism $h_{r,a}\colon S^n \to S_r^n(a)$, $x \mapsto rx + a$ and $r_b\colon \mathbb{R}^{n+1} \smallsetminus b \to S^n$, $x \mapsto N(x-b)$. The ***winding number*** of $f\colon S_r^n(a) \to \mathbb{R}^{n+1} \smallsetminus b$ about b is defined as the degree of $r_b \circ f \circ h_{r,a}$.

10.5.6 Local degree. Let $K \subset S^n$ be compact and different from S^n and let U be an open neighbourhood of K. Then we have an excision isomorphism $H_n(U, U \smallsetminus K) \cong H_n(S^n, S^n \smallsetminus K)$. For a continuous map $f\colon S^n \to S^n$ we let $K = f^{-1}(p)$. Consider the diagram

$$\begin{array}{ccc}
H_n(S^n) & \xrightarrow{\;f_*\;} & H_n(S^n) \\
\downarrow & & \downarrow{\scriptstyle (1)} \\
H_n(S^n, S^n \smallsetminus K) & \xrightarrow{\;f_*\;} & H_n(S^n, S^n \smallsetminus p) \\
\uparrow{\scriptstyle \cong} & & \uparrow{\scriptstyle =} \\
H_n(U, U \smallsetminus K) & \xrightarrow{\;f_*^U\;} & H_n(S^n, S^n \smallsetminus p)
\end{array}$$

with the restriction f^U of f. The exact sequence of the pair $(S^n, S^n \smallsetminus p)$ shows that (1) is an isomorphism ($n \geq 1$). Let $z \in H_n(S^n)$ be a generator and $z(U, K)$ its image in $H_n(U, U \smallsetminus K)$. Commutativity of the diagram shows $f_*^U z_{U,K} = d(f) z_{S^n,p}$. Hence the degree only depends on the restriction f^U. For any compact set L with $f(L) = \{p\}$ and open neighbourhood W of L we define the (partial) degree $d(f, L)$ by $f_*^W z(W, L) = d(f, L) z(S^n, p)$; it is independent of the choice of W.

(10.5.7) Lemma. *Suppose $U = U_1 \cup U_2$ is the disjoint union of open sets U_j. Set $K_j = U_j \cap K$. The inclusions induce the additivity isomorphism*

$$\langle i_1, i_2 \rangle \colon H_n(U_1, U_1 \smallsetminus K_1) \oplus H_n(U_2, U_2 \smallsetminus K_2) \to H_n(U, U \smallsetminus K).$$

Then the relation $z(U, K) = i_1 z(U_1, K_1) + i_2 z(U_2, K_2)$ holds.

Proof. Consider the diagram with $M = S^n$:

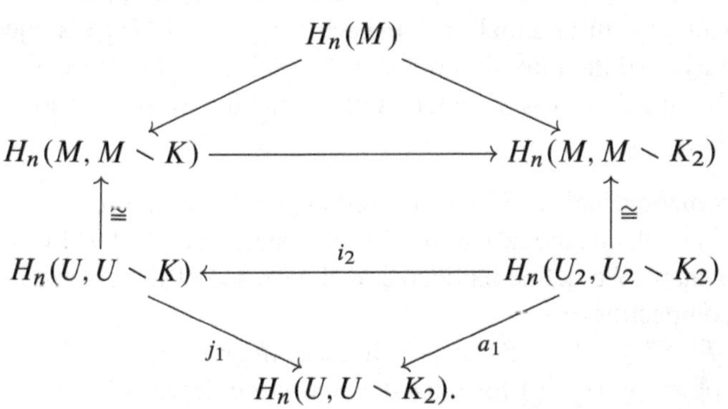

There exist x_1, x_2 such that $z(U, K) = i_1 x_1 + i_2 x_2$. We compute

$$j_1 z(U, K) = z(U, K_2) = a_1 z(U_2, K_2) = j_1 i_2 x_2 = a_1 x_2.$$

This proves $x_2 = z(U_2, K_2)$. □

As a consequence of this lemma we obtain the additivity of the degree $d(f) = d(f, K_1) + d(f, K_2)$. Suppose K is a finite set, then we can choose U as a disjoint union of open sets $U_x, x \in K$ such that $U_x \cap K = \{x\}$. In that case $\bigoplus_{x \in K} H_n(U_x, U_x \smallsetminus x) \cong H_n(U, U \smallsetminus K)$; and $H_n(U_x, U_x \smallsetminus x) \cong \mathbb{Z}$, by excision and $H_n(D^n, S^{n-1}) \cong \mathbb{Z}$. We call the integer $d(f, x)$ defined by $f_* z_{U_x, x} = d(f, x) z_{S^n, p}$ the ***local degree*** of f at p. With this notation we therefore have $d(f) = \sum_{x \in K} d(f, x)$. ◇

Let $U \subset \mathbb{R}^n$ be an open neighbourhood of the origin. Then we have an excision isomorphism $h_k(U, U \smallsetminus 0) \cong h_k(\mathbb{R}^n, \mathbb{R}^n \smallsetminus 0)$. Suppose $g \colon U \to \mathbb{R}^n$ is a map with the properties: (1) Continuously differentiable (a C^1-map); (2) $g^{-1}(0) = \{0\}$; (3) the differential $Dg(0)$ is invertible. Under these conditions we show:

(10.5.8) Proposition. *The induced map*

$$h_k(\mathbb{R}^n, \mathbb{R}^n \smallsetminus 0) \cong h_k(U, U \smallsetminus 0) \xrightarrow{g_*} h_k(\mathbb{R}^n, \mathbb{R}^n \smallsetminus 0)$$

is multiplication by the sign of the determinant of the differential $Dg(0)$.

Proof. There exist continuous maps $h_i : U \to \mathbb{R}^n$ with $g(x) = \sum_{i=1}^n x_i h_i(x)$ and $Dg(0)(x_1, \dots, x_n) = \sum_{i=1}^n x_i h_i(0)$. We define a homotopy $h_t : (U, U \smallsetminus 0) \to (\mathbb{R}^n, \mathbb{R}^n \smallsetminus 0)$ from g to $Dg(0)$

$$h_t(x) = \sum_{i=1}^n x_i h_i(tx) = \begin{cases} t^{-1} g(tx), & t > 0, \\ Dg(0), & t = 0. \end{cases}$$

Now we use (10.5.5). □

(10.5.9) Example. Let $f : S^n \to S^n$ be continuously differentiable and p a regular value of f, i.e., the differential of f in each point $x \in f^{-1}(p)$ is bijective. Then $d(f, x) = \pm 1$, and the plus-sign holds, if the differential respects the orientation at x. For the proof express f in terms of orientation preserving local charts and apply (10.5.8). ◇

If S is homeomorphic to S^n and f a self map of S we choose a homeomorphism $h : S^n \to S$ and define the degree of f as the degree of $h^{-1} f h$. Let $g : \mathbb{R}^n \to \mathbb{R}^n$ be a proper map. We define its degree as the degree of the extension of g to the one-point compactification.

A map $f : S^n \times S^n \to S^n$, $n \geq 1$, has a ***bi-degree*** $(a, b) \in \mathbb{Z} \times \mathbb{Z}$ where a is the degree of $x \mapsto f(x, y)$ for a fixed y and b the degree of $y \mapsto f(x, y)$ for a fixed x.

Problems

1. If $f : S^n \to S^n$ is not surjective, then f is null homotopic and hence $d(f) = 0$.
2. Suppose $f(x) \neq -x$ for each $x \in S^n$, then $h(x,t) = tf(x) + (1-t)x \neq 0$ for $t \in [0,1]$. We compose with $N : x \mapsto x/\|x\|$ and obtain in $F(x,t) = Nh(x,t)$ a homotopy from the identity to f. If always $f(x) \neq x$, then $G(x,t) = N(-tx + (1-t)f(x))$ is a homotopy from f to the antipodal map. Thus if $d(f) \neq \pm 1$, there exists x such that $f(x) = x$ or $f(x) = -x$.
3. A permutation λ of $(t_0, \dots, t_n)$ induces an affine homeomorphism of $(\Delta^n, \partial \Delta^n)$. The induced homomorphism in $h_k(\Delta^n, \partial \Delta^n)$ is the multiplication with the sign of the permutation λ. The same holds for the linear permutation map l_λ induced by λ on the vector space $N = \{(t_0, \dots, t_n) \mid \sum_i t_i = 0\}$ and $h_k(N, N \smallsetminus 0)$. One can compute the determinant of l_λ by using the decomposition $\mathbb{R}^{n+1} = N \oplus D$ with the diagonal $D = \{(t, \dots, t) \mid t \in \mathbb{R}\}$.
4. The map $S^i \to S^i$, $(y,t) \mapsto (2ty, 2t^2 - 1)$, $y \in \mathbb{R}^i$, $t \in \mathbb{R}$ has degree $1 + (-1)^{i+1}$. The point $(0, \dots, 0, 1)$ is a regular value.
5. Consider a complex polynomial as self-map of the Riemann sphere $\mathbb{C}P^1 \cong S^2$. Then the homological degree is the algebraic degree of the polynomial. A quotient $f = p/q$ of two complex polynomials (without common divisor) p of degree m and q of degree n can be considered as a self-map of $\mathbb{C}P^1$. Show that the homological degree is $a = \max(m, n)$. In homogeneous coordinates f can be written as $[z, w] \mapsto [w^a p(z/w), w^a q(z/w)]$. Suppose

$c \in \mathbb{C}$ is such that $z \mapsto p(z) - cq(z)$ has a pairwise different zeros. Then $[c, 1]$ is a regular value of f.

6. Consider S^3 as the topological group of quaternions of norm 1. Determine the degree of $z \mapsto z^k, k \in \mathbb{Z}$.

7. The map $S^n \times S^n \to S^n$, $(x, \xi) \mapsto \xi - 2\langle x, \xi\rangle x$ has bi-degree $(1 + (-1)^{n-1}, -1)$. If $\xi = (1, 0, \dots, 0) = e_0$, then $-e_0$ is a regular value of $x \mapsto e_0 - 2\langle e_0, x\rangle x$. (Here $\langle -, -\rangle$ is the standard inner product.)

8. Let $a, b, p, q \in \mathbb{N}$ with $ap - bq = 1$. Then $f: \mathbb{C}^2 \to \mathbb{C}^2$, $(x, y) \mapsto (x^p \bar{y}^q, x^b + y^a)$ is proper and has degree 1. The point $(1, 0)$ is a regular value.

10.6 The Theorem of Borsuk and Ulam

We describe another classical result which uses the homotopy notion in the presence of a symmetry. As a rather striking application to a problem in combinatorics we present the proof of Greene [74] for the determination of the chromatic number of the Kneser graphs.

We have the antipodal symmetry $x \mapsto -x$ on the Euclidean spaces. A map $f: A \to B$ which is equivariant with respect to this symmetry, i.e., which satisfies $f(-x) = -f(x)$, is called ***antipodal*** or ***odd***; here A and B are subsets of Euclidean spaces that are invariant with respect to the antipodal symmetry. The additional presence of the symmetry has remarkable consequences: Classical theorems known under the name of Borsuk–Ulam theorems and Lusternik–Schnirelmann theorems. The basic result has a number of equivalent formulations. A proof of statement (5) in the following theorem is a consequence of the more general theorem (10.6.3).

(10.6.1) Theorem. *The following assertions are equivalent:*

(1) *Let $f: S^n \to \mathbb{R}^n$ be continuous. Then there exists $x \in S^n$ such that $f(x) = f(-x)$.*
(2) *Let $f: S^n \to \mathbb{R}^n$ be antipodal. Then there exists $x \in S^n$ such that $f(x) = 0$.*
(3) *There does not exist an antipodal map $f: S^n \to S^{n-1}$.*
(4) *There does not exist a continuous map $f: D^n \to S^{n-1}$ which restricts to an antipodal map on the boundary.*
(5) *An antipodal map $S^{n-1} \to S^{n-1}$ is not null homotopic.*
(6) *Suppose $S^n = F_1 \cup F_2 \cup \dots \cup F_{n+1}$ with non-empty closed sets F_j. Then at least one of the sets F_j contains an antipodal pair of points.*
(7) *Let $S^n = A_1 \cup A_2 \cup \dots \cup A_{n+1}$ and assume that each A_j is either open or closed. Then at least one of the A_j contains an antipodal pair.*

Proof. (1) $\Rightarrow$ (2). By (1) there exists x with $f(x) = f(-x)$. Since f is antipodal, $f(x) = -f(x)$ and hence $f(x) = 0$.

(2) $\Rightarrow$ (3). The existence of an antipodal map contradicts (2).

(3) $\Rightarrow$ (4). Let $D^n_\pm = \{(x_0, \dots, x_n) \in S^n \mid \pm x_n \geq 0\}$. The projection $h\colon D^n_+ \to D^{n-1}$ onto the first $n-1$ coordinates is a homeomorphism which is the identity on the boundary. Suppose $f\colon D^n \to S^{n-1}$ is antipodal on the boundary. Define $g\colon S^n \to S^{n-1}$ by

$$g(x) = \begin{cases} fh(x), & x \in D^n_+, \\ -fh(-x), & x \in D^n_-. \end{cases}$$

If $x \in S^{n-1} = D^n_+ \cap D^n_-$, then $h(x) = x$, $h(-x) = -x$ and $fh(x) = f(x) = -f(-x) = -fh(-x)$. Hence g is well-defined and continuous. One verifies that g is antipodal.

(4) $\Leftrightarrow$ (5). If an antipodal map $S^{n-1} \to S^{n-1}$ were null homotopic, then we could extend this map to D^n, contradicting (4). Conversely, if a map of type (4) would exist, then the restriction to S^{n-1} would contradict (5).

(1) $\Rightarrow$ (6). We consider the function

$$f\colon S^n \to \mathbb{R}^n, \quad x \mapsto (d(x, F_1), \dots, d(x, F_n))$$

defined with the Euclidean distance d. By (1), there exists x such that $f(x) = f(-x) = y$. If the i-th component of y is zero, then $d(x, F_i) = d(-x, F_i) = 0$ and therefore $x, -x \in F_i$ since F_i is closed. If all coordinates of y are non-zero, then x and $-x$ are not contained in $\bigcup_{i=1}^n F_i$, so they are contained in F_{n+1}.

(6) $\Rightarrow$ (3). There exists a closed covering $F_1, \dots, F_{n+1}$ of S^{n-1} such that no F_i contains an antipodal pair, e.g., project the standard simplex onto the sphere and take the images of the faces. Suppose $f\colon S^n \to S^{n-1}$ is antipodal. Then the covering by the $f^{-1}(F_i)$ contradicts (6).

(3) $\Rightarrow$ (2). If $f(x) \neq 0$ for all x, then $x \mapsto \|f(x)\|^{-1} f(x)$ is an antipodal map $S^n \to S^{n-1}$.

(2) $\Rightarrow$ (1). Given $f\colon S^n \to \mathbb{R}^n$. Then $g(x) = f(x) - f(-x)$ is antipodal; $g(x) = 0$ implies $f(x) = f(-x)$.

(6) $\Rightarrow$ (7). Suppose for the moment that the A_j are open. Then we can find a closed shrinking and apply (6). In the general case let $A_1, \dots, A_j$ be closed and $U_{j+1}, \dots, U_{n+1}$ open. Suppose there are no antipodal pairs in the U_j. Thicken the A_i to open ε-neighbourhoods $U_\varepsilon(A_i)$. Let $\varepsilon = n^{-1}$. By the case of an open covering we find an antipodal pair $(x_n, -x_n)$ in some $U_\varepsilon(A_i)$. By passing to a subsequence we find an $i \leq j$ and

$$\lim_{n \to \infty} d(x_n, A_i) = \lim_{n \to \infty} d(-x_n, A_i) = 0.$$

A convergent subsequence yields an antipodal pair in A_i.

(7) $\Rightarrow$ (3). As in the case (6) $\Rightarrow$ (3). □

Since the identity is antipodal, we see that (10.6.1) implies the retraction theorem, see (6.6.1). Part (1) shows that $\mathbb{R}^n$ does not contain a subset homeomorphic to S^n.

(10.6.2) Lemma. *Let $F: S^n \to S^n$ be an odd map. Then F is homotopic as odd map to a map g such that $g(S^i) \subset S^i$ for $0 \le i \le n$.*

Proof. Let $p\colon S^n \to \mathbb{R}P^n$ be the quotient map and $f: \mathbb{R}P^n \to \mathbb{R}P^n$ be induced by F on the orbit space. Choose a homotopy f_t from $f = f_0$ to a cellular map, i.e., $f_1(\mathbb{R}P^i) \subset \mathbb{R}P^i$ for $0 \le i \le n$. Lift the homotopy $f_t p = h_t$ to a homotopy $H_t\colon S^n \to S^n$ with initial condition f. Then H_t is an odd map and $H_1 = g$ has the desired property. □

The following theorem implies statement (5) of (10.6.1). See also (17.9.9) and Problem 7 in Section 18.8.

(10.6.3) Theorem. *An odd map has odd degree.*

Proof. The proof is by induction on the dimension of the sphere. Let $f: S^n \to S^n$ be an odd map. By (10.6.2) we can deform f as odd map into a map g such that $g(S^{n-1}) \subset S^{n-1}$. The induction is now a consequence of (10.6.4). □

(10.6.4) Proposition. *Let $f: S^n \to S^n$ be an odd map such that $f(S^{n-1}) \subset S^{n-1}$. Then we have the degree $D(f)$ of f and the degree $d(f)$ of its restriction to S^{n-1}. These degrees have the same parity.*

Proof. We study the diagram in the proof of (10.5.2) more closely for singular homology with coefficients in $\mathbb{Z}$. We fix a generator $z \in \tilde{H}_{n-1}(S^{n-1})$ and define other generators by

$$\partial_\pm z_\pm = z, \quad s_\pm z_\pm = w_\pm, \quad l_\pm v_\pm = w_\pm.$$

We set $i_\mp z_\pm = u_\pm$. The Sum Lemma tells us that $u_\pm$ is a $\mathbb{Z}$-basis of the group $H_n(S^n, S^{n-1})$. Let $T: S^n \to S^n$ be the antipodal map. Suppose given $f: S^n \to S^n$ such that $f(S^{n-1}) \subset S^{n-1}$ and $Tf = fT$. Then we have two degrees $D(f)$ and $d(f)$ defined by

$$f_*(v_+) = D(f)v_+, \quad f_*(z) = d(f)z.$$

Since $u_\pm$ is a $\mathbb{Z}$-basis we can write $f_*(u_+) = au_+ + bu_-$. Using this notation, we show

$$d(f) = a + b, \quad D(f) = a - b.$$

Hence $d(f)$ and $D(f)$ are congruent modulo 2. From $f_*z = f_*\partial u_+ = \partial f_* u_+ = \partial(au_+ + bu_-) = (a+b)z$ we obtain the first assertion $d(f) = a+b$. Naturality of the boundary operator $\partial_- T_* = T_* \partial_+$ and $T_* z = (-1)^n z$ imply $T_* z_+ = (-1)^n z_-$ and $T_* u_+ = (-1)^n u_-$. We conclude

$$\begin{aligned} f_* u_- &= (-1)^n f_* T_* u_+ = (-1)^n T_* f_* u_+ \\ &= (-1)^n T_*(au_+ + bu_-) = au_- + bu_+. \end{aligned}$$

The exactness of (η, ∂) shows that the image of η is generated by $u_+ - u_-$, hence $\eta(v_+) = \varepsilon(u_+ - u_-)$ with $\varepsilon = \pm 1$. The computation

$$j_+u_+ = s_+i_-z_+ = l_+ = j_+\eta v_+ = \varepsilon j_+(u_+ - u_-) = \varepsilon j_+u_+$$

shows $\varepsilon = 1$. The computation

$$\begin{aligned} D(f)(u_+ - u_+) &= D(f)\eta v_+ = \eta f_* v_+ = f_*\eta v_+ \\ &= f_*(u_+ - u_-) = (a-b)(u_+ - u_-) \end{aligned}$$

finally yields the second assertion $D(f) = a - b$. □

(10.6.5) Example. The map $f : S^1 \to S^1, z \mapsto z^{2k+1}$ satisfies the hypothesis of (10.6.4). We know already that $D(f) = 2k+1$ and $d(f) = 1$, hence $a = k+1$ and $b = -k$. Let d_+ denote the singular 1-simplex represented by the path from 1 to -1 and d_- the 1-simplex running from -1 to 1 (both counter-clockwise). Then $d_+ + d_-$ is a cycle and $v_+ = [d_+ + d_-]$ is a natural choice of a generator. Then d_- represents u_+, w_+, and z_+; and $z = [1] - [-1]$. By considering the simplex fd_-, the relation $f_*u_+ = [fd_-] = (k+1)[d_-] + k[d_+]$ becomes apparent. For $k = 1$ say, fd_- runs counter-clockwise from -1 to 1 to -1 to 1.

Let $f : S^1 \to S^1$ be any self-map which commutes with the antipodal map. We can multiply f by a constant such that the new map g satisfies $g(1) = 1$. From (10.6.4) we see that g and hence f has odd degree, since g has degree 1 on S^0. ◇

We now apply the Borsuk–Ulam theorem to a problem in combinatorics: The determination of the chromatic number of the so-called Kneser graphs.

We begin by explaining the problem. Let $[n] = \{1, \dots, n\}$ and denote by N_k the set of subsets of $[n]$ with k elements.

A ***graph*** consists of a set E of vertices and a set K of edges. Each edge has two boundary points, and they are identified with one or two points in E. (In other terms: A graph is a 1-dimensional CW-complex.) The ***Kneser graph*** $KG_{n,k}$ has $E = N_k$. Vertices F_1, F_2 are connected by an edge if they represent disjoint subsets of $[n]$.

Let $\mathcal{C} = [k]$, and call the elements of $\mathcal{C}$ colours. A ***k-colouring*** of a graph (E, K) is a map $f : E \to \mathcal{C}$ such that $f(e_1) \neq f(e_2)$ whenever e_1 and e_2 are connected by an edge. The ***chromatic number*** of a graph (E, K) is the smallest k such that there exists a k-colouring. The following result was conjectured by Martin Kneser (1955) [105]. This conjecture was proved by Lovász [114] with topological methods. The following ingenious proof was given by Greene [74].

(10.6.6) Theorem. *Let $k > 0$ and $n \leq 2k - 1$. Then $KG_{n,k}$ has the chromatic number $n - 2k + 2$.*

Proof. An explicit construction shows that the chromatic number is at most $n-2k+2$. We associate to a set F with k elements the colour $\varphi(F) = \min(\min(F),$

$n-2k+2$). Suppose $\varphi(F)=\varphi(F')=i<n-2k+2$. Then these sets are not disjoint, since they contain the element i. If their colour is $n-2k+2$, then they are contained in $\{n-2k+2,\dots,n\}$, and then they cannot be disjoint.

Now we come to the topological part. Let $d=2n-2k+1$. Choose a set $X\subset S^d$ with n elements and such that no hyperplane (through the origin) contains more than d points of X. We consider the subsets of X with cardinality k as the vertices of the Kneser graph $KG_{n,k}$. Suppose there exists a colouring with $d=n-2k+1$ elements and we choose one. Let

$$A_i=\{x\in S^d\mid H(x)=\{y|\langle x,y\rangle>0\}\text{ contains a }k\text{-tuple in }X\text{ with colour }i\}$$

and $A_{d+1}=S^d\smallsetminus(A_1\cup\dots\cup A_d)$. The sets $A_1,\dots,A_d$ are open and A_{d+1} is closed. By (10.6.1), one of the sets A_i contains an antipodal pair $x,-x$.

If $i\le d$, then we have two disjoint k-tuples with colour i, one in $H(x)$ the other one in $H(-x)$. This contradicts the definition of a colouring.

Let $i=d+1$. Then, by definition of $A_1,\dots,A_d$, the half-space $H(x)$ contains at most $k-1$ points of X, and similarly for $H(-x)$. The set $S^d\smallsetminus(H(x)\cup H(-x))$ is contained in a hyperplane and contains at least $n-2(k-1)=d+1$ points, and this contradicts the choice of X. □

Problems

1. Let n be odd. Then $\pm\,\mathrm{id}\colon S^n\to S^n$ are homotopic as odd maps.

2. Let $f,g\colon S^n\to S^n$ be odd maps with the same degree. Then they are homotopic as odd maps.

3. Let $d_0,d_1,\dots,d_n$ be a family of odd integers with $d_0=\pm1$. There exists an odd map $f\colon S^n\to S^n$ such that $f(S^i)\subset S^i$ and the map $S^i\to S^i$ induced by f has degree d_i.

10.7 Mayer–Vietoris Sequences

We derive further exact sequences for homology from the axioms, the so-called Mayer–Vietoris sequences.

Let h_* be a homology theory. Let $(X;A,B)$ be a triad, i.e., $A,B\subset X=A\cup B$. The triad $(X;A,B)$ is said to be ***excisive*** for the homology theory if the inclusion induces an (excision) isomorphism $h_*(A,A\cap B)\cong h_*(X,B)$. The condition is actually symmetric in A and B. We write $AB=A\cap B$.

(10.7.1) Proposition. *The following are equivalent:*

(1) *$(A\cup B;A,B)$ is excisive.*

(2) *$(A\cup B;B,A)$ is excisive.*

(3) *$\iota\colon h_*(A,AB)\oplus h_*(B,AB)\to h_*(A\cup B,AB)$ is an isomorphism.*

(4) *$\pi\colon h_*(A\cup B,AB)\to h_*(A\cup B,A)\oplus h_*(A\cup B,B)$ is an isomorphism.*

Proof. Apply the Sum Lemma (11.1.2) to the diagram

$$\begin{array}{ccccc} h_*(A, A\cap B) & & \longrightarrow & & h_*(A\cup B, B) \\ & \searrow & & \nearrow & \\ & & h_*(A\cup B, A\cap B) & & \\ & \nearrow & & \searrow & \\ h_*(B, A\cap B) & & \longrightarrow & & h_*(A\cup B, A). \end{array}$$

The morphisms are induced by the inclusions. □

The ***boundary operator*** Δ of an excisive triad is defined by

$$\Delta\colon h_n(X) \to h_n(X,B) \xleftarrow{\cong} h_n(A, AB) \xrightarrow{\partial} h_{n-1}(AB).$$

This operator is part of the ***Mayer–Vietoris sequence*** (= MVS) of the triad.

(10.7.2) Theorem. *Let $(A\cup B; A, B)$ be an excisive triad and $C\subset AB$. Then the sequence*

$$\dots \xrightarrow{\Delta} h_n(AB, C) \xrightarrow{(1)} h_n(A,C)\oplus h_n(B,C) \xrightarrow{(2)} h_n(A\cup B, C)$$
$$\xrightarrow{\Delta} h_{n-1}(AB) \xrightarrow{(1)} \cdots$$

is exact. The inclusions $i^A\colon AB\subset A$ and $i^B\colon AB\subset B$ yield the first map $x\mapsto(-i^A_*x, i^B_*x)$; and the inclusions $j^A\colon A\subset A\cup B$ and $j^B\colon B\subset A\cup B$ yield the second map $(a,b)\mapsto j^A_*a + j^B_*b$. If $C=\{*\}$ is a point, we obtain the MVS for reduced homology groups.

There exists another relative MVS. Let $(A\cup B; A, B)$ be excisive and $A\cup B\subset X$. We define a ***boundary operator*** Δ by

$$\Delta\colon h_n(X, A\cup B) \xrightarrow{\partial} h_{n-1}(A\cup B, A) \cong h_{n-1}(B, AB) \to h_{n-1}(X, AB).$$

(10.7.3) Theorem. *The sequence*

$$\cdots \xrightarrow{\Delta} h_n(X, AB) \xrightarrow{(1)} h_n(X,A)\oplus h_n(X,B) \xrightarrow{(2)} h_n(X, A\cup B)$$
$$\xrightarrow{\Delta} h_{n-1}(X, AB) \xrightarrow{(1)} \cdots$$

is exact. The maps (1) *and* (2) *are defined as in* (10.7.2).

Proof. The homology sequences of the triples of $(B, A\cap B, C)$ and (X, A, C) yield a commutative diagram (a "ladder")

$$\begin{array}{ccccccccc} \cdots \to & h_n(AB, C) & \to & h_n(A, C) & \to & h_n(A, AB) & \to & h_{n-1}(AB, C) & \to\cdots \\ & \downarrow & & \downarrow & & \downarrow & & \downarrow & \\ \cdots \to & h_n(B, C) & \to & h_n(X, C) & \to & h_n(X, B) & \to & h_{n-1}(B, C) & \to\cdots. \end{array}$$

We apply (10.7.4) to this diagram and obtain (10.7.2). There exists a similar diagram which compares the sequences of the triples (X, A, AB) and $(X, A \cup B, B)$, and (10.7.4) yields now (10.7.3). □

(10.7.4) Lemma. *Suppose the following diagram of abelian groups and homomorphisms is commutative and has exact rows.*

$$\begin{array}{ccccccccc} \cdots \to & A_i & \xrightarrow{f_i} & B_i & \xrightarrow{g_i} & C_i & \xrightarrow{h_i} & A_{i-1} & \to \cdots \\ & \downarrow{\scriptstyle a_i} & & \downarrow{\scriptstyle b_i} & & \downarrow{\scriptstyle c_i} & & \downarrow{\scriptstyle a_{i-1}} & \\ \cdots \to & A'_i & \xrightarrow{f'_i} & B'_i & \xrightarrow{g'_i} & C'_i & \xrightarrow{h'_i} & A'_{i-1} & \to \cdots . \end{array}$$

Assume moreover that the c_i are isomorphisms. Then the sequence

$$\cdots \to A_i \xrightarrow{(-f_i, a_i)} B_i \oplus A'_i \xrightarrow{\langle b_i, f'_i \rangle} B'_i \xrightarrow{h_i c_i^{-1} g'_i} A_{i-1} \to \cdots .$$

is exact ([17, p. 433]). □

We use abbreviations of the type $I \times X = IX$, $\partial I \times X = \partial I X$, $0 \times X = 0X$. We associate to a triad $(X; A, B)$ the subspace $N = N(A, B) = 0A \cup IAB \cup 1B$ of $I \times X$. Let $p \colon N(A, B) \to X$ be the projection onto the second factor.

(10.7.5) Proposition. *The following are equivalent:*

(1) *The triad $(X; A, B)$ is excisive.*

(2) *$p_* \colon h_*(N(A, B)) \to h_*(X)$ is an isomorphism.*

Proof. We have isomorphisms

$$h_*(A, AB) \oplus h_*(B, AB) \cong h_*(0A + 1B, 0AB + 1AB) \cong h_*(N, IAB),$$

by additivity, excision and h-equivalence. It transforms p_* into the map ι of item (3) in (10.7.1). Hence (1) and (2) are equivalent. □

The excision axiom says that the triad is excisive if $X = A^\circ \cup B^\circ$. The auxiliary space $N(A, B)$ allows us to transfer the problem into homotopy theory: A triad is excisive for each homology theory, if $p \colon N \to X$ is an h-equivalence. Recall from Section 3.3:

(10.7.6) Proposition. *Suppose the covering A, B of X is numerable. Then p is an h-equivalence.* □

We now give a second proof of the MVS; it uses the homotopy axiom but not lemma (10.7.4). Let $(X; A, B)$ be a triad. Via excision and h-invariance we see that the inclusion induces an isomorphism

$$h_n((I, \partial I) \times AB) \cong h_n(N, 0 \times A \cup 1 \times B).$$

We rewrite the exact sequence of the pair $(N, 0 \times A \cup 1 \times B)$. Using the suspension isomorphism and the additivity, we obtain an exact sequence

$$\cdots \to h_n(A) \oplus h_n(B) \to h_n(N) \to h_{n-1}(AB) \to \cdots .$$

If the triad is excisive, we can use (10.7.5) and replace $h_*(N)$ by $h_*(X)$. It is an exercise to compare the boundary operators of the two constructions of the MVS.

There exists a more general MVS for pairs of spaces. It comprises the previously discussed cases.

(10.7.7) Theorem. *Let* $(A; A_0, A_1) \subset (X; X_0, X_1)$ *be two excisive triads. Set* $X_{01} = X_0 \cap X_1$, $A_{01} = A_0 \cap A_1$. *Then there exists an exact Mayer–Vietoris sequence of the form*

$$\cdots \to h_n(X_{01}, A_{01}) \to h_n(X_0, A_0) \oplus h_n(X_1, A_1) \to h_n(X, A) \to \cdots .$$

Proof. Let $N(X, A) = 0X_0 \cup IA_{01} \cup 1X_1$. The sequence in question arises from a rewriting of the exact sequence of the triple $(N(X), N(X, A), N(A))$. We consider three typical terms.

(1) $p_*\colon h_*(N(X), N(A)) \cong h_*(X, A)$, by (10.7.5) and the hypotheses.

(2) The inclusions $(X_j, A_j) \cong \{j\} \times (X_j, A_j) \to (N(X, A), N(A))$ induce an isomorphism

$$h_*(X_0, A_0) \oplus h_*(X_1, A_1) \to h_*(N(X, A), N(A)).$$

For the proof one excises $[1/3, 2/3] \times A_{01}$ and then uses an h-equivalence and additivity.

(3) The group $h_*(N(X), N(X, A))$ is isomorphic to $h_*(IX_{01}, \partial IX_{01} \cup IA_{01})$ via inclusion, and the latter via suspension isomorphic to $h_{*-1}(X_{01}, A_{01})$. For the proof one replaces $N(X, A)$ by the thickened space

$$0X_0 \cup [0, 1/4]X_{01} \cup IA_{01} \cup [3/4, 1]X_{01} \cup 1X_1.$$

Then one can excise $0X_0$ and $1X_1$ and use suitable h-equivalences.

It remains to identify the morphisms in the resulting sequence. The map $h_n(N(X, A), N(A)) \to h_n(N(X), N(A))$ becomes

$$\langle j_*^0, j_*^1 \rangle\colon h_n(X_0, A_0) \oplus h_n(X_1, A_1) \to h_n(X, A).$$

The map $\partial\colon h_{n+1}(N(X), N(X, A)) \to h_n(N(X, A), N(A)$ becomes, with our definition of the suspension isomorphism,

$$(-i_*^0, i_*^1)\colon h_n(X_{01}, A_{01}) \to h_n(X_0, A_0) \oplus h_n(X_1, A_1).$$

The boundary operator Δ of the generalized MV-sequence becomes, in the special cases $X = X_0 = X_1 = X_{01}$ and $A = A_0 = A_1 = A_{01} \subset X_{01}$, the same as in the previously discussed algebraic derivation of the MV-sequences. □

(10.7.8) Example. Let $i^1, i^2 \colon S^n \to S^n \times S^n$, $x \mapsto (x, y_0)$, resp. $x \mapsto (x_0, y)$. Then $\langle i^1_*, i^2_* \rangle \colon H_n(S^n) \oplus H_n(S^n) \to H_n(S^n \times S^n)$ is an isomorphism ($n \geq 1$) with inverse $(\mathrm{pr}^1_*, \mathrm{pr}^2_*)$. We fix a generator $z \in H_n(S^n)$ and write $z_j = i^j_* z$. Then (z_1, z_2) is a $\mathbb{Z}$-basis of $H_n(S^n \times S^n)$. Let $\alpha = (\alpha_1, \alpha_2) \colon S^n \times S^n \to S^n \times S^n$ be a map with bi-degree (a, b) of α_1 and bi-degree (c, d) of α_2. Then $\alpha_*(z_1) = az_1 + cz_2$ and $\alpha_*(z_2) = bz_1 + dz_2$.

Construct a space, a $(2n+1)$-manifold, X by identifying in $D^{n+1} \times S^n + D^{n+1} \times S^n$ the point $(x, y) \in S^n \times S^n$ in the first summand with $\alpha(x, y)$ in the second summand via a homeomorphism $\alpha = (\alpha_1, \alpha_2)$ of $S^n \times S^n$ as above. The two summands are embedded as X_1 and X_2 into X. We use the MV-sequence of $(X; X_1, X_2)$ to determine the integral homology of X. Let us consider a portion of this sequence

$$H_n(S^n \times S^n) \xrightarrow{j} H_n(D^{n+1} \times S^n) \oplus H_n(D^{n+1} \times S^n) \to H_n(X).$$

We use the $\mathbb{Z}$-basis (z_1, z_2) as above. The inclusion $S^n \to D^{n+1} \times S^n$, $x \mapsto (0, x)$ give us as image of z the generators u_1, u_2 in the summands $H_n(D^{n+1} \times S^n)$. The image of the basis elements under j is seen to be $j(z_1) = cu_2$, $j(z_2) = u_1 + du_2$ (we do not use the minus sign for the second summand). We conclude for $c \neq 0$ that $H_n(X)$ is the cyclic group of order $|c|$; the other homology groups of X are in this case $H_0(X) \cong \mathbb{Z} \cong H_{2n+1}(X)$ and $H_j(X) = 0$ for $j \neq 0, n, 2n+1$. We leave the case $c = 0$ to the reader. ◇

Problems

1. Let $\mathbb{R}^n$ be the union of two open sets U and V.

(i) If U and V are path connected, then $U \cap V$ is path connected.

(ii) Suppose two of the sets $\pi_0(U), \pi_0(V), \pi_0(U \cap V)$ are finite, then the third set is also finite and the relation

$$|\pi_0(U \cap V)| - (|\pi_0(U)| + |\pi_0(V)|) + |\pi_0(U \cup V)| = 0$$

holds.

(iii) Suppose $x, y \subset U \cap V$ can be connected by a path in U and in V. Then they can be connected by a path in $U \cap V$.

Can you prove these assertions without the use of homology directly from the definition of path components?

2. Let the real projective plane P be presented as the union of a Möbius band M and a disk D, glued together along the common boundary S^1. Determine the groups and homomorphisms in the MV-sequence of $(P; M, D)$. Do the same for the Klein bottle $(K; M, M)$. (Singular homology with arbitrary coefficients.)

3. Let $(X_1, \dots, X_n)$ be an open covering of X and $(Y_1, \dots, Y_n)$ be an open covering of Y. Let $f \colon X \to Y$ be a map such that $f(X_i) \subset Y_i$. Suppose that the restriction $\bigcap_{a \in A} X_a \to \bigcap_{a \in A} Y_a$ of f induces a homology isomorphism for each $\emptyset \neq A \subset \{1, \dots, n\}$. Then f

induces a homology isomorphism.
4. Suppose $AB \subset A$ and $AB \subset B$ are closed cofibrations. Then $p\colon N(A, B) \to A \cup B$ is an h-equivalence.
5. The boundary operators in (10.7.2) and (10.7.3) which result when we interchange the roles of A and B differ from the original ones by -1. (Apply the Hexagon Lemma to the two boundary operators.)
6. The triad $(X \times \partial I \cup A \times I; X \times 0 \cup A \times I, X \times 1 \cup A \times I)$ is always excisive. (Excision of $X \times 0 \cup A \times [0, 1/2[$ and h-equivalence.)
7. Verify the assertions about the morphisms in the sequence (10.7.7).

10.8 Colimits

The additivity axiom for a homology theory expresses a certain compatibility of homology and colimits (namely sums). We show that this axiom has consequences for other colimits.

Let $(X_\bullet, f_\bullet)$ be a sequence $X_1 \xrightarrow{f^1} X_2 \xrightarrow{f^2} X_3 \xrightarrow{f^3} \cdots$ of continuous maps f^j. Recall that a colimit (a direct limit) of this sequence consists of a space X and continuous maps $j^k\colon X_k \to X$ with the following universal property:

(1) $j^{k+1} f^k = j^k$.

(2) If $a^k\colon X_k \to Y$ is a family of maps such that $a^{k+1} f^k = a^k$, then there exists a unique map $a\colon X \to Y$ such that $a j^k = a^k$.

(This definition can be used in any category.) Let us write

$$\operatorname{colim}(X_\bullet, f_\bullet) = \operatorname{colim}(X_k)$$

for the colimit. In the case that the $f^k\colon X_k \subset X_{k+1}$ are inclusions, we can take as colimit the union $X = \bigcup_i X_i$ together with the colimit topology: $U \subset X$ open if and only if $U \cap X_n$ open in X_n for each n.

Colimits are in general not suitable for the purpose of homotopy theory, one has to weaken the universal property "up to homotopy". We will construct a so-called homotopy colimit. Colimits of sequences allow a special and simpler treatment than general colimits. A model of a homotopy colimit in the case of sequences is the ***telescope***. We identify in $\coprod_i X_i \times [i, i+1]$ the point $(x_i, i+1)$ with $(f^i(x_i), i+1)$ for $x_i \in X_i$. Denote the result by

$$T = T(X_\bullet, f_\bullet) = \operatorname{hocolim}(X_\bullet, f_\bullet).$$

We have injections $j^k\colon X_k \to T, x \mapsto (x, k)$ and a homotopy $\kappa^k\colon j^{k+1} f^k \simeq j^k$, a linear homotopy in $X_k \times [k, k+1]$. Thus the telescope T consists of the mapping cylinders of the maps f^i glued together. The data define the ***homotopy colimit*** of the sequence.

Given maps $a^k\colon X_k \to Y$ and homotopies $h^k\colon X_k \times [k, k+1] \to Y$ from $a^{k+1} f^k$ to a^k. Then there exists a map $a\colon T \to Y$ such that $j^k a = a^k$, and the composition of the canonical map $X_k \times [k, k+1] \to T$ with a is h^k.

We have subspaces $T_k \subset T$, the image of $\coprod_{i=1}^{k-1} X_i \times [i, i+1] + X_k \times \{k\}$. The canonical inclusion $\iota^k\colon X_k \to T_k$ is a homotopy equivalence (compare the analogous situation for a mapping cylinder).

In homology we have the equality $j_*^k = j_*^{k+1} f_*^k\colon h_n(X_k) \to h_n(X_{k+1}) \to h_n(T)$. By the universal property of the colimit of groups we therefore obtain a homomorphism

$$\iota\colon \operatorname{colim} h_n(X_k) \to h_n(T(X_\bullet, f_\bullet)).$$

(10.8.1) Theorem. *In an additive homology theory ι is an isomorphism.*

Proof. We recall an algebraic construction of the colimit $A_1 \xrightarrow{a_1} A_2 \xrightarrow{a_2} \cdots$ of abelian groups. Consider

$$\textstyle\bigoplus_{k \geq 1} A_k \to \bigoplus_{k \geq 1} A_k, \quad (x_k) \mapsto (x_{k+1} - a_k(x_k)).$$

The cokernel is the colimit, together with the canonical maps (inclusion of the j-th summand composed with the projection) $A_j \to \bigoplus_k A_k \to \operatorname{colim} A_k$. We therefore need a computation of $h_n(T)$ which has this form. We cover T by the subspaces

$$A = T \smallsetminus \textstyle\bigcup_{i \geq 1} X_{2i} \times \{2i + \frac{1}{2}\}, \quad B = T \smallsetminus \bigcup_{i \geq 1} X_{2i-1} \times \{2i - \frac{1}{2}\}.$$

(10.8.2) Lemma. *The inclusions*

$$\coprod_{i \geq 1} X_{2i} \times \{2i\} \to A, \quad \coprod_{i \geq 1} X_{2i-1} \times \{2i-1\} \to B, \quad \coprod_{i \geq 1} X_i \times \{i\} \to A \cap B$$

are h-equivalences, and (A, B) is a numerable covering of T. □

Because of this lemma we have a Mayer–Vietoris sequence

$$\begin{array}{ccccc} h_n(A \cap B) & \longrightarrow & h_n(A) \oplus h_n(B) & \longrightarrow & h_n(T) \\ \cong\uparrow & & \cong\uparrow & & \\ \bigoplus h_n(X_j) & \xrightarrow{\alpha} & \bigoplus_{j \equiv 0(2)} h_n(X_j) \oplus \bigoplus_{j \equiv 1(2)} h_n(X_j). & & \end{array}$$

The map α has the form

$$\alpha(x_{2i}) = (x_{2i}, f_*^{2i}(x_{2i})) \quad \text{and} \quad \alpha(x_{2i+1}) = (f_*^{2i+1}(x_{2i+1}), x_{2i+1})$$

for $x_j \in h_n(X_j)$. We see that α is injective; therefore we can obtain $h_n(T)$ as cokernel of α. The automorphism $(x_i) \mapsto ((-1)^i x_i)$ transforms α into the map which was used in the algebraic definition of the colimit. □

For applications we have to find conditions under which the homotopy colimit is h-equivalent to the colimit. We consider the case that the $f^k\colon X_k \to X_{k+1}$ are inclusions, and denote the colimit by $X = \bigcup_k X_k$. We change the definition of the telescope slightly and consider it now as the subspace

$$T = \bigcup_k X_k \times [k, k+1] \subset X \times [0, \infty[\,.$$

The topology of T may be different, but the proof of (10.8.1) works equally well in this case. The projection onto the first factor yields $p\colon T \to X$.

(10.8.3) Example. Let $\lambda\colon X \to [1, \infty[$ be a function such that

$$s = (\mathrm{id}, \lambda)\colon X \to X \times [1, \infty[$$

has an image in T. Then $s\colon X \to T$ is a section of p. The composition sp is homotopic to the identity by the homotopy $((x,u),t) \mapsto (x, (1-t)u + t\lambda(x))$. This is a homotopy over X, hence p is shrinkable. The property required by λ amounts to $\lambda(x) < i \Rightarrow x \in X_{i-1}$ for each i.

Let $(U_i \mid i \in \mathbb{N})$ be a numerable covering of X with locally finite numeration (τ_i) (see the chapter on partitions of unity). Set $X_k = \bigcup_{i=1}^k U_i$ and $\lambda = \sum_{i=1}^\infty i\tau_i$. Then $\lambda(x) < i$ implies $x \in X_{i-1}$. ◇

(10.8.4) Proposition. *Suppose the inclusions $X_k \subset X_{k+1}$ are cofibrations. Then $T \subset X \times [1, \infty[$ is a deformation retract.*

Proof. Since $X_k \subset X$ is a cofibration, there exists by (5.1.3) a homotopy

$$h_t^k\colon X \times [1, \infty[\to X \times [1, \infty[\quad \text{rel } Y_k = X_k \times [1, \infty[\cup X \times [k+1, \infty[$$

from the identity to a retraction $X \times [0, \infty[\to Y_k$. The retraction R_l acts as the identity on $X_k \times [1, \infty[$ for $l > k$, and therefore the infinite composition $R^j = \cdots \circ R_{j+2} \circ R_{j+1} \circ R_j$ is a well-defined continuous map. From h^j we obtain a homotopy $R^j \simeq R^{j+1}$ relative to Y_j. We can concatenate these homotopies and obtain a homotopy from the retraction R^1 to the identity relative to T. □

Problems

1. Let T be a subring. Find a system of homomorphisms $\mathbb{Z} \to \mathbb{Z} \to \cdots$ such that the colimit is T.

2. Let $S^n \to S^n \to S^n \to \cdots$ be a sequence of maps where each map has degree two. Let X be the homotopy colimit. Show that $\pi_n(X) \cong \mathbb{Z}[1/2]$, the ring of rational numbers with denominators a power of two. What system of maps between S^n would yield a homotopy colimit Y such that $\pi_n(Y) \cong \mathbb{Q}$?

3. Let X be a CW-complex and T the telescope of the skeleton filtration. Then the inclusion $T \subset X \times [1, \infty[$ induces isomorphisms of homotopy groups and is therefore a homotopy equivalence. One can also apply (10.8.4) in this case.

10.9 Suspension

Recall the homological suspension isomorphism 10.2.5. We use abbreviations of the type $I \times A = IA$, $\partial I \times A = \partial IA$, $0 \times X = 0A$. We set $k_n(A,B) = h_n((I,\partial I) \times (A,B))$. The $k_*(-)$ are the data for a new homology theory. The boundary operator of this homology theory is defined for a triple (A,B,C) by

$$\tilde{\partial}\colon h_{n+1}(IA, \partial IA \cup IB) \xrightarrow{\partial} h_n(\partial IA \cup IB, \partial IA \cup IC) \xleftarrow{\cong} h_n(IB, \partial IB \cup IC).$$

In order to work with this definition we use

(10.9.1) Lemma. *For each triple (A,B,C) the triad $(\partial IA \cup IB; IB, \partial IA \cup IC)$ is excisive.*

Proof. The inclusion induces an isomorphism

$$h_n(\partial IA, \partial IB) \to h_n(\partial IA \cup IC, \partial IB \cup IC),$$

excise $\frac{1}{2} \times C$ and use a homotopy equivalence. If we use this also for $B = C$ we conclude that $h_n(\partial IA \cup IC, \partial IB \cup IC) \to h_n(\partial IA \cup IB, IB)$ is an isomorphism. □

The exact sequence of the triple $(IA, \partial IA \cup IB, \partial IA)$ is transformed with the isomorphism $k_n(B) = h_n(IB, \partial IB) \cong h_n(IB \cup \partial IA, \partial IA)$ into the exact sequence of (A,B) for the k_*-groups. Let $U \subset B \subset A$ and $\bar{U} \subset B^\circ$. The excision isomorphism for the k_*-theory claims that

$$h_n(IA \smallsetminus U), I(B \smallsetminus U) \cup \partial I(A \smallsetminus U)) \to h_n(IA, IB \cup \partial IA)$$

is an isomorphism. This is a consequence of $\overline{IU} = I\bar{U} \subset (IB \cup \partial IA)^\circ$ and the usual excision isomorphism. The isomorphisms $\sigma^{(A,B)}\colon h_n(A,B) \to k_{n+1}(A,B)$ defined in (10.2.5) form a natural transformation. The next proposition says that they are natural transformations of homology theories of degree 1.

(10.9.2) Proposition. *For each triple (A,B,C) the diagram*

$$\begin{array}{ccc} h_{n+1}(A,B) & \xrightarrow{\partial} & h_n(B,C) \\ \downarrow{\scriptstyle \sigma^{(A,B)}} & & \downarrow{\scriptstyle -\sigma^{(B,C)}} \\ k_{n+2}(A,B) & \xrightarrow{\tilde{\partial}} & k_{n+1}(B,C) \end{array}$$

is commutative.

Proof. By naturality it suffices to consider the case $C = \emptyset$. The Hexagon Lemma shows $\alpha = -\beta$ for the maps

$$\begin{aligned}\alpha\colon h_{n+1}(IA, IB \cup \partial IA) &\to h_n(IB \cup \partial IA, \partial IA)\\ &\cong h_n(IB \cup 0A, 0A \cup 1B) \to h_{n-1}(0A \cup 1B, 0A),\\ \beta\colon h_{n+1}(IA, IB \cup \partial IA) &\to h_n(IB \cup \partial IA, IB \cup 0A)\\ &\cong h_n(\partial IA, 0A \cup 1B) \to h_{n-1}(0A \cup 1B, 0A),\end{aligned}$$

and the center group $h_n(IB \cup \partial IA, 1B \cup 0A)$. Let j be the isomorphism

$$j\colon h_{n-1}(1B \cup 0A, 0A) \xleftarrow{\cong} h_{n-1}(1B) \cong h_{n-1}(B).$$

By diagram chasing one verifies $\sigma^{(B)} j\alpha = \tilde{\partial}$ and $j\beta\sigma^{(A,B)} = \partial$. □

(10.9.3) Lemma. *Let (A, B, C) be a triple. Then we have an isomorphism*

$$\langle \iota_1, \iota_0, \iota\rangle\colon h_n(A, B) \oplus h_n(A, B) \oplus h_n(IB, \partial IB \cup IC) \to h_n(\partial IA \cup IB, \partial IB \cup IC).$$

Here ι is induced by the inclusion, and ι_ν by $a \mapsto (\nu, a)$.

Proof. This is a consequence of (10.9.1) and (10.7.1). □

(10.9.4) Proposition. *For each triple (A, B, C) the diagram*

$$\begin{array}{ccc} h_n(A, B) & \xrightarrow{\sigma^{(A,B)}} & h_{n+1}(IA, \partial IA \cup IB) \\ \downarrow{\scriptstyle\alpha} & & \downarrow{\scriptstyle\partial} \\ h_n(A, B) \oplus h_n(A, B) \oplus h_n(IB, \partial IB \cup IC) & \underset{\cong}{\xrightarrow{\beta}} & h_n(\partial IA \cup IB, \partial IB \cup IC) \end{array}$$

is commutative; here $\alpha(x) = (x, -x, -\sigma^{(B,C)}\partial x)$, and the isomorphism β is taken from (10.9.3).

Proof. The assertion about the third component of α follows from (10.9.2). The other components require a little diagram chasing. For the verification it is helpful to use the inverse isomorphism of β given by the procedure of (10.7.1). The minus sign in the second component is due to the fact that the suspension isomorphism changes the sign if we interchange the roles of 0, 1, see 10.2.5. □

Chapter 11

Homological Algebra

In this chapter we collect a number of algebraic definitions and results which are used in homology theory. Reading of this chapter is absolutely essential, but it only serves practical purposes and is not really designed for independent study. "Homological Algebra" is also the name of a mathematical field – and the reader may wish to look into the appropriate textbooks.

The main topics are diagrams and exact sequences, chain complexes, derived functors, universal coefficients and Künneth theorems. We point out that one can imitate a lot of homotopy theory in the realm of chain complexes. It may be helpful to compare this somewhat simpler theory with the geometric homotopy theory.

11.1 Diagrams

Let R be a commutative ring and denote by R-MOD the category of left R-modules and R-linear maps. (The category ABEL of abelian groups can be identified with $\mathbb{Z}$-MOD.) Recall that a sequence of R-modules and R-linear maps $\cdots \to A_{i+1} \xrightarrow{a_{i+1}} A_i \xrightarrow{a_i} A_{i-1} \to \cdots$ is exact at A_i if $\operatorname{Im}(a_{i+1}) = \operatorname{Ker}(a_i)$ and exact if it is exact at each A_i. The language of exact sequences is a convenient way to talk about a variety of algebraic situations.

(1) $0 \to A \xrightarrow{a} B$ exact $\Leftrightarrow$ a injective. (We also use $A \rightarrowtail B$ for an injective homomorphism.)

(2) $B \xrightarrow{b} C \to 0$ exact $\Leftrightarrow$ b surjective. (We also use $B \twoheadrightarrow C$ for a surjective homomorphism.)

(3) $0 \to A \xrightarrow{a} B \xrightarrow{b} C \to 0$ exact $\Leftrightarrow$ a is injective, b is surjective, b induces an isomorphism of the cokernel of a with C.

(4) Let $A \overset{(1)}{\to} B \overset{(2)}{\to} C \overset{(3)}{\to} D$ be exact. Then the following are equivalent:

$$(1) \text{ surjective} \quad \Leftrightarrow \quad (2) \text{ zero} \quad \Leftrightarrow \quad (3) \text{ injective}.$$

An exact sequence of the form (3) is called a ***short exact sequence***. It sometimes happens that in a longer exact sequence every third morphism has the property (1) (or (2), (3)). Then the sequence can be decomposed into short exact sequences. Note that the exact homotopy sequences and the exact homology sequences have "period" 3.

A family $(M_j \mid j \in J)$ of modules has a direct sum $\bigoplus_{j\in J} M_j$, the sum in the category R-MOD, and a product $\prod_{j\in J} M_j$, the product in the category

R-MOD. The underlying set of the product is the Cartesian product of the underlying sets, consisting of all families $(x_j \in M_j \mid j \in J)$; the R-module structure is given by pointwise addition and scalar multiplication. The canonical projection $p_k \colon \prod_j M_j \to M_k$ onto the k-th factor is part of the categorical product structure. The sum $\bigoplus_j M_j$ is a submodule of the product and consists of all families (x_j) where all but a finite number of x_j are zero. We have the canonical injection $i_k \colon M_k \to \bigoplus_j M_j$, defined by $p_k i_k = \mathrm{id}$ and $p_k i_l = 0$ for $k \neq l$. If $f_j \colon M_j \to N$ is a family of R-linear maps, then

$$\langle f_j \rangle \colon \bigoplus M_j \to N$$

denotes the morphism determined by $\langle f_j \rangle \circ i_k = f_k$. If $g_j \colon L \to M_j$ is a family of R-linear maps, then

$$(g_j) \colon L \to \textstyle\prod_j M_j$$

denotes the morphism determined by $p_k \circ (g_j) = g_k$.

Let $(A_j \mid j \in J)$ be a family of submodules of M. Then $\sum_j A_j$ denotes the submodule generated by $\bigcup_j A_j$. We say that M is the internal direct sum of the A_j if the map

$$\textstyle\bigoplus_j A_j \to M, \quad (a_j) \mapsto \sum_j a_j$$

is an isomorphism. In that case we also write $M = \bigoplus_j A_j$. A submodule $A \subset M$ is called a direct summand of M if there exists a complement of A, i.e., a submodule B such that M is the internal direct sum of A and B.

We assume known the structure theory of finitely generated abelian groups. An element of finite order in an abelian group A is called a torsion element. The torsion elements form a subgroup, the torsion subgroup $T(A)$. The torsion subgroup of $A/T(A)$ is trivial. If the group is finitely generated, then the torsion subgroup has a complement F, and F is a free abelian group. The cardinality of a basis of F is called the rank of A. A finitely generated torsion group is the direct sum of cyclic groups of prime power order $\mathbb{Z}/(p^k)$, and the number of factors isomorphic to $\mathbb{Z}/(p^k)$ is uniquely determined by the group. A similar structure theory exists for finitely generated modules over principal ideal domains.

A linear map $p \colon M \to M$ with the property $p \circ p = p$ is called a projection operator on M.

(11.1.1) Splitting Lemma. *Let* $0 \to E \xrightarrow{f} F \xrightarrow{g} G \to 0$ *be a short exact sequence of modules. Then the following assertions are equivalent:*

(1) *The image of f is a direct summand of F.*

(2) *There exists a homomorphism $r \colon F \to E$ such that $rf = \mathrm{id}$.*

(3) *There exists a homomorphism $s \colon G \to F$ such that $gs = \mathrm{id}$.*

If (1)–(3) *holds, we say the sequence* ***splits***. *We then call r and s* ***splittings***, *r a retraction of f, s a section of g. In case* (2) *fr is a projection operator, hence we* *have* $F = \mathrm{Im}(fr) \oplus \mathrm{Ker}(fr) = \mathrm{Im}(f) \oplus \mathrm{Ker}(r)$. *In case* (3) sg *is a projection operator, hence we have* $F = \mathrm{Im}(sg) \oplus \mathrm{Ker}(sg) = \mathrm{Im}(s) \oplus \mathrm{Ker}(g)$. □

(11.1.2) Sum Lemma. *Suppose given a commutative diagram in* R-MOD:

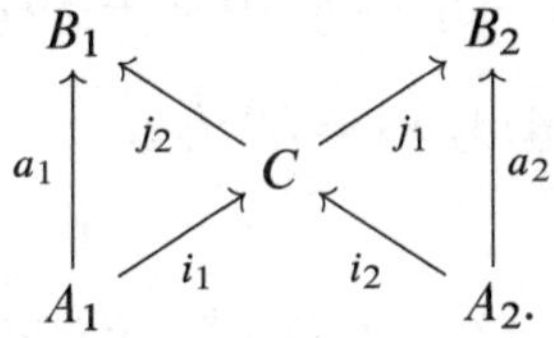

Assume $j_k i_k = 0$ *for* $k = 1, 2$.

(1) *If the* a_k *are isomorphisms and* (i_2, j_2) *is exact, then*

$$\langle i_1, i_2 \rangle \colon A_1 \oplus A_2 \to C \quad \textit{and} \quad (j_2, j_1) \colon C \to B_2 \oplus B_1$$

are isomorphisms and (i_1, j_1) *is exact.*

(2) *If* $\langle i_1, i_2 \rangle$ *is an isomorphism and* (i_2, j_2) *is short exact, then* a_1 *is an isomorphism* (j_1, a_2 *are not needed*). *If* (j_2, j_1) *is an isomorphism and* (i_1, j_1) *is short exact, then* a_1 *is an isomorphism* (i_2, a_2 *are not needed*).

Proof. (1) The hypothesis implies $(j_2, j_1) \circ \langle i_1, i_2 \rangle = a_1 \oplus a_2$. We show that $\langle i_1, i_2 \rangle$ is surjective. Given $c \in C$ we have $j_2(c - i_1 a_1^{-1} j_2(c)) = 0$, by commutativity. Hence there exists by exactness $x_2 \in A_2$ such that $c - i_1 a_1^{-1} j_2(c) = i_2(x_2)$, i.e., c is contained in the image of $\langle i_1, i_2 \rangle$.

Let $j_1(c) = 0$ and write $c = i_1 x_1 + i_2 x_2$. Then $0 = j_1(c) = j_1 i_1 x_1 + j_1 i_2 x_2 = j_1 i_2 x_2 = a_2(x_2)$, hence $x_2 = 0$ and $c \in \mathrm{Im}(i_1)$.

(2) Exercise. □

(11.1.3) Hexagon Lemma. *Given a commutative diagram of abelian groups.*

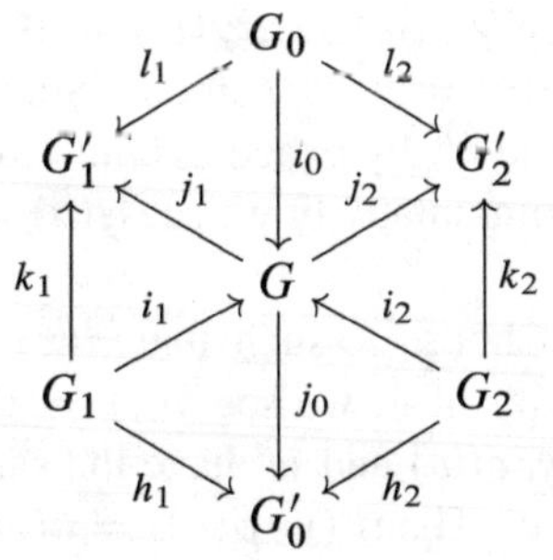

Suppose that k_1, k_2 *are isomorphisms,* (i_1, j_2) *exact,* (i_2, j_1) *exact, and* $j_0 i_0 = 0$. *Then* $h_1 k_1^{-1} l_1 = -h_2 k_2^{-1} l_2$.

Proof. The part i, j, k satisfies the hypothesis of the Sum Lemma (11.1.2). Given $x \in G_0$ there exist $x_j \in G_j$ such that $i_0 x = i_1 x_1 + i_2 x_2$. We compute

$$\begin{aligned} 0 = j_0 i_0 x &= j_0 i_1 x_1 + j_0 i_2 x_2 = h_1 x_1 + h_2 x_2, \\ l_1 x = j_1 i_0 x &= j_1 i_1 x + j_1 i_2 x = j_1 i_1 x = k_1 x_1, \end{aligned}$$

hence $x_1 = k_1^{-1} l_1 x$ and similarly $x_2 = k_2^{-1} l_2 x$. □

(11.1.4) Five Lemma. *Given a commutative diagram of groups and homomorphisms with exact rows:*

$$\begin{array}{ccccccccc} A & \xrightarrow{\alpha} & B & \xrightarrow{\beta} & C & \xrightarrow{\gamma} & D & \xrightarrow{\delta} & E \\ \downarrow{\scriptstyle a} & & \downarrow{\scriptstyle b} & & \downarrow{\scriptstyle c} & & \downarrow{\scriptstyle d} & & \downarrow{\scriptstyle e} \\ A' & \xrightarrow{\alpha'} & B' & \xrightarrow{\beta'} & C' & \xrightarrow{\gamma'} & D' & \xrightarrow{\delta'} & E'. \end{array}$$

(1) *a surjective, b, d injective $\Rightarrow$ c injective. (Here the E-part of the diagram is not needed.)*

(2) *b, d surjective, e injective $\Rightarrow$ c surjective. (Here the A-part of the diagram is not needed.)*

(3) *a surjective, b, d bijective, e injective $\Rightarrow$ c bijective.*

Proof. For another proof see (11.2.7). We give here a direct proof by the "method" called diagram chasing. One refers to diagram chasing whenever the proof (chasing elements through the diagram) does not really require a mathematical idea, only careful patience.

(1) Let $c(w) = 0$. Then $\gamma' c(w) = d\gamma(w) = 0$, and injectivity of d shows $\gamma(w) = 0$. By exactness, $\beta(v) = w$ for some v. Since $\beta' b(v) = c\beta(v) = 0$, we have $\alpha'(u') = b(v)$ for some u', by exactness, and $a(u) = u'$ by surjectivity of a. By injectivity of b and commutativity we see $\alpha(u) = v$ and hence by exactness $w = \beta(v) = 0$.

(2) Given $w' \in C'$. Choose x such that $d(x) = \gamma'(c')$. By exactness, commutativity, and injectivity of e, we see $\delta(x) = 0$ and hence $\gamma(w) = x$ for some w. By commutativity, $c(w)$ and w' have the same image under γ'. Hence $w' = c(w) \cdot \beta'(v')$ for some v'. Then $c(w \cdot \beta(v)) = c(w) \cdot c\beta(v) = c(w) \cdot \beta' b(v) = c(w) \cdot \beta'(v') = w'$, i.e., w' is contained in the image of c.

(3) A consequence of (1) and (2). □

Problems

1. Let p be a projection operator on M. Then $1-p$ is a projection operator. The equalities $\mathrm{Im}(1-p) = \mathrm{Ker}(p)$ and $\mathrm{Ker}(1-p) = \mathrm{Im}(p)$ hold. Moreover $M = \mathrm{Im}(p) \oplus \mathrm{Im}(1-p)$. The submodule A of M is a direct summand if and only if there exists a projection operator with image A.

2. Let $(A_j \mid j \in J)$ be a finite family of modules. Suppose given linear maps $i^k\colon A_k \to A$ and $p^l\colon A \to A_l$ such that $p^k i^k = \mathrm{id}$ and $p^k i^l = 0$ for $k \neq l$ (we write $p^k i^l = \delta^{kl}$ in this case). Then $(p^k) \circ \langle i^k \rangle = \mathrm{id}$ and $\langle i^k \rangle \circ (p^k) = \sum_j i^j p^j$ is a projection operator. Hence the following are equivalent: (1) $\langle i^k \rangle$ is an isomorphism. (2) (p^k) is an isomorphism. (3) $\sum_j i^j p^j = \mathrm{id}$.

3. Let p be a prime number. Determine the number of subgroups of $\mathbb{Z}/(p^k) \oplus \mathbb{Z}/(p^l)$.

4. Consider the group $\mathbb{Z}/(6) \oplus \mathbb{Z}$. Determine the subgroups of index $2, 3, 4, 5, 6$. Determine the number of complements of the torsion subgroup.

5. Let A be a finitely generated abelian group. Then $A \otimes_{\mathbb{Z}} \mathbb{Q}$ is a $\mathbb{Q}$-vector space. Show that its dimension is the rank of A.

6. Let M_j, $j \in J$ be submodules of M. The following assertions are equivalent:

(1) $\sum_j M_j$ is the direct sum of the M_j.

(2) For each $i \in J$, $M_i \cap \sum_{j, j \neq i} M_j = \{0\}$.

(3) Suppose $\sum_j x_j = 0$, $x_j \in M_j$, almost all $x_j = 0$, then $x_j = 0$ for each $j \in J$.

11.2 Exact Sequences

We start with a commutative diagram of modules.

$$\begin{array}{ccccc} A & \xrightarrow{a} & B & \xrightarrow{b} & C \\ \downarrow{\scriptstyle \alpha} & & \downarrow{\scriptstyle \beta} & & \downarrow{\scriptstyle \gamma} \\ A' & \xrightarrow{a'} & B' & \xrightarrow{b'} & C' \end{array}$$

It yields two derived diagrams (Ke = kernel, Ko = cokernel, Im = Image).

$$\begin{array}{ccccc} \mathrm{Ke}(\alpha) & \xrightarrow{a} & \mathrm{Ke}(\beta) & \xrightarrow{b} & \mathrm{Ke}(\gamma) \\ \downarrow{\scriptstyle (1)} & & \downarrow{\scriptstyle =} & & \uparrow{\scriptstyle (2)} \\ \mathrm{Ke}(a'\alpha) & \xrightarrow{a} & \mathrm{Ke}(\beta) & \xrightarrow{b} & \mathrm{Ke}(\gamma) \cap \mathrm{Im}(b) \end{array}$$

$$\begin{array}{ccccc} \mathrm{Ko}(\alpha) & \xrightarrow{a'} & \mathrm{Ko}(\beta) & \xrightarrow{b'} & \mathrm{Ko}(\gamma) \\ \downarrow{\scriptstyle (3)} & & \uparrow{\scriptstyle =} & & \uparrow{\scriptstyle (4)} \\ \dfrac{A'}{\mathrm{Im}(\alpha) + \mathrm{Ke}(a')} & \xrightarrow{a'} & \dfrac{B'}{\mathrm{Im}(\beta)} & \xrightarrow{b'} & \dfrac{C'}{\mathrm{Im}(\gamma b)} \end{array}$$

The morphisms named a, b, a', b' are induced by the original morphisms with the same name by applying them to representatives. (1) and (2) are inclusions, (3) and (4) are quotients.

(11.2.1) Proposition. *Let (a, b) and (a', b') be exact. Then a* ***connecting morphism***

$$\delta\colon \mathrm{Ke}(\gamma) \cap \mathrm{Im}(b) \to \frac{A'}{\mathrm{Im}(\alpha) + \mathrm{Ke}(a')}$$

is defined by the correspondence $(a')^{-1}\beta b^{-1}$.

Proof. For $z \in \mathrm{Ke}(\gamma) \cap \mathrm{Im}(b)$ there exists $y \in B$ such that $b(y) = z$; since $z \in \mathrm{Ke}(\gamma)$ and $\gamma b = b'\beta$ we have $\beta(y) \in \mathrm{Ke}(b')$; since $\mathrm{Ke}(b') \subset \mathrm{Im}(a')$, there exists $x' \in A'$ such that $a'(x') = \beta(y)$. We set $\delta(z) = x'$ and show that this assignment is well-defined. If $\tilde{y} \in B$, $b(\tilde{y}) = z$, then $b(y - \tilde{y}) = 0$; since $\mathrm{Ke}(b) \subset \mathrm{Im}(a)$, there exists $x \in A$ such that $a(x) = y - \tilde{y}$. We have $\beta(y) - \beta(\tilde{y}) = a'\alpha(x)$, because of $a'\alpha = \beta a$, and with $a'(\tilde{x}') = \beta(\tilde{y})$ we obtain $a'(x' - \tilde{x}' - \alpha(x)) = 0$, i.e., $x' \equiv \tilde{x}' \bmod \mathrm{Im}(\alpha) + \mathrm{Ke}(a')$. □

We add further hypotheses to the original diagram and list the consequences for the derived diagrams. We leave the verification of (11.2.2), (11.2.3), (11.2.4), (11.2.5) to the reader.

11.2.2 If a' is injective, then (1) and (3) are bijective. If b is surjective, then (2) and (4) are bijective. ◇

11.2.3 Let (a', b') be exact. Then

$$\frac{A'}{\mathrm{Im}(\alpha) + \mathrm{Ke}(a')} \xrightarrow{a'} \frac{B'}{\mathrm{Im}(\beta)} \xrightarrow{b'} \frac{C'}{\mathrm{Im}(\gamma b)}$$

is exact. If, moreover, b is surjective, then (4) is bijective and therefore

$$\mathrm{Ko}(\alpha) \xrightarrow{a'} \mathrm{Ko}(\beta) \xrightarrow{b'} \mathrm{Ko}(\gamma)$$

is exact. If b' is surjective, then the derived b' is surjective too. ◇

11.2.4 Let (a, b) be exact. Then

$$\mathrm{Ke}(a'\alpha) \xrightarrow{a} \mathrm{Ke}(\beta) \xrightarrow{b} \mathrm{Ke}(\gamma) \cap \mathrm{Im}(b)$$

is exact. If, moreover, a' is injective, then (1) is bijective and therefore

$$\mathrm{Ke}(\alpha) \xrightarrow{a} \mathrm{Ke}(\beta) \xrightarrow{b} \mathrm{Ke}(\gamma)$$

is exact. If a is injective, then the derived a is injective too. ◇

11.2.5 Let (a, b) and (a', b') be exact. Then, as we have seen, δ is defined. Under these assumptions the bottom lines of the derived diagrams together with δ yield an exact sequence. (See the special case (11.2.6).) ◇

(11.2.6) Kernel–Cokernel Lemma. *If in the original diagram a' is injective and b surjective, then* (1), (2), (3), *and* (4) *are bijective and the kernel-cokernel-sequence*

$$\operatorname{Ke}(\alpha) \xrightarrow{a} \operatorname{Ke}(\beta) \xrightarrow{b} \operatorname{Ke}(\gamma) \xrightarrow{\delta} \operatorname{Ko}(\alpha) \xrightarrow{a'} \operatorname{Ko}(\beta) \xrightarrow{b'} \operatorname{Ko}(\gamma)$$

is exact.

Proof. We show the exactness at places involving δ; the other cases have already been dealt with. The relations $\delta b = 0$ and $a'\delta = 0$ hold by construction.

If the class of x' is contained in the kernel of a', then there exists y such that $a'(x') = \beta(y)$. Hence $z = b(y) \in \operatorname{Ke}(\gamma)$ by commutativity, and $\delta(z) = x'$.

Suppose $z \in \operatorname{Ke}(\gamma)$ is contained in the kernel of δ. Then there exists y such that $z = b(y)$, $\beta(y) = a'(x')$ and $\beta(z) = \alpha(x) \in \operatorname{Im}(\alpha)$. Then $b(y - a(x)) = z$ and $\beta(y - a(x)) = \beta(y) - \beta a(x) = \beta(y) - a'\alpha(x) = \beta(y) - a'(x') = 0$. Hence $y - a(x)$ is a pre-image of z. □

We now relate the Kernel–Cokernel Lemma to the Five Lemma (11.2.7); see also (11.1.4). Given a commutative five-term diagram of modules and homomorphisms with exact rows.

$$\begin{array}{ccccccccc} A & \xrightarrow{\alpha} & B & \xrightarrow{\beta} & C & \xrightarrow{\gamma} & D & \xrightarrow{\delta} & E \\ \downarrow{\scriptstyle a} & & \downarrow{\scriptstyle b} & & \downarrow{\scriptstyle c} & & \downarrow{\scriptstyle d} & & \downarrow{\scriptstyle e} \\ A' & \xrightarrow{\alpha'} & B' & \xrightarrow{\beta'} & C' & \xrightarrow{\gamma'} & D' & \xrightarrow{\delta'} & E' \end{array}$$

We have three derived diagrams.

$$\begin{array}{ccccccc} 0 & \longrightarrow & \operatorname{Ke}(\delta) & \longrightarrow & D & \xrightarrow{\delta} & E \\ & & \downarrow{\scriptstyle \tilde{d}} & & \downarrow{\scriptstyle d} & & \downarrow{\scriptstyle e} \\ 0 & \longrightarrow & \operatorname{Ke}(\delta') & \longrightarrow & D' & \xrightarrow{\delta'} & E', \end{array} \qquad \begin{array}{ccccccc} A & \xrightarrow{\alpha} & B & \longrightarrow & \operatorname{Ko}(\alpha) & \longrightarrow & 0 \\ \downarrow{\scriptstyle a} & & \downarrow{\scriptstyle b} & & \downarrow{\scriptstyle \tilde{b}} & & \\ A' & \xrightarrow{\alpha'} & B' & \longrightarrow & \operatorname{Ko}(\alpha') & \longrightarrow & 0, \end{array}$$

$$\begin{array}{ccccccccc} 0 & \longrightarrow & \operatorname{Ko}(\alpha) & \xrightarrow{\beta} & C & \xrightarrow{\gamma} & \operatorname{Ke}(\delta) & \longrightarrow & 0 \\ & & \downarrow{\scriptstyle \tilde{b}} & & \downarrow{\scriptstyle c} & & \downarrow{\scriptstyle \tilde{d}} & & \\ 0 & \longrightarrow & \operatorname{Ko}(\alpha') & \xrightarrow{\beta'} & C' & \xrightarrow{\gamma'} & \operatorname{Ke}(\delta') & \longrightarrow & 0. \end{array}$$

The rows of the first two diagrams are exact for trivial reasons. The exactness of the rows of the original diagram implies that the third diagram has exact rows. From the considerations so far we obtain the exact sequences

$$\mathrm{Ke}(e) \cap \mathrm{Im}(\delta) \to \mathrm{Ko}(\tilde{d}) \to \mathrm{Ko}(d),$$

$$\mathrm{Ke}(b) \to \mathrm{Ke}(\tilde{b}) \to A'/(\mathrm{Im}(a) + \mathrm{Ke}(\alpha')),$$

$$0 \to \mathrm{Ke}(\tilde{b}) \to \mathrm{Ke}(c) \to \mathrm{Ke}(\tilde{d}) \to \mathrm{Ko}(\tilde{b}) \to \mathrm{Ko}(c) \to \mathrm{Ko}(\tilde{d}) \to 0.$$

This yields:

(11.2.7) Five Lemma. *Given a five-term diagram as above. Then the following holds:*

(1) $\mathrm{Ke}(\tilde{b}) = 0, \mathrm{Ke}(\tilde{d}) = 0 \Rightarrow \mathrm{Ke}(c) = 0.$

(2) $\mathrm{Ko}(\tilde{b}) = 0, \mathrm{Ko}(\tilde{d}) = 0 \Rightarrow \mathrm{Ko}(c) = 0.$

(3) $\mathrm{Ke}(b) = 0, A'/(\mathrm{Im}(a) + \mathrm{Ke}(\alpha')) = 0 \Rightarrow \mathrm{Ke}(\tilde{b}) = 0.$

(4) $\mathrm{Ke}(e) \cap \mathrm{Im}(\delta) = 0, \mathrm{Ko}(d) = 0 \Rightarrow \mathrm{Ko}(\tilde{d}) = 0.$

(5) *a surjective, b, d injective $\Rightarrow$ c injective. (Here the E-part of the diagram is not needed.)*

(6) *b, d surjective, e injective $\Rightarrow$ c surjective. (Here the A-part of the diagram is not needed.)*

(7) *a surjective, b, d bijective, e injective $\Rightarrow$ c bijective.* □

Problems

1. Given homomorphisms $f\colon A \to B$ and $g\colon B \to C$ between R-modules. Then there is a natural exact sequence

$$0 \to \mathrm{Ke}(f) \to \mathrm{Ke}(gf) \to \mathrm{Ke}(g) \to \mathrm{Ko}(f) \to \mathrm{Ko}(gf) \to \mathrm{Ko}(g) \to 0.$$

The connection to the previous considerations: The commutative diagram with exact rows

$$\begin{array}{ccccccccc} 0 & \longrightarrow & A & \xrightarrow{(1,f)} & A \oplus B & \xrightarrow{\langle -f,1 \rangle} & B & \longrightarrow & 0 \\ & & \downarrow{\scriptstyle f} & & \downarrow{\scriptstyle gf \oplus 1} & & \downarrow{\scriptstyle g} & & \\ 0 & \longrightarrow & B & \xrightarrow{(g,1)} & C \oplus B & \xrightarrow{\langle -1,g \rangle} & C & \longrightarrow & 0 \end{array}$$

can be viewed as an exact sequence of chain complexes. Its homology sequence (11.3.2) is the desired sequence, if we identify the kernel and cokernel of $gf \oplus 1$ with the corresponding modules for gf.

Describe the morphisms of the sequence and give also a direct proof.

11.3 Chain Complexes

The algebraic terminology of chain complexes arose from the definition of homology groups. Since then it also has become of independent interest in algebra (homological algebra). The construction of (singular) homology proceeds in two stages: First one associates to a space a so-called chain complex. Then the chain complex yields, by algebra, the homology groups. The category of chain complexes and chain maps has an associated homotopy structure.

We work in this section with the category R-MOD of left modules over some fixed ring R. A family $A_* = (A_n \mid n \in \mathbb{Z})$ of modules A_n is called a ***$\mathbb{Z}$-graded module***. We call A_n the component of degree or dimension n. One sometimes considers the direct sum $\bigoplus_{n\in\mathbb{Z}} A_n$; then the elements in A_n are said to be homogeneous of degree n. Typical examples are polynomial rings; if $k[x, y]$ is the polynomial ring in two indeterminates x, y of degree 1 say, then the homogeneous polynomials of degree n are spanned by $x^i y^{n-i}$ for $0 \le i \le n$, and in this manner we consider $k[x, y]$ as a graded k-module (actually a graded algebra, as defined later). One can also consider formal power series; this would correspond to taking the product $\prod_{n\in\mathbb{Z}} A_n$ instead of the sum.

Let A_* and B_* be $\mathbb{Z}$-graded modules. A family $f_n : A_n \to B_{n+k}$ of homomorphisms is called a ***morphism of degree k*** between the graded modules.

A sequence $C_\bullet = (C_n, \partial_n \mid n \in \mathbb{Z})$ of modules C_n and homomorphisms $\partial_n : C_n \to C_{n-1}$, called ***boundary operators*** or ***differentials***, is said to be a ***chain complex***, if for each $n \in \mathbb{Z}$ the boundary relation $\partial_{n-1} \circ \partial_n = 0$ holds. We associate to a chain complex $C_\bullet$ the modules

$$\begin{aligned} Z_n &= Z_n(C_\bullet) = \operatorname{Ker}(\partial_n : C_n \to C_{n-1}), \\ B_n &= B_n(C_\bullet) = \operatorname{Im}(\partial_{n+1} : C_{n+1} \to C_n), \\ H_n &= H_n(C_\bullet) = Z_n / B_n. \end{aligned}$$

We call C_n (Z_n, B_n) the module of ***n-chains*** (***n-cycles***, ***n-boundaries***) and H_n the n-th ***homology module*** of the chain complex. (The boundary relation $\partial\partial = 0$ implies $B_n \subset Z_n$, and therefore H_n is defined.) Two n-chains whose difference is a boundary are said to be ***homologous***. Often, in particular in the case $R = \mathbb{Z}$, we talk about ***homology groups***.

Let $C_\bullet = (C_n, c_n)$ and $D_\bullet = (D_n, d_n)$ be chain complexes. A ***chain map*** $f_\bullet : C_\bullet \to D_\bullet$ is a sequence of homomorphisms $f_n : C_n \to D_n$ which satisfy the commutation rules $d_n \circ f_n = f_{n-1} \circ c_n$. A chain map induces (by restriction and passage to the factor groups) homomorphisms of the cycles, boundaries, and homology groups

$$\begin{aligned} Z_n(f_\bullet) &: Z_n(C_\bullet) \to Z_n(D_\bullet), \\ B_n(f_\bullet) &: B_n(C_\bullet) \to B_n(D_\bullet), \\ f_* = H_n(f_\bullet) &: H_n(C_\bullet) \to H_n(D_\bullet). \end{aligned}$$

A (short) ***exact sequence of chain complexes***

$$0 \to C' \xrightarrow{f} C \xrightarrow{g} C'' \to 0$$

consists of chain maps f and g such that $0 \to C'_n \xrightarrow{f_n} C_n \xrightarrow{g_n} C''_n \to 0$ is exact for each n.

We certainly have the induced morphisms $H_n(f)$ and $H_n(g)$. Moreover, there exists a ***connecting morphism*** $\partial_n \colon H_n(C'') \to H_{n-1}(C')$, also called ***boundary operator***, which is induced by the correspondence $f_{n-1}^{-1} \circ d_n \circ g_n^{-1}$.

$$\begin{array}{ccc} & C_n & \xrightarrow{g_n} C''_n \ni z'' \\ & \downarrow d_n & \\ z' \in C'_{n-1} \xrightarrow{f_{n-1}} & C_{n-1} & \end{array}$$

(11.3.1) Lemma. *For a cycle $z'' \in C''_n$ with pre-image z under g_n the relation $g_{n-1} d_n z = d''_n g_n z = d''_n z'' = 0$ and exactness shows that there exists z' with $d_n(z) = f_{n-1}(z')$. The assignment $z'' \mapsto z'$ induces a well-defined homomorphism $\partial_n \colon H_n(C'') \to H_{n-1}(C')$.*

Proof. The relation $f_{n-2} d'_{n-1} z' = d_{n-1} f_{n-1} z' = d_{n-1} d_n z = 0$ and the injectivity of f_{n-2} show that z' is a cycle. If we choose another pre-image $z + f_n w'$ of z'', then we have to replace z' by $z' + d'_n w'$, so that the homology class of z' is well-defined. Finally, if we change z'' by a boundary, we can replace z by the addition of a boundary and hence $d_n z$ does not change. □

(11.3.2) Proposition. *The sequence*

$$\cdots \to H_n(C') \xrightarrow{f_*} H_n(C) \xrightarrow{g_*} H_n(C'') \xrightarrow{\partial_n} H_{n-1}(C') \to \cdots$$

is exact.

Proof. The boundary operator $d_n \colon C_n \to C_{n-1}$ induces a homomorphism

$$d_n \colon K_n = C_n / B_n \to Z_{n-1},$$

and its kernel and cokernel are H_n and H_{n-1}. By (11.2.3) and (11.2.4) the rows of the next diagram are exact.

$$\begin{array}{ccccccc} & & K'_n & \xrightarrow{f_n} & K_n & \xrightarrow{g_n} & K''_n \longrightarrow 0 \\ & & \downarrow d'_n & & \downarrow d_n & & \downarrow d''_n \\ 0 & \longrightarrow & Z'_{n-1} & \xrightarrow{f_{n-1}} & Z_{n-1} & \xrightarrow{g_{n-1}} & Z''_{n-1} \end{array}$$

The associated sequence (11.2.6) $H'_n \to H_n \to H''_n \xrightarrow{\partial} H'_{n-1} \to H_{n-1} \to H''_{n-1}$ is the exact homology sequence. □

Let $f, g\colon C = (C_n, c_n) \to D = (D_n, d_n)$ be chain maps. A ***chain homotopy*** s from f to g is a sequence $s_n\colon C_n \to D_{n+1}$ of homomorphisms which satisfy

$$d_{n+1} \circ s_n + s_{n-1} \circ c_n = g_n - f_n.$$

(This definition has two explanations; firstly, one can define "chain homotopy" in analogy to the topological definition by using the chain complex analogue of the unit interval; secondly, it codifies the boundary relation of a geometric homotopy.) We call f and g ***homotopic*** or ***chain homotopic***, if there exists a chain homotopy s from f to g, in symbols $s\colon f \simeq g$. "Chain homotopic" is an equivalence relation on the set of chain maps $C \to D$; the data $s\colon f \simeq g$ and $t\colon g \simeq h$ imply $(s_n + t_n)\colon f \simeq h$. This relation is also compatible with composition; if $s\colon f \simeq f'\colon C \to D$ and $t\colon g \simeq g'\colon D \to E$, then $(g_{n+1}s_n)\colon gf \simeq gf'$ and $(t_n f_n)\colon gf \simeq g'f$. We call $f\colon C \to D$ a ***chain equivalence***, if there exists a chain map $g\colon D \to C$ and chain homotopies $fg \simeq \mathrm{id}$ and $gf \simeq \mathrm{id}$.

(11.3.3) Proposition. *Chain homotopic maps induce the same morphisms between the homology groups.*

Proof. Let $x \in C_n$ be a cycle. The homotopy relation $g_n(x) - f_n(x) = d_{n+1}s_n(x)$ shows that $f_n(x)$ and $g_n(x)$ are homologous. □

11.4 Cochain complexes

Let $C_\bullet = (C_n, \partial_n)$ be a chain complex of R-modules. Let G be another R-module. We apply the functor $\mathrm{Hom}_R(-, R)$ to $C_\bullet$ and obtain a chain complex $C^\bullet = (C^n, \delta^n)$ of R-modules with $C^n = \mathrm{Hom}_R(C_n, R)$ and the R-linear map

$$\delta^n\colon C^n = \mathrm{Hom}_R(C_n, R) \to \mathrm{Hom}_R(C_{n+1}, R) = C^{n+1}$$

defined by $\delta^n(\varphi) = (-1)^{n+1}\varphi \circ \partial_{n+1}$ for $\varphi \in \mathrm{Hom}(C_n, R)$.

For the choice of this sign see 11.7.4. The reader will find different choices of signs in the literature. Other choices will not effect the cohomology functors. But there seems to be an agreement that our choice is the best one when it comes to products.

Now some "co" terminology. A ***cochain complex*** $C^\bullet = (C^n, \delta^n)$ is a $\mathbb{Z}$-graded module $(C^n \mid n \in \mathbb{Z})$ together with homomorphisms $\delta^n\colon C^n \to C^{n+1}$, called ***coboundary operators*** or ***differentials***[1], such that $\delta^{n+1}\delta^n = 0$. We set

$$Z^n = \mathrm{Ker}\,\delta^n, \quad B^n = \mathrm{Im}\,\delta^{n-1}, \quad H^n = Z^n/B^n$$

and call C^n, Z^n, B^n the module of ***n-cochains***, ***n-cocycles***, ***n-coboundaries***, and H^n the n-th ***cohomology module*** of the cochain complex.

[1]An important cochain complex arises from the exterior differentiation of differential forms. So one should not use a "co" word here.

11.5 Natural Chain Maps and Homotopies

Let $\mathcal{C}$ be an arbitrary category and CH_+ the category of chain complexes (C_n, c_n) of abelian groups with $C_n = 0$ for $n < 0$ and chain maps. A functor $F_*\colon \mathcal{C} \to \mathrm{CH}_+$ consists of a family of functors $F_n\colon \mathcal{C} \to \mathbb{Z}\text{-MOD}$ and natural transformations $d_n^F\colon F_n \to F_{n-1}$ such that $d_{n-1}^F \circ d_n^F = 0$. A natural transformation $\varphi_*\colon F_* \to G_*$ between such functors is a family of natural transformations $\varphi_n\colon F_n \to G_n$ such that $d_n^G \varphi_n = \varphi_{n-1} d_n^F$. A natural chain homotopy $s_*\colon \varphi_* \simeq \psi_*$ from φ_* to ψ_* is a family $s_n\colon F_n \to G_{n+1}$ of natural transformations such that

$$d_{n+1}^G \circ s_n + s_{n-1} \circ d_n^F = \psi_n - \varphi_n.$$

A functor $F_n\colon \mathcal{C} \to \mathbb{Z}\text{-MOD}$ is called free if there exists a family $((B_{n,j}, b_{n,j}) \mid j \in J(n))$ of objects $B_{n,j}$ of $\mathcal{C}$ (called models) and elements $b_{n,j} \in F_n(B_{n,j})$ such that

$$F_n(f)(b_{n,j}), \quad j \in J(n), \ f \in \mathrm{Hom}_{\mathcal{C}}(B_{n,j}, X)$$

is for each object X of $\mathcal{C}$ a $\mathbb{Z}$-basis of $F_n(X)$. A natural transformation $\varphi_n\colon F_n \to G_n$ from a free functor F_n into another functor G_n is then determined by the values $\varphi_n(b_{n,j})$ and the family of these values can be fixed arbitrarily in order to obtain a natural transformation. We call F_* free if each F_n is free. We call G_* acyclic (with respect to the families of models for F_*) if the homology groups $H_n(G_*(B_{n,j})) = 0$ for $n > 0$ and each model $B_{n,j}$.

(11.5.1) Theorem. *Let F_* be a free and G_* be an acyclic functor. For each natural transformation $\overline{\varphi}\colon H_0 \circ F_0 \to H_0 \circ G_0$ there exists a natural transformation $\varphi_*\colon F_* \to G_*$ which induces $\overline{\varphi}$. Any two natural transformations φ and ψ with this property are naturally chain homotopic* ([57]).

Proof. We specify a natural transformation φ_0 by the condition that $\varphi(b_{0,j})$ represents the homology class $\overline{\varphi}[b_{0,j}]$. Let now $\varphi_i\colon F_i \to G_i$ be natural transformations $(0 \le i < n)$ such that $d_i^G \varphi_i = \varphi_{i-1} d_i^F$ for $0 < i < n$. Consider the elements $\varphi_{n-1} d_n^F b_{n,j} \in G_{n-1}(B_{n,j})$. For $n = 1$ this element represents 0 in H_0, by the construction of φ_0. For $n > 1$ we see from the induction hypothesis that $d_{n-1}^G \varphi_{n-1} d_n^F b_{n,j} = \varphi_{n-2} d_{n-1}^F d_n^F b_{n,j} = 0$. Since G_* is acyclic we find $g_{n,j} \in G_n(B_{n,j})$ such that $d_n^G g_{n,j} = \varphi_{n-1} d_n^F b_{n,j}$. We specify a natural transformation φ by the conditions $\varphi(b_{n,j}) = g_{n,j}$. This transformation then satisfies $d_n^G \varphi_n = \varphi_{n-1} d_n^F$. This finishes the induction step.

Let now φ_* and ψ_* be given. Then $\psi_0(b_{0,j}) - \varphi_0(b_{0,j}) = d_1^G c_{0,j}$ for some $c_{0,j}$, since $\psi_0(b_{0,j})$ and $\varphi_0(b_{0,j})$ represent the same homology class. We define the transformation $s_0\colon F_0 \to G_1$ by the condition $s_0(b_{0,j}) = c_{0,j}$. Suppose now that $s_n\colon F_n \to G_{n+1}$ are given such that $d_{i+1}^G s_i + s_{i-1} d_i^F = \psi_i - \varphi_i$ for $0 \le i < n$

(and $s_{-1} = 0$). We compute with the induction hypothesis

$$\begin{aligned} &d_n^G(\psi_n - \varphi_n - s_{n-1}d_n^F) \\ &\quad = \psi_{n-1}d_n^F - \varphi_{n-1}d_n^F - (\psi_{n-1}d_n^F - \varphi_{n-1}d_n^F - s_{n-2}d_{n-1}^F)d_n^F = 0. \end{aligned}$$

Thus, by acyclicity, we can choose $c_{n,j} \in G_{n+1}(B_{n,j})$ such that

$$d_{n+1}^G c_{n,j} = (\psi_n - \varphi_n - s_{n-1}d_n^F)(b_{n,j}).$$

We now specify a natural transformation $s_n \colon F_n \to G_{n+1}$ by $s_n(b_{n,j}) = c_{n,j}$. It then has the required property $d_{n+1}^G s_n = \psi_n - \varphi_n - s_{n-1}d_n^F$. □

Problems

1. Let $F_0 \leftarrow F_1 \leftarrow \cdots$ be a chain complex of free R-modules F_i and $D_0 \leftarrow D_1 \leftarrow \cdots$ an exact sequence of R-modules. A chain map $(\varphi \colon F_i \to D_i \mid i \in \mathbb{N}_0)$ induces a homomorphism $H_0(\varphi_*) \colon H_0(F_*) \to H_0(D_*)$. Given a homomorphism $\alpha \colon H_0(F_*) \to H_0(D_0)$ there exists up to chain homotopy a unique chain map (φ_i) such that $H_0(\varphi_*) = \alpha$. This can be obtained as a special case of (11.5.1).

The reader should now study the notion of a projective module (one definition is: direct summand of a free module) and then show that a similar result holds if the F_i are only assumed to be projective.

An exact sequence of the form $0 \leftarrow A \leftarrow P_0 \leftarrow P_1 \leftarrow \cdots$ with projective modules P_i is called a ***projective resolution*** of the module A. The result stated at the beginning says that projective resolutions are unique up to chain equivalence. (***Fundamental Lemma*** of homological algebra). Each module has a free resolution.

11.6 Chain Equivalences

A chain map which induces an isomorphism of homology groups is under certain circumstances a chain equivalence. This is one of the results of this section.

The notion of a chain homotopy can be used to develop a homotopy theory of chain complexes in analogy to the topological homotopy theory.

We have the null complex; the chain groups are zero in each dimension. A chain complex is called ***contractible*** if it is chain equivalent to the null complex, or equivalently, if the identity is chain homotopic to the null map. A chain complex is said to be ***acyclic*** if its homology groups are zero.

Let $f \colon (K, d^K) \to (L, d^L)$ be a chain map. We construct a new chain complex Cf, the ***mapping cone*** of f, by

$$(Cf)_n = L_n \oplus K_{n-1}, \quad d^{Cf}(y, x) = (d^L y + fx, -d^K x).$$

This can also be written in matrix form

$$\begin{pmatrix} y \\ x \end{pmatrix} \mapsto \begin{pmatrix} d^L & f \\ 0 & -d^K \end{pmatrix} \begin{pmatrix} y \\ x \end{pmatrix}.$$

The ***suspension*** ΣK of K is defined by $(\Sigma K)_n = K_{n-1}$ and $d^{\Sigma K} = -d^K$. The canonical injection and projection yield an exact sequence of chain complexes $0 \to L \to Cf \to \Sigma K \to 0$. Associated is an exact sequence (11.3.2), and the boundary morphism $\partial\colon H_{n+1}(\Sigma K) \to H_n(L)$ equals $H_n(f)$, if we use the canonical identifications $H_{n+1}(\Sigma K) \cong H_n(K)$. The next result shows a typical difference between the topological and the algebraic homotopy theory.

(11.6.1) Theorem. *Let Cf be contractible. Then f is a chain equivalence.*

Proof. The inclusion $\iota\colon L \to Cf$, $y \mapsto (y, 0)$ is null homotopic, since Cf is contractible. Let $s\colon \iota \simeq 0$ be a null homotopy. We write $s(y) = (\gamma(y), g(y)) \in L \oplus K$ (without notation for the dimensions). The condition $\partial s + s\partial = \iota$ then reads

$$(\partial\gamma y + fgy + \gamma\partial y, -\partial gy + g\partial y) = (y, 0),$$

i.e., $\partial g = g\partial$ and $\partial\gamma + \gamma\partial = \mathrm{id} - fg$. Hence g is a chain map, and because of the γ-relation, a right homotopy inverse of f.

The projection $\kappa\colon Cf \to \Sigma K$ is likewise null homotopic. Let $t\colon \kappa \simeq 0$ be a null homotopy. We write $t(y, x) = h(y) + \eta(x)$. The equality $\partial t + t\partial = \kappa$ then means

$$-\partial hy + h\partial y - \partial\eta x + \eta\partial x + hfx = x,$$

hence $\partial h = h\partial$ and $\partial\eta + \eta\partial = hf - \mathrm{id}$. Therefore h is a chain map and a left homotopy inverse of f. □

(11.6.2) Proposition. *Let K be acyclic and suppose that $Z_n \subset K_n$ is always a direct summand. Then K is contractible.*

Proof. We have the exact sequence $0 \to Z_n \to K_n \xrightarrow{\partial} B_{n-1} \to 0$, and since K is acyclic we conclude $Z_n = B_n$. Moreover there exists $t_{n-1}\colon B_{n-1} \to K_n$ with $\partial t_{n-1} = \mathrm{id}$, since Z_n is a direct summand of K_n. We therefore have a direct decomposition $K_n = B_n \oplus t_{n-1}B_{n-1}$. We define $s\colon K_n \to K_{n+1}$ by $s|B_n = t_n$ and $s_n|t_{n-1}B_{n-1} = 0$. With these definitions one verifies separately on B_n as well as on $t_{n-1}B_{n-1}$ that $\partial s + s\partial$ is the identity, i.e., s is a null homotopy of the identity. □

(11.6.3) Theorem. *Let $f\colon K \to L$ be a chain map between chain complexes which consist of free modules over a principal ideal domain R. If f induces isomorphisms $f_*\colon H_*(K) \cong H_*(L)$, then f is a chain equivalence.*

Proof. The exact homology sequence and the hypothesis imply that Cf is acyclic. A submodule of a free R-module is free. Hence the boundary groups of the complex Cf are free, and therefore the exact sequence $0 \to Z_n \to Cf_n \to B_{n-1} \to 0$ splits. Now we apply (11.6.1) and (11.6.2), in order to see that f is a chain equivalence. □

In the topological applications we often have to work with large chain complexes. In some situations it is useful to replace them by smaller chain equivalent complexes. A graded R-module $A = (A_n)$ is said to be of finite type if the modules A_n are finitely generated R-modules.

(11.6.4) Proposition. *Let R be a principal ideal domain. Let $C = (C_n)$ be a chain complex of free R-modules such that its homology groups are finitely generated. Then there exists a free chain complex D of finite type which is chain equivalent to C.*

Proof. Let F_n be a finitely generated submodule of $Z_n(C)$ which is mapped onto $H_n(C)$ under the quotient map $Z_n(C) \to H_n(C)$, and denote by G_n the kernel of the epimorphism $F_n \to H_n(C)$. Define a chain complex $D = (D_n, d_n)$ by $D_n = F_n \oplus G_{n-1}$ and $d_n(x, y) = (y, 0)$. Then D is a free chain complex of finite type and $H_n(D) = F_n/G_n \cong H_n(C)$. Since G_n is a free submodule of $B_n(C)$ we can choose for each n a homomorphism $\varphi_n : G_n \to C_{n+1}$ such that $c_{n+1}\varphi_n(y) = y$ for each $y \in G_n$. Define $\psi_n : D_n = F_n \oplus G_{n-1} \to C_n$, $(x, y) \mapsto x + \varphi_{n-1}(y)$. One verifies that $\psi = (\psi_n)$ is a chain map which induces an isomorphism of homology groups. By (11.6.3), ψ is a chain equivalence. □

11.7 Linear Algebra of Chain Complexes

We work in the category R-MOD for a commutative ring R.

11.7.1 Graded modules. Let $A_\bullet = (A_n)$ and $B_\bullet = (B_n)$ be $\mathbb{Z}$-graded left R-modules over a commutative ring R. The tensor product $A_\bullet \otimes B_\bullet$ is the module with $\bigoplus_{p+q=n} A_p \otimes_R B_q$ as entry in degree n. If $f : A_\bullet \to A'_\bullet$ and $g : B_\bullet \to B'_\bullet$ are morphisms of some degree, then their tensor product $f \otimes g$ is defined by

$$(f \otimes g)(a \otimes b) = (-1)^{|g||a|} f(a) \otimes g(b).$$

Here $|a|$ denotes the degree of a. The formula for the tensor product obeys the (heuristic) "graded sign rule": Whenever entities of degree x and y are interchanged, then the sign $(-1)^{xy}$ appears. The tensor product of objects and of morphisms is associative and compatible with composition (in the graded sense)

$$(f \otimes g) \circ (f' \otimes g') = (-1)^{|g||f'|} ff' \otimes gg'$$

(sign rule). This composition is associative, as it should be. When we use the degree as upper index (e.g., in cohomology), then the agreement $A^k = A_{-k}$ is sometimes suitable. ◇

11.7.2 Graded algebras. A $\mathbb{Z}$-graded R-algebra $A^\bullet$ is a $\mathbb{Z}$-graded R-module $(A^n \mid n \in \mathbb{Z})$ together with a family of R-linear maps

$$A^i \otimes_R A^j \to A^{i+j}, \quad x \otimes y \mapsto x \cdot y.$$

The algebra is associative, if always $x \cdot (y \cdot z) = (x \cdot y) \cdot z$ holds, and commutative, if always $x \cdot y = (-1)^{|x||y|} y \cdot x$ holds (sign rule). A unit element $1 \in A^0$ of the algebra satisfies $1 \cdot x = x = x \cdot 1$. Let $M^\bullet = (M^n)$ be a $\mathbb{Z}$-graded R-module. A family

$$A^i \otimes M^j \to M^{i+j}, \quad a \otimes x \mapsto a \cdot x$$

of R-linear maps is the structure of an $A^\bullet$-module on $M^\bullet$, provided the associativity $a \cdot (b \cdot x) = (a \cdot b) \cdot x$ holds for $a, b \in A$ and $x \in M$. If A has a unit element, then the module is unital, provided $1 \cdot x = x$ always holds. Let $A^\bullet$ and $B^\bullet$ be $\mathbb{Z}$-graded algebras. Their tensor product $A \otimes B$ is the tensor product of the underlying graded modules $(A \otimes B)^n = \bigoplus_{i+j=n} A^i \otimes B^j$ together with the multiplication $(a \otimes b) \cdot (a' \otimes b') = (-1)^{|b||a'|} aa' \otimes bb'$ (sign rule). If A and B are associative, then $A \otimes B$ is associative. If both have a unit element 1, then $1 \otimes 1$ is a unit element for the tensor product. If both algebras are commutative, then their tensor product is commutative. The tensor product of graded algebras is an associative functor. ◇

11.7.3 Tensor product of chain complexes. Let $(A_\bullet, d_A)$ and $(B_\bullet, d_B)$ be chain complexes. Then the graded module $A_\bullet \otimes B_\bullet$ is a chain complex with boundary operator $d = d_A \otimes 1 + 1 \otimes d_B$. Here we have to take the sign rule into account, i.e.,

$$d(a \otimes b) = d_A a \otimes b + (-1)^{|a|} a \otimes d_B b.$$

One verifies $dd = 0$, using this sign rule. Passage to homology induces

$$H_p(A_\bullet) \otimes H_q(B_\bullet) \to H_{p+q}(A_\bullet \otimes B_\bullet), \quad [a] \otimes [b] \mapsto [a \otimes b].$$

The tensor product of chain complexes is associative. ◇

11.7.4 Dual chain complex. We regard the ground ring R as a trivial chain complex with R in degree 0 and zero modules otherwise. Let (A_n, ∂) be a chain complex. We define the dual graded R-module by $A^*_{-n} = \operatorname{Hom}_R(A_n, R)$. We require a boundary operator $\delta\colon A^*_{-n} \to A^*_{-n-1}$ on the dual module such that the evaluation $\varepsilon\colon A^*_\bullet \otimes A_\bullet \to R$

$$\varepsilon\colon A^*_{-n} \otimes A_n \to R, \quad \varphi \otimes a \mapsto \varphi(a),$$

$\varepsilon = 0$ otherwise, becomes a chain map. This condition, $\varepsilon(\varphi \otimes a) = 0$ and 11.7.3 yield for $\varphi \otimes a \in A^*_{-n-1} \otimes A_n$

$$\begin{aligned} 0 = d\varepsilon(\varphi \otimes a) &= \varepsilon(d(\varphi \otimes a) \\ &= \varepsilon(\delta\varphi \otimes a + (-1)^{|\varphi|} \varphi \otimes \partial a \\ &= (\delta\varphi)(a) + (-1)^{|\varphi|} \varphi(\partial a), \end{aligned}$$

i.e., we have to define $\delta\varphi = (-1)^{|\varphi|+1} \varphi \circ \partial$. ◇

11.7.5 Hom-Complex. For graded modules $A_\bullet$ and $B_\bullet$ we let $\mathrm{Hom}(A_\bullet, B_\bullet)$ be the module with $\prod_{a\in\mathbb{Z}} \mathrm{Hom}(A_a, B_{a+n})$ as component in degree n. On this Hom-module we use the boundary operator

$$d(f_i) = (\partial \circ f_i) - ((-1)^n f_i \circ \partial)$$

for $(f_i\colon A_i \to B_{i+n})$, i.e., the a-component $\mathrm{pr}_a(df) \in \mathrm{Hom}(A_a, B_{a+n-1})$ for $f = (f_a) \in \mathrm{Hom}(A_\bullet, B_\bullet)_n$ is defined to be

$$\mathrm{pr}_a(df) = \partial \circ f_a - (-1)^n g_{a-1} \circ \partial.$$

One verifies $dd = 0$. This definition generalizes our convention about the dual module. ◇

11.7.6 Canonical maps. The following canonical maps from linear algebra are chain maps.

(1) The composition

$$\mathrm{Hom}(B, C) \otimes \mathrm{Hom}(A, B) \to \mathrm{Hom}(A, B), \quad (f_i) \otimes (g_j) \mapsto (f_{l+|g|} \circ g_l).$$

(2) The adjunction

$$\Phi\colon \mathrm{Hom}(A\otimes B, C) \to \mathrm{Hom}(A, \mathrm{Hom}(B, C)), \quad \Phi(f_i)(x)(y) = f_{|x|+|y|}(x\otimes y).$$

(3) The tautological map

$$\gamma\colon \mathrm{Hom}(C, C') \otimes \mathrm{Hom}(D, D') \to \mathrm{Hom}(C \otimes D, C' \otimes D')$$

with $\gamma(f \otimes g)(x \otimes y) = (-1)^{|g||x|} f(x) \otimes g(y)$ (sign rule).

(4) The trace map $\lambda\colon A^* \otimes B \to \mathrm{Hom}(A, B)$, $\lambda(\varphi \otimes b)(a) = (-1)^{|a||b|}\varphi(a)b$. ◇

Problems

1. Tensor product is compatible with chain homotopy. Let $s\colon f \simeq g\colon C \to C'$ be a chain homotopy. Then $s \otimes \mathrm{id}\colon f \otimes \mathrm{id} \simeq g \otimes \mathrm{id}\colon C \otimes D \to C' \otimes D$ is a chain homotopy.

2. A chain complex model of the unit interval is the chain complex $I_\bullet$ with two non-zero groups $I_1 \cong R$ with basis e, $I_0 \cong R \oplus R$ with basis e_0, e_1 and boundary operator $d(e) = e_1 - e_0$ (in the topological context: the cellular chain complex of the unit interval). We use this model to define chain homotopies with the cylinder $I_\bullet \otimes C$. Note

$$C_n \oplus C_n \oplus C_{n-1} \cong (I_\bullet \otimes C)_n, \quad (x_1, x_0, y) \mapsto e_1 \otimes x_1 + e_0 \otimes x_0 + e \otimes y.$$

A chain map $h\colon I_\bullet \otimes C \to D$ consists, via these isomorphisms, of homomorphisms $h^t_n\colon C_n \to D_n$ and $s_n\colon C_n \to D_{n+1}$. The $h^t_\bullet$ are chain maps ($t = 0, 1$) and $ds_n(y) = h^1_n(y) - h^0_n(y) - s_{n-1}cy$, i.e., $s_\bullet\colon h^1_\bullet \simeq h^0_\bullet$ is a chain homotopy in our previous definition.

3. Imitate the topological definition of the mapping cone and define the mapping cone of a chain map $f\colon C \to D$ as a quotient of $I_\bullet \otimes C \oplus D$. The n-th chain group is then canonically isomorphic to $C_{n-1} \oplus D_n$ and the resulting boundary operator is the one we defined in the section on chain equivalences. Consider also the mapping cylinder from this view-point.

11.8 The Functors Tor and Ext

Let R be a principal ideal ring. We work in the category R-MOD; this comprises the category of abelian groups ($\mathbb{Z}$-modules). An exact sequence $0 \to F_1 \to F_0 \to A \to 0$ with free modules F_1, F_0 is a ***free resolution*** of A. Since submodules of free modules are free, it suffices to require that F_0 is free. Let $F(A)$ denote the free R-module generated by the set A. Denote the basis element of $F(A)$ which belongs to $a \in A$ by $[a]$. We have a surjective homomorphism $p \colon F(A) \to A$, $\sum n_a[a] \mapsto \sum n_a a$. Let $K(A)$ denote its kernel. The exact sequence

$$0 \to K(A) \xrightarrow{i} F(A) \xrightarrow{p} A \to 0$$

will be called the ***standard resolution*** of A. We take the tensor product (over R) of this sequence with a module G, denote the kernel of $i \otimes 1$ by $\operatorname{Tor}^R(A, G) = \operatorname{Tor}(A, G)$ and call it the ***torsion product*** of A, G.

We now derive some elementary properties of torsion products. We show that $\operatorname{Tor}(A, G)$ can be determined from any free resolution, and we make $\operatorname{Tor}(-,-)$ into a functor in two variables. In the next lemma we compare free resolutions.

(11.8.1) Lemma. *Given a homomorphism $f \colon A \to A'$ and free resolution $\mathcal{F}$ and $\mathcal{F}'$ of A and A', there exists a commutative diagram*

$$\begin{array}{ccccccccccc}
0 & \longrightarrow & F_1 & \xrightarrow{i} & F_0 & \xrightarrow{p} & A & \longrightarrow & 0 & & \mathcal{F} \\
 & & \downarrow f_1 & \swarrow s & \downarrow f_0 & & \downarrow f & & & & \\
0 & \longrightarrow & F_1' & \xrightarrow[i']{} & F_0' & \xrightarrow[p']{} & A' & \longrightarrow & 0 & & \mathcal{F}'
\end{array}$$

(without s). If $(\bar{f}_1, \bar{f}_0)$ is another choice of homomorphisms making the diagram commutative, then there exists a homomorphism $s \colon F_0 \to F_1'$ with $f_0 - f_0' = i's$ and $f_1 - f_1' = si$.

Proof. Let (x_k) be a basis of F_0. Choose $x_k' \in F_0'$ such that $p'(x_k') = fp(x_k)$. Define f_0 by $f_0(x_k) = x_k'$. Then $fp = p'f_0$. Since $p'f_0 i = 0$, there exists by exactness of $\mathcal{F}'$ a unique f_1 such that $f_0 i = i'f_1$. Since $p'(f_0 - f_0') = fp - fp = 0$, the elements $(f_0 - f_0')(x_k)$ are contained in the kernel of p'. Hence we have $(f_0 - f_o')(x_k) = i'(y_k)$ for suitable y_k. We define s by $s(x_k) = y_k$. From $i'(f_1 - f_1') = f_0 i - f_0' i = i'si$ and the injectivity of i' we conclude $f_1 - f_1' = si$. □

We take the tensor product $\otimes G$ of the diagram in (11.8.1). The homomorphism $f_1 \otimes 1$ induces a homomorphism $\operatorname{Ker}(i \otimes 1) \to \operatorname{Ker}(i' \otimes 1)$ and $f_1 - f_1' = si$ shows that this homomorphism does not depend on the choice of (f_1, f_0). Let us denote this homomorphism by $T(f; \mathcal{F}, \mathcal{F}')$. If $g \colon A' \to A''$ is given and $\mathcal{F}''$ a free resolution of A'', then $T(g; \mathcal{F}', \mathcal{F}'') \circ T(f; \mathcal{F}, \mathcal{F}') = T(gf; \mathcal{F}, \mathcal{F}'')$. This implies that an isomorphism f induces an isomorphism $T(f; \mathcal{F}, \mathcal{F}')$. In particular each free resolution yields a unique isomorphism $\operatorname{Ker}(i \otimes 1) \cong \operatorname{Tor}(A; G)$, if we compare $\mathcal{F}$ with the standard resolution. The standard resolution is functorial in A. This fact is used to make $\operatorname{Tor}(-, G)$ into a functor. It is clear that a homomorphism $G \to G'$ induces a homomorphisms $\operatorname{Tor}(A, G) \to \operatorname{Tor}(A, G')$. Hence Tor is also a functor in the variable G (and the two functor structures commute).

If we view $0 \to F_1 \to F_0 \to 0$ in (11.8.1) as a chain complex, then (f_1, f_0) is a chain map and s yields a chain homotopy between (f_1, f_0) and (f_1', f_0').

(11.8.2) Proposition. *Elementary properties of torsion groups in the category of abelian groups are:*

(1) *Let A be a free abelian group. Then* $\operatorname{Tor}(A, G) = 0$.
(2) $\operatorname{Tor}(\mathbb{Z}/n, G) \cong \{g \in G \mid ng = 0\} \subset G$.
(3) *If G is torsion free, then* $\operatorname{Tor}(\mathbb{Z}/n, G) = 0$.
(4) $\operatorname{Tor}(\mathbb{Z}/m, \mathbb{Z}/n) \cong \mathbb{Z}/d$ *with d the greatest common divisor of m, n.*
(5) *A direct sum decomposition $A \cong A_1 \oplus A_2$ induces a direct sum decomposition* $\operatorname{Tor}(A, G) \cong \operatorname{Tor}(A_1, G) \oplus \operatorname{Tor}(A_2, G)$.

Proof. (1) $0 \to 0 \to A \to A \to 0$ is a free resolution. (2) Use the free resolution $0 \to \mathbb{Z} \xrightarrow{n} \mathbb{Z} \to \mathbb{Z}/n \to 0$. (3) and (4) are consequences of (2). In order to verify (5), use the direct sum of free resolutions. □

We can also work with a resolution of the other variable. Let $Q_1 \rightarrowtail Q_0 \twoheadrightarrow B$ be a free resolution and define $\operatorname{Tor}'(A, B) = \operatorname{Ker}(A \otimes Q_1 \to A \otimes Q_0)$.

(11.8.3) Proposition. *There exists a canonical isomorphism*

$$\operatorname{Tor}(A, B) \cong \operatorname{Tor}'(A, B).$$

Proof. Let $P_1 \to P_0 \to A$ be a free resolution. From the resolutions of A and B we obtain a commutative diagram:

$$\begin{array}{ccccccc}
 & & & & & & \operatorname{Tor}(A, B) \\
 & & & & & & \downarrow \\
 & & P_1 \otimes Q_1 & \longrightarrow & P_1 \otimes Q_0 & \longrightarrow & P_1 \otimes B \\
 & & \downarrow \alpha & & \downarrow \beta & & \downarrow \gamma \\
 & & P_0 \otimes Q_1 & \longrightarrow & P_0 \otimes Q_0 & \longrightarrow & P_0 \otimes B \\
 & & \downarrow & & \downarrow & & \downarrow \\
\operatorname{Tor}'(A, B) & \rightarrowtail & A \otimes Q_1 & \longrightarrow & A \otimes Q_0 & \longrightarrow & A \otimes B.
\end{array}$$

The Kernel–Cokernel Lemma (11.2.6) yields an isomorphism δ of $\mathrm{Tor}(A, B) = \mathrm{Ker}(\gamma)$ with the submodule $\mathrm{Tor}'(A, B)$ of $\mathrm{Coker}\,\alpha$. □

Interchanging the tensor factors yields an isomorphism $\mathrm{Tor}(B, A) \cong \mathrm{Tor}'(A, B)$. We combine this with (11.8.3) and see that the isomorphisms (11.8.2) also hold if we interchange the variables. It is now no longer necessary to use the notation Tor′.

The functor Ext is defined in analogy to the functor Tor, the tensor product is replaced by the Hom-functor.

Let R be a principal ideal domain and $0 \to K(A) \xrightarrow{i} F(A) \xrightarrow{p} A \to 0$ the standard free resolution of A as above. We apply the functor $\mathrm{Hom}_R(-, B)$ to this sequence. The cokernel of $i^*\colon \mathrm{Hom}(F(A), B) \to \mathrm{Hom}(K(A), B)$ is defined to be $\mathrm{Ext}^R(A, B) = \mathrm{Ext}(A, B)$. We show that $\mathrm{Ext}(A, B)$ can be determined from any free resolution. We start with a diagram as in (11.8.1) and obtain a well-defined homomorphism $\mathrm{Coker}(\mathrm{Hom}(i, B)) \to \mathrm{Coker}(\mathrm{Hom}(i', B))$; in particular we obtain an isomorphism $\mathrm{Ext}(A, B) \cong \mathrm{Coker}(\mathrm{Hom}(i, B))$.

(11.8.4) Proposition. *Elementary properties of* Ext *in the category of abelian groups are:*

(1) $\mathrm{Ext}(A, B) = 0$ *for a free abelian group* A.

(2) $\mathrm{Ext}(\mathbb{Z}/n, B) \cong B/nB$.

(3) $\mathrm{Ext}(\mathbb{Z}/n, B) = 0$ *for* $B = \mathbb{Q}, \mathbb{Q}/\mathbb{Z}, \mathbb{R}$.

(4) $\mathrm{Ext}(\mathbb{Z}/m, \mathbb{Z}/n) \cong \mathbb{Z}/(m, n)$.

(5) $\mathrm{Ext}(A_1 \oplus A_2, B) \cong \mathrm{Ext}(A_1, B) \oplus \mathrm{Ext}(A_2, B)$. □

The foregoing develops what we need in this text. We should at least mention the general case. Let $0 \leftarrow C \leftarrow P_0 \leftarrow P_1 \leftarrow \cdots$ be a projective resolution of the R-module C and let A be another R-module. We apply $\mathrm{Hom}(-, A)$ to the chain complex P_* and obtain a cochain complex $\mathrm{Hom}(P_*, A)$; its i-th cohomology group $(i \ge 1)$ is denoted $\mathrm{Ext}^i_R(C, A)$. Since projective resolutions are unique up to chain equivalence, the Ext^i_R-groups are unique up to isomorphism. For principal ideal domains only Ext_1 occurs, since we have resolution of length 1. The notation Ext has its origin in the notion of extensions of modules. An exact sequence

$$0 \to A \to B_{n-1} \to \cdots \to B_1 \to B_0 \to C \to 0$$

is called an ***n*-fold extension of *A* by *C***. One can obtain $\mathrm{Ext}^n_R(C, A)$ as certain congruence classes of n-fold extension of A by C, see [120, Chapter III]. Write $E \rightsquigarrow E'$ if there exists a commutative diagram

$$\begin{array}{ccccccccccccc}
E\colon & 0 & \to & C & \to & B_{n-1} & \to \cdots \to & B_0 & \to & A & \to & 0 \\
 & & & \downarrow = & & \downarrow & & \downarrow & & \downarrow = & & \\
E'\colon & 0 & \to & C' & \to & B'_{n-1} & \to \cdots \to & B'_0 & \to & A & \to & 0.
\end{array}$$

The congruence relation is generated by $\rightsquigarrow$.

Problems

1. Suppose $\mathrm{Tor}(A, \mathbb{Z}/p) = 0$ for each prime p. Then the abelian group A is torsion free.
2. The kernel of $A \to A \otimes_{\mathbb{Z}} \mathbb{Q}$, $a \mapsto a \otimes 1$ is the torsion subgroup of A.
3. Does there exist a non-trivial abelian group A such that $A \otimes \mathbb{F} = 0$ for each field $\mathbb{F}$?

11.9 Universal Coefficients

We still work in R-MOD for a principal ideal domain R. Let $C = (C_n, c_n)$ be a chain complex of modules. Then $C \otimes G = (C_n \otimes G, c_n \otimes 1)$ is again a chain complex.

(11.9.1) Proposition (Universal coefficients). *Let C be a chain complex of free modules. Then there exists an exact sequence*

$$0 \to H_q(C) \otimes G \xrightarrow{\alpha} H_q(C \otimes G) \xrightarrow{\beta} \mathrm{Tor}(H_{q-1}(C), G) \to 0.$$

The sequence is natural in C and G and splits. The homomorphism α sends $[z] \otimes g$ for a cycle z to the homology class $[z \otimes g]$.

Proof. The sequence $0 \to Z_n \to C_n \xrightarrow{c_n} B_{n-1} \to 0$ is exact; B_{n-1} is a submodule of C_{n-1} and hence free. Therefore the sequence splits and the induced sequence

$$0 \to Z_n \otimes G \to C_n \otimes G \to B_{n-1} \otimes G \to 0$$

is again a split exact sequence. We consider the totality of these sequences as an exact sequence of chain complexes, the Z- and the B-complex have trivial boundary operator. Associated to this short exact sequence of chain complexes is a long exact homology sequence of the form

$$B_n \otimes G \xrightarrow{i \otimes 1} Z_n \otimes G \longrightarrow H_n(C \otimes G) \longrightarrow B_{n-1} \otimes G \xrightarrow{i \otimes 1} Z_{n-1} \otimes G.$$

One verifies that the boundary operator (11.3.1) of the homology sequence is $i \otimes 1$, where $i : B_n \subset Z_n$. The sequence $B_n \otimes G \to Z_n \otimes G \to H_n \otimes G \to 0$ is exact, hence the cokernel of $i \otimes 1$ is $H_n(C) \otimes G$, and the resulting map $H_n(C) \otimes G \to H_n(C \otimes G)$ is α. The kernel of $B_{n-1} \otimes G \to Z_{n-1} \otimes G$ is $\mathrm{Tor}(H_{n-1}(C), G)$, because $0 \to B_{n-1} \to Z_{n-1} \to H_{n-1}(C) \to 0$ is a free resolution. Let $r : C_n \to Z_n$ be a splitting of $Z_n \subset C_n$. Then

$$Z_n(C \otimes G) \subset C_n \otimes G \xrightarrow{r \otimes 1} Z_n \otimes G \to H_n(C) \otimes G$$

maps $B_n(C \otimes G)$ to zero and induces $\rho : H_n(C \otimes G) \to H_n(C) \otimes G$ with $\rho\alpha = \mathrm{id}$, i.e., a splitting of the universal coefficient sequence. $\Box$

Let again $C = (C_n, c_n)$ be a chain complex with free R-modules C_n. We obtain the cochain complex with cochain groups $\mathrm{Hom}(C_n, G)$ and cohomology groups $H^n(C; G)$.

(11.9.2) Proposition (Universal coefficients). *There exists an exact sequence*

$$0 \to \mathrm{Ext}(H_{n-1}(C), G) \to H^n(C; G) \xrightarrow{\alpha} \mathrm{Hom}(H_n(C), G) \to 0.$$

The map α sends the cohomology class of the cocycle $\varphi\colon C_n \to G$ to the homomorphism $H_n(C) \to G$, $[c] \mapsto \varphi(c)$. The sequence is natural with respect to chain maps (variable C) and module homomorphisms (variable G). The sequence splits, and the splitting is natural in G but not in C.

Proof. Again we start with the split exact sequence $0 \to Z_n \to C_n \to B_{n-1} \to 0$ and the induced exact sequence

$$0 \leftarrow \mathrm{Hom}(Z_n, G) \leftarrow \mathrm{Hom}(C_n, G) \leftarrow \mathrm{Hom}(B_{n-1}, G) \leftarrow 0.$$

We consider the totality of these sequences as an exact sequence of cochain complexes, the Z- and the B-complex have trivial coboundary operator. Associated to this short exact sequence of cochain complexes is a long exact cohomology sequence of the form

$$\cdots \leftarrow \mathrm{Hom}(B_n, G) \xleftarrow{d^n} \mathrm{Hom}(Z_n, G) \leftarrow H^n(C; G) \leftarrow \mathrm{Hom}(B_{n-1}, G) \leftarrow \cdots$$

which induces a short exact sequence

$$(4) \qquad 0 \leftarrow \mathrm{Ker}\, d^n \xleftarrow{\alpha} H^n(C; G) \leftarrow \mathrm{Coker}\, d^{n-1} \leftarrow 0.$$

We need:

(11.9.3) Lemma. *The formal coboundary operator d^n (without the additional sign introduced earlier!) is the homomorphism induced by $i\colon B_n \to Z_n$.*

Proof. Let $\varphi\colon Z_n \to G$ be given. Then $d^n(\varphi)$ is obtained as follows: Extend φ to $\tilde\varphi\colon C_n \to G$. Apply δ and find a pre-image of $\delta(\tilde\varphi) = \tilde\varphi c_{n+1}$ in $\mathrm{Hom}(B_n, G)$. One verifies that φi is a pre-image. □

From the exact sequence $0 \to B_n \to Z_n \to H_n(C) \to 0$ we obtain the exact sequence

$$\mathrm{Hom}(B_n, G) \xleftarrow{i^*} \mathrm{Hom}(Z_n, G) \leftarrow \mathrm{Hom}(H_n(C), G) \leftarrow 0.$$

We use it to identify the $\mathrm{Ker}\, i^*$ with $\mathrm{Hom}(H_n(C), G)$. One verifies that α is as claimed in the statement (11.9.2). From the free presentation and the definition of Ext we thus obtain the exact sequence of the theorem. The naturality of this sequence is a consequence of the construction. It remains to verify the splitting. We choose a splitting $r\colon C_n \to Z_n$ of the inclusion $Z_n \subset C_n$. Now consider the diagram

$$\begin{array}{ccccccc} 0 \longrightarrow & Z^n(\mathrm{Hom}(C,G)) & \xrightarrow{\subset} & \mathrm{Hom}(C_n,G) & \xrightarrow{\delta} & \mathrm{Hom}(C_{n+1},G) \\ & & & \downarrow{\scriptstyle r^*} & & \downarrow \\ 0 \longrightarrow & \mathrm{Hom}(H_n(C),G) & \longrightarrow & \mathrm{Hom}(Z_n,G) & \xrightarrow{i^*} & \mathrm{Hom}(B_n,G). \end{array}$$

If $\varphi \in \operatorname{Ker} i^*$, then $r^*(\varphi) = \varphi \circ r \in \operatorname{Ker} \delta$. The splitting is induced by $\operatorname{Ker} i^* \to Z^n(\mathrm{Hom}(C,G))$, $\varphi \mapsto \varphi t$. □

Without going into the definition of Ext we see from the discussion:

(11.9.4) Proposition. *Suppose $H_{n-1}(C)$ is a free R-module. Then the homomorphism $\alpha\colon H^n(C;G) \to \mathrm{Hom}(H_n(C),G)$ in (11.9.2) is an isomorphism.*

Proof. The sequence $0 \to B_{n-1} \to Z_{n-1} \to H_{n-1} \to 0$ splits and therefore the cokernel of d^{n-1} is zero. □

Given a cochain complex $C^\bullet = (C^q, \delta^q)$ we can view it as a chain complex $C_\bullet = (C_q, \partial_q)$ by a shift of indices: We set $C_q = C^{-q}$ and we define $\partial_q\colon C_q \to C_{q-1}$ as $\delta^{-q}\colon C^{-q} \to C^{-q+1}$. We can now rewrite (11.9.1):

(11.9.5) Proposition. *Let $C^\bullet$ be a cochain complex of free R-modules. Then we have a split exact sequence*

$$0 \to H^q(C^\bullet) \otimes G \to H^q(C^\bullet \otimes G) \to \mathrm{Tor}(H^{q+1}(C^\bullet), G) \to 0. \qquad \square$$

Let now $C_\bullet$ be a chain complex of free modules. We apply (11.9.5) to the dual cochain complex with $C^q = \mathrm{Hom}(C_q, R)$ and cohomology groups $H^q(C;R)$.

(11.9.6) Proposition. *Let C be a free chain complex and G be a module such that either $H_*(C)$ is of finite type or G is finitely generated. Then there exists a natural exact sequence*

$$0 \to H^p(C) \otimes G \to H^q(C;G) \to \mathrm{Tor}(H^{q+1}(C),G) \to 0$$

and this sequence splits.

Proof. If G is finitely generated we have a canonical isomorphism of the form $\mathrm{Hom}(C,R) \otimes G \cong \mathrm{Hom}(C,G)$; we use this isomorphism in (11.9.5). If $H_*(C)$ is of finite type we replace C by a chain equivalent complex C' of finite type (see (11.6.4)). In that case we have again a canonical isomorphism $\mathrm{Hom}(C',R) \otimes G \cong \mathrm{Hom}(C',G)$. We apply now (11.9.5) to C'. □

(11.9.7) Proposition. *Let $f: C \to D$ be a chain map between complexes of free abelian groups. Suppose that for each field $\mathbb{F}$ the map $f \otimes \mathbb{F}$ induces isomorphisms of homology groups. Then f is a chain equivalence.*

Proof. Let $C(f)$ denote the mapping cone of f. The hypothesis implies that $H_*(C(f) \otimes \mathbb{F}) = 0$. We use the universal coefficient sequence. It implies that $\mathrm{Tor}(H_*(C(f)), \mathbb{Z}/p) = 0$ for each prime p. Hence $H_*(C(f))$ is torsion-free. From $H_*(C(f)) \otimes \mathbb{Q}$ we conclude that $H_*(C(f))$ is a torsion group. Hence $H_*(C(f)) = 0$. Now we use (11.6.3). □

11.10 The Künneth Formula

Let C and D be chain complexes of R-modules over a principal ideal domain R. We have the tensor product chain complex $C \otimes_R D$ and the associated homomorphism

$$\alpha\colon H_i(C) \otimes_R H_j(D) \to H_{i+j}(C \otimes_R D), \quad [x] \otimes [y] \mapsto [x \otimes y].$$

We use the notation $*$ for Tor^R. The next theorem and its proof generalizes the universal coefficient formula (11.9.1).

(11.10.1) Theorem (Künneth formula). *Suppose C consists of free R-modules. Then there exists an exact sequence*

$$0 \to \bigoplus_{i+j=n} H_i(C) \otimes_R H_j(D) \to H_n(C \otimes_R D) \to \bigoplus_{i+j=n-1} H_i(C) * H_j(D) \to 0.$$

If also D is a free complex, then the sequence splits.

Proof. We consider the graded modules $Z(C)$ and $B(C)$ of cycles and boundaries as chain complexes with trivial boundary. Since $Z(C)$ is free, we have the equalities (canonical isomorphisms)

$$(Z(C) \otimes Z(D))_n = \mathrm{Ker}(1 \otimes \partial\colon (Z(C) \otimes D)_n \to (Z(C) \otimes D)_{n-1})$$

and

$$(Z(C) \otimes B(D))_n = \mathrm{Im}(1 \otimes \partial\colon (Z(C) \otimes D)_{n+1} \to (Z(C) \otimes D)_n),$$

and they imply $H(Z(C) \otimes D) \cong Z(C) \otimes H(D)$ (homology commutes with the tensor product by a free module). In a similar manner we obtain an isomorphism $H(B(C) \otimes D) \cong B(C) \otimes H(D)$. We form the tensor product of the free resolution of chain complexes

$$0 \to B(C) \xrightarrow{i} Z(C) \to H(C) \to 0$$

with $H(D)$. We obtain the following exact sequence, referred to as (†), with injective morphism (1) and surjective morphism (2)

$$\begin{array}{ccccccc} H(C) * H(D) & \xrightarrow{(1)} & B(C) \otimes H(D) & \xrightarrow{i\otimes 1} & Z(C) \otimes H(D) & \xrightarrow{(2)} & H(C) \otimes H(D) \\ & & \downarrow\cong & & \downarrow\cong & & \\ & & H(B(C) \otimes D) & \xrightarrow{(i\otimes 1)_*} & H(Z(C) \otimes D). & & \end{array}$$

Let us use the notation $(A[-1])_n = A_{n-1}$ for a graded object A. We tensor the exact sequence of chain complexes $0 \to Z(C) \to C \to B(C)[-1] \to 0$ with D and obtain an exact sequence

$$0 \to Z(C) \otimes D \to C \otimes D \to (B(C) \otimes D)[-1] \to 0.$$

Its exact homology sequence has the form

$$\begin{aligned} \ldots \to H(B(C) \otimes D) &\xrightarrow{(1)} H(Z(C) \otimes D) \to H(C \otimes D) \\ \to H(B(C) \otimes D)[-1] &\xrightarrow{(1)} H(Z(C) \otimes D)[-1] \to \cdots . \end{aligned}$$

One verifies that (1) is the map $(i \otimes 1)_*$. Hence we obtain the exact sequence

$$0 \to \operatorname{Coker}(i_*) \to H(C \otimes D) \to \operatorname{Ker}(i_*)[-1] \to 0$$

which yields, together with the sequence (†), the exact sequence of the theorem.

Choose retractions $r\colon C_n \to Z_n(C)$ and $s\colon D_n \to Z_n(D)$. Then $(C \otimes D)_n \to H(C) \otimes H(D)$, $c \otimes d \mapsto [r(c)] \otimes [s(d)]$ sends the boundaries of $(C \otimes D)_n$ to zero and induces a retraction $\rho\colon H_n(C \otimes D) \to (H(C) \otimes H(D))_n$ of α. □

As in the case of the universal coefficient theorem we can rewrite (11.10.1) in terms of cochain complexes. Under suitable finiteness conditions we can then apply the result to the dual complex of a chain complex and obtain:

(11.10.2) Theorem (Künneth formula). *Let C and D be free chain complexes such that $H_*(C)$ or $H_*(D)$ is of finite type. Then there exists a functorial exact sequence*

$$0 \to \bigoplus_{i+j=n} H^i(C) \otimes H^j(D) \to H^n(C \otimes D) \to \bigoplus_{i+j=n+1} H^i(C) * H^j(D) \to 0$$

and this sequence splits. □

Chapter 12

Cellular Homology

In this chapter we finally show that ordinary homology theory is determined on the category of cell complexes by the axioms of Eilenberg and Steenrod. From the axioms one constructs the cellular chain complex of a CW-complex. This chain complex depends on the skeletal filtration, and the boundary operators of the chain complex are determined by the so-called incidence numbers; these are mapping degrees derived from the attaching maps. The main theorem then says that the algebraic homology groups of the cellular chain complex are isomorphic to the homology groups of the homology theory (if it satisfies the dimension axiom). From this fact one obtains immediately qualitative results and explicit computations of homology groups. Thus if X has $k(n)$ n-cells, then $H_n(X;\mathbb{Z})$ is a subquotient of the free abelian group of rank $k(n)$. A finite cell complex has finitely generated homology groups. We deduce that the combinatorial Euler characteristic is a homotopy invariant that can be computed from the homology groups.

In the case of a simplicial complex we show that singular homology is isomorphic to the classical combinatorial simplicial homology. In this context, simplicial homology is a special case of cellular homology.

12.1 Cellular Chain Complexes

Let h_* be an additive homology theory. Let X be obtained from A by attaching n-cells via $(\Phi,\varphi)\colon \coprod_{e\in E}(D^n_e,S^{n-1}_e)\to(X,A)$. The characteristic map of the cell e is denoted by (Φ^e,φ^e). The index e distinguishes different copies.

(12.1.1) Proposition. *The induced map*

$$\Phi^n = \langle \Phi^e_* \rangle\colon \bigoplus_e h_*(D^n_e,S^{n-1}_e)\to h_*(X,A)$$

is an isomorphism.

Proof. By (10.4.6), $\Phi_*\colon h_*(\coprod_e(D^n_e,S^{n-1}_e))\to h_*(X,A)$ is an isomorphism. Now apply the additivity isomorphism $\bigoplus_e h_*(D^n_e,S^{n-1}_e)\cong h_*(\coprod_e(D^n_e,S^{n-1}_e)$ and compose it with Φ_*. □

The isomorphism inverse to Φ^n is obtained as follows. Given $z\in h_k(X,A)$. We use the inclusion $p^e\colon(X,A)\subset(X,X\smallsetminus e)$ and the relative homeomorphism $\Phi^e\colon(D^n,S^{n-1})\to(X,X\smallsetminus e)$. Let $z_e\in h_k(D^n_e,S^{n-1}_e)$ denote the image of z under

$$z\in h_k(X,A)\xrightarrow{p^e_*}h_k(X,X\smallsetminus e)\xleftarrow{\Phi^e_*}h_k(D^n_e,S^{n-1}_e)\ni z_e.$$

Then $z \mapsto (z_e \mid e \in E)$ is inverse to Φ^n.

Let X be a CW-complex. The boundary operator $\partial \colon h_{k+1}(X^{n+1}, X^n) \to h_k(X^n, X^{n-1})$ of the triple (X^{n+1}, X^n, X^{n-1}) is transformed via the isomorphisms (12.1.1) into a matrix of linear maps

$$m(e, f) \colon h_{k+1}(D_f^{n+1}, S_f^n) \to h_k(D_e^n, S_e^{n-1})$$

for each pair (f, e) of an $(n+1)$-cell f and an n-cell e (as always in linear algebra). Let $\iota^{e,f}$ be the composition

$$S_f^n \xrightarrow{\varphi^f} X^n \xrightarrow{q^e} X^n/(X^n \smallsetminus e) \xleftarrow{\Phi^e} D^n/S^{n-1}.$$

If we compose $\iota^{e,f}$ with an h-equivalence $\kappa^n \colon D^n/S^{n-1} \to S^n$, then $\kappa^n \iota^{e,f}$ has as a self-map of S^n a degree $d(e, f)$. We call $d(e, f)$ the ***incidence number*** of the pair (f, e) of cells. The case $n = 0$ is special, so let us consider it separately. Note that D^0/S^{-1} is the point $D^0 = \{0\}$ together with a disjoint base point $\{*\}$. Let κ^0 be given by $\kappa^0(0) = +1$ and $\kappa^0(*) = -1$. We have two 0-cells $\varphi^f(\pm 1) = e_\pm$ (they could coincide). With these conventions $d(f, e_\pm) = \pm 1$.

In the following considerations we use different notation $\partial, \partial', \partial''$ for the boundary operators.

(12.1.2) Proposition. *The diagram*

$$\begin{array}{ccc}
h_{k+1}(D_f^{n+1}, S_f^n) & \xrightarrow{m(e,f)} & h_k(D_e^n, S_e^{n-1}) \\
\downarrow{\scriptstyle \partial''} & & \downarrow{\scriptstyle p_*} \\
\tilde h_k(S_f^n) & \xrightarrow{\iota_*^{e,f}} & \tilde h_k(D_e^n/S_e^{n-1})
\end{array}$$

is commutative.

Proof. Consider the diagram

$$\begin{array}{ccccccc}
h_{k+1}(X^{n+1}, X^n) & \xrightarrow{\partial'} & \tilde h_k(X^n) & \xrightarrow{j} & h_k(X^n, X^{n-1}) & \xrightarrow{p_*} & \tilde h_k(X^n/X^{n-1}) \\
\uparrow{\scriptstyle \Phi_*^f} & & \uparrow{\scriptstyle \varphi_*^f} & & \downarrow{\scriptstyle p_*^e} & & \downarrow \\
h_{k+1}(D^{n+1}, S^n) & \xrightarrow{\partial''} & \tilde h_k(S^n) & & h_k(X^n, X^n \smallsetminus e) & \xrightarrow{p_*} & \tilde h_k(X^n/X^n \smallsetminus e) \\
 & & & & \uparrow{\scriptstyle \Phi_*^e} & & \uparrow{\scriptstyle \Phi_*^e} \\
 & & & & h_k(D^n, S^{n-1}) & \xrightarrow{p_*} & \tilde h_k(D^n/S^{n-1}).
\end{array}$$

Given $x \in h_{k+1}(D^{n+1}, S^n)$. Then $p_* m(e, f)x$ is, by definition of $m(e, f)$, the image of x in $\tilde h_k(D^n/S^{n-1})$. Now use the commutativity of the diagram. □

(12.1.3) Corollary. *Let $\sigma\colon h_k(D^n, S^{n-1}) \to h_{k+1}(D^{n+1}, S^n)$ be a suspension isomorphism. Then $m(e,f)\circ\sigma$ is the multiplication by $d(e,f)$, provided the relation $\partial''\circ\sigma = \kappa^n_*\circ p_*$ holds.* □

We now write the isomorphism (12.1.1) in a different form. We use an iterated suspension isomorphism $\sigma^n\colon h_{k-n} \to h_k(D^n, S^{n-1})$ in each summand. Let $C_n(X)$ denote the free abelian group on the n-cells of (X,A). Elements in $C_n(X)\otimes_{\mathbb{Z}} h_{k-n}$ will be written as finite formal sums $\sum_e e\otimes u_e$ where $u_e\in h_{k-n}$; the elements in $C_n(X)\otimes h_{k-n}$ are called ***cellular n-chains with coefficients in h_{k-n}***. We thus have constructed an isomorphism

$$\zeta_n\colon C_n(X)\otimes_{\mathbb{Z}} h_{k-n} \to h_k(X,A), \quad \textstyle\sum_e e\otimes u_e \mapsto \sum_e \Phi^e_*\sigma^n(u_e).$$

The matrix of incidence numbers provides us with the $\mathbb{Z}$-linear map

$$M(n)\colon C_{n+1}(X) \to C_n(X), \quad f\mapsto \textstyle\sum_e d(e,f)e.$$

The sum is finite: $d(e,f)$ can only be non-zero if the image of φ^f intersects e (property (W3) of a Whitehead complex). From the preceding discussion we obtain:

(12.1.4) Proposition. *Suppose σ and κ are chosen such that the relation* (12.1.3) *holds. Then the diagram*

$$\begin{array}{ccc} h_{k+1}(X^{n+1},X^n) & \xrightarrow{\ \partial\ } & h_k(X^n,X^{n-1}) \\ \uparrow{\scriptstyle \zeta_{n+1}} & & \uparrow{\scriptstyle \zeta_n} \\ C_{n+1}(X)\otimes h_{k-n} & \xrightarrow{M(n)\otimes \mathrm{id}} & C_n(X)\otimes h_{k-n} \end{array}$$

is commutative. □

The composition of the boundary operators (belonging to the appropriate triples)

$$h_{m+1}(X^{n+1},X^n) \xrightarrow{\partial} h_m(X^n,X^{n-1}) \xrightarrow{\partial} h_{m-1}(X^{n-1},X^{n-2})$$

is zero, because the part $h_m(X^n)\to h_m(X^n,X^{n-1})\to h_{m-1}(X^{n-1})$ of the exact sequence of the pair (X^n,X^{n-1}) is "contained" in this composition. We set $h_{n,k}(X) = h_{n+k}(X^n,X^{n-1})$. Thus the groups $(h_{n,k}(X)\mid n\in\mathbb{Z})$ together with the boundary operators just considered form a chain complex $h_{\bullet,k}(X)$.

(12.1.5) Proposition. *The product $M(n-1)M(n)$ of two adjacent incidence matrices is zero. The cellular chain groups $C_n(X)$ together with the homomorphisms $M(n)\colon C(n)\to C(n-1)$ form a chain complex $C_\bullet(X)$. This chain complex is called the* ***cellular chain complex*** *of X.*

Proof. The relation $M(n-1)M(n) = 0$ follows from (12.1.4) applied to the chain complex $H_{\bullet,0}(X)$ obtained from singular homology with coefficients in $\mathbb{Z}$. □

The cellular chain complex has its algebraically defined homology groups. In the next section we prove that in the case of an ordinary homology theory the algebraic homology groups of the cellular chain complex are naturally isomorphic to the homology groups of the theory. We should point out that the algebraic homology groups of the chain complexes $h_{\bullet,k}(X)$ only depend on the space and the coefficients of the homology theory, so are essentially independent of the theory. Nevertheless, they can be used to obtain further information about general homology theories – this is the topic of the so-called spectral sequences [130].

The definition of incidence numbers uses characteristic maps and a homotopy equivalence κ. These data are not part of the structure of a CW-complex so that the incidence numbers are not completely determined by the CW-complex. The choice of a characteristic map determines, as one says, an orientation of the cell. If $\Phi, \Psi\colon (D^n, S^{n-1}) \to (X^n, X^{n-1})$ are two characteristic maps of a cell e, then

$$\Psi^{-1}\Phi\colon D^n/S^{n-1} \to X^n/X^n \smallsetminus e \leftarrow D^n/S^{n-1}$$

is a homeomorphism and hence has degree ± 1. One concludes that the incidence numbers are defined up to sign by the CW-complex.

(12.1.6) Proposition. *A cellular map $f\colon X \to Y$ induces a chain map with components $f_*\colon h_m(X^n, X^{n-1}) \to h_m(Y^n, Y^{n-1})$. Homotopic cellular maps induce chain homotopic maps.*

Proof. The first assertion is clear. Let $f, g\colon X \to Y$ be cellular maps and let $\varphi\colon X \times I \to Y$, $f \simeq g$ be a homotopy between them. By the cellular approximation theorem we can assume that φ is cellular, i.e., $\varphi((X \times I)^n) \subset Y^n$. Note that $(X \times I)^n = X^n \times \partial I \cup X^{n-1} \times I$. We define a chain homotopy as the composition

$$s_n\colon h_m(X^n, X^{n-1}) \xrightarrow{\sigma} h_{m+1}((X^n, X^{n-1}) \times (I, \partial I)) \xrightarrow{\varphi_*} h_{m+1}(Y^{n+1}, Y^n).$$

In order to verify the relation $\partial s_n = g_* - f_* - s_{n-1}\partial$ we apply (10.9.4) to $(A, B, C) = (X^n, X^{n-1}, X^{n-2})$ and compose with φ_*. □

(12.1.7) Proposition. *Let $\rho\colon k_*(-) \to l_*(-)$ be a natural transformation between additive homology theories such that ρ induces isomorphisms of the coefficient groups $\rho\colon k_n(P) \cong l_n(P)$, $n \in \mathbb{Z}$, P a point. Then ρ is an isomorphism $k_*(X) \to l_*(X)$ for each CW-complex X.*

Proof. Since ρ is compatible with the suspension isomorphism we see from (12.1.1) that $\rho\colon k_*(X^n, X^{n-1}) \cong l_*(X^n, X^{n-1})$. Now one uses the exact homology sequences and the Five Lemma to prove by induction on n that ρ is an isomorphism for n-dimensional complexes. For the general case one uses (10.8.1). □

Problems

1. The map $x \mapsto (2\sqrt{1-\|x\|^2}x, 2\|x\|^2 - 1)$ induces a homeomorphism κ^n. Let σ be the suspension isomorphism (10.2.5). Then commutativity holds in (12.1.3). For the proof show that $S^n \to S^n/D^n_+ \cong D^n_-/S^{n-1} \xrightarrow{r} D^n/S^{n-1} \xrightarrow{\kappa^n} S^n$, with the projection r which deletes the last coordinate, has degree 1.
2. Prove $M(n-1)M(n) = 0$ without using homology by homotopy theoretic methods.

12.2 Cellular Homology equals Homology

Let $H_*(-) = H_*(-;G)$ be an ordinary additive homology theory with coefficients in G (not necessarily singular homology). The cellular chain complex $C_\bullet(X) = C_\bullet(X;G)$ of a CW-complex X with respect to this theory has its algebraically defined homology groups. It is a remarkable and important fact that these algebraic homology groups are naturally isomorphic to the homology groups of the space X. This result says that the homology groups are computable from the combinatorial data (the incidence matrices) of the cellular complex.

(12.2.1) Theorem. *The n-th homology group of the cellular chain complex $C_\bullet(X)$ is naturally isomorphic to $H_n(X)$.*

Proof. We show that the isomorphism is induced by the correspondence

$$H_n(X^n, X^{n-1}) \leftarrow H_n(X^n) \to H_n(X).$$

We divide the proof into several steps.

(1) A basic input is $H_k(X^n, X^{n-1}) = 0$ for $k \neq 0$; this follows from our determination of the cellular chain groups in (12.1.1) and the dimension axiom.

(2) $H_k(X^n) = 0$ for $k > n$. Proof by induction on n. The result is clear for X^0 by the dimension axiom. Let $k > n+1$. We have the exact sequence $H_k(X^n) \to H_k(X^{n+1}) \to H_k(X^{n+1}, X^n)$. The first group is zero by induction, the third by (1).

(3) Since $H_{n-1}(X^{n-2}) = 0$, the map $H_{n-1}(X^{n-1}) \to H_{n-1}(X^{n-1}, X^{n-2})$ is injective. Hence the cycle group Z_n of the cellular chain complex is the kernel of $\partial\colon H_n(X^n, X^{n-1}) \to H_{n-1}(X^{n-1})$.

(4) The exact sequence $0 \to H_n(X^n) \to H_n(X^n, X^{n-1}) \to H_{n-1}(X^{n-1})$ induces an isomorphism $(b)\colon H_n(X^n) \cong Z_n$.

(5) $H_k(X, X^n) = 0$ for $k \le n$. One shows by induction on t that the groups $H_k(X^{n+t}, X^n)$ are zero for $t \ge 0$ and $k \le n$. We know that for an additive theory the canonical map $\operatorname{colim}_t H_k(X^{n+t}, X^n) \to H_k(X, X^n)$ is an isomorphism (see (10.8.1) and (10.8.4)). For singular homology one can also use that a singular chain has compact support and that a compact subset of X is contained in some skeleton X^m.

(6) The map $H_n(X^{n+1}) \to H_n(X)$ is an isomorphism. This follows from the exact sequence of the pair (X, X^{n+1}) and (5).

(7) The diagram

$$\begin{array}{ccccccc}
H_{n+1}(X^{n+1}, X^n) & \xrightarrow{\partial} & Z_n & \longrightarrow & Z_n/B_n & \longrightarrow & 0 \\
\uparrow{=} & & \uparrow{(b)\,\cong} & & \uparrow{(a)\,\cong} & & \\
H_{n+1}(X^{n+1}, X^n) & \xrightarrow{\partial} & H_n(X^n) & \longrightarrow & H_n(X^{n+1}) & \longrightarrow & 0 \\
 & & \downarrow & \swarrow{\cong} & & & \\
 & & H_n(X) & & & &
\end{array}$$

shows us that we have an induced isomorphism (a) (Five Lemma). □

(12.2.2) Corollary. *Suppose X has a finite number of n-cells; then $H_n(X;\mathbb{Z})$ is a finitely generated abelian group. Let X be n-dimensional; then $H_k(X;G) = 0$ for $k > n$.* □

(12.2.3) Example (Real projective space). The diagram

$$\begin{array}{ccc}
S^{i-1} & \longrightarrow & \mathbb{R}P^{i-1} \\
\downarrow & & \downarrow \\
D^i & \xrightarrow{\Phi} & \mathbb{R}P^i
\end{array}$$

with attaching map $\Phi\colon x \mapsto [x, \sqrt{1-\|x\|^2}]$ is a pushout. The incidence map with the homomorphism κ^{i-1} of Problem 1 in the previous section is computed to be $S^{i-1} \to S^{i-1}, (y,t) \mapsto (2ty, 2t^2-1)$ of degree $1+(-1)^i$. This yields the cellular chain complex

$$C_0 \xleftarrow{0} C_1 \xleftarrow{2} C_2 \xleftarrow{0} \cdots$$

with $C_i = \mathbb{Z}$ for $0 \le i \le n$ and boundary operator, alternatively, the zero morphism and the multiplication by 2. The cellular chain complex with coefficients in the abelian group G is of the same type (using the canonical identification $\mathbb{Z} \otimes G \cong G$). Let ${}_2G = \{g \in G \mid 2g = 0\}$. We obtain the cellular homology

$$H_j(\mathbb{R}P^n; G) \cong \begin{cases} G, & j = 0, \\ G/2G, & 0 < j = 2k-1 < n, \\ {}_2G, & 0 < j = 2k \le n, \\ G & n = 2k-1. \end{cases}$$

Similarly for $\mathbb{R}P^\infty$. ◇

(12.2.4) Example. The sphere S^n has a CW-decomposition with two i-cells in each dimension i, $0 \le i \le n$. The attaching diagram is

$$\begin{array}{ccc} S^{n-1} + S^{n-1} & \longrightarrow & S^{n-1} \\ \downarrow & & \downarrow \\ D^n + D^n & \xrightarrow{\langle \Phi_-, \Phi_+ \rangle} & S^n \end{array}$$

with $\Phi_+(x) = (x, \sqrt{1 - \|x\|^2})$ and $\Phi_-(x) = -\Phi_+(x)$. This attaching map is G-equivariant, if the cyclic group $G = \{1, t \mid t^2 = 1\}$ acts on the spheres by the antipodal map and on the left column by permutation of the summands. The equivariant chain groups are therefore isomorphic to the group ring $\mathbb{Z}G = \mathbb{Z} \cdot 1 \oplus \mathbb{Z} \cdot t$. In order to determine the equivariant boundary operator we use the fact that we know already the homology of this chain complex. If we add the homology groups in dimension 0 and n we obtain an exact sequence

$$0 \to \mathbb{Z}(\varepsilon_n) \xrightarrow{\eta} \mathbb{Z}G \to \mathbb{Z}G \to \cdots \to \mathbb{Z}G \xrightarrow{\varepsilon} \mathbb{Z} \to 0.$$

The map ε sends $1, t$ to 1. The kernel is generated by $1 - t$. From the geometry we see that the first boundary operator sends the generator $1 \in C_1$ represented by a suitably oriented 1-cell to $\pm(1 - t)$. We orient the cell such that the plus-sign holds. Then d_1 is the multiplication by $1 - t$. The kernel of d_1 is thus generated by $1 + t$. We can again orient the 2-cells such that d_2 is multiplication by $1 + t$. If we continue in this manner, we see that $d_k = 1 - t$ for k odd, and $d_k = 1 + t$ for k even. The homology module $H_n(S^n) = \mathbb{Z}(\varepsilon_n)$ carries the t-action $\varepsilon_n = (-1)^{n+1}$, the degree of the antipodal map. One can, of course, determine the boundary operator by a computation of degrees. We leave this as an exercise. Similar results hold for S^∞. ◇

Let X and Y be CW-complexes. The product inherits a cell decomposition. The cross product induces an isomorphism

$$\bigoplus_{k+l=n} H_k(X^k, X^{k-1}) \otimes H_l(Y^l, Y^{l-1}) \to H_n((X \times Y)^n, (X \times Y)^{n-1}).$$

With a careful choice of cell orientations these isomorphisms combine to an isomorphism $C_*(X) \otimes C_*(Y) \cong C_*(X \times Y)$ of cellular chain complexes.

12.3 Simplicial Complexes

We describe the classical combinatorial definition of homology groups of polyhedra. These groups are isomorphic to the singular groups for this class of spaces. The combinatorial homology groups of a finite polyhedron are finitely generated abelian groups and they are zero above the dimension of the polyhedron. This finite generation is not at all clear from the definition of the singular groups.

Recall that a ***simplicial complex*** $K = (E, S)$ consists of a set E of ***vertices*** and a set S of finite subsets of E. A set $s \in S$ with $q+1$ elements is called a ***q-simplex*** of K. We require the following axioms:

(1) A one-point subset of E is a simplex in S.

(2) $s \in S$ and $\emptyset \neq t \subset s$ imply $t \in S$.

An ordering of a p-simplex is a bijection $\{0, 1, \dots, p\} \to s$. An ***ordering*** of K is a partial order on E which induces a total ordering on each simplex. We write $s = \langle v_0, \dots, v_p \rangle$, if the vertices of s satisfy $v_0 < v_1 < \dots < v_p$ in the given partial ordering. Let $C_p(K)$ denote the free abelian group with basis the set of p-simplices. Its elements are called the simplicial p-chains of K. Now fix an ordering of K and define a boundary operator

$$\partial\colon C_p(K) \to C_{p-1}(K), \quad \langle v_0, \dots, v_p \rangle \mapsto \textstyle\sum_{i=0}^{p} (-1)^i \langle v_0, \dots, \widehat{v_i}, \dots, v_p \rangle.$$

The symbol $\widehat{v_i}$ means that this v_i is to be omitted from the string of vertices. The boundary relation $\partial\partial$ holds (we set $C_p(K) = 0$ for $p \leq -1$). We denote the p-th homology group of this chain complex by $H_p(K)$. This is the classical combinatorial homology group.

A simplicial complex K has a geometric realization $|K|$. An ordered simplex $s = \langle v_0, \dots, v_p \rangle$ has an associated singular simplex

$$\Phi^s\colon \Delta^p \to |K|, \quad (t_0, \dots, t_p) \mapsto \textstyle\sum t_j v_j.$$

We extend $s \mapsto \Phi^s$ by linearity to a homomorphism $\rho_p\colon C_p(K) \to S_p(|K|)$. The boundary operators are arranged so that $\rho = (\rho_p)$ is a chain map.

(12.3.1) Theorem. *ρ induces isomorphisms $R_p\colon H_p(K) \cong H_p(|K|)$.*

Proof. We write $X = |K|$. Let $S(p)$ be the set of p-simplices. The characteristic maps $\Phi^s\colon (\Delta^p, \partial\Delta^p) \to (X^p, X^{p-1})$ yield an isomorphism (12.1.1),

$$\Phi^p\colon \textstyle\bigoplus_{s \in S(p)} H_p(\Delta^p_s, \partial\Delta^p_s) \to H_p(X^p, X^{p-1}).$$

The identity of Δ^p represents a generator ι_p of $H_p(\Delta^p, \partial\Delta^p)$. Let x_s be its image under Φ^s_*. Then $(x_s \mid s \in S(p))$ is a $\mathbb{Z}$-basis of $H_s(X^p, X^{p-1})$. If we express $x \in H_p(X^p, X^{p-1})$ in terms of this basis, $x = \sum_s n_s x_s$, then n_s is determined by the image $n_s \iota_p$ of x under

$$x \in H_p(X^p, X^{p-1}) \to H_p(X^p, X^p \smallsetminus e_s) \xleftarrow{\Phi^s_*} H_p(\Delta^p, \partial\Delta^p) \ni n_s \iota_p.$$

Here e_s is the open simplex which belongs to s. Let $s(i)$ denote the i-th face of Δ^p and $x_{s(i)} \in H_{p-1}(X^{p-1}, X^{p-2})$ the corresponding basis element. We claim $\partial x_s = \sum_{i=0}^{p} (-1)^i x_{s(i)}$. It is clear for geometric reasons that the expression of ∂x_s

in terms of the basis $(x_t \mid t \in S(p-1))$ can have a non-zero coefficient only for the $x_{s(i)}$. The coefficient of $x_{s(i)}$ is seen from the commutative diagram

$$\begin{array}{ccc}
H_p(\Delta^p, \partial\Delta^p) & \xrightarrow{\Phi^s_*} & H_p(X^p, X^{p-1}) \\
\big\downarrow \partial & & \big\downarrow \partial \\
H_{p-1}(\partial\Delta^p, \partial\Delta^p \smallsetminus s(i)^\circ) & \xrightarrow{\Phi^s_*} & H_{p-1}(X^{p-1}, X^{p-2}) \\
\big\uparrow (d^p_i)_* & & \big\downarrow \\
H_{p-1}(\Delta^{p-1}, \partial\Delta^{p-1}) & \xrightarrow{\Phi^{s(i)}} & H^{p-1}(X^{p-1}, X^{p-1} \smallsetminus e_{s(i)}).
\end{array}$$

Note $\Phi^s d^p_i = \Phi^{s(i)}$. The left column sends ι_p to $(-1)^i \iota_{p-1}$. We have constructed so far an isomorphism of $C_\bullet(K)$ with the cellular chain complex $C_\bullet(|K|)$ of $|K|$. Let $P_p\colon H_p(C_\bullet(K)) \to H_p(C_\bullet(|K|))$ be the induced isomorphism. Let $Q_p\colon H_p(C_\bullet(|K|)) \to H_p(|K|)$ be the isomorphism in the proof of (12.2.1). Tracing through the definitions one verifies $R_p = Q_p P_p$. Hence R_p is the composition of two isomorphisms. □

An interesting consequence of (12.3.1) is that $\rho\colon C_\bullet(K) \to S_\bullet(|K|)$ is a chain equivalence. Thus, for a finite complex K, the singular complex of $|K|$ is chain equivalent to a chain complex of finitely generated free abelian groups, zero above the dimension of K.

(12.3.2) Example. A circle S^1 can be triangulated by a regular n-gon with vertices $\{e_0, \dots, e_{n-1}\}$ and ordered simplices $s_i = \langle e_i, e_{i+1}\rangle$, $0 \le i \le n-1$ and mod n notation $e_n = e_0$. The cellular chain complex is given by $\partial\langle e_i, e_{i+1}\rangle = \langle e_i\rangle - \langle e_{i+1}\rangle$. The sum $z = \sum_{i=0}^{n-1} s_i$ is a 1-cycle. Let $\iota\colon \mathbb{Z} \to C_1(K)$, $1 \mapsto z$ and $\varepsilon\colon C_0(K) \to \mathbb{Z}$, $e_j \mapsto 1$. Then the sequence

$$0 \to \mathbb{Z} \xrightarrow{\iota} C_1(K) \xrightarrow{\partial} C_0(K) \xrightarrow{\varepsilon} \mathbb{Z} \to 0$$

is exact. Hence ι induces an isomorphism $H_1(C) \cong \mathbb{Z}$. ◇

Problems

1. Let K be the tetrahedral simplicial complex; it consists of $E = \{0, 1, 2, 3\}$, and all subsets are simplices. Verify $H_i(K) = 0$ for $n > 0$. Generalize to an n-simplex.

12.4 The Euler Characteristic

Let X be a finite CW-complex and $f_i(X)$ the number of its i-cells. The ***combinatorial Euler characteristic*** of X is the alternating sum

$$\chi(X) = \textstyle\sum_{i \ge 0} (-1)^i f_i(X).$$

The fundamental and surprising property of this number is its topological invariance, in fact its homotopy invariance – it does not depend on the cellular decomposition of the space. The origin is the famous result of Euler which says that in the case $X = S^2$ the value $\chi(X)$ always equals 2 ([61], [62], [64]).

We prepare the investigation of the Euler characteristic by an algebraic result about chain complexes. Let $\mathcal{M}$ be a category of R-modules. An ***additive invariant*** for $\mathcal{M}$ with values in the abelian group A assigns to each module M in $\mathcal{M}$ an element $\lambda(M) \in A$ such that for each exact sequence $0 \to M_0 \to M_1 \to M_2 \to 0$ in $\mathcal{M}$ the additivity

$$(1) \qquad \lambda(M_0) - \lambda(M_1) + \lambda(M_2) = 0$$

holds. For the zero-module M we have $\lambda(M) = 0$ since there exists an exact sequence $0 \to M \to M \to M \to 0$. We consider only categories which contain with a module also its submodules and its quotient modules as well as all exact sequences between its objects. Let

$$C_* \colon 0 \longrightarrow C_k \xrightarrow{\partial_k} C_{k-1} \to \cdots \to C_1 \xrightarrow{\partial_1} C_0 \xrightarrow{\partial_0} 0$$

be a chain complex in this category. Then its homology groups $H_i(C_*)$ are also contained in this category.

(12.4.1) Proposition. *Let λ be an additive invariant for $\mathcal{M}$. Then for each chain complex C_* in $\mathcal{M}$ as above the following equality holds:*

$$\textstyle\sum_{i=0}^k (-1)^i \lambda(C_i) = \sum_{i=0}^k (-1)^i \lambda(H_i(C_*)).$$

Proof. Induction on the length k of C_*. We set $H_i = H_i(C_*)$, $B_i = \operatorname{Im} \partial_{i+1}$, $Z_i = \operatorname{Ker} \partial_i$. For $k = 1$ there exist, by definition of homology groups, exact sequences

$$0 \to B_0 \to C_0 \to H_0 \to 0, \quad 0 \to H_1 \to C_1 \to B_0 \to 0.$$

We apply the additivity (1) to both sequences and thereby obtain $\lambda(H_0) - \lambda(H_1) = \lambda(C_0) - \lambda(C_1)$. For the induction step we consider the sequences

$$C'_* \colon 0 \to C_{k-1} \to \cdots \to C_0 \to 0,$$

$$0 \to H_k \to C_k \to B_{k-1} \to 0, \quad 0 \to B_{k-1} \to Z_{k-1} \to H_{k-1} \to 0;$$

the last two are exact and the first one is a chain complex. The homology groups of the chain complex are, for $k \geq 2$,

$$H_i(C'_*) = H_i(C_*), \quad 0 \leq i \leq k-2, \quad H_{k-1}(C'_*) = Z_{k-1}.$$

We apply the induction hypothesis to C'_* and (1) to the other sequences. We obtain the desired result by eliminating $\lambda(B_{k-1})$ and $\lambda(Z_{k-1})$. □

The relation of the combinatorial Euler characteristic to homology groups goes back to Henri Poincaré [150], [152]. The i-th ***Betti number***, named after Enrico Betti [20], $b_i(X)$ of X is the rank of $H_i(X;\mathbb{Z})$, i.e., the cardinality of a basis of its free abelian part, or equivalently, the dimension of the $\mathbb{Q}$-vector space $H_i(X;\mathbb{Z}) \otimes \mathbb{Q} \cong H_i(X;\mathbb{Q})$. The result of Poincaré says:

(12.4.2) Theorem. *For each finite CW-complex X the combinatorial Euler characteristic equals the homological Euler characteristic $\sum_{i\ge 0}(-1)^i b_i(X)$.*

Proof. For finitely generated abelian groups $A \mapsto \operatorname{rank} A$ is an additive invariant. We apply (12.4.1) to the cellular chain complex $C(X)$ of X and observe that $\operatorname{rank} C_i(X) = f_i(X)$. □

If λ is an additive invariant for $\mathcal{M}$ and C_* a chain complex of finite length in $\mathcal{M}$, then we call $\chi(C_*) = \sum_{i\ge 0}(-1)^i\lambda(C_i) = \sum_{i\ge 0}(-1)^i\lambda(H_i(C_*))$ the ***Euler characteristic*** of C_* with respect to λ.

(12.4.3) Proposition. *Let*

$$0 \leftarrow H_0' \leftarrow H_0 \leftarrow H_0'' \leftarrow H_1' \leftarrow H_1 \leftarrow H_1'' \leftarrow H_2' \leftarrow H_2 \leftarrow \cdots$$

be an exact sequence of modules in $\mathcal{M}$ which consists eventually of zero-modules. Let $\chi(H_*) = \sum_{i\ge 0}(-1)^i\lambda(H_i)$ and similarly for H' and H''. Then

$$\chi(H_*') - \chi(H_*) + \chi(H_*'') = 0.$$

Proof. Apply (12.4.1) to the given exact sequence, considered as chain complex, and order the terms according to H, H', and H''. □

One can define the Euler characteristic by homological methods for spaces which are not necessarily finite CW-complexes. There are several possibilities depending on the homology theory being used.

Let R be a principal ideal domain. We call (X,A) of finite R-type if the groups $H_i(X,A;R)$ are finitely generated R-modules and only finitely many of them are non-zero. In that case we have the associated ***homological Euler characteristic***

$$\chi(X,A;R) = \textstyle\sum_{i\ge 0}(-1)^i \operatorname{rank}_R H_i(X,A;R).$$

(12.4.4) Proposition. *If (X,A) is of finite $\mathbb{Z}$-type, then it is of finite R-type and the equality $\chi(X,A;\mathbb{Z}) = \chi(X,A;R)$ holds.*

Proof. If (X,A) is of finite $\mathbb{Z}$-type, then the singular complex $S_\bullet(X,A)$ is chain equivalent to a chain complex $D_\bullet$ of finitely generated free abelian groups with only finitely many of the D_n non-zero (see (11.6.4)). Therefore

$$\begin{aligned}\chi(X,A;R) &= \textstyle\sum_i(-1)^i \operatorname{rank}_R H_i(D_\bullet \otimes R) = \sum_i(-1)^i\operatorname{rank}_R(D_i \otimes R)\\ &= \textstyle\sum_i(-1)^i\operatorname{rank}_{\mathbb{Z}}(D_i) = \chi(X,A;\mathbb{Z}),\end{aligned}$$

by (12.4.1), and some elementary algebra. □

Proposition (12.4.3) has the following consequence. Suppose two of the spaces $A, X, (X, A)$ are of finite R-type. Then the third is of finite R-type and the additivity relation

$$\chi(A; R) + \chi(X, A; R) = \chi(X; R)$$

holds. Let A_0, A_1 be subspaces of X with MV-sequence, then

$$\chi(A_0; R) + \chi(A_1; R) = \chi(A_0 \cup A_1; R) + \chi(A_0 \cap A_1; R)$$

provided the spaces involved are of finite R-type. Similarly in the relative case. Let (X, A) and (Y, B) be of finite R-type. Then the Künneth formula is used to show that the product is of finite R-type and the product formula

$$\chi(X, A; R) \cdot \chi(Y, B; R) = \chi((X, A) \times (Y, B); R)$$

holds. These relations should be clear for finite CW-complex and the combinatorial Euler characteristic by counting cells.

For the more general case of Lefschetz invariants and fixed point indices see [51], [52], [109], [116].

12.5 Euler Characteristic of Surfaces

We report about the classical classification of surfaces and relate this to the Euler characteristic. For details of the combinatorial or differentiable classification see e.g., [167], [80], [123]. See also the chapter about manifolds.

Let F_1 and F_2 be connected surfaces. The connected sum $F_1 \# F_2$ of these surfaces is obtained as follows. Let $D_j \subset F_j$ be homeomorphic to the disk D^2 with boundary S_j. In the topological sum $F_1 \smallsetminus D_1^\circ + F_2 \smallsetminus D_2^\circ$ we identify $x \in S_1$ with $\varphi(x) \in S_2$ via a homeomorphism $\varphi\colon S_1 \to S_2$. The additivity of the Euler characteristic is used to show

$$\chi(F_1) - 1 + \chi(F_2) - 1 = \chi(F_1 \# F_2),$$

i.e., the assignment $F \mapsto \chi(F) - 2$ is additive with respect to the connected sum. Let mF denote the m-fold connected sum of F with itself. We have the standard surfaces sphere S^2, torus T, and projective plane P. The Euler characteristics are

$$\chi(S^2) = 2, \quad \chi(mT) = 2 - 2m, \quad \chi(nP) = 2 - n.$$

If F is a compact surface with k boundary components, then we can attach k disks D^2 along the components in order to obtain a closed surface F^*. By additivity $\chi(F^*) = \chi(F) + k$. Connected surfaces F_j with the same number of boundary components are homeomorphic if and only if the associated surfaces F_j^* are homeomorphic.

(12.5.1) Theorem. *A closed connected surface is homeomorphic to exactly one of the surfaces S^2, mT with $m \geq 1$, nP with $n \geq 1$. The nP are the non-orientable surfaces.*

The homeomorphism type of a closed orientable surface is determined by the orientation behaviour and the Euler characteristic. The homeomorphism type of a compact connected surface with boundary is determined by the orientation behaviour, the Euler characteristic and the number of boundary components.

The sphere has **genus** *0, mT has genus m and nP has genus n.* □

12.5.2 Platonic solids. A convex polyhedron is called regular if each vertex is the end point of the same number of edges, say m, and each 2-dimensional face has the same number of boundary edges, say n. If E is the number of vertices, K the number of edges and F the number of 2-faces, then $mE = 2K$ and $nF = 2K$. We insert this into the Euler relation $E + F = K + 2$, divide by $2K$, and obtain

$$\frac{1}{m} + \frac{1}{n} = \frac{1}{K} + \frac{1}{2}.$$

We have $m \geq 3$, $n \geq 3$. The equation has only the solutions which are displayed in the next table.

m	n	K	solid	E	F
3	3	6	tetrahedron	4	4
4	3	12	octahedron	6	8
3	4	12	cube	8	6
3	5	30	dodecahedron	20	12
5	3	30	icosahedron	12	20

12.5.3 Lines in the projective plane. Let $G_1, \dots, G_n$ be lines in the projective plane P. We consider the resulting cells decomposition of P. Let t_r be the number of points which are incident with r lines. We have the Euler characteristic relation $f_0 - f_1 + f_2 = 1$ where f_i is the number of i-cells. Thus $f_0 = t_2 + t_3 + \cdots$. From an r-fold intersection point there start $2r$ edges. The sum over the vertices yields $f_1 = 2t_2 + 3t_3 + 4t_4 + \cdots$. Let p_n denote the numbers of n-gons, then $f_2 = \sum p_s$, $2f_1 = \sum sp_s$. We insert these relations into the Euler characteristic relation and obtain

$$\textstyle\sum_{r\geq 2}(3-r)t_r + \sum_{s\geq 2}(3-s)p_s = 3f_0 - f_1 + 3f_2 - 2f_1 = 3.$$

We now assume that not all lines are incident with a single point; then we do not have 2-gons. From $2f_1 \geq 3f_2$ and then $f_1 \leq 3(f_0 - 1)$ we conclude

$$t_2 \geq 3 + \textstyle\sum_{r\geq 4}(r-3)t_r.$$

Thus there always exist at least three double points. ◇

(12.5.4) Proposition. *Let X be a Hausdorff space and $p\colon X \to Y$ a local homeomorphism onto a connected space. Then p is a covering with finitely many leaves if and only if p is proper.*

Proof. (1) Suppose p is proper. For $n \in \mathbb{N}$ let $Y_n = \{y \in Y \mid n \le |p^{-1}(y)|\}$. We show that Y_n is open and closed. Since Y is connected, either $Y_n = \emptyset$ or $Y_n = Y$. The inclusion $Y_n \supset Y_{n+1}$ shows that there is a largest n such that $Y_n = Y$ and $Y_{n+1} = \emptyset$. Hence the fibres have the cardinality n.

Let $p^{-1}(y) = \{x_1, \dots, x_m\}$. Since X is a Hausdorff space and p a local homeomorphism, there exist open pairwise disjoint sets $U_i \ni x_i$ which are mapped homeomorphically under p onto the same open set $V \ni y$. Hence each fibre $p^{-1}(z)$, $z \in V$ has at least cardinality n. If $y \in Y_n$, then $m \ge n$ and hence $z \in Y_n$, i.e., $V \subset Y_n$. This shows that Y_n is open.

Let $p^{-1}(y) = \{x_1, \dots x_t\}$, $t < n$, i.e., $y \notin Y_n$. Let again the $U_i \ni x_i$ be open with homeomorphic image $V \in Y$. Since p is closed, being a proper map, the set $C = (X \smallsetminus (U_1 \cup \dots \cup U_t))$ is closed in Y. This set does not contain y. Hence $Y \smallsetminus C = W$ is an open neighbourhood of y and

$$p^{-1}(W) = p^{-1}(Y) \smallsetminus p^{-1}p(X \smallsetminus (U_1 \cup \dots \cup U_t)) \subset U_1 \cup \dots \cup U_t.$$

This shows $|p^{-1}(z)| \le t$ for each $z \in W$, and the complement of Y_n is seen to be open.

We now know that all fibres of p have the same cardinality, and since p is a local homeomorphism it must be a covering.

(2) Suppose conversely that p is an n-fold covering. We have to show that p is closed. A projection $\mathrm{pr}\colon B \times F \to B$ with a finite discrete set F as fibre is closed. Now we use (1.5.4). □

A continuous map $p\colon X \to Y$ between surfaces is called a ***ramified covering*** if for each $x \in X$ there exist centered charts (U, φ, U') about x and (V, ψ, V') about $y = p(x)$ with $p(U) \subset V$ such that

$$\psi\varphi^{-1}\colon \varphi(U) = U' \to \mathbb{C}, \quad z \mapsto z^n$$

with $n \in \mathbb{N}$. We call $n - 1$ the ***ramification index*** of p at x. In the case that $n = 1$ we say that p is unramified at x and for $n > 1$ we call x a ***ramification point***.

(12.5.5) Proposition. *Let $p\colon X \to Y$ be a ramified covering between compact connected surfaces. Let V be the image under p of the ramification points. Then $p\colon X \smallsetminus p^{-1}(p) \to Y \smallsetminus V$ is a covering with finitely many leaves.*

Proof. For each $y \in Y$ the set $p^{-1}(y) \subset X$ is closed and hence compact. The pre-images $p^{-1}(y)$ in a ramified covering are always discrete, hence finite. The set V is also discrete and hence finite. The map p is, as a continuous map between compact Hausdorff spaces, closed. Thus we have shown that the map in question is proper. Now we use (12.5.4). □

(12.5.6) Proposition (Riemann–Hurwitz). *Let $p: X \to Y$ be a ramified covering between compact connected surfaces. Let $P_1, \dots, P_r \in X$ be the ramification points with ramification index $v(P_j)$. Let n be the cardinality of the general fibre. Then for the Euler characteristics the relation*

$$\chi(X) = n\chi(Y) - \sum_{j=1}^{r} v(P_j)$$

holds.

Proof. Let $Q_1, \dots, Q_s$ be the images of the ramification points. Choose pairwise disjoint neighbourhoods $D_1, \dots, D_s \subset Y$ where D_j is homeomorphic to a disk. Then

$$p: X_0 = X \smallsetminus \bigcup_{j=1}^{s} p^{-1}(D_j^\circ) \to Y \smallsetminus \bigcup_{j=1} D_j^\circ = Y_0$$

is an n-fold covering (see (12.5.4)). We use the relation $\chi(X_0) = n\chi(Y_0)$ for n-fold coverings. If C is a finite set in a surface X, then $\chi(X \smallsetminus C) = \chi(X) - |C|$. Thus we see

$$\chi(X) - \sum_{j=1}^{s} |p^{-1}(Q_j)| = n(\chi(Y) - s).$$

Moreover

$$ns - \sum_{j=1}^{s} |p^{-1}(Q_j)| = \sum_{i=1}^{r} v(P_i),$$

since for $p^{-1}(Q_j) = \{P_1^j, \dots, P_{n(j)}^j\}$ the relation $\sum_{t=1}^{n(j)} (v(P_t^j) + 1) = n$ holds. □

An interesting application of the Riemann–Hurwitz formula concerns actions of finite groups on surfaces. Let F be a compact connected orientable surface and $G \times F \to F$ an effective orientation preserving action. We assume that this action has the following properties:

(1) The isotropy group G_x of each point $x \in F$ is cyclic.

(2) There exist about each point x a centered chart $\varphi: U \to \mathbb{R}^2$ such that U is G_x-invariant and φ transforms the G_x-action on U into a representation on $\mathbb{R}^2$, i.e., a suitable generator of G_x acts on $\mathbb{R}^2$ as rotation about an angle $2\pi/|G_x|$.

In this case the orbit map $p: F \to F/G$ is a ramified covering, F/G is orientable, and the ramification points are the points with non-trivial isotropy group. One can show that each orientation preserving action has the properties (1) and (2). Examples are actions of a finite group $G \subset SO(3)$ on S^2 by matrix multiplication and of a finite group $G \subset GL_2(\mathbb{Z})$ on the torus $T = \mathbb{R}^2/\mathbb{Z}^2$ by matrix multiplication.

The ramified coverings which arise as orbit maps from an action are of a more special type. If $x \in F$ is a ramification point, then so is each point in $p^{-1}px$, and these points have the same ramification index, since points in the same orbit have conjugate isotropy groups. Let $C_1, \dots, C_r$ be the orbits with non-trivial isotropy group and let n_j denote the order of the isotropy group of $x \in C_i$; hence $|C_i|n_i = |G|$. The Riemann–Hurwitz formula yields in this case:

12.5.7 Riemann–Hurwitz formula for group actions.

$$\chi(F) = |G|\big(\chi(F/G) - \textstyle\sum_{j=1}^{r}(1 - 1/n_j)\big).$$

In the case of a free action $r = 0$ and there is no sum. ◇

12.5.8 Actions on spheres. Let $F = S^2$ and $|G| \geq 2$. Since $\chi(S^2) = 2$ we see that $\chi(F/G) \leq 0$ is not compatible with 12.5.7, hence $\chi(F/G) = 2$ and the orbit space is again a sphere. We also see that $r \leq 3$ and $r = 0, 1$ are not possible. For $r = 2$ we have $2/|G| = 1/n_1 + 1/n_2$, $2 = |C_1 + |C_2$. Hence there are two fixed points (example: rotation about an axis). For $r = 3$ one verifies that

$$1 + \frac{2}{|G|} = \frac{1}{n_1} + \frac{1}{n_2} + \frac{1}{n_3}$$

has the solutions (for $n_1 \geq n_2 \geq n_3$) displayed in the next table.

n_1	n_2	n_3	$\|G\|$
$\|G\|/2$	2	2	$\|G\|$
3	3	2	12
4	3	2	24
5	3	2	60

Examples are the standard actions of subgroups of SO(3), namely D_{2n} (dihedral), A_4 (tetrahedral), S_4 (octahedral), A_5 (icosahedral). Up to homeomorphism there are no other actions. ◇

12.5.9 Action on the torus. Let $F = T = S^1 \times S^1$ be the torus, $\chi(F) = 0$. The Riemann–Hurwitz formula shows that for $r \geq 1$ we must have $\chi(F/G) = 2$. The cases $r \geq 5$ and $r = 1, 2$ are impossible. For $r = 4$ we must have $n_1 = n_2 = n_3 = n_4 = 2$ and $G = \mathbb{Z}/2$. For $r = 3$ the solutions of 12.5.7 are displayed in a table.

n_1	n_2	n_3
3	3	3
2	3	6
2	4	4

Consider the matrices in $SL_2(\mathbb{Z})$

$$A = \begin{pmatrix} 0 & 1 \\ -1 & 1 \end{pmatrix}, \quad \begin{pmatrix} -1 & 1 \\ -1 & 0 \end{pmatrix}.$$

The cyclic groups generated by A, A^2, A^3 realize cases 2 and 1 of the table and the case $r = 4$ above. The matrix B realizes case 3 of the table. ◇

Problems

1. Let G act effectively on a closed orientable surface F of genus 2 preserving the orientation. Then $|G|$ divides 48 or 10. There exist groups of orders 48 and 10 which act on a surface of genus 2. The group of order 48 has a central subgroup C of order 2 and G/C is the octahedral group S_4 acting on the sphere F/C. Study the solutions of 12.5.7 and determine the groups which can act on F. Use covering space theory and work towards a topological classification of the actions.

2. The nicest models of surfaces are of course Riemann surfaces. Here we assume known the construction of a compact Riemann surface from a polynomial equation in two variables. The equation $y^2 = f(x)$ with $2g+2$ branch points defines a surface of genus g. Such curves are called hyper-elliptic ($g \geq 2$). It is known that all surfaces of genus 2 are ***hyper-elliptic***. A hyper-elliptic surface always has the ***hyper-elliptic involution*** $I(x,y) = (x,-y)$. Here are some examples. Let us write $e(a) = \exp(2\pi i a)$.

(1) $y^2 = x(x^2-1)(x^2-4)$ has a $\mathbb{Z}/4$-action generated by $A(x,y) = (-x, e(1/4)y)$. Note $A^2 = I$.

(2) $y^2 = (x^3-1)/(x^3-8)$ has a $\mathbb{Z}/3$-action generated by $B(x,y) = (e(1/3)x, y)$. Since B commutes with I, we obtain an action of $\mathbb{Z}/6$.

(3) $y^2 = (x^3-1)/(x^3+1)$ has a $\mathbb{Z}/6$-action generated by $C(x,y) = (e(1/6)x, 1/y)$. Since C commutes with I, we obtain an action of $\mathbb{Z}/6 \oplus \mathbb{Z}/2$. It has an action of $\mathbb{Z}/4$ generated by $D(x,y) = (1/x, e(1/4)y)$. Note $D^2 = I$. The actions C and D do not commute, in fact $CD = DC^5$. Thus we obtain an action of a group F which is an extension

$$1 \to \mathbb{Z}/2 \to F \to D_{12} \to 1$$

where $\mathbb{Z}/2$ is generated by I and D_{12} denotes the dihedral group of order 12.

(4) $y^2 = x(x^4-1)$ has the following automorphisms (see also [121, p. 94])

$$G(x,y) = (e(1/4)x, e(1/8)y), \quad G^8 = \mathrm{id}, \quad G^4 = I,$$
$$H(x,y) = (1/x, e(1/4)y/x^3), \quad H^4 = \mathrm{id}, \quad H^2 = I,$$
$$K(x,y) = (-(x-i)/(x+i), 2\sqrt{2}e(1/8)y/(x+i)), \quad K^3 = I.$$

The elements G, H, K generate a group of order 48. If we quotient out the central hyper-elliptic involution we obtain the octahedral group of order 24 acting on the sphere. Thus there also exists an action of a group of order 24 such that the quotient by the hyper-elliptic involution is the tetrahedral group (and not the dihedral group D_{12}, as in (3)).

(5) $y^2 = x^5-1$ has an action of $\mathbb{Z}/5$ generated by $J(x,y) = (e(1/5)x, y)$. It commutes with I and gives an action of $\mathbb{Z}/10$.

3. By an analysis of 12.5.7 one can show that, for an effective action of G on a closed orientable surface of genus $g \geq 2$, the inequality $|G| \leq 84(g-1)$ holds. There exists a group of order 168 which acts on a surface of genus 3 [63, p. 242].

Chapter 13
Partitions of Unity in Homotopy Theory

Partitions of unity and numerable coverings of a space are useful tools in order to obtain global results from local data. (A related concept is that of a paracompact space.) We present some notions about partitions of unity in the context of point-set topology. Then we use them to show that, roughly, local homotopy equivalences are global ones and a map is a fibration if it is locally a fibration (see the precise statements in (13.3.1) and (13.4.1)). We apply the results to prove a theorem of Dold about fibrewise homotopy equivalences (see (13.3.4)). Conceptually, partitions of unity are used to relate the homotopy colimit of a covering to the colimit of the covering; see (13.2.4) for a result of this type. There are many other results of this type in the literature. This chapter only can serve as an introduction to this topic.

13.1 Partitions of Unity

Let $t\colon X \to \mathbb{R}$ be continuous. The closure of $t^{-1}(\mathbb{R} \smallsetminus 0)$ is the ***support*** $\operatorname{supp}(t)$ of t. A family $T = (t_j\colon X \to \mathbb{R} \mid j \in J)$ of continuous functions is said to be ***locally finite*** if the family of supports $(\operatorname{supp}(t_j) \mid j \in J)$ is locally finite. We call T ***point finite*** if $\{j \in J \mid t_j(x) \neq 0\}$ is a finite set for each $x \in X$. We call a locally finite T a ***partition of unity*** if the t_j assume only non-negative values and if for each $x \in X$ we have $\sum_{j\in J} t_j(x) = 1$. A covering $\mathcal{U} = (U_j \mid j \in J)$ is ***numerable*** if there exists a partition of unity T such that $\operatorname{supp}(t_j) \subset U_j$ holds for each $j \in J$; the family T is then called a ***numeration*** of $\mathcal{U}$ or a partition of unity ***subordinate*** to $\mathcal{U}$.

(13.1.1) Theorem. *A locally finite open covering of a normal space is numerable.*

Proof. Let $U = (U_j \mid j \in J)$ be a locally finite open covering of the normal space X and $V = (V_j \mid j \in J)$ a shrinking of U and $W = (W_j \mid j \in J)$ a shrinking of V. By the theorem of Urysohn, there exist continuous functions $\tau_j\colon X \to [0,1]$ which assume the value 1 on W_j and the value 0 on the complement of V_j. The function $\tau = \sum_{j\in J} \tau_j\colon X \to [0,1]$ is well-defined and continuous, since by local finiteness of V, in a suitable neighbourhood of a point only a finite number of τ_j are non-zero. We set $f_j(x) = \tau_j(x)\cdot\tau(x)^{-1}$. The functions $(f_j \mid j \in J)$ are a numeration of U. □

(13.1.2) Lemma. *Let the covering $V = (V_k \mid k \in K)$ be a refinement of the covering $U = (U_j \mid j \in J)$. If V is numerable, then also U is numerable.*

Proof. Let $(f_k \mid k \in K)$ be a numeration of V. For each $k \in K$ choose $a(k) \in J$ with $V_k \subset U_{a(k)}$. This defines a map $a\colon K \to J$. We set $g_j(x) = \sum_{k,a(k)=j} f_k(x)$; this is the zero function if the sum is empty. Then g_j is continuous; the support of g_j is contained in the union of the supports of the f_k with $a(k) = j$ and is therefore contained in U_j. Moreover, the sum of the g_j is 1. The family $(g_j \mid j \in J)$ is locally finite: If W is an open neighbourhood of x which meets only a finite number of supports $\mathrm{supp}(f_k)$, $k \in E \subset K$, E finite, then W meets only the supports of the g_j with $j \in a(E)$. □

(13.1.3) Theorem. *Each open covering of a paracompact space is numerable.*

Proof. Let $U = (U_j \mid j \in J)$ be an open covering of the paracompact space X and let $V = (V_k \mid k \in K)$ be a locally finite refinement. Since X is normal, there exists a numeration $(f_k \mid k \in K)$ of V. Now apply the previous lemma. □

(13.1.4) Lemma. *Let $(f_j\colon X \to [0,\infty[\mid j \in J)$ be a family of continuous functions such that $U = (f_j^{-1}]0,\infty[\mid j \in J)$ is a locally finite covering of X. Then U is numerable and has, in particular, a shrinking.*

Proof. Since U is locally finite, $f\colon x \mapsto \max(f_j(x) \mid j \in J)$ is continuous and nowhere zero. We set $g_j(x) = f_j(x)f(x)^{-1}$. Then

$$t_j\colon X \to [0,1], \quad x \mapsto \max(2g_j(x) - 1, 0)$$

is continuous. Since $t_j(x) > 0$ if and only if $g_j(x) > 1/2$, we have the inclusions $\mathrm{supp}(t_j) \subset g_j^{-1}[1/4,\infty[\subset f_j^{-1}]0,\infty[$. For $x \in X$ and $i \in J$ with $f_i(x) = \max(f_j(x))$ we have $t_i(x) = 1$. Hence the supports of the t_j form a locally finite covering of X, and the functions $x \mapsto t_i(x)/t(x)$, $t(x) = \sum_{j\in J} t_j(x)$ are a numeration of U. □

(13.1.5) Theorem. *Let $\mathcal{U} = (U_j \mid j \in J)$ be a covering of the space X. The following assertions are equivalent:*

(1) *$\mathcal{U}$ is numerable.*

(2) *There exists a family $(s_{a,n}\colon X \to [0,\infty[\mid a \in A,\ n \in \mathbb{N}) = S$ of continuous functions $s_{a,n}$ with the properties:*

   (a) *S, i.e., $(s_{a,n}^{-1}]0,\infty[)$, refines $\mathcal{U}$.*

   (b) *For each n the family $(s_{a,n}^{-1}]0,\infty[\mid a \in A)$ is locally finite.*

   (c) *For each $x \in X$ there exists (a,n) such that $s_{a,n}(x) > 0$.*

Proof. (1) ⇒ (2) is clear.

(2) ⇒ (1). $(s_{a,n})$ is, by assumption, a countable union of locally finite families. From these data we construct a locally finite family. By replacing $s_{a,n}$ with $s_{a,n}/(1 + s_{a,n})$ we can assume that $s_{a,n}$ has an image contained in $[0,1]$. Let

$$q_r(x) = \sum_{a\in A,\, i<r} s_{a,i}(x), \quad r \geq 1$$

where $q_r(x) = 0$ for $r = 1$. (The sum is finite for each $x \in X$.) Then q_r and

$$p_{a,r}(x) = \max(0, s_{a,r}(x) - rq_r(x))$$

are continuous. Let $x \in X$; then there exists $s_{a,k}$ with $s_{a,k}(x) \neq 0$; we choose such a function with minimal k; then $q_k(x) = 0$, $p_{a,k}(x) = s_{a,k}(x)$. Therefore the sets $p_{a,k}^{-1}]0,1]$ also cover X. Choose $N \in \mathbb{N}$ such that $N > k$ and $s_{a,k}(x) > \frac{1}{N}$. Then $q_N(x) > \frac{1}{N}$, and therefore $Nq_N(y) > 1$ for all y in a suitable neighbourhood of x. In this neighbourhood, all $p_{a,r}$ with $r \geq N$ vanish. Hence

$$(p_{a,n}^{-1}]0,1] \mid a \in A,\ n \in \mathbb{N})$$

is a locally finite covering of X which refines $\mathcal{U}$. We finish the proof by an application of the previous lemma and (13.1.2). □

(13.1.6) Theorem. *Let $(U_j \mid j \in J)$ be a numerable covering of $B \times [0,1]$. Then there exists a numerable covering $(V_k \mid k \in K)$ of B and a family $(\epsilon(k) \mid k \in K)$ of positive real numbers such that, for $t_1, t_2 \in [0,1]$, $t_1 < t_2$ and $|t_1 - t_2| < \epsilon(k)$, there exist a $j \in J$ with $V_k \times [t_1, t_2] \subset U_j$.*

Proof. Let $(t_j \mid j \in J)$ be a numeration of (U_j). For each r-tuple $k = (j_1, \dots, j_r) \in J^r$, define a continuous map

$$v_k \colon B \to I, \quad x \mapsto \prod_{i=1}^{r} \min\left(t_{j_i}(x,s) \mid s \in \left[\tfrac{i-1}{r+1}, \tfrac{i+1}{r+1}\right]\right).$$

Let $K = \bigcup_{r=1}^{\infty} J^r$. We show that the $V_k = v_k^{-1}]0,1]$ and $\epsilon(k) = \frac{1}{2r}$ for $k = (j_1, \dots, j_r)$ satisfy the requirements of the theorem. Namely if $|t_1 - t_2| < \frac{1}{2r}$, there exists i with $[t_1, t_2] \subset [\frac{i-1}{r+1}, \frac{i+1}{r+1}]$ and hence $V_k \times [t_1, t_2] \subset U_{j_i}$.

We show that (V_k) is a covering. Let $x \in B$ be given. Each point (x,t) has an open neighbourhood of the form $U(x,t) \times V(x,t)$ which is contained in a suitable set $W(i) = t_i^{-1}]0,1]$ and meets only a finite number of the $W(j)$. Suppose $V(x,t_1), \dots, V(x,t_n)$ cover the interval $I = [0,1]$; let $\frac{2}{r+1}$ be a Lebesgue number of this covering. We set $U = U(x,t_1) \cap \dots \cap U(x,t_n)$. Each set $U \times [\frac{i-1}{r+1}, \frac{i+1}{r+1}]$ is then contained in a suitable $W(j_i)$. Hence x is contained in V_k, $k = (j_1, \dots, j_r)$.

There are only a finite number of $j \in J$ for which $W(j) \cap (U \times I) \neq \emptyset$. Since $v_k(x) \neq 0$ implies the relation $W(j_i) \cap \{x\} \times I \neq \emptyset$, the family $(V_k \mid k \in J^r)$ is locally finite for r fixed. The existence of a numeration for $(V_k \mid k \in K)$ follows now from theorem (13.1.5). □

A family of continuous maps $(t_j : X \to [0,1] \mid j \in J)$ is called a ***generalized partition of unity*** if for each $x \in X$ the family $(t_j(x) \mid j \in J)$ is summable with sum 1.

(13.1.7) Lemma. *Let $(t_j \mid j \in J)$ be a generalized partition of unity. Then $(t_j^{-1}]0,1] \mid j \in J)$ is a numerable covering.*

Proof. Summability of $(t_j(a))$ means: For each $\varepsilon > 0$ there exists a finite set $E \subset J$ such that for all finite sets $F \supset E$ the inequality $|1 - \sum_{j \in F} t_j(a)| > 1 - \varepsilon$ holds. In that case $V = \{x \mid \sum_{j \in E} t_j(x) > 1 - \varepsilon\}$ is an open neighbourhood of a. If $k \notin E$, $x \in V$ and $t_k(x) > \varepsilon$, then $t_k(x) + \sum_{j \in E} t_j(x) > 1$. This is impossible. Hence for each $a \in X$ there exists an open neighbourhood $V(a)$ such that only a finite number of functions t_j have a value greater than ε on $V(a)$. Let $s_{j,n}(x) = \max(t_j(x) - n^{-1}, 0)$ for $j \in J$ and $n \in \mathbb{N}$. By what we have just shown, the $s_{j,n}$ are locally finite for fixed n. The claim now follows from (13.1.5). □

It is a useful fact that arbitrary partitions of unity can be reduced to countable ones. The method of proof is inspired by the barycentric subdivision of a simplicial complex. Let $\mathcal{U} = (U_j \mid j \in J)$ be a covering of the space Z with subordinate partition of unity $T = (t_j \mid j \in J)$. For each finite set $E \subset J$ we set $q_E(z) = \max(0, \min_{i \in E} t_i(z) - \max_{j \notin E} t_j(z))$. The function q_E is continuous, since T is locally finite. From this definition one verifies:

(13.1.8) Lemma. *If $q_E(x) \neq 0 \neq q_F(x)$, then either $E \subset F$ or $F \subset E$. The family $(|E| q_E \mid E \subset J$ finite$)$ is a locally finite partition of unity.* □

(13.1.9) Corollary. *Let $U(E) = q_E^{-1}$. If $U(E) \cap U(F) \neq \emptyset$ and $|E| = |F|$, then $E = F$. We set $U\langle n \rangle = \coprod_{|E|=n} U(E)$ and define $\tau_n \colon X \to [0,1]$ as $\tau_n = \sum_{|E|=n} |E| q_E$. This is a continuous function with $\tau^{-1}]0,1] = U\langle n \rangle$.* □

Suppose the functions t_j are non-zero. Let N be the nerve of the covering $(t_j^{-1}]0,1] \mid j \in J)$. Then the nerve of the covering $(q_E^{-1}]0,1] \mid E \subset J$ finite$)$ is the barycentric subdivision of N.

13.2 The Homotopy Colimit of a Covering

Let $K = (V, S)$ be a simplicial complex. We consider it as a category: The objects are the simplices, the morphisms are the inclusions of simplices. A contravariant functor $X \colon K \to \mathrm{TOP}$ is called a ***simplicial K-diagram*** (in TOP). It associates to each simplex s a space X_s and to each inclusion $t \subset s$ a continuous map $r_t^s \colon X_s \to X_t$. We also have a covariant functor $\Delta \colon K \to \mathrm{TOP}$ which $\Delta(s) = \{\sum_{v \in s} t_v v \mid t_v \in I, \sum t_v = 1\}$, and for an inclusion $t \subset s$ we have the canonical inclusion $i_t^s \colon \Delta(t) \to \Delta(s)$. The ***geometric realization*** $|X|$ of a K-diagram X is the quotient of $\coprod_s X_s \times \Delta(s)$ with respect to the relation

$$X_s \times \Delta(s) \ni (x, i_t^s(a) \sim (r_t^s(x), a)) \in X_t \times \Delta(t).$$

Restriction to the n-skeleton K^n yields a functor $X^n \colon K^n \to \mathrm{TOP}$. In $|X|$ we have the subspace $|X|^n$ which is the image of the $X_s \times \Delta(s)$ with $\dim s \leq n$. Since $|X^n|$ is a quotient of the sum of these products we obtain a continuous map $|X^n| \to |X|^n$.

(13.2.1) Proposition. *The space $|X|$ is the colimit of the subspaces $|X|^n$. The canonical map $|X^n| \to |X|^n$ is a homeomorphism. There exists a canonical pushout diagram*

$$\begin{array}{ccc} \coprod_{s,\dim s=n} X_s \times \partial\Delta(s) & \xrightarrow{\varphi^n} & |X^{n-1}| \\ \downarrow & & \downarrow \\ \coprod_{s,\dim s=n} X_s \times \Delta(s) & \xrightarrow{\Phi^n} & |X^n|. \end{array}$$

The attaching map φ^n is defined as follows: $\partial\Delta(s)$ is the union of the $i_t^s\Delta(t)$, where the $t \subset s$ have one element less than s. The map φ^n is defined on $X_s \times i_t^s\Delta(s)$ by $r_t^s \times (i_t^s)^{-1}$ composed with the canonical map into $|X^{n-1}|$. □

Let $\mathcal{U} = \{U_j \mid j \in J\}$ be a covering of a space X. For each finite non-empty $E \subset J$ we write $U_E = \bigcap_{j\in} U_j$. We define a subspace $C(\mathcal{U})$ of $X \times \prod_{j\in J} I_j$, $I_j = I$, as the set of families $y = (x; t_j)$ such that:

(1) Only a finite number of the t_j are non-zero.

(2) $\sum_j t_j = 1$.

(3) If $J(y) = \{j \in J \mid t_j \neq 0\}$ then $x \in U_{J(y)}$.

We have coordinate maps $\mathrm{pr} = \mathrm{pr}^C : C(\mathcal{U}) \to X$, $(x; t_j) \mapsto x$ and $t_i : C(\mathcal{U}) \to I$, $(x; t_j) \mapsto t_i$. They are restrictions of the product projections and therefore continuous. The t_j form a point-finite partition of unity on $C(\mathcal{U})$. We view $C(\mathcal{U})$ via pr^C as a space over X.

We define a second space $B(\mathcal{U})$ with the same underlying set but with a new topology. Recall the nerve $N(\mathcal{U})$ of the covering $\mathcal{U}$. We have the simplicial $N(\mathcal{U})$-diagram which associates to a simplex E of the nerve the space U_E and to an inclusion $F \subset E$ of simplices the inclusion $r_F^E : U_E \to U_F$. The space $B(\mathcal{U})$ is the geometric realization of this $N(\mathcal{U})$-diagram. Thus $B(\mathcal{U})$ is the quotient space of $\coprod_E U_E \times \Delta(E)$ by the relation

$$U_E \times \Delta(E) \ni (x, d_F^E(a)) \sim (i_F^E(x), a) \in U_F \times \Delta(F).$$

The sum is taken over the finite non-empty subsets E of J. Let $\Delta(E)^\circ$ be the interior of $\Delta(E)$ and $\partial\Delta(E)$ its boundary. Then each element of $B(\mathcal{U})$ has a unique representative of the form $(x; t) \in U_E \times \Delta(E)^\circ$ for a unique E. We can interpret this element as an element of $C(\mathcal{U})$, and in this manner we obtain a bijection of sets $\rho\colon B(\mathcal{U}) \to C(\mathcal{U})$. This map is continuous, since the canonical maps $U_E \times \Delta(E) \to C(\mathcal{U})$ are continuous. The space $B(\mathcal{U})$ has a projection $\mathrm{pr} = \mathrm{pr}^B$ onto X and ρ is a map over X.

(13.2.2) Proposition. *The map ρ is a homotopy equivalence over X.*

Proof. We construct a map $\pi : C(\mathcal{U}) \to B(\mathcal{U})$. For this purpose we choose a locally finite partition of unity (τ_j) subordinate to the open covering $t_j^{-1}]0, 1]$ of $C(\mathcal{U})$.

Then we define

$$\pi \colon y = (x;t_j) \mapsto (x;\tau_j(y)) = z.$$

The map is well-defined and continuous: Let $j \in J(z)$, i.e., $0 \neq t_j(z) = \tau_j(y)$ hence $J(z) \subset J(y)$ and $x \in J(z)$; this shows that $z \in B(\mathcal{U})$. Let $W \subset C(\mathcal{U})$ be an open set such that $J(W) = \{j \mid \tau_j|W \neq 0\}$ is finite. Let $y = (x;t_j) \in W$. Then $J(W) \subset J(y)$, therefore π factors on W through a map $W \to U_{J(W)} \times \Delta(J(W))$, and this shows the continuity.

A homotopy $\rho\pi \simeq \mathrm{id}_{C(\mathcal{U})}$ is defined by

$$(y,t) = ((x;t_j),t) \mapsto (x;tt_j + (1-t)\tau_j).$$

This assignment is clearly well-defined and continuous. A homotopy $\pi\rho \simeq \mathrm{id}$ is defined by the same formula $(y;t) \mapsto (x;tt_j(y) + (1-t)\tau_j\rho(y))$. In order to verify the continuity, we let again W be as above, but now considered as a subset of $B(\mathcal{U})$. We consider the composition with $X_E \times \Delta(E) \to B(\mathcal{U})$. The formula for the homotopy on the pre-image of W has an image in $X_E \times \Delta(E)$. □

Let $B(\mathcal{U})^n$ be the subspace of $B(\mathcal{U})$ which is the image of the $U_E \times \Delta(E)$ with $|E| \leq n+1$. We state (13.2.1) for the special case at hand.

(13.2.3) Proposition. *$B(\mathcal{U})$ is the colimit of the sequence of subspaces $B(\mathcal{U})^n$. The canonical map $\coprod_{\dim E \leq n} U_E \times \Delta(E) \to B(\mathcal{U})^n$ is a quotient map. The inclusion $B^{n-1} \subset B^n$ is obtained via a pushout diagram*

$$\begin{array}{ccc} \coprod_{\dim E = n} U_E \times \partial\Delta(E) & \xrightarrow{k_n} & B(\mathcal{U})^{n-1} \\ \downarrow & & \downarrow \\ \coprod_{\dim E = n} U_E \times \Delta(E) & \xrightarrow{K_n} & B(\mathcal{U})^n. \end{array}$$

The map k_n is defined on $X_E \times \partial\Delta(E)$ as follows: Let $F \subset E$ be a proper subset. Then k_n is defined on $X_E \times \Delta(F)$ by $X_E \times \Delta(F) \subset X_F \times \Delta(F) \to B^{n-1}$. □

(13.2.4) Proposition. *Let $\mathcal{U}$ be numerable. Then the projections $C(\mathcal{U}) \to X$ and $B(\mathcal{U}) \to X$ are shrinkable.*

Proof. Let $(\tau_j \mid j \in J)$ a numeration of $\mathcal{U}$. Then $x \mapsto (x;\tau_j(x))$ is a section s of pr^C and $((x;t_j),t) \mapsto (x;tt_j + (1-t)\tau_j(x))$ is a homotopy from $s\,\mathrm{pr}^C$ to the identity over X. Thus pr^C is shrinkable, and (13.2.2) shows that also pr^B is shrinkable. □

For some applications we need a barycentric subdivision of $B(\mathcal{U})$. Recall that we have the barycentric subdivision $N'(\mathcal{U})$ of the nerve of $\mathcal{U}$. An n-simplex of $N'(\mathcal{U})$ is an ordered set $\tau = (E_0 \subsetneq E_1 \subsetneq \cdots \subsetneq E_n)$ such that $U_{E_n} \neq \emptyset$. We write $q(\tau) = E_n$. We have the $N'(\mathcal{U})$-diagram which associates to τ the space

$X_{q(\tau)}$ and to $\sigma \subset \tau$ the inclusion $X_{q(\tau)} \subset X_{q(\sigma)}$. Let $B'(\mathcal{U})$ denote the geometric realization of this $N'(\mathcal{U})$-diagram. Since the simplices are ordered, we can replace $\Delta(\tau)$ by the standard simplex $\Delta[n]$ spanned by $[n] = \{0, 1, \dots, n\}$.

(13.2.5) Remark. In the case of the barycentric subdivision the pushout diagram in (13.2.3) reads as follows:

$$\begin{array}{ccc} \coprod_{\tau \in A_n} X_{q(\tau)} \times \partial \Delta^n & \xrightarrow{k_n} & B'(\mathcal{U})^{n-1} \\ \big\downarrow{\scriptstyle j_n} & & \big\downarrow{\scriptstyle J_n} \\ \coprod_{\tau \in A_n} X_{q(\tau)} \times \Delta^n & \xrightarrow{K_n} & B'(\mathcal{U})^n. \end{array}$$

The sum is over the set $A_n = \{(\sigma_0, \dots, \sigma_n) \mid \sigma_0 \subsetneq \dots \subsetneq \sigma_n,\ \sigma_n \subset J \text{ finite}\}$, and $q(\sigma_0, \dots, \sigma_n) = \sigma_n$. ◇

13.3 Homotopy Equivalences

The main result (13.3.1) of this section asserts that being a homotopy equivalence is in a certain sense a local property.

(13.3.1) Theorem. *Let $p \colon X \to B$ and $q \colon Y \to B$ be spaces over B and $f \colon X \to Y$ a map over B. Let $\mathcal{X} = (X_j \mid j \in J)$ be a covering of X and $\mathcal{Y} = (Y_j \mid j \in J)$ a covering of Y. Let $f(X_j) \subset Y_j$ and assume that for each finite $E \subset J$ the map $_E \colon X_E \to Y_E$ induced by f is a homotopy equivalence over B. Then the induced map $B(f) \colon B(\mathcal{X}) \to B(\mathcal{Y})$ is a homotopy equivalence over B. Thus if the coverings $\mathcal{X}$ and $\mathcal{Y}$ are numerable, then f is a homotopy equivalence over B.*

Proof. From (5.3.4) and (13.2.3) we prove inductively that the induced maps $B(\mathcal{X})^n \to B(\mathcal{Y})^n$ are h-equivalences. Now we use (5.2.9), in order to show that Bf is an h-equivalence. In the case of numerable coverings we also use (13.2.4). □

(13.3.2) Remark. In the situation of (13.3.1) we can conclude that f is a homotopy equivalence, if the projections $p_X \colon B(\mathcal{X}) \to X$ and p_Y are homotopy equivalences. ◇

(13.3.3) Theorem. *Let $p \colon X \to B$ and $q \colon Y \to B$ be spaces over B and $f \colon X \to Y$ a map over B. Let $(U_j \mid j \in J)$ be a numerable covering of B. Let $f_j \colon p^{-1}(U_j) \to q^{-1}(U_j)$ be the map induced by f over U_j. Suppose each f_j is a fibrewise homotopy equivalence. Then f is a fibrewise homotopy equivalence.*

Proof. The hypothesis implies that f is a fibrewise homotopy equivalence over each set $V \subset V_j$. We can therefore apply (13.3.1). □

We say that a covering $(U_j \mid j \in J)$ of B is ***null homotopic*** if every inclusion $U_j \subset B$ is null homotopic.

(13.3.4) Theorem. *Let $f : X \to Y$ be a map over B from $p : X \to B$ to $q : Y \to B$. Assume that p and q are fibrations. Suppose B has a numerable null homotopic covering $(V_j \mid j \in J)$ and that each path component of B contains a point b such that f is a homotopy equivalence over b. The f is a fibrewise homotopy equivalence.*

Proof. Let $V_j \subset B$ be homotopic to the constant map $V_j \to \{b(j)\}$. We can assume that $f_{b(j)} : X_{b(j)} \to Y_{b(j)}$ is an h-equivalence. By the homotopy theorem for fibrations we obtain a homotopy commutative diagram of maps over V_j

$$\begin{array}{ccc} p^{-1}(V_j) & \xrightarrow{f_j} & q^{-1}(V_j) \\ \downarrow{\scriptstyle (1)} & & \downarrow{\scriptstyle (2)} \\ V_j \times X_{b(j)} & \xrightarrow{\mathrm{id} \times f_{b(j)}} & V_j \times Y_{b(j)} \end{array}$$

with fibrewise homotopy equivalences (1) and (2). Hence f_j is a fibrewise equivalence and we can apply (13.3.3) □

Problems

1. For each $j \in J$ we let $C(\mathcal{U})_j = \mathrm{pr}^{-1}(U_j)$ and similarly for B. Then the partial projection maps $\mathrm{pr}_j^C : C(\mathcal{U})_j \to U_j$ and $\mathrm{pr}_j^B : B(\mathcal{U})_j \to U_j$ are shrinkable.
2. If the coverings in (13.3.1) are open, then $f_* : [Z, X] \to [Z, Y]$ is for each paracompact Z a bijection. The canonical projection $p : B(\mathcal{X}) \to X$ induces for a paracompact space Z a bijection $p_* : [Z, B(\mathcal{X})] \to [Z, X]$.

13.4 Fibrations

(13.4.1) Theorem. *Let $\mathcal{V} = (B_j \mid j \in J)$ be a covering of B and $p : E \to B$ a continuous map. Assume that the map $p_j : p^{-1}(B_j) \to B_j$ induced by p is for each $j \in J$ a fibration. If the covering $\mathcal{V}$ is numerable, then p is a fibration. If the covering $\mathcal{V}$ is open, then p has the* HLP *for paracompact spaces.*

Proof. We have to solve a homotopy lifting problem (left diagram)

$$\begin{array}{ccc} X & \xrightarrow{a} & E \\ \downarrow{\scriptstyle i} & & \downarrow{\scriptstyle p} \\ X \times I & \xrightarrow{h} & B \end{array} \qquad\qquad \begin{array}{ccc} Y & \longrightarrow & E \\ \downarrow{\scriptstyle q} & & \downarrow{\scriptstyle p} \\ X \times I & \xrightarrow{h} & B. \end{array}$$

We form the pullback of p along h (right diagram). The initial condition a yields a section s_0 of q over $X \times 0$. A lifting of h with this initial condition amounts to a section s of q which extends s_0. We pull back the numerable covering of B to a numerable covering of $X \times I$. There exists a numerable covering $\mathcal{U} = (U_k \mid k \in K)$ of X such that q is a fibration over the sets $U_k \times I$ (Problem 1). We begin by constructing a lifting $t\colon B(\mathcal{U}) \times I \to E$ of pr^B which extends the lifting t_0 over $B(\mathcal{U}) \times 0$ determined by s_0. The lifting is constructed inductively from partial liftings t_n over $B(\mathcal{U})^n \times I$. The induction step is again based on the pushout diagram (13.2.3), now multiplied by I. The extension of the lifting t_n amounts to solving a lifting problem of the type

$$\begin{array}{ccc} \coprod U_E \times (\partial\Delta(E) \times I \cup \Delta(E) \times 0) & \longrightarrow & Y \\ \downarrow & \nearrow & \downarrow{\scriptstyle q} \\ \coprod U_E \times \Delta(E) \times I & \longrightarrow & X \times I \end{array}$$

and this is possible (by (5.5.3)), because $U_E \times \Delta(E) \times I$ is mapped into a subset over which q is a fibration. If the covering $\mathcal{U}$ is numerable we compose it with a section of pr^B and obtain the desired extension of s_0. □

(13.4.2) Theorem. *Let $p\colon Y \to X$ be a continuous map. Let $\mathcal{Y} = (Y_j \mid j \in J)$ be a family of subsets of Y and $\mathcal{X} = (X_j \mid j \in J)$ a numerable covering of X. Assume that $p(Y_j) \subset X_j$ and that for finite $E \subset J$ the map $p_E\colon Y_E \to X_E$ induced by p is shrinkable. Then p has a section.* (*Note that $\mathcal{Y}$ is not assumed to be a covering of Y.*)

Proof. We work with the barycentric subdivision $B'(\mathcal{X})$. We show the existence of a map $s\colon B'(\mathcal{X}) \to Y$ such that $ps = \mathrm{pr}^B$. The proof does not use the numerability of the covering. We construct inductively maps $s^n\colon B'(\mathcal{X})^n \to Y$ with the appropriate properties and an additional property which makes the induction work.

The map $B^0 = \coprod_E X_E \to Y$ is given as follows: We choose sections $X_E \to Y_E$ of p_E and compose them with the inclusion $Y_E \subset Y$.

Suppose s^{n-1} is given. We want to extend

$$s^{n-1}k_n\colon \coprod (X_{q(\tau)} \times \partial\Delta[n]) \to E$$

over $\coprod (X_{q(\tau)} \times \Delta[n])$. If $\tau = (E_0, \dots, E_n)$ we impose the additional hypothesis that the image of $X_{q(\tau)} \times \partial\Delta[n]$ under $s^{n-1}k_n$ is contained in Y_{E_0}. The construction of s^0 agrees with this requirement. Under this additional hypothesis we have a commutative diagram

$$\begin{array}{ccc} X_{q(\tau)} \times \partial\Delta[n] & \xrightarrow{s^{n-1}k_n} & Y_{E_0} \\ \downarrow & & \downarrow \\ X_{q(\tau)} & \xrightarrow{\subset} & X_{E_0}. \end{array}$$

From (5.5.3) we see that $s^{n-1}k_n$ can be extended over $\coprod X_{q(\tau)} \times \Delta[n]$. With an extension we construct s^n via the pushout (13.2.5). We show that s^n satisfies the additional hypothesis. Given $\tau = (E_0, \dots, E_{n+1})$ we describe

$$k_{n+1}: X_{q(\tau)} \times \partial\Delta[n+1] \to B(\mathcal{U})^n.$$

Let $d_i : \Delta[n] \to \Delta[n]_i$ be the standard map onto the i-th face of $\Delta[n+1]$ with inverse homeomorphism e_i. Let $\partial_i : X_{q(\tau)} \to X_{q(\varepsilon_i \tau)}$ be the inclusion where $\varepsilon_i \tau = (E_0, \dots, E_{i-1}, E_{i+1}, \dots, E_{n+1})$. The restriction of k_{n+1} to the subset $X_{q(\tau)} \times \Delta[n+1]_i$ is $K_n(\partial_i \times e_i)$. By construction of s^n the image of $s^n K_n(\partial_i \times e_i)$ is contained in X_{E_0} (for $i > 0$) or X_{E_1} (for $i = 0$). But $X_{E_1} \subset X_{E_0}$, hence s^n has the desired property.

If $\mathcal{X}$ is numerable, then pr has a section t and st is a section of p. □

(13.4.3) Theorem. *Let $p: X \to B$ be a continuous map and $\mathcal{X} = (X_j \mid j \in J)$ a numerable covering of X. Assume that for each finite $E \subset J$ the restriction $p_E : X_E \to B$ is a fibration (an h-fibration, shrinkable). Then p is a fibration (an h-fibration, shrinkable).*

Proof. Recall that for a fibration $p: X \to B$ the canonical map $r: X^I \to W(p)$ is shrinkable (see (5.6.5)), and that p is a fibration if this map has a section (see (5.5.1)). If the p_E are fibrations, then the $r_E : X_E^I \to W(p_E)$ are shrinkable. The $W(p_j)$ form a numerable covering of $W(p)$. Theorem (13.4.2) shows that r has a section, hence p is a fibration.

Assume that the p_E are shrinkable, i.e., homotopy equivalences over B. We apply (13.3.1) and see that p is shrinkable.

Assume that the p_E are h-fibrations. The map p is an h-fibration if and only if the canonical map $b: W(p) \to X$ is a homotopy equivalence over B (this can be taken as a definition of an h-fibration). The $W(p_j)$ form a numerable covering of $W(p)$ and the $b_E : W(p_E) \to X_E$ are homotopy equivalences over B, since p_E are h-fibrations. Thus we are in a position where (13.3.1) can be applied. □

The hypothesis of (13.4.3) is satisfied if the X_E are either empty or contractible.

Problems

1. If $q: M \to N \times [a,b]$ is a fibration over $N \times [a,c]$ and $N \times [c,b]$, then q is a fibration.
2. Let $\mathcal{X} = (X_j \mid j \in J)$ be a numerable covering of X. If the spaces X_E have the homotopy type of a CW-complex, then X has the homotopy type of a CW-complex.
3. Let $p: E \to B$ be an h-fibration. Suppose B and each fibre $p^{-1}(b)$ have the homotopy type of a CW-complex. Then E has the homotopy type of a CW-complex.

Chapter 14
Bundles

Bundles (also called fibre bundles) are one of the main objects and tools in topology and geometry. They are locally trivial maps with some additional structure. A basic example in geometry is the tangent bundle of a smooth manifold and its associated principal bundle. They codify the global information that is contained in the transitions functions (coordinate changes).

The classification of bundles is reduced to a homotopy problem. This is achieved via universal bundles and classifying spaces. We construct for each topological group G the universal G-principal bundle $EG \to BG$ over the so-called classifying space BG. The isomorphism classes of numerable bundles over X are then in bijection with the homotopy set $[X, BG]$.

The classification of vector bundles is equivalent to the classification of their associated principal bundles. A similar equivalence holds between n-fold covering spaces and principal bundles for the symmetric group S_n. This leads to a different setting for the classification of coverings.

From the set of (complex) vector bundles over a space X and their linear algebra one constructs the Grothendieck ring $K(X)$. The famous Bott periodicity theorem in one of its formulations is used to make the functor $K(X)$ part of a cohomology theory, the so-called topological K-theory. Unfortunately lack of space prevents us from developing this very important aspect.

Classifying spaces and universal bundles have other uses, and the reader may search in the literature for information.

The cohomology ring $H^*(BG)$ of the classifying space BG of a discrete group G is also called the cohomology of the group. There is a purely algebraic theory which deals with such objects.

If X is a G-space, one can form the associated bundle $EG \times_G X \to BG$. This bundle over BG can be interpreted as an invariant of the transformation group X. The cohomology of $EG \times_G X$ is a module over the cohomology ring $H^*(BG)$ (Borel-cohomology). The module structure contains some information about the transformation group X, e.g., about its fixed point set (see [7], [43]).

14.1 Principal Bundles

Let G be a topological group. In the general theory we use multiplicative notation and denote the unit element of G by e. Let $r\colon E \times G \to E$, $(x, g) \mapsto xg$ be a continuous right action of G on E, and $p\colon E \to B$ a continuous map. The pair (p, r) is called a (right) ***G*-principal bundle** if the following axioms hold:

(1) For $x \in E$ and $g \in G$ we have $p(xg) = p(x)$.

(2) For each $b \in B$ there exists an open neighbourhood U of b in B and a G-homeomorphism $\varphi\colon p^{-1}(U) \to U \times G$ which is a trivialization of p over U with typical fibre G. Here G acts on $U \times G$ by $((u,h),g) \mapsto (u,hg)$.

If we talk about a G-principal bundle $p\colon E \to B$, we understand a given action of G on E. From the axioms we see that G acts freely on E. The map p factors through the orbit map $q\colon E \to E/G$ and induces a continuous bijection $h\colon E/G \to B$. Since q and p are open maps, hence quotient maps, h is a homeomorphism. Thus G-principal bundles can be identified with suitable free right G-spaces. In contrast to an arbitrary locally trivial map with typical fibre G, the local trivializations in a principal bundle have to be compatible with the group action. In a similar manner we define left principal bundles.

A G-principal bundle with a discrete group G is called a ***G-principal covering***. The continuity of the action r is in this case equivalent to the continuity of all right translations $r_g\colon E \to E$, $x \mapsto xg$. This is due to the fact that $E \times G$ is homeomorphic to the topological sum $\amalg_{g \in G} E \times \{g\}$, if G is discrete.

Let $E \times G \to E$ be a free action and set $C(E) = \{(x,xg) \mid x \in E,\ g \in G\}$. We call $t = t_E : C(E) \to G$, $(x,xg) \mapsto g$ the ***translation map*** of the action.

(14.1.1) Lemma. *Let $p\colon E \to E/G$ be locally trivial. Then the translation map is continuous.*

Proof. Let $W = p^{-1}(U) \subset E$ be a G-stable open set which admits a trivialization $\psi\colon U \times G \to W$. The pre-image of $(W \times W) \cap C(E)$ under $\psi \times \psi$ is $\{(u,g,u,h) \mid u \in U,\ g,h \in G\}$, and $t \circ (\psi \times \psi)$ is the continuous map $(u,g,u,h) \mapsto g^{-1}h$. □

A free G-action on E is called ***weakly proper*** if the translation map is continuous. It is called ***proper*** if, in addition, $C(E)$ is closed in $E \times E$.

(14.1.2) Proposition. *A free action of G on E is weakly proper if and only if $\theta'\colon E \times G \to C(E)$, $(x,g) \mapsto (x,xg)$ is a homeomorphism.*

Proof. The map $\psi\colon C(E) \to E \times G$, $(x,y) \mapsto (x, t_E(x,y))$ is a set-theoretical inverse of θ'. It is continuous if and only if t_E is continuous. □

Let E carry a free right G-action and F a left G-action. We have a commutative diagram

$$\begin{array}{ccc} E \times F & \xrightarrow{\mathrm{pr}_1} & E \\ \downarrow{\scriptstyle P} & & \downarrow{\scriptstyle p} \\ E \times_G F & \xrightarrow{q} & E/G \end{array}$$

with orbit maps P and p and $q = \mathrm{pr}_1/G$.

(14.1.3) Proposition. *A free right G action on E is weakly proper if and only if for each left G-space F the diagram is a topological pullback.*

Proof. We compare the diagram with the canonical pullback

$$\begin{array}{ccc} X & \xrightarrow{\;b\;} & E \\ \big\downarrow{\scriptstyle a} & & \big\downarrow{\scriptstyle p} \\ E \times_G F & \xrightarrow{\;q\;} & E/G \end{array}$$

with $X = \{((x,f),y) \in (E \times_G F) \times E \mid p(x) = p(y)\}$ and $a = \mathrm{pr}_1$, $b = \mathrm{pr}_2$. There exists a unique map $\lambda \colon E \times F \to X$ such that $b\lambda = \mathrm{pr}_1$, $a\lambda = P$, i.e., $\lambda(x,f) = ((x,f),x)$. The diagram in question is a pullback if and only if λ is a homeomorphism. Suppose this is the case for the left G-space G. The homeomorphism $E \times_G G \to E$, $(x,g) \mapsto xg$ transforms q into p, X into $C(E)$ and λ into $(x,g) \mapsto (xg,x)$. The latter is, in different notation, θ'. Hence θ' is a homeomorphism if the diagram is a pullback for $F = G$.

Conversely, let the action be weakly proper. The map

$$\tilde{\mu} \colon \tilde{X} = \{((x,f),y) \mid p(x) = p(y)\} \to E \times F, \quad ((x,f),y) \mapsto (y, t(x,y)^{-1} f)$$

is continuous. One verifies that $\tilde{\mu}$ induces a map $\mu \colon X \to E \times F$. The equalities

$$\mu\lambda(x,f) = \mu((x,f),x) = (x, t^{-1}(x,x)f) = (x,f)$$

show that μ is an inverse of λ. □

(14.1.4) Proposition. *Let G act freely and weakly properly on E. The sections of $q \colon E \times_G F \to E/G$ correspond bijectively to the maps $f \colon E \to F$ with the property $f(xg) = g^{-1} f(x)$; here we assign to f the section $s_f \colon x \mapsto (x, f(x))$.*

Proof. It should be clear that s_f is a continuous section. Conversely, let $s \colon B \to E \times_G F$ be a continuous section. We use the pullback diagram displayed before (14.1.3). It yields an induced section σ of pr_1 which is determined by the conditions $\mathrm{pr}_1 \circ \sigma = \mathrm{id}$ and $P \circ \sigma = s \circ p$. Let $f = \mathrm{pr}_2 \circ \sigma \colon E \to F$. Then $\mathrm{pr}_1 \sigma(xg) = xg = \mathrm{pr}_1(\sigma(x)g)$, since σ is a section and pr_1 a G-map. (The right action on $E \times F$ is $(e,f,g) \mapsto (eg, g^{-1}f)$.) The equalities $P\sigma(xg) = sp(xg) = sp(x) = P\sigma(x) = P(\sigma(x)g)$ hold, since σ is an induced section and P the orbit map. Since $\sigma(xg)$ and $\sigma(x)g$ have the same image under P and pr_1, these elements are equal; we now apply the G-map pr_2 and obtain finally $f(xg) = g^{-1} f(x)$. □

(14.1.5) Proposition. *Let the free G-action on E be weakly proper. Then the orbit map $p \colon E \to E/G = B$ is isomorphic to* $\mathrm{pr} \colon B \times G \to B$, *if and only if p has a section.*

Proof. Let s be a section of p. Then $B \times G \to E$, $(b, g) \mapsto s(b)g$ and $E \to B \times G$, $x \mapsto (p(x), t(spx, x))$ are inverse G-homeomorphisms, compatible with the projections to B. Conversely, pr has a section and hence also the isomorphic map p. □

(14.1.6) Proposition. *Let X and Y be free G-spaces and $\Phi\colon X \to Y$ a G-map. If $\varphi = \Phi/G$ is a homeomorphism and Y weakly proper, then Φ is a homeomorphism.*

Proof. X is weakly proper, since the translation map of X is obtained from the translation map of Y by composition with $\Phi \times \Phi$. We have to find an inverse $\Psi\colon Y \to X$. By (14.1.4), it corresponds to a section of $\pi_Y\colon (Y \times X)/G \to Y/G$. We have the section $s\colon x \mapsto (x, \Phi(x))$ of $\pi_X\colon (X \times Y)/G \to X/G$. Let ψ be the inverse of φ. With the interchange map $\tau\colon (X \times Y)/G \to (Y \times X)/G$ we form $\sigma = \tau \circ s \circ \psi$. One verifies that σ is a section of π_Y. □

Let a commutative diagram below with principal G-bundles p and q be given, and let F be a G-map. Then F or (F, f) is called a ***bundle map***.

$$\begin{array}{ccc} Y & \xrightarrow{F} & X \\ \downarrow{\scriptstyle q} & & \downarrow{\scriptstyle p} \\ C & \xrightarrow{f} & B \end{array}$$

If f is a homeomorphism, then F is a homeomorphism (see (14.1.6)). If f is the identity, then F is called a ***bundle isomorphism***.

Given a principal bundle $p\colon X \to B$ and a map $f\colon C \to B$, we have a pullback diagram as above with $Y = \{(c, x) \mid f(c) = p(x)\} \subset C \times X$. The maps q and F are the restrictions of the projections onto the factors. The G-action on Y is $(c, x)g = (c, xg)$. If p is trivial over V, then q is trivial over $f^{-1}(V)$. Therefore q is a principal bundle, called the bundle ***induced*** from p by f. Also F is a bundle map. From the universal property of a pullback we see, that the bundle map diagram above is a pullback.

(14.1.7) Proposition. *Let U be a right G-space. The following are equivalent:*

(1) *There exists a G map $f\colon U \to G$.*
(2) *There exists a subset $A \subset U$ such that $m\colon A \times G \to U$, $(a, g) \mapsto ag$ is a homeomorphism.*
(3) *The orbit map $p\colon U \to U/G$ is G-homeomorphic over U/G to the projection $\mathrm{pr}\colon U/G \times G \to U/G$.*
(4) *U is a free G-space, $p\colon U \to U/G$ has a section, and t_U is continuous.*

Proof. (1) $\Rightarrow$ (2). Let $A = f^{-1}(e)$ and $v\colon U \to A$, $x \mapsto x \cdot f(x)^{-1}$. Then $(v, f)\colon U \to A \times G$ is an inverse of m.

(2) $\Rightarrow$ (3). The G-homeomorphism m induces a homeomorphism m/G of the orbit spaces. We have the homeomorphism $\varepsilon\colon A \to (A \times G)/G, a \mapsto (a, e)$. With these data $m \circ (\varepsilon \times \mathrm{id}) \circ m/G = \mathrm{pr}$.

(3) $\Rightarrow$ (4). If $\varphi\colon U/G \times G \to U$ is a G-homeomorphism over U/G, then the G-action is free and $x \mapsto \varphi(x, e)$ is a section of p. The translation map of $U/G \times G$ is continuous and hence, via φ, also t_U.

(4) $\Rightarrow$ (1). Let $s\colon U/G \to G$ be a section. Then $U \to G, u \mapsto t_U(sp(u), u)$ is a G-map. □

A right G-space U is called ***trivial*** if there exists a continuous G-map $f\colon U \to G$ into the G-space G with right translation action. A right G-space is called ***locally trivial*** if it has an open covering by trivial G-subspaces.

(14.1.8) Proposition. *The total space E of a G-principal bundle is locally trivial. If E is locally trivial, then $E \to E/G$ is a G-principal bundle.* □

14.1.9 Hopf fibrations. Consider $S^{2n-1} \subset \mathbb{C}^n$ as a free S^1-space with action induced from scalar multiplication. Let U_j be the subset of points $z = (z_k)$ with $z_j \neq 0$. The map $z \mapsto z_j |z_j|^{-1}$ shows that U_j is a trivial S^1-space. The orbit space of this action is $\mathbb{C}P^{n-1}$. The S^1-principal bundle $p\colon S^{2n-1} \to \mathbb{C}P^{n-1}$, i.e., the orbit map, is called a Hopf fibration. There is a similar $\mathbb{Z}/2$-principal bundle $S^{n-1} \to \mathbb{R}P^{n-1}$ onto the real projective space and an S^3-principal bundle $S^{4n-1} \to \mathbb{H}P^{n-1}$ onto the quaternionic projective space. ◇

(14.1.10) Proposition. *Let $f\colon X \to Y$ be a G-map and $p_Y\colon Y \to Y/G$ a G-principal bundle. Then the diagram*

$$\begin{array}{ccc} X & \xrightarrow{f} & Y \\ \downarrow{\scriptstyle p_X} & & \downarrow{\scriptstyle p_Y} \\ X/G & \xrightarrow{f/G} & Y/G \end{array}$$

is a pullback.

Proof. Let $U \subset Y$ be a G-set with a G-map $h\colon U \to G$. Then $h \circ f\colon f^{-1}(U) \to U \to G$ is a G-map. Hence p_X is a G-principal bundle. The diagram is therefore a bundle map and hence a pullback. □

We say, a map $f\colon X \to Y$ has ***local sections*** if each $y \in Y$ has an open neighbourhood V and a section $s\colon V \to X$ of f over V; the latter means $fs(v) = v$ for all $v \in V$.

(14.1.11) Proposition. *Let G be a topological group and H a subgroup. The quotient map $q\colon G \to G/H$ is an H-principal bundle if and only if q has a section over some neighbourhood of the unit coset.*

Proof. A locally trivial map has local sections. Conversely, let $s: U \to G$ be a section of q over U. The map t_G is continuous. For each $k \in G$ we have the H-equivariant map $kq^{-1}(U) \to H$, $kg \mapsto s(q(g))^{-1}g$. We thus can apply (14.1.8) and (14.1.7). □

Actions with the properties of the next proposition were called earlier properly discontinuous.

(14.1.12) Proposition. *Let $E \times G \to E$, $(x, g) \mapsto xg$ be a free right action of the discrete group G. The following assertions are equivalent:*

(1) *The orbit map $p: E \to E/G$ is a G-principal covering.*

(2) *Each $x \in E$ has a neighbourhood U, such that $U \cap Ug = \emptyset$ for each $g \neq e$.*

(3) *The set $t^{-1}(e)$ is open in C.*

(4) *The map t is continuous.*

Proof. (1) ⇒ (4) holds by (14.1.1).

(4) ⇒ (3). The set $\{e\} \subset G$ is open, since G is discrete.

(3) ⇒ (2). We have $t(x, x) = e$. Since $t^{-1}(e)$ is open, there exists an open neighbourhood U of x in E such that $(U \times U) \cap C \subset t^{-1}(e)$. Let $U \cap Ug \neq \emptyset$, say $v = ug$ for $u, v \in U$. Then $(u, v) = (u, ug) \in (U \times U) \cap C$, hence $t(u, ug) = g = e$.

(2) ⇒ (1). Let U be open. Then $U \times G \to UG$, $(u, g) \mapsto ug$ is a G-homeomorphism, hence UG an open trivial G-subspace. □

(14.1.13) Example. Let G be a closed discrete subgroup of the topological group E. Then the action $G \times E \to E$, $(g, x) \mapsto gx$ is free and has property (4) of the previous proposition. Examples are $\mathbb{Z} \subset \mathbb{R}$ or $\mathbb{Z} \subset \mathbb{C}$. ◇

(14.1.14) Example. The map $g: \mathbb{R} \to S^1$, $t \mapsto \exp(2\pi i t)$ has kernel $\mathbb{Z}$. Therefore there exists a bijective map $f: \mathbb{R}/\mathbb{Z} \to S^1$ such that $fp = g$. Since g is an open map, g is a quotient map and therefore f a homeomorphism. By the previous example, g is therefore a covering. Similarly $\exp: \mathbb{C} \to \mathbb{C}^*$ is seen to be a covering. ◇

(14.1.15) Example. Let G be a Lie group and H a closed subgroup. Then the quotient map $p: G \to G/H$ is an H-principal bundle. In the chapter on differentiable manifolds we show that G/H carries the structure of a smooth manifold such that p is a submersion. A submersion has always smooth local sections. ◇

We construct locally trivial bundles from principal bundles. Let $p: E \to B$ be a right G-principal bundle and F a left G-space. The projection $E \times F \to E$ induces, via passage to orbit spaces, $q: E \times_G F \to B$. The map q is locally trivial with typical fibre F. A bundle chart $\varphi\colon p^{-1}(U) \to U \times G$ of p yields a bundle chart

$$q^{-1}(U) = p^{-1}(U) \times_G F \xrightarrow{\varphi \times_G \mathrm{id}} (U \times G) \times_G F \cong U \times F$$

for q. We call q the ***associated fibre bundle*** with typical fibre F, and G is said to be the ***structure group*** of this fibre bundle. The structure group contains additional information: The local trivializations have the property that the transition functions are given by homeomorphisms of the fibre which arise from an action of an element of the group G.

Let $p\colon Y \to B$ be a right G-principal bundle. It may happen that there exists a right H-principal bundle $q\colon X \to B$ for a subgroup $H \subset G$ and a G-homeomorphism $\rho\colon X \times_H G \to Y$ over B. In that case (q, ρ) is called a ***reduction of the structure*** of p. One can consider more generally a similar problem for homomorphisms $\alpha\colon H \to G$.

(14.1.16) Example. Let $E \to B$ be a G-principal bundle and $H \subset G$ a subgroup. Then $E \times_G G/H \to E/H$, $(x, gH) \mapsto xgH$ is a homeomorphism. Therefore $E/H \to E/G$, $xH \mapsto xG$ is isomorphic to the associated bundle $E \times_G G/H \to E/G$. From a subgroup $K \subset H \subset G$ we obtain in this manner $G/K \to G/H$ as a bundle with structure group H and fibre H/K, if $G \to G/H$ has local sections.

If X is a G-space and $H \lhd G$, then G/H acts on X/H by $(xH, gH) \mapsto xgH$. The quotient map $X/H \to X/G$ induces a homeomorphism $(X/H)/(G/H) \cong X/G$. In particular, $E/H \to E/G$ becomes in this manner a G/H-principal bundle. ◇

One can "topologize" various algebraic notions in analogy to the passage from groups to topological groups. An important notion in this respect is that of a small category.

A (small) ***topological category*** C consists of an object space $\mathrm{Ob}(C)$ and a morphism space $\mathrm{Mor}(C)$ such that the structure data

$$\begin{array}{lll} s\colon \mathrm{Mor}(C) \to \mathrm{Ob}(C) & & \text{(source)}, \\ r\colon \mathrm{Mor}(C) \to \mathrm{Ob}(C) & & \text{(range)}, \\ i\colon \mathrm{Ob}(C) \to \mathrm{Mor}(C) & & \text{(identity)}, \\ c\colon \mathrm{Mor}_2(C) \to \mathrm{Mor}(C) & & \text{(composition)} \end{array}$$

are continuous. Here

$$\mathrm{Mor}_2(C) = \{(\beta, \alpha) \in \mathrm{Mor}(C) \times \mathrm{Mor}(C) \mid s(\beta) = r(\alpha)\}$$

carries the subspace topology of $\mathrm{Mor}(C) \times \mathrm{Mor}(C)$. For these data the usual axioms of a category hold. In a groupoid each morphism has an inverse. In a ***topological***

groupoid we require in addition that passage to the inverse is a continuous map $\mathrm{Mor}(C) \to \mathrm{Mor}(C)$.

A topological group can be considered as a topological groupoid with object space a point. Principal bundles yield a parametrized version.

(14.1.17) Example. Let E be a free right G-space with a weakly proper action. Let $p: E \to B$ be a map which factors over the orbit map $q: E \to E/G$ and induces a homeomorphism $E/G \cong B$. We construct a topological groupoid with object space B. The product $p \times p: E \times E \to B \times B$ factors over the orbit map $Q: E \times E \to (E \times E)/G$, $(a,b) \mapsto [a,b]$ of the diagonal action and induces

$$(s,r): (E \times E)/G \to B \times B.$$

We define $(E \times E)/G$ as the morphism space of our category and s, r as source, range. The diagonal of E induces $i: B \cong E/G \to (E \times E)/G$, and this is defined to be the identity. We define the composition by

$$[a,b] \circ [x,y] = [x, b \cdot t(y,a)^{-1}]$$

with the translation map $t = t_E$ of E. One verifies that composition is associative and continuous. (The space Mor_2 is a quotient space of $E \times C(E) \times E$.) The morphism $[b,a]$ is inverse to $[a,b]$, hence the inverse is continuous and we have obtained a topological groupoid. ◇

Problems

1. A free action of a finite group G on a Hausdorff space E is proper.

2. The action $\mathbb{R}^2 \smallsetminus 0 \times \mathbb{R}^* \to \mathbb{R}^2 \smallsetminus 0$, $((x,y),t) \mapsto (tx, t^{-1}y)$ is a non-trivial $\mathbb{R}^*$-principal bundle. Determine the orbits and the orbit space.

3. Let E be a space with a free right G action. Then the translation map t_E is continuous if and only if the pullback of p along p is a trivial G-space.

4. The continuous maps $E \times_G F_1 \to E \times_G F_2$ over E/G correspond via $(x,u) \mapsto (x, \alpha(x,u))$ to the continuous maps $\alpha: E \times F_1 \to F_2$ with the equivariance condition $g\alpha(x,u) = \alpha(xg^{-1}gu)$.

If E is connected and F_1, F_2 discrete, then α does not depend on $x \in E$ and has the form $\beta \circ \mathrm{pr}$ with a uniquely determined equivariant map $\beta: F_1 \to F_2$.

5. The groupoid (14.1.17) has further properties. There exists at least one morphism between any two objects. The map $(s,r): (E \times E)/G \to B \times B$ is open.

6. Reconsider in the light of (14.1.17) the topological groupoid that was used in the determination of the fundamental groupoid of S^1.

14.2 Vector Bundles

Vector bundles are, roughly speaking, continuous families of vector spaces. Suppose given a continuous map $p: X \to B$ and the structure of an n-dimensional

$\mathbb{R}$-vector space on each fibre $X_b = p^{-1}(b)$. A ***bundle chart*** or a ***trivialization*** over the open basic set $U \subset B$ for these data is a homeomorphism $\varphi\colon p^{-1}(U) \to U \times \mathbb{R}^n$ over U which is fibrewise linear. A set of bundle charts is a ***bundle atlas***, if their basic domains cover B. The data $p\colon X \to B$ together with the vector space structures on the fibres are an ***n-dimensional real vector bundle*** over the space B if a bundle atlas exists. Thus a vector bundle is in particular a locally trivial map. In a similar manner one defines complex vector bundles or quaternionic vector bundles. A vector bundle $\xi\colon E(\xi) \to B$ is called ***numerable***, if there exists a numerable covering $\mathcal{U}$ of B such that ξ is trivial over the members $U \in \mathcal{U}$. (This notion is also used for other types of locally trivial bundles.) A bundle has ***finite type***, if it has a finite bundle atlas.

Let (U, φ) and (V, ψ) be bundle charts for p. Then the ***transition map*** is

$$\psi\varphi^{-1}\colon (U \cap V) \times \mathbb{R}^n \to (U \cap V) \times \mathbb{R}^n, \quad (x, v) \mapsto (x, g_x(v))$$

with $g_x \in \mathrm{GL}_n(\mathbb{R})$. The assignment $g\colon U \cap V \to \mathrm{GL}_n(\mathbb{R})$, $x \mapsto g_x$ is continuous.

A bundle atlas is said to be ***orienting***, if the g_x have positive determinant. If an orienting atlas exists, then the bundle is ***orientable***. An ***orientation*** of a vector bundle $p\colon E \to B$ consists of a vector space orientation of each fibre $p^{-1}(b)$ with the property: For each $x \in B$ there exists a bundle chart (U, φ) about x such that φ transports for each $b \in U$ the given orientation on $p^{-1}(b)$ into the standard orientation of $\mathbb{R}^n$. A chart with this property is called ***positive*** with respect to the given orientation. The positive charts form an orienting atlas, and for each orienting atlas there exists a unique orientation such that its charts are positive with respect to the orientation. A complex vector space has a canonical orientation. If one uses this orientation in each fibre, then the bundle, considered as a real bundle, is an oriented bundle.

Let $\xi\colon E(\xi) \to B$ and $\eta\colon E(\eta) \to C$ be real vector bundles. A ***bundle morphism*** $\xi \to \eta$ over $\varphi\colon B \to C$ is a commutative diagram

$$\begin{array}{ccc} E(\xi) & \xrightarrow{\Phi} & E(\eta) \\ \downarrow{\scriptstyle \xi} & & \downarrow{\scriptstyle \eta} \\ B & \xrightarrow{\varphi} & C \end{array}$$

with a map Φ which is fibrewise linear. If Φ is bijective on fibres, then we call the bundle morphism a ***bundle map***. Thus we have categories of vector bundles with bundle morphisms or bundle maps as morphisms. The trivial n-dimensional bundle is the product bundle $\mathrm{pr}\colon B \times \mathbb{R}^n \to B$. More generally, we call a bundle trivial if it is isomorphic to the product bundle.

(14.2.1) Proposition. *A bundle map over the identity is a bundle isomorphism.*

Proof. We have to show that the inverse of Φ is continuous. Via bundle charts this can be reduced to a bundle map between trivial bundles

$$U \times \mathbb{R}^n \to U \times \mathbb{R}^n, \quad (u, v) \mapsto (u, g_u(v)).$$

In that case $(u, v) \mapsto (u, g_u^{-1}(v))$ is a continuous inverse, since $u \mapsto g_u^{-1}$ is continuous. □

(14.2.2) Proposition. *If the previous diagram is a pullback in* TOP *and η a vector bundle, then there exists a unique structure of a vector bundle on ξ such that the diagram is a bundle map.*

Proof. Since Φ is bijective on fibres, we define the vector space structures in such a way that Φ becomes fibrewise linear. It remains to show the existence of bundle charts.

If η is the product bundle, then ξ can be taken as product bundle. If η has a bundle chart over V, then ξ has a bundle chart over $\varphi^{-1}(V)$, by transitivity of pullbacks. □

We call ξ in (14.2.2) the bundle ***induced*** from η along φ and write occasionally $\xi = \varphi^*\eta$ in this situation. The previous considerations show that a bundle map is a pullback. (Compare the analogous situation for principal bundles.)

A bundle morphism over $\mathrm{id}(B)$ has for each $b \in B$ a rank, the rank of the linear map between the fibres over b. It is then clear what we mean by a bundle morphism of constant rank.

A subset $E' \subset E$ of an n-dimensional real bundle $p\colon E \to B$ is a k-dimensional ***subbundle*** of p, if there exists an atlas of bundle charts (called ***adapted charts***) $\varphi\colon p^{-1}(U) \to U \times \mathbb{R}^n$ such that $\varphi(E' \cap p^{-1}(U)) = U \times (\mathbb{R}^k \times 0)$. The restriction $p\colon E' \to B$ is then a vector bundle.

A vector bundle is to be considered as a continuous family of vector spaces. We can apply constructions and notions from linear algebra to these vector spaces. We begin with kernels, cokernels, and images.

(14.2.3) Proposition. *Let $\alpha\colon \xi_1 \to \xi_2$ be a bundle morphism over B of constant rank and α_b the induced linear map on the fibres over b. Then the following hold:*

(1) $\operatorname{Ker} \alpha = \bigcup_{b \in B} \operatorname{Ker}(\alpha_b) \subset E(\xi_1)$ *is a subbundle of* ξ_1.
(2) $\operatorname{Im}(\alpha) = \bigcup_{b \in B} \operatorname{Im}(\alpha_b) \subset E(\xi_2)$ *is a subbundle of* ξ_2.
(3) *Suppose* $\operatorname{Coker}(\alpha) = \bigcup_{b \in B} E(\xi_2)_b / \operatorname{Im}(\alpha_b)$ *carries the quotient topology from* $E(\xi_2)$. *Then, with the canonical projection onto* B, $\operatorname{Coker}(\alpha)$ *is a vector bundle.*

Proof. The problem in all three cases is the existence of bundle charts. This is a local problem. Therefore it suffices to consider morphisms

$$\alpha\colon B \times \mathbb{R}^m \to B \times \mathbb{R}^n, \quad (b, v) \mapsto (b, \alpha_b(v))$$

between trivial bundles.

We write $K_x = \mathrm{Ker}(\alpha_x)$ and $L_x = \mathrm{Im}(\alpha_x)$. We fix $b \in B$ and choose complements $\mathbb{R}^m = K \oplus K'$ and $\mathbb{R}^n = L \oplus L'$ for $K = K_b$ and $L = L_b$. Let $q\colon \mathbb{R}^m \to K$ and $p\colon \mathbb{R}^n \to L$ be the projections with $\mathrm{Ker}(q) = K'$ and $\mathrm{Ker}(p) = L'$. Then

$$\gamma_x\colon \mathbb{R}^m \oplus L' \to \mathbb{R}^n \oplus K, \quad (v, w) \mapsto (\alpha_x(v) + w, q(v))$$

is an isomorphism for $x = b$, hence also an isomorphism for x in a neighbourhood of b. Thus let us assume without essential restriction that γ_x is always an isomorphism. Since α_x has constant rank k, we conclude that $K_x \cap K' = 0$ and $L_x \cap L' = 0$. This fact is used to verify that

$$B \times \mathbb{R}^m \to B \times (L \times K), \quad (x, v) \mapsto (x, p\alpha_x(v), q(v))$$

$$B \times (K' \times L') \to B \times \mathbb{R}^n, \quad (x, v, w) \mapsto (x, \alpha_x(v) + w)$$

are fibrewise linear homeomorphisms. The first one maps $\bigcup_{x \in B}\{x\} \times K_x$ onto $B \times (0 \times K)$ and the second one $B \times (K' \times 0)$ onto $\bigcup_{x \in B}\{x\} \times L_x$. Moreover, the second induces a bijection of $B \times (0 \times L')$ with $\bigcup_{x \in B}\{x\} \times \mathbb{R}^n/L_x$. Thus we have verified (1)–(3) in the local situation. □

14.2.4 Tangent bundle of the sphere. Let

$$TS^n = \{(x, v) \mid \langle x, v \rangle = 0\} \subset S^n \times \mathbb{R}^{n+1}$$

with the projection $p\colon TS^n \to S^n$ onto the first factor. The fibre $p^{-1}(x)$ is the orthogonal complement of x in $\mathbb{R}^{n+1}$. These data define the tangent bundle of the sphere. One can apply (14.2.3) to the family $\alpha_x\colon \mathbb{R}^{n+1} \to \mathbb{R}^{n+1}$, $v \mapsto \langle x, v \rangle x$ of orthogonal projections.

Recall the stereographic projections $\varphi_\pm\colon S^n \smallsetminus \{\pm e_{n+1}\} \to \mathbb{R}^n$. The differential of $\varphi_- \circ \varphi_+^{-1}\colon \mathbb{R}^n \smallsetminus 0 \to \mathbb{R}^n \smallsetminus 0$ at x is the linear map

$$\mathbb{R}^n \to \mathbb{R}^n, \quad \xi \mapsto \frac{\|x\|^2\xi - 2\langle x, \xi \rangle x}{\|x\|^4}.$$

For $\|x\| = 1$ we obtain the reflection $\xi \mapsto \xi - 2\langle x, \xi \rangle x$ at the hyperplane orthogonal to x. Let $U_\pm = p^{-1}(S^n \smallsetminus \{e_{n+1}\})$. The differential of $\varphi_\pm$ yields a homeomorphism which is fibrewise linear and the next diagram commutes:

$$\begin{array}{ccc} U_\pm & \xrightarrow{D\varphi_\pm} & \mathbb{R}^n \times \mathbb{R}^n \\ \downarrow{\scriptstyle p} & & \downarrow{\scriptstyle \mathrm{pr}_1} \\ S^n \smallsetminus \{\pm e_{n+1}\} & \xrightarrow{\varphi_\pm} & \mathbb{R}^n. \end{array}$$

If we identify in $\mathbb{R}^n \times \mathbb{R}^n + \mathbb{R}^n \times \mathbb{R}^n$ the point $(x,v) \in (\mathbb{R}^n \smallsetminus 0) \times \mathbb{R}^n$ in the first summand with $\left(\frac{x}{\|x\|^2}, \frac{\|x\|^2\xi - 2\langle x,\xi\rangle x}{\|x\|^4}\right)$ in the second summand then we obtain a vector bundle over $S = (\mathbb{R}^n \smallsetminus 0 + \mathbb{R}^n \smallsetminus 0)/x \sim x\|x\|^{-2}$ which is isomorphic to $p\colon TS^n \to S^n$. We can simplify the situation by identifying in $D^n \times \mathbb{R}^n + D^n \times \mathbb{R}^n$ the point $(x,v) \in S^{n-1} \times \mathbb{R}^n$ in the first summand with $(x, v - 2\langle x,v\rangle x)$ in the second summand.

In the tangent bundle we have the subspace of tangent vectors of length 1. In our case this is the space $\{(u,v) \in S^n \times S^n \mid \langle u,v\rangle = 1\}$, the Stiefel manifold $V_2(\mathbb{R}^{n+1})$. We can obtain it from $D^n \times S^{n-1} + D^n \times S^{n-1}$ by the identification $(x,v) \sim (x, v - 2\langle x,v\rangle v)$. For even n we obtain from (10.7.8) the integral homology of this Stiefel manifold.

For "smallest" structure groups of the tangent bundle of S^n see [36]. The vector field problem [3] is a special case of this problem. ◇

Important vector bundles in geometry are the tangent bundles of differentiable manifolds and the normal bundles of immersed manifolds.

14.2.5 Tautological bundles. Let V be an n-dimensional real vector space and $G_k(V)$ the Grassmann manifold of the k-dimensional subspaces of V. We set

$$E_k(V) = \{(x,v) \mid x \in G_k(V), v \in x\} \subset G_k(V) \times V.$$

We have the projections $\gamma_k(V) = \gamma_k\colon E_k(V) \to G_k(V)$, $(x,v) \mapsto x$, and the fibre over the element $x \in G_k(V)$ is the subspace x. For this reason we call this bundle the ***tautological bundle***. It remains to verify that γ_k is locally trivial. For this purpose we recall the $\mathrm{O}(k)$-principal bundle $p\colon S_k(\mathbb{R}^n) \to G_k(\mathbb{R}^n)$ from the Stiefel manifold to the Grassmann manifold. The map

$$S_k(\mathbb{R}^n) \times_{\mathrm{O}(k)} \mathbb{R}^k \to E_k(\mathbb{R}^n),$$
$$((v_1,\dots,v_k),(\lambda_1,\dots,\lambda_k)) \mapsto (\langle v_1,\dots,v_k\rangle, \textstyle\sum \lambda_j v_j)$$

is a fibrewise linear homeomorphism; it describes the tautological bundle as an associated fibre bundle.

Here is a different argument. Suppose x is spanned by $(v_1,\dots,v_k) \in S_k(\mathbb{R}^n)$; then $p_x\colon \mathbb{R}^n \to \mathbb{R}^n$, $v \mapsto \sum_{j=1}^k \langle v, v_j\rangle v_j$ is the orthogonal projection onto x. It depends continuously on $(v_1,\dots,v_k)$ and induces a continuous map $G_k(\mathbb{R}^n) \to \mathrm{Hom}(\mathbb{R}^n,\mathbb{R}^n)$, $x \mapsto p_x$, and

$$G_k(\mathbb{R}^n) \times \mathbb{R}^n \to G_k(\mathbb{R}^n) \times \mathbb{R}^n, \quad (x,v) \mapsto (x, p_x(v))$$

is a bundle morphism of constant rank with image $E_k(\mathbb{R}^n)$. Now one can use (14.2.3).

There exist analogous complex tautological bundles over the complex Grassmannians. ◇

14.2.6 Line bundles over $\mathbb{C}P^n$. Let $H\colon \mathbb{C}^{n+1} \smallsetminus 0 \to \mathbb{C}P^n$ be the defining $\mathbb{C}^*$-principal bundle. Let $\mathbb{C}(k)$ be the one-dimensional complex $\mathbb{C}^*$-representation $\mathbb{C}^* \times \mathbb{C} \to \mathbb{C}$, $(\lambda, z) \mapsto \lambda^k z$. We obtain the associated complex line bundle

$$H(k) = (\mathbb{C}^{n+1} \smallsetminus 0) \times_{\mathbb{C}^*} \mathbb{C}(-k) \to \mathbb{C}P^n.$$

Thus $H(k)$ is the quotient of $(\mathbb{C}^{n+1} \smallsetminus 0) \times \mathbb{C}$ under the equivalence relation $(z,u) \sim (\lambda z, \lambda^k u)$ for $\lambda \in \mathbb{C}^*$. We also have the S^1-principal bundle $S^{2n+1} \to \mathbb{C}^n$ (the Hopf bundle). The inclusion $S^{2n+1} \times \mathbb{C} \to (\mathbb{C}^{n+1} \smallsetminus 0) \times \mathbb{C}$ induces a homeomorphism

$$S^{2n+1} \times_{S^1} \mathbb{C}(-k) \to (\mathbb{C}^{n+1} \smallsetminus 0) \times_{\mathbb{C}^*} \mathbb{C}(-k).$$

The inverse homeomorphism is induced by $(z,u) \mapsto (\|z\|^{-1} z, \|z\|^{-k} u)$. The unit sphere bundle is $S^{2n+1} \times_{S^1} S^1(-k)$. The assignment $z \mapsto (z,1)$ induces a homeomorphism

$$S^{2n+1}/C_{|k|} \to S^{2n+1} \times_{S^1} S^1(-k).$$

Here $C_m \subset S^1$ is the subgroup of order m (roots of unity).

The bundles $H(k)$ over $\mathbb{C}P^n$ exist for $1 \le n \le \infty$. We call the bundle $H(1)$ the ***canonical complex line bundle***. For $n = \infty$ it will serve as a universal one-dimensional vector bundle.

The tautological bundle over $\mathbb{C}P^n = G_1(\mathbb{C}^{n+1})$ is $H(-1)$, since

$$(\mathbb{C}^{n+1} \smallsetminus 0) \times_{\mathbb{C}^*} \mathbb{C}(1) \to E_1(\mathbb{C}^{n+1}), \quad (x,u) \mapsto ([x], ux)$$

is an isomorphism.

The sections of $H(k)$ correspond to the functions $f\colon \mathbb{C}^{n+1} \smallsetminus 0 \to \mathbb{C}$ with the property $f(\lambda z) = \lambda^k f(z)$, they are homogenous functions of degree k. This is the reason to define $H(k)$ with $\mathbb{C}(-k)$. ◇

Let $q\colon E \to B$ be a right G-principal bundle and V an n-dimensional representation of G. Then the associated bundle $p\colon E \times_G V \to B$ is an n-dimensional vector bundle. A bundle chart $\varphi\colon q^{-1}(U) \to U \times G$ for q induces a bundle chart $q^{-1}(U) \times_G V \to U \times V$ for p, and the vector space structure on the fibres of p is uniquely determined by the requirement that the bundle charts are fibrewise linear.

We now show that vector bundles are always associated to principal bundles. Let $p\colon X \to B$ be an n-dimensional real vector bundle. Let $E_b = \mathrm{Iso}(\mathbb{R}^n, X_b)$ be the space of linear isomorphisms. The group $G = \mathrm{GL}_n(\mathbb{R}) = \mathrm{Iso}(\mathbb{R}^n, \mathbb{R}^n) = \mathrm{Aut}(\mathbb{R}^n)$ acts freely and transitively on E_b from the right by composition of linear maps. We have the set map

$$q\colon E(\xi) = E = \textstyle\coprod_{b \in B} E_b \to B, \quad E_b \to \{b\}$$

with fibrewise $\mathrm{GL}_n(\mathbb{R})$-action just explained. If $\varphi\colon p^{-1}(U) \to U \times \mathbb{R}^n$, $\varphi_b\colon X_b \to \mathbb{R}^n$ is a bundle chart of p, we define

$$\tilde{\varphi}\colon q^{-1}(U) = \textstyle\coprod_{b \in U} E_b \to U \times \mathrm{Iso}(\mathbb{R}^n, \mathbb{R}^n), \quad \alpha \in E_b \mapsto (b, \varphi_b \circ \alpha)$$

to be a bundle chart for q. The transition function for two such charts has the form

$$(U \cap V) \times \mathrm{Aut}(\mathbb{R}^n) \to (U \cap V) \times \mathrm{Aut}(\mathbb{R}^n), \quad (b, \gamma) \mapsto (b, \psi_b \varphi_b^{-1} \gamma).$$

This map is continuous, because $b \mapsto \psi_b \varphi_b^{-1}$ is continuous. Therefore there exists a unique topology on E in which the sets $q^{-1}(U)$ are open and the charts $\tilde{\varphi}$ homeomorphisms. The fibrewise $\mathrm{GL}_n(\mathbb{R})$-action on E now becomes continuous and $\tilde{\varphi}$ is equivariant. This shows $q: E \to B$ to be a $\mathrm{GL}_n(\mathbb{R})$-principal bundle. The evaluation $\mathrm{Iso}(\mathbb{R}^n, X_b) \times \mathbb{R}^n \to X_b$, $(f, u) \mapsto f(u)$ induces an isomorphism $E(\xi) \times_{\mathrm{GL}_n(\mathbb{R})} \mathbb{R}^n \cong X(\xi)$ of vector bundles.

(14.2.7) Theorem. *The assignment which associated to a* $\mathrm{GL}_n(\mathbb{R})$*-principal bundle* $E \to B$ *the vector bundle* $E \times_{\mathrm{GL}_n(\mathbb{R})} \mathbb{R}^n \to B$ *is an equivalence of the category of* $\mathrm{GL}_n(\mathbb{R})$*-principal bundles with the category of* n*-dimensional real vector bundles; the morphisms are in both cases the bundle maps.*

Proof. The construction above shows that each vector bundle is, up to isomorphism, in the image of this functor. The construction also associates to each bundle map $(F, f): \xi \to \eta$ between vector bundles a bundle map between principal bundles which is given on fibres by

$$E(\xi)_b = \mathrm{Iso}(\mathbb{R}^n, X(\xi)_b) \to \mathrm{Iso}(\mathbb{R}^n, X(\eta)_{f(b)}) = E(\eta)_{f(b)},$$

and these isomorphisms are compatible with the original bundle maps, i.e., they constitute a natural isomorphism. Therefore the functor is surjective on morphism sets between two given objects. The injectivity is a consequence of the fact that a G-map $E(\xi)_b \to E(\eta)_{f(b)}$ is determined by the associated linear map $E(\xi)_b \times_G \mathbb{R}^n \to E(\eta)_{f(b)} \times_G \mathbb{R}^n$ (where $G = \mathrm{GL}_n(\mathbb{R})$). □

Problems

1. Determine an $\mathrm{O}(n)$-principal bundle such that the associated vector bundle is the tangent bundle of S^n.
2. Consider in $\mathbb{C}^{n+1}$ the set of points $(z_0, \dots, z_n)$ which satisfy the equations

$$z_0^2 + z_1^2 + \cdots + z_n^2 = 0, \quad |z_0|^2 + |z_1|^2 + \cdots + |z_n|^2 = 2.$$

Show that this space is homeomorphic to the Stiefel manifold $V_2(\mathbb{R}^{n+1})$. Show that $V_2(\mathbb{R}^3)$ is the projective space $\mathbb{R}P^3$. Thus we have embedded this space into S^5.
3. An n-dimensional real vector bundle is trivial if and only if it has n continuous sections which are everywhere linearly independent.
4. The bundle $H(k)$ over $\mathbb{C}P^1$ is obtained from two trivial bundles $\mathrm{pr}_1: \mathbb{C} \times \mathbb{C} \to \mathbb{C}$ by gluing over $\mathbb{C}^* \times \mathbb{C}$ with the transition maps $(z, \omega) \mapsto (z^{-1}, z^{-k}\omega)$.
5. The complex tangent bundle of $\mathbb{C}P^1$ is $H(2)$.
6. A bundle morphism $J: \xi \to \xi$ of a real vector bundle which satisfies $J^2(x) = -x$ for

each $x \in E(\xi)$ is called a ***complex structure*** on ξ. If we define in each fibre the multiplication by $i \in \mathbb{C}$ as the map J, then ξ becomes a complex vector bundle. One has to verify the local triviality as a complex bundle.
7. An oriented real vector bundle is associated to a $\mathrm{GL}_n(\mathbb{R})_+$-principal bundle. It is constructed as above using the orientation preserving isomorphisms $\mathrm{Iso}_+(\mathbb{R}^n, \xi^{-1}(b))$. There exists an equivalence of categories analogous to (14.2.7).

A reduction of the structure group from $\mathrm{GL}_n(\mathbb{R})$ to $\mathrm{GL}_n(\mathbb{R})_+$ corresponds to the choice of an orientation of the vector bundle. The reductions of a $\mathrm{GL}_n(\mathbb{R})$-bundle $E \to B$ correspond to the sections of $E/\mathrm{GL}_n(\mathbb{R})_+ \to B$; the latter is a twofold covering, the ***orientation covering.***
8. The simplest non-trivial vector bundle is a line bundle over S^1. Its total space X can be viewed as the (open) Möbius band [140, Werke II, p. 484]. A formal definition is $X = S^1 \times_G \mathbb{R}$, where $G = \{\pm 1\}$ acts on S^1 and $\mathbb{R}$ by $(\lambda, z) \mapsto \lambda z$. It is also associated to the G-principal bundle $q\colon S^1 \to S^1$, $z \mapsto z^2$ (where $S^1 \subset \mathbb{C}$). Suppose the bundle were trivial. Then there would exist a nowhere vanishing section $s\colon S^1 \to E$, hence a map $\sigma\colon S^1 \to \mathbb{R} \setminus \{0\}$ satisfying $\sigma(-z) = -\sigma(z)$. The latter contradicts the intermediate value theorem of calculus.

In the same manner one constructs for each $n \geq 1$ a non-trivial line bundle $S^n \times_G \mathbb{R}^1 \to \mathbb{R}P^n = S^n/G$.
9. Let $f\colon X \to G/H$ be a G-map and $A = f^{-1}(eH)$. Then A is a left H-space and we have a bijective G-map $F\colon G \times_H A \to X$, $(g, x) \mapsto gx$. The map F is a homeomorphism if $G \to G/H$ has local sections. If p is a vector bundle and G acts by bundle automorphisms, then A is an H-representation.

14.3 The Homotopy Theorem

A locally trivial bundle is called ***numerable*** if it is trivial over the members of a numerable covering of the base space. We show that homotopic maps induce isomorphic bundles. We begin with the universal situation of a homotopy.

(14.3.1) Theorem. *Let $p\colon E \to B \times I$ be a numerable, locally trivial bundle with typical fibre F. Then there exists a bundle map $R\colon E \to E$ over $r\colon B \times I \to B \times I$, $(b, t) \mapsto (b, 1)$ which is the identity on $E|B \times 1$ and the morphism (R, r) is a pullback.*

Proof. (1) By (13.1.6), (3.1.4), and (13.1.8) we choose a numerable countable covering $(U_j \mid j \in \mathbb{N})$ of B such that p is trivial over $U_j \times I$. We then choose a numeration (t_j) of (U_j). Let $t(x) = \max(t_j(x))$ and set $u_j(x) = t_j(x)/t(x)$. Then the support of u_j is contained in U_j and $\max\{u_j(x) \mid j \in \mathbb{N}\} = 1$ holds. Let

$$r_j\colon B \times I \to B \times I, \quad (x, t) \mapsto (x, \max(u_j(x), t)).$$

We define over r_j a bundle map $R_j\colon E \to E$: It is the identity in the complement of $p^{-1}(U_j \times I)$, and over $U_j \times I$ a trivialization $U_j \times I \times G \to E|U_j \times I$ transforms it into

$$(x, t, g) \mapsto (x, \max(u_j(x), t), g).$$

Then R_j is the identity on $E|B \times 1$. From the construction we see that (R_j, r_j) is a pullback. The desired bundle map R is the composition of the R_j according to the ordering of $\mathbb{N}$. This is sensible, since for each $x \in E$ only a finite number of $R_j(x)$ are different from x. The condition $\max\{u_j(x) \mid j \in J\} = 1$ shows that R is a map over r. $\square$

If we apply the previous proof to principal bundles (to vector bundles), then (R, r) is a bundle map in the corresponding category of bundles.

Let $p\colon E \to B \times I$ be as in (14.3.1) and denote by $p_t\colon E_t \to B \times t \cong B$ its restriction to $B \times t$. We obtain from (14.3.1) a pullback $(R, r)\colon p \to p_1$. The map r induces from p the product bundle $p_1 \times \mathrm{id}\colon E_1 \times I \to B \times I$. We conclude that there exists an isomorphism $E \cong E_1 \times I$ of bundles which is the identity $E_1 = E_1 \times 1$ over $B \times 1$.

(14.3.2) Theorem. *Under the assumptions of* (14.3.1) *the bundles E_0 and E_1 are isomorphic.*

Proof. We have bundle maps $E_0 = E|B \times 0 \subset E \cong E_1 \times I \xrightarrow{\mathrm{pr}} E_1$. $\square$

14.3.3 Homotopy theorem. Let $q\colon E \to C$ be a numerable G-principal bundle and $h\colon B \times I \to C$ a homotopy. Then the bundles induced from p along h_0 and h_1 are isomorphic. A similar statement holds for vector bundles.

Proof. This follows from the previous theorem, since $h_j^* q = (h^* q)_j$. $\square$

14.3.4 Homotopy lifting. Let

$$\begin{array}{ccc} X & \xrightarrow{\Phi} & Y \\ \downarrow{\scriptstyle p} & & \downarrow{\scriptstyle q} \\ B & \xrightarrow{\varphi} & C \end{array}$$

be a bundle map between numerable G-principal bundles. Let $h\colon B \times I \to C$ be a homotopy with $h_0 = \varphi$. Then there exists a homotopy of bundle maps $H\colon X \times I \to Y$ with $H_0 = \Phi$ and $q \circ H = h \circ (p \times \mathrm{id})$.

Proof. There exists a diagram

$$\begin{array}{ccccc} X & \xrightarrow{\iota} & Z & \xrightarrow{\tilde{h}} & Y \\ \downarrow{\scriptstyle p} & & \downarrow{\scriptstyle Q} & & \downarrow{\scriptstyle q} \\ B & \xrightarrow{i_0} & B \times I & \xrightarrow{h} & C \end{array}$$

with two pullback squares and $hi_0 = \varphi$ and $\tilde{h}\iota = \Phi$. There exists an isomorphism $\alpha\colon X \times I \to Z$ of $p \times \mathrm{id}$ with Q such that $\alpha i_0 = \iota$. The desired homotopy is $H = \tilde{h} \circ \alpha$. □

(14.3.5) Theorem. *Let $q\colon X \to C$ be a numerable locally trivial map.Then q is a fibration.*

Proof. Given a homotopy $h\colon B \times I \to C$ and an initial condition $a\colon B \to X$. We pull back the bundle along h. The initial condition gives a section of the pullback bundle over $B \times 0$. We have seen that the bundle over a product $B \times I$ is isomorphic to a product bundle, and in a product the section has an obvious extension to $B \times I$. We have remarked earlier that the extendibility of the section is equivalent to finding a lifting of the homotopy with given initial condition. □

14.4 Universal Bundles. Classifying Spaces

We denote by $\mathcal{B}(B, G)$ the set of isomorphism classes of numerable G-principal bundles over B. (This is a set!) A continuous map $f\colon B \to C$ induces via pullback a well-defined map $\mathcal{B}(f) = f^*\colon \mathcal{B}(C, G) \to \mathcal{B}(B, G)$. We thus obtain a homotopy invariant functor $\mathcal{B}(-, G)$.

Let $p_G\colon EG \to BG$ be a numerable G-principal bundle and $[B, BG]$ the set of homotopy classes $B \to BG$. Since homotopic maps induce isomorphic bundles, we obtain a well-defined map

$$\iota_B\colon [B, BG] \to \mathcal{B}(B, G), \quad [f] \mapsto [f^* p_G].$$

The ι_B constitute a natural transformation.

We call the total space EG ***universal*** if each numerable free G-space E has up to G-homotopy a unique G-map $E \to EG$. (Thus EG is a terminal object in the appropriate homotopy category.) The corresponding bundle $p_G\colon EG \to BG$ is also called universal.

Let $\xi\colon E(\xi) \to B$ be a numerable G-principal bundle. Then there exists a G-map $\Phi\colon E(\xi) \to EG$ and an induced map $\bar{\Phi}\colon B \to BG$; and G-homotopic maps induce homotopic maps between the base spaces. We assign to ξ the class $[\bar{\Phi}] \in [B, BG]$. Isomorphic bundles yield the same homotopy class. Thus we obtain a well-defined map $\kappa_B\colon \mathcal{B}(B, G) \to [B, BG]$, and the κ_B constitute a natural transformation. The compositions $\iota_B \kappa_B$ and $\kappa_B \iota_B$ are the identity.

If $p'\colon E'G \to B'G$ is another universal bundle, then there exist bundle maps $\beta\colon EG \to E'G$, $\gamma\colon E'G \to EG$. The compositions $\beta\gamma$ and $\gamma\beta$ are homotopic to the identity as bundle maps. In particular, the spaces BG and $B'G$ are homotopy equivalent. The space BG is called a ***classifying space*** of the group G. A map $k\colon B \to BG$ which induces from $EG \to BG$ a given bundle $q\colon E \to B$ is called a ***classifying map*** of the bundle q. Hence:

(14.4.1) Theorem (Classification theorem). *We assign to each isomorphism class of numerable G-principal bundles the homotopy class of a classifying map and obtain a well-defined bijection $\mathcal{B}(G, B) \cong [B, BG]$. The inverse assigns to $k\colon B \to BG$ the bundle induced by k from the universal bundle.* □

(14.4.2) Theorem. *There exist universal G-principal bundles.*

The proof of the theorem will be given in three steps.

(1) Construction of the space EG (14.4.3).

(2) Proof that any two G-maps $E \to EG$ are G-homotopic (see (14.4.4)).

(3) Proof that each numerable G-space E admits a G-map (see (14.4.5)).

14.4.3 The Milnor space. We present a construction of the universal bundle which is due to Milnor [131]. It uses the notion of a join of a family of spaces. Let $(X_j \mid j \in J)$ be a family of spaces X_j. The ***join***

$$X = \bigstar_{j \in J} X_j$$

is defined as follows. The elements of X are represented by families

$$(t_j x_j \mid j \in J), \quad t_j \in [0,1], \quad x_j \in X_j, \quad \textstyle\sum_{j \in J} t_j = 1$$

in which only a finite number of t_j are different from zero. The families $(t_j x_j)$ and $(u_j y_j)$ represent the same element of X if and only if

(1) $t_j = u_j$ for each $j \in J$,

(2) $x_j = y_j$ whenever $t_j \neq 0$.

The notation $t_j x_j$ is short-hand for the pair (t_j, x_j). This is suggestive, since we can replace $0x_j$ by $0y_j$ for arbitrary x_j and y_j in X_j. We therefore have coordinate maps

$$t_j\colon X \to [0,1],\ (t_i x_i) \mapsto t_j, \quad p_j\colon t_j^{-1}]0,1] \to X_j,\ (t_i x_i) \mapsto x_j.$$

The Milnor topology on X shall be the coarsest topology for which all t_j and p_j are continuous. This topology is characterized by the following universal property: A map $f: Y \to X$ from any space Y is continuous if and only if the maps $t_j f: Y \to [0,1]$ and $p_j f\colon f^{-1}t_j^{-1}]0,1] \to X_j$ are continuous. For a finite number of spaces we use the notation $X_1 \star \cdots \star X_n$ for their join.

If the spaces X_j are right G-spaces, then $((t_j x_j), g) \mapsto (t_j x_j g)$ defines a continuous action of G on X. Continuity is verified with the universal property of the join topology. The Milnor space is

$$EG = G \star G \star G \star \cdots,$$

a join of a countably infinite number of copies of G. We write $BG = EG/G$ for the orbit space and $p\colon EG \to BG$ for the orbit map.

It remains to show that $EG \to BG$ is numerable. The coordinate functions t_j are G-invariant and induce therefore functions τ_j on BG. The τ_j are a point-finite partition of unity subordinate to the open covering by the V_j/G, $V_j = t_j^{-1}]0,1]$. The bundle is trivial over V_j/G, since we have, by construction, G-maps $p_j : V_j \to G$.◇

(14.4.4) Proposition. *Let E be a G-space. Any two G-maps $f, g: E \to EG$ are G-homotopic.*

Proof. We consider the coordinate form of $f(x)$ and $g(x)$,

$$(t_1(x)f_1(x), t_2(x)f_2(x), \dots) \quad \text{and} \quad (u_1(x)g_1(x), u_2(x)g_2(x), \dots),$$

and show that f and g are G-homotopic to maps with coordinate form

$$(t_1(x)f_1(x), 0, t_2(x)f_2(x), 0, \dots) \quad \text{and} \quad (0, u_1(x)g_1(x), 0, u_2(x)g_2(x), \dots)$$

where 0 denotes an element of the form $0 \cdot y$. In order to achieve this, for f say, we construct a homotopy in an infinite number of steps. The first step has in the homotopy parameter t the form

$$(t_1 f_1, t t_2 f_2, (1-t)t_2 f_2, t t_3 f_3, (1-t)t_3 f_3, \dots).$$

It removes the first zero in the final result just stated. We now iterate this process appropriately. We obtain the desired homotopy by using the first step on the interval $[0, \frac{1}{2}]$, the second step on the interval $[\frac{1}{2}, \frac{3}{4}]$, and so on. The total homotopy is continuous, since in each coordinate place only a finite number of homotopies are relevant.

Having arrived at the two forms above, they are now connected by the homotopy $((1-t)t_1 f_1, t u_1 g_1, (1-t)t_2 f_2, t u_2 g_2, \dots)$ in the parameter t. □

(14.4.5) Proposition. *Let E be a G-space. Let $(U_n \mid n \in \mathbb{N})$ be an open covering by G-trivial sets. Suppose there exists a point-finite partition of unity $(v_n \mid n \in \mathbb{N})$ by G-invariant functions subordinate to the covering (U_n). Then there exists a G-map $\varphi: E \to EG$.*

A numerable free G-space E admits a G-map $E \to EG$.

Proof. By definition of a G-trivial space, there exist G-maps $\varphi_j : U_j \to G$. The desired map φ is now given by $\varphi(z) = (v_1(z)\varphi_1(z), v_2(z)\varphi_2(z), \dots)$. It is continuous, by the universal property of the Milnor topology.

In order to apply the last result to the general case, we reduce arbitrary partitions of unity to countable ones (see (13.1.8)). □

(14.4.6) Proposition. *The space EG is contractible.*

Proof. We have already seen that there exists a homotopy of the identity to the map $(t_j g_j) \mapsto (t_1 g_1, 0, t_2 g_2, t_3 g_3, \dots)$. The latter map has the null homotopy $((1-t)t_1 g_1, te, (1-t)t_2 g_2, \dots)$. □

(14.4.7) Example. The locally trivial map $p\colon EG \to BG$ is a fibration with contractible total space by (14.3.5). We also have the path fibration $P \to BG$ with contractible total space and fibre ΩBG. We can turn a homotopy equivalence $EG \to P$ into a fibrewise map, and this map is then a fibrewise homotopy equivalence. Hence we have a homotopy equivalence $\Omega BG \simeq G$. The exact homotopy sequence then yields an isomorphism $\partial\colon \pi_n(BG) \cong \pi_{n-1}(G)$. ◇

(14.4.8) Example. For a discrete group, BG is an Eilenberg–Mac Lane space of type $K(G,1)$. The space BS^1 is an Eilenberg–Mac Lane space $K(\mathbb{Z},2)$. Models for $B\mathbb{Z}/2$ and BS^1 are $\mathbb{R}P^\infty$ and $\mathbb{C}P^\infty$, respectively. ◇

A continuous homomorphism $\alpha\colon K \to L$ induces the map

$$E(\alpha)\colon EK \to EL, \quad (t_i k_i) \mapsto (t_i \alpha(k_i))$$

which is compatible with the projections to the classifying spaces. We obtain an induced map $B(\alpha)\colon BK \to BL$. In this manner B becomes a functor from the category of topological groups into TOP.

(14.4.9) Proposition. *An inner automorphism $\alpha\colon K \to K$, $k \mapsto uku^{-1}$ induces a map $B(\alpha)$ which is homotopic to the identity.*

Proof. The map $(t_i k_i) \mapsto (t_i u k_i)$ is a K-map and therefore K-homotopic to the identity. The assignment $(t_i k_i) \mapsto (t_i u k_i u^{-1})$ induces the same map between the orbit spaces. □

(14.4.10) Proposition. *Let X be a free numerable G-space. Then the join $E = X \star X \star \cdots$ is a universal G-space.*

Proof. As in the proof of 14.4.3 we see that any two G-maps into E are G-homotopic. Since X is numerable, so is E. □

(14.4.11) Corollary. *Let H be a subgroup of G. Assume that G is numerable as H-space. Then EG is, considered as H-space, universal.* □

The next theorem characterizes universal bundles so that we need not rely on a special construction.

(14.4.12) Theorem. *A numerable G-principal bundle $q\colon E \to B$ is universal if and only if E is contractible (as a space without group action).*

Proof. We know already that Milnor's space EG is contractible. If p is universal, the G-space E is G-homotopy equivalent to EG and hence contractible.

Conversely, assume that E is contractible. Then the associated fibre bundle $E \times_G EG \to B$ has a contractible fibre EG and is therefore shrinkable (use (13.3.3)). Hence it has a section and any two sections are homotopic as sections. A section corresponds to a bundle map $\alpha\colon E \to EG$ (see (14.1.4). For the same reason there exists a bundle map $\beta\colon EG \to E$. By 14.4.3, $\alpha\beta$ is homotopic to the identity as a bundle map. In order to see that $\beta\alpha$ is homotopic $\mathrm{id}(E)$, we use that sections are homotopic. □

We compare classifying spaces of different groups and discuss the functorial properties of classifying spaces. Let $\alpha\colon K \to L$ be a continuous homomorphism between topological groups. We denote by ${}_\alpha L$ the K-space

$$K \times L \to L, \quad (k,l) \mapsto \alpha(k) \cdot l.$$

The associated bundle $E(K) \times_K {}_\alpha L \to B(K)$ inherits a right L-action and is an L-principal bundle. It has a classifying map $B(\alpha)\colon B(K) \to B(L)$. For the Milnor bundle the homotopy class is the same as the one already defined. If $\beta\colon L \to M$ is a further homomorphism, then the relation $B(\beta)B(\alpha) \simeq B(\beta\alpha)$ is easily verified.

Let $i\colon H \subset G$ be the inclusion of a subgroup. We restrict the G-action to H and obtain a free and contractible H-space $\mathrm{res}_H\, EG$. If $G \to G/H$ is a numerable H-principal bundle, then $\mathrm{res}_H\, EG$ is numerable as H-space; hence we have in this case in $\mathrm{res}_H EG \to (\mathrm{res}_H EG)/H$ as model for $EH \to BH$. We then obtain, because of $EG \times_G H \cong EG/H$, a map

$$Bi\colon BH = (EG)/H \to (EG)/G = BG,$$

which is a fibre bundle with fibre G/H. If G/H is contractible, then Bi is a numerable fibration with contractible fibre, hence a homotopy equivalence. This situation occurs for the inclusions $\mathrm{O}(n) \to \mathrm{GL}_n(\mathbb{R})$ and $\mathrm{U}(n) \to \mathrm{GL}_n(\mathbb{C})$, and in general for the inclusion $K \subset G$ of a maximal compact subgroup K of a connected Lie group G [84, p. 180].

(14.4.13) Proposition. *The inclusions of subgroups induce homotopy equivalences* $B\mathrm{O}(n) \to B\mathrm{GL}_n(\mathbb{R})$ *and* $B\mathrm{U}(n) \to B\mathrm{GL}_n(\mathbb{C})$. □

Let H be a normal subgroup of G. Then $E(G/H) \times E(G)$ is a numerable free G-space; hence $(E(G/H) \times EG)/G$ is a model for BG. (In general, for each G-space X which is contractible, the product $X \times EG$ is another model for EG.) With this model and the orbit map of the projection $E(G/H) \times EG \to E(G/H)$ we obtain a map $p\colon BG \to B(G/H)$ which is a fibre bundle with structure group G/H and fibre BH. In this case $BH = EG/H$ with induced G/H-action. The map p and the inclusion $i\colon BH \to BG$ of a fibre are induced maps of the type $B\rho$ for the cases $i\colon H \subset G$ and $\rho\colon G \to G/H$. Therefore we have a fibre bundle

$$BH \xrightarrow{Bi} BG \xrightarrow{Bp} B(G/H).$$

Principal bundles for a discrete group G are covering spaces; this holds in particular for the universal bundle $EG \to BG$. Since EG is simply connected, it is also the universal covering of BG. Thus two notions of "universal" meet. What is the relation between these concepts?

Let B be a pathwise connected space with universal covering $p\colon E \to B$, a right π-principal covering for $\pi = \pi_1(B)$. Let $\varphi\colon \pi \to G$ be a homomorphism. We have as before the right G-principal covering $E \times_{\pi\,\varphi} G$. If φ and ψ are conjugate homomorphisms, i.e., if $g\varphi(a)g^{-1} = \psi(a)$ for a $g \in G$, then

$$E \times_{\pi\,\varphi} G \to E \times_{\pi\,\psi} G, \quad (x,h) \mapsto (x, gh)$$

is an isomorphism of G-principal coverings. The assignment $\varphi \mapsto E \times_{\pi\varphi} G$ is a map $\alpha\colon \mathrm{Hom}(\pi, G)_c \to \mathcal{B}(G, B)$ from the set of conjugacy classes of homomorphisms (index c) to the set of isomorphism classes of G-principal bundles over B.

(14.4.14) Proposition. *The map α is a bijection.* □

(14.4.15) Example. Let G be discrete and abelian. Then conjugate homomorphisms are equal. If all coverings of B are numerable (say B paracompact), then $\mathcal{B}(B, G) = [B, BG]$. A bijection

$$\alpha\colon \mathrm{Hom}(\pi, G) \cong [B, BG] = H^1(B; G)$$

is obtained from (14.4.14). For the last equality note that BG is an Eilenberg–Mac Lane space and represents the first cohomology. ◇

(14.4.16) Example. The fibration $\mathrm{U}(n)/\mathrm{U}(n-1) \to B\mathrm{U}(n-1) \to B\mathrm{U}(n)$ and $\mathrm{U}(n)/\mathrm{U}(n-1) \cong S^{2n-1}$ show that the map $i(n)\colon B\mathrm{U}(n-1) \to B\mathrm{U}(n)$ induced by the inclusion $\mathrm{U}(n-1) \subset \mathrm{U}(n)$ is $(2n-1)$-connected. The induced map $i(n)_*\colon [X, B(n-1)] \to [X, B\mathrm{U}(n)]$ is therefore bijective (surjective) for a CW-complex X of dimension $\dim X < 2n - 1$ ($\dim X \le 2n-1$). So if $\dim X \le 2n-2$, then a k-dimensional complex vector bundle ξ over X is isomorphic to $\eta \oplus (k-n)\varepsilon$ for a unique isomorphism class η of an n-dimensional bundle η. ◇

Problems

1. Work out a proof of (14.4.14).
2. The canonical diagram

$$\begin{array}{ccc} K & \xrightarrow{\ \simeq\ } & \Omega BK \\ \big\downarrow{\scriptstyle\alpha} & & \big\downarrow{\scriptstyle\Omega B(\alpha)} \\ L & \xrightarrow{\ \simeq\ } & \Omega BL \end{array}$$

is homotopy commutative. See (14.4.7).

3. The abelianized group $\pi_1(B)$ is isomorphic to the homology group $H_1(B;\mathbb{Z})$. Therefore we can write (14.4.15) in the form $\alpha\colon \mathrm{Hom}(H_1(B), G) \cong H^1(B; G)$. The classifying map $f\colon B \to BG$ of a G-principal bundle $q\colon X \to B$ induces homomorphisms $f_*\colon \pi_1(X) \to$

$\pi_1(B)$ and $f_*\colon H_1(B) \to H_1(BG)$. If G is abelian and discrete, then $G \cong \pi_1(BG) \cong H_1(BG)$. We thus obtain a map

$$\beta\colon [B, BG] \to \mathrm{Hom}(H_1(B), G).$$

Under the hypotheses of (14.4.15), β is inverse to α.

4. We give an example of a non-numerable bundle. The equation $xz + y^2 = 1$ yields a hyperboloid Q in $\mathbb{R}^3$. The action of the additive group $\mathbb{R}$ on $\mathbb{R}^3$

$$c \cdot (x, y, z) = (x, y + cx, z - 2cy - c^2 x),$$

is free on Q, and Q becomes an $\mathbb{R}$-principal bundle. Numerable $\mathbb{R}$-principal bundles are trivial, since $\mathbb{R}$ is contractible. The bundle Q is non-trivial. The orbit space is the non-Hausdorff line with two origins. If the bundle were trivial it would have a section, and this would imply that the orbit space is separated.

5. The join $S^m \star S^n$ is homeomorphic to S^{m+n+1}, $(t_1 z_1, t_2 z_2) \mapsto (\sqrt{t_1} z_1, \sqrt{t_2} z_2)$ is a homeomorphism. The join of k copies S^1 is homeomorphic to S^{2k-1}. A suitable homeomorphism respects the S^1-action, if we let S^1 act on S^{2k-1} by scalar multiplication $(\lambda, v) \mapsto \lambda v$. The Milnor construction thus yields in this case the Hopf bundle $S^{2k-1} \to \mathbb{C}P^{k-1}$.

6. A suitable isomorphism $\pi_n(BG) \cong \pi_{n-1}(G)$ has the following interpretation. Let $p\colon E \to S^n$ be a G-principal bundle. Write $S^n = D_+ \cup D_-$, $S^{n-1} = D_+ \cap D_-$ as usual. Then $p|D_+$ and $p|D_-$ are trivial. Choose trivializations $t_\pm\colon p^{-1}(D_\pm) \to D_\pm \times G$. They differ over S^{n-1} by an automorphism

$$S^{n-1} \times G \to S^{n-1} \times G, \quad (x, g) \mapsto (x, \alpha_x(g))$$

of principal bundles. Hence $\alpha_x(g) = \alpha_x(e)g$, and $x \mapsto \alpha_x(e)$ represents an element in $\pi_{n-1}(G)$ which corresponds under the isomorphism in question to the classifying map of p.

7. The canonical map $S^\infty \to \mathbb{C}P^\infty$ is an S^1-principal bundle with contractible total space. Hence $\mathbb{C}P^\infty$ is a model for BS^1. This space is also an Eilenberg–Mac Lane space of type $K(\mathbb{Z}, 2)$. In a similar manner one has $B(\mathbb{Z}/2) = \mathbb{R}P^\infty = K(\mathbb{Z}/2, 1)$.

8. Suppose the X_j are Hausdorff spaces. Then their join is a Hausdorff space.

9. The map $(X_1 \star X_2) \star X_3 \to X_1 \star X_2 \star X_3$ which sends $(u_1(t_1x_1, t_2x_2), u_2x_3)$ to $(u_1t_1x_1, u_1t_2x_2, u_2x_3)$ is a homeomorphism. Discuss in general the associativity of the join.

10. The join of a family $(X_j \mid j \in J)$ is a subspace of the product $\prod_{j\in J} CX_j$ of cones, when the cone $CX = I \times X/0 \times X$ is given the Milnor topology with coordinate functions $t\colon (x,t) \mapsto t$ and $t^{-1}]0,1] \to X$, $(x,t) \mapsto x$ (and not the quotient topology as previously used in homotopy theory).

11. As a set, $X_0 \star \cdots \star X_n$ is a quotient of $X_0 \times \cdots \times X_n \times \Delta[n]$. If the X_j are compact Hausdorff spaces, then the join carries the quotient topology. In general the two topologies yield h-equivalent spaces.

12. A theorem of type (14.2.7) can be proved in certain other situations. Let $p\colon E \to B$ be a right $S(n)$-principal bundle for the symmetric group $S(n)$ and $F(n) = \{1, \dots, n\}$ the standard left $S(n)$-set. Then $p_n: E \times_{S(n)} F(n) \to B$ is an n-fold covering. The assignment $p \mapsto p_n$ is part of an equivalence between the category of $S(n)$-principal bundles and n-fold coverings. Isomorphism classes of n-fold numerable coverings of B are therefore classified by $[B, BS(n)]$.

13. Use the previous problem in order to classify n-fold coverings of $S^1 \vee S^1$ via the bijection to $[S^1 \vee S^1, BS(n)]$.

14.5 Algebra of Vector Bundles

Let $\xi: E(\xi) \to B$ and $\eta: E(\eta) \to C$ be vector bundles over the same field. The product $\xi \times \eta: E(\xi) \times E(\eta) \to B \times C$ is a vector bundle. Let $B = C$; we pull back the product bundle along the diagonal $d: B \to B \times B$, $b \mapsto (b, b)$ and obtain $d^*(\xi \times \eta) = \xi \oplus \eta$, the ***Whitney sum*** of ξ and η. The fibre of $\xi \oplus \eta$ over b is the direct sum of the fibres $\xi_b \oplus \eta_b$. A bundle η is called an ***inverse*** of ξ, if $\xi \oplus \eta$ is isomorphic to a trivial bundle.

A ***Riemannian metric*** on a real vector bundle ξ is a continuous map $s: E(\xi \oplus \xi) \to \mathbb{R}$ which is on each fibre an inner product $s(x)$ on $E(\xi)_x$. If ξ has a Riemannian metric and if $\alpha: \eta \to \xi$ is a fibrewise injective bundle morphism over B, then $\mathrm{Coker}(\alpha)$ is isomorphic to the fibrewise orthogonal complement ζ of $\mathrm{Im}(\alpha)$ (this is a subbundle). We have therefore an isomorphism $\xi \cong \eta \oplus \zeta$. Similarly for complex bundles and hermitian metrics.

(14.5.1) Proposition. *A numerable vector bundle has a Riemannian metric.*

Proof. Let $\mathcal{U}$ be a covering of B with numeration $(\tau_U \mid U \in \mathcal{U})$. A trivial bundle certainly has a Riemannian metric; so let s_U be a metric on $\xi|U$. Then $\sum_U \tau_U s_U$ is a Riemannian metric on ξ. The sum is short-hand notation for the inner product $\sum_U \tau_U(x) s_U(x)$ on the fibre over x, and we agree that "zero times undefined = zero". □

(14.5.2) Proposition. *A bundle $\xi: E(\xi) \to B$ has an inverse if and only if it is numerable of finite type.*

Proof. Let $\varphi_j: \xi^{-1}(U_j) \to U_j \times \mathbb{R}^n$ be bundle charts and (τ_j) a numeration of (U_j), $(j \in J$, J finite). Then

$$\alpha: E(\xi) \to B \times \textstyle\bigoplus_{j \in J} \mathbb{R}^n, \quad x \mapsto (\xi(x); \tau_j(\xi(x)) \operatorname{pr}_2 \varphi_j(x) \mid j \in J)$$

is a fibrewise injective bundle morphism into a trivial bundle. The orthogonal complement of the image of α is an inverse of ξ.

Let ξ be a k-dimensional bundle with inverse η. From an isomorphism of $\xi \oplus \eta$ to a trivial bundle we obtain a map $f: E(\xi) \subset E(\xi \oplus \eta) \cong B \times \mathbb{R}^n \xrightarrow{\mathrm{pr}} \mathbb{R}^n$ which is injective on each fibre. Given a map with this property we get a bundle map $\xi \to \gamma_k(\mathbb{R}^n)$ into the tautological bundle by mapping $x \in \xi_b$ to $(f(\xi_b), f(x)) \in$

$E_k(\mathbb{R}^n) \subset G_k(\mathbb{R}^n) \times \mathbb{R}^n$. (Verify that $b \mapsto f(\xi_b)$ is continuous.) This map is called the ***Gauss map*** of ξ. The bundle γ_k is numerable of finite type, as a bundle over a compact Hausdorff space, hence the induced bundle ξ has the same properties. □

Standard constructions of linear algebra can be applied fibrewise to vector bundles. Examples are:

V^*	dual space of V	ξ^*	dual bundle of ξ
$V \oplus W$	direct sum	$\xi \oplus \eta$	Whitney sum
$V \otimes W$	tensor product	$\xi \otimes \eta$	tensor product
$\Lambda^i V$	i-th exterior power	$\Lambda^i \xi$	i-th exterior power
$\mathrm{Hom}(V, W)$	homomorphisms	$\mathrm{Hom}(\xi, \eta)$	homomorphism bundle

Canonical isomorphisms between algebraic constructions yield canonical isomorphisms for the corresponding vector bundles. Examples are:

$$(\xi \oplus \eta) \otimes \zeta \cong (\xi \otimes \zeta) \oplus (\eta \otimes \zeta)$$
$$\mathrm{Hom}(\xi, \eta) \cong \xi^* \otimes \eta$$
$$\Lambda^k(\xi \oplus \eta) \cong \textstyle\bigoplus_{i+j=k}(\Lambda^i \xi \otimes \Lambda^j \eta).$$

In the last isomorphism $\Lambda^0 \xi$ is the trivial one-dimensional bundle and $\Lambda^1 \xi \cong \xi$. In order to prove such statements, one has to use that the constructions of linear algebra are in an appropriate sense continuous. It suffices to consider an example, say the tensor product. Let $\xi \colon E(\xi) \to B$ and $\eta \colon E(\eta) \to B$ be real vector bundles. The total space of $\xi \otimes \eta$ has the underlying set

$$\textstyle\bigcup_{b \in B}(\xi_b \otimes \eta_b) = E(\xi \otimes \eta),$$

the disjoint union of the tensor products of the fibres. Let $\varphi \colon \xi^{-1}(U) \to U \times \mathbb{R}^m$ be a bundle chart of ξ and $\psi \colon \eta^{-1}(U) \to U \times \mathbb{R}^n$ a chart of η. Then a bundle chart for $\xi \otimes \eta$ over U should be

$$\gamma \colon \textstyle\bigcup_{b \in U}(\xi_b \otimes \eta_b) \to U_j \times (\mathbb{R}^m \otimes \mathbb{R}^n),$$

the fibre $\xi_b \otimes \eta_b$ is mapped by the tensor product of the linear maps φ over b and ψ over b. At this point it is now important to observe that the transition maps of such charts are homeomorphisms. Therefore there exists a unique topology on $E(\xi \otimes \eta)$ such that the sources of the γ are open and the γ are homeomorphisms. In this manner we have obtained the data of the bundle $\xi \otimes \eta$. In dealing with tensor products one has to distinguish $\otimes_{\mathbb{R}}$ for real bundles and $\otimes_{\mathbb{C}}$ for complex bundles.

If we start with bundles $\xi \colon E(\xi) \to B$ and $\eta \colon E(\eta) \to C$ we obtain in a similar manner a bundle $\xi \hat{\otimes} \eta$ over $B \times C$ with fibres $\xi_b \otimes \eta_c$. It is called the exterior tensor product. Let $B = C$ and let $d \colon B \times B$ be the diagonal; then $d^*(\xi \hat{\otimes} \eta) = \xi \otimes \eta$. Let $p \colon B \times C \to B$ and $q \colon B \times C \to C$ be the projections; then $\xi \hat{\otimes} \eta = p^* \xi \otimes q^* \eta$.

(14.5.3) Example. Let $p\colon E \to B$ be an S^1-principal bundle and $\xi\colon E(\xi) = E \times_{S^1} \mathbb{C} \to B$ the associated complex line bundle. Then p is the unit-sphere bundle of ξ. Let $C_m \subset S^1$ be the cyclic subgroup of order m. The m-fold tensor product $\xi^{\otimes m} = \xi \otimes \dots \otimes \xi$ is $E \times_{S^1} \mathbb{C}(m)$, where S^1 acts on $\mathbb{C}$ by $(\lambda, z) \mapsto \lambda^m z$. The unit-sphere bundle of $\eta^{\otimes m}$ is $E/C_m \to B$. If we use the model $S^\infty \to BS^1$ for the universal S^1-principal bundle, we obtain the canonical map $BC_m \to BS^1$ as the sphere bundle of the m-fold tensor product of the universal line bundle. ◇

(14.5.4) Example. Let $p\colon E \to B$ be a right G-principal bundle. Let V, W be complex G-representations. Let $p_V\colon E \times_G V \to B$ be the associated complex vector bundle. Then there are canonical isomorphisms $p_V \oplus p_W \cong p_{V \oplus W}$ and $p_V \otimes p_W \cong p_{V \otimes W}$. For the bundles $H(k)$ over $\mathbb{C}P^n$ the relations $H(k) \otimes_{\mathbb{C}} H(l) \cong H(k+l)$ hold. ◇

14.5.5 Complex vector bundles over $S^2 = \mathbb{C}P^1$. We have the line bundles $H(k)$ over $\mathbb{C}P^1$ for $k \in \mathbb{Z}$. The total space is $H(k) = (\mathbb{C}^2 \smallsetminus 0) \times_{\mathbb{C}^*} \mathbb{C}$ with equivalence relation $((z_0, z_1), u) \sim ((\lambda z_0, \lambda z_1), \lambda^k u)$. Set $\eta = H(1)$; then $\eta^n \cong H(n)$. Let ξ be an arbitrary line bundle over $\mathbb{C}P^1$. We have the charts $\varphi_0\colon \mathbb{C} \to U_0 = \{[z_0, z_1] \mid z_0 \neq 0\}$ and $\varphi_1\colon \mathbb{C} \to U_1\{[z_0, z_1] \mid z_1 \neq 0\}$. We pull back ξ along φ_j and obtain a trivial bundle. Let $\Phi_j\colon \mathbb{C} \times \mathbb{C} \to \xi | U_j$ be a trivialization. Then $\Phi_1^{-1}\Phi_0\colon \mathbb{C}^* \times \mathbb{C} \to \mathbb{C}^* \times \mathbb{C}$ has the form $(z, u) \mapsto (z^{-1}, a_z \cdot u)$ for some map $a\colon \mathbb{C}^* \to \mathbb{C}^*$. The map a is homotopic to a map $z \mapsto z^{-k}$. We use a homotopy in order to construct a bundle over $\mathbb{C}P^1 \times [0,1]$ which is over $\mathbb{C}P^1 \times 0$ given by the gluing $(z, u) \mapsto (z^{-1}, a_z u)$ and over $\mathbb{C}P^1 \times 1$ by the gluing $(z, u) \mapsto (z^{-1}, z^{-k} u)$. The latter gives $H(k)$. By the homotopy theorem we see that ξ is isomorphic to $H(k)$. Let $\mathcal{B}$ denote the set of isomorphism classes of complex line bundles over $\mathbb{C}P^1$ with tensor product as composition law (see Problem 4). We have just seen that $\kappa\colon \mathbb{Z} \to \mathcal{B}$, $k \mapsto H(k)$ is a surjective homomorphism. We know that $\mathcal{B} \cong [\mathbb{C}P^1, \mathbb{C}P^\infty] \cong \pi_2(\mathbb{C}P^\infty) \cong \mathbb{Z}$. Therefore κ has a trivial kernel, because otherwise $\mathcal{B}$ would be a finite cyclic group. Altogether we have seen that the $H(k)$ represent the isomorphism classes of complex line bundles.

Now we use (14.4.16) and see that a k-dimensional bundle ($k \geq 1$) is isomorphic to $H(n) \oplus (k-1)\varepsilon$ for a unique $n \in \mathbb{Z}$. Bundles over $\mathbb{C}P^1$ have a cancellation property: An isomorphism $\xi \oplus \zeta \cong \eta \oplus \zeta$ implies $\xi \cong \eta$; this is again a consequence of (14.4.16).

For the bundles $H(k)$ over $\mathbb{C}P^1$ the relations $H(k) \oplus H(l) \cong H(k+l) \oplus \varepsilon$ hold. In order to prove this relation, we construct an isomorphism of $\eta \oplus \eta^{-1}$ to the trivial bundle. We write the bundle in the form $S^3 \times_{S^1} (\mathbb{C}(-1) \oplus \mathbb{C}(1))$. A fibrewise map to $\mathbb{C}^2$ is given by

$$((z_0, z_1), (u_0, u_1)) \mapsto (z_0 u_0 - \overline{z}_1 u_1, z_1 u_0 + \overline{z}_0 u_1).$$

Observe that the matrix with rows $(z_0, -\overline{z}_1)$, $(z_1, \overline{z}_0)$ is unitary. For u_0 the image is the tautological bundle $H(-1)$, for u_1 the orthogonal complement $H(1)$. We show by induction $\eta^k + (k-1)\varepsilon = k\eta$. This relation is clear for $k = 1$ and follows for $k = 2$ from $\eta \oplus \eta^{-1} = 2\varepsilon$. Multiply $\eta^k + (k-1)\varepsilon = k\eta$ by η, add $k\varepsilon$ and cancel $(k-1)\eta$; the desired relation for $k+1$ drops out. Suppose $k, l \in \mathbb{N}$. Then

$$\eta^k \oplus \eta^l \oplus (k-1+l-1)\varepsilon = (k+l)\eta = \eta^{k+l} \oplus (k+l-1)\varepsilon,$$

and cancellation of $(k-1+l-1)\varepsilon$ gives $\eta^k \oplus \eta^l = \eta^{k+l} \oplus \varepsilon$. We multiply this relation by η^{-k}, η^{-l}, or $\eta^{-(k+l)}$ in order to verify the remaining cases. ◇

Problems

1. Let ξ and η be vector bundles over B. An orientation of ξ and η induces an orientation of $\xi \oplus \eta$, fibrewise the sum orientation of the vector spaces. If two of the bundles ξ, η, and $\xi \oplus \eta$ are orientable, then the third is orientable.

2. In a bundle with Riemannian metric the fibrewise orthogonal complement of a subbundle is a subbundle.

3. Let $p\colon E \to B$ be an n-dimensional bundle with Riemannian metric. Then there exists a bundle atlas such that the transition maps have an image in the orthogonal group $\mathrm{O}(n)$. The structure group is therefore reducible to $\mathrm{O}(n)$. If the bundle is orientable, then the structure group is reducible to $\mathrm{SO}(n)$.

4. Let ξ and η be complex line bundles over B. Then $\xi \otimes_{\mathbb{C}} \eta$ is again a line bundle. The bundle $\xi \otimes_{\mathbb{C}} \xi^*$ is trivial; the assignments $\xi_b \otimes \xi_b^* \to \mathbb{C}$, $(x, \lambda) \mapsto \lambda(x)$ are an isomorphism to the trivial bundle. The isomorphism classes of complex line bundles are an abelian group with composition law the tensor product.

5. Let X be a normal space and $Y \subset X$ a closed subset. A section $s\colon Y \to E|Y$ over Y of a numerable vector bundle $\xi\colon E \to X$ has an extension to a section over X.

6. Let $p\colon E \to X$ and $q\colon F \to X$ be vector bundles. The bundle morphisms $E \to F$ correspond to the sections of $\mathrm{Hom}(E, F) \to X$.

7. Let $p\colon E \to X$ and $q\colon F \to X$ be numerable bundles over the normal space X. If $f\colon E|Y \to F|Y$ is an isomorphism over the closed set Y, then there exists an open neighbourhood U of Y and an isomorphism $f\colon E|U \to F|U$ which extends f over Y.

8. The map $\mathbb{C}P^a \times \mathbb{C}P^b \to \mathbb{C}P^{a+b}$, $([x_i], [y_j]) \mapsto [z_k]$ with $x_t = \sum_{i+j=t} x_i y_j$ induces from $H(1)$ the exterior tensor product $H(1)\hat{\otimes}H(1)$. In the case $a = b = \infty$ the map is associative and defines the structure of an H-space. It induces on $[B, \mathbb{C}P^\infty]$ the group structure on the set of line bundles given by the tensor product.

9. Determine two-dimensional real bundles over $\mathbb{R}P^2$ [44, p. 434].

14.6 Grothendieck Rings of Vector Bundles

Denote by $V(X)$ the set of isomorphism classes of complex vector bundles over X. The Whitney sum and the tensor product induce on $V(X)$ two associative and commutative composition laws (addition $+$, multiplication $\cdot$), and the distributive law holds. Addition has a zero element, the 0-dimensional bundle; multiplication has a unit element, the 1-dimensional trivial bundle.

A commutative monoid M is a set together with an associative and commutative composition law $+$ with zero element. A ***universal group*** for M is a homomorphism $\kappa\colon M \to K(M)$ into an abelian group $K(M)$ such that each monoid-homomorphism $\varphi\colon M \to A$ into an abelian group A has a unique factorization $\Phi \circ \kappa = \varphi$ with a homomorphism $\Phi\colon K(M) \to A$. A monoid-homomorphism $f\colon M \to N$ induces a homomorphism $K(f)\colon K(M) \to K(N)$. Let $N \subset M$ be a submonoid. We define an equivalence relation on M by

$$x \sim y \quad \Leftrightarrow \quad \text{there exist } a, b \in N \text{ such that } x + a = y + b.$$

Let $p\colon M \to M/N, x \mapsto [x]$ denote the quotient map onto the set of equivalence classes. We obtain by $[x]+[y] = [x+y]$ a well-defined composition law on M/N which is a monoid structure.

In the product monoid $M \times M$ we have the diagonal submonoid $D(M) = \{(m,m)\}$. We set $K(M) = (M \times M)/D(M)$ and $\kappa(x) = [x,0]$. Then $K(M)$ is an abelian group and κ a universal homomorphism. Since $[x,0]+[0,x] = [x,x] = 0$, we see $[0,x] = -\kappa(x)$. The elements of $K(M)$ are formal differences $x-y$, $x, y \in M$, $\kappa(x)-\kappa(y) = [x,y]$, and $x-y = x'-y'$ if and only if $x+y'+z = x'+y+z$ holds as equality in M for some $z \in M$.

We apply these concepts to $M = V(X)$ and write $K(X) = K(V(X))$. The tensor product induces a bi-additive map $V(X) \times V(X) \to V(X)$. It induces a bi-additive map in the K-groups:

(14.6.1) Proposition. *Let A, B, C be abelian monoids and $m\colon A \times B \to C$ be a bi-additive map. The there exists a unique bi-additive map $K(m)$ such that the diagram*

$$\begin{array}{ccc} A \times B & \xrightarrow{\quad m \quad} & C \\ \big\downarrow{\scriptstyle \kappa_A \times \kappa_B} & & \big\downarrow{\scriptstyle \kappa_C} \\ K(A) \times K(B) & \xrightarrow{\ K(m)\ } & K(C) \end{array}$$

is commutative. □

The bi-additive map induced by the tensor product is written as multiplication, and $K(X)$ becomes in this way a commutative ring. This ring is often called the ***Grothendieck ring*** of complex vector bundles. In general, the universal groups $K(M)$ are called Grothendieck groups.

We can apply the same construction to real vector bundles and obtain the Grothendieck ring $KO(X)$. Other notations for these objects are $KO(X) = K_{\mathbb{R}}(X)$ and $K(X) = KU(X) = K_{\mathbb{C}}(X)$.

Pullback of bundles along $f : X \to Y$ induces a ring homomorphism $K(f) = f^*\colon K(Y) \to K(X)$ and similarly for KO. Homotopic maps induce the same homomorphism.

(14.6.2) Example. $K(S^2)$ is free abelian as an additive group with basis 1 and η. The multiplicative structure is determined by $\eta^2 = 2\eta - 1$. This is a consequence of 14.5.5. ◇

The inclusions $\mathrm{U}(n) \to \mathrm{U}(n+1)$, $A \mapsto \left(\begin{smallmatrix} A & 0 \\ 0 & 1 \end{smallmatrix}\right)$ are used to define $\mathrm{U} = \operatorname{colim} \mathrm{U}(n)$, a topological group with the colimit topology. The inclusion of groups $\mathrm{U}(n) \to \mathrm{U}$ induces $B\mathrm{U}(n) \to B\mathrm{U}$. If we compose a classifying map $X \to B\mathrm{U}(n)$ with this map, we call the result $X \to B\mathrm{U}$ the ***stable classifying map***. Bundles ξ and $\xi \oplus a\varepsilon$ are called ***stably equivalent***, and they have the same stable classifying map (up to homotopy).

(14.6.3) Proposition. *Let X be a path connected compact Hausdorff space. Then there exists a natural bijection $K(X) \cong [X, \mathbb{Z} \times B\mathrm{U}]$. Here $\mathbb{Z}$ carries the discrete topology.*

Proof. Let $[\xi] - [\eta] \in K(X)$. A bundle η over a compact Hausdorff space has an inverse bundle η^- (see (14.5.2)). Hence $[\xi \oplus \eta^-] - [\eta \oplus \eta^-] \in K(X)$ is the same element. Therefore each element in $K(X)$ can be written in the form $[\xi] - n$. Suppose $[\xi] - n = [\eta] - m$. Then $\xi \oplus m\varepsilon \oplus \zeta \cong \eta \oplus n\varepsilon \oplus \zeta$ for some ζ. We add an inverse of ζ and arrive at a relation of the form $\xi \oplus a\varepsilon \cong \eta \oplus b\varepsilon$, i.e., ξ and η are stably equivalent. The homotopy class of a stable classifying map $k_\xi \colon X \to B\mathrm{U}$ is therefore uniquely determined by the element $[\xi] - n$. We define $\kappa \colon K(X) \to [X, \mathbb{Z} \times B\mathrm{U}]$ by sending $[\xi] - n$ to $k_\xi \colon X \to (\dim \xi - n) \times B\mathrm{U}$.

Conversely, let $f \colon X \to \mathbb{Z} \times B\mathrm{U}$ be given. Since X is path connected, the image is contained in some $k \times B\mathrm{U}$. The compactness of X is used to verify that f admits a factorization $X \to B\mathrm{U}(n) \to B\mathrm{U}$. We obtain a well-defined inverse map $[X, \mathbb{Z} \times B\mathrm{U}] \to K(X)$, if we assign to $f \colon X \to B\mathrm{U}(n) \to B\mathrm{U}$ the element $[f_n^* \gamma_n] - [n - k]$. □

For more general spaces the Grothendieck ring $K(X)$ can differ substantially from the homotopy group $[X, \mathbb{Z} \times B\mathrm{U}]$, e.g., for $X = \mathbb{C}P^\infty$. The latter is a kind of completion of the Grothendieck ring.

The (exterior) tensor product of bundles yields a ring homomorphism

$$K(X) \otimes_{\mathbb{Z}} K(Y) \to K(X \times Y).$$

A fundamental result is the periodicity theorem of Bott. One of its formulations is: For compact spaces X the tensor product yields isomorphisms

$$K(X) \otimes K(S^2) \cong K(X \times S^2), \quad KO(X) \otimes KO(S^8) \cong KO(X \times S^8).$$

Starting from this isomorphism one constructs the cohomology theories which are called K-theories. For an introduction see [15], [9], [10], [12], [11], [13], [14], [102]. For the Bott periodicity see also [6], [106], [19].

Chapter 15
Manifolds

This chapter contains an introduction to some concepts and results of differential topology. For more details see [30], [44], [107]. We restrict attention to those parts which are used in the proof of the so-called Pontrjagin–Thom theorem in the chapter on bordism theory. We do not summarize the results here, since the table of contents should give enough information.

15.1 Differentiable Manifolds

A topological space X is ***n-dimensional locally Euclidean*** if each $x \in X$ has an open neighbourhood U which is homeomorphic to an open subset V of $\mathbb{R}^n$. A homeomorphism $h\colon U \to V$ is a ***chart*** or ***local coordinate system*** of X about x with ***chart domain*** U. The inverse $h^{-1}\colon V \to U$ is a ***local parametrization*** of X about x. If $h(x) = 0$, we say that h and h^{-1} are ***centered*** at x. A set of charts is an ***atlas*** for X if their domains cover X. If X is n-dimensional locally Euclidean, we call n the ***dimension*** of X and write $\dim X = n$. The dimension is well-defined, by invariance of dimension.

An ***n-dimensional manifold*** or just ***n-manifold*** is an n-dimensional locally Euclidean Hausdorff space with countable basis for its topology. Hence manifolds are locally compact. A ***surface*** is a 2-manifold. A 0-manifold is a discrete space with at most a countably infinite number of points. The notation M^n is used to indicate that $n = \dim M$.

Suppose (U_1, h_1, V_1) and (U_2, h_2, V_2) are charts of an n-manifold. Then we have the associated ***coordinate change*** or ***transition function***

$$h_2 h_1^{-1}\colon h_1(U_1 \cap U_2) \to h_2(U_1 \cap U_2),$$

a homeomorphism between open subsets of Euclidean spaces.

Recall: A map $f\colon U \to V$ between open subsets of Euclidean spaces ($U \subset \mathbb{R}^n, V \subset \mathbb{R}^m$) is a C^k-map if it is k-times continuously differentiable in the ordinary sense of analysis ($1 \le k \le \infty$). A continuous map is also called a C^0-map. A C^k-map $f\colon U \to V$ has a differential $Df(x)\colon \mathbb{R}^n \to \mathbb{R}^m$ at $x \in U$.

If the coordinate changes $h_2 h_1^{-1}$ and $h_1 h_2^{-1}$ are C^k-maps, we call the charts (U_1, h_1, V_1) and (U_2, h_2, V_2) ***C^k-related*** ($1 \le k \le \infty$). An atlas is a C^k-atlas if any two of its charts are C^k-related. We call C^∞-maps ***smooth*** or just ***differentiable***; similarly, we talk about a smooth or differentiable atlas.

(15.1.1) Proposition. *Let $\mathcal{A}$ be a smooth atlas for M. The totality of charts which are smoothly related to all charts of $\mathcal{A}$ is a smooth atlas $D(\mathcal{A})$. If $\mathcal{A}$ and $\mathcal{B}$ are smooth atlases, then $\mathcal{A} \cup \mathcal{B}$ is a smooth atlas if and only if $D(\mathcal{A}) = D(\mathcal{B})$. The atlas $D(\mathcal{A})$ is the uniquely determined maximal smooth atlas which contains $\mathcal{A}$.* □

A ***differential structure*** on the n-manifold M is a maximal smooth atlas $\mathcal{D}$ for M. The pair $(M, \mathcal{D})$ is called a ***smooth*** or ***differentiable manifold***. A maximal atlas serves just the purpose of this definition. Usually we work with a smaller atlas which then generates a unique differential structure. Usually we omit the differential structure from the notation; the charts of $\mathcal{D}$ are then called the charts of the differentiable manifold M.

Let M and N be smooth manifolds. A map $f : M \to N$ is ***smooth*** if it is continuous and for each $x \in M$ there exist charts (U, h, U') about x and (V, k, V') about $f(x)$ such that kfh^{-1} is smooth. We call kfh^{-1} the ***expression of f in local coordinates***. The map f is smooth if it is differentiable at each point. The composition of smooth maps is smooth. Thus we have the category of smooth manifolds and smooth maps. A ***diffeomorphism*** is a smooth map which has a smooth inverse. Manifolds M and N are ***diffeomorphic*** if there exists a diffeomorphism $f : M \to N$.

Smooth manifolds M and N have a ***product*** in the category of smooth manifolds. The charts of the form $(U \times V, f \times g, U' \times V')$ for charts (U, f, U') of M and (V, g, V') of N define a smooth structure on $M \times N$. The projections onto the factors are smooth. The canonical isomorphisms $\mathbb{R}^m \times \mathbb{R}^n = \mathbb{R}^{m+n}$ are diffeomorphisms.

A subset N of an n-manifold M is a ***k-dimensional submanifold*** of M if the following holds: For each $x \in N$ there exists a chart $h : U \to U'$ of M about x such that $h(U \cap N) = U' \cap (\mathbb{R}^k \times 0)$. A chart with this property is called ***adapted*** to N. The difference $n - k$ is the ***codimension*** of N in M. (The subspace $\mathbb{R}^k \times 0$ of $\mathbb{R}^n$ may be replaced by any k-dimensional linear or affine subspace if this is convenient.) If we identify $\mathbb{R}^k \times 0 = \mathbb{R}^k$, then $(U \cap N, h, U' \cap \mathbb{R}^k)$ is a chart of N. If M is smooth, we call N a ***smooth submanifold*** of M if there exists about each point an adapted chart from the differential structure of M. The totality of charts $(U \cap N, h, U' \cap \mathbb{R}^k)$ which arise from adapted smooth charts of M is then a smooth atlas for N. In this way, a differentiable submanifold becomes a smooth manifold, and the inclusion $N \subset M$ is a smooth map. A smooth map $f : N \to M$ is a ***smooth embedding*** if $f(N) \subset M$ is a smooth submanifold and $f : N \to f(N)$ a diffeomorphism.

The spheres are manifolds which need an atlas with at least two charts. We have the atlas with two charts (U_N, φ_N) and (U_S, φ_S) coming from the stereographic projection (see (2.3.2)). The coordinate transformation is $\varphi_S \circ \varphi_N^{-1}(y) = \|y\|^{-2} y$. The differential of the coordinate transformation at x is $\xi \mapsto (\|x\|^2 \xi - 2\langle x, \xi \rangle x) \cdot \|x\|^{-4}$. For $\|x\| = 1$ we obtain the reflection $\xi \mapsto \xi - 2\langle x, \xi \rangle x$ in a hyperplane.

We now want to construct charts for the projective space $\mathbb{R}P^n$. The subset $U_i = \{[x_0, \dots, x_n] \mid x_i \neq 0\}$ is open. The assignment

$$\varphi_i : U_i \to \mathbb{R}^n, \quad [x_0, \dots, x_n] \mapsto x_i^{-1}(x_0, \dots, x_{i-1}, x_{i+1}, \dots, x_n)$$

is a homeomorphism. These charts are smoothly related.

Charts for $\mathbb{C}P^n$ can be defined by the same formulas. Note that $\mathbb{C}P^n$ has dimension $2n$ as a smooth manifold. (It is n-dimensional when viewed as a so-called complex manifold.)

(15.1.2) Proposition. *Let M be an n-manifold and $\mathcal{U} = (U_j \mid j \in J)$ an open covering of M. Then there exist charts $((V_k, h_k, B_k) \mid k \in \mathbb{N})$ of M with the following properties:*

(1) *Each V_k is contained in some member of $\mathcal{U}$.*

(2) $B_k = U_3(0) = \{x \in \mathbb{R}^n \mid \|x\| < 3\}$.

(3) *The family $(V_k \mid k \in \mathbb{N})$ is a locally finite covering of M.*

In particular, each open cover has a locally finite refinement, i.e., manifolds are paracompact. If M is smooth, there exists a smooth partition of unity $(\sigma_k \mid k \in \mathbb{N})$ subordinate to (V_k). There also exists a smooth partition of unity $(\alpha_j \mid j \in J)$ such that the support of α_j is contained in U_j and at most a countable number of the α_j are non-zero.

Proof. The space M is a locally compact Hausdorff space with a countable basis. Therefore there exists an exhaustion

$$M_0 \subset M_1 \subset M_2 \subset \cdots \subset M = \textstyle\bigcup_{i=0}^{\infty} M_i$$

by open sets M_i such that $\overline{M}_i$ is compact and contained in M_{i+1}. Hence $K_i = \overline{M}_{i+1} \smallsetminus M_i$ is compact. For each i we can find a finite number of charts (V_ν, h_ν, B_ν), $B_\nu = U_3(0)$, such that $V_\nu \subset U_j$ for some j and such that the $h_\nu^{-1} U_1(0)$ cover K_i and such that $V_\nu \subset M_{i+2} \smallsetminus \overline{M}_{i-1}$ ($M_{-1} = \emptyset$). Then the V_ν form a locally finite, countable covering of M, now denoted $(V_k, h_k, B_k \mid k \in \mathbb{N})$.

The function $\lambda\colon \mathbb{R} \to \mathbb{R}$, $\lambda(t) = 0$ for $t \le 0$, $\lambda(t) = \exp(-1/t)$ for $t > 0$, is a C^∞-function. For $\varepsilon > 0$, the function $\varphi_\varepsilon(t) = \lambda(t)(\lambda(t) + \lambda(\varepsilon - t))^{-1}$ is C^∞ and satisfies $0 \le \varphi_\varepsilon \le 1$, $\varphi_\varepsilon(t) = 0 \Leftrightarrow t \le 0$, $\varphi_\varepsilon(t) = 1 \Leftrightarrow t \ge \varepsilon$. Finally, $\psi\colon \mathbb{R}^n \to \mathbb{R}$, $x \mapsto \varphi_\varepsilon(\|x\| - r)$ is a C^∞ map which satisfies $0 \le \psi(x) \le 1$, $\psi(x) = 1 \Leftrightarrow x \in U_r(0)$, $\psi(x) = 0 \Leftrightarrow \|x\| \ge r + \varepsilon$.

We use these functions ψ for $r = 1$ and $\varepsilon = 1$ and define ψ_i by $\psi \circ h_i$ on V_i and as zero on the complement. Then the $\sigma_k = s^{-1}\psi_k$ with $s = \sum_{j=1}^{\infty} \psi_j$ yield a smooth, locally finite partition of unity subordinate to $(V_k \mid k \in \mathbb{N})$.

The last statement follows from (13.1.2). □

Let C_0 and C_1 be closed disjoint subsets of the smooth manifold M. Then there exists a smooth function $\varphi\colon M \to [0,1]$ such that $\varphi(C_j) \subset \{j\}$; apply the previous proposition to the covering by the $U_j = M \smallsetminus C_j$.

Let A be a closed subset of the smooth manifold M and U an open neighbourhood of A in M. Let $f: U \to [0,1]$ be smooth. Then there exists a smooth function $F: M \to [0,1]$ such that $F|A = f|A$. For the proof choose a partition of unity (φ_0, φ_1) subordinate to $(U, M \setminus A)$. Then set $F(x) = \varphi_0(x) f(x)$ for $x \in U$ and $F(x) = 0$ otherwise.

(15.1.3) Proposition. *Let M be a closed submanifold of N. A smooth function $f: M \to \mathbb{R}$ has a smooth extension $F: N \to \mathbb{R}$.*

Proof. From the definition of a submanifold we obtain for each $p \in M$ an open neighbourhood U of p in N and a smooth retraction $r: U \to U \cap M$. Hence we can find an open covering $(U_j \mid j \in J)$ of M in N and smooth extensions $f_j: U_j \to \mathbb{R}$ of $f|U_j \cap M$. Let $U_0 = N \setminus M$ be an additional open set and $f_0 = 0$. Let $(\alpha_j \mid j \in J \cup \{0\})$ be a subordinate smooth partition of unity and set $F(x) = \sum_{j \in J} \alpha_j(x) f_j(x)$, where a summand is defined to be zero if $f_j(x)$ is not defined. □

(15.1.4) Proposition. *Let M be a smooth manifold. There exists a smooth proper function $f: M \to \mathbb{R}$.*

Proof. A function between Hausdorff spaces is proper if the pre-image of a compact set is compact. We choose a countable partition of unity $(\tau_k \mid k \in \mathbb{N})$ such that the functions τ_k have compact support. Then we set $f = \sum_{k=1}^{\infty} k \cdot \tau_k : M \to \mathbb{R}$. If $x \notin \bigcup_{j=1}^{n} \operatorname{supp}(\tau_j)$, then $1 = \sum_{j \geq 1} \tau_j(x) = \sum_{j>n} \tau_j(x)$ and therefore $f(x) = \sum_{j>n} j\tau_j(x) > n$. Hence $f^{-1}[-n,n]$ is contained in $\bigcup_{j=1}^{n} \operatorname{supp}(\tau_j)$ and therefore compact. □

In working with submanifolds we often use, without further notice, the following facts. Let M be a smooth manifold and $A \subset M$. Then A is a submanifold if and only if each $a \in A$ has an open neighbourhood U such that $A \cap U$ is a submanifold of U. (Being a submanifold is a local property.) Let $f: N_1 \to N_2$ be a diffeomorphism. Then $M_1 \subset N_1$ is a submanifold if and only if $f(M_1) = M_2 \subset N_2$ is a submanifold. (Being a submanifold is invariant under diffeomorphisms.)

Important objects in mathematics are the group objects in the smooth category. A ***Lie group*** consists of a smooth manifold G and a group structure on G such that the group multiplication and the passage to the inverse are smooth maps. The fundamental examples are the classical matrix groups. A basic result in this context says that a closed subgroup of a Lie group is a submanifold and with the induced structure a Lie group [84], [29].

Problems

1. The gluing procedure (1.3.7) can be adapted to the smooth category. The maps g_i^j are assumed to be diffeomorphisms, and the result will be a locally Euclidean space. Again one has to take care that the result will become a Hausdorff space.

2. Let E be an n-dimensional real vector space $0 < r < n$. We define charts for the Grassmann manifold $G_r(E)$ of r-dimensional subspaces of E. Let K be a subspace of codimension r in E. Consider the set of complements in K

$$U(K) = \{F \in G_r(E) \mid F \oplus K = E\}.$$

The sets are the chart domains. Let $P(K) = \{p \in \mathrm{Hom}(E,E) \mid p^2 = p, p(E) = K\}$ be the set of projections with image K. Then $P(K) \to U(K)$, $p \mapsto \mathrm{Ker}(p)$ is a bijection. The set $P(K)$ is an affine space for the vector space $\mathrm{Hom}(E/K, K)$. Let $j : K \subset E$ and let $q : E \to E/K$ be the quotient map. Then

$$\mathrm{Hom}(E/K, K) \times P(K) \to P(K), \quad (\varphi, p) \mapsto p + j\varphi q$$

is a transitive free action. We choose a base point $p_0 \in P(K)$ in this affine space and obtain a bijection

$$U(K) \leftarrow P(K) \to \mathrm{Hom}(E/K, K), \quad \mathrm{Ker}(p) \leftarrow\!\!\shortmid p \mapsto p - p_0.$$

The bijections are the charts for a smooth structure.

3. $\{(x,y,z) \subset \mathbb{R}^3 \mid z^2x^3 + 3zx^2 + 3x - zy^2 - 2y = 1\}$ is a smooth submanifold of $\mathbb{R}^3$ diffeomorphic to $\mathbb{R}^2$. If one considers the set of solutions $(x,y,z) \in \mathbb{C}^2$, then one obtains a smooth complex submanifold of $\mathbb{C}^3$ which is contractible but not homeomorphic to $\mathbb{C}^2$ (see [47]).

15.2 Tangent Spaces and Differentials

We associate to each point p of a smooth m-manifold M an m-dimensional real vector space $T_p(M)$, the ***tangent space*** of M at the point p, and to each smooth map $f : M \to N$ a linear map $T_p f : T_p(M) \to T_{f(p)}(N)$, the ***differential*** of f at p, such that the functor properties hold (***chain rule***)

$$T_p(gf) = T_{f(p)}g \circ T_p f, \quad T_p(\mathrm{id}) = \mathrm{id}.$$

The elements of $T_p(M)$ are the ***tangent vectors*** of M at p.

Since there exist many different constructions of tangent spaces, we define them by a universal property.

A ***tangent space*** of the m-dimensional smooth manifold M at p consists of an m-dimensional vector space $T_p(M)$ together with an isomorphism $i_k : T_p M \to \mathbb{R}^m$ for each chart $k = (U, \varphi, U')$ about p such that for any two such charts k and $l = (V, \psi, V')$ the isomorphism $i_l i_k^{-1}$ is the differential of the coordinate change $\psi\varphi^{-1}$ at $\varphi(p)$. If $(T'_p M, i'_k)$ is another tangent space, then $\iota_p = i_k^{-1} \circ i'_k : T'_p M \to T_p M$ is independent of the choice of k. Thus a tangent space is determined, up to unique isomorphism, by the universal property. If we fix a chart k, an *arbitrary* m-dimensional vector space $T_p M$, and an isomorphism $i_k : T_p M \to \mathbb{R}^m$, then there exists a unique tangent space with underlying vector space $T_p M$ and isomorphism i_k; this follows from the chain rule of calculus. Often we talk about the tangent space T_pM and understand a suitable isomorphism $i_k : T_pM \to \mathbb{R}^m$ as structure datum.

Let $f : M^m \to N^n$ be a smooth map. Choose charts $k = (U, \varphi, U')$ about $p \in M$ and $l = (V, \psi, V')$ about $f(p) \in N$. There exists a unique linear map $T_p f$ which makes the diagram

$$\begin{array}{ccc} T_pM & \xrightarrow{T_p f} & T_{f(p)}N \\ \uparrow{\scriptstyle i_k} & & \uparrow{\scriptstyle i_l} \\ \mathbb{R}^m & \xrightarrow{D(\psi f \varphi^{-1})} & \mathbb{R}^n \end{array}$$

commutative; the morphism at the bottom is the differential of $\psi f \varphi^{-1}$ at $\varphi(p)$. Again by the chain rule, $T_p f$ is independent of the choice of k and l. Differentials, defined in this manner, satisfy the chain rule. This definition is also compatible with the universal maps ι_p for different choices of tangent spaces $T_p f \circ \iota_p = \iota_{f(p)} \circ T'_p f$.

In abstract terms: Make a choice of $T_p(M)$ for each pair $p \in M$. Then the T_pM and the $T_p f$ constitute a functor from the category of pointed smooth manifolds and pointed smooth maps to the category of real vector spaces. Different choices of tangent spaces yield isomorphic functors.

The purpose of tangent spaces is to allow the definition of differentials. The actual vector spaces are adapted to the situation at hand and can serve other geometric purposes (e.g., they can consist of geometric tangent vectors).

We call a smooth map f an ***immersion*** if each differential $T_x f$ is injective and a ***submersion*** if each differential $T_x f$ is surjective. The point $x \in M$ is a ***regular*** point of f if $T_x f$ is surjective. A point $y \in N$ is a ***regular value*** of f if each $x \in f^{-1}(y)$ is a regular point, and otherwise a ***singular value***. If $f^{-1}(y) = \emptyset$, then y is also called a regular value.

(15.2.1) Rank theorem. *Let $f : M \to N$ be a smooth map from an m-manifold into an n-manifold.*

(1) *If $T_a f$ is bijective, then there exist open neighbourhoods U of a and V of $f(a)$, such that f induces a diffeomorphism $f : U \to V$.*

(2) *If $T_a f$ is injective, then there exist open neighbourhoods U of a, V of $f(a)$, W of $0 \in \mathbb{R}^{n-m}$ and a diffeomorphism $F : U \times W \to V$ such that $F(x, 0) = f(x)$ for $x \in U$.*

(3) *If $T_a f$ is surjective, then there exist open neighbourhoods U of a, V of $f(a)$, W of $0 \in \mathbb{R}^{m-n}$ and a diffeomorphism $F : U \to V \times W$ such that $\mathrm{pr}_V\, F(x) = f(x)$ for $x \in U$ with the projection $\mathrm{pr}_V : V \times W \to V$.*

(4) *Suppose $T_x f$ has rank r for all $x \in M$. Then for each $a \in M$ there exist open neighbourhoods U of a, V of $f(a)$ and diffeomorphisms $\varphi : U \to U'$, $\psi : V \to V'$ onto open sets $U' \subset \mathbb{R}^m$, $V' \subset \mathbb{R}^n$ such that $f(U) \subset V$ and $\psi f \varphi^{-1}(x_1, \dots, x_m) = (x_1, \dots, x_r, 0, \dots, 0)$ for all $(x_1, \dots, x_m) \in U'$.*

Proof. The assertions are of a local nature. Therefore we can, via local charts, reduce to the case that M and N are open subsets of Euclidean spaces. Then these assertions are known from calculus. □

(15.2.2) Proposition. *Let y be a regular value of the smooth map $f: M \to N$. Then $P = f^{-1}(y)$ is a smooth submanifold of M. For each $x \in P$, we can identify $T_x P$ with the kernel of $T_x f$.*

Proof. Let $x \in P$. The rank theorem (15.2.1) says that f is in suitable local coordinates about x and $f(x)$ a surjective linear map; hence P is locally a submanifold.

The differential of a constant map is zero. Hence $T_x P$ is contained in the kernel of $T_x f$. For reasons of dimension, the spaces coincide. □

(15.2.3) Example. The differentials of the projections onto the factors yield an isomorphism $T_{(x,y)}(M \times N) \cong T_x(M) \times T_y(N)$ which we use as an identification. With these identifications, $T_{(x,y)}(f \times g) = T_x f \times T_y g$ for smooth maps f and g. Let $h: M \times N \to P$ be a smooth map. Then $T_{(x,y)}h$, being a linear map, is determined by the restrictions to $T_x M$ and to $T_y N$, hence can be computed from the differentials of the partial maps $h_1: a \mapsto h(a, y)$ and $h_2: b \mapsto h(x, b)$ via $T_{(x,y)}h(u, v) = T_x h_1(u) + T_y h_2(v)$. ◇

(15.2.4) Proposition. *Suppose $f: M \to N$ is an immersion which induces a homeomorphism $M \to f(M)$. Then f is a smooth embedding.*

Proof. We first show that $f(M)$ is a smooth submanifold of N of the same dimension as M. It suffices to verify this locally.

Choose U, V, W and F according to (15.2.1). Since U is open and $M \to f(M)$ a homeomorphism, the set $f(U)$ is open in $f(M)$. Therefore $f(U) = f(M) \cap P$, with some open set $P \subset N$. The set $R = V \cap P$ is an open neighbourhood of b in N, and $R \cap f(M) = f(U)$ holds by construction. It suffices to show that $f(U)$ is a submanifold of R. We set $Q = F^{-1}R$, and have a diffeomorphism $F: Q \to R$ which maps $U \times 0$ bijectively onto $f(U)$. Since $U \times 0$ is a submanifold of $U \times W$, we see that $f(U)$ is a submanifold.

By assumption, $f: M \to f(M)$ has a continuous inverse. This inverse is smooth, since $f: M \to f(M)$ has an injective differential, hence bijective for dimensional reasons, and is therefore a local diffeomorphism. □

(15.2.5) Proposition. *Let $f: M \to N$ be a surjective submersion and $g: N \to P$ a set map between smooth manifolds. If gf is smooth, then g is smooth.*

Proof. Let $f(x) = y$. By the rank theorem, there exist chart domains U about x and V about y such that $f(U) = V$ and $f: U \to V$ has, in suitable local coordinates, the form $(x_1, \dots, x_m) \mapsto (x_1, \dots, x_n)$. Hence there exists a smooth map $s: V \to U$ such that $fs(z) = z$ for all $z \in V$. Then $g(z) = gfs(z)$, and gfs is smooth. (The map s is called a ***local section*** of f about y.) □

It is an important fact of analysis that most values are regular. A set $A \subset N$ in the n-manifold N is said to have (Lebesgue) ***measure zero*** if for each chart (U, h, V) of N the subset $h(U \cap A)$ has measure zero in $\mathbb{R}^n$. A subset of $\mathbb{R}^n$ has measure zero if it can be covered by a countable number of cubes with arbitrarily small total volume. We use the fact that a diffeomorphism (in fact a C^1-map) sends sets of measure zero to sets of measure zero. An open (non-empty) subset of $\mathbb{R}^n$ does not have measure zero. The next theorem is a basic result for differential topology; in order to save space we refer for its proof to the literature [136], [30], [177].

(15.2.6) Theorem (Sard). *The set of singular values of a smooth map has measure zero, and the set of regular values is dense.* □

Problems

1. An injective immersion of a compact manifold is a smooth embedding.
2. Let $f: M \to N$ be a smooth map which induces a homeomorphism $M \to f(M)$. If the differential of f has constant rank, then f is a smooth embedding. By the rank theorem, f has to be an immersion, since f is injective.
3. Let M be a smooth m-manifold and $N \subset M$. The following assertions are equivalent: (1) N is a k-dimensional smooth submanifold of M. (2) For each $a \in N$ there exist an open neighbourhood U of a in M and a smooth map $f: U \to \mathbb{R}^{m-k}$ such that the differential $Df(u)$ has rank $m-k$ for all $u \in U$ and such that $N \cap U = f^{-1}(0)$. (Submanifolds are locally solution sets of "regular" equations.)
4. $f: \mathbb{R}^{n+1} \to \mathbb{R}$, $(x_0, \dots, x_n) \mapsto \sum x_i^2 = \|x\|^2$ has, away from the origin, a non-zero differential. The sphere $S^n(c) = f^{-1}(c^2) = \{x \in \mathbb{R}^{n+1} \mid c = \|x\|\}$ of radius $c > 0$ is therefore a smooth submanifold of $\mathbb{R}^{n+1}$. From Proposition (15.2.2) we obtain $T_x S^n(c) = \{v \in \mathbb{R}^{n+1} \mid x \perp v\}$.
5. Let $M(m,n)$ be the vector space of real (m,n)-matrices and $M(m,n;k)$ for $0 \le k \le \min(m,n)$ the subset of matrices of rank k. Then $M(m,n;k)$ is a smooth submanifold of $M(m,n)$ of dimension $k(m+n-k)$.
6. The subset $S_k(\mathbb{R}^n) = \{(x_1, \dots, x_k) \mid x_i \in \mathbb{R}^n; x_1, \dots, x_k$ linearly independent$\}$ of the k-fold product of $\mathbb{R}^n$ is called the ***Stiefel manifold*** of k-frames in $\mathbb{R}^n$. It can be identified with $M(k,n;k)$ and carries this structure of a smooth manifold.
7. The group $\mathrm{O}(n)$ of orthogonal (n,n)-matrices is a smooth submanifold of the vector space $M_n(\mathbb{R})$ of real (n,n)-matrices. Let $S_n(\mathbb{R})$ be the subspace of symmetric matrices. The map $f: M_n(\mathbb{R}) \to S_n(\mathbb{R})$, $B \mapsto B^t \cdot B$ is smooth, $\mathrm{O}(n) = f^{-1}(E)$, and f has surjective differential at each point $A \in \mathrm{O}(n)$. The derivative at $s = 0$ of $s \mapsto (A^t + sX^t)(A + sX)$ is $A^t \cdot X + X^t \cdot A$; the differential of f at A is the linear map $M_n(\mathbb{R}) \to S_n(\mathbb{R})$, $X \mapsto A^t \cdot X + X^t \cdot A$. It is surjective, since the symmetric matrix S is the image of $X = \frac{1}{2}AS$. From (15.2.2) we obtain

$$T_A\mathrm{O}(n) = \{X \in M_n(\mathbb{R}) \mid A^t \cdot X + X^t \cdot A = 0\},$$

and in particular for the unit matrix E, $T_E\mathrm{O}(n) = \{X \in M_n(\mathbb{R}) \mid A^t + A = 0\}$, the space of skew-symmetric matrices. A local parametrization of $\mathrm{O}(n)$ about E can be obtained from the exponential map $T_E\mathrm{O}(n) \to \mathrm{O}(n)$, $X \mapsto \exp X = \sum_0^\infty X^k/k!$. Group multiplication

and passage to the inverse are smooth maps.
8. Make a similar analysis of the unitary group $U(n)$.
9. The Stiefel manifolds have an orthogonal version which generalizes the orthogonal group, the ***Stiefel manifold of orthonormal k-frames***. Let $V_k(\mathbb{R}^n)$ be the set of orthonormal k-tuples $(v_1,\dots,v_k)$, $v_j \in \mathbb{R}^n$. If we write v_j as row vector, then $V_k(\mathbb{R}^n)$ is a subset of the vector space $M = M(k,n;\mathbb{R})$ of real (k,n)-matrices. Let $S = S_k(\mathbb{R})$ again be the vector space of symmetric (k,k)-matrices. Then $f\colon M \to S$, $A \mapsto A\cdot A^t$ has the pre-image $f^{-1}(E) = V_k(\mathbb{R}^n)$. The differential of f at A is the linear map $X \mapsto XA^t + AX^t$ and it is surjective. Hence E is a regular value. The dimension of $V_k(\mathbb{R}^n)$ is $(n-k)k + \frac{1}{2}k(k-1)$.
10. The defining map $\mathbb{R}^{n+1}\smallsetminus 0 \to \mathbb{R}P^n$ is a submersion. Its restriction to S^n is a submersion and an immersion (a 2-fold regular covering).
11. The graph of a smooth function $f\colon \mathbb{R}^n \to \mathbb{R}$ is a smooth submanifold of $\mathbb{R}^{n+1}$.
12. Let Y be a smooth submanifold of Z and $X \subset Y$. Then X is a smooth submanifold of Y if and only if it is a smooth submanifold of Z. If X is a smooth submanifold, then there exists about each point $x \in X$ a chart (U,φ,V) of Z such that $\varphi(U\cap X)$ as well as $\varphi(U \cap Y)$ are intersections of V with linear subspaces. (Charts ***adapted*** to $X \subset Y \subset Z$. Similarly for inclusions of submanifolds $X_1 \subset X_2 \subset \dots \subset X_r$.)
13. Let $\Lambda^k(\mathbb{R}^n)$ be the k-th exterior power of $\mathbb{R}^n$. The action of $O(n)$ on $\mathbb{R}^n$ induces an action on $\Lambda^k(\mathbb{R}^n)$, a smooth representation. If we assign to a basis $x(1),\dots,x(k)$ of a k-dimensional subspace the element $x(1)\wedge\dots\wedge x(k) \in \Lambda^k(\mathbb{R}^n)$, we obtain a well-defined, injective, $O(n)$-equivariant map $j\colon G_k(\mathbb{R}^n) \to P(\Lambda^k\mathbb{R}^n)$ (***Plücker coordinates***). The image of j is a smooth submanifold of $P(\Lambda^k\mathbb{R}^n)$, i.e., j is an embedding of the Grassmann manifold $G_k(\mathbb{R}^n)$.
14. The ***Segre embedding*** is the smooth embedding

$$\mathbb{R}P^m \times \mathbb{R}P^n \to \mathbb{R}P^{(m+1)(n+1)-1}, \quad ([x_i],[y_j]) \mapsto [x_i y_j].$$

For $m = n = 1$ the image is the quadric $\{[s_0,s_1,s_2,s_3] \mid s_0s_3 - s_1s_2 = 0\}$.
15. Let $h\colon \mathbb{R}^{n+1}\times\mathbb{R}^{n+1} \to \mathbb{R}^{n+k+1}$ be a symmetric bilinear form such that $x \neq 0$, $y \neq 0$ implies $h(x,y) \neq 0$. Let $g\colon S^n \to S^{n+k}$, $x \mapsto h(x,x)/|h(x,x)|$. If $g(x) = g(y)$, hence $h(x,x) = t^2 h(y,y)$ with some $t \in \mathbb{R}$, then $h(x+ty, x-ty) = 0$ and therefore $x + ty = 0$ or $x - ty = 0$. Hence g induces a smooth embedding $\mathbb{R}P^n \to S^{n+k}$. The bilinear form $h(x_0,\dots,x_n,y_0,\dots,y_n) = (z_0,\dots,z_{2n})$ with $z_k = \sum_{i+j=k} x_i y_j$ yields an embedding $\mathbb{R}P^n \to S^{2n}$ [89], [95].
16. Remove a point from $S^1 \times S^1$ and show (heuristically) that the result has an immersion into $\mathbb{R}^2$. (Removing a point is the same as removing a big 2-cell!).

15.3 Smooth Transformation Groups

Let G be a Lie group and M a smooth manifold. We consider smooth action $G \times M \to M$ of G on M. The left translations $l_g\colon M \to M$, $x \mapsto gx$ are then diffeomorphisms. The map $\beta\colon G \to M$, $g \mapsto gx$ is a smooth G-map with image the orbit $B = Gx$ through x. We have an induced bijective G-equivariant set map $\gamma\colon G/G_x \to B$. The map β has constant rank; this follows from the equivariance. If $L_g\colon G \to G$ and $l_g\colon M \to M$ denote the left translations by g, then $l_g\beta = \beta L_g$, and since L_g and l_g are diffeomorphisms, we see that $T_e\beta$ and $T_g\beta$ have the same rank.

(15.3.1) Proposition. *Suppose the orbit $B = Gx$ is a smooth submanifold of M. Then:*

(1) *$\beta\colon G \to B$ is a submersion.*

(2) *$G_x = \beta^{-1}(x)$ is a closed Lie subgroup of G.*

(3) *There exists a unique smooth structure on G/G_x such that the quotient map $G \to G/G_x$ is a submersion. The induced map $\gamma\colon G/G_x \to B$ is a diffeomorphism.*

Proof. If β would have somewhere a rank less than the dimension of B, the rank would always be less than the dimension, by equivariance. This contradicts the theorem of Sard. We transport via γ the smooth structure from B to G/G_x. The smooth structure is unique, since $G \to G/G_x$ is a submersion. The pre-image G_x of a regular value is a closed submanifold. □

The previous proposition gives us G_x as a closed Lie subgroup. We need not use the general theorem about closed subgroups being Lie subgroups.

(15.3.2) Example. The action of $\mathrm{SO}(n)$ on S^{n-1} by matrix multiplication is smooth. We obtain a resulting equivariant diffeomorphism $S^{n-1} \cong \mathrm{SO}(n)/\mathrm{SO}(n-1)$. In a similar manner we obtain equivariant diffeomorphisms $S^{2n-1} \cong \mathrm{U}(n)/\mathrm{U}(n-1) \cong \mathrm{SU}(n)/\mathrm{SU}(n-1)$. ◇

(15.3.3) Theorem. *Let M be a smooth n-manifold. Let $C \subset M \times M$ be the graph of an equivalence relation R on M, i.e., $C = \{(x,y) \mid x \sim y\}$. Then the following are equivalent:*

(1) *The set of equivalence classes $N = M/R$ carries the structure of a smooth manifold such that the quotient map $p\colon M \to N$ is a submersion.*

(2) *C is a closed submanifold of $M \times M$ and $\mathrm{pr}_1\colon C \to M$ is a submersion.* □

(15.3.4) Theorem. *Let M be a smooth G-manifold with free, proper action of the Lie group G. Then the orbit space M/G carries a smooth structure and the orbit map $p\colon M \to M/G$ is a submersion.*

Proof. We verify the hypothesis of the quotient theorem (15.3.3). We have to show that C is a closed submanifold. The set C is homeomorphic to the image of the map $\Theta\colon G \times M \to M \times M$, $(g,x) \mapsto (x, gx)$, since the action is proper. We show that Θ is a smooth embedding. It suffices to show that Θ is an immersion (see (15.2.4)). The differential

$$T_{(g,x)}\Theta\colon T_gG \times T_xM \to T_xM \times T_{gx}M$$

will be decomposed according to the two factors

$$T_{(g,x)}\Theta(u,v) = T_g\Theta(?,x)u + T_x\Theta(g,?)v.$$

The first component of $T_g\Theta(?,x)u$ is zero, since the first component of the partial map is constant. Thus if $T_{(g,x)}(u,v) = 0$, the component of $T_x\Theta(g,?)v$ in T_xM is zero; but this component is v. It remains to show that $T_g f\colon T_gG \to T_{gx}M$ is injective for $f\colon G \to M$, $g \mapsto gx$. Since the action is free, the map f is injective; and f has constant rank, by equivariance. An injective map of constant rank has injective differential, by the rank theorem. Thus we have verified the first hypothesis of (15.3.3). The second one holds, since $\mathrm{pr}_1 \circ \Theta = \mathrm{pr}_2$ shows that pr_1 is a submersion. □

(15.3.5) Example. The cyclic group $G = \mathbb{Z}/m \subset S^1$ acts on $\mathbb{C}^n$ by

$$\mathbb{Z}/m \times \mathbb{C}^n \to \mathbb{C}^n, \quad (\zeta,(z_1,\dots,z_n)) \mapsto (\zeta^{r_1}z_1,\dots,\zeta^{r_n}z_n)$$

where $r_j \in \mathbb{Z}$. This action is a smooth representation. Suppose the integers r_j are coprime to m. The induced action on the unit sphere is then a free G-manifold $S(r_1,\dots,r_n)$; the orbit manifold $L(r_1,\dots,r_n)$ is called a (generalized) ***lens space***. ◇

(15.3.6) Example. Let H be a closed Lie subgroup of the Lie group G. The H-action on G by left translation is smooth and proper. The orbit space $H\backslash G$ carries a smooth structure such that the quotient map $G \to H\backslash G$ is a submersion. The G-action on $H\backslash G$ is smooth. One can consider the projective spaces, Stiefel manifolds and Grassmann manifolds as homogeneous spaces from this view-point. ◇

(15.3.7) Theorem. *Let M be a smooth G-manifold. Then:*

(1) *An orbit $C \subset M$ is a smooth submanifold if and only if it is a locally closed subset.*

(2) *If the orbit C is locally closed and $x \in C$, then there exists a unique smooth structure on G/G_x such that the orbit map $G \to G/G_x$ is a submersion. The map $G/G_x \to C$, $gG_x \mapsto gx$ is a diffeomorphism. The G-action on G/G_x is smooth.*

(3) *If the action is proper, then* (1) *and* (2) *hold for each orbit.*

Proof. (1) $\beta\colon G \to C$, $g \mapsto gx$ has constant rank by equivariance. Hence there exists an open neighbourhood of e in G such that $\beta(U)$ is a submanifold of M. Since C is locally closed in the locally compact space M, the set C is locally compact and therefore $\beta\colon G \to C$ is an open map (see (1.8.6)). Hence there exists an open set W in M such that $C \cap W = \beta(U)$. Therefore C is a submanifold in a neighbourhood of x and, by equivariance, also globally a submanifold.

(2) Since C is locally closed, the submanifold C has a smooth structure. The map β has constant rank and is therefore a submersion. We now transport the smooth structure from C to G/G_x.

(3) The orbits of a proper action are closed. □

15.4 Manifolds with Boundary

We now extend the notion of a manifold to that of a manifold with boundary. A typical example is the n-dimensional disk $D^n = \{x \in \mathbb{R}^n \mid \|x\| \le 1\}$. Other examples are half-spaces. Let $\lambda \colon \mathbb{R}^n \to \mathbb{R}$ be a non-zero linear form. We use the corresponding ***half-space*** $H(\lambda) = \{x \in \mathbb{R}^n \mid \lambda(x) \ge 0\}$. Its ***boundary*** $\partial H(\lambda)$ is the kernel of λ. Typical half-spaces are $\mathbb{R}^n_\pm = \{(x_1, \dots, x_n) \in \mathbb{R}^n \mid \pm x_1 \ge 0\}$. If $A \subset \mathbb{R}^m$ is any subset, we call $f \colon A \to \mathbb{R}^n$ ***differentiable*** if for each $a \in A$ there exists an open neighbourhood U of a in $\mathbb{R}^m$ and a differentiable map $F \colon U \to \mathbb{R}^n$ such that $F|U \cap A = f|U \cap A$. We only apply this definition to open subsets A of half-spaces. In that case, the differential of F at $a \in A$ is independent of the choice of the extension F and will be denoted $Df(a)$.

Let $n \ge 1$ be an integer. An ***n-dimensional manifold with boundary*** or ***∂-manifold*** is a Hausdorff space M with countable basis such that each point has an open neighbourhood which is homeomorphic to an open subset in a half-space of $\mathbb{R}^n$. A homeomorphism $h \colon U \to V$, U open in M, V open in $H(\lambda)$ is called a ***chart*** about $x \in U$ with ***chart domain*** U. With this notion of chart we can define the notions: C^k-related, atlas, differentiable structure. An ***n-dimensional smooth manifold with boundary*** is therefore an n-dimensional manifold M with boundary together with a (maximal) smooth C^∞-atlas on M.

Let M be a manifold with boundary. Its ***boundary*** ∂M is the following subset: The point x is contained in ∂M if and only if there exists a chart (U, h, V) about x such that $V \subset H(\lambda)$ and $h(x) \in \partial H(\lambda)$. The complement $M \smallsetminus \partial M$ is called the ***interior*** $\mathrm{In}(M)$ of M. The following lemma shows that specifying a boundary point does not depend on the choice of the chart (invariance of the boundary).

(15.4.1) Lemma. *Let $\varphi \colon V \to W$ be a diffeomorphism between open subsets $V \subset H(\lambda)$ and $W \subset H(\mu)$ of half-spaces in $\mathbb{R}^n$. Then $\varphi(V \cap \partial H(\lambda)) = W \cap \partial H(\mu)$.* □

(15.4.2) Proposition. *Let M be an n-dimensional smooth manifold with boundary. Then either $\partial M = \emptyset$ or ∂M is an $(n-1)$-dimensional smooth manifold. The set $M \smallsetminus \partial M$ is a smooth n-dimensional manifold with empty boundary.* □

The boundary of a manifold can be empty. Sometimes it is convenient to view the empty set as an n-dimensional manifold. If $\partial M = \emptyset$, we call M a manifold without boundary. This coincides then with the notion introduced in the first section. In order to stress the absence of a boundary, we call a compact manifold without boundary a ***closed manifold***.

A map $f\colon M \to N$ between smooth manifolds with boundary is called ***smooth*** if it is continuous and C^∞-differentiable in local coordinates. ***Tangent spaces*** and the ***differential*** are defined as for manifolds without boundary.

Let $x \in \partial M$ and $k = (U, h, V)$ be a chart about x with V open in $\mathbb{R}^n_-$. Then the pair (k, v), $v \in \mathbb{R}^n$ represents a vector w in the tangent space $T_x M$. We say that w is ***pointing outwards*** (***pointing inwards, tangential***) to ∂M if $v_1 > 0$ ($v_1 < 0$, $v_1 = 0$, respectively). One verifies that this disjunction is independent of the choice of charts.

(15.4.3) Proposition. *The inclusion $j\colon \partial M \subset M$ is smooth and the differential $T_x j\colon T_x(\partial M) \to T_x M$ is injective. Its image consists of the vectors tangential to ∂M. We consider $T_x j$ as an inclusion.* □

The notion of a submanifold can have different meanings for manifolds with boundary. We define therefore submanifolds of type I and type II.

Let M be a smooth n-manifold with boundary. A subset $N \subset M$ is called a k-dimensional smooth ***submanifold*** (of type I) if the following holds: For each $x \in N$ there exists a chart (U, h, V), $V \subset \mathbb{R}^n_-$ open, of M about x such that $h(U \cap N) = V \cap (\mathbb{R}^k \times 0)$. Such charts of M are ***adapted*** to N. The set $V \cap (\mathbb{R}^k \times 0) \subset \mathbb{R}^k_- \times 0 = \mathbb{R}^k_-$ is open in $\mathbb{R}^k_-$. A diffeomorphism onto a submanifold of type I is an embedding of type I. From this definition we draw the following conclusions.

(15.4.4) Proposition. *Let $N \subset M$ be a smooth submanifold of type* I. *The restrictions $h\colon U \cap N \to h(U \cap N)$ of the charts (U, h, V) adapted to N form a smooth atlas for N which makes N into a smooth manifold with boundary. The relation $N \cap \partial M = \partial N$ holds, and ∂N is a submanifold of ∂M.* □

Let M be a smooth n-manifold without boundary. A subset $N \subset M$ is a k-dimensional smooth ***submanifold*** (of type II) if the following holds: For each $x \in N$ there exists a chart (U, h, V) of M about x such that $h(U \cap N) = V \cap (\mathbb{R}^k_- \times 0)$. Such charts are ***adapted*** to N.

The intersection of D^n with $\mathbb{R}^k \times 0$ is a submanifold of type I ($k < n$). The subset D^n is a submanifold of type II of $\mathbb{R}^n$. The next two propositions provide a general means for the construction of such submanifolds.

(15.4.5) Proposition. *Let M be a smooth n-manifold with boundary. Let $f\colon M \to \mathbb{R}$ be smooth with regular value* 0. *Then $f^{-1}[0, \infty[$ is a smooth submanifold of type* II *of M with boundary $f^{-1}(0)$.*

Proof. We have to show that for each $x \in f^{-1}[0, \infty[$ there exists a chart which is adapted to this set. If $f(x) > 0$, then x is contained in the open submanifold $f^{-1}]0, \infty[$; hence the required charts exist. Let therefore $f(x) = 0$. By the rank theorem (15.2.1), f has in suitable local coordinates the form $(x_1, \dots, x_n) \mapsto x_1$. From this fact one easily obtains the adapted charts. □

(15.4.6) Proposition. *Let $f: M \to N$ be smooth and $y \in f(M) \cap (N \smallsetminus \partial N)$ be a regular value of f and $f|\partial M$. Then $P = f^{-1}(y)$ is a smooth submanifold of type* I *of M with $\partial P = (f|\partial M)^{-1}(y) = \partial M \cap P$.*

Proof. Being a submanifold of type I is a local property and invariant under diffeomorphisms. Therefore it suffices to consider a local situation. Let therefore U be open in $\mathbb{R}^m_-$ and $f: U \to \mathbb{R}^n$ a smooth map which has $0 \in \mathbb{R}^n$ as regular value for f and $f|\partial U$ $(n \geq 1, m > n)$.

We know already that $f^{-1}(0) \cap \mathrm{In}(U)$ is a smooth submanifold of $\mathrm{In}(U)$. It remains to show that there exist adapted charts about points $x \in \partial U$. Since x is a regular point of $f|\partial U$, the Jacobi matrix $(D_i f_j(x) \mid 2 \leq i \leq m,\ 1 \leq j \leq n)$ has rank n. By interchange of the coordinates $x_2, \dots, x_m$ we can assume that the matrix

$$(D_i f_j(x) \mid m - n + 1 \leq i \leq m,\ 1 \leq j \leq n)$$

has rank n. (This interchange is a diffeomorphism and therefore harmless.) Under this assumption, $\varphi: U \to \mathbb{R}^m_-$, $u \mapsto (u_1, \dots, u_{m-n}, f_1(u), \dots, f_n(u))$ has bijective differential at x and therefore yields, by part (1) of the rank theorem applied to an extension of f to an open set in $\mathbb{R}^m$, an adapted chart about x. □

If only one of the two manifolds M and N has a non-empty boundary, say M, then we define $M \times N$ as the manifold with boundary which has as charts the products of charts for M and N. In that case $\partial(M \times N) = \partial M \times N$. If both M and N have a boundary, then there appear "corners" along $\partial M \times \partial N$; later we shall explain how to define a differentiable structure on the product in this case.

Problems

1. The map

$$D_n(+) = \{(x,t) \mid t > 0, \|x\|^2 + t^2 \leq 1\} \to]-1, 0] \times U_1(0),\ (x,t) \mapsto \Big(\frac{t}{\sqrt{1 - \|x\|^2}} - 1, x\Big)$$

is an adapted chart for $S^{n-1} = \partial D^n \subset D^n$.

2. Let B be a ∂-manifold. A smooth function $f: \partial B \to \mathbb{R}$ has a smooth extension to B. A smooth function $g: A \to \mathbb{R}$ from a closed submanifold A of type I or of type II of B has a smooth extension to B.

3. Verify the invariance of the boundary for topological manifolds (use local homology groups).

4. A ∂-manifold M is connected if and only if $M \smallsetminus \partial M$ is connected.

5. Let M be a ∂-manifold. There exists a smooth function $f: M \to [0, \infty[$ such that $f(\partial M) = \{0\}$ and $T_x f \neq 0$ for each $x \in \partial M$.

6. Let $f: M \to \mathbb{R}^k$ be an injective immersion of a compact ∂-manifold. Then the image is a submanifold of type II.

7. Verify that "pointing inwards" is well-defined, i.e., independent of the choice of charts.

8. Unfortunately is not quite trivial to classify smooth 1-dimensional manifolds by just starting from the definitions. The reader may try to show that a connected 1-manifold without boundary is diffeomorphic to $\mathbb{R}^1$ or S^1; and a ∂-manifold is diffeomorphic to $[0, 1]$ or $[0, 1[$.

15.5 Orientation

Let V be an n-dimensional real vector space. We call ordered bases $b_1, \dots, b_n$ and $c_1, \dots, c_n$ of V positively related if the determinant of the transition matrix is positive. This relation is an equivalence relation on the set of bases with two equivalence classes. An equivalence class is an ***orientation*** of V. We specify orientations by their representatives. The ***standard orientation*** of $\mathbb{R}^n$ is given by the standard basis $e_1, \dots, e_n$, the rows of the unit matrix. Let W be a complex vector space with complex basis $w_1, \dots, w_n$. Then $w_1, iw_1, \dots, w_n, iw_n$ defines an orientation of the underlying real vector space which is independent of the choice of the basis. This is the orientation ***induced by the complex structure***. Let $u_1, \dots, u_m$ be a basis of U and $w_1, \dots, w_n$ a basis of W. In a direct sum $U \oplus W$ we define the ***sum orientation*** by $u_1, \dots, u_m, w_1, \dots, w_n$. If $o(V)$ is an orientation of V, we denote the ***opposite orientation*** (the occidentation) by $-o(V)$. A linear isomorphism $f : U \to V$ between oriented vector spaces is called ***orientation preserving*** or ***positive*** if for the orientation $u_1, \dots, u_n$ of U the images $f(u_1), \dots, f(u_n)$ yield the given orientation of V.

Let M be a smooth n-manifold with or without boundary. We call two charts ***positively related*** if the Jacobi matrix of the coordinate change has always positive determinant. An atlas is called ***orienting*** if any two of its charts are positively related. We call M ***orientable***, if M has an orienting atlas. An orientation of a manifold is represented by an orienting atlas; and two such define the same orientation if their union contains only positively related charts. If M is oriented by an orienting atlas, we call a chart ***positive*** with respect to the given orientation if it is positively related to all charts of the orienting atlas. These definitions apply to manifolds of positive dimension. An orientation of a zero-dimensional manifold M is a function $\epsilon \colon M \to \{\pm 1\}$.

Let M be an oriented n-manifold. There is an induced orientation on each of its tangent spaces $T_x M$. It is specified by the requirement that a positive chart (U, h, V) induces a positive isomorphism $T_x h \colon T_x M \to T_{h(x)} V = \mathbb{R}^n$ with respect to the standard orientation of $\mathbb{R}^n$. We can specify an orientation of M by the corresponding orientations of the tangent spaces.

If M and N are oriented manifolds, the ***product orientation*** on $M \times N$ is specified by declaring the products $(U \times V, k \times l, U' \times V')$ of positive charts (U, k, U') of M and (V, l, V') of N as positive. The canonical isomorphism $T_{(x,y)}(M \times N) \cong T_x M \oplus T_y N$ is then compatible with the sum orientation of vector spaces. If N is a point, then the canonical identification $M \times N \cong M$ is orientation preserving if and only if $\epsilon(N) = 1$. If M is oriented, then we denote the manifold with the opposite orientation by $-M$.

Let M be an oriented manifold with boundary. For $x \in \partial M$ we have a direct decomposition $T_x(M) = N_x \oplus T_x(\partial M)$. Let $n_x \in N_x$ be pointing outwards. The ***boundary orientation*** of $T_x(\partial M)$ is defined by that orientation $v_1, \dots, v_{n-1}$ for which $n_x, v_1, \dots, v_{n-1}$ is the given orientation of $T_x(M)$. These orientations correspond to the ***boundary orientation*** of ∂M; one verifies that the restriction of positive charts for M yields an orienting atlas for ∂M.

In $\mathbb{R}^n_-$, the boundary $\partial \mathbb{R}^n_- = 0 \times \mathbb{R}^{n-1}$ inherits the orientation defined by $e_2, \dots, e_n$. Thus positive charts have to use $\mathbb{R}^n_-$.

Let $D^2 \subset \mathbb{R}^2$ carry the standard orientation of $\mathbb{R}^2$. Consider S^1 as boundary of D^2 and give it the boundary orientation. An orienting vector in $T_x S^1$ is then the velocity vector of a counter-clockwise rotation. This orientation of S^1 is commonly known as the positive orientation. In general if $M \subset \mathbb{R}^2$ is a two-dimensional submanifold with boundary with orientation induced from the standard orientation of $\mathbb{R}^2$, then the boundary orientation of the curve ∂M is the velocity vector of a movement such that M lies "to the left".

Let M be an oriented manifold with boundary and N an oriented manifold without boundary. Then product and boundary orientation are related as follows

$$o(\partial(M \times N)) = o(\partial M \times N), \quad o(\partial(N \times M)) = (-1)^{\dim N} o(N \times \partial M).$$

The unit interval $I = [0,1]$ is furnished with the standard orientation of $\mathbb{R}$. Since the outward pointing vector in 0 yields the negative orientation, we specify the orientation of ∂I by $\epsilon(0) = -1$, $\epsilon(1) = 1$. We have $\partial(I \times M) = 0 \times M \cup 1 \times M$. The boundary orientation of $0 \times M \cong M$ is opposite to the original one and the boundary orientation of $1 \times M \cong M$ is the original one, if $I \times M$ carries the product orientation. We express these facts by the suggestive formula $\partial(I \times M) = 1 \times M - 0 \times M$. (These conventions suggest that homotopies should be defined with the cylinder $I \times X$.)

A diffeomorphism $f: M \to N$ between oriented manifolds ***respects the orientation*** if $T_x f$ is for each $x \in M$ orientation preserving. If M is connected, then f respects or reverses the orientation.

Problems

1. Show that a 1-manifold is orientable.

2. Let $f: M \to N$ be a smooth map and let A be the pre-image of a regular value $y \in N$. Suppose M is orientable, then A is orientable.

We specify an orientation as follows. Let M and N be oriented. We have an exact sequence $0 \to T_a A \xrightarrow{(1)} T_a M \xrightarrow{(2)} T_y N \to 0$, with inclusion (1) and differential $T_a f$ at (2). This orients $T_a A$ as follows: Let $v_1, \dots, v_k$ be a basis of $T_a A$, $w_1, \dots, w_l$ a basis of $T_y N$, and $u_1, \dots, u_l$ be pre-images in $T_a M$; then $v_1, \dots, v_k, u_1, \dots, u_l$ is required to be the given orientation of $T_a M$. These orientations induce an orientation of A. This orientation

of A is called the ***pre-image orientation***.
3. Let $f: \mathbb{R}^n \to \mathbb{R}$, $(x_i) \mapsto \sum x_i^2$ and $S^{n-1} = f^{-1}(1)$. Then the pre-image orientation coincides with the boundary orientation with respect to $S^{n-1} \subset D^n$.

15.6 Tangent Bundle. Normal Bundle

The notions and concepts of bundle theory can now be adapted to the smooth category. A smooth bundle $p: E \to B$ has a smooth bundle projection p and the bundle charts are assumed to be smooth. A smooth subbundle of a smooth vector bundle has to be defined by smooth bundle charts. Let $\alpha: \xi_1 \to \xi_2$ be a smooth bundle morphism of constant rank; then $\operatorname{Ker} \alpha$ and $\operatorname{Im} \alpha$ are smooth subbundles. The proof of (14.2.3) can also be used in this situation. A smooth vector bundle has a smooth Riemannian metric; for the existence proof one uses a smooth partition of unity and proceeds as in (14.5.1). Let ξ be a smooth subbundle of the smooth vector bundle η with Riemannian metric; then the orthogonal complement of ξ in η is a smooth subbundle.

Let M be a smooth n-manifold. Denote by TM the disjoint union of the tangent spaces $T_p(M)$, $p \in M$. We write a point of $T_p(M) \subset TM$ in the form (p, v) with $v \in T_p(M)$, for emphasis. We have the projection $\pi_M: TM \to M$, $(p, v) \mapsto p$. Each chart $k = (U, h, V)$ of M yields a bijection

$$\varphi_k: TU = \textstyle\bigcup_{p \in U} T_p(M) \to U \times \mathbb{R}^n, \quad (p, v) \mapsto (p, i_k(v)).$$

Here i_k is the morphism which is part of the definition of a tangent space. The map φ_k is a map over U and linear on fibres. The next theorem is a consequence of the general gluing procedure.

(15.6.1) Theorem. *There exists a unique structure of a smooth manifold on TM such that the TU are open subsets and the φ_k are diffeomorphisms. The projection π_M carries the structure of an n-dimensional vector bundle for which the φ_k are bundle charts.* □

The vector bundle $\pi_M: TM \to M$ is called the ***tangent bundle*** of M. A smooth map $f: M \to N$ induces a smooth fibrewise map $Tf: TM \to TN$, $(p, v) \mapsto (f(p), T_p f(v))$.

(15.6.2) Proposition. *Let $M \subset \mathbb{R}^q$ be a smooth n-dimensional submanifold. Then*

$$TM = \{(x, v) \mid x \in M, v \in T_x(M)\} \subset M \times \mathbb{R}^q$$

is a smooth subbundle of the trivial bundle $M \times \mathbb{R}^q \to M$.

Proof. Let (U, φ, V) be a chart of $\mathbb{R}^q$ adapted to M. Then the diffeomorphism

$$U \times \mathbb{R}^q \to U \times \mathbb{R}^q, \quad (x, v) \mapsto (x, T_x\varphi(v))$$

restricted to $U \cap M$ is a bundle chart adapted to TM. □

We can apply (15.6.2) to $S^n \subset \mathbb{R}^{n+1}$ and obtain the model of the tangent bundle of S^n, already used at other occasions.

Let the Lie group G act smoothly on M. We have an induced action

$$G \times TM \to TM, \quad (g, v) \mapsto (Tl_g)v.$$

This action is again smooth and the bundle projection is equivariant, i.e., $TM \to M$ is a smooth G-vector bundle.

(15.6.3) Proposition. *Let $\xi\colon E \to M$ be a smooth G-vector bundle. Suppose the action on M is free and proper. Then the orbit map $E/G \to M/G$ is a smooth vector bundle. We have an induced bundle map $\xi \to \xi/G$.* □

The differential $Tp\colon TM \to T(M/G)$ of the orbit map p is a bundle morphism which factors over the orbit map $TM \to (TM)/G$ and induces a bundle morphism $q\colon (TM)/G \to T(M/G)$ over M/G. The map is fibrewise surjective. If G is discrete, then M and M/G have the same dimension, hence q is an isomorphism.

(15.6.4) Proposition. *For a free, proper, smooth action of the discrete group G on M we have a bundle isomorphism $(TM)/G \cong T(M/G)$ induced by the orbit map $M \to M/G$.* □

(15.6.5) Example. We have a bundle isomorphism $TS^n \oplus \varepsilon \cong (n+1)\varepsilon$. If $G = \mathbb{Z}/2$ acts on TS^n via the differential of the antipodal map and trivially on ε, then the said isomorphism transforms the action into $S^n \times \mathbb{R}^{n+1} \to S^n \times \mathbb{R}^{n+1}$, $(x, v) \mapsto (-x, -v)$. We pass to the orbit spaces and obtain an isomorphism $T(\mathbb{R}P^n) \oplus \varepsilon \cong (n+1)\eta$ with the tautological line bundle η over $\mathbb{R}P^n$. ◇

In the general case the map $q\colon (TM)/G \to T(M/G)$ has a kernel, a bundle $K \to M/G$ with fibre dimension $\dim G$. See [44, IX.6] for details.

(15.6.6) Example. The defining map $\mathbb{C}^{n+1} \smallsetminus 0 \to (\mathbb{C}^{n+1} \smallsetminus 0)/\mathbb{C}^* = \mathbb{C}P^n$ yields a surjective bundle map $q\colon T(\mathbb{C}^{n+1} \smallsetminus 0)/\mathbb{C}^* \to T(\mathbb{C}P^n)$. The source of q is the $(n+1)$-fold Whitney sum $(n+1)\eta$ where $E(\eta)$ is the quotient of $(\mathbb{C}^{n+1} \smallsetminus 0) \times \mathbb{C}$ with respect to $(z, x) \sim (\lambda z, \lambda x)$ for $\lambda \in \mathbb{C}^*$ and $(z, x) \in (\mathbb{C}^{n+1} \smallsetminus 0) \times \mathbb{C}$. The kernel bundle of q is trivial: We have a canonical section of $(n+1)\eta$

$$\mathbb{C}P^n \to ((\mathbb{C}^{n+1} \smallsetminus 0) \times \mathbb{C}^{n+1})/\mathbb{C}^*, \quad [z] \mapsto (z, z)/\sim,$$

and the subbundle generated by this section is contained in the kernel of q. Hence the complex tangent bundle of $\mathbb{C}P^n$ satisfies $T(\mathbb{C}P^n) \oplus \varepsilon \cong (n+1)\eta$. For η see $H(1)$ in (14.2.6). ◇

Let $p\colon E \to M$ be a smooth vector bundle. Then E is a smooth manifold and we can ask for its tangent bundle.

(15.6.7) Proposition. *There exists a canonical exact sequence*

$$0 \to p^*E \xrightarrow{\alpha} TE \xrightarrow{\beta} p^*TM \to 0$$

of vector bundles over E, written in terms of total spaces.

Proof. The differential of p is a bundle morphism $Tp\colon TE \to TM$, and it induces a bundle morphism $\beta\colon TE \to p^*TM$ which is fibrewise surjective, since p is a submersion. We consider the total space of $p^*E \to E$ as $E \oplus E$ and the projection onto the first summand is the bundle projection. Let $(v, w) \in E_x \oplus E_x$. We define $\alpha(v, w)$ as the derivative of the curve $t \mapsto v + tw$ at $t = 0$. The bundle morphism α has an image contained in the kernel of β and is fibrewise injective. Thus, for reasons of dimension, the sequence is exact. □

(15.6.8) Remark. We restrict the exact sequence given in (15.6.7) to the zero section $i\colon M \subset E$. Since $pi = \mathrm{id}$ we obtain an exact sequence

$$0 \to E \xrightarrow{\alpha} TE|M \xrightarrow{\beta} TM \to 0$$

of vector bundles over M. For $w \in E_x$, $x \in M$ the vector $\alpha(w) \in T_xE$ is the derivative at $t = 0$ of the curve $t \mapsto tw \in E_x$. The bundle map β has the right inverse $ti\colon TM \to TE|M$. We therefore obtain a canonical isomorphism $(\alpha, Ti)\colon E \oplus TM \cong TE|M$. We have written this in a formal manner. The geometric meaning is that T_xE splits into the tangent vectors in direction of the fibre and the tangent vectors to the submanifold M. Since the tangent space of a vector space is canonically identified with the vector space, we can consider α as an inclusion. ◇

Let $f\colon M \to N$ be an immersion. Then Tf is fibrewise injective. We pull back TN along f and obtain a fibrewise injective bundle morphism $i\colon TM \to f^*TN|M$. The quotient bundle is called the ***normal bundle*** of the immersion. In the case of a submanifold $M \subset N$ the normal bundle $\nu(M, N)$ of M in N is the quotient bundle of $TN|M$ by TM. If we give TN a smooth Riemannian metric, then we can take the orthogonal complement of TM as a model for the normal bundle. The normal bundle of $S^n \subset \mathbb{R}^{n+1}$ is the trivial bundle.

We will show that the total space of the normal bundle of an embedding $M \subset N$ describes a neighbourhood of M in N. We introduce some related terminology. Let $\nu\colon E(\nu) \to M$ denote the smooth normal bundle. A ***tubular map*** is a smooth map $t\colon E(\nu) \to N$ with the following properties:

(1) It is the inclusion $M \to N$ when restricted to the zero section.

(2) It embeds an open neighbourhood U of the zero section onto an open neighbourhood V of M in N.

(3) The differential of t, restricted to $TE(\nu)|M$, is a bundle morphism

$$\tau\colon TE(\nu)|M \to TN|M.$$

We compose with the inclusion (15.6.8)

$$\alpha\colon E(\nu) \to E(\nu) \oplus TM \cong TE(\nu)|M$$

and the projection

$$\pi\colon TN|M \to E(\nu) = (TN|M)/TM.$$

We require that $\pi\tau\alpha$ is the identity. If we use another model of the normal bundle given by an isomorphism $\iota\colon E(\nu) \to (TN|M)/TM$, then we require $\pi\tau\alpha = \iota$.

The purpose of (3) is to exclude bundle automorphisms.

(15.6.9) Remark (Shrinking). Let $t\colon E(\nu) \to N$ be a tubular map for a submanifold M with embedding $t\colon U \xrightarrow{\cong} V$ as above. There exists a smooth function $\varepsilon\colon M \to]0,\infty[$ such that

$$E_\varepsilon(\nu) = \{y \in E(\nu)_x \mid \|y\| < \varepsilon(x)\} \subset U.$$

(Here $E(\nu)$ carries a smooth Riemannian metric, and $\|y\|$ is the corresponding norm.) The set $E_\varepsilon(\nu)$ is an open neighbourhood of the zero section. The assignment

$$h\colon E(\nu) \to E_\varepsilon(\nu), \quad y \mapsto \frac{\varepsilon(x)y}{\sqrt{\varepsilon(x)^2 + \|y\|^2}}, \quad y \in E(\nu)_x$$

is a fibrewise diffeomorphism and $t \circ h$ is a tubular map which embeds $E(\nu)$ onto an open neighbourhood of M in N. ◇

The image U of a tubular map $t\colon E(\nu) \to N$ which embeds $E(\nu)$ is called a ***tubular neighbourhood*** of M in N.

Let M be an m-dimensional smooth submanifold $M \subset \mathbb{R}^n$ of codimension k. We take $N(M) = \{(x,v) \mid x \in M, v \perp T_xM\} \subset M \times \mathbb{R}^n$ as the normal bundle of $M \subset \mathbb{R}^n$.

(15.6.10) Proposition. *The set*

$$N(M) = \{(x,v) \mid x \in M, v \in N_xM\} \subset M \times \mathbb{R}^q$$

together with the projection onto M is a smooth subbundle of the trivial bundle $M \times \mathbb{R}^q \to M$.

Proof. Let $A\colon \mathbb{R}^n \to \mathbb{R}^k$ be a linear map. Its transpose A^t with respect to the standard inner product is defined by $\langle Av, w\rangle = \langle v, A^t w\rangle$. If A is surjective, then A^t is injective, and the relation image $(A^t) = (\text{kernel } A)^\perp$ holds; moreover $A \cdot A^t \in \mathrm{GL}_k(\mathbb{R})$.

We define M locally as the solution set: Suppose $U \subset \mathbb{R}^n$ is open, $\varphi\colon U \to \mathbb{R}^k$ a submersion, and $\varphi^{-1}(0) = U \cap M = W$. We set $N(M) \cap (W \times \mathbb{R}^n) = N(W)$. The smooth maps

$$\Phi\colon W \times \mathbb{R}^n \to W \times \mathbb{R}^k, \quad (x,v) \mapsto (x, T_x\varphi(v)),$$
$$\Psi\colon W \times \mathbb{R}^k \to W \times \mathbb{R}^n, \quad (x,v) \mapsto (x, (T_x\varphi)^t(v))$$

satisfy

$$N(W) = \mathrm{Im}\,\Psi, \quad T(W) = \mathrm{Ker}\,\Phi.$$

The composition $\Phi\Psi$ is a diffeomorphism: it has the form $(w,v) \mapsto (w, g_w(v))$ with a smooth map $W \to \mathrm{GL}_k(\mathbb{R})$, $w \mapsto g_w$ and therefore $(w,v) \mapsto (w, g_w^{-1}(v))$ is a smooth inverse. Hence Ψ is a smooth embedding with image $N(W)$ and $\Psi^{-1}|N(W)$ is a smooth bundle chart. Now use (14.2.3). □

(15.6.11) Proposition. *The map $a\colon N(M) \to \mathbb{R}^n$, $(x,v) \mapsto x + v$ is a tubular map for $M \subset \mathbb{R}^n$.*

Proof. We show that a has a bijective differential at each point $(x,0) \in N(M)$. Let $N_xM = T_xM^\perp$. Since $M \subset \mathbb{R}^n$ we consider T_xM as a subspace of $\mathbb{R}^n$. Then $T_{(x,0)}N(M)$ is the subspace $T_xM \times N_xM \subset T_{(x,0)}(M \times \mathbb{R}^n) = T_xM \times \mathbb{R}^n$. The differential $T_{(x,0)}a$ is the identity on each of the subspaces T_xM and N_xM. Therefore we can consider this differential as the map $(u,v) \mapsto u + v$, i.e., essentially as the identity.

It is now a general topological fact (15.6.13) that a embeds an open neighbourhood of the zero section. Finally it is not difficult to verify property (3) of a tubular map. □

(15.6.12) Corollary. *If we transport the bundle projection $N(M) \to M$ via the embedding a we obtain a smooth retraction $r\colon U \to M$ of an open neighbourhood U of $M \subset \mathbb{R}^n$.* □

(15.6.13) Theorem. *Let $f\colon X \to Y$ be a local homeomorphism. Let $A \subset X$ and $f\colon A \to f(A) = B$ be a homeomorphism. Suppose that each neighbourhood of B in Y contains a paracompact neighbourhood. Then there exists an open neighbourhood U of A in X which is mapped homeomorphically under f onto an open neighbourhood V of B in Y (see* [30, p. 125]*).* □

For embeddings of compact manifolds and their tubular maps one can apply another argument as in the following proposition.

(15.6.14) Proposition. *Let $\Phi\colon X \to Y$ be a continuous map of a locally compact space into a Hausdorff space. Let Φ be injective on the compact set $A \subset X$. Suppose that each $a \in A$ has a neighbourhood U_a in X on which Φ is injective. Then there exists a compact neighbourhood V of A in X on which Φ is an embedding.*

Proof. The coincidence set $K = \{(x, y) \in X \times X \mid \Phi(x) = \Phi(y)\}$ is closed in $X \times X$, since Y is a Hausdorff space. Let $D(B)$ be the diagonal of $B \subset X$. If Φ is injective on U_a, then $(U_a \times U_a) \cap K = D(U_a)$. Hence $D(U_a)$ and $U = \bigcup_{a \in A} D(U_a)$ is open in K and therefore $W = X \times X \smallsetminus (K \smallsetminus U)$ open in $X \times X$. By assumption, $A \times A$ is contained in W. Since $A \times A$ is compact and X locally compact, there exists a compact neighbourhood V of A such that $V \times V \subset W$. Then $\Phi|V$ is injective and, being a map from a compact space into a Hausdorff space, an embedding. □

(15.6.15) Proposition. *A submanifold $M \subset N$ has a tubular map.*

Proof. We fix an embedding of $N \subset \mathbb{R}^n$. By (15.6.12) there exists an open neighbourhood W of V in $\mathbb{R}^n$ and a smooth retraction $r\colon W \to V$. The standard inner product on $\mathbb{R}^n$ induces a Riemannian metric on TN. We use as normal bundle for $M \subset N$ the model

$$E = \{(x, v) \in M \times \mathbb{R}^n \mid v \in (T_x M)^\perp \cap T_x N\}.$$

Again we use the map $f\colon E \to \mathbb{R}^n$, $(x, v) \mapsto x + v$ and set $U = f^{-1}(W)$. Then U is an open neighbourhood of the zero section of E. The map $g = rf\colon U \to N$ is the inclusion when restricted to the zero section. We claim that the differential of g at points of the zero section is the identity, if we use the identification $T_{(x,0)}E = T_x M \oplus E_x = T_x N$. On the summand $T_x M$ the differential $T_{(x,0)}g$ is obviously the inclusion $T_x M \subset T_x V$. For $(x, v) \in E_x$ the curve $t \mapsto (x, tv)$ in E has (x, v) as derivative at $t = 0$. Therefore we have to determine the derivative of $t \mapsto r(x + tv)$ at $t = 0$. The differential of r at $(x, 0)$ is the orthogonal projection $\mathbb{R}^n \to T_x N$, if we use the retraction r in (15.6.12). The chain rule tells us that the derivative of $t \mapsto r(x + tv)$ at $t = 0$ is v. We now apply again (15.6.13). One verifies property (3) of a tubular map. □

15.7 Embeddings

This section is devoted to the embedding theorem of Whitney:

(15.7.1) Theorem. *A smooth n-manifold has an embedding as a closed submanifold of $\mathbb{R}^{2n+1}$.*

We begin by showing that a compact n-manifold has an embedding into some Euclidean space. Let $f\colon M \to \mathbb{R}^t$ be a smooth map from an n-manifold M. Let $(U_j, \phi_j, U_3(0))$, $j \in \{1, \dots, k\}$ be a finite number of charts of M (see (15.1.2) for the definition of $U_3(0)$). Choose a smooth function $\tau\colon \mathbb{R}^n \to [0, 1]$ such that $\tau(x) = 0$ for $\|x\| \geq 2$ and $\tau(x) = 1$ for $\|x\| \leq 1$. Define $\sigma_j\colon M \to \mathbb{R}$ by $\sigma_j(x) = 0$ for $x \notin U_j$ and by $\sigma_j(x) = \tau\phi_j(x)$ for $x \in U_j$; then σ_j is a smooth function on M. With the help of these functions we define

$$\Phi\colon M \to \mathbb{R}^t \times (\mathbb{R} \times \mathbb{R}^n) \times \cdots \times (\mathbb{R} \times \mathbb{R}^n) = \mathbb{R}^t \times \mathbb{R}^N$$

$$\Phi(x) = (f(x); \sigma_1(x), \sigma_1(x)\phi_1(x); \dots; \sigma_k(x), \sigma_k(x)\phi_k(x)),$$

(k factors $\mathbb{R} \times \mathbb{R}^n$), where $\sigma_j(x)\phi_j(x)$ should be zero if $\phi_j(x)$ is not defined. The differential of this map has the rank n on $W_j = \phi_j^{-1}(U_1(0))$, as $\Phi(W_j) \subset V_j = \{(z; a_1, x_1; \dots; a_k, x_k) \mid a_j \neq 0\}$, and the composition of $\Phi|W_j$ with $V_j \to \mathbb{R}^n$, $(z; a_1, x_1; \dots) \mapsto a_j^{-1}x_j$ is ϕ_j. By construction, Φ is injective on $W = \bigcup_{j=1}^k W_j$, since $\Phi(a) = \Phi(b)$ implies $\sigma_j(a) = \sigma_j(b)$ for each j, and then $\phi_i(a) = \phi_i(b)$ holds for some i. Moreover, Φ is equal to f composed with $\mathbb{R}^t \subset \mathbb{R}^t \times \mathbb{R}^N$ on the complement of the $\phi_j^{-1}U_2(0)$. Hence if f is an (injective) immersion on the open set U, then Φ is an (injective) immersion on $U \cup W$. In particular, if M is compact, we can apply this argument to an arbitrary map f and $M = W$. Thus we have shown:

(15.7.2) Note. *A compact smooth manifold has a smooth embedding into some Euclidean space.* □

We now try to lower the embedding dimension by applying a suitable parallel projcction.

Let $\mathbb{R}^{q-1} = \mathbb{R}^{q-1} \times 0 \subset \mathbb{R}^q$. For $v \in \mathbb{R}^q \setminus \mathbb{R}^{q-1}$ we consider the projection $p_v \colon \mathbb{R}^q \to \mathbb{R}^{q-1}$ with direction v, i.e., for $x = x_0 + \lambda v$ with $x_0 \in \mathbb{R}^{q-1}$ and $\lambda \in \mathbb{R}$ we set $p_v(x) = x_0$. In the sequel we only use vectors $v \in S^{q-1}$. Let $M \subset \mathbb{R}^q$. We remove the diagonal D and consider $\sigma \colon M \times M \setminus D \to S^{q-1}$, $(x, y) \mapsto N(x - y) = (x - y)/\|x - y\|$.

(15.7.3) Note. *$\varphi_v = p_v|M$ is injective if and only if v is not contained in the image of σ.*

Proof. The equality $\varphi_v(x) = \varphi_v(y)$, $x \neq y$ and $x = x_0 + \lambda v$, $y = y_0 + \mu v$ imply $x - y = (\lambda - \mu)v \neq 0$, hence $v = \pm N(x - y)$. Note $\sigma(x, y) = -\sigma(y, x)$. □

Let now M be a smooth n-manifold in $\mathbb{R}^q$. We use the bundle of unit vectors

$$STM = \{(x, v) \mid v \in T_x M,\ \|v\| = 1\} \subset M \times S^{q-1}$$

and its projection to the second factor $\tau = \mathrm{pr}_2\,|STM \colon STM \to S^{q-1}$. The function $(x, v) \mapsto \|v\|^2$ on $TM \subset \mathbb{R}^q \times \mathbb{R}^q$ has 1 as regular value with pre-image STM, hence STM is a smooth submanifold of the tangent bundle TM.

(15.7.4) Note. *φ_v is an immersion if and only if v is not contained in the image of τ.*

Proof. The map φ_v is an immersion if for each $x \in M$ the kernel of $T_x p_v$ has trivial intersection with $T_x M$. The differential of p_v is again p_v. Hence $0 \neq z = p_v(z) + \lambda v \in T_x M$ is contained in the kernel of $T_x p_v$ if and only if $z = \lambda v$ and hence v is a unit vector in $T_x M$. □

(15.7.5) Theorem. *Let M be a smooth compact n-manifold. Let $f: M \to \mathbb{R}^{2n+1}$ be a smooth map which is an embedding on a neighbourhood of a compact subset $A \subset M$. Then there exists for each $\varepsilon > 0$ an embedding $g: M \to \mathbb{R}^{2n+1}$ which coincides on A with f and satisfies $\|f(x) - g(x)\| < \varepsilon$ for $x \in M$.*

Proof. Suppose f embeds the open neighbourhood U of A and let $V \subset U$ be a compact neighbourhood of A. We apply the construction in the beginning of this section with chart domains U_j which are contained in $M \setminus V$ and such that the sets W_j cover $M \smallsetminus U$. Then Φ is an embedding on some neighbourhood of $M \setminus U$ and

$$\Phi: M \to \mathbb{R}^{2n+1} \oplus \mathbb{R}^N = \mathbb{R}^q, \quad x \mapsto (f(x), \Psi(x))$$

is an embedding which coincides on V with f (up to composition with the inclusion $\mathbb{R}^{2n+1} \subset \mathbb{R}^q$). For $2n < q - 1$ the image of σ is nowhere dense and for $2n - 1 < q - 1$ the image of τ is nowhere dense (theorem of Sard). Therefore in each neighbourhood of $w \in S^{q-1}$ there exist vectors v such that $p_v \circ \Phi = \Phi_v$ is an injective immersion, hence an embedding since M is compact. By construction, Φ_v coincides on V with f. If necessary, we replace Ψ with $s\Psi$ (with small s) such that $\|f(x) - \Phi(x)\| \le \varepsilon/2$ holds. We can write f as composition of Φ with projections $\mathbb{R}^q \to \mathbb{R}^{q-1} \to \cdots \to \mathbb{R}^{2n+1}$ along the unit vectors $(0, \dots, 1)$. Sufficiently small perturbations of these projections applied to Φ yield a map g such that $\|f(x) - g(x)\| < \varepsilon$, and, by the theorem of Sard, we find among these projections those for which g is an embedding. □

The preceding considerations show that we need one dimension less for immersions.

(15.7.6) Theorem. *Let $f: M \to \mathbb{R}^{2n}$ be a smooth map from a compact n-manifold. Then there exists for each $\varepsilon > 0$ an immersion $h: M \to \mathbb{R}^{2n}$ such that $\|h(x) - f(x)\| < \varepsilon$ for $x \in M$. If $f: M \to \mathbb{R}^{2n+1}$ is a smooth embedding, then the vectors $v \in S^{2n}$ for which the projection $p_v \circ f: M \to \mathbb{R}^{2n}$ is an immersion are dense in S^{2n}.* □

Let $f: M \to \mathbb{R}$ be a smooth proper function from an n-manifold without boundary. Let $t \in \mathbb{R}$ be a regular value and set $A = f^{-1}(t)$. The manifold A is compact. There exists an open neighbourhood U of A in M and a smooth retraction $r: U \to A$.

(15.7.7) Proposition. *There exists an $\varepsilon > 0$ and open neighbourhood $V \subset U$ of A such that $(r, f): V \to A \times]t - \varepsilon, t + \varepsilon[$ is a diffeomorphism.*

Proof. The map $(r, f): U \to A \times \mathbb{R}$ has bijective differential at points of A. Hence there exists an open neighbourhood $W \subset U$ of A such that (r, f) embeds W onto an open neighbourhood of $A \times \{t\}$ in $A \times \mathbb{R}$. Since f is proper, each neighbourhood W of A contains a set of the form $V = f^{-1}]t - \varepsilon, t + \varepsilon[$. The restriction of (r, f) to V has the required properties. □

In a similar manner one shows that a proper submersion is locally trivial (theorem of Ehresmann).

We now show that a non-compact n-manifold M has an embedding into $\mathbb{R}^{2n+1}$ as a closed subset. For this purpose we choose a proper smooth function $f: M \to \mathbb{R}_+$. We then choose a sequence $(t_k \mid k \in \mathbb{N})$ of regular values of f such that $t_k < t_{k+1}$ and $\lim_k t_k = \infty$. Let $A_k = f^{-1}(t_k)$ and $M_k = f^{-1}[t_k, t_{k+1}]$. Choose $\varepsilon_k > 0$ small enough such that the intervals $J_k =]t_k - \varepsilon_k, t_k + \varepsilon_k[$ are disjoint and such that we have diffeomorphisms $f^{-1}(J_k) \cong A_k \times J_k$ of the type (15.7.7). We then use (15.7.7) in order to find embeddings $\Phi_k: f^{-1}(J_k) \to \mathbb{R}^{2n} \times J_k$ which have f as their second component. We then use the method of (15.7.5) to find an embedding $M_k \to \mathbb{R}^{2n} \times [t_k, t_{k+1}]$ which extends the embeddings Φ_k and Φ_{k+1} in a neighbourhood of $M_k + M_{k+1}$. All these embeddings fit together and yield an embedding of M as a closed subset of $\mathbb{R}^{2n+1}$.

A ***collar*** of a smooth ∂-manifold M is a diffeomorphism $\kappa: \partial M \times [0,1[\to M$ onto an open neighbourhood U of ∂M in M such that $\kappa(x,0) = x$. Instead of $[0,1[$ one can also use $\mathbb{R}_\pm$.

(15.7.8) Proposition. *A smooth ∂-manifold M has a collar.*

Proof. There exists an open neighbourhood U of ∂M in M and a smooth retraction $r: U \to \partial M$. Choose a smooth function $f: M \to \mathbb{R}_+$ such that $f(\partial M) = \{0\}$ and the derivative of f at each point $x \in \partial M$ is non-zero. Then $(r, f): U \to \partial M \times \mathbb{R}_+$ has bijective differential along ∂M. Therefore this map embeds an open neighbourhood V of ∂M onto an open neighbourhood W of $\partial M \times 0$. Now choose a smooth function $\varepsilon: \partial M \to \mathbb{R}_+$ such that $\{x\} \times [0, \varepsilon(x)[\subset W$ for each $x \in \partial M$. Then compose $\partial M \times [0,1[\to \partial M \times \mathbb{R}_+$, $(x,s) \mapsto (x, \varepsilon(x)s)$ with the inverse of the diffeomorphism $V \to W$. □

(15.7.9) Theorem. *A compact smooth n-manifold B with boundary M has a smooth embedding of type* I *into D^{2n+1}.*

Proof. Let $j: M \to S^{2n}$ be an embedding. Choose a collar $k: M \times [0,1[\to U$ onto the open neighbourhood U of M in B, and let $l = (l_1, l_2)$ be its inverse. We use the collar to extend j to $f: B \to D^{2n+1}$

$$f(x) = \begin{cases} \max(0, 1 - 2l_2(x))\, j(l_1(x)), & x \in U, \\ 0, & x \notin U. \end{cases}$$

Then f is a smooth embedding on $k(M \times [0, \frac{1}{2}[)$. As in the proof of (15.7.4) we approximate f by a smooth embedding $g: B \to D^{2n+1}$ which coincides with f on $k(M \times [0, \frac{1}{4}[)$ and which maps $B \smallsetminus M$ into the interior of D^{2n+1}. The image of g is then a submanifold of type I of D^{2n+1}. □

15.8 Approximation

Let M and N be smooth manifolds and $A \subset M$ a closed subset. We assume that $N \subset \mathbb{R}^p$ is a submanifold and we give N the metric induced by this embedding.

(15.8.1) Theorem. *Let $f: M \to N$ be continuous and $f|A$ smooth. Let $\delta: M \to]0,\infty[$ be continuous. Then there exists a smooth map $g: M \to N$ which coincides on A with f and satisfies $\|g(x) - f(x)\| < \delta(x)$ for $x \in M$.*

Proof. We start with the special case $N = \mathbb{R}$. The fact that f is smooth at $x \in A$ means, by definition, that there exists an open neighbourhood U_x of x and a smooth function $f_x: U_x \to \mathbb{R}$ which coincides on $U_x \cap A$ with f. Having chosen f_x, we shrink U_x, such that for $y \in U_x$ the inequality $\|f_x(y) - f(y)\| < \delta(y)$ holds.

Fix now $x \in M \smallsetminus A$. We choose an open neighbourhood U_x of x in $M \smallsetminus A$ such that for $y \in U_x$ the inequality $\|f(y) - f(x)\| < \delta(y)$ holds. We define $f_x: U_x \to \mathbb{R}$ in this case by $f_x(y) = f(x)$.

Let $(\tau_x \mid x \in M)$ be a smooth partition of unity subordinate to $(U_x \mid x \in M)$. The function $g(y) = \sum_{x \in M} \tau_x(y) f_x(y)$ now has the required property.

From the case $N = \mathbb{R}$ one immediately obtains a similar result for $N = \mathbb{R}^p$. The general case will now be reduced to the special case $N = \mathbb{R}^p$. For this purpose we choose an open neighbourhood U of N in $\mathbb{R}^p$ together with a smooth retraction $r: U \to N$. We show in a moment:

(15.8.2) Lemma. *There exists a continuous function $\varepsilon: M \to]0,\infty[$ with the properties:*

(1) *$U_x = U_{\varepsilon(x)}(f(x)) \subset U$ for each $x \in M$.*

(2) *For each $x \in M$ the diameter of $r(U_x)$ is smaller than $\delta(x)$.*

Assuming this lemma, we apply (15.8.1) to $N = \mathbb{R}^p$ and ε instead of δ. This provides us with a map $g_1: M \to \mathbb{R}^p$ which has an image contained in U. Then $g = r \circ g_1$ has the required properties. □

Proof. We first consider the situation locally. Let $x \in M$ be fixed. Choose $\gamma(x) > 0$ and a neighbourhood W_x of x such that $\delta(y) \geq 2\gamma(x)$ for $y \in W_x$. Let

$$V_x = r^{-1}(U_{\gamma(x)/2}(f(x)) \cap N).$$

The distance $\eta(x) = d(f(x), \mathbb{R}^p \smallsetminus V_x)$ is greater than zero. We shrink W_x to a neighbourhood Z_x such that $\|f(x) - f(y)\| < \frac{1}{4}\eta(x)$ for $y \in Z_x$.

The function $f|Z_x$ satisfies the lemma with the constant function $\varepsilon = \varepsilon_x: y \mapsto \frac{1}{4}\eta(x)$. In order to see this, let $y \in Z_x$ and $\|z - f(y)\| < \frac{1}{4}\eta(x)$, i.e., $z \in U_y$. Then, by the triangle inequality, $\|z - f(x)\| < \frac{1}{2}\eta(x)$, and hence, by our choice of $\eta(x)$,

$$z \in V_x \subset U, \quad r(z) \in U_{\gamma(x)/2}(f(x)).$$

If $z_1, z_2 \in U_y$, then the triangle inequality yields $\|r(z_1)-r(z_2)\| < \gamma(x) \le \frac{1}{2}\delta(y)$. Therefore the diameter of $r(U_y)$ is smaller than $\delta(y)$.

After this local consideration we choose a partition of unity $(\tau_x \mid x \in M)$ subordinate to $(Z_x \mid x \in M)$. Then we define $\varepsilon\colon M \to]0,\infty[$ as $\varepsilon(x) = \sum_{a\in M} \frac{1}{4}\tau_a(x)\eta(a)$. This function has the required properties. □

(15.8.3) Proposition. *Let $f\colon M \to N$ be continuous. For each continuous map $\delta\colon M \to]0,\infty[$ there exists a continuous map $\varepsilon\colon M \to]0,\infty[$ with the following property: Each continuous map $g\colon M \to N$ with $\|g(x) - f(x)\| < \varepsilon(x)$ and $f|A = g|A$ is homotopic to f by a homotopy $F\colon M \times [0,1] \to N$ such that $F(a,t) = f(a)$ for $(a,t) \in A \times [0,1]$ and $\|F(x,t) - f(x)\| < \delta(x)$ for $(x,t) \in M \times [0,1]$.*

Proof. We choose $r\colon U \to N$ and $\varepsilon\colon M \to]0,\infty[$ as in (15.8.1) and (15.8.2). For $(x,t) \in M \times [0,1]$ we set $H(x,t) = t\cdot g(x) + (1-t)\cdot f(x) \in U_{\varepsilon(x)}(f(x))$. The composition $F(x,t) = rH(x,t)$ is then a homotopy with the required properties. □

(15.8.4) Theorem. (1) *Let $f\colon M \to N$ be continuous and $f|A$ smooth. Then f is homotopic relative to A to a smooth map. If f is proper and N closed in $\mathbb{R}^p$, then f is properly homotopic relative to A to a smooth map.*

(2) *Let $f_0, f_1\colon M \to N$ be smooth maps. Let $f_t\colon M \to N$ be a homotopy which is smooth on $B = M\times[0,\varepsilon[\,\cup M\times]1-\varepsilon,1]\cup A\times[0,1]$. Then there exists a smooth homotopy g_t from f_0 to f_1 which coincides on $A\times[0,1]$ with f. If f_t is a proper homotopy and N closed in $\mathbb{R}^p$, then g_t can be chosen as a proper homotopy.*

Proof. (1) We choose δ and ε according to (15.8.3) and apply (15.8.1). Then (15.8.3) yields a suitable homotopy. If f is proper, δ bounded, and if $\|g(x) - f(x)\| < \delta(x)$ holds, then g is proper.

(2) We now consider $M\times]0,1[$ instead of M and its intersection with B instead of A and proceed as in (1). □

15.9 Transversality

Let $f\colon A \to M$ and $g\colon B \to M$ be smooth maps. We form the pullback diagram

$$\begin{array}{ccc} C & \xrightarrow{F} & B \\ \downarrow{\scriptstyle G} & & \downarrow{\scriptstyle g} \\ A & \xrightarrow{f} & M \end{array}$$

with $C = \{(a,b) \mid f(a) = g(b)\} \subset A \times B$. If $g\colon B \subset M$, then we identify C with $f^{-1}(B)$. If also $f\colon A \subset M$, then $f^{-1}(B) = A \cap B$. The space C can also be obtained as the pre-image of the diagonal of $M \times M$ under $f \times g$. The maps f and g are said to be ***transverse in*** $(a,b) \in C$ if

$$T_a f(T_a A) + T_b g(T_b B) = T_y M,$$

$y = f(a) = g(b)$. They are called ***transverse*** if this condition is satisfied for all points of C. If $g\colon B \subset M$ is the inclusion of a submanifold and $f(a) = b$, then we say that f is ***transverse*** to B in a if

$$T_a f(T_a M) + T_b B = T_b M$$

holds. If this holds for each $a \in f^{-1}(B)$, then f is called ***transverse*** to B. We also use this terminology if C is empty, i.e., we also call f and g transverse in this case. In the case that $\dim A + \dim B < \dim M$, the transversality condition cannot hold. Therefore f and g are then transverse if and only if C is empty. A submersion f is transverse to every g.

In the special case $B = \{b\}$ the map f is transverse to B if and only if b is a regular value of f. We reduce the general situation to this case.

We use a little linear algebra: Let $a\colon U \to V$ be a linear map and $W \subset V$ a linear subspace; then $a(U) + W = V$ if and only if the composition of a with the canonical projection $p\colon V \to V/W$ is surjective.

Let $B \subset M$ be a smooth submanifold. Let $b \in B$ and suppose $p\colon Y \to \mathbb{R}^k$ is a smooth map with regular value 0, defined on an open neighbourhood Y of b in M such that $B \cap Y = p^{-1}(0)$. Then:

(15.9.1) Note. *$f: A \to M$ is transverse to B in $a \in f^{-1}(B \cap Y)$ if and only if a is a regular point of $p \circ f : f^{-1}(Y) \to Y \to \mathbb{R}^k$.*

Proof. The space $T_b B$ for $b = f(a)$ is the kernel of $T_b p$. The composition of $T_a f : T_a A \to T_b M / T_b B$ with the isomorphism $T_b M/T_b B \cong T_0 \mathbb{R}^k$ induced by $T_b\colon T_b M \to T_0\mathbb{R}^k$ is $T_a(p \circ f)$. Now we apply the above remark from linear algebra. □

(15.9.2) Proposition. *Let $f : A \to M$ and $f|\partial A$ be smooth and transverse to the submanifold B of M of codimension k. Suppose B and M have empty boundary. Then $C = f^{-1}(B)$ is empty or a submanifold of type* I *of A of codimension k. The equality $T_a C = (T_a f)^{-1}(T_{f(a)} B)$ holds.* □

Let, in the situation of the last proposition, $\nu(C, A)$ and $\nu(B, M)$ be the normal bundles. Then Tf induces a smooth bundle map $\nu(C, A) \to \nu(B, M)$; for, by definition of transversality, $T_a f : T_a A / T_a C \to T_{f(a)}/T_{f(a)} B$ is surjective and then bijective for reasons of dimension.

From (15.9.1) we see that transversality is an "open condition": If $f : A \to M$ is transverse in a to B, then it is transverse in all points of a suitable neighbourhood of a, since a similar statement holds for regular points.

(15.9.3) Proposition. *Let $f: A \to M$ and $g: B \to M$ be smooth and let $y = f(a) = g(b)$. Then f and g are transverse in (a,b) if and only if $f \times g$ is transverse in (a,b) to the diagonal of $M \times M$.*

Proof. Let $U = T_a f(T_a A)$, $V = T_b g(T_b B)$, $W = T_y M$. The statement amounts to: $U + V = W$ and $(U \oplus V) + D(W) = W \oplus W$ are equivalent relations ($D(W)$ diagonal). By a small argument from linear algebra one verifies this equivalence. □

(15.9.4) Corollary. *Suppose f and g are transverse. Then C is a smooth submanifold of $A \times B$. Let $c = (a,b) \in C$. We have a diagram*

$$\begin{array}{ccc} T_c C & \xrightarrow{TF} & T_b B \\ \Big\downarrow{\scriptstyle TG} & & \Big\downarrow{\scriptstyle Tg} \\ T_a A & \xrightarrow{Tf} & T_y M. \end{array}$$

It is bi-cartesian, i.e., $\langle Tf, Tg \rangle$ is surjective and the kernel is $T_c C$. Therefore the diagram induces an isomorphism of the cokernels of TG and Tg (and similarly of TF and Tf).

(15.9.5) Corollary. *Let a commutative diagram of smooth maps be given,*

$$\begin{array}{ccccc} & & C & \xrightarrow{F} & B \\ & & \Big\downarrow{\scriptstyle G} & & \Big\downarrow{\scriptstyle g} \\ Z & \xrightarrow{h} & A & \xrightarrow{f} & M. \end{array}$$

Let f be transverse to g and C as above. Then h is transverse to G if and only if fh is transverse to g.

Proof. The proof uses the isomorphisms of cokernels in (15.9.4). □

(15.9.6) Corollary. *We apply* (15.9.5) *to the diagram*

$$\begin{array}{ccccc} & & M & \longrightarrow & \{s\} \\ & & \Big\downarrow{\scriptstyle i_s} & & \Big\downarrow \\ W & \xrightarrow{f} & M \times S & \xrightarrow{\mathrm{pr}} & S \end{array}$$

and obtain: f is transverse to $i_s: x \mapsto (x,s)$ if and only if s is a regular value of $\mathrm{pr} \circ f$. □

Let $F: M \times S \to N$ be smooth and $Z \subset N$ a smooth submanifold. Suppose S, Z, and N have no boundary. For $s \in S$ we set $F_s: M \to N$, $x \mapsto F(x,s)$. We consider F as a parametrized family of maps F_s. Then:

(15.9.7) Theorem. *Suppose $F: M \times S \to N$ and $\partial F = F|(\partial M \times S)$ are transverse to Z. Then for almost all $s \in S$ the maps F_s and ∂F_s are both transverse to Z.*

Proof. By (15.9.2), $W = F^{-1}(Z)$ is a submanifold of $M \times S$ with boundary $\partial W = W \cap \partial(M \times S)$. Let $\pi: M \times S \to S$ be the projection. The theorem of Sard yields the claim if we can show: If $s \in S$ is a regular value of $\pi: W \to S$, then F_s is transverse to Z, and if $s \in S$ is a regular value of $\partial\pi: \partial W \to S$, then ∂F_s is transverse to Z. But this follows from (15.9.6). □

(15.9.8) Theorem. *Let $f: M \to N$ be a smooth map and $Z \subset N$ a submanifold. Suppose Z and N have no boundary. Let $C \subset M$ be closed. Suppose f is transverse to Z in points of C and ∂f transverse to Z in points of $\partial M \cap C$. Then there exists a smooth map $g: M \to N$ which is homotopic to f, coincides on C with f and is on M and ∂M transverse to Z.*

Proof. We begin with the case $C = \emptyset$. We use the following facts: N is diffeomorphic to a submanifold of some $\mathbb{R}^k$; there exists an open neighbourhood U of N in $\mathbb{R}^k$ and a submersion $r: U \to N$ with $r|N = \text{id}$. Let $S = E^k \subset \mathbb{R}^k$ be the open unit disk and set

$$F: M \times S \to N, \quad (x, s) \mapsto r(f(x) + \varepsilon(x)s).$$

Here $\varepsilon: M \to]0, \infty[$ is a smooth function for which this definition of F makes sense. We have $F(x, 0) = f(x)$. We claim: F and ∂F are submersions. For the proof we consider for fixed x the map

$$S \to U_\varepsilon(f(x)), \quad s \mapsto f(x) + \varepsilon(x)s;$$

it is the restriction of an affine automorphism of $\mathbb{R}^k$ and hence a submersion. The composition with r is then a submersion too. Therefore F and ∂F are submersions, since already the restrictions to the $\{x\} \times S$ are submersions.

By (15.9.7), for almost all $s \in S$ the maps F_s and ∂F_s are transverse to Z. A homotopy from F_s to f is $M \times I \to N$, $(x, t) \mapsto F(x, st)$.

Let now C be arbitrary. There exists an open neighbourhood W of C in M such that f is transverse to Z on W and ∂f transverse to Z on $W \cap \partial M$. We choose a set V which satisfies $C \subset V^\circ \subset \overline{V} \subset W^\circ$ and a smooth function $\tau: M \to [0, 1]$ such that $M \setminus W \subset \tau^{-1}(1)$, $V \subset \tau^{-1}(0)$. Moreover we set $\sigma = \tau^2$. Then $T_x\sigma = 0$, whenever $\tau(x) = 0$. We now modify the map F from the first part of the proof

$$G: M \times S \to N, \quad (x, s) \mapsto F(x, \sigma(x)s)$$

and claim: G is transverse to Z. For the proof we choose $(x, s) \in G^{-1}(Z)$. Suppose, to begin with, that $\sigma(x) \neq 0$. Then $S \to N$, $t \mapsto G(x, t)$ is, as a composition of a diffeomorphism $t \mapsto \sigma(x)t$ with the submersion $t \mapsto F(x, t)$, also a submersion and therefore G is regular at (x, s) and hence transverse to Z.

Suppose now that $\sigma(x) = 0$. We compute $T_{(x,s)}G$ at $(v, w) \in T_x M \times T_s S = T_x M \times \mathbb{R}^k$. Let

$$m\colon M \times S \to M \times S, \quad (x, s) \mapsto (x, \sigma(x)s).$$

Then

$$T_{(x,s)}m(v, w) = (v, \sigma(x)w + T_x\sigma(v)s).$$

The chain rule, applied to $G = F \circ m$, yields

$$T_{(x,s)}G(v, w) = T_{m(x,s)}F \circ T_{(x,s)}m(v, w) = T_{(x,0)}F(v, 0) = T_x f(v),$$

since $\sigma(x) = 0$, $T_x\sigma = 0$ and $F(x, 0) = f(x)$. Since $\sigma(x) = 0$, by choice of W and τ, f is transverse to Z in x, hence – since $T_{(x,s)}G$ and $T_x f$ have the same image – also G is transverse to Z in (x, s). A similar argument is applied to ∂G. Then one finishes the proof as in the case $C = \emptyset$. □

15.10 Gluing along Boundaries

We use collars in order to define a smooth structure if we glue manifolds with boundaries along pieces of the boundary. Another use of collars is the definition of a smooth structure on the product of two manifolds with boundary (smoothing of corners).

15.10.1 Gluing along boundaries. Let M_1 and M_2 be ∂-manifolds. Let $N_i \subset \partial M_i$ be a union of components of ∂M_i and let $\varphi\colon N_1 \to N_2$ be a diffeomorphism. We denote by $M = M_1 \cup_\varphi M_2$ the space which is obtained from $M_1 + M_2$ by the identification of $x \in N_1$ with $\varphi(x) \in N_2$. The image of M_i in M is again denoted by M_i. Then $M_i \subset M$ is closed and $M_i \smallsetminus N_i \subset M$ open. We define a smooth structure on M. For this purpose we choose collars $k_i : \mathbb{R}_- \times N_i \to M_i$ with open image $U_i \subset M_i$. The map

$$k\colon \mathbb{R} \times N_1 \to M, \quad (t, x) \mapsto \begin{cases} k_1(t, x), & t \le 0, \\ k_2(-t, \varphi(x)), & t \ge 0, \end{cases}$$

is an embedding with image $U = U_1 \cup_\varphi U_2$. We define a smooth structure (depending on k) by the requirement that $M_i \smallsetminus N_i \to M$ and k are smooth embeddings. This is possible since the structures agree on $(M_i \smallsetminus N_i) \cap U$. ◇

15.10.2 Products. Let M_1 and M_2 be smooth ∂-manifolds. We impose a canonical smooth structure on $M_1 \times M_2 \smallsetminus (\partial M_1 \times \partial M_2)$ by using products of charts for M_i as charts. We now choose collars $k_i : \mathbb{R}_- \times \partial M_i \to M_i$ and consider the composition λ,

$$\begin{array}{ccc} \mathbb{R}^2_- \times \partial M_1 \times \partial M_2 & \xrightarrow{\pi \times \mathrm{id}} & \mathbb{R}_- \times \mathbb{R}_- \times \partial M_1 \times \partial M_2 \\ \downarrow \lambda & & \downarrow (1) \\ M_1 \times M_2 & \xleftarrow{k_1 \times k_2} & (\mathbb{R}_- \times \partial M_1) \times (\mathbb{R}_- \times \partial M_2). \end{array}$$

Here $\pi : \mathbb{R}^2_- \to \mathbb{R}^1_- \times \mathbb{R}^1_-$, $(r, \varphi) \mapsto (r, \frac{1}{2}\varphi + \frac{3\pi}{4})$, written in polar coordinates (r, φ), and (1) interchanges the second and third factor. There exists a unique smooth structure on $M_1 \times M_2$ such that $M_1 \times M_2 \smallsetminus (\partial M_1 \times \partial M_2) \subset M_1 \times M_2$ and λ are diffeomorphisms onto open parts of $M_1 \times M_2$. ◇

15.10.3 Boundary pieces. Let B and C be smooth n-manifolds with boundary. Let M be a smooth $(n-1)$-manifold with boundary and suppose that

$$\varphi_B : M \to \partial B, \quad \varphi_C : M \to \partial C$$

are smooth embeddings. We identify in $B + C$ the points $\varphi_B(m)$ with $\varphi_C(m)$ for each $m \in M$. The result D carries a smooth structure with the following properties:

(1) $B \smallsetminus \varphi_B(M) \subset D$ is a smooth submanifold.

(2) $C \smallsetminus \varphi_C(M) \subset D$ is a smooth submanifold.

(3) $\iota : M \to D, m \mapsto \varphi_B(m) \sim \varphi_C(m)$ is a smooth embedding as a submanifold of type I.

(4) The boundary of D is diffeomorphic to the gluing of $\partial B \smallsetminus \varphi_B(M)^\circ$ with $\partial C \smallsetminus \varphi_C(M)^\circ$ via $\varphi_B(m) \sim \varphi_C(m)$, $m \in \partial M$.

The assertions (1) and (2) are understood with respect to the canonical embeddings $B \subset D \supset C$. We have to define charts about the points of $\iota(M)$, since the conditions (1) and (2) specify what happens about the remaining points. For points of $\iota(M \smallsetminus \partial M)$ we use collars of B and C and proceed as in 15.10.1. For $\iota(\partial M)$ we use the following device.

Choose collars $\kappa_B : \mathbb{R}_- \times \partial B \to B$ and $\kappa : \mathbb{R}_- \times \partial M \to M$ and an embedding $\tau_B : \mathbb{R} \times \partial M \to \partial B$ such that the next diagram commutes:

$$\begin{array}{ccc} \mathbb{R} \times \partial M & \xrightarrow{\tau_B} & \partial B \\ \uparrow \cup & & \uparrow \varphi_B \\ \mathbb{R}_- \times \partial M & \xrightarrow{\kappa} & M. \end{array}$$

Here τ_B can essentially be considered as a tubular map, the normal bundle of $\varphi(\partial M)$ in ∂B is trivial. And κ is "half" of this normal bundle.

Then we form $\Phi_B = \kappa_B \circ (\mathrm{id} \times \tau_B) : \mathbb{R}_- \times \mathbb{R} \times \partial M \to B$. For C we choose in a similar manner κ_C and τ_C, but we require $\varphi_C \circ \kappa^- = \tau_C$ where $\kappa^-(m, t) =$

$\kappa(m, -t)$. Then we define Φ_C from κ_C and τ_C. The smooth structure in a neighbourhood of $\iota(\partial M)$ is now defined by the requirement that $\alpha\colon \mathbb{R}_- \times \mathbb{R} \times \partial M \to D$ is a smooth embedding where

$$\alpha(r, \psi, m) = \begin{cases} \Phi_B(r, 2\psi - \pi/2, m), & \frac{\pi}{2} \le \psi \le \pi, \\ \Phi_C(r, 2\psi - 3\pi/2, m), & \pi \le \psi \le \frac{3\pi}{2}, \end{cases}$$

with the usual polar coordinates (r, ψ) in $\mathbb{R}_- \times \mathbb{R}$. ◇

15.10.4 Connected sum. Let M_1 and M_2 be n-manifolds. We choose smooth embeddings $s_i\colon D^n \to M_i$ into the interiors of the manifolds. In $M_1 \smallsetminus s_1(E^n) + M_2 \smallsetminus s_2(E^n)$ we identify $s_1(x)$ with $s_2(x)$ for $x \in S^{n-1}$. The result is a smooth manifold (15.10.1). We call it the ***connected sum*** $M_1 \# M_2$ of M_1 and M_2. Suppose M_1, M_2 are oriented connected manifolds, assume that s_1 preserves the orientation and s_2 reverses it. Then $M_1 \# M_2$ carries an orientation such that the $M_i \smallsetminus s_i(E^n)$ are oriented submanifolds. One can show by isotopy theory that the oriented diffeomorphism type is in this case independent of the choice of the s_i. ◇

15.10.5 Attaching handles. Let M be an n-manifold with boundary. Furthermore, let $s\colon S^{k-1} \times D^{n-k} \to \partial M$ be an embedding and identify in $M + D^k \times D^{n-k}$ the points $s(x)$ and x. The result carries a smooth structure (15.10.3) and is said to be obtained by attaching a k-handle to M.

Attaching a 0-handle is the disjoint sum with D^n. Attaching an n-handle means that a "hole" with boundary S^{n-1} is closed by inserting a disk. A fundamental result asserts that each (smooth) manifold can be obtained by successive attaching of handles. A proof uses the so-called Morse theory (see e.g., [134], [137]). A handle decomposition of a manifold replaces a cellular decomposition, the advantage is that the handles are themselves n-dimensional manifolds. ◇

15.10.6 Elementary surgery. If M' arises from M by attaching a k-handle, then $\partial M'$ is obtained from ∂M by a process called ***elementary surgery***. Let $h\colon S^{k-1} \times D^{n-k} \to X$ be an embedding into an $(n-1)$-manifold with image U. Then $X \smallsetminus U^\circ$ has a piece of the boundary which is via h diffeomorphic to $S^{k-1} \times S^{n-k-1}$. We glue the boundary of $D^k \times S^{n-k-1}$ with h; in symbols

$$X' = (X \smallsetminus U^\circ) \cup_h D^k \times S^{n-k-1}.$$

The transition from X to X' is called ***elementary surgery of index k*** at X via h. The method of surgery is very useful for the construction of manifolds with prescribed topological properties. See [191], [162], [108] to get an impression of surgery theory. ◇

Problems

1. The subsets of $S^{m+n+1} \subset \mathbb{R}^{m+1} \times \mathbb{R}^{n+1}$

$$D_1 = \{(x,y) \mid \|x\|^2 \geq \tfrac{1}{2}, \|y\|^2 \leq \tfrac{1}{2}\}, \quad D_2 = \{(x,y) \mid \|x\|^2 \leq \tfrac{1}{2}, \|y\|^2 \geq \tfrac{1}{2}\}$$

are diffeomorphic to $D_1 \cong S^m \times D^{n+1}$, $D_2 \cong D^{m+1} \times S^n$. They are smooth submanifolds with boundary of S^{m+n+1}. Hence S^{m+n+1} can be obtained from $S^m \times D^{n+1}$ and $D^{m+1} \times S^n$ by identifying the common boundary $S^m \times S^n$ with the identity. A diffeomorphism $D_1 \to S^m \times D^{n+1}$ is $(z,w) \mapsto (\|z\|^{-1}z, \sqrt{2}w)$.

2. Let M be a manifold with non-empty boundary. Identify two copies along the boundary with the identity. The result is the ***double*** $D(M)$ of M. Show that $D(M)$ for a compact M is the boundary of some compact manifold. (Hint: Rotate M about ∂M about 180 degrees.)

3. Show $M \# S^n \cong M$ for each n-manifold M.

4. Study the classification of closed connected surfaces. The orientable surfaces are S^2 and connected sums of tori $T = S^1 \times S^1$. The non-orientable ones are connected sums of projective planes $P = \mathbb{R}P^2$. The relation $T \# P = P \# P \# P$ holds. The connected sum with T is classically also called attaching of a handle.

Chapter 16
Homology of Manifolds

The singular homology groups of a cell complex vanish above its dimension. It is an obvious question whether the same holds for a manifold. It is certainly technically complicated to produce a cell decomposition of a manifold and also an artificial structure. Locally the manifold looks like a Euclidean space, so there arises no problem locally. The Mayer–Vietoris sequences can be used to paste local information, and we use this technique to prove the vanishing theorem.

The homology groups of an n-manifold M in dimension n also have special properties. They can be used to define and construct homological orientations of a manifold. A local orientation about $x \in M$ is a generator of the local homology group $H_n(M, M \smallsetminus x; \mathbb{Z}) \cong \mathbb{Z}$. In the case of a surface, the two generators correspond to "clockwise" and "counter-clockwise". If you pick a local orientation, then you can transport it along paths, and this defines a functor from the fundamental groupoid and hence a twofold covering. If the covering is trivial, then the manifold is called orientable, and otherwise (as in the case of a Möbius-band) non-orientable.

Our first aim in this chapter will be to construct the orientation covering and use it to define orientations as compatible families of local orientations.

In the case of a closed compact connected manifold we can define a global homological orientation to be a generator of $H_n(M; \mathbb{Z})$; we show that this group is either zero or $\mathbb{Z}$. In the setting of a triangulation of a manifold, the generator is the sum of the n-dimensional simplices, oriented in a coherent manner. In a non-orientable manifold it is impossible to orient the simplices coherently; but in that case their sum still gives a generator in $H_n(M; \mathbb{Z}/2)$, since $\mathbb{Z}/2$-coefficients mean that we can ignore orientations.

Once we have global orientations, we can define the degree of a map between oriented manifolds. This is analogous to the case of spheres already studied.

16.1 Local Homology Groups

Let $h_*(-)$ be a homology theory and M an n-dimensional manifold. Groups of the type $h_k(M, M \smallsetminus x)$ are called ***local homology groups***. Let $\varphi\colon U \to \mathbb{R}^n$ be a chart of M centered at x. We excise $M \smallsetminus U$ and obtain an isomorphism

$$h_k(M, M \smallsetminus x) \cong h_k(U, U \smallsetminus x) \xrightarrow{\varphi_*} h_k(\mathbb{R}^n, \mathbb{R}^n \smallsetminus 0).$$

For singular homology with coefficients in G we see that $H_n(M, M \smallsetminus x; G) \cong G$, and the other local homology groups are zero. Let R be a commutative ring. Then

$H_n(M, M \smallsetminus x; R) \cong R$ is a free R-module of rank 1. A generator, corresponding to a unit of R, is called a ***local R-orientation*** of M about x. We assemble the totality of local homology groups into a covering.

Let $K \subset L \subset M$. The homomorphism $r^L_K \colon h_k(M, M \smallsetminus L) \to h_k(M, M \smallsetminus K)$, induced by the inclusion, is called restriction. We write r^L_x in the case that $K = \{x\}$.

(16.1.1) Lemma. *Each neighbourhood W of x contains an open neighbourhood U of x such that the restriction r^U_y is for each $y \in U$ an isomorphism.*

Proof. Choose a chart $\varphi \colon V \to \mathbb{R}^n$ with $V \subset W$ centered at x. Set $U = \varphi^{-1}(E^n)$, $E^n = \{x \in \mathbb{R}^n \mid \|x\| < 1\}$. We have a commutative diagram

$$\begin{CD} h_k(M, M \smallsetminus U) @<{(1)}<< h_k(V, V \smallsetminus U) \\ @V{r^U_y}VV @VV{(3)}V \\ h_k(M, M \smallsetminus y) @<{(2)}<< h_k(V, V \smallsetminus y), \end{CD}$$

with morphisms induced by inclusion. The maps (1) and (2) are excisions, and (3) is an isomorphism, because $V \smallsetminus U \subset V \smallsetminus y$ is for each $y \in U$ an h-equivalence (see Problem 1). □

We construct a covering $\omega \colon h_k(M, M \smallsetminus \bullet) \to M$. As a set

$$h_k(M, M \smallsetminus \bullet) = \textstyle\coprod_{x \in M} h_k(M, M \smallsetminus x),$$

and $h_k(M, M \smallsetminus x)$ is the fibre of ω over x (with discrete topology). Let U be an open neighbourhood of x such that r^U_y is an isomorphism for each $y \in U$. We define bundle charts

$$\varphi_{x,U} \colon U \times h_k(M, M \smallsetminus x) \to \omega^{-1}(U), \quad (y, a) \mapsto r^U_y (r^U_x)^{-1}(a).$$

We give $h_k(M, M \smallsetminus \bullet)$ the topology which makes $\varphi_{x,U}$ a homeomorphism onto an open subset. We have to show that the transition maps

$$\varphi^{-1}_{y,V} \varphi_{x,U} \colon (U \cap V) \times h_k(M, M \smallsetminus x) \to (U \cap V) \times h_k(M, M \smallsetminus y)$$

are continuous. Given $z \in U \cap V$, choose $z \in W \subset U \cap V$ such that r^W_w is an isomorphism for each $w \in W$. Consider now the diagram

$$\begin{array}{ccccc} h_k(M, M \smallsetminus x) & \xleftarrow{r^U_x} & h_k(M, M \smallsetminus U) & \xrightarrow{r^U_w} & h_k(M, M \smallsetminus w) \\ & & \downarrow{\scriptstyle r^U_W} & \nearrow{\scriptstyle r^W_w} & \uparrow{\scriptstyle r^V_w} \\ & & h_k(M, M \smallsetminus W) & \xleftarrow[r^V_W]{} & h_k(M, M \smallsetminus V) \\ & & & & \downarrow{\scriptstyle r^V_y} \\ & & & & h_k(M, M \smallsetminus y). \end{array}$$

It shows $\varphi_{y,V}^{-1}\varphi_{x,U} = r_y^V(r_W^V)^{-1}r_W^U(r_x^U)^{-1}$. Hence the second component of $\varphi_{y,V}^{-1}\varphi_{x,U}$ is independent of $w \in W$, and this shows the continuity of the transition map.

We take advantage of the fact that the fibres are groups. For $A \subset M$ we denote by $\Gamma(A)$ the set of continuous (= locally constant) sections over A of $\omega\colon h_k(M, M \smallsetminus \bullet) \to M$. If s and t are sections, we can define $(s+t)(a) = s(a)+t(a)$. One uses the bundle charts to see that $s+t$ is again continuous. Hence $\Gamma(A)$ is an abelian group. We denote by $\Gamma_c(A) \subset \Gamma(A)$ the subgroup of sections with compact support, i.e., of sections which have values zero away from a compact set.

(16.1.2) Proposition. *Let $z \in h_k(M, M \smallsetminus U)$. Then $y \mapsto r_y^U z$ is a continuous section of ω.*

Proof. The bundle chart $\varphi_{x,U}$ transforms the constant section $y \mapsto (y, r_x^U z)$ into the section $y \mapsto r_y^U z$. □

Problems

1. $S^{n-1} \subset \mathbb{R}^n \smallsetminus E^n$ and $\mathbb{R}^n \smallsetminus e \to S^{n-1}$, $y \mapsto (y-e)/\|y-e\|$ are h-equivalences ($e \in E^n$). The map $S^{n-1} \to S^{n-1}$, $y \mapsto (y-e)/\|y-e\|$ is homotopic to the identity. These facts imply that $\mathbb{R}^n \smallsetminus E \subset \mathbb{R}^n \smallsetminus e$ is an h-equivalence.

2. Let M be an n-dimensional manifold with boundary ∂M. Show that $x \in \partial M$ if and only if $H_n(M, M \smallsetminus x) = 0$. From this homological characterization of boundary points one obtains: Let $f\colon M \to M$ be a homeomorphism. Then $f(\partial M) = \partial M$ and $f(M \smallsetminus \partial M) = M \smallsetminus \partial M$.

3. Let $\varphi\colon (C,0) \to (D,0)$ be a homeomorphism between open neighbourhoods of 0 in $\mathbb{R}^n$. Then $\varphi_*\colon H_n(C, C \smallsetminus 0; R) \to H_n(D, D \smallsetminus 0; R)$ is multiplication by ± 1.

16.2 Homological Orientations

Let M be an n-manifold and $A \subset M$. An ***R*-orientation of *M* along *A*** is a section $s \in \Gamma(A; R)$ of $\omega\colon H_n(M, M \smallsetminus \bullet; R) \to M$ such that $s(a) \in H_n(M, M \smallsetminus a; R) \cong R$ is for each $a \in A$ a generator of this group. Thus s combines the local orientations in a continuous manner. In the case that $A = M$, we talk about an ***R*-orientation** of M, and for $R = \mathbb{Z}$ we just talk about orientations. If an orientation exists, we call M (homologically) orientable. If M is orientable along A and $B \subset A$, then M is orientable along B.

(16.2.1) Note. *Let $\mathrm{Ori}(M) \subset H_n(M, M \smallsetminus \bullet; \mathbb{Z})$ be the subset of all generators of all fibres. Then the restriction $\mathrm{Ori}(M) \to M$ of ω is a 2-fold covering of M, called the **orientation covering** of M.* □

(16.2.2) Proposition. *The following are equivalent:*

(1) *M is orientable.*

(2) *M is orientable along compact subsets.*

(3) *The orientation covering is trivial.*

(4) *The covering $\omega\colon H_n(M, M \smallsetminus \bullet; \mathbb{Z}) \to M$ is trivial.*

Proof. (1) $\Rightarrow$ (2). Special case.

(2) $\Rightarrow$ (3). The orientation covering is trivial if and only if the covering over each component is trivial. Therefore let M be connected. Then a 2-fold covering $\tilde{M} \to M$ is trivial if and only if $\tilde{M}$ is not connected, hence the components of $\tilde{M}$ are also coverings.

Suppose $\mathrm{Ori}(M) \to M$ is non-trivial. Since $\mathrm{Ori}(M)$ is then connected, there exists a path in $\mathrm{Ori}(M)$ between the two points of a given fibre. The image S of such a path is compact and connected, and the covering is non-trivial over S, since we can connect two points of a fibre in it. By the assumption (2), the orientation covering is trivial over the compact set S, hence it has a section over S. Contradiction.

(3) $\Rightarrow$ (4). Let s be a section of the orientation covering. Then $M \times \mathbb{Z} \to H_n(M, M \smallsetminus \bullet; \mathbb{Z})$, $(x, k) \mapsto ks(x)$ is a trivialization of ω: It is a map over M, bijective on fibres, continuous, and a morphism between coverings.

(4) $\Rightarrow$ (1). If ω is trivial, then it has a section with values in the set of generators. □

(16.2.3) Note. $\mathrm{Ori}(M) \to M$ *is a twofold principal covering with automorphism group $C = \{1, t \mid t^2 = 1\}$. Let t act on G as multiplication by -1. Then the associated covering* $\mathrm{Ori(M)} \times_C G$ *is isomorphic to $H_n(M, M \smallsetminus \bullet; G)$.*

Proof. The map

$$\mathrm{Ori}(M) \times G \to H_n(M, M \smallsetminus \bullet; \mathbb{Z}) \otimes G \cong H_n(M, M \smallsetminus \bullet; G), \quad (u, g) \mapsto u \otimes g$$

induces the isomorphism. (The isomorphism is the fibrewise isomorphism $H_n(M, M \smallsetminus x; \mathbb{Z}) \otimes G \cong H_n(M, M \smallsetminus x; G)$ from the universal coefficient formula.) □

(16.2.4) Remark. The sections $\Gamma(A; G)$ of ω over A correspond bijectively to the continuous maps $\lambda\colon \mathrm{Ori}(M)|A \to G$ with the property $\lambda \circ t = -\lambda$. This is a general fact about sections of associated bundles. ◇

Problems

1. Let M be a smooth n-manifold with an orienting atlas. Then there exists a unique homological $\mathbb{Z}$-orientation such that the local orientations in $H_n(M, M \smallsetminus x; \mathbb{Z})$ are mapped via positive charts to a standard generator of $H_n(\mathbb{R}^n, \mathbb{R}^n \smallsetminus 0; \mathbb{Z})$. Conversely, if M is $\mathbb{Z}$-oriented, then M has an orienting atlas which produces the given $\mathbb{Z}$-orientation.

2. Every manifold has a unique $\mathbb{Z}/2$-orientation.

16.3 Homology in the Dimension of the Manifold

Let M be an n-manifold and $A \subset M$ a closed subset. We use in this section singular homology with coefficients in the abelian group G and sometimes suppress G in the notation.

(16.3.1) Proposition. *For each $\alpha \in H_n(M, M \smallsetminus A; G)$ the section*

$$J^A(\alpha)\colon A \to H_n(M, M \smallsetminus \bullet; G), \quad x \mapsto r_x^A(\alpha)$$

of ω (over A) is continuous and has compact support.

Proof. Let the chain $c \in S_n(M;G)$ represent the homology class α. There exists a compact set K such that c is a chain in K. Let $x \in A \smallsetminus K$. Then the image of c under

$$S_n(K;G) \to S_n(M;G) \to S_n(M,K;G) \to S_n(M, M \smallsetminus x; G)$$

is zero. Since this image represents $r_x^A(\alpha)$, the support of $J^A(\alpha)$ is contained in $A \cap K$.

The continuity is a general fact (16.1.2). □

From (16.3.1) we obtain a homomorphism

$$J^A\colon H_n(M, M \smallsetminus A; G) \to \Gamma_c(A; G), \quad \alpha \mapsto (x \mapsto r_x^A(\alpha)).$$

(16.3.2) Theorem. *Let $A \subset M$ be closed.*

(1) *Then $H_i(M, M \smallsetminus A) = 0$ for $i > n$.*

(2) *The homomorphism $J^A\colon H_n(M, M \smallsetminus A) \to \Gamma_c(A)$ is an isomorphism.*

Proof. Let $D(A,1)$ and $D(A,2)$ stand for the statement that (1) and (2) holds for the subset A, respectively. We use the fact that J^A is a natural transformation between contravariant functors on the category of closed subsets of M and their inclusions. The proof is a kind of induction over the complexity of A. It will be divided into several steps.

(1) $D(A,j)$, $D(B,j)$, $D(A \cap B, j)$ imply $D(A \cup B, j)$. For the proof we use the relative Mayer–Vietoris sequence for $(M \smallsetminus A \cap B; M \smallsetminus A, M \smallsetminus B)$ and an analogous sequence for sections. This leads us to consider the diagram

$$\begin{array}{ccc}
H_{n+1}(M, M \smallsetminus (A \cap B)) & \xrightarrow[\cong]{} & 0 \\
\downarrow & & \downarrow \\
H_n(M, M \smallsetminus (A \cup B)) & \xrightarrow{J^{A \cup B}} & \Gamma_c(A \cup B) \\
\downarrow & & \downarrow \\
H_n(M, M \smallsetminus A) \oplus H_n(M, M \smallsetminus B) & \xrightarrow[\cong]{J^A \oplus J^B} & \Gamma_c(A) \oplus \Gamma_c(B) \\
\downarrow & & \downarrow \\
H_n(M, M \smallsetminus (A \cap B)) & \xrightarrow[\cong]{J^{A \cap B}} & \Gamma_c(A \cap B).
\end{array}$$

The Five Lemma and the hypotheses now show $D(A \cup B, 2)$. The Mayer–Vietoris sequence alone yields $D(A \cup B, 1)$.

(2) $D(A, j)$ holds for compact convex subsets A in a chart domain U, i.e., $\varphi(A) = B$ is compact convex for a suitable chart $\varphi\colon U \to \mathbb{R}^n$. For the proof we show that for $x \in A$ the restriction r_x^A is an isomorphism. By an appropriate choice of φ we can assume $0 \in B \subset E^n$. Then we have a commutative diagram

$$\begin{array}{ccccccc} H_i(D^n, S^{n-1}) & \xrightarrow{(1)} & H_i(\mathbb{R}^n, \mathbb{R}^n \smallsetminus B) & \xrightarrow[\cong]{\varphi_*} & H_i(U, U \smallsetminus A) & \xrightarrow{(3)} & H_i(M, M \smallsetminus A) \\ \downarrow = & & \downarrow & & \downarrow & & \downarrow r_x^A \\ H_i(D^n, S^{n-1}) & \xrightarrow{(2)} & H_i(\mathbb{R}^n, \mathbb{R}^n \smallsetminus 0) & \xrightarrow[\cong]{\varphi_*} & H_i(U, U \smallsetminus x) & \xrightarrow{(4)} & H_i(M, M \smallsetminus x). \end{array}$$

The maps (3) and (4) are excisions. The maps (1) and (2) are isomorphisms, because $S^{n-1} \subset \mathbb{R}^n \smallsetminus B$ is an h-equivalence. The isomorphism r_x^A shows, firstly, that $D(A, 1)$ holds; and, secondly, $D(A, 2)$, since a section of a covering over a connected set is determined by a single value.

(3) Suppose $A \subset U$, $\varphi\colon U \to \mathbb{R}^n$ a chart, $A = K_1 \cup \dots \cup K_r$, $\varphi(K_i)$ compact convex. We show $D(A, j)$ by induction on r. Let $B = K_1 \cup \dots \cup K_{r-1}$ and $C = K_r$. Then B and $B \cap C$ are unions of $r - 1$ sets of type (2), hence $D(B, j)$ and $D(B \cap C, j)$ holds by induction. Now use (1).

(4) Let $K \subset U$, $\varphi\colon U \to \mathbb{R}^n$ a chart, K compact. Let $K \subset W \subset U$, W open. Then there exists a neighbourhood V of K inside W of type (3). In this case J^V is an isomorphism. The restrictions r_K^V induce canonical maps from the colimits

$$\begin{array}{ccc} \operatorname{colim}_V H_i(M, M \smallsetminus V) & \xrightarrow{(*)} & H_i(M, M \smallsetminus K) \\ \downarrow \operatorname{colim}_V J^V & & \downarrow J^K \\ \operatorname{colim}_V \Gamma_c(V) & \xrightarrow{(**)} & \Gamma_c(K), \end{array}$$

where the colimit is taken over the directed set of neighbourhoods V of K of type (3). What does this isomorphism statement mean in explicit terms? Firstly, an element x_K in the image has the form $r_K^V x_V$ for a suitable V; and secondly, if x_V and x_W have the same image x_K, then they become equal under a restriction to a suitable smaller neighbourhood. From this description it is then easy to verify that J^K is indeed an isomorphism. Suppose $(*)$ and $(**)$ are isomorphisms. Then we obtain $D(K, 1)$, and the isomorphisms J^V yield $D(K, 2)$.

The isomorphism $(*)$ holds already for the singular chain groups. It uses the fact that a chain has compact support; if the support is contained in $M \smallsetminus K$, then already in $M \smallsetminus V$ for a suitable neighbourhood V of K.

$(**)$ is an isomorphism: See Problem 1.

(5) $D(K, j)$ holds for arbitrary compact subsets K, for K is a union of a finite number of sets of type (4). Then we can use induction as in case (3).

(6) Let $K = K_1 \cup K_2 \cup K_3 \cup \cdots$ with compact K_i. Suppose there are pairwise disjoint open neighbourhoods U_i of K_i. By additivity of homology groups and section groups, J^K is the direct sum of the J^{K_i}.

(7) Let A finally be an arbitrary closed subset. Since M is locally compact with countable basis, there exists an exhaustion $M = \cup K_i$, $K_1 \subset K_2 \subset \cdots$ by compact sets K_i such that $K_i \subset K_{i+1}^\circ$. Set $A_i = A \cap (K_i \smallsetminus K_{i-1}^\circ)$, $K_0 = \emptyset$, $B = \bigcup_{i=2n} A_i$, $C = \bigcup_{i=2n+1} A_i$. Then $D(B, j)$, $D(C, j)$, and $D(B \cap C, j)$ hold by (6); the hypothesis of (6) follows from the fact that a manifold is a normal space. Now $D(A, j)$ holds by (1), since $A = B \cup C$. □

(16.3.3) Theorem. *Suppose A is a closed connected subset of M. Then:*

(1) *$H_n(M, M \smallsetminus A; G) = 0$, if A is not compact.*

(2) *$H_n(M, M \smallsetminus A; G) \cong G$, if M is R-orientable along A and A is compact. Moreover $H_n(M, M \smallsetminus A; G) \to H_n(M, M \smallsetminus x; G)$ is an isomorphism for each $x \in A$.*

(3) *$H_n(M, M \smallsetminus A; \mathbb{Z}) \cong {}_2G = \{g \in G \mid 2g = 0\}$, if M is not orientable along A and A is compact.*

Proof. (1) Since A is connected, a section in $\Gamma(A; G)$ is determined by its value at a single point. If this value is non-zero, then the section is non-zero everywhere. Therefore there do not exist non-zero sections with compact support over a non-compact A, and (16.3.2) shows $H_n(M, M \smallsetminus A; G) = 0$.

(2) Let A be compact. Then $H_n(M, M \smallsetminus A; G) \cong \Gamma(A; G)$. Again a section is determined by its value at a single point. We have a commutative diagram

$$\begin{array}{ccc} H_n(M, M \smallsetminus A; G) & \xrightarrow{\cong} & \Gamma(A; G) \\ \downarrow r_x^A & & \downarrow b \\ H_n(M, M \smallsetminus x; G) & \xrightarrow{\cong} & \Gamma(\{x\}; G). \end{array}$$

If M is orientable along A, then there exists in $\Gamma(A)$ an element such that its value at x is a generator. Hence b is an isomorphism and therefore also r_x^A.

(3) A section in $\Gamma(A; G)$ corresponds to a continuous map $\lambda\colon \mathrm{Ori}(M)|A \to G$ with $\lambda t = -\lambda$. If M is not orientable along A, then $\mathrm{Ori}(M)|A$ is connected and therefore λ constant. The relation $\lambda t = -\lambda$ shows that the value of λ is contained in ${}_2G$. In this case $r_x^A\colon H_n(M, M \smallsetminus A; G) \to H_n(M, M \smallsetminus x; G) \cong G$ is injective and has image ${}_2G$. □

Theorem (16.3.3) can be considered as a duality result, since it relates an assertion about A with an assertion about $M \smallsetminus A$.

(16.3.4) Proposition. *Let M be an n-manifold and $A \subset M$ a closed connected subset. Then the torsion subgroup of $H_{n-1}(M, M \smallsetminus A; \mathbb{Z})$ is of order 2 if A is compact and M non-orientable along A, and is zero otherwise.*

Proof. Let $q \in \mathbb{N}$ and suppose M is orientable along the compact set A. Then

$$\begin{aligned} \mathbb{Z}/q &\cong H_n(M, M \smallsetminus A; \mathbb{Z}/q) \\ &\cong H_n(M, M \smallsetminus A; \mathbb{Z}) \otimes \mathbb{Z}/q \oplus H_{n-1}(M, M \smallsetminus A; \mathbb{Z}) * \mathbb{Z}/q \\ &\cong \mathbb{Z}/q \oplus H_{n-1}(M, M \smallsetminus A; \mathbb{Z}) * \mathbb{Z}/q. \end{aligned}$$

We have used: (16.3.3); universal coefficient theorem; again (16.3.3). This implies that $H_{n-1}(M, M \smallsetminus A; \mathbb{Z}) * \mathbb{Z}/q = 0$. Similarly for non-compact A or q odd

$$0 \cong H_n(M, M \smallsetminus A; \mathbb{Z}/q) \cong H_{n-1}(M, M \smallsetminus A; \mathbb{Z}) * \mathbb{Z}/q.$$

Since $\operatorname{Tor}(G, \mathbb{Z}/q) \cong \{g \in G \mid qg = 0\}$, this shows that $H_{n-1}(M, M \smallsetminus A; \mathbb{Z})$ has no q-torsion in these cases. If A is compact and M non-orientable along A, then

$$\mathbb{Z}/2 \cong H_n(M, M \smallsetminus A; \mathbb{Z}/4) \cong H_{n-1}(M, M \smallsetminus A; \mathbb{Z}) * \mathbb{Z}/4$$

by (16.3.3) and the universal coefficient formula. Since we know already that the group in question has no odd torsion, we conclude that there exists a single non-zero element of finite order and the order is 2. □

Problems

1. Let s be a section over the compact set K. For each $x \in K$ there exists an open neighbourhood $U(x)$ and an extension s_x of $s|U(x) \cap K$. Cover K by $U(x_1), \dots, U(x_r)$. Let $W = \{y \mid s_{x_i}(y) = s_{x_j}(x) \text{ if } y \in U(x_i) \cap U(x_j)\}$ and define $s(y)$, $y \in W$ as the common value. Show that W is open. Let s, s' be sections over V which agree on K. Show that they agree in a smaller neighbourhood $V_1 \subset V$ of K. (These assertions hold for sections of coverings.)

16.4 Fundamental Class and Degree

The next theorem is a special case of (16.3.3).

(16.4.1) Theorem. *Let M be a compact connected n-manifold. Then one of the following assertions holds:*

(1) *M is orientable, $H_n(M) \cong \mathbb{Z}$, and for each $x \in M$ the restriction $H_n(M) \to H_n(M, M \smallsetminus x)$ is an isomorphism.*

(2) *M is non-orientable and $H_n(M) = 0$.* □

Under the hypothesis of (16.4.1), the orientations of M correspond to the generators of $H_n(M)$. A generator will be called ***fundamental class*** or homological orientation of the orientable manifold.

We now use fundamental classes in order to define the degree; we proceed as in the special case of S^n. Let M and N be compact oriented n-manifolds. Let

N be connected and suppose M has components $M_1, \dots, M_r$. Then we have fundamental classes $z(M_j)$ for the M_j and $z(M) \in H_n(M) \cong \bigoplus_j H_n(M_j)$ is the sum of the $z(M_j)$. For a continuous map $f: M \to N$ we define its degree $d(f) \in \mathbb{Z}$ via the relation $f_* z(M) = d(f) z(N)$. From this definition we see immediately the following properties of the degree:

(1) The degree is a homotopy invariant.

(2) $d(f \circ g) = d(f) d(g)$.

(3) A homotopy equivalence has degree ± 1.

(4) If $M = M_1 + M_2$, then $d(f) = d(f|M_1) + d(f|M_2)$.

(5) If we pass in M or N to the opposite orientation, then the degree changes the sign.

We now come to the computation of the degree in terms of local data of the map. Let M and N be connected and set $K = f^{-1}(p)$. Let U be an open neighbourhood of K in M. In the commutative diagram

$$\begin{array}{ccccc}
z(M) \in & H_n(M) & \xrightarrow{f_*} & H_n(N) & \ni z(N) \\
\Big\downarrow & \downarrow & & \downarrow \cong & \Big\downarrow \\
& H_n(M, M \smallsetminus K) & \xrightarrow{f_*} & H_n(N, N \smallsetminus p) & \\
& \uparrow \cong & & \uparrow = & \\
z(U, K) \in & H_n(U, U \smallsetminus K) & \xrightarrow{f_*^U} & H_n(N, N \smallsetminus p) & \ni z(N, p)
\end{array}$$

we have $f_*^U z(U, K) = d(f) z(N, p)$. Thus the degree only depends on the restriction f^U of f to U. One can now extend the earlier investigations of self maps of S^n to this more general case. The additivity of the degree is proved in exactly the same manner. Let K be finite. Choose $U = \bigcup_{x \subset K} U_x$ where the U_x are pair-wise disjoint open neighbourhoods of x. We then have

$$\textstyle\bigoplus_{x \in K} H_n(U_x, U_x \smallsetminus x) \cong H_n(U, U \smallsetminus K), \quad H_n(U_x, U_x \smallsetminus x) \cong \mathbb{Z}.$$

The image $z(U_x, x)$ of $z(M)$ is a generator, the local orientation determined by the fundamental class $z(M)$. The local degree $d(f, x)$ of f about x is defined by $f_* z(U_x, x) = d(f, x) z(N, p)$. The additivity yields $d(f) = \sum_{x \in K} d(f, x)$.

(16.4.2) Remark. Let f be a C^1-map in a neighbourhood of x; this shall mean the following. There exist charts $\varphi: U_x \to \mathbb{R}^n$ centered at x and $\psi: V \to \mathbb{R}^n$ centered at p such that $f(U_x) \subset V$ and $g = \psi f \varphi^{-1}$ is a C^1-map. We can suppose that the charts preserve the local orientations; this shall mean for φ that $\varphi_*: H_n(U_x, U_x \smallsetminus x) \to H_n(\mathbb{R}^n, \mathbb{R}^n \smallsetminus 0)$ sends $z(U_x, x)$ to the standard generator. Such charts are called ***positive*** with respect to the given orientations. Suppose now in addition that the differential of g at x is regular. Then $d(f, x)$ is the sign of the determinant of the Differential $Dg(0)$. ◇

(16.4.3) Proposition. *Let M be a connected, oriented, closed n-manifold. Then there exists for each $k \in \mathbb{Z}$ a map $f: M^n \to S^n$ of degree k.*

Proof. If $f: M \to S^n$ has degree a and $g: S^n \to S^n$ degree b, then gf has degree ab. Thus it suffices to realize a degree ± 1. Let $\varphi: D^n \to M$ be an embedding. Then we have a map $f: M \to D^n/S^{n-1}$ which is the inverse of φ on $U = \varphi(E^n)$ and sends $M \smallsetminus U$ to the base point. This map has degree ± 1. □

Let the manifolds M and N be oriented by the fundamental classes $z_M \in H_m(M)$ and $z_N \in H_n(N)$. Then the homology product $z_M \times z_N$ is a fundamental class for $M \times N$, called the ***product orientation***.

Problems

1. Let $p: M \to N$ be a covering of n-manifolds. Then the pullback of $\mathrm{Ori}(N) \to N$ along p is $\mathrm{Ori}(M) \to M$.
2. Let $p: M \to N$ be a G-principal covering between connected n manifolds with orientable M. Then N is orientable if and only if G acts by orientation-preserving homeomorphisms.
3. The manifold $\mathrm{Ori}(M)$ is always orientable.
4. $\mathbb{R}P^n$ is orientable if and only if n is odd.
5. Let M be a closed oriented connected n-manifold. Suppose that M carries a CW-decomposition with k-skeleton M_k. The inclusion induces an injective map $H_n(M) \to H_n(M_n, M_{n-1})$. The fundamental class is therefore represented by a cellular chain in $H_n(M_n, M_{n-1})$. If we orient the n-cells in accordance with the local orientations of the manifold, then the fundamental class chain is the sum of the n-cells. This is the classical interpretation of the fundamental class of a triangulated manifold. A similar assertion holds for unoriented manifolds and coefficients in $\mathbb{Z}/2$ and manifolds with boundary (to be considered in the next section).

In a sense, a similar assertion should hold for non-compact manifolds; but the cellular chain would have to be an infinite sum. Therefore a fundamental class has to be defined via an inverse limit.

6. Let M be a closed connected n-manifold. Then $H_n(M;\mathbb{Z}/2) \cong \mathbb{Z}/2$ and the restrictions $r_x^M: H_n(M;\mathbb{Z}/2) \to H_n(M, M \smallsetminus x;\mathbb{Z}/2)$ are isomorphisms.
7. Let $f: M \to N$ be a map between closed connected n-manifold. Then one can define the ***degree modulo 2*** $d_2(f) \in \mathbb{Z}/2$; it is zero (one) if $f_*: H_n(M;\mathbb{Z}/2) \to H_n(N;\mathbb{Z}/2)$ is the zero map (an isomorphism). If this degree is non-zero, then f is surjective. If the manifolds are oriented, then $d(f) \bmod 2 = d_2(f)$.
8. Let G be a compact connected Lie group and let T be a maximal torus of G. The map $q: G/T \times T \to G$, $(g,t) \mapsto gtg^{-1}$ has degree $|W|$. Here $W = NT/T$ is the Weyl group. Since q has non-zero degree, this map is surjective (see [29, IV.1]).
9. Let G be a compact connected Lie group and T a maximal torus of G. The degree of $f: G \to G$, $g \mapsto g^k$ has degree k^r, $r = \dim T$. Let $c \in T$ be an element such that the powers of c are dense in T, then $|f^{-1}(c)| = k^r$, $f^{-1}(c) \subset T$, and c is a regular value of f ([90]).
10. Let $f: M \to N$ be a proper map between oriented connected n-manifolds. Define the degree of f.

16.5 Manifolds with Boundary

Let M be an n-dimensional manifold with boundary. We call $z \in H_n(M, \partial M)$ a ***fundamental class*** if for each $x \in M \smallsetminus \partial M$ the restriction of z is a generator in $H_n(M, M \smallsetminus x)$.

(16.5.1) Theorem. *Let M be a compact connected n-manifold with non-empty boundary. Then one of the following assertions hold:*

(1) *$H_n(M, \partial M) \cong \mathbb{Z}$, and a generator of this group is a fundamental class. The image of a fundamental class under $\partial\colon H_n(M, \partial M) \to H_{n-1}(\partial M)$ is a fundamental class. The interior $M \smallsetminus \partial M$ is orientable.*

(2) *$H_n(M, \partial M) = 0$, and $M \smallsetminus \partial M$ is not orientable.*

Proof. Let $\kappa\colon [0, \infty[\, \times \partial M \to U$ be a collar of M, i.e., a homeomorphism onto an open neighbourhood U of ∂M such that $\kappa(0, x) = x$ for $x \in \partial M$. For simplicity of notation we identify U with $[0, \infty[\, \times \partial M$ via κ; similarly for subsets of U. In this sense $\partial M = 0 \times \partial M$. We have isomorphisms

$$H_n(M, \partial M) \cong H_n(M, [0, 1[\, \times \partial M) \cong H_n(M \smallsetminus \partial M,]0, 1[\, \times \partial M) \cong \Gamma(A).$$

The first one by h-equivalence; the second one by excision; the third one uses the closed set

$$A = M \smallsetminus ([0, 1[\, \times \partial M) \subset M \smallsetminus \partial M$$

and (16.3.2). The set A is connected, hence $\Gamma(A) \cong \mathbb{Z}$ or $\Gamma(A) \cong 0$. If $\Gamma(A) \cong \mathbb{Z}$, then $M \smallsetminus \partial M$ is orientable along A. Instead of A we can argue with the complement of $[0, \varepsilon[\, \times \partial M$. Since each compact subset of $M \smallsetminus \partial M$ is contained in some such complement, we see that $M \smallsetminus \partial M$ is orientable along compact subsets, hence orientable (see (16.2.2)). The isomorphism $H_n(M \smallsetminus \partial M,]0, 1[\, \times \partial M) \cong \Gamma(A)$ says that there exists an element $z \in H_n(M \smallsetminus \partial M,]0, 1[\, \times \partial M)$ which restricts to a generator of $H_n(M \smallsetminus \partial M, M \smallsetminus \partial M \smallsetminus x)$ for each $x \in A$. For the corresponding element $z \in H_n(M, \partial M)$ a similar assertion holds for each $x \in M \smallsetminus \partial M$, i.e., z is a fundamental class (move around x within the collar).

It remains to show that ∂z is a fundamental class. The lower part of the diagram (for $x \in]0, 1[\, \times \partial M$)

$$\begin{array}{ccccc}
H_{n-1}(\partial M) & \xrightarrow{\cong} & H_{n-1}(\partial M \cup A, A) & \xleftarrow{\cong} & H_{n-1}(\partial I \times \partial M, 1 \times \partial M) \\
\uparrow \partial & & \uparrow \partial\ \cong & & \uparrow \partial\ \cong \\
H_n(M, \partial M) & \longrightarrow & H_n(M, \partial M \cup A) & \xleftarrow{\cong} & H_n(I \times \partial M, \partial I \times \partial M) \\
& \searrow & \downarrow & & \downarrow \\
& & H_n(M, M \smallsetminus x) & \xleftarrow{\cong} & H_n(I \times \partial M, I \times \partial M \smallsetminus x)
\end{array}$$

shows that z yields a fundamental class in $H_n(I \times \partial M, \partial I \times \partial M)$. The upper part shows that this fundamental class corresponds to a fundamental class in $H_{n-1}(\partial M)$, since fundamental classes are characterized by the fact that they are generators (for each component of ∂M). □

(16.5.2) Example. Suppose the n-manifold M is the boundary of the compact orientable $(n+1)$-manifold B. We have the fundamental classes $z(B) \in H_{n+1}(B, \partial B)$ and $z(M) = \partial z_B \in H_n(M)$. Let $f: M \to N$ be a map which has an extension $F: B \to N$, then the degree of f (if defined) is zero, $d(f) = 0$, for we have $f_* z(M) = f_* \partial z(B) = F_* i_* \partial z(B) = 0$, since $i_* \partial = 0$ as consecutive morphisms in the exact homology sequence of the pair (B, M). We call maps $f_\nu : M_\nu \to N$ ***orientable bordant*** if there exists a compact oriented manifold B with oriented boundary $\partial B = M_1 - M_2$ and an extension $F: B \to N$ of $\langle f_1, f_2 \rangle : M_1 + M_2 \to N$. The minus sign in $\partial B = M_1 - M_2$ means $\partial z(B) = z(M_1) - z(M_2)$. Under these assumptions we have $d(f_1) = d(f_2)$. This fact is called the ***bordism invariance*** of the degree; it generalizes the homotopy invariance. ◇

Problems

1. Let M be a compact connected n-manifold with boundary. Then $H_n(M, \partial M; \mathbb{Z}/2)$ is isomorphic to $\mathbb{Z}/2$; the non-zero element is a $\mathbb{Z}/2$-fundamental class $z(M; \mathbb{Z}/2)$. The restriction to $H_n(M, M \smallsetminus x; \mathbb{Z}/2)$ is for each $x \in M \smallsetminus \partial M$ an isomorphism and $\partial z(M; \mathbb{Z}/2)$ is a $\mathbb{Z}/2$-fundamental class for ∂M.

2. Show that the degree $d_2(f)$ is a bordism invariant.

16.6 Winding and Linking Numbers

Let M be a closed connected oriented n-manifold. Let $f: M \to \mathbb{R}^{n+1}$ and $a \notin \mathrm{Im}(f)$. The ***winding number*** $W(f, a)$ of f with respect to a is the degree of the map

$$p_{f,a} = p_a : M \to S^n, \quad x \mapsto N(f(x) - a)$$

where $N: \mathbb{R}^{n+1} \smallsetminus 0 \to S^n$, $x \mapsto \|x\|^{-1} x$. If f_t is a homotopy with $a \notin \mathrm{Im}(f_t)$ for each t, then $W(f_0, a) = W(f_t, a)$.

(16.6.1) Theorem. *Let M be the oriented boundary of the compact smooth oriented manifold B. Let $F: B \to \mathbb{R}^{n+1}$ be smooth with regular value 0 and assume $0 \notin f(M)$. Then*

$$W(f, 0) = \textstyle\sum_{x \in P} \varepsilon(F, x), \quad P = F^{-1}(0), \; f = F|\partial B$$

where $\varepsilon(F, x) \in \{\pm 1\}$ is the orientation behaviour of the differential $T_x F : T_B \to T_0(\mathbb{R}^{n+1})$.

Proof. Let $D(x) \subset B \smallsetminus \partial B, x \in P$ be small disjoint disks about x. Then $G = N \circ F$ is defined on $C = B \sum \bigcup_{x \in P} D(x)$, and by the bordism invariance of the degree $d(G|\partial B) = \sum_{x \in P} d(G|\partial D(x))$. By (16.4.2), $d(G|\partial D(x)) = \varepsilon(F, x)$. □

Let M and N be oriented closed submanifolds of $\mathbb{R}^{k+1}$ of dimensions m and n with $k = m + n$. Let $M \times N$ carry the product orientation. The degree of the map

$$f_{M,N} = f : M \times N \to S^k, \quad (x, y) \mapsto N(x - y)$$

is the ***linking number*** $L(M, N)$ of the pair (M, N). More generally, if the maps $\mu \colon M^m \to \mathbb{R}^{k+1}$ and $\nu \colon N^n \to \mathbb{R}^{k+1}$ have disjoint images, then the degree of $(x, y) \mapsto N(\mu(x) - \nu(y))$ is the linking number of (μ, ν).

Problems

1. Let the n-manifold M be the oriented boundary of the smooth connected compact manifold B. Suppose $f \colon M \to S^n$ has degree zero. Then f can be extended to B.
2. Show $L(M, N) = (-1)^{m+n+1} L(N, M)$.
3. Let $f, g \colon \mathbb{R} \to \mathbb{R}^3$ be smooth embeddings with disjoint closed images. Define a linking number for the pair (f, g) and justify the definition.

Chapter 17
Cohomology

The axioms for a cohomology theory are analogous to the axioms of a homology theory. Now we consider contravariant functors. The reader should compare the two definitions, also with respect to notation. One advantage of cohomology is an additional internal product structure (called cup product) which will be explained in subsequent sections. The product structure suggests to view the family $(h^n(X) \mid n \in \mathbb{Z})$ as a single object; the product then furnishes it with the structure of a ring (graded algebra). Apart from the additional information in the product structure, the ring structure is also notationally convenient (for instance, a polynomial ring has a better description than its additive group without using the multiplicative structure).

Singular cohomology is obtained from the singular chain complex by an application of the Hom-functor. We present an explicit definition of the cup product in singular cohomology (Alexander–Whitney). In more abstract terms the product can also be obtained from the Eilenberg–Zilber chain equivalences as in the case of the homology product.

We use the product structure to prove a powerful theorem (Leray–Hirsch) which says roughly that the cohomology of the total space of a fibration is a free module over the cohomology ring of the base, provided the fibre is a free module and a basis of the fibre-cohomology can be lifted to the total space. In the case of a topological product, the result is a special case of the Künneth theorem if the cohomology of the fibre is free, since the Ext-groups vanish in that case. One interesting application is to vector bundles; the resulting so-called Thom isomorphism can be considered as a twisted suspension isomorphism in that the suspension is replaced by a sphere bundle. As a specific example we determine the cohomology rings of the projective spaces.

17.1 Axiomatic Cohomology

17.1.1 The axioms. A ***cohomology theory*** for pairs of spaces with values in the category of R-modules consists of a family $(h^n \mid n \in \mathbb{Z})$ of contravariant functors $h^n\colon \mathrm{TOP}(2) \to R\text{-}\mathrm{MOD}$ and a family $(\delta^n \mid n \in \mathbb{Z})$ of natural transformations $\delta^n\colon h^{n-1} \circ \kappa \to h^n$. These data are required to satisfy the following axioms.

(1) **Homotopy invariance.** Homotopic maps f_0 and f_1 between pairs of spaces induce the same homomorphism, $h^n(f_0) = h^n(f_1)$.

(2) **Exact sequence.** For each pair (X, A) the sequence

$$\cdots \to h^{n-1}(A,\emptyset) \xrightarrow{\delta} h^n(X,A) \to h^n(X,\emptyset) \to h^n(A,\emptyset) \xrightarrow{\delta} \cdots$$

is exact. The undecorated arrows are induced by the inclusions.

(3) **Excision.** Let (X, A) be a pair and $U \subset A$ such that $\bar{U} \subset A^\circ$. Then the inclusion $(X \smallsetminus U, A \smallsetminus U) \to (X, A)$ induces an excision isomorphism $h^n(X, A) \cong h^n(X \smallsetminus U, A \smallsetminus U)$.

We call $h^n(X, A)$ the n-th ***cohomology group*** of (X, A). The δ^n are called the ***coboundary operators***. We write $h^n(X, \emptyset) = h^n(X)$ and $h^n(f) = f^*$. Occasionally we refer to the homomorphisms $i^*\colon h^n(X) \to h^n(A)$ induced by an inclusion $i\colon A \subset X$ as ***restriction***. The groups $h^n(P) \cong h^n$ for a point P are said to be the ***coefficient groups*** of the theory (compatible family of isomorphisms to a given module h^n). In the case that $h^n(P) = 0$ for $n \neq 0$, we talk about an ordinary or classical cohomology theory and say that the theory satisfies the ***dimension axiom***. The notation $h^n(X, A) = H^n(X, A; G)$ stands for an ordinary cohomology theory with a given isomorphism $h^0(P) \cong G$.

The cohomology theory is ***additive*** if

$$h^n\left(\textstyle\coprod_j X_j, \coprod_j A_j\right) \to \textstyle\prod_j h^n(X_j, A_j), \quad x \mapsto (h^n(i_j)(x))$$

is always an isomorphism (i_j the inclusion of the j-th summand). For finite J the additivity isomorphism follows from the other axioms. ◇

Several formal consequences of the homology axioms have analogues in cohomology and the proofs are similar. We mention some of them.

We begin with the ***exact sequence of a triple*** (X, A, B). The coboundary operator is in this case defined by $\delta\colon h^{n-1}(A, B) \to h^{n-1}(A) \to h^n(X, A)$. The first map is induced by the inclusion and the second map is the given coboundary operator. For each triple (X, A, B) the sequence

$$\cdots \to h^{n-1}(A,B) \xrightarrow{\delta} h^n(X,A) \to h^n(X,B) \to h^n(A,B) \xrightarrow{\delta} \cdots$$

is exact. The undecorated arrows are restrictions.

17.1.2 Suspension. The suspension isomorphism σ is defined by the commutative diagram

$$\begin{array}{ccc} h^n(Y,B) & \xrightarrow{\cong} & h^n(0\times Y, 0\times B) \\ \downarrow{\scriptstyle\sigma} & & \uparrow{\scriptstyle\cong} \\ h^{n+1}((I,\partial I)\times(Y,B)) & \xleftarrow[\cong]{\delta} & h^n(\partial I\times Y\cup I\times B, 1\times Y\cup I\times B). \end{array}$$

For homology we used a definition with the roles of 0, 1 interchanged. ◇

17.1.3 Reduced cohomology. The reduced cohomology groups of a non-empty space X are defined as $\tilde{h}^n(X) = \operatorname{coker}(p^*\colon h^n(P) \to h^n(X))$ where $p\colon X \to P$ denotes the unique map to a point. The functors $\tilde{h}^n(-)$ are homotopy invariant. For a pointed space $(X, *)$ we have the canonical split exact sequence

$$0 \to h^n(X, *) \xrightarrow{j} h^n(X) \xrightarrow{i} h^n(*) \to 0.$$

The restriction j induces an isomorphism $h^n(X, *) \cong \tilde{h}^n(X)$ and we have isomorphisms

$$\langle j, p^* \rangle\colon h^n(X, *) \oplus h^n \cong h^n(X), \quad (q, i)\colon h^n(X) \cong \tilde{h}^n(X) \oplus h^n$$

with the quotient map q. The coboundary operator $\delta^n : h^{n-1}(A) \to h(X, A)$ factors over the quotient map $q\colon h^{n-1}(A) \to \tilde{h}^{n-1}(A)$. Passing to quotients yields the exact sequence

$$\cdots \to \tilde{h}^{n-1}(A) \xrightarrow{\delta} h(X, A) \to \tilde{h}^n(X) \to \tilde{h}^n(A) \xrightarrow{\delta} \cdots$$

for the reduced groups. ◇

17.1.4 Mayer–Vietoris sequence. A triad $(X; A, B)$ is ***excisive*** for the cohomology theory if the inclusion induces an isomorphism $h^*(A \cup B, A) \cong h^*(B, A \cap B)$. This property can be characterized in different ways as in the case of a homology theory, see (10.7.1) and (10.7.5). In particular the property is symmetric in A, B.

We have exact Mayer–Vietoris sequences for excisive triads. As in the case of a homology theory one can derive some MV-sequences by diagram chasing. For the general case of two excisive triads we use a method which we developed in the case of homology theory; the MV-sequence was obtained as the exact sequence of a triad of auxiliary spaces by some rewriting (see (10.7.6)). This procedure also works for cohomology.

Let $(A; A_0, A_1) \subset (X; X_0, X_1)$ be excisive triads. Set $X_{01} = X_0 \cap X_1$ and $A_{01} = A_0 \cap A_1$. Then there exists an exact Mayer–Vietoris sequence of the following form

$$\begin{aligned}\cdots \longleftarrow h^n(X_{01}, A_{01}) &\xleftarrow{(1)} h^n(X_0, A_0) \oplus h^n(X_1, A_1) \xleftarrow{(2)} h^n(X, A)\\ &\xleftarrow{\Delta} h^{n-1}(X_{01}, A_{01}) \longleftarrow \cdots .\end{aligned}$$

The map (1) is $(x_0, x_1) \mapsto i_0^* x_0 - i_1^* x_1$ with the inclusions i_ν; the components of (2) are the restrictions. The connecting morphism in the case $A = A_0 = A_1$ is the composition

$$\Delta\colon h^{n-1}(X_{01}, A) \xrightarrow{\delta} h^n(X_0, X_{01}) \cong h^n(X, X_1) \to h^n(X, A)$$

and the connecting morphism in the case $X = X_0 = X_1$ is

$$\Delta\colon h^n(X, A_{01}) \to h^n(A_0, A_{01}) \cong h^n(A, A_1) \xrightarrow{\delta} h^{n+1}(X, A). \qquad \diamond$$

17.1.5 Limits. We have seen that an additive homology theory is compatible with colimits. The situation for cohomology and limits is more complicated.

Let $G_\bullet : G_1 \xleftarrow{p_1} G_2 \xleftarrow{p_2} G_3 \xleftarrow{p_3} \cdots$ be a sequence of groups and homomorphisms. The group $\prod_{i\in\mathbb{N}} G_i$ acts on the set $\prod_{i\in\mathbb{N}} G_i$

$$(g_1, g_2, \dots)\cdot(h_1, h_2, \dots) = (g_1 h_1 p_1(g_2)^{-1}, g_2 h_2 p_2(g_3)^{-1}, \dots).$$

The orbit set is denoted $\lim^1(G_\bullet) = \lim^1(G_i, p_i) = \lim^1(G_i)$. A direct consequence of this definition is:

Suppose the groups G_i are abelian. Then we have an exact sequence

$$0 \to \lim(G_i, p_i) \to \textstyle\prod_{i\in\mathbb{N}} G_i \xrightarrow{d} \prod_{i\in\mathbb{N}} G_i \to \lim^1(G_i, p_i) \to 0.$$

Here $d(g_1, g_2, \dots) = (g_1 - p_1(g_2), g_2 - p_2(g_3), \dots)$ and lim is the limit of the sequence. As a consequence of the ker-coker-sequence we obtain:

A short exact sequence

$$0 \to (G_i', p_i') \to (G_i, p_i) \to (G_i'', p_i'') \to 0$$

of inverse systems induces an exact sequence

$$\begin{aligned} 0 &\to \lim{}^0(G_i') \to \lim{}^0(G_i) \to \lim{}^0(G_i'') \\ &\to \lim{}^1(G_i') \to \lim{}^1(G_i) \to \lim{}^1(G_i'') \to 0. \end{aligned}$$

Let now an additive cohomology theory by given. We apply the Mayer–Vietoris sequence to the telescope T of a sequence $X_1 \subset X_2 \subset \cdots$ of spaces with colimit X. The result is [133]:

(17.1.6) Proposition. *There exists an exact sequence*

$$0 \to \lim{}^1(h^{n-1}(X_i)) \to h^n(T) \to \lim(h^n(X_i)) \to 0.$$

Proof. We use the MV-sequence of the triad $(T; A, B)$ as in (10.8.2). It has the form

$$\cdots \to h^n(T) \to h^n(A) \oplus h^n(B) \xrightarrow{\beta^n} h^n(A \cap B) \to \cdots$$

and yields the short exact sequence

$$0 \to \operatorname{Coker}(\beta^{n-1}) \to h^n(T) \to \operatorname{Ker}(\beta^n) \to 0.$$

Thus we have to determine the kernel and the cokernel. By additivity of the cohomology theory we obtain

$$h^n(A) \cong \prod_{i \equiv 0(2)} h^n(X_i), \quad h^n(B) \cong \prod_{i \equiv 1(2)} h^n(X_i), \quad h^n(A \cap B) \cong \prod_{i \in \mathbb{N}} h^n(X_i).$$

These isomorphisms transform β^n into

$$(x_1, x_2, x_3, \dots) \mapsto (-x_1 + f_1^*(x_2), x_2 - f_2^*(x_3), \dots).$$

The isomorphism $(a_i) \mapsto ((-1)^i a_i)$ transforms this map finally into the map d in the definition of lim and $\lim^1$. □

(17.1.7) Proposition. *Suppose the homomorphisms p_i between the abelian groups G_i are surjective. Then* $\lim^1(G_i, p_i) = 0$.

Proof. Let $g = (g_1, g_2, \dots) \in \prod_{i \in \mathbb{N}} G_i$. We have to show that g is contained in the image of d, i.e., we have to solve the equations $g_i = x_i - p_i(x_{i+1})$ with suitable $x_i \in G_i$. This is done inductively. □

Problems

1. The system (G_i, p_i) satisfies the ***Mittag-Leffler condition*** (= ML) if for each i there exists j such that for $k \geq j$ the equality $\mathrm{Im}(G_{i+k} \to G_i) = \mathrm{Im}(G_{i+j} \to G_i)$ holds. If (G_i, p_i) satisfies ML, then $\lim^1(G_i, p_i) = 0$. Thus if the groups G_i are finite, then $\lim^1(G_i, p_i) = 0$. If the G_i are countable and if $\lim^1(G_i, p_i) = 0$, then the system satisfies ML ([44, p. 154]).
2. Imitate the earlier investigation of cellular homology and show that $H^*(X)$ can be determined from a cellular cochain complex which arises from a cellular decomposition of X.
3. Let $\rho\colon k^*(-) \to l^*(-)$ be a natural transformation between additive cohomology theories which induces isomorphisms of the coefficient groups. Show that ρ is an isomorphism for each CW-complex.

17.2 Multiplicative Cohomology Theories

Let h^* be a cohomology theory with values in R-MOD. A ***multiplicative structure*** on this theory consists of a family of R-linear maps ($m, n \in \mathbb{Z}$)

$$h^m(X, A) \otimes_R h^n(X, B) \to h^{m+n}(X, A \cup B), \quad x \otimes y \mapsto x \cup y,$$

defined for suitable triads $(X; A, B)$, and in any case for excisive (A, B) in X. We call $x \cup y$ the ***cup product*** of x, y. The products are always defined if A or B is

empty or if $A = B$. In this section, tensor products will be taken over R. The cup product maps are required to satisfy the following axioms.
(1) **Naturality.** For maps of triads $f: (X; A, B) \to (X'; A', B')$ the commutativity $f^*(x \cup y) = f^*x \cup f^*y$ holds.
(2) **Stability.** Let (A, B) be excisive in X. We use the restriction morphism $\iota_A: h^j(X, B) \to h^j(A, A \cap B)$ and the coboundary operator $\delta_A: h^r(A, A \cap B) \cong h^r(A \cup B, B) \xrightarrow{\delta} h^{r+1}(X; A \cup B)$. The diagram

$$\begin{array}{ccccc} h^i(A) \otimes h^j(X, B) & \xrightarrow{1 \otimes \iota_A} & h^i(A) \otimes h^j(A, A \cap B) & \xrightarrow{\cup} & h^{i+j}(A, A \cap B) \\ \downarrow{\scriptstyle \delta \otimes 1} & & & & \downarrow{\scriptstyle \delta_A} \\ h^{i+1}(X, A) \otimes h^j(X, B) & & \xrightarrow{\cup} & & h^{i+j+1}(X, A \cup B) \end{array}$$

is commutative.
(3) **Stability.** Let (A, B) be excisive in X. We use the restriction morphism $\iota_B: h^i(X, A) \to h^i(B, A \cap B)$ and the coboundary operator $\delta_B: h^r(B, A \cap B) \cong h^r(A \cup B, A) \xrightarrow{\delta} h^{r+1}(X, A \cup B)$. The diagram

$$\begin{array}{ccccc} h^i(X, A) \otimes h^j(B) & \xrightarrow{\iota_B \otimes 1} & h^i(B, A \cap B) \otimes h^j(B) & \xrightarrow{\cup} & h^{i+j}(B, A \cap B) \\ \downarrow{\scriptstyle 1 \otimes \delta} & & & & \downarrow{\scriptstyle \delta_B} \\ h^i(X, A) \otimes h^{j+1}(X, B) & & \xrightarrow{\cup} & & h^{i+j+1}(X, A \cup B) \end{array}$$

is commutative up to the sign $(-1)^i$.
(4) **Unit element.** There is given a unit $1 \in h^0(P)$, P a fixed point, as additional structure datum. It induces $1 = 1_X = p^*(1) \in h^0(X)$, $p: X \to P$. Then $1 \cup x = x \cup 1 = x$ for each $x \in h^m(X, A)$.
(5) **Associativity.** $(x \cup y) \cup z = x \cup (y \cup z)$.
(6) **Commutativity.** $x \cup y = (-1)^{|x||y|} y \cup x$.

One can also consider situations where (6) or (5) do not hold. This is the reason for requiring (2) and (3) separately. For (5) it is required that the products are defined. For the convenience of the reader we also display the properties in a table and refer to the detailed description above.

$$
\begin{aligned}
f^*(x \cup y) &= f^*x \cup f^*y \\
\delta(a) \cup x &= \delta_A(a \cup \iota_A x) \\
x \cup \delta(b) &= (-1)^{|x|}\delta_B(\iota_B x \cup b) \\
1 \cup x &= x = x \cup 1 \\
(x \cup y) \cup z &= x \cup (y \cup z) \\
x \cup y &= (-1)^{|x||y|} y \cup x
\end{aligned}
$$

The cup product defines on $h^*(X, A)$ the structure of a $\mathbb{Z}$-graded associative and commutative algebra. If $A = \emptyset$ this algebra has a unit element $1_X \in h^0(X)$. Moreover $h^*(X, A)$ becomes a unital graded left $h^*(X)$-module. These structures have the obvious naturality properties which follow from the axioms, e.g., a map $f: X \to Y$ induces a unital algebra homomorphism $f^*: h^*(Y) \to h^*(X)$.

In particular the graded module $h^* = (h^n)$ of the coefficient groups becomes a unital commutative graded algebra. We make $h^*(X, A)$ into a unital left h^*-module via $a \cdot x = p^*(a) \cup x$ with $p: X \to P$ the unique map to a point. Morphisms induced by continuous maps are then h^*-linear. This is a particular case of the cross product introduced later.

We list some consequences of naturality and stability. Let $y \in h^n(X, B)$ be fixed. Right multiplication by y yields a morphism of degree n from the exact sequence of the pair into the exact sequence of the triple $(X, A \cup B, B)$. This means: We have a commutative diagram

$$
\begin{array}{ccccc}
h^m(X) & \longrightarrow & h^m(A) & \xrightarrow{\delta} & h^{m+1}(X, A) \\
\downarrow{\scriptstyle r_y} & & \downarrow{\scriptstyle r_y} & & \downarrow{\scriptstyle r_y} \\
h^{m+n}(X, B) & \longrightarrow & h^{m+n}(A \cup B, B) & \xrightarrow{\delta} & h^{m+n+1}(X, A \cup B).
\end{array}
$$

The r_y in the middle is defined by multiplication of $x \in h^m(A)$ with the restriction of y along $h^n(X, B) \to h^n(A, A \cap B)$ and then using $h^n(A, A \cap B) \cong h^n(A \cup B, B)$.

Let $y \in h^n(X)$ be fixed. For each pair (A, B) in X we obtain a product $r_y: h^k(A, B) \to h^{k+n}(A, B)$ by right multiplication with the restriction of y along $h^n(X) \to h^n(A)$. The following diagram commutes:

$$
\begin{array}{ccccc}
h^m(A) & \longrightarrow & h^m(B) & \xrightarrow{\delta} & h^{m+1}(A, B) \\
\downarrow{\scriptstyle r_y} & & \downarrow{\scriptstyle r_y} & & \downarrow{\scriptstyle r_y} \\
h^{m+n}(A) & \longrightarrow & h^{m+n}(B) & \xrightarrow{\delta} & h^{m+n+1}(A, B).
\end{array}
$$

If (A, B) is excisive, then the coboundary operators of the MV-sequences

$$\Delta: h^{m-1}(A \cap B, C) \to h^m(A, A \cap B) \cong h^m(A \cup B, B) \to h^m(A \cup B, C),$$

$$\Delta\colon h^m(X, A \cap B) \to h^m(A, A \cap B) \cong h^m(A \cup B, B) \to h^{m+1}(X, A \cap B)$$

commute with r_y. Also in the general case 17.1.4, the boundary operator commutes with products:

(17.2.1) Proposition. *Let $(A; A_0, A_1) \subset (X; X_0, X_1)$ be excisive triads. Let $X^0 \subset X$ and assume that also $(A \cup X^0; A_0 \cup X_0^0, A_1 \cup A_1^0)$ is excisive for $X_\nu^0 = X_\nu \cap X^0$. Then the diagram*

$$\begin{array}{ccc} h^r(X_{01}, A_{01}) & \xrightarrow{r_y} & h^{r+|y|}(X_{01}, A_{01} \cup X_{01}^0) \\ \downarrow{\scriptstyle\Delta} & & \downarrow{\scriptstyle\Delta} \\ h^{r+1}(X, A) & \xrightarrow{r_y} & h^{r+1+|y|}(X, A \cup X^0) \end{array}$$

commutes. Here r_y is right multiplication by $y \in h(X, X^0)$ (bottom) and multiplication with the restriction of y to $h(X_{01}, X_{01}^0)$ (top).

Proof. The coboundary operators are defined via suspension and appropriate induced morphisms. The commutativity of the diagram then amounts to the naturality of the cup product and a compatibility (17.2.2) with the suspension. □

(17.2.2) Proposition. *The diagram*

$$\begin{array}{ccc} h(IY, IB \cup \partial IY) \otimes h(IY, IY^0) & \xrightarrow{\cup} & h(IY, I(B \cup Y^0) \cup \partial IY) \\ \uparrow{\scriptstyle\sigma\otimes \mathrm{pr}^*} & & \uparrow{\scriptstyle\sigma} \\ h(Y, B) \otimes h(Y, Y^0) & \xrightarrow{\cup} & h(Y, B \cup Y^0) \end{array}$$

commutes. (Notation: $IY = I \times Y$, $\partial IY = \partial I \times Y$ etc.)

Proof. We use the associated cross product and (17.3.1), (17.3.3) in the computation $(e^1 \times x) \cup (1_I \times y) = (e^1 \cup 1_I) \times (x \cup y)$. □

(17.2.3) Example. Let $(X, *)$ be a pointed space. The ring homomorphism $i\colon h^*(X) \to h^*$ has as kernel the two-sided h^*-ideal $h^*(X, *)$. We have the isomorphism $h^*(X, *) \oplus h^* \cong h^*(X)$, $(a, b) \mapsto ja + p^*b$, see 17.1.3. This isomorphism transforms the cup product on $h^*(X)$ into the product

$$(a_1, b_1) \cdot (a_2, b_2) = (a_1 \cup a_2 + b_1 a_2 + a_1 b_2, b_1 b_2). \qquad \diamond$$

(17.2.4) Example. The commutativity $x \cup x = (-1)^{|x||x|} x \cup x$ has for $|x| \equiv 1 \bmod 2$ the consequence $2(x \cup x) = 0$. Hence $x \cup x = 0$, if multiplication by 2 is injective. ◇

(17.2.5) Example. Let $X = A_1 \cup \cdots \cup A_k$. Suppose $a_i \in h^{n(i)}(X)$ are elements in the kernel of the restriction $h^{n(i)}(X) \to h^{n(i)}(A_i)$. Let $b_i \in h^{n(i)}(X, A_i)$ be a pre-image of a_i. Then $a_1 \cup \cdots \cup a_k = 0$, because this element is the image of $b_1 \cup \cdots \cup b_k \in h^n(X, \bigcup_i A_i) = 0$. This argument requires that the relative products are defined.

This simple consequence of the existence of products has interesting geometric consequences. The projective space $\mathbb{C}P^n$ has a covering by $n+1$ affine (hence contractible) open sets $U_j = \{[z_i] \mid z_j \neq 0\}$. Therefore $a_0 \cup \cdots \cup a_n = 0$ for elements $a_j \in h^{n(j)}(\mathbb{C}P^n, *)$. Later we show the existence of an element $c \in H^2(\mathbb{C}P^n; \mathbb{Z})$ with $c^n \neq 0$. Hence $\mathbb{C}P^n$ cannot be covered by n contractible open sets (more generally, by open sets U such that $U \to \mathbb{C}P^n$ is null homotopic). An analogous result holds for $\mathbb{R}P^n$. ◇

(17.2.6) Example. The argument of the preceding example shows that products in $h^*(S^n, *)$, $n \geq 1$ are trivial, $a_1 \cup a_2 = 0$.

More generally, let X be a well-pointed space. Then products in $h^*(\Sigma X, *)$ are trivial. It suffices to prove this fact for the unreduced suspension $\Sigma' X$; but this space has a covering by two contractible open sets (cones).

Additively, the cohomology groups only depend on the stable homotopy type. The product structure contains more subtle information. ◇

(17.2.7) Example. Let $p\colon E \to B$ be any map. Then we make $h^*(E)$ into a graded right $h^*(B)$- module by the definition $y \cdot x = y \cup p^*x$ for $x \in h^*(B)$ and $y \in h^*(E)$. If $f\colon X \to Y$ is a morphism from $p\colon X \to B$ to $q\colon Y \to B$ in TOP_B, then $f^*\colon h^*(Y) \to h^*(X)$ is $h^*(B)$-linear. The same device works for pairs of spaces over B. The coboundary operator is then also $h^*(B)$-linear, $\delta(y \cdot x) = \delta y \cdot x$. ◇

17.3 External Products

A multiplicative structure on a cohomology theory h^* of ***external products*** consists of a family of R-linear maps ($m, n \in \mathbb{Z}$)

$$h^m(X, A) \otimes_R h^n(Y, B) \to h^{m+n}((X, A) \times (Y, B)), \quad x \otimes y \mapsto x \times y,$$

defined for a suitable class of pairs (X, A) and (Y, B) and in any case if the pair $(X \times B, A \times Y)$ is excisive in $X \times Y$. The products are defined if A or B is empty. These maps are required to satisfy the following axioms.

(1) **Naturality.** For continuous maps $f\colon (X, A) \to (X', A')$ and $g\colon (Y, B) \to (Y', B')$ we have $(f \times g)^*(x \times y) = f^*x \times g^*y$.
(2) **Stability.** Let $(X \times B, A \times Y)$ be excisive. For $x \in h^m(A)$ and $y \in h^n(Y, B)$ the relation $\delta x \times y = \delta'(x \times y)$ holds. Here δ' is the composition of the excision

isomorphism with δ for $k = m + n$

$$\delta' \colon h^k(A \times (Y, B)) \cong h^k(A \times Y \cup X \times B, X \times B) \xrightarrow{\delta} h^{k+1}((X, A) \times (Y, B)).$$

(3) **Stability.** Let $(X \times B, A \times Y)$ be excisive. For $x \in h^m(X, A)$ and $y \in h^n(B)$ the relation $x \times \delta y = (-1)^{|x|} \delta''(x \times y)$ holds. Here δ'' is a similar composition of excision with δ for $k = m + n$

$$\delta'' \colon h^k((X, A) \times B) \cong h^k(X \times B \cup A \times Y, A \times Y) \xrightarrow{\delta} H^{k+1}((X, A) \times (Y, B)).$$

(4) **Unit element.** There is given $1 \in h^0(P)$ as a further structure datum. It satisfies $1 \times x = x \times 1 = x$ (with respect to $P \times X = X \times P = X$).
(5) **Associativity.** $(x \times y) \times z = x \times (y \times z)$.
(6) **Commutativity.** Let $\tau \colon X \times Y \to Y \times X$, $(x, y) \mapsto (y, x)$. Then for $x \in h^m(X, A)$ and $y \in h^n(Y, B)$ the relation $\tau^*(x \times y) = (-1)^{|x||y|} y \times x$ holds.

Also in this case we display the properties in a table and refer to the detailed explanation above.

$$\begin{aligned}
(f \times g)^*(x \times y) &= f^*x \times g^*y \\
\delta x \times y &= \delta'(x \times y) \\
x \times \delta y &= (-1)^{|x|} \delta''(x \times y) \\
1 \times x &= x = x \times 1 \\
(x \times y) \times z &= x \times (y \times z) \\
\tau^*(x \times y) &= (-1)^{|x||y|} y \times x
\end{aligned}$$

The term "stability" usually refers to compatibility with suspension. We explain this for the present setup. As a first application we show that the suspension isomorphism is given by multiplication with a standard element. Let $e^1 \in h^1(I, \partial I)$ be the image of $1 \in h^0$ under

$$1 \in h^0 \cong h^0(0) \xleftarrow{\cong} h^0(\partial I, 1) \xrightarrow{\delta} h^1(I, \partial I) \ni e^1.$$

(17.3.1) Proposition. $e^1 \times y = \sigma(y)$.

Proof. Consider the diagram

$$\begin{array}{ccccc}
h^m(Y, B) & \xrightarrow{\tilde{1}\times} & h^m(\partial I Y, \partial I B) & \longrightarrow & h^m(0Y, 0B) \\
\downarrow & & \uparrow \alpha \, \cong & & \uparrow \beta \, \cong \\
h^{m+1}((I, \partial I) \times (Y, B)) & \xleftarrow{\delta} & h^m(\partial I Y \cup I B, I B) & \longleftarrow & h^m(\partial I Y \cup I B, 1Y \cup I B).
\end{array}$$

Let $\tilde{1} \in h^0(\partial I)$ be the image of $1 \in h^0(0)$ under $h^0(0) \cong h^0(\partial I, 1) \to h^0(\partial I)$. Then $e^1 = \delta(\tilde{1})$. Stability (2) of the cross product shows $e^1 \times y = \delta\alpha^{-1}(\tilde{1} \times y)$. The maps α and β are isomorphisms by excision and h-equivalence. The outer diagram path is $y \mapsto \sigma(y)$. □

Suppose we have a cup product. We construct an associated external product. Let $p\colon (X,A) \times Y \to (X,A)$ and $q\colon X \times (Y,B) \to (Y,B)$ be the projections onto the factors. We define for $x \in h^m(X,A)$ and $y \in h^n(Y,B)$ the product $x \times y = p^*y \cup q^*y$.

Conversely, suppose an external product is given. We define an associated $\cup$-product. Let $d\colon (X, A \cup B) \to (X,A) \times (X,B)$ be the diagonal. Then we set $x \cup y = d^*(x \times y)$.

(17.3.2) Proposition. *The $\times$-product associated to a $\cup$-product satisfies the axioms of an external product. The $\cup$-product associated to a $\times$-product satisfies the axioms of an internal product. The processes $\times \leadsto \cup$ and $\cup \leadsto \times$ are inverse to each other.* □

(17.3.3) Proposition. *Let $x_i \in h^*(X, A_i)$ and $y_i \in h^*(Y, B_i)$. Then*

$$(x_1 \times y_1) \cup (x_2 \times y_2) = (-1)^{|x_2||y_1|}(x_1 \cup x_2) \times (y_1 \cup y_2).$$

In particular $h^*(X) \otimes h^*(Y) \to h^*(X \times Y)$, $x \otimes y \mapsto x \times y$ is a homomorphism of unital graded algebras. □

(17.3.4) Proposition. *Let $s \in h^n(S^n, *) \subset h^n(S^n)$ be the element which corresponds to $1 \in h^0$ under a suspension isomorphism. Then for each space F the map*

$$h^k(F) \oplus h^{k-n}(F) \to h^k(F \times S^n), \quad (a,b) \mapsto a \times 1 + b \times s$$

is an isomorphism. These isomorphisms show that $h^*(F \times S^n)$ is a free graded left $h^*(F)$-module with homogeneous basis 1×1, $1 \times s$.

Proof. We start with the isomorphism

$$h^{k-n}(F) \to h^k(I^n \times F, \partial I^n \times F), \quad x \mapsto e^n \times x$$

where $e^n = e^1 \times \cdots \times e^1$, see (17.3.1). Let $s \in h^n(S^n, *)$ correspond to e^n under the isomorphism $h^n(S^n, *) \cong h^n(I^n/\partial I^n, *) \cong h^n(I^n, \partial I^n)$. Then, by naturality,

$$h^{k-n}(F) \to h^k(S^n \times F, * \times F), \quad b \mapsto s \times b$$

is an isomorphism. We now use the split exact sequence

$$0 \to h^k(S^n \times F, * \times F) \to h^k(S^n \times F) \to h^k(F) \to 0$$

with splitting $h^k(F) \to h^k(S^n \times F)$, $a \mapsto 1 \times a = \mathrm{pr}^*(a)$ in order to see that $(a,b) \mapsto 1 \times a + s \times b$ is an isomorphism.

Under this isomorphism we have the following expression of the product structure

$$(a + sb)(a' + sb') = aa' + s(ba' + (-1)^{|a|n} ab')$$

since $s^2 = 0$.

We use commutativity to show that $(a,b) \mapsto a \times 1 + b \times s$ is an isomorphism as claimed. If we use $1 = 1_F \times 1_{S^n}$ and $\overline{s} = 1_F \times s$ as basis elements for the left $h^*(F)$-module $h^*(F \times S^n)$, then the $h^*(F)$ algebra is seen to be the graded exterior algebra $h^*(F)[s]/(s^2)$. □

Let B be a CW-complex. The ***skeleton filtration*** $(F^k h^i(B) \mid k \in \mathbb{N})$ on $h^i(B)$ is defined by $F^k h^i(B) = \mathrm{Ker}(h^i(B) \to h^i(B^{k-1}))$.

(17.3.5) Proposition. *The skeleton filtration is multiplicative: If $a \in F^k h^i(B)$ and $b \in F^l h^j(B)$, then $a \cup b \in F^{k+l} h^{i+j}(B)$.*

Proof. Choose pre-images $a' \in h^i(B, B^{k-1})$, $b' \in h^l(B, B^{l-1})$. Then $a' \times b' \in h^{i+j}(B \times B, B^{k-1} \times B \cup B \times B^{l-1})$. The product $a \cup b$ is the image of $a \times b$ under the diagonal d^*. A cellular approximation $d' \colon B \to B \times B$ of the diagonal d sends B^{k+l-1} into $(B \times B)^{k+l-1}$ and the latter is contained in $B^{k-1} \times B \cup B \times B^{l-1}$. Naturality of the $\times$-product now shows that $a \cup b$ is contained in the image of $h^{i+j}(B, B^{k+l-1}) \to h^{i+j}(B)$. □

(17.3.6) Corollary. *Let $n \in h^0(B)$ be contained in $F^1 h^0(B)$. Then $n^k \in F^k h^0(B)$. Thus if B is finite-dimensional, n is nilpotent and $1 + n$ a unit.* □

Problems

1. Supply the proofs for (17.3.2) and (17.3.3).
2. Determine the algebras $h^*(S^{n(1)} \times \cdots \times S^{n(k)})$ as graded h^*-algebras (graded exterior algebra).

17.4 Singular Cohomology

The singular cohomology theory is constructed from the singular chain complexes by a purely algebraic process. The algebraic dual of the singular chain complex is again a chain complex, and its homology groups are called cohomology groups. It is customary to use a "co" terminology in this context.

Let $C_\bullet = (C_n, \partial_n)$ be a chain complex of R-modules. Let G be another R-module. We apply the functor $\mathrm{Hom}_R(-, G)$ to $C_\bullet$ and obtain a chain complex $C^\bullet = (C^n, \delta^n)$ of R-modules with $C^n = \mathrm{Hom}_R(C_n, G)$ and the R-linear map

$$\delta^n \colon C^n = \mathrm{Hom}_R(C_n, G) \to \mathrm{Hom}_R(C_{n+1}, G) = C^{n+1}$$

defined by $\delta^n(\varphi) = (-1)^{n+1}\varphi \circ \partial_{n+1}$ for $\varphi \in \mathrm{Hom}(C_n, G)$.

For the choice of this sign see 11.7.4. The reader will find different choices of signs in the literature. Other choices will not effect the cohomology functors. But there seems to be an agreement that our choice is the best one when it comes to products.

For a pair of spaces (X, A) we have the singular chain complex $S_\bullet(X, A)$. With an abelian group G we set

$$S^n(X, A; G) = \mathrm{Hom}(S_n(X, A), G)$$

and use the coboundary operator above. Here we are using $\mathrm{Hom}_{\mathbb{Z}}$. Elements in $S^n(X, A; G)$ are functions which associate to each singular simplex $\sigma\colon \Delta^n \to X$ an element of G and the value 0 to simplices with image in A. If G is an R-module, then the set of these functions becomes an R-module by pointwise addition and scalar multiplication. A continuous map $f\colon (X, A) \to (Y, B)$ induces homomorphisms

$$f^\# = S^n(f)\colon S^n(Y, B; G) \to S^n(X, A; G), \quad (f^\#\varphi)(\sigma) = \varphi(f\sigma)$$

which are compatible with the coboundary operators. In this manner we obtain a contravariant functor from TOP(2) into the category of cochain complexes of R-modules. The n-th cohomology group of $S^\bullet(X, A; G)$ is denoted $H^n(X, A; G)$ and called the n-th ***singular cohomology module*** of (X, A) with coefficients in G. We often write $H^n(X, A; \mathbb{Z}) = H^n(X, A)$ and talk about integral cohomology, in this case ($G = \mathbb{Z}/(p)$ mod-p cohomology). Dualization of the split exact sequence $0 \to S_\bullet(A) \to S_\bullet(X) \to S_\bullet(X, A) \to 0$ yields again an exact sequence $0 \to S^\bullet(X, A; G) \to S^\bullet(X; G) \to S^\bullet(A; G) \to 0$. It induces a long exact cohomology sequence

$$\cdots \to H^{n-1}(A; G) \xrightarrow{\delta} H^n(X, A; G) \to H^n(X; G) \to H^n(A; G) \to \cdots.$$

(17.4.1) Remark. We recall the definition of the coboundary operator δ for the present situation. Let $\varphi\colon S_{n-1}(A) \to G$ be a cocycle, i.e., $\varphi \circ \partial = 0$. Extend φ in an arbitrary manner to a function $\tilde{\varphi}\colon S_{n-1}(X) \to G$. The element $\delta(\tilde{\varphi}) = (-1)^n\tilde{\varphi} \circ \partial\colon S_n(X) \to G$ vanishes on $S_n(A)$, since its restriction to $S_n(A)$ is $\varphi \circ \partial$. Therefore $\delta(\tilde{\varphi})$ yields an element $\psi \in S^n(X, A; G)$. The coboundary operator is then defined by the assignment $[\varphi] \mapsto [\psi]$. ◇

So far we have defined the data of a cohomology theory. If we apply $\mathrm{Hom}(-, G)$ to a chain homotopy we obtain a cochain homotopy; this yields the homotopy invariance. The excision axiom holds, as the chain equivalence $S_\bullet(X \smallsetminus U, A \smallsetminus U) \simeq S_\bullet(X, A)$ induces a chain equivalence $S^\bullet(X, A; G) \simeq S^\bullet(X \smallsetminus U, A \smallsetminus U; G)$.

There exist several algebraic relations among homology and cohomology. The first one comes from the evaluation of the Hom-complex. Let us use coefficients in a commutative ring R. The evaluations

$$\mathrm{Hom}(S_n(X, A), R) \otimes S_n(X, A; R) \to R, \quad \varphi \otimes (c \otimes r) \mapsto \varphi(c)r$$

induces a pairing (sometimes called ***Kronecker pairing***)

$$H^n(X,A;R)\otimes H_n(X,A;R)\to R,\quad x\otimes y\mapsto \langle x,y\rangle.$$

This is due to the fact that the evaluations combine to a chain map, see 11.7.4.

(17.4.2) Proposition. *For $f\colon (X,A)\to(Y,B)$, $y\in H^n(Y,B;R)$, and $x\in H_n(X,A;R)$ we have*

$$\langle f^*(y),x\rangle = \langle y, f_*(x)\rangle.$$

For $a\in H^{n-1}(A;R)$, $b\in H_n(X,A;R)$ we have

$$\langle \delta a,b\rangle + (-1)^{n-1}\langle a,\partial b\rangle = 0.$$

Proof. We verify the second relation. Let $a=[\varphi]$, $\varphi\in \mathrm{Hom}(S_{n-1}(A),\mathbb{Z})$. Let $\tilde\varphi\colon S_{n-1}(X)\to\mathbb{Z}$ be an extension of φ. Then $\delta(a)$ is represented by the homomorphism $(-1)^n\tilde\varphi\partial\colon S_n(X)\to\mathbb{Z}$ (see (17.4.1)). Let $y=[c]$, $c\in S_n(X)$. Then $\langle\delta x,y\rangle=(-1)^n\tilde\varphi\partial(c)$. From $\partial(c)\in S_{n-1}(A)$ we conclude $\tilde\varphi(\partial(c))=\varphi(\partial c)=\langle x,\partial y\rangle$. □

The canonical generator $e_1\in H_1(I,\partial I)$ is represented by the singular simplex $s\colon\Delta^1\to I$, $(t_0,t_1)\mapsto t_1$. Let $[x]\in H^0(X)$ denote the element represented by the cochain which assumes the value 1 on $x\in X$ and 0 otherwise.

(17.4.3) Proposition. *Let e^1 be the generator which is the image of $[0]\in H^0(0)$ under $H^0(0)\leftarrow H^0(\partial I,1)\xrightarrow{\delta} H^1(I,\partial I)$. Then $\langle e^1,e_1\rangle=1$.* □

The singular cohomology groups $H^n(X,A;G)$ can be computed from the homology groups of (X,A). This is done via the ***universal coefficient formula***. We have developed the relevant algebra in (11.9.2). The application to topology starts with the chain complex $C=S(X,A;R)=S(X,A)\otimes_{\mathbb{Z}} R$ of singular chains with coefficients in a principal ideal domain R. It is a complex of free R-modules. Note that $\mathrm{Hom}_R(S_n(X,A)\otimes_{\mathbb{Z}} R,G)\cong \mathrm{Hom}_{\mathbb{Z}}(S_n(X,A),G)$ where in the second group G is considered as abelian group ($=\mathbb{Z}$-module). Then (11.9.2) yields the ***universal coefficient formula for singular cohomology***.

(17.4.4) Theorem. *For each pair of spaces (X,A) and each R-module G there exists an exact sequence*

$$0\to \mathrm{Ext}(H_{n-1}(X,A;R),G)\to H^n(X,A;G)\to \mathrm{Hom}(H_n(X,A;R),G)\to 0.$$

The sequence is natural in (X,A) and in G and splits. In particular, we have isomorphisms $H^0(X;G)\cong\mathrm{Hom}(H_0(X),G)\cong \mathrm{Map}(\pi_0(X),G)$. □

The statement that the splitting is not natural means that, although the term in the middle is the direct sum of the adjacent terms, this is not a direct sum of functors. There is some additional information that cannot be obtained directly from the homology functors. The topological version of (11.9.6) is:

(17.4.5) Theorem. *Let R be a principal ideal domain. Assume that either $H_*(X, A; R)$ is of finite type or the R-module G is finitely generated. Then there is a functorial exact sequence*

$$0 \to H^n(X, A; R) \otimes G \to H^n(X, A; G) \to \mathrm{Tor}(H^{n+1}(X, A; R), G) \to 0.$$

The sequence splits. □

(17.4.6) Proposition. *Let M be a closed, connected, non-orientable n-manifold. Then $H^n(M; \mathbb{Z}) \cong \mathbb{Z}/2$.*

Proof. Since M is non-orientable, $H_n(M; \mathbb{Z}) = 0$. Theorem (17.4.4) for $R = G = \mathbb{Z}$ then shows $H^n(M; \mathbb{Z}) \cong \mathrm{Ext}(H_{n-1}(M; \mathbb{Z}), \mathbb{Z})$. From (16.3.4) we know that $H_{n-1}(M; \mathbb{Z}) \cong \mathbb{Z}/2 \oplus F$ with a finitely generated free abelian group F. □

Problems

1. Let $0 \to G_1 \to G_2 \to G_3 \to 0$ be an exact sequence of abelian groups. It induces a short exact sequence of cochain complexes

$$0 \to \mathrm{Hom}(S(X), G_1) \to \mathrm{Hom}(S(X), G_2) \to \mathrm{Hom}(S(X), G_3) \to 0$$

and an associated long exact sequence of cohomology groups. The coboundary operator

$$\beta\colon H^n(X; G_3) \to H^{n+1}(X; G_1)$$

is a natural transformation of functors $H^n(-; G_3) \to H^{n+1}(-; G_1)$ and called a ***Bockstein operator***. A typical and interesting case arises from the exact sequence $0 \to \mathbb{Z}/p \to \mathbb{Z}/p^2 \to \mathbb{Z}/p \to 0$.

2. If the functor $\mathrm{Hom}(-, G)$ preserves exact sequences, then no Ext-term appears in the universal coefficient formula. Examples are

$$H^n(X, A; \mathbb{Q}) \cong \mathrm{Hom}_{\mathbb{Z}}(H_n(X, A), \mathbb{Q}) \cong \mathrm{Hom}_{\mathbb{Q}}(H_n(X, A; \mathbb{Q}), \mathbb{Q})$$

and $H^n(X; \mathbb{Z}/p) \cong \mathrm{Hom}_{\mathbb{Z}/p}(H_n(X; \mathbb{Z}/p), \mathbb{Z}/p)$ for the prime field $\mathbb{Z}/p$ as coefficient ring.

17.5 Eilenberg–Mac Lane Spaces and Cohomology

The representability theorem (8.6.10) of Brown can be used to find a natural isomorphism $\lambda\colon [-, K(A, n)] \to H^n(-; A)$ on the homotopy category of CW-complexes. In this section we construct this isomorphism and give some applications. A natural transformation λ is determined by its value on the identity of $K(A, n)$, and this value can be prescribed arbitrarily. Thus we have to find a suitable element $\iota_n \in H^n(K(A, n); A)$.

Let us begin with the special case $n = 0$. As a model for $K(A, 0)$ we take the abelian group A with discrete topology. Since A is discrete, a map $X \to A$

is continuous if and only if it is locally constant. Moreover, all homotopies are constant. Therefore $[X, K(A,0)] = [X, A]$ is the group of locally constant maps. A locally constant map is constant on a path component. Therefore the surjection $q\colon X \to \pi_0(X)$ induces an injective homomorphism

$$\lambda^0(X)\colon [X, A] \to \mathrm{Map}(\pi_0(X), A) \cong \mathrm{Hom}(H_0(X), A) \cong H^0(X; A).$$

The last isomorphism is the one that appears in the universal coefficient formula. If X is locally path connected, then the homomorphism induced by q is an isomorphism. Hence we have obtained a natural isomorphism on the category of locally path connected spaces, in particular on the category of CW-complexes. If X is connected but not path connected, then $\lambda^0(X)$ is not an isomorphism.

Let now $n \geq 1$ and write $K = K_n = K(A, n)$. Consider the composition

$$\iota_n \in H^n(K; A) \overset{(1)}{\cong} \mathrm{Hom}(H_n(K; \mathbb{Z}), A) \overset{(2)}{\cong} \mathrm{Hom}(\pi_n(K), A) \overset{(3)}{\cong} \mathrm{Hom}(A, A) \ni \mathrm{id}.$$

The isomorphism (1) is the universal coefficient isomorphism. The isomorphism (2) is induced by the Hurewicz isomorphism $h\colon \pi_n(K, *) \to H_n(K; \mathbb{Z})$, see (20.1.1). It sends $[f] \in [S(n), K]^0 = \pi_n(K)$ to $f_*(z_n)$ where $z_n \in H_n(S(n); \mathbb{Z})$ is a suitable generator. The isomorphism (3) is induced by a fixed polarization $\rho\colon A \cong \pi_n(K)$. We define ι_n as the element which corresponds to the identity of A. Let λ^n be the natural transformation which is determined by the condition $\lambda^n[\mathrm{id}] = \iota_n$. Note that category theory does not tell us yet that the $\lambda^n(X)\colon [X, K] \to H^n(X; A)$ are homomorphisms of abelian groups.

Let us compare λ^{n-1} and λ^n. It suffices to consider connected CW-complexes and pointed homotopy classes. The diagram with the structure map $e(n)\colon \Sigma K_{n-1} \to K_n$ of the spectrum

$$\begin{array}{ccccc} [X, K_{n-1}]^0 & \xrightarrow{\Sigma} & [\Sigma X, \Sigma K_{n-1}]^0 & \xrightarrow{e(n)_*} & [\Sigma X, K_n]^0 \\ \downarrow{\scriptstyle \lambda^{n-1}} & & & & \downarrow{\scriptstyle \lambda^n} \\ \tilde{H}^{n-1}(X; A) & & \xrightarrow{\sigma} & & \tilde{H}^n(\Sigma X; A) \end{array}$$

is commutative up to sign, provided $e(n)^*(\iota_n) = \lambda^n(e(n)) = \pm\sigma(\iota_{n-1})$. The morphism $\lambda^n(\Sigma X)$ is a homomorphism, since the group structures are also induced by the cogroup structure of ΣX. In order to check the commutativity one has to arrange and prove several things: (1) The Hurewicz homomorphisms commute with suspensions. (2) The structure map $e(n)$ and the polarizations satisfy $e(n)_* \circ \Sigma \circ \rho_{n-1} = \rho_n\colon A \to \pi_n(K_n)$. (3) The homomorphisms from the universal coefficient formula commute up to sign with suspension. Since the sign is not important for the moment, we do not go into details.

(17.5.1) Theorem. *The transformation λ^n is an isomorphism on the category of pointed CW-complexes.*

Proof. We work with connected pointed CW-complexes. Both functors have value 0 for $S^m, m \neq n$. The reader is asked to trace through the definitions and verify that $\lambda^n(S^n)$ is an isomorphism. By additivity, $\lambda^n(X)$ is an isomorphism for X a wedge of spheres. Now one uses the cofibre sequence

$$\bigvee S^{k-1} \to X^{k-1} \to X^k \to X^k/X^{k-1} \to \Sigma X^{k-1}$$

and applies the functors $[-, K_n]^0$ and $H^n(-; A)$. We use induction on k. Then λ^n is an isomorphism for $\bigvee S^{k-1}$, X^{k-1}, and X^k/X^{k-1}. The Five Lemma implies that $\lambda^n(X^k)$ and hence also $\lambda^n(\Sigma X^{k-1})$ are surjective. By another application of the Five Lemma we see that $\lambda^n(X^k)$ is also injective. This settles the case of finite-dimensional CW-complexex. The general case follows from the fact that both functors yield an isomorphism when applied to $X^{n+1} \subset X$. □

A CW-complex $K(\mathbb{Z}, n)$ can be obtained by attaching cells of dimension $\geq n+2$ to S^n. The cellular approximation theorem (8.5.4) tells us that the inclusion $i^n \colon S^n \subset K(\mathbb{Z}, n)$ induces for each CW-complex X of dimension at most n a bijection

$$i^n_* \colon [X, S^n] \xrightarrow{\cong} [X, K(\mathbb{Z}, n)].$$

We combine this with the isomorphism λ and obtain a bijection

$$[X, S^n] \xrightarrow{\cong} H^n(X; \mathbb{Z}).$$

It sends a class $[f] \in [X, S^n]$ to the image of $1 \in H^n(S^n; \mathbb{Z})$ under $f^* \colon H^n(S^n) \to H^n(X)$. Here we use the isomorphism $A \cong H^n(S^n; A)$ which is (for an arbitrary abelian group A) defined as the composition

$$H^n(S^n; A) \cong \mathrm{Hom}(H_n(S^n), A) \cong \mathrm{Hom}(\pi_n(S^n), A) \cong \mathrm{Hom}(\mathbb{Z}, A) \cong A.$$

Again we have used universal coefficients and the Hurewicz isomorphism.

(17.5.2) Proposition (Hopf). *Let $M = X$ be a closed connected n-manifold which has an n-dimensional CW-decomposition. Suppose M is oriented by a fundamental class z_M. Then we have an isomorphism*

$$H^n(M; \mathbb{Z}) \cong \mathrm{Hom}(H_n(M), \mathbb{Z}) \cong \mathbb{Z}$$

where the second isomorphism sends α to $\alpha(z_M)$. The isomorphism $[M, S^n] \cong H^n(M; \mathbb{Z}) \cong \mathbb{Z}$ maps the class $[f]$ to the degree $d(f)$ of f. □

(17.5.3) Proposition (Hopf). *Let M be a closed, connected and non-orientable n-manifold with a CW-decomposition. Then $H^n(M; \mathbb{Z}) \cong \mathbb{Z}/2$; hence we have a bijection $[M, S^n] \cong \mathbb{Z}/2$. It sends $[f]$ to the degree $d_2(f)$ modulo 2 of f.*

Proof. The universal coefficient theorem and $H_n(M;\mathbb{Z}) = 0$ show

$$\operatorname{Ext}(H_{n-1}(M), \mathbb{Z}/2) \cong H^n(M;\mathbb{Z}/2).$$

We use again that $H_{n-1}(M;\mathbb{Z}) \cong F \oplus \mathbb{Z}/2$ with a finitely generated free abelian group F. This proves $H^n(M;\mathbb{Z}/2) \cong \mathbb{Z}/2$. Naturality of the universal coefficient sequences is used to show that the canonical map $H^n(M;\mathbb{Z}) \to H^n(M;\mathbb{Z}/2)$ is an isomorphism. The commutative diagram

$$\begin{array}{ccc} H^n(S^n;\mathbb{Z}/2) & \xrightarrow{f^*} & H^n(M;\mathbb{Z}/2) \\ \downarrow & & \downarrow \\ \operatorname{Hom}(H_n(S^n;\mathbb{Z}/2),\mathbb{Z}/2) & \longrightarrow & \operatorname{Hom}(H_n(M;\mathbb{Z}/2),\mathbb{Z}/2) \end{array}$$

is used to show the assertion about the degree. □

Problems

1. λ is not always an isomorphism: $A = \mathbb{Z}$, $n = 1$ and the pseudo-circle.

17.6 The Cup Product in Singular Cohomology

Let R be a commutative ring. The cup product in singular cohomology with coefficients in R arises from a cup product on the cochain level

$$S^k(X;R) \otimes S^l(X;R) \to S^{k+l}(X;R), \quad \varphi \otimes \psi \mapsto \varphi \cup \psi.$$

It is defined by

$$(1) \qquad (\varphi \cup \psi)(\sigma) = (-1)^{|\varphi||\psi|}\varphi(\sigma|[e_0,\dots,e_k])\psi(\sigma|[e_k,\dots,e_{k+l}]).$$

Here $\sigma\colon \Delta^{k+l} = [e_0,\dots,e_{k+l}] \to X$ is a singular $(k+l)$-simplex. Let $[v_0,\dots,v_n]$ be an affine n-simplex and $\tau\colon [v_0,\dots,v_n] \to X$ a continuous map. We denote by $\tau|[v_0,\dots,v_n]$ the singular simplex obtained from the composition of τ with the map $\Delta^n \to [v_0,\dots,v_n]$, $e_j \mapsto v_j$. This explains the notation in (1).

(17.6.1) Proposition. *The cup product is a chain map*

$$S^\bullet(X;R) \otimes S^\bullet(X;R) \to S^\bullet(X;R),$$

i.e., the following relation holds:

$$\delta(\varphi \cup \psi) = \delta\varphi \cup \psi + (-1)^{|\varphi|}\varphi \cup \delta\psi.$$

Proof. From the definition we compute $(\delta\varphi \cup \psi)(\sigma)$ to be

$$(-1)^{(|\varphi|+1)|\psi|}\sum_{i=0}^{k}(-1)^{i+|\varphi|+1}\varphi(\sigma|[e_0,\dots,\widehat{e_i},\dots,e_{k+1}])\psi(\sigma|[e_{k+1},\dots,e_{k+l+1}])$$

and $(-1)^{|\varphi|}(\varphi \cup \delta\psi)(\sigma)$ to be

$$(-1)^{|\varphi|(|\psi|+1)}\sum_{i=k}^{k+l+1}(-1)^{i+|\psi|+1}\varphi(\sigma|[e_0,\dots,e_k])\psi(\sigma|[e_k,\dots,\widehat{e_i},\dots,e_{k+l+1}]).$$

If we add the two sums, the last term of the first sum and the first term of the second sum cancel, the remaining terms yield $(-1)^{|\varphi|+|\psi|+1}(\varphi \cup \psi)(\partial\sigma)$, and this equals $(\delta(\varphi \cup \psi))(\sigma)$. □

We extend the cup product to the relative case. Suppose $\varphi \in S^k(X, A; R)$ and $\varphi \in S^l(X, B; R)$ are given. This means: φ vanishes on singular simplices $\Delta^k \to A$ and ψ vanishes on simplices $\Delta^l \to B$. The relation (17.6.1) then shows that $\varphi \cup \psi$ vanishes on the submodule $S_{k+l}(A + B; R) \subset S_{k+l}(X; R)$ generated by simplices in A and B. The pair (A, B) is excisive for singular homology if $S_\bullet(A+B) = S_\bullet(A)+S_\bullet(B) \subset S_\bullet(A \cup B)$ is a chain equivalence. The dual maps $S^\bullet(A \cup B; R) \to S^\bullet(A + B; R)$ and $S^\bullet(X, A \cup B; R) \to S^\bullet(X; A + B; R)$ are then chain equivalences. The last module consists of the cochains which vanish on $S_\bullet(A + B)$; let $H^*(X; A + B; R)$ denote the corresponding cohomology group. Thus we obtain a cup product (again a chain map)

$$S^\bullet(X, A; R) \otimes S^\bullet(X, B; R) \to S^\bullet(X; A + B; R).$$

We pass to cohomology, obtain $H^*(X, A; R) \otimes H^*(X, B; R) \to H^*(X; A+B; R)$ and in the case of an excisive pair a cup product

$$H^*(X, A; R) \otimes H^*(X, B; R) \to H^*(X, A \cup B; R).$$

We now prove that the cup product satisfies the axioms of Section 2. From the definition (1) we see that naturality and associativity hold on the cochain level. The unit element $1_X \in H^0(X; R)$ is represented by the cochain which assumes the value 1 on each 0-simplex. Hence it acts as unit element on the cochain level. In the relative case the associativity holds for the representing cocycles in the group $H^*(X; A + B + C; R)$. In order that the products are defined one needs that the pairs (A, B), (B, C), $(A \cup B, C)$ and $(A, B \cup C)$ are excisive.

Commutativity does not hold on the cochain level. We use: The homomorphisms

$$\rho\colon S_n(X) \mapsto S_n(X), \quad \sigma \mapsto \varepsilon_n\overline{\sigma}_n$$

with $\varepsilon_n = (-1)^{(n+1)n/2}$ and $\overline{\sigma} = \sigma|[e_n, e_{n-1}, \dots, e_0]$ form a natural chain map which is naturally chain homotopic to the identity (see (9.3.5) and Problem 1 in that section). The cochain map $\rho^{\#}$ induced by ρ satisfies

$$\rho^{\#}\varphi \cup \rho^{\#}\psi = (-1)^{|\varphi||\psi|}\rho^{\#}(\varphi \cup \psi).$$

Since $\rho^{\#}$ induces the identity on cohomology, the commutativity relation follows.

The stability relation (2) is a consequence of the commutativity of the next diagram (which exists without excisiveness). Coefficients are in R.

$$\begin{array}{ccccc} H^i(A) \otimes H^j(X,B) & \longrightarrow & H^i(A) \otimes H^j(A, A, A \cap B) & \xrightarrow{\cup} & H^{i+j}(A, A \cap B) \\ \downarrow{\scriptstyle \delta \otimes 1} & & & & \downarrow{\scriptstyle \delta'} \\ H^{i+1}(X,A) \otimes H^j(X,B) & & \xrightarrow{\qquad\qquad \cup \qquad\qquad} & & H^{i+j+1}(X; A+B) \end{array}$$

where δ' is the composition of $H^{i+j}(A, A \cap B) \xleftarrow{\cong} H^{i+j}(A + B; B)$, which is induced by the algebraic isomorphism

$$S_{\bullet}(A)/S_{\bullet}(A \cap B) \cong S_{\bullet}(A+B)/S_{\bullet}(B),$$

and $\delta \colon H^{i+j}(A + B; B) \to H^{i+j+1}(X; A + B)$. In order to verify the commutativity, one has to recall the construction of δ, see (17.4.1). Suppose $[\varphi] \in H^i(A)$ and $[\psi] \in H^j(X, A)$ are given. Let $\tilde{\varphi} \in S^i(X)$ be an extension of φ; then $\delta\tilde{\varphi}$ vanishes on $S_i(A)$, and the cochain $\delta\tilde{\varphi} \in S^{i+1}(X, A)$ represents $\delta[\varphi]$. The image of $[\varphi] \otimes [\psi]$ along the down-right path is represented by $\delta\tilde{\varphi} \cup \psi$, and one verifies that the image along the right-down path is represented by $\delta(\tilde{\varphi} \cup \psi)$. Since ψ is a cocycle, the representing elements coincide. A similar verification can be carried out for stability (3).

One can deal with products from the view-point of Eilenberg–Zilber transformations. We have the tautological chain map

$$S^{\bullet}(X; R) \otimes S^{\bullet}(Y; R) \to \operatorname{Hom}(S_{\bullet} \otimes S_{\bullet}, R \otimes R).$$

We compose it with the ring multiplication $R \otimes R \to R$ and an Eilenberg–Zilber transformation $S_{\bullet}(X \times Y) \to S_{\bullet}(X) \otimes S_{\bullet}(Y)$ and obtain a chain map

$$S^{\bullet}(X; R) \otimes S^{\bullet}(Y; R) \to S^{\bullet}(X \times Y; R), \quad f \otimes g \mapsto f \times g,$$

a $\times$-product on the cochain level. If the pair $(A \times Y, X \times B)$ is excisive, we obtain a $\times$-product

$$S^{\bullet}(X, A; R) \otimes S^{\bullet}(Y, B; R) \to S^{\bullet}((X, A) \times (Y, B); R).$$

Our previous explicit definition of the cup product arises in this manner from the Alexander–Whitney equivalence and the related approximation of the diagonal. We can now apply the algebraic Künneth theorem for cohomology to the singular cochain complexes and obtain:

(17.6.2) Theorem. *Let R be a principal ideal domain and $H_*(Y,B;R)$ of finite type. Assume that the pair $(A\times Y, X\times B)$ is excisive. Then there exists a natural short exact sequence*

$$0 \to \bigoplus\nolimits_{i+j=n} H^i(X,A;R)\otimes H^j(Y,B;R) \to H^n((X,A)\times(Y,B);R)$$
$$\to \bigoplus\nolimits_{i+j=n+1} H^i(X,A;R) * H^j(Y,B;R)$$

and this sequence splits. □

Problems

1. We have defined the cup product for simplicity by explicit formulas. One can use instead Eilenberg–Zilber morphisms. Let us consider the absolute case. Consider the composition

$$S^q(X;R)\otimes S^q(Y;R) = \mathrm{Hom}(S_pX,R)\otimes \mathrm{Hom}(S_q(Y),R)$$
$$\to \mathrm{Hom}(S_p(X)\otimes S_q(Y),R) \to \mathrm{Hom}(S_{p+q}(X\times Y),R)$$

where the first morphism is the tautological map and the second induced by an Eilenberg–Zilber morphism $S_\bullet(X\times Y)\to S_\bullet(X)\otimes S_\bullet(Y)$. Then this composition induces the $\times$-product in cohomology. The cup product in the case $X=Y$ is obtained by composition with $S_\bullet(X)\to S_\bullet(X\times X)$ induced by the diagonal. Instead we can go directly from $\mathrm{Hom}(S_p(X)\otimes S_q(X),R)$ to $\mathrm{Hom}(S_{p+q}(X),R)$ by an approximation of the diagonal. Our previous definition used the Alexander–Whitney diagonal.

17.7 Fibration over Spheres

Let $p\colon X\to S^n$ be a fibration ($n\ge 2$). We write as usual $S^n = D^n_+\cup D^n_-$ and $S^{n-1} = D^n_+\cap D^n_-$. Let $b_0\in S^{n-1}$ be a base point. We set $X_\pm = p^{-1}(D^n_\pm)$, $X_0 = p^{-1}(S^{n-1})$, and $F = p^{-1}(b_0)$. From the homotopy theorem of fibrations we obtain the following result.

(17.7.1) Proposition. *There exist h-equivalences $\varphi_\pm\colon D^n_\pm\times X\to X_\pm$ over $D^n_\pm$ such that $\varphi_\pm(b_0,y) = y$ for $y\in F$. The h-inverses $\psi_\pm$ also satisfy $\psi_\pm(y) = (b_0,y)$ and the fibrewise homotopies of $\psi_\pm\varphi_\pm$ and $\varphi_\pm\psi_\pm$ to the identity are constant on F. Since $X_0\subset X_\pm\subset X$ are closed cofibrations, we have a Mayer–Vietoris sequence for (X_+,X_-).* □

We use the data of (17.7.1) in order to rewrite the MV-sequence. We work with a multiplicative cohomology theory. The embedding $j\colon F\to D^n_\pm\times F$, $y\mapsto(b_0,y)$ is an h-equivalence. Therefore we have isomorphisms

$$i^*_\pm\colon h^k(X_\pm)\xrightarrow[\cong]{\varphi^*_\pm} h^k(D^n_\pm\times F)\xrightarrow[\cong]{j^*} h^k(F).$$

The restriction of φ_+ gives us another isomorphism

$$\varphi_+^* \colon h^k(X_0) \xrightarrow{\cong} h^k(S^{n-1} \times F).$$

We insert these isomorphisms into the MV-sequence of (X_+, X_-)

$$\begin{array}{ccccc} h^k(X) & \longrightarrow & h^k(X_+) \oplus h^k(X_-) & \longrightarrow & h^k(X_0) \\ \downarrow = & & \downarrow i_+^* \oplus i_-^* & & \downarrow \varphi_+^* \\ h^k(X) & \xrightarrow{(1)} & h^k(F) \oplus h^k(F) & \xrightarrow{(2)} & h^k(S^{n-1} \times F). \end{array}$$

The two components of (1) equal i^* where $i \colon F \subset X$. The first component of (2) is induced by the projection $\mathrm{pr} \colon S^{n-1} \times F \to F$. We write $\psi_- \circ \varphi_+$ in the form $(s, y) \mapsto (s, \alpha(s, y))$. Then the second component of (2) is $-\alpha^*$. Both maps yield the identity when composed with $j \colon F \to S^{n-1} \times F$, $y \mapsto (b_0, y)$. The product structure provides us with an isomorphism

$$h^k(F) \oplus h^{k-n+1}(F) \to h^k(S^{n-1} \times F), \quad (a, b) \mapsto 1 \times a + s \times b.$$

We also use this isomorphism to change the MV-sequence. We set

$$\alpha^*(x) = 1 \times x - s \times \Theta(x), \quad \Theta(x) \in h^{k-n+1}(F).$$

The relation $j^*(1 \times a - s \times b) = a$ shows that $\alpha^*(x)$ has the displayed form.

(17.7.2) Theorem (Wang sequence). *There exists an exact sequence*

$$\cdots \to h^k(X) \xrightarrow{i^*} h^k(F) \xrightarrow{\Theta} h^{k-n+1}(F) \to h^{k-n+1}(X) \to \cdots.$$

The map Θ is a derivation, i.e., $\Theta(x \cup y) = \Theta(x) \cup y + (-1)^{|x|(n-1)} x \cup \Theta y$.

Proof. We start with the modified MV-sequence

$$\cdots \to h^k(X) \xrightarrow{(1)} h^k(F) \oplus h^k(F) \xrightarrow{(2)} h^k(F) \oplus h^{k-n+1}(F) \to \cdots.$$

The morphism (1) is as before, and (2) has the form $(a, b) \mapsto (a - b, \Theta(b))$. Then we form the quotients with respect to the left $h^k(F)$ summands in order to obtain the stated exact sequence.

Since α^* is a homomorphism and $s^2 = 0$, we obtain

$$\begin{aligned} \alpha^*(xy) &= 1 \times xy + s \times \Theta(xy) \\ \alpha^*(x)\alpha^*(y) &= (1 \times x + s \times \Theta(x))(1 \times y + s \times \Theta(y)) \\ &= 1 \times xy + s \times \Theta(x) \cdot y + (-1)^{|x||s|} s \times x \cdot \Theta(y). \end{aligned}$$

This proves the derivation property of Θ. □

We can, of course, also consider the MV-sequence in homology. It assumes after an analogous rewriting the form

$$\cdots \to H_q(F) \xrightarrow{i_*} H_q(X) \to H_{q-n}(F) \xrightarrow{\Theta_*} H_{q-1}(F) \to \cdots.$$

Problems

1. As an example for the use of the Wang sequence compute the integral cohomology ring $H^*(\Omega S^{n+1})$ of the loop space of S^{n+1}. Use the path fibration $\Omega S^{n+1} \xrightarrow{i} P \xrightarrow{p} S^{n+1}$ with contractible P.

If $n = 0$, then ΩS^1 is h-equivalent to the discrete space $\mathbb{Z}$. So let $n \geq 1$. Since P is contractible, the Wang sequence yields $H_q(\Omega S^{n+1}) \cong H_{q-n}(\Omega S^{n+1})$ and therefore $H_k(\Omega S^{n+1}) \cong \mathbb{Z}$ for $k \equiv 0 \bmod (n)$ and $\cong 0$ otherwise. Similarly for cohomology $H^k(\Omega S^{n+1}) \cong \mathbb{Z}$ for $k \equiv 0 \bmod (n)$ and zero otherwise. Using the isomorphism Θ we define inductively elements $z_0 = 1$ and $\Theta z_k = z_{k-1}$ for $k \geq 1$.

Let n be even. Then $k!z_k = z_1^k$ for $k \geq 1$. For the proof use induction over k and the derivation property of Θ.

The relation above yields the multiplication rule

$$z_k z_l = \binom{k+l}{k} z_{k+l}.$$

A multiplicative structure of this type is called a ***polynomial ring with divided powers***. With coefficient ring $\mathbb{Q}$ one obtains a polynomial ring $H^*(\Omega S^{n+1};\mathbb{Q}) \cong \mathbb{Q}[z_1]$.

Let n be odd. Then $z_1 z_{2k} = z_{2k+1}$, $z_1 z_{2k+1} = 0$, and $z_2^k = k!z_{2k}$.

Again use induction and the derivation property. Since $z_1^2 = -z_1^2$, one has $z_1^2 = 0$. Then $\Theta(z_1 z_{2k}) = \Theta(z_1) z_{2k} - z_1\Theta(z_{2k}) = z_{2k} - z_1 z_{2k-1} = z_{2k} = \Theta(z_{2k+1})$, hence $z_1 z_{2k} = z_{2k+1}$, since Θ is an isomorphism. Next compute $z_1 z_{2k+1} = z_1(z_1 z_{2k}) = z_1^2 z_{2k} = 0$. For the last formula use that $\Theta \circ \Theta$ is a derivation of even degree which maps z_{2k} to z_{2k-2}. The induction runs then as for even n. The elements z_{2k} generate a polynomial algebra with divided powers and z_1 generates an exterior algebra.

17.8 The Theorem of Leray and Hirsch

The theorem of Leray and Hirsch determines the additive structure of the cohomology of the total space of a fibration as the tensor product of the cohomology of the base and the fibre. We work with singular cohomology with coefficients in the ring R. A relative fibration

$$(F, F') \to (E, E') \to B$$

consists of a fibration $p\colon E \to B$ such that the restriction $p'\colon E' \to B$ to the subspace E' of E is also a fibration. The fibres of p and p' over a base point $* \in B$ are F and F'. The case $E' = \emptyset$ and hence $F' = \emptyset$ is allowed. We assume that B is path connected.

(17.8.1) Theorem (Leray–Hirsch). *Let $(F, F') \xrightarrow{i} (E, E') \xrightarrow{p} B$ be a relative fibration. Assume that $H^n(F, F')$ is for each n a finitely generated free R-module.*

Let $c_j \in H^*(E, E')$ be a family of elements such that the restrictions $i^*(c_j)$ form an R-basis of $H^*(F, F')$. Then

$$L\colon H^*(B) \otimes H^*(F, F') \to H^*(E, E'), \quad b \otimes i^*(c_j) \mapsto p^*(b) \cup c_j$$

is an isomorphism of R-modules. Thus $H^*(E)$ is a free graded $H^*(B)$-module with basis $\{c_j\}$.

We explain the statement of the theorem. The source of L is a direct sum of modules $H^k(B) \otimes H^l(F, F')$. The elements i^*c_μ which are contained in $H^l(F, F')$ are a finite R-basis of the R-module $H^l(F, F')$. A basic property of the tensor product says that each element has a unique expression of the form

$$\textstyle\sum_\mu b_\mu \otimes i^*(c_\mu), \quad b_\mu \in H^k(B).$$

By the conventions about tensor products of graded modules, L is a map of degree zero between graded modules.

Proof. Let $A \subset B$ be a subspace. We have the restricted fibrations $(F, F') \to (E|A, E'|A) \to A$ with $E|A = p^{-1}(A)$ and $E'|A = E' \cap E|A$. The elements c_j yield by restriction elements $c_j|A \in H^*(E|A, E'|A)$ which again restrict to a basis of $H^*(F, F')$.

We first prove the theorem for CW-complexes B by induction over the skeleta B^n. If $B^0 = \{*\}$, then L has the form $H^0(B^0) \otimes H^k(F, F') \to H^k(F, F')$ and it is an isomorphism by the unit element property of the cup product.

Suppose the theorem holds for the $(n-1)$-skeleton B^{n-1}. We write $B^n = U \cup V$, where U is obtained from B^n by deleting a point in each open n-cell and V is the union of the open n-cells. We use the MV-sequence of U, V and $E|U, E|V$ and obtain a commutative diagram

$$\begin{array}{ccc}
H^*(U \cup V) \otimes M^* & \xrightarrow{L_{U\cup V}} & H^*(E|U \cup V, E'|U \cup V) \\
\downarrow & & \downarrow \\
H^*(U) \otimes M^* \oplus H^*(V) \otimes M^* & \xrightarrow{L_U \oplus L_V} & H^*(E|U, E'|U) \oplus H^*(E|V, E'|V) \\
\downarrow & & \downarrow \\
H^*(U \cap V) \otimes M^* & \xrightarrow{L_{U\cap V}} & H^*(E|U \cap V, E'|U \cap V).
\end{array}$$

The left column is the tensor product of the MV-sequence for U, V with the graded module $M^* = H^*(F, F')$. It is exact, since the tensor product with a free module preserves exactness. We show that L_U, L_V and $L_{U\cap V}$ are isomorphisms. The Five Lemma then shows that $L_{U\cup V}$ is an isomorphism. This finishes the induction step.

Case U. We have the commutative diagram

$$\begin{array}{ccc} H^*(U) \otimes H^*(F,F') & \xrightarrow{L_U} & H^*(E|U, E'|U) \\ \downarrow & & \downarrow \\ H^*(B^{n-1}) \otimes H^*(F,F') & \longrightarrow & H^*(E|B^{n-1}, E'|B^{n-1}). \end{array}$$

Since $B^{n-1} \subset U$, $E|B^{n-1} \subset E|U$, and $E'|B^{n-1} \subset E'|U$ are deformation retracts, the vertical maps are isomorphisms. We use the induction hypothesis and see that L_U is an isomorphism.

Case V. The set V is the disjoint union of the open n-cells $V = \coprod_j e_j^n$. We obtain a commutative diagram

$$\begin{array}{ccc} H^*(V) \otimes H^*(F,F') & \xrightarrow{L_V} & H^*(E|V, E'|V) \\ {\scriptstyle (1)}\downarrow\cong & & {\scriptstyle (2)}\downarrow\cong \\ (\prod H^*(e_j^n)) \otimes H^*(F,F') & & \prod H^*(E|e_j^n, E'|e_j^n) \\ {\scriptstyle (3)}\downarrow & & \downarrow = \\ \prod \big(H^*(e_j^n) \otimes H^*(F,F')\big) & \xrightarrow{\prod L_{e_j^n}} & \prod H^*(E|e_j^n, E'|e_j^n). \end{array}$$

(1) and (2) are isomorphisms by the additivity of cohomology. The map (3) is the direct sum of homomorphisms of the type $(\prod M_j) \otimes N \to \prod(M_j \otimes N)$ with a finitely generated free module N and other modules M_j. In a situation like this the tensor product commutes with the product. Hence (3) is an isomorphism. The homomorphisms $L(e_j^n)$ are isomorphisms, since e_j^n is pointed contractible.

Case $U \cap V$. We combine the arguments of the two previous cases. By additivity and finite generation we reduce to the case of $U \cap e_j^n$, a cell with a point deleted. This space has the $(n-1)$-sphere as a deformation retract. By induction, the theorem holds for an $(n-1)$-sphere.

From the finite skeleta we now pass to arbitrary CW-complexes via the lim–lim^1-sequence (17.1.6). For general base spaces B we pull back the fibration along a CW-approximation. □

(17.8.2) Example. Consider the product fibration $p = \mathrm{pr}_B \colon B \times (F,F') \to B$. Let $H^*(F,F')$ be a free R-module with homogeneous basis $(d_j \mid j \in J)$, finite in each dimension. Let $c_j = \mathrm{pr}_F^* d_j$. Then $i^* c_j = d_j$. Therefore (17.8.1) says in this case that

$$H^*(B) \otimes H^*(F,F') \to H^*(B \times (F,F')), \quad x \otimes y \mapsto x \times y$$

is an isomorphism (of graded algebras). This is a special case of the Künneth formula. ◇

(17.8.3) Remark. The methods of proof for (17.8.1) (induction, Mayer–Vietoris sequences) gives also the following result. Suppose $H^k(F, F') = 0$ for $k < n$. Then $H^k(E, E') = 0$ for $k < n$. ◇

Let now h^* be an arbitrary additive and multiplicative cohomology theory and $(F, F') \to (E, E') \to B$ a relative fibration over a CW-complex. We prove a Leray- Hirsch theorem in this more general situation. We assume now that there is given a finite number of elements $t_j \in h^{n(j)}(E, E')$ such that the restrictions $t_j|b \in h^{n(j)}(E_b, E'_b)$ to each fibre over b are a basis of the graded h^*-module $h^*(E_b, E'_b)$. Under these assumptions:

(17.8.4) Theorem (Leray–Hirsch). *$h^*(E, E')$ is a free left $h^*(B)$-module with basis (t_j).*

Proof. Let us denote by $h^*(C)\langle t \rangle$ the free graded $h^*(C)$-module with (formal) basis t_j in degree $n(j)$. We have the $h^*(C)$-linear map of degree zero

$$\varphi(C)\colon h^*(C)\langle t \rangle \to h^*(E|C, E'|C)$$

which sends t_j to $t_j|C$. These maps are natural in the variable $C \subset B$. We view $h^*(-)\langle t \rangle$ as a cohomology theory, a direct sum of the theories $h^*(-)$ with shifted degrees. Thus we have Mayer–Vietoris sequences for this theory. If U and V are open in B, we have a commutative diagram of MV-sequences.

$$\begin{array}{ccc} h^*(U \cup V)\langle t \rangle & \xrightarrow{\varphi(U\cup V)} & h^*(E|U \cup V, E'|U \cup V) \\ \downarrow & & \downarrow \\ h^*(U)\langle t \rangle \oplus h^*(V)\langle t \rangle & \xrightarrow{\varphi(U)\oplus\varphi(V)} & h^*(E|U, E'|U) \oplus h^*(E|V, E'|V) \\ \downarrow & & \downarrow \\ h^*(U \cap V)\langle t \rangle & \xrightarrow{\varphi(U\cap V)} & h^*(E|U \cap V, E'|U \cap V) \end{array}$$

We use this diagram as in the proof of (17.8.1). We need for the inductive proof that $\varphi(e)$ is an isomorphism for an open cell e. This follows from two facts:

(1) $\varphi(P)$ is an isomorphism for a point $P = \{b\} \subset B$, by our assumption about the t_j.

(2) $(E_P, E'_P) \to (E|e, E'|e)$ induces an isomorphism in cohomology, since $E_P \subset E|e$ is a homotopy equivalence by the homotopy theorem for fibrations.

The finiteness of the set $\{t_j\}$ is used for the compatibility of products and finite sums. The passage from the skeleta of B to B uses again (17.1.6). □

There is a similar application of (17.8.4) as we explained in (17.8.2).

17.9 The Thom Isomorphism

We again work with a cohomology theory which is additive and multiplicative. Under the obvious finiteness conditions (e.g., finite CW-complexes) additivity is not needed.

Let $(p, p')\colon (E, E') \to B$ be a relative fibration over a CW-complex. A ***Thom class*** for p is an element $t = t(p) \in h^n(E, E')$ such that the restriction to each fibre $t_b \in h^n(F_b, F'_b)$ is a basis of the h^*-module $h^*(F_b, F'_b)$. We apply the theorem of Leray–Hirsch (17.8.4) and obtain:

(17.9.1) Theorem (Thom isomorphism). *The Thom homomorphism*

$$\Phi\colon h^k(B) \to h^{k+n}(E, E'), \quad b \mapsto p^*(b) \cup t$$

is an isomorphism. □

Let us further assume that p induces an isomorphism $p^*\colon H^*(B) \to H^*(E)$. We use the Thom isomorphism and the isomorphism p^* in order to rewrite the exact sequence of the pair (E, E'); we set $\Delta = \Phi^{-1}\delta$.

$$\begin{array}{ccccccc}
h^{k-1}(E') & \xrightarrow{\delta} & h^k(E, E') & \xrightarrow{j} & h^k(E) & \longrightarrow & h^k(E') \\
& \searrow^{\Delta} & \cong \uparrow \Phi & & \uparrow p^* & \nearrow_{(p')^*} & \\
& & h^{k-n}(B) & \xrightarrow{J} & h^k(B) & &
\end{array}$$

Let $e = e(p) \in H^n(B)$ be the image of t under $h^n(E, E') \xrightarrow{j} h^n(E) \xleftarrow{p^*} h^n(B)$. We call e the ***Euler class*** of p with respect to t. From the definitions we verify that J is the cup product with e, i.e., $J(x) = x \cup e$.

(17.9.2) Theorem (Gysin sequence). *Let $(E, E') \to B$ be a relative fibration as above such that $p^*\colon h^*(B) \cong h^*(E)$ with Thom class t and associated Euler class $e \in h^n(B)$. Then we have an exact Gysin sequence*

$$\cdots \to h^{k-1}(E') \to h^{k-n}(B) \xrightarrow{(1)} h^k(B) \xrightarrow{(2)} h^k(E') \to h^{k-n+1}(B) \to \cdots.$$

(1) *is the cup product $x \mapsto x \cup e$ and* (2) *is induced by p'.* □

We discuss the existence of Thom classes for singular cohomology $H^*(-; R)$. In this case it is not necessary to assume that B is a CW-complex (see (17.8.1)). We have for each path $w\colon I \to B$ from b to c a fibre transport $w^\#$ defined as follows: Let $q\colon (X, X') \to I$ be the pullback of $p\colon (E, E') \to B$ along w. Then we have isomorphisms induced by the inclusions

$$w^\#\colon H^n(F_c, F'_c) \xleftarrow{\cong} H^n(X, X') \xrightarrow{\cong} H^n(F_b, F'_b).$$

The homomorphism $w^\#$ only depends on the class of w in the fundamental groupoid. In this manner we obtain a transport functor "fibre cohomology".

(17.9.3) Proposition. *We assume that $H^*(F_b, F'_b; R)$ is a free R-module with a basis element in $H^n(F_b, F'_b; R)$. A Thom class exists if and only if the transport functor is trivial.*

Proof. Let t be a Thom class and $w: I \to B$ a path from b to c. Then $w^\# \circ i_c^* = i_b^*$, where i_b denotes the inclusion of the fibre over b. Thus $w^\#$ sends the restricted Thom class to the restricted Thom class and is therefore independent of the path. (In general, the transport is trivial on the image of the i_c^*.)

Let now the transport functor be trivial. Then we fix a basis element in a particular fibre $H^n(E_b, E'_b)$ and transport it to any other fibre uniquely (B path connected). A Thom class $t_C \in H^n(E|C, E'|C)$ for $C \subset B$ is called distinguished, if the restriction to each fibre is the specified basis element. This requirement determines t_C. By a MV-argument and (17.8.3) we prove by induction that $H^k(E|B^n, E'|B^n) = 0$ for $k < n$ and that a distinguished Thom class exists. Then we pass to the limit and to general base spaces as in the proof of (17.8.1). □

The preceding considerations can be applied to vector bundles. Let $\xi: E \to B$ be a real n-dimensional vector bundle and E^0 the complement of the zero section. Then for each fibre $H^n(E_b, E_b^0) \cong R$ and ξ is a homotopy equivalence. A Thom class $t(\xi) \in H^n(E, E^0; R)$ is called an R-orientation of ξ. If it exists, we have a Thom isomorphism and a Gysin sequence. We discuss the existence of Thom classes and its relation to the geometric orientations.

(17.9.4) Theorem. *There exists a Thom class of ξ with respect to singular cohomology $H^*(-;\mathbb{Z})$ if and only if the bundle is orientable. The Thom classes with respect to $H^*(-;\mathbb{Z})$ correspond bijectively to orientations.*

Proof. Let us consider bundles over CW-complexes. Let t be a Thom class. Consider a bundle chart $\varphi: U \times \mathbb{R}^n \to \xi^{-1}(U)$ over a path connected open U. The image of $t|U$ in $H^n(U \times (\mathbb{R}^n, \mathbb{R}^n \smallsetminus 0)) \cong H^0(U)$ under φ^* and a canonical suspension isomorphism is an element $\varepsilon(U)$ which restricts to $\varepsilon(u) = \pm 1$ for each point $u \in U$, and $u \mapsto \varepsilon(u)$ is constant, since U is path connected. We can therefore change the bundle chart by an automorphism of $\mathbb{R}^n$ such that $\varepsilon(u) = 1$ for each u. Bundle charts with this property yield an orienting bundle atlas.

Conversely, suppose ξ has an orienting atlas. Let $\varphi: U \times \mathbb{R}^n \to \xi^{-1}(U)$ be a positive chart. From a canonical Thom class for $U \times \mathbb{R}^n$ we obtain via φ a local Thom class t_U for $\xi^{-1}(U)$. Two such local Thom classes restrict to the same Thom class over the intersection of the basic domains, since the atlas is orienting. We can now paste these local classes by the Mayer–Vietoris technique in order to obtain a global Thom class. □

As a canonical generator of $H^n(\mathbb{R}^n, \mathbb{R}^n \smallsetminus 0; \mathbb{Z})$ we take the element $e^{(n)}$ satisfying the Kronecker pairing relation $\langle e^{(n)}, e_n \rangle = 1$ where $e_n \in H_n(\mathbb{R}^n, \mathbb{R}^n \smallsetminus 0; \mathbb{Z})$ is the n-fold product $e_1 \times \cdots \times e_1$ of the canonical generator $e_1 \in H_1(\mathbb{R}^1, \mathbb{R}^1 \smallsetminus 0; \mathbb{Z})$. If a bundle ξ is oriented and $\varphi_b \colon \mathbb{R}^n \to \xi_b$ a positive isomorphism, then we require $\varphi_b^* t(\xi) = e^{(n)}$ for its associated Thom class $t(\xi)$.

Let $t(\eta) \in h^n(E, E^0)$ be a Thom class of $\eta \colon E \to B$ and $e(\eta) \in h^n(B)$ the associated Euler class, defined as the restriction of $t(\eta)$ to the zero section

$$t(\eta) \in h^n(E, E^0) \to h^n(E) \xrightarrow{s^*} h^n(B) \ni e(\eta).$$

Here is a geometric property of the Euler class:

(17.9.5) Proposition. *Suppose η has a section which is nowhere zero. Then the Euler class is zero.*

Proof. Let $s' \colon B \to E^0$ be a map such that $\eta \circ s' = \mathrm{id}$. The section s' is homotopic to the zero section by a linear homotopy in each fibre. Therefore $e(\eta)$ is the image of $t(\eta)$ under a map

$$h^n(E, E^0) \to h^n(E) \to h^n(E^0) \xrightarrow{s'} h^n(B)$$

and therefore zero. □

The Thom classes and the Euler classes have certain naturality properties. Let $f^* \colon \xi \to \eta$ be a bundle map. If $t(\eta)$ is a Thom class, then $f^* t(\eta)$ is a Thom class for ξ and $f^* e(\eta)$ is the corresponding Euler class. If $\xi \colon X \to B$ and $\eta \colon Y \to C$ are bundles with Thom classes, then the $\times$-product $t(\xi) \times t(\eta)$ is a Thom class for $\xi \times \eta$ and $e(\xi) \times e(\eta)$ is the corresponding Euler class.

In general, Thom classes are not unique. Let us consider the case of a trivial bundle $\xi = \mathrm{pr}_2 \colon \mathbb{R}^n \times B \to B$. It has a canonical Thom class $\mathrm{pr}_2^* e^n$. If $t(\xi)$ is an arbitrary Thom class, then it corresponds under the suspension isomorphism $h^0(B) \to h^n(\mathbb{R}^n \times B, \mathbb{R}^n_0 \times B)$ to an element $v(\xi)$ with the property that its restriction to each point $b \in B$ is the element $\pm 1 \in h^0(b)$. Under reasonable conditions, an element with this property (call it a point-wise unit) is a (global) unit in the ring $h^0(B)$. We call a Thom class for a numerable bundle ***strict*** if the restrictions to the sets of a numberable covering correspond under bundle charts and suspension isomorphism to a unit in h^0.

(17.9.6) Proposition. *Let $\mathcal{U}$ be a numerable covering of X. Let $\varepsilon \in h^0(X)$ be an element such that its restriction to each $U \in \mathcal{U}$ is a unit. Then ε is a unit.*

Proof. Let $X = U \cup V$ and assume that (U, V) is excisive. Let $\varepsilon | U = \varepsilon_U$ and $\varepsilon | V = \varepsilon_V$ be a unit. Let η_U, η_V be inverse to $\varepsilon_U, \varepsilon_V$. Then η_U and η_V have the same restriction to $U \cap V$. By the exactness of the MV-sequence there exists

$\eta \in h^0(X)$ with $\eta|U = \eta_U, \eta|V = \eta_V$. Then $x = \varepsilon\eta - 1$ has restriction 0 in U and V. Let $x_U \in h^0(X,U)$ be a pre-image of x and similarly $x_V \in h^0(X,V)$. Then $x_U x_V = 0$ and hence $x^2 = 0$. The relation $1 = \varepsilon\eta(2 - \varepsilon\eta)$ shows that ε is a unit. Restrictions of units are units. By additivity, if X is the disjoint union of $\mathcal{U}$ and $\varepsilon|U$ is a unit for each $U \in \mathcal{U}$, then ε is a unit. We finish the proof as in the proof of (17.9.7). □

The Thom isomorphism is a generalized (twisted) suspension isomorphism. It is given by the product with a Thom class. Let $\xi \colon E \to B$ be an n-dimensional real vector bundle and $t(\xi) \in h^n(E, E^0)$ a Thom class with respect to a given multiplicative cohomology theory. The ***Thom homomorphism*** is the map

$$\Phi(\xi)\colon h^k(B,A) \to h^{k+n}(E(\xi), E^0(\xi) \cup E(\xi_A)), \quad x \mapsto x \cdot t(\xi) = \xi^*(x) \cup t(\xi)$$

where $\xi^*\colon h^k(B,A) \to H^k(E(\xi), E^0(\xi))$ is the homomorphism induced by ξ. The Thom homomorphism defines on $h^*(E,E^0)$ the structure of a left graded $h^*(B)$-module. The Thom homomorphism is natural with respect to bundle maps. Let

$$\begin{array}{ccc} E(\xi) & \xrightarrow{F} & E(\eta) \\ \downarrow{\scriptstyle \xi} & & \downarrow{\scriptstyle \eta} \\ B & \xrightarrow{f} & C \end{array}$$

be a bundle map. Let $t(\eta)$ be a Thom class. We use $t(\xi) = F^* t(\eta)$ as the Thom class for ξ. Then the diagram

$$\begin{array}{ccc} h^k(C,D) & \xrightarrow{\Phi(\xi)} & h^{k+n}(E(\eta), E^0(\eta) \cup E(\eta_D)) \\ \downarrow{\scriptstyle f^*} & & \downarrow{\scriptstyle F^*} \\ h^k(B,A) & \xrightarrow{\Phi(\eta)} & h^{k+n}(E(\eta), E^0(\eta) \cup E(\eta_A)) \end{array}$$

is commutative. We assume that $f\colon (B,A) \to (C,D)$ is a map of pairs.

The Thom homomorphism is also compatible with the boundary operators. Let $t(\xi_A)$ be the restriction of $t(\xi)$. Then the diagram

$$\begin{array}{ccc} h^k(A) & \xrightarrow{\quad\delta\quad} & h^{k+1}(B,A) \\ \downarrow{\scriptstyle \Phi(\xi_A)} & & \\ h^{k+n}(E(\xi_A), E^0(\xi_A)) & & \downarrow{\scriptstyle \Phi(\xi)} \\ \uparrow{\scriptstyle \cong} & & \\ h^{k+n}(E^0(\xi) \cup E(\xi_A), E^0(\xi_A)) & \xrightarrow{\delta} & h^{k+n+1}(E(\xi), E^0(\xi) \cup E(\xi_A)) \end{array}$$

is commutative.

The Thom homomorphisms are also compatible with the morphisms in the MV-sequence. We now consider the Thom homomorphism under a different hypothesis.

(17.9.7) Theorem. *The Thom homomorphism of a numerable bundle with strict Thom class is an isomorphism.*

Proof. Let ξ be a numerable bundle of finite type. By hypothesis, B has a finite numerable covering $U_1, \dots, U_t$ such that the bundle is trivial over each U_j and the Thom class is strict over U_j. We prove the assertion by induction over t. For $t = 1$ it holds by the definition of a strict Thom class. For the induction step consider $C = U_1 \cup \dots \cup U_{t-1}$ and $D = U_t$. By induction, the Thom homomorphism is an isomorphism for C, D, and $C \cap D$. Now we use that the Thom homomorphism is compatible with the MV-sequence associated to C, D. By the Five Lemma we see that $\Phi(\xi)$ is an isomorphism over $C \cup D$.

Suppose the bundle is numerable over a numerable covering $\mathcal{U}$. Assume that for each $U \in \mathcal{U}$ the Thom class $t(\xi_U)$ is strict. In that case $\Phi(\xi_U)$ is an isomorphism. For each $V \subset U$ the Thom class $t(\xi_V)$ is also strict. There exists a numerable covering $(U_n \mid n \in \mathbb{N})$ such that $\xi | U_n$ is numerable of finite type with strict Thom class. Let $(\sigma_n \mid n \in \mathbb{N})$ be a numeration of (u_n). Set $f\colon B \to [0, \infty[$, $f(x) = \sum_n n\sigma_n(x)$. If $x \notin \bigcup_{j=1}^n \operatorname{supp}(\sigma_j)$, then $1 = \sum_{j \ge 1} \sigma_j(x) = \sum_{j>n} \sigma_j(x)$ and therefore

$$f(x) = \textstyle\sum_{j>n} j\sigma_j(x) \ge (n+1) \sum_{j>n} \sigma_j(x) = n + 1.$$

Hence $f^{-1}[0, n]$ is contained in $\bigcup_{j=1}^n \operatorname{supp}(\sigma_j)$. Hence $f^{-1}]r, s[$ is always contained in a finite number of V_j and therefore the bundle over such a set is numerable of finite type. The sets $C_n = f^{-1}]2n-1, 2n+1[$ are open and disjoint. Over C_n the bundle is of finite type. By additivity, we have for $C = \bigcup C_n$ the isomorphism $h^k(E|C, E^0|C) \cong \prod h^k(E|C_n, E^0|C_n)$. The Thom classes over C_n yield a unique Thom class over C. Now we use the same argument for $D_n = f^{-1}]2n, 2n+2[$, $D = \bigcup D_n$ and $C \cap D = \bigcup f^{-1}]n, n+1[$ and then apply the MV-argument to C, D. □

(17.9.8) Example. Let $\gamma_1(n)\colon H(1) \to \mathbb{C}P^n$ be the canonical line bundle introduced in (14.2.6). A complex vector bundle has a canonical orientation and an associated Thom class (17.9.4). Let $c \in H^2(\mathbb{C}P^n)$ be the Euler class of $\gamma_1(n)$. The associated sphere bundle is the Hopf fibration $S^{2n+1} \to \mathbb{C}P^n$. Since $H^k(S^{2n+1}) = 0$ for $0 < k < 2n+1$ the multiplications by the Euler class $c \in H^2(\mathbb{C}P^n)$ are isomorphisms

$$\mathbb{Z} \cong H^0(\mathbb{C}P^n) \cong H^2(\mathbb{C}P^n) \cong \dots \cong H^{2n}(\mathbb{C}P^n)$$

and similarly $H^k(\mathbb{C}P^n) = 0$ for odd k. We obtain the structure of the cohomology ring

$$H^*(\mathbb{C}P^n) \cong \mathbb{Z}[c]/(c^{n+1}).$$

In the infinite case we obtain $H^*(\mathbb{C}P^\infty; R) \cong R[c]$ where R is an arbitrary commutative ring.

A similar argument for the real projective space yields for the cohomology ring $H^*(\mathbb{R}P^n;\mathbb{Z}/2) \cong \mathbb{Z}/2[w]/(w^{n+1})$ with $w \in H^1(\mathbb{R}P^n;\mathbb{Z}/2)$ and for the infinite projective space $H^*(\mathbb{R}P^\infty;\mathbb{Z}/2) \cong \mathbb{Z}/2[w]$. ◇

(17.9.9) Example. The structure of the cohomology ring of $\mathbb{R}P^n$ can be used to give another proof of the Borsuk–Ulam theorem: There does not exist an odd map $F\colon S^n \to S^{n-1}$.

For the proof let $n \geq 3$. (We can assume this after suspension.) Suppose there exists an odd map F. It induces a map $f\colon \mathbb{R}P^n \to \mathbb{R}P^{n-1}$ of the orbit spaces. Let $v\colon I \to S^n$ be a path from x to $-x$. Composed with the orbit map $p_n\colon S^n \to \mathbb{R}P^n$ we obtain a loop $p_n v$ that generates $\pi_1(\mathbb{R}P^n) \cong \mathbb{Z}/2$. The path $u = Fv$ from $F(x)$ to $F(-x) = -F(x)$ yields a loop that generates $\pi_1(\mathbb{R}P^{n-1})$. Hence $f_*\colon \pi_1(\mathbb{R}P^n) \to \pi_1(\mathbb{R}P^{n-1})$ is an isomorphism. This fact implies (universal coefficients) that f^* is an isomorphism in $H^1(-;\mathbb{Z}/2)$. Since $w^n = 0$ in $H^*(\mathbb{R}P^{n-1};\mathbb{Z}/2)$ but $w^n \neq 0$ in $H^*(\mathbb{R}P^n;\mathbb{Z}/2)$, we have arrived at a contradiction. ◇

Problems

1. A point-wise unit is a unit under one of the following conditions: (1) For singular cohomology $H^*(-;R)$. (2) B has a numerable null homotopic covering. (3) B is a CW-complex.
2. Prove the Thom isomorphism for vector bundles over general spaces and for singular cohomology.
3. Let $\xi\colon E(\xi) \to B$ and $\eta\colon E(\eta) \to B$ be vector bundles with Thom classes $t(\xi)$ and $t(\eta)$. Define a relative Thom homomorphism as the composition of $x \mapsto x \times t(\xi)$,

$$h^k(E(\eta), E^0(\eta)\cup E(\eta_A)) \to h^{k+n}(E(\eta)\times E(\xi), (E^0(\eta)\cup E(\eta_A))\times E(\xi)\cup E(\eta)\times E^0(\xi))$$

with the map induced by

$$\begin{aligned}&(E(\eta \oplus \xi), E^0(\eta \oplus \xi) \cup E(\eta_A \oplus \xi_A))\\ &\quad \to (E(\eta) \times E(\xi), (E^0(\eta) \cup E(\eta_A) \times E(\xi) \cup E(\eta) \times E^0(\xi)),\end{aligned}$$

a kind of diagonal, on each fibre given by $(b, v, w) \mapsto ((b, v), (b, w))$. This is the previously defined map in the case that $\dim \eta = 0$. The product $t(\eta) \times t(\xi)$ is a Thom class and also its restriction $t(\eta \oplus \xi)$ to the diagonal. Using this Thom class one has the transitivity of the Thom homomorphism $\Phi(\eta)\Phi(\xi) = \Phi(\eta \oplus \xi)$.
4. Given $i\colon X \to Y, r\colon Y \to X$ such that $ri = \mathrm{id}$ (a retract). Let $\xi\colon E \to X$ be a bundle over X and $r^*\xi = \eta$ the induced bundle. Let $t(\xi)$ be a Thom class and $t(\eta)$ its pullback. If $\Phi(\eta)$ is an isomorphism, then $\Phi(\xi)$ is an isomorphism.
5. Let $C_m \subset S^1$ be the cyclic subgroup of m-th roots of unity. A model for the canonical map $p_m\colon BC_m \to BS^1$ is the sphere bundle of the m-fold tensor product $\eta^m = \eta \otimes \cdots \otimes \eta$ of the canonical (universal) complex line bundle over BS^1.
6. Let R be a commutative ring. Then $H^*(BS^1; R) \cong R[c]$ where c is the Euler class of η.

7. We use coefficients in the ring R. The Gysin sequence of $p_m\colon BC_m \to BS^1$ splits into short exact sequences

$$0 \to H^{2k-1}(BC_m) \to H^{2k-2}(BS^1) \xrightarrow{\cup mc} H^{2k}(BS^1) \xrightarrow{p_m^*} H^{2k}(BC_m) \to 0.$$

This implies $H^{2k}(BC_m) \cong R/mR$, $H^{2k-1}(BC_m) \cong {}_mR$ for $k > 0$, where ${}_mR$ is the m-torsion $\langle x \in R \mid mx = 0 \rangle$ of R. In even dimensions we have the multiplicative isomorphism $H^{2*}(BS^1)/(mc) \cong H^{2*}(BC_m)$ induced by p_m.

8. The sphere bundle of the canonical bundle $EC_m \times_{C_m} \mathbb{C} \to BC_m$ has a contractible total space. Therefore the Gysin sequence of this bundle shows that the cup product $\cup t\colon H^j(BC_m; R) \to H^{j+2}(BC_m; R)$ is an isomorphism for $j > 0$; here $t = p_m^* c$.

9. The cup product

$$H^1(BC_m; R) \times H^1(BC_m; R) \to H^2(BC_m; R)$$

is the R-bilinear form

$${}_mR \times {}_mR \to R/mR, \quad (u, v) \mapsto m(m-1)/2 \cdot uv.$$

Here one has to take the product of $u, v \in {}_mR \subset R$ and reduce it modulo m. Thus if m is odd, this product is zero; and if m is even it is $(u, v) \mapsto m/2 \cdot uv$.

Chapter 18
Duality

We have already given an introduction to duality theory from the view point of homotopy theory. In this chapter we present the more classical duality theory based on product structures in homology and cohomology. Since we did not introduce products for spectral homology and cohomology we will not directly relate the two approaches of duality in this book.

Duality theory has several aspects. There is, firstly, the classical Poincaré duality theorem. It states that for a closed orientable n-dimensional manifold the groups $H^k(M)$ and $H_{n-k}(M)$ are isomorphic. A consequence is that the cup product pairing $H^*(M) \otimes H^{n-*}(M) \to H^n(M)$ is a regular bilinear form (say with field coefficients). This quadratic structure of a manifold is a basic ingredient in the classification theory (surgery theory). The cup product pairing for a manifold has in the context of homology an interpretation as intersection. Therefore the bilinear cup product form is called the intersection form. In the case of a triangulated manifold there exists the so-called dual cell decomposition, and the simplicial chain complex is isomorphic to the cellular cochain complex of dual cells; this is a very strong form of a combinatorial duality theorem [167].

The second aspect relates the cohomology of a closed subset $K \subset \mathbb{R}^n$ with the homology of the complement $\mathbb{R}^n \smallsetminus K$ (Alexander duality). This type of duality is in fact a phenomenon of stable homotopy theory as we have explained earlier.

Both types of duality are related. In this chapter we prove in the axiomatic context of generalized cohomology theories a theorem which compares the cohomology of pairs (K, L) of compact subsets of an oriented manifold M with the homology of the dual pair $(M \smallsetminus L, M \smallsetminus K)$. The duality isomorphism is constructed with the cap product by the fundamental class. We construct the cap product for singular theory.

18.1 The Cap Product

The ***cap product*** relates singular homology and cohomology with coefficients in the ring R. Let M and N be left R-modules. The cap product consists of a family of R-linear maps

$$H^k(X, A; M) \otimes H_n(X, A \cup B; N) \to H_{n-k}(X, B; M \otimes_R N), \quad x \otimes y \mapsto x \cap y$$

and is defined for excisive pairs (A, B) in X. (Compare the definition of the cup product for singular cohomology.) If a linear map $\lambda\colon M \otimes N \to P$ is given, we compose with the induced map; then $x \cap y \in H_{n-k}(X, B; P)$. This device is typically applied in the cases $M = R$ and $\lambda\colon R \otimes N \to N$ is an R-module structure, or $M = N = \Lambda$ is an R-algebra and $\lambda\colon \Lambda \otimes \Lambda \to \Lambda$ is the multiplication.

We first define a cap product for chains and cochains

$$S^k(X; M) \otimes S_n(X; N) \to S_{n-k}(X; M \otimes N), \quad \varphi \otimes c \mapsto \varphi \cap c.$$

Given $\varphi \in S^p(X; M)$ and $\sigma\colon \Delta^{p+q} = [e_0, \dots, e_{p+q}] \to X$, we set

$$\varphi \cap (\sigma \otimes b) = (-1)^{pq} (\varphi(\sigma | [e_q, \dots, e_{p+q}] \otimes b) \sigma | [e_0, \dots, e_q]$$

and extend linearly. (Compare in this context the definition of the cup product.)

From this definition one verifies the following properties.

(1) Let $f\colon X \to Y$ be continuous. Then $f_\#(f^\# \varphi \cap c) = \varphi \cap f_\# c$.

(2) $\partial(\varphi \cap c) = \delta\varphi \cap c + (-1)^{|\varphi|} \cap \partial c$.

(3) $(\varphi \cup \psi) \cap c = \varphi \cap (\psi \cap c)$.

(4) $1 \cap c = c$.

Case (3) needs conventions about the coefficients. It can be applied in the case that $\varphi \in S^p(X; R), \psi \in S^q(X; \Lambda)$ and $c \in S_n(X; \Lambda)$ for an R-algebra Λ. In case (4) we assume that $1 \in S^0(X; R)$ is the cocycle which send a 0-simplex to 1 and $c \in S_n(X; N)$ for an R-module N.

We now extend the definition to relative groups. If $\varphi \in S^p(X, A; M) \subset S^p(X; M)$ and $c \in S_{p+q}(A; N) + S_{p+q}(B; N)$, then $\varphi \cap c \in S_q(B; M \otimes N)$. Thus we have an induced cap product

$$S^p(X, A; M) \otimes \frac{S_{p+q}(X; N)}{S_{p+q}(A; N) + S_{p+q}(B; N)} \to \frac{S_q(X; M \otimes N)}{S_q(B; M \otimes N)}.$$

Let A, B be excisive. We use the chain equivalence $S_\bullet(A) + S_\bullet(B) \to S_\bullet(A \cup B)$. After passing to cohomology we obtain the cap product as stated in the beginning. We list the

18.1.1 Properties of the cap product.

(1) For $f\colon (X; A, B) \to (X'; A', B')$, $x' \in H^p(X', A'; M)$, and for $u \in H_{p+q}(X, A \cup B; N)$ the relation $f_*(f^* x' \cap u) = x' \cap f_* u$ holds.

(2) Let A, B be excisive, $j_B\colon (B, A \cap B) \to (X, A)$ the inclusion and

$$\partial_B\colon H_{p+q}(X, A \cup B) \xrightarrow{\partial} H_{p+q-1}(A \cup B, A) \xleftarrow{\cong} H_{p+q-1}(B, A \cap B).$$

Then for $x \in H^p(X, A; M)$, $y \in H_{p+q}(X, A \cup B; N)$,

$$j_B^* x \cap \partial_B y = (-1)^p \partial(x \cap y) \in H_{q-1}(B; M \otimes N).$$

(3) Let A, B be excisive, $j_A\colon (A, A \cap B) \to (X, B)$ the inclusion and

$$\partial_A\colon H_{p+q}(X, A \cup B) \xrightarrow{\partial} H_{p+q-1}(A \cup B, B) \xleftarrow{\cong} H_{p+q-1}(A, A \cap B).$$

Then for $x \in H^p(A; M)$, $y \in H_{p+q}(X, A \cup B; N)$,

$$j_{A*}(x \cap \partial_A y) = (-1)^{p+1}\delta x \cap y \in H_{q-1}(X, B; M \otimes N).$$

(4) $1 \cap x = x$, $1 \in H^0(X)$, $x \in H_n(X, B)$.

(5) $(x \cup y) \cap z = x \cap (y \cap z) \in H_{n-p-q}(X, C; \Lambda)$ for $x \in H^*(X, A; R)$, $y \in H^*(X, B; \Lambda)$, $z \in H_*(X, A \cup B \cup C; \Lambda)$.

(6) Let $\varepsilon\colon H_0(X; M \otimes N) \to M \otimes N$ denote the augmentation. For $x \in H^p(X, A; M)$, $y \in H_p(X, A; N)$,

$$\varepsilon(x \cap y) = \langle x, y \rangle$$

where $\langle -, - \rangle$ is the Kronecker pairing. □

We display again the properties in a table and refer to the detailed description above.

$$\begin{aligned} f_*(f^*x' \cap u) &= x' \cap f_*u \\ j_B^*x \cap \partial_B y &= (-1)^{|x|}\partial(x \cap y) \\ (j_A)_*(x \cap \partial_A y) &= (-1)^{|x|+1}\delta x \cap y \\ 1 \cap x &= x \\ (x \cup y) \cap z &= x \cap (y \cap z) \\ \varepsilon(x \cap y) &= \langle x, y \rangle \end{aligned}$$

We use the algebra of the cap product and deduce the homological Thom isomorphism from the cohomological one.

(18.1.2) Theorem. *Let $\xi\colon E \to B$ be an oriented n-dimensional real vector bundle with Thom class $t \in H^n(E, E^0; \mathbb{Z})$. Then*

$$t\cap\colon H_{n+k}(E, E^0; N) \to H_k(E; N)$$

is an isomorphism.

Proof. Let $z \in S^n(E, E^0)$ be a cocycle which represents t. Then the family

$$S_{n+k}(E, E^0; N) \to S_k(E; N), \quad x \mapsto z \cap x$$

is a chain map of degree $-n$. This chain map is obtained from the corresponding one for $N = \mathbb{Z}$ by taking the tensor product with N. It suffices to show that the integral chain map induces an isomorphism of homology groups, and for this purpose it suffices to show that for coefficients in a field $N = \mathbb{F}$ an isomorphism is induced (see (11.9.7)). The diagram

$$\begin{array}{ccc} H^k(E;\mathbb{F}) & \xrightarrow{\ \cup t\ } & H^{k+n}(E,E^0;\mathbb{F}) \\ \cong\downarrow\alpha & & \cong\downarrow\alpha \\ \operatorname{Hom}(H_k(E;\mathbb{F}),\mathbb{F}) & \xrightarrow{(t\cap)^*} & \operatorname{Hom}(H_{k+n}(E,E^0;\mathbb{F}),\mathbb{F}) \end{array}$$

is commutative (by property (6) in 18.1.1) where α is the isomorphism of the universal coefficient theorem. Since $\cup t$ is an isomorphism, we conclude that $t\cap$ is an isomorphism. □

Problems

1. The cap product for an excisive pair (A, B) in X is induced by the following chain map (coefficient group $\mathbb{Z}$):

$$\begin{aligned} S^\bullet(X,A)\otimes S_\bullet(X,A\cup B) &\xleftarrow{1\otimes\iota} S^\bullet(X,A)\otimes S_\bullet(X)/(S_\bullet(A)+S_\bullet(B)) \\ &\xrightarrow{1\otimes D} S^\bullet(X,A)\otimes S_\bullet(X,B)\otimes S_\bullet(X,A) \\ &\xrightarrow{1\otimes\tau} S^\bullet(X,A)\otimes S_\bullet(X,A)\otimes S_\bullet(X,B) \\ &\xrightarrow{\varepsilon} \mathbb{Z}\otimes S_\bullet(X,B)\cong S_\bullet(X,B). \end{aligned}$$

D is an approximation of the diagonal, τ the graded interchange map, and ε the evaluation. The explicit form above is obtained from the Alexander–Whitney map D.

2. $(x\times y)\cap(a\times b) = (-1)^{|y||a|}(x\cap a)\times(y\cap b)$.

3. From the cap product one obtains the ***slant product*** $x\otimes u\mapsto x\backslash u$ which makes the following diagram commutative:

$$\begin{array}{ccc} H^q(X,A)\otimes H_n((X,A)\times(Y,B)) & \xrightarrow{\ \backslash\ } & H_{n-q}(Y,B) \\ \downarrow \mathrm{pr}^*\otimes 1 & & \uparrow \mathrm{pr}_* \\ H^q((X,A)\times Y)\otimes H_n((X,A)\times(Y,B)) & \xrightarrow{\ \cap\ } & H_{n-q}(X\times(Y,B)). \end{array}$$

The properties (1)–(5) of the cap product can be translated into properties of the slant product, and the cap product can be deduced from the slant product. (This is analogous to the $\cup$- and $\times$-product.)

18.2 Duality Pairings

We use the properties of the cap product in an axiomatic context. Let h^* be a cohomology theory and k_*, h_* homology theories with values in R-MOD. A

duality pairing (a cap product) between these theories consists of a family of linear maps

$$h^p(X, A) \otimes k_{p+q}(X, A \cup B) \to h_q(X, B), \quad x \otimes y \mapsto x \cap y$$

defined for pairs (A, B) which are excisive for the theories involved. They have the following properties:

(1) **Naturality.** For $f\colon (X; A, B) \to (X'; A', B')$ the relation $f_*(f^*x' \cap u) = x' \cap f_*u$ holds.
(2) **Stability.** Let A, B be excisive. Define the mappings j_B and ∂_B as in (18.1.1). Then $j_B^*x \cap \partial_B y = (-1)^{|x|}\partial(x \cap y)$.
(3) **Stability.** Let A, B be excisive. Define the mappings j_A and ∂_A as in (18.1.1). Then $(j_A)_*(x \cap \partial_A y) = (-1)^{|x|+1}\delta x \cap y$.
(4) **Unit element.** There is given a unit element $1 \in k_0(P)$. The homomorphism $h^k(P) \to h_{-k}(P)$, $x \mapsto x \cap 1$ is assumed to be an isomorphism (P a point).

(In the following investigations we deal for simplicity of notation only with the case $h_* = k_*$.) Note that we do not assume given a multiplicative structure for the cohomology and homology theories.

As a first consequence of the axioms we state the compatibility of the cap product with the suspension isomorphisms.

(18.2.1) Proposition. *The following diagrams are commutative:*

$$\begin{array}{ccc}
h^p(X, A) \otimes h_{p+q}(X, A \cup B) & \xrightarrow{\ \cap\ } & h_q(X, B) \\
\downarrow{\scriptstyle \mathrm{pr}^* \otimes \sigma} & & \downarrow{\scriptstyle (-1)^p \sigma} \\
h^p(IX, IA) \otimes h_{p+q+1}(IX, \partial IX \cup IA \cup IB) & \xrightarrow{\ \cap\ } & h_{q+1}(IX, \partial IX \cup IB),
\end{array}$$

$$\begin{array}{ccc}
h^p(X, A) \otimes h_{p+q}(X, A \cup B) & \xrightarrow{\ \cap\ } & h_q(X, B) \\
\downarrow{\scriptstyle \sigma \otimes \sigma} & & \uparrow{\scriptstyle (-1)^p \mathrm{pr}_*} \\
h^{p+1}(IX, \partial IX \cup IA) \otimes h_{p+q+1}(IX, \partial IX \cup IA \cup IB) & \xrightarrow{\ \cap\ } & h_q(IX, IB).
\end{array}$$

(For the second diagram one should recall our conventions about the suspension isomorphisms, they were different for homology and cohomology. Again we use notations like $IX = I \times X$.)

Proof. We consider the first diagram in the case that $A = \emptyset$. The proof is based on the next diagram.

$$\begin{array}{ccc}
h^p(1X) \otimes h_{p+q}(1X, 1B) & \xrightarrow{\ \cap\ } & h_q(1X, 1B) \\
\uparrow {\scriptstyle j^* \otimes j_*^{-1}} & & \uparrow {\scriptstyle j_*^{-1}} \\
h^p(\partial IX \cup IB) \otimes h_{p+q}(\partial IX \cup IB, IB) & \xrightarrow{\ \cap\ } & h_q(\partial IX \cup IB, IB) \\
\uparrow {\scriptstyle 1 \otimes i_*} & & \uparrow {\scriptstyle i_*} \\
h^p(\partial IX \cup IB) \otimes h_{p+q}(\partial IX \cup IB) & \xrightarrow{\ \cap\ } & h_q(\partial IX \cup IB) \\
\uparrow {\scriptstyle k^* \otimes \partial} & & {\scriptstyle \partial} \uparrow \\
h^p(IX) \otimes h_{p+q+1}(IX, \partial IX \cup IB) & \xrightarrow{\ \cap\ } & h_{q+1}(IX, \partial IX \cup IB)
\end{array}$$

The maps i, j, k are inclusions. The first and the second square commute by naturality. The third square is $(-1)^p$-commutative by stability (2). □

(18.2.2) Proposition. *Let $e_n \in h_n(\mathbb{R}^n, \mathbb{R}^n \smallsetminus 0)$ be obtained from $1 \in h_0(P)$ under an iterated suspension isomorphism. Then*

$$h^k(\mathbb{R}^n) \to h_{n-k}(\mathbb{R}^n, \mathbb{R}^n \smallsetminus 0), \quad x \mapsto x \cap e_n$$

is for $k \in \mathbb{Z}$ and $n \geq 1$ an isomorphism.

Proof. This follows by induction on n. One uses the first diagram in (18.2.1) and an analogous suspension isomorphism with $(\mathbb{R}, \mathbb{R} \smallsetminus 0)$ in place of $(I, \partial I)$. □

(18.2.3) Proposition. *Let (U, V) and (U', V') be pairs of open subsets in the space $X = U \cup U'$. Let $\xi \in h_n(U \cup U', V \cup V')$ be a fixed element. From it we produce elements α and β via*

$$\xi \in h_n(X, V \cup V') \to h_n(X, V \cup U') \xleftarrow{\cong} h_n(U, V \cup UU') \ni \alpha,$$

$$\xi \in h_n(X, V \cup V') \to h_n(X, U \cup V') \xleftarrow{\cong} h_n(U', V' \cup UU') \ni \beta.$$

Then the diagram

$$\begin{array}{ccccc}
h^{k-1}(U, V) & \longrightarrow & h^{k-1}(UU', VU') & \xrightarrow{\ \delta\ } & h^k(U', UU') \\
\downarrow {\scriptstyle \cap \alpha} & & & & \downarrow {\scriptstyle \cap \beta} \\
h_{n-k+1}(U, UU') & \xrightarrow{\ \partial\ } & h_{n-k+1}(UU', UV') & \longrightarrow & h_{n-k}(U', V')
\end{array}$$

is commutative. (We again have used notations like $UU' = U \cap U'$.)

Proof. We use naturality and stability (2) and show that the down-right path of the diagram is $(-1)^{k-1}$ times the map

$$h^{k-1}(U,V) \to h^{k-1}(UU',VU') \xrightarrow{\cap\alpha_1} h_{n-k}(UU',UV') \to h_{n-k}(U',V')$$

and the right-down path $(-1)^k$ times the analogous map where α_1 is replaced by β_1; the element α_1 is obtained from α via the morphism

$$\begin{aligned}\alpha \in h_n(U, V \cup UU') &\xrightarrow{\partial} h_{n-1}(V \cup UU', V) \xleftarrow{\cong} h_{n-1}(UU',VU') \\ &\to h_{n-1}(UU',(UV') \cup (VU')) \ni \alpha_1\end{aligned}$$

and β_1 from β via the analogous composition in which the primed and unprimed spaces are interchanged. Thus it remains to show $\alpha_1 = -\beta_1$. This is essentially a consequence of the Hexagon Lemma. One of the outer paths in the hexagon is given by the composition

$$\begin{aligned}h_n(U \cup U', V \cup V') &\longrightarrow h_n(U \cup U', U \cup V') \\ &\xleftarrow{\cong} h_n(U' \cup V, (V \cup U')(U \cup V')) \\ &\xrightarrow{\partial} h_{n-1}((U \cup V')(V \cup U'), V \cup V')\end{aligned}$$

and the other path is obtained by interchanging the primed and unprimed objects. The center of the hexagon is $h_n(U \cup U', (U \cup V')(V \cup U'))$. We then compose the outer paths of the hexagon with the excision

$$h_{n-1}(UU', UV' \cup VU') \to h_{n-1}((U \cup V')(V \cup U'), V \cup V');$$

then ξ is mapped along the paths to α_1 and β_1, respectively; this follows from the original definition of the elements by a little rewriting. The displayed morphism yields β_1. □

18.3 The Duality Theorem

For the statement of the duality theorem we need two ingredients: A homological orientation of a manifold and a duality homomorphism. We begin with the former.

Let M be an n-dimensional manifold. For $K \subset L \subset M$ we write

$$r_K^L : h_*(M, M \smallsetminus L) \to h_*(M, M \smallsetminus K)$$

for the homomorphism induced by the inclusion, and r_x^L in the case that $K = \{x\}$. An element $o_L \in h_n(M, M \smallsetminus L)$ is said to be a ***homological orientation*** along L if for each $y \in L$ and each chart $\varphi\colon U \to \mathbb{R}^n$ centered at y the image of o_L under

$$h_n(M, M \smallsetminus L) \xrightarrow{r_y^L} h_n(M, M \smallsetminus y) \xleftarrow{\cong} h_n(U, U \smallsetminus y) \xrightarrow{\varphi_*} h_n(\mathbb{R}^n, \mathbb{R}^n \smallsetminus 0)$$

is $\pm e_n$ where e_n is the element which arises from $1 \in h_0$ under suspension. A family $(o_K \mid K \subset M$ compact) is called ***coherent*** if for each compact pair $K \subset L$ the restriction relation $r^L_K o_L = o_K$ holds. A coherent family (o_K) of orientations is a ***(homological) orientation*** of M. If M is compact, then $K = M$ is allowed and an orientation is determined by the element $o_M \in h_n(M)$, called the ***fundamental class*** of M.

In order to state the duality theorem we need the definition of a duality homomorphism. We fix a homological orientation (o_K) of M. Given closed sets $L \subset K \subset M$ and open sets $V \subset U \subset M$ such that $L \subset V, K \subset U$. We fix an element $z \in h_n(M, M \smallsetminus K)$. From $o_K = z$ we obtain z^{UV}_{KL} via

$$\begin{array}{ccc} z \in h_n(M, M \smallsetminus K) & \longrightarrow & h_n(M, (M \smallsetminus K) \cup V) \\ & & \uparrow {\scriptstyle (\#)}\ \cong \\ & & z^{UV}_{KL} \in h_n(U \smallsetminus L, (U \smallsetminus K) \cup (V \smallsetminus L)). \end{array}$$

The morphism (#) is an excision, since $M \smallsetminus (U \smallsetminus L) = (M \smallsetminus U) \cup L$ (closed) is contained in $(M \smallsetminus K) \cup V$ (open).

From z^{UV}_{KL} we obtain the homomorphism D^{UV}_{KL} via the commutative diagram

$$\begin{array}{ccc} h^k(U, V) & \longrightarrow & h^k(U \smallsetminus L, V \smallsetminus L) \\ \downarrow D^{UV}_{KL} & & \downarrow \cap z^{UV}_{KL} \\ h_{n-k}(M \smallsetminus L, M \smallsetminus K) & \xleftarrow{\cong} & h_{n-k}(U \smallsetminus L, U \smallsetminus K). \end{array}$$

We state some naturality properties of these data. They are easy consequences of the naturality of the cap product.

(18.3.1) Lemma. *Let* $(K, L) \subset (U', V') \subset (U, V)$ *and*

$$i\colon (U' \smallsetminus L, U' \smallsetminus K, V' \smallsetminus L) \subset (U \smallsetminus L, U \smallsetminus K, V \smallsetminus L).$$

Then $i_* z^{U'V'}_{KL} = z^{UV}_{KL}$ *and* $D^{UV}_{KL} = D^{U'V'}_{KL} \circ i^*$. □

(18.3.2) Lemma. *Let* $(K', L') \subset (K, L) \subset (U, V)$ *and*

$$j\colon (U \smallsetminus L, U \smallsetminus K, V \smallsetminus L) \subset (U \smallsetminus L', U \smallsetminus K', V \smallsetminus L').$$

Then $j_* z^{UV}_{KL} = z^{UV}_{K'L'}$ *and* $j^* \circ D^{UV}_{K'L'} = D^{UV}_{KL}$. □

The naturality (18.3.1) allows us to pass to the colimit over the neighbourhoods (U, V) of (K, L) in M. We obtain a ***duality homomorphism***

$$D_{KL}\colon \check{h}^k(K, L) = \operatorname{colim}_{UV} h^k(U, V) \to h_{n-k}(M \smallsetminus L, M \smallsetminus K).$$

We explain this with some remarks about colimits. An element $x \in h^k(U,V)$ represents an element of the colimit. Two elements $x \in h^k(U,V)$ and $x' \in h^k(U',V')$ represent the same element if and only if there exists a neighbourhood (U'',V'') with $U'' \subset U \cap U'$, $V'' \subset V \cap V'$ such that x and x' have the same restriction in $h^k(U'',V'')$. Thus we have canonical homomorphisms $l_{UV}\colon h^k(U,V) \to \check{h}^k(K,L)$. Via these homomorphisms the colimit is characterized by a universal property: If $\lambda_{UV}\colon h^k(U,V) \to h$ is a family of homomorphisms such that $\lambda_{U'V'} \circ i = \lambda_{UV}$ for the restrictions $i\colon h^k(U,V) \to h^k(U',V')$, then there exists a unique homomorphism $\lambda\colon \check{h}^k(K,L) \to h$ such that $\lambda l_{UV} = \lambda_{UV}$. The restrictions $h^k(U,V) \to h^k(K,L)$ are compatible in this sense, and we obtain a canonical homomorphism $\check{h}^k(K,L) \to h^k(K,L)$. In sufficiently regular situations this homomorphism is an isomorphism; we explain this later.

(18.3.3) Duality theorem. *Let M be an oriented manifold. Then the duality homomorphism D_{KL} is, for each compact pair (K,L) in M, an isomorphism.*

We postpone the proof and discuss some of its applications. Let M be compact and $[M] \in h_n(M)$ a fundamental class. In the case $(K,L) = (M,\emptyset)$ we have $\check{h}^k(M,\emptyset) = h^k(M)$ and D_{KL} is the cap product with $[M]$. Thus we obtain as a special case:

(18.3.4) Poincaré duality theorem. *Suppose the compact n-manifold is oriented by the fundamental class $[M] \in h_n(M)$. Then*

$$h^k(M) \to h_{n-k}(M), \quad x \mapsto x \cap [M]$$

is an isomorphism. □

A duality pairing exists for the singular theory

$$H^p(X,A;G) \otimes H_{p+q}(X,A \cup B;R) \to H_q(X,B;G)$$

for commutative rings R and R-modules G. The Euclidean space $\mathbb{R}^n$ is orientable for $H_*(-;\mathbb{Z})$. Thus we have:

(18.3.5) Alexander duality theorem. *For a compact pair (K,L) in $\mathbb{R}^n$*

$$\check{H}^k(K,L;G) \cong H_{n-k}(\mathbb{R}^n \smallsetminus K, \mathbb{R}^n \smallsetminus L; G).$$

A similar isomorphism exists for S^n in place of $\mathbb{R}^n$. □

(18.3.6) Example. We generalize the Jordan separation theorem. Let M be a connected and orientable (with respect to $H_*(-;\mathbb{Z})$) n-manifold. Suppose that $H_1(M;R) = 0$. Let $A \subset M$ be compact, $A \neq M$. Then $\check{H}^{n-1}(A;R)$ is a free R-module, and $|\pi_0(M \smallsetminus A)| = 1 + \operatorname{rank} \check{H}^{n-1}(A;R)$.

By duality $\check{H}^{n-1}(A;R) \cong H_1(M, M \smallsetminus A; R)$. The hypothesis shows

$$\partial\colon H_1(M, M \smallsetminus A; R) \cong \tilde{H}_0(M \smallsetminus A; R),$$

and the latter is a free R-module of rank $|\pi_0(M \smallsetminus A)| - 1$. ◇

(18.3.7) Example. $H^2(\mathbb{R}P^2;\mathbb{Z}) \cong \mathbb{Z}/2$. This is not a free $\mathbb{Z}$-module. Hence the projective plane cannot be embedded into S^3. (A similar proof shows that $\mathbb{R}P^{2n}$ has no embedding into S^{2n+1}.) ◇

(18.3.8) Remark. From Alexander duality and the Thom isomorphism one can deduce Poincaré duality. Let $M \subset \mathbb{R}^{n+t}$ be a smooth closed submanifold of dimension n. Suppose we have an Alexander duality isomorphism $h_k(M) \cong h^{n+t-k}(\mathbb{R}^{n+t}, \mathbb{R}^{n+t} \smallsetminus M)$. Let $\tau\colon E(\nu) \to U$ be a tubular map. We use τ and excision and obtain $h^{n+t-k}(\mathbb{R}^{n+t}, \mathbb{R}^{n+t} \smallsetminus M) \cong h^{n+t-k}(E(\nu), E^0(\nu))$. Suppose the normal bundle ν is oriented by a Thom class. Then we have a Thom-isomorphism $h^{n+t-k}(E(\nu), E^0(\nu)) \cong h^{n-k}(M)$. Altogether we obtain an isomorphism $h_k(M) \cong h^{n-k}(M)$ of Poincaré duality type. A similar device works if we start from an isomorphism $h^k(M) \cong h_{n+k-t}(\mathbb{R}^{n+t}, \mathbb{R}^{n+t} \smallsetminus M)$ and use a homological Thom isomorphism. This approach would be used if one starts with homotopical duality as a foundation stone. ◇

Problems

1. Let $D \subset \mathbb{R}^2$ be connected and open. The following are equivalent: (1) D is homeomorphic to $\mathbb{R}^2$. (2) D is simply connected. (3) $H_1(D;\mathbb{Z}) = 0$. (4) $H^1(D;\mathbb{Z}) = 0$. (5) $\mathbb{R}^2 \smallsetminus D$ is connected. (6) The boundary of D is connected. (7) If $J \subset D$ is a Jordan curve, then D contains the interior of J. [44, p. 394]

2. Let $i\colon S^n \to K(\mathbb{Z}, n)$ be an inclusion of a subcomplex which induces an isomorphism of π_n. For each compact subset $K \subset \mathbb{R}^{n+1}$ the induced map $i_*\colon [K, S^n] \to [K, K(\mathbb{Z}, n)]$ is bijective.

3. Use cohomology $H^n(X;\mathbb{Z}) = [X, K(\mathbb{Z}, n)]$ defined with an Eilenberg–Mac Lane complex $K(\mathbb{Z}, n)$. Then for a compact subset X in a Euclidean space $\check{H}^n(X;\mathbb{Z}) \cong H^n(X;\mathbb{Z})$. Similar isomorphisms hold for the stable cohomotopy groups.

4. $\mathbb{R}^n$ is orientable for each homology theory. Let $K \subset D(r) = \{x \mid \|x\| \le r\}$ be compact. Define o_K as the image of the canonical class under $h_n(\mathbb{R}^n, \mathbb{R}^n \smallsetminus 0) \leftarrow h_n(\mathbb{R}^n, \mathbb{R}^n \smallsetminus D(r)) \to h_n(\mathbb{R}^n, \mathbb{R}^n \smallsetminus K)$.

18.4 Euclidean Neighbourhood Retracts

For applications it is interesting to compare $\check{H}^k$ with H^k.

A space X is called a ***Euclidean neighbourhood retract* (*ENR*)** if there exists an embedding $j: X \to \mathbb{R}^n$, an open neighbourhood U of $j(X)$ in $\mathbb{R}^n$ and a retraction $r: U \to X$, i.e., $rj = \mathrm{id}(X)$.

Let $X \subset \mathbb{R}^n$ be a retract of an open set U, then X is closed in U and hence locally compact: Let $r: U \to X$ be a retraction; then X is the coincidence set of $r: U \to U$ and $\mathrm{id}: U \to U$. Recall that a locally compact set Y in a Hausdorff space Z is locally closed, i.e., has the form $Y = \overline{Y} \cap W$ for an open set W in Z.

(18.4.1) Proposition. *Let $X \subset \mathbb{R}^n$ be a retract of an open neighbourhood. Let Z be a metric space and $Y \subset Z$ homeomorphic to X. Then Y is a retract of an open neighbourhood V of Y in Z.*

Proof. Let $f: X \to Y$ be a homeomorphism and $r: U \to X$ a retraction. Then Y is locally compact and we can write $Y = \overline{Y} \cap W$ with an open $W \subset Z$. Then Y is closed in W. Since W is a normal space, the map $Y \xrightarrow{f^{-1}} X \to \mathbb{R}^n$ has a continuous extension $h: W \to \mathbb{R}^n$ by the Tietze extension theorem. Let $V = h^{-1}(U)$. Then $frh: V \to Y$ is a retraction of $Y \subset V$. □

(18.4.2) Proposition. *Let $X \subset \mathbb{R}^n$ be locally compact. Then there exists an embedding of X into $\mathbb{R}^{n+1}$ as a closed subset.*

Proof. Let $U \subset \mathbb{R}^n$ be open ($U \neq \mathbb{R}^n$). Then

$$j: U \to \mathbb{R}^n \times \mathbb{R}, \quad x \mapsto (x, d(x, \mathbb{R}^n \smallsetminus U)^{-1})$$

is an embedding. The image of j is closed , since

$$j(U) = \{(x,t) \mid t \cdot d(x, \mathbb{R}^n \smallsetminus U) = 1\}.$$

We can assume that $X \subset U$ is closed; then $j(X)$ is closed in $j(U)$, hence closed in $\mathbb{R}^{n+1}$. □

(18.4.3) Proposition. *Let X be an ENR. Suppose $f_0, f_1: Y \to X$ are maps which coincide on a subset $B \subset Y$. Then there exists a neighbourhood W of B in Y and a homotopy h: $f_0|W \simeq f_1|W$ relative to B.*

Proof. Let $X \xrightarrow{i} U \xrightarrow{r} X$ be a presentation as a retract with $U \subset \mathbb{R}^n$ open. Let

$$W = \{y \mid (1-t)if_0(y) + tif_1(y) \in U \ \text{ for all } t \in I\}.$$

Then certainly $B \subset W$. Since

$$\lambda: Y \times I \to \mathbb{R}^n, \quad (y,t) \mapsto (1-t)if_0(y) + tif_1(y)$$

is continuous, $\lambda^{-1}(U)$ is open. If $\{y\} \times I \subset \lambda^{-1}(U)$, then there exists an open neighbourhood U_y of y such that $U_y \times I \subset \lambda^{-1}(U)$. Hence W is open in Y. A suitable homotopy h is now obtained as the restriction of λ to $W \times I$. □

(18.4.4) Remark. Suppose $B \subset X$ are Euclidean neighbourhood retracts. Then there exists a retraction $r\colon V \to B$ from an open neighbourhood V of B in X. There further exists a neighbourhood $W \subset V$ of B in X such that $j : W \subset V$ and $i \circ (r|W)\colon W \to B \subset V$ are homotopic relative to B.

An ENR is locally contractible: Each neighbourhood V of x contains a neighbourhood W of x such that $W \subset V$ is homotopic to $W \to \{x\} \subset V$ relative to $\{x\}$. ◇

(18.4.5) Lemma. *Let the Hausdorff space $X = X_1 \cup \cdots \cup X_r$ be a union of locally compact open subsets X_j which are homeomorphic to a subset of a Euclidean space. Then X is homeomorphic to a closed subset of a Euclidean space.*

Proof. There exist embeddings $h_i : X_i \to \mathbb{R}^{m(i)}$ as a closed subset. We extend h_i to a continuous map $k_i : X \to S^{m(i)} = \mathbb{R}^{m(i)} \cup \{\infty\}$ by $k_i(X \smallsetminus X_i) = \{\infty\}$ (if $X \neq X_i$). The product $(k_i)\colon X \to \prod_{i=1}^{r} S^{m(i)}$ is an embedding. The product of the spheres can be embedded into $\prod_i \mathbb{R}^{m(i)+1}$, and then we can apply (18.4.2) if necessary. □

(18.4.6) Theorem. *Let the Hausdorff space $X = X_0 \cup \cdots \cup X_r$ be a union of open subsets X_i which are ENR's. Then X is an ENR.*

Proof. Induction on r. It suffices to consider $X = X_0 \cup X_1$. By (18.4.5) we can assume that X is a closed subset of some $\mathbb{R}^n$. Let $r_i : U_i \to X_i$ be retractions (see (18.4.1)). Set

$$U_{01} = r_0^{-1}(X_0 \cap X_1) \cap r_1^{-1}(X_0 \cap X_1).$$

Then $r_0, r_1\colon U_{01} \to X_0 \cap X_1$ are retractions of a neighbourhood. The open subset $X_0 \cap X_1$ of the ENR X_0 is an ENR. Hence there exists $X_0 \cap X_1 \subset V_{01} \subset U_{01}$ such that r_0, r_1 are homotopic on V_{01} relative to $X_0 \cap X_1$ by a homotopy r_t (see (18.4.3)). Let $V_0 \subset U_0$, $V_1 \subset U_1$ be open neighbourhoods of $X \smallsetminus X_1, X \smallsetminus X_0$ such that $\bar{V}_0 \cap \bar{V}_1 = \emptyset$. Choose a continuous function $\tau\colon \mathbb{R}^n \to [0,1]$ such that $\tau(V_0) = 0$ and $\tau(V_1) = 1$. Let $V = V_0 \cup V_{01} \cup V_1$. Then $\rho\colon V \to X$, defined as $\rho|V_0 = r_0|V_0$, $\rho|V_1 = r_1|V_1$, $\rho(x) = r_{\tau(x)}(x)$ for $x \in V_{01}$, is a suitable retraction. □

(18.4.7) Corollary. *A compact manifold is an ENR.* □

(18.4.8) Remark. Since an ENR is dominated by a CW-complex it has the homotopy type of a CW-complex. A compact ENR is dominated by a finite CW-complex; therefore its singular homology groups with coefficients in $\mathbb{Z}$ are finitely generated abelian groups. This holds in particular for compact manifolds. ◇

(18.4.9) Proposition. *Let K be a compact ENR in an n-manifold M. Then the canonical map $\rho\colon \check{H}^k(K) \to H^k(K)$ is an isomorphism.*

Proof. We use that M is a metrizable space or, at least, that open subsets are normal. Then K is a retract $r\colon U \to K$ of an open neighbourhood U of K in M. Suppose $x \in H^k(K)$; then $r^*(x)$ represents an element $\check{x} \in \check{H}^k(K)$ with $\rho(\check{x}) = x$. This shows that ρ is surjective. Let $x_U \in H^k(U)$ represent an element in the kernel of ρ.

Suppose there exists a neighbourhood $V \subset U$ of K and a homotopy from $j\colon V \subset U$ to $ir\colon V \to K \to U$. Then we have the situation

$$\begin{array}{ccccc} x_U & H^k(U) & \xrightarrow{\;i^*\;} & H^k(K) & \xrightarrow{\;r^*\;} H^k(V), \qquad H^k(U) \xrightarrow{\;j^*\;} H^k(V) \\ \big\downarrow & \big\downarrow{\scriptstyle j^*} & & \nearrow & \\ x_V & H^k(V) & & & \end{array}$$

$$x_U \mapsto x_V, \qquad j^*\colon H^k(U) \to H^k(V), \qquad H^k(V) \to H^k(K) \xrightarrow{\;r^*\;} H^k(V).$$

Since $i^*(x_U) = 0$, by assumption, we see that $x_V = 0$; but x_V represents the same element as x_U.

The homotopy exists by (18.4.3) if also U is an ENR. If we choose U as the union of a finite number of sets which are homeomorphic to open subsets of $\mathbb{R}^n$, then we can apply (18.4.6). □

Let X be an ENR and $X \subset \mathbb{R}^n$. Since the canonical homomorphism $\check{h}^*(X) \to h^*(X)$ is an isomorphism, $\check{h}^*(X)$ does not depend on the embedding $X \subset \mathbb{R}^n$.

Let M be a compact n-manifold. Then $H_k(M;\mathbb{Z})$ and $H^k(M;\mathbb{Z})$ are finitely generated abelian groups. For each field $\mathbb{F}$ the groups $H_k(M;\mathbb{F})$ are finite-dimensional vector spaces. The Euler characteristic $\chi(M;\mathbb{F})$ is independent of $\mathbb{F}$ and equal to the Euler characteristic $\chi(M)$.

Let $K \subset M$ be a compact ENR. Then

$$H^{n-i}(K;\mathbb{F}_2) \cong \check{H}^{n-i}(K;\mathbb{F}_2) \cong H_i(M, M \smallsetminus K;\mathbb{F}_2)$$

and these are finite-dimensional vector spaces over the prime field $\mathbb{F}_2$.

(18.4.10) Proposition. *$H_*(K;\mathbb{F}_2)$ is finite-dimensional if and only if the same holds for $H_*(M \smallsetminus K;\mathbb{F}_2)$. If finiteness holds, then for the Euler characteristic χ_2 the relation*

$$\chi_2(M) = \chi_2(M \smallsetminus K) + (-1)^n \chi_2(K)$$

holds. (*Note that $\chi_2(K) = \chi(K)$.*)

Proof. The first statement follows from the exact homology sequence of the pair $(M, M \smallsetminus K)$. It also yields $\chi_2(M) = \chi_2(M \smallsetminus K) + \chi_2(M, M \smallsetminus K)$. The equality $\chi_2(M, M \smallsetminus K) = (-1)^n \chi_2(K)$ is obtained, if we insert the consequence $\dim H_i(M, M \smallsetminus K;\mathbb{F}_2) = \dim H^{n-i}(K;\mathbb{F}_2)$ of the duality into the homological definition of the Euler characteristic. □

(18.4.11) Corollary. *Let M be a closed manifold of odd dimension. Then $\chi(M) = 0$. If $K \subset M$ is a compact ENR, then $\chi(K) = \chi(M \setminus K)$.* □

Problems

1. Let F be a compact, connected, non-orientable surface. The universal coefficient formula and $H_1(F;\mathbb{Z}) \cong \mathbb{Z}^{g-1} \oplus \mathbb{Z}/2$ show $H^2(F;\mathbb{Z}) \cong \mathbb{Z}/2$. Therefore F cannot be embedded into S^3.

2. Let $S \subset \mathbb{R}^2$ be the pseudo-circle. Show, heuristically, that the pseudo-circle has a system of neighbourhoods $U_1 \supset U_2 \supset \cdots$ with $U_i \cong S^1 \times [0,1]$ and $\cap U_n = S$.

Then

$$\check{H}^1(S;\mathbb{Z}) \cong H_1(\mathbb{R}^2, \mathbb{R}^2 \setminus S;\mathbb{Z}) \cong \tilde{H}_0(\mathbb{R} \setminus S;\mathbb{Z}) \cong \mathbb{Z},$$

the last isomorphism because $\mathbb{R}^2 \setminus S$ has two path components. By the universal coefficient formula $H^1(S;\mathbb{Z}) \cong \mathrm{Hom}(H_1(S),\mathbb{Z})$. The singular homology group $H_1(S;\mathbb{Z})$ is zero, a singular 1-chain is always contained in a contractible subset. In fact, S has the weak homotopy type of a point. This shows that singular theory is the wrong one for spaces like S.

3. Let $F \subset \mathbb{R}^3$ be a connected orientable compact surface. Then $\mathbb{R}^3 \setminus F$ has two path components (interior and exterior).

4. Let $M = \mathbb{R}^{n+1}$ and $S \subset \mathbb{R}^{n+1}$ be homeomorphic to S^n. Then S is an ENR and $\check{H}^n(S;\mathbb{Z}) \cong H^n(S;\mathbb{Z}) \cong \mathbb{Z}$. From the duality theorem one obtains that $\mathbb{R}^{n+1} \setminus S$ has two path components.

18.5 Proof of the Duality Theorem

We have to collect some formal properties of the groups $\check{h}$ and of the duality homomorphisms D. We want the $\check{h}^k(K,L)$ to be part of functors from the category $\mathcal{K}(M)$ of compact pairs in M and inclusions.

Let $(K',L') \subset (K,L)$. The induced map $\check{h}^k(K,L) \to \check{h}^k(K',L')$ sends an element represented by $x \in h^k(U,V)$ to the element which is represented by the same x. This makes the $\check{h}^k$ into functors on $\mathcal{K}(M)$ and the canonical maps $\check{h}^k \to h^k$ into natural transformations. We define a coboundary operator $\check{\delta}\colon \check{h}^k(L) \to \check{h}^{k+1}(K,L)$ as follows. Let $V \supset L$ be open. Choose $U \supset V$ as an open neighbourhood of K. Then we map representing elements via $\delta\colon h^k(V) \to h^{k+1}(U,V)$. This process yields a well-defined $\check{\delta}$ and the diagram

$$\begin{array}{ccc} \check{h}^k(L) & \xrightarrow{\check{\delta}} & \check{h}^{k+1}(K,L) \\ \downarrow & & \downarrow \\ h^k(K) & \xrightarrow{\delta} & h^{k+1}(K,L) \end{array}$$

is commutative. From (18.3.2) we obtain by passage to the colimit:

(18.5.1) Lemma. *The D_{KL} yield a natural transformation, i.e., the diagram*

$$\begin{array}{ccc} \check{h}^k(K,L) & \xrightarrow{D_{KL}} & h_{n-k}(M \smallsetminus L, M \smallsetminus K) \\ \downarrow & & \downarrow \\ \check{h}^k(K',L') & \xrightarrow{D_{K'L'}} & h_{n-k}(M \smallsetminus L', M \smallsetminus K') \end{array}$$

is commutative for each inclusion $(K',L') \subset (K,L)$. □

(18.5.2) Lemma. *The sequence*

$$\cdots \to \check{h}^k(K) \longrightarrow \check{h}^k(L) \xrightarrow{\check{\delta}} \check{h}^{k+1}(K,L) \to \cdots$$

is exact. Similarly for triples of compact subsets.

Proof. This is a special case of the general fact that a colimit over a directed set of exact sequences is again exact. A direct verification from the definitions and the exact sequences for the representing elements is not difficult. An example should suffice. Suppose $\check{\delta}(x) = 0$ and let x be represented by $x_1 \in h^k(V)$. We use the representing element $\delta(x_1) \in h^{k+1}(U,V)$ for $\check{\delta}(x)$. Since $\check{\delta}(x)$ is zero, x_1 is contained in the kernel of some restriction $h^{k+1}(U,V) \to h^{k+1}(U',V')$. Another representative of x is the restriction $x_2 \in h^k(V')$ of x_1. By exactness, x_2 has a pre-image in $h^k(U')$, and it represents a pre-image of x in $\check{h}^k(K)$. □

(18.5.3) Lemma. *Each compact pair K, L is excisive for $\check{h}^k$, i.e.,*

$$\check{h}^k(K \cup L, K) \to \check{h}^k(L, K \cap L)$$

is an isomorphism.

Proof. This is a consequence of the isomorphisms $h^k(U \cup V, U) \cong h^k(V, U \cap V)$ for open neighbourhoods $U \supset K$, $V \supset L$. □

(18.5.4) Corollary. *For each compact pair there exist an exact MV-sequence*

$$\cdots \to \check{h}^k(K \cup L) \to \check{h}^k(K) \oplus \check{h}^k(L) \to \check{h}^k(K \cap L) \xrightarrow{\delta} \cdots.$$

Proof. The MV-sequence is constructed by algebra from suitable exact sequences using (18.5.2) and (18.5.3). □

(18.5.5) Lemma. *For each pair (K,L) the diagram*

$$\begin{array}{ccc} \check{h}^k(L) & \xrightarrow{D_L} & h_{n-k}(M, M \smallsetminus L) \\ \downarrow{\check{\delta}} & & \downarrow{\partial} \\ \check{h}^{k+1}(K,L) & \xrightarrow{D_{KL}} & h_{n-k-1}(M \smallsetminus L, M \smallsetminus K) \end{array}$$

is commutative.

Proof. By passage to the colimit this is a consequence of the commutativity of the diagrams

$$\begin{array}{ccc}
h^{k-1}(V) & \xrightarrow{\quad\delta\quad} & h^k(U,V) \\
\Big\downarrow D^{V\emptyset}_{L\emptyset} & & \Big\downarrow D^{UV}_{KL} \\
h_{n-k+1}(M, M \smallsetminus L) & \xrightarrow{\partial} & h_{n-k}(M \smallsetminus L, M \smallsetminus K).
\end{array}$$

In order to verify this commutativity we apply (18.2.3) to the sets

$$(U, V, U', V') = (V, \emptyset, U \smallsetminus L, U \smallsetminus K).$$

The element ξ arises from o_K via the excision $h_n(U, U \smallsetminus K) \cong h_n(M, M \smallsetminus K)$. One now verifies from the definitions that the element α is $z^{V\emptyset}_{L\emptyset}$ and β becomes z^{UV}_{KL}. These data yield the commutative diagram

$$\begin{array}{ccccc}
h^{k-1}(V) & & \xrightarrow{\qquad\qquad\delta\qquad\qquad} & & h^k(U,V) \\
\Big\downarrow = & & & & \Big\downarrow \\
h^{k-1}(V) & \longrightarrow & h^{k-1}(V \smallsetminus L) & \xrightarrow{\delta} & h^k(U \smallsetminus L, V \smallsetminus L) \\
\Big\downarrow \cap z^{V\emptyset}_{L\emptyset} & & & & \Big\downarrow \cap z^{UV}_{KL} \\
h^{n-k+1}(V, V \smallsetminus L) & \xrightarrow{\partial} & h_{n-k}(V \smallsetminus L, V \smallsetminus K) & \longrightarrow & h_{n-k}(U \smallsetminus L, U \smallsetminus K) \\
\Big\downarrow & & & & \Big\downarrow \\
h_{n-k+1}(M, M \smallsetminus L) & & \xrightarrow{\qquad\qquad\partial\qquad\qquad} & & h_{n-k}(M \smallsetminus L, M \smallsetminus K).
\end{array}$$

The upper and lower rectangles commute by naturality of ∂ and δ. □

For the proof of the duality theorem we note that it suffices to consider the special case $D_K \colon \check{h}^k(K) \to h_{n-k}(M, M \smallsetminus K)$, by the Five Lemma and the previous results. The proof is based on the following principle.

(18.5.6) Theorem. *Let $D(K)$ be an assertion about compact subsets in M. Suppose:*

(1) *$D(K)$ holds for sets K in a chart domain which are mapped onto a convex subset of $\mathbb{R}^n$ under the coordinate map.*
(2) *If $D(K)$, $D(L)$, and $D(K \cap L)$ hold, then also $D(K \cup L)$ holds.*
(3) *Let $K_1 \supset K_2 \supset \cdots$, $K = \bigcap K_i$. If $D(K_i)$ holds for each i, then $D(K)$ holds.*

Under these assumptions, $D(K)$ holds for all compact K.

Proof. Since an intersection of convex sets is convex, (1) and (2) yield by induction on t that $D(K_1 \cup \dots \cup K_t)$ holds for compact subsets K_i of type (1) which are contained in the same chart domain.

If K is a compact set in a chart domain, then K is the intersection of a sequence $K_1 \supset K_2 \supset \cdots$ where each K_i is a finite union of compact convex sets.

Each compact set K is a finite union of compact sets in chart domains. Again $D(K)$ follows by induction from (2). □

We now verify (1)–(3) of (18.5.6) in the case that $D(K)$ is the assertion: D_K is an isomorphism.

(2) The duality homomorphisms D yield a morphism of the MV-sequence for K, L into the MV-sequence of the complements; this follows from the fact that the (co-)boundary operators of the MV-sequences are defined from induced morphisms and ordinary (co-)boundary operators. Now use the Five Lemma.

(1) Let $\varphi\colon U \to \mathbb{R}^n$ be a chart, $K \subset U$ and $\varphi(K)$ convex. We begin with the special case of a point K and $\varphi(K) = \{0\}$. We have a commutative diagram

$$\begin{array}{ccccc}
h^k(\mathbb{R}^n) & \xrightarrow{\varphi^*} & h^k(U) & \xrightarrow{\cong} & \check{h}^k(K) \\
\downarrow {\scriptstyle \cap \varphi_*(z_K^U)} & & \downarrow {\scriptstyle \cap z_K^U} & & \downarrow {\scriptstyle D_K} \\
h_{n-k}(\mathbb{R}^n, \mathbb{R}^n \smallsetminus 0) & \xleftarrow[\varphi_*]{} & h_{n-k}(U, U \smallsetminus K) & \xrightarrow[\cong]{} & h_{n-k}(M, M \smallsetminus K).
\end{array}$$

The right square commutes by definition of D_K; here z_K^U is the image of the orientation under the restriction $h_n(M, M \smallsetminus K) \to h_n(U, U \smallsetminus K)$. The left square commutes by naturality of the cap product. By definition of the orientation, $\varphi_*(z_K^U) = \pm e_n$. The fact that $\cap e_n$ is an isomorphism follows from the compatibility with suspension and the unit element axiom of the pairing. Hence $D(K)$ holds for a point K.

Let now K be arbitrary and $P \subset K$ a point. From naturality we see that $D(K)$ holds, if $\check{h}^k(K) \to \check{h}^k(P)$ and $h_{n-k}(M, M \smallsetminus K) \to h_{n-k}(M, M \smallsetminus P)$ are isomorphisms.

The set $X = \varphi(K)$, being compact convex, is the intersection of a sequence of open neighbourhoods $U_1 \supset U_2 \supset \cdots$ which are contractible onto P, and each neighbourhood of X contains eventually all U_j. Hence the restriction $h^k(U_j) \to h^k(U_{j+1})$ are isomorphisms, and we see $h^k(U_j) \cong \check{h}^k(X) \cong h^k(X)$. This shows the first isomorphism. The second isomorphism is verified by standard methods (excision, h-equivalence).

(3) We show that the canonical maps

$$\operatorname{colim}_i \check{h}^k(K_i) \to \check{h}^k(K), \quad \operatorname{colim}_i h_{n-k}(M, M \smallsetminus K_i) \to h_{n-k}(M, M \smallsetminus K)$$

are isomorphisms.

The first isomorphism is an immediate consequence of the colim-definition. Given $x \in \check{h}^k(K)$ represented by $y \in h^k(U)$. There exists i such that $K_i \subset U$. Hence y represents an element in $\check{h}^k(K_i)$. This shows surjectivity, and injectivity is shown by a similar argument.

The second isomorphism is easily seen for singular homology, if one uses that singular chains have compact carrier. In the general case one uses that additive homology theories commute with colimits.

18.6 Manifolds with Boundary

We now treat duality for manifolds with boundary.

(18.6.1) Theorem. *Let M be a compact n-manifold with boundary ∂M, oriented by a fundamental class $[M] \in H_n(M, \partial M; \mathbb{Z})$. Then*

$$\begin{aligned} H^p(M;G) &\to H_{n-p}(M,\partial M;G), & x &\mapsto x \cap [M], \\ H^p(M,\partial M;G) &\to H_{n-p}(M;G), & x &\mapsto x \cap [M], \end{aligned}$$

are isomorphisms for each coefficient group G.

Proof. By naturality and stability of the cap product the following diagram commutes up to sign (coefficients are G); $[\partial M] = \partial[M]$ is a fundamental class:

$$\begin{array}{ccccccc} H^p(M,\partial M) & \xrightarrow{j^*} & H^p(M) & \xrightarrow{i^*} & H^p(\partial M) & \xrightarrow{\delta} & H^{p+1}(M,\partial M) \\ \downarrow{\scriptstyle \cap[M]} & & \downarrow{\scriptstyle \cap[M]} & & \downarrow{\scriptstyle \cap[\partial M]} & & \downarrow{\scriptstyle \cap[M]} \\ H_{n-p}(M) & \xrightarrow{j_*} & H_{n-p}(M,\partial M) & \xrightarrow{\partial} & H_{n-p-1}(\partial M) & \xrightarrow{i_*} & H_{n-p-1}(M). \end{array}$$

We know already that $\cap [\partial M]$ is an isomorphism. Therefore it suffices to show that the left-most vertical map is an isomorphism. We reduce the problem to the duality already proved. We use the non-compact auxiliary manifold

$$P = M \cup (\partial M \times [0,1[),$$

which is obtained by the identification $x \sim (x,0)$ for $x \in \partial M$. We also use the subspaces

$$\begin{aligned} M(t) &= M \cup (\partial M \times [0,t]), & 0 \le t < 1, \\ P(t) &= M \cup (\partial M \times [0,t[), & 0 < t \le 1. \end{aligned}$$

The $P(t)$ are a cofinal system of open neighbourhoods of $M = M(0)$ in P. The $M(t)$ are a compact exhaustion of P. The inclusions $M \subset P(t) \subset M(t)$ are h-equivalences. Let $i(t)\colon (M,\partial M) \subset (P(t), P(t) \smallsetminus M^\circ)$. The diagram

$$\begin{array}{ccc} H^p(M) & \xleftarrow{\;i(t)^*\;} & H^p(P(t)) \\ \downarrow{\scriptstyle \cap[M]} & & \downarrow{\scriptstyle \cap i(t)_*[M]} \\ H_{n-p}(M,\partial M) & \xrightarrow{\;i(t)_*\;} & H_{n-p}(P(t), P(t) \smallsetminus M^\circ) \end{array}$$

is commutative (naturality of the cap product). The map $i(t)^*$ is an isomorphism, since $i(t)$ is an h-equivalence; the map $i(t)_*$ is an excision. Thus it suffices to show that $\cap i(t)_*[M]$ is an isomorphism. We use the duality theorem for P. We have isomorphisms

$$H_n(P, P \smallsetminus M(t)) \cong H_n(P, P \smallsetminus M) \cong H_n(P, P \smallsetminus M^\circ) \cong H_n(M, \partial M).$$

Let $z(t) \in H_n(P, P \smallsetminus M(t))$ correspond to the fundamental class $[M]$. One verifies that $z(t)$ is an orientation along $M(t)$. Since the $M(t)$ form a compact exhaustion, the coherent family of the $z(t)$ yields an orientation of P. Let $w(t) \in H_n(P(t), P(t) \smallsetminus M)$ and $v(t) \in H_n(P(t), P(t) \smallsetminus M^\circ)$ correspond to the fundamental class under

$$H_n(P(t), P(t) \smallsetminus M) \cong H_n(P(t), P(t) \smallsetminus M^\circ) \cong H_n(M, \partial M).$$

By definition of the duality homomorphism, $D_M\colon \check{H}^p(M) \to H_{n-p}(P, P \smallsetminus M)$ is the colimit of the maps

$$H^p(P(t)) \xrightarrow{\cap w(t)} H_{n-p}(P(t), P(t) \smallsetminus M) \xrightarrow{\cong} H_{n-p}(P, P \smallsetminus M).$$

Since $H^p(P(t)) \to H^p(M)$ is an isomorphism, the canonical maps $H^p(P(t)) \to \check{H}^p(M) \to H^p(M)$ are isomorphisms. Since D_M is an isomorphism, so is $\cap w(t)$. The diagram

$$\begin{array}{ccc} H^p(P(t)) & \xleftarrow{\;i^*\;} & H^p(P(t)) \\ \downarrow{\scriptstyle \cap w(t)} & & \downarrow{\scriptstyle \cap v(t)} \\ H_{n-p}(P(t), P(t) \smallsetminus M) & \xrightarrow{\;i_*\;} & H_{n-p}(P(t), P(t) \smallsetminus M^\circ) \end{array}$$

is commutative by naturality of the cap product and $v(t) = i_* w(t)$. Since $v(t) = i(t)_*[M]$, the map $\cap i(t)_*[M]$ is an isomorphism.

□

(18.6.2) Proposition. *Let B be a compact $(n+1)$-manifold with boundary $\partial B = M$. Then $\chi(M) = (1 + (-1)^n)\chi(B)$. In particular $\chi(M)$ is always even.*

Proof. Let $M = B \cup (\partial B \times [0,1[)$. Then B is a compact deformation retract of M and $M \smallsetminus B \cong \partial B \times]0,1[\simeq \partial B$. Hence $\chi(B) = \chi(M) = \chi(M \smallsetminus B) + (-1)^{n+1}\chi(B) = \chi(\partial B) - (-1)^n \chi(B)$. □

(18.6.3) Example. $\mathbb{R}P^{2n}$ is not the boundary of a compact manifold, since $\chi(\mathbb{R}P^{2n})$ is odd. The same holds for an arbitrary finite product of even-dimensional real projective spaces.

Problems

1. Suppose $\partial M = A + B$ is a decomposition into two closed submanifolds. The diagram

$$\begin{array}{ccccccc} \cdots \longrightarrow & H^p(M,A) & \longrightarrow & H^p(M) & \longrightarrow & H^p(A) & \longrightarrow \cdots \\ & \downarrow{\scriptstyle \cap[M]} & & \downarrow{\scriptstyle \cap[M]} & & \downarrow{\scriptstyle \cap[\partial M]} & \\ \cdots \longrightarrow & H_{n-p}(M,B) & \longrightarrow & H_{n-p}(M,\partial M) & \longrightarrow & H_{n-p-1}(\partial M,A) & \longrightarrow \cdots \end{array}$$

is commutative up to sign. Hence $\cap[M]\colon H^p(M,A) \to H_{n-p}(M,B)$ is an isomorphism.

18.7 The Intersection Form. Signature

Let M be a closed n-manifold oriented by a fundamental class $[M] \in H_n(M;\mathbb{K})$, coefficients in a field $\mathbb{K}$. The ***evaluation on the fundamental class*** is $H^n(M) \to \mathbb{K}$, $x \mapsto x[M] = \langle x, [M]\rangle$. We can also write this as the composition $H^n(M) \cong \operatorname{Hom}(H_n(M),\mathbb{K}) \to \mathbb{K}$ where we evaluate a homomorphism on $[M]$. The canonical map $\varepsilon\colon H_0(M) \to \mathbb{K}$ allows us to write $\langle x, [M]\rangle = \varepsilon(x \cap [M])$.

(18.7.1) Proposition. *The bilinear form*

$$H^k(M) \times H^{n-k}(M) \xrightarrow{\ \cup\ } H^n(M) \xrightarrow{\langle -,[M]\rangle} \mathbb{K}$$

is regular. We write this form also as $(x,y) \mapsto x \odot y$.

Proof. We use the rule $(x \cup y) \cap [M] = x \cap (y \cap [M])$. It gives us the commutative diagram

$$\begin{array}{ccc} H^k(M) \times H^{n-k} & \xrightarrow{\ \cup\ } & H^n(M) \\ \cong\downarrow{\scriptstyle \mathrm{id}\times\cap[M]} & & \downarrow{\scriptstyle \cap[M]} \\ H^k(M) \times H_k(M) & \xrightarrow{\ \cap\ } & H_0(M) \\ \cong\downarrow{\scriptstyle \alpha\times\mathrm{id}} & & \downarrow{\scriptstyle \varepsilon} \\ \operatorname{Hom}(H_k(M),\mathbb{K}) \times H_k(M) & \xrightarrow{\ \mathrm{eval}\ } & \mathbb{K}. \end{array}$$

The bilinear form in question is isomorphic to the Hom-evaluation, and the latter is for each finite-dimensional vector space a regular form. □

We take now $\mathbb{R} = \mathbb{K}$ as coefficients and assume $n = 4t$. In that case

$$H^{2t}(M) \times H^{2t}(M) \to \mathbb{R}, \quad (x, y) \mapsto x \odot y$$

is a regular symmetric bilinear form. Recall from linear algebra: Let (V, β) be a real vector space together with a symmetric bilinear form β. Then V has a decomposition $V = V_+ \oplus V_- + V_0$ such that the form is positive definite on V_+, negative definite on V_- and zero on V_0. By Sylvester's theorem the dimensions of V_+ and V_- are determined by β. The integer $\dim V_+ - \dim V_-$ is called the signature of β. We apply this to the intersection form and call

$$\sigma(M) = \dim H^{2t}(M)_+ - \dim H^{2t}(M)_-$$

the ***signature*** of the closed oriented $4t$-manifold M. We also set $\sigma(M) = 0$, if the dimension of M is not divisible by 4. If $-M$ denotes the manifold with the opposite orientation, then one has $\sigma(-M) = -\sigma(M)$. If $M = M_1 + M_2$ then $H^{2t}(M) = H^{2t}(M_1) + H^{2t}(M_2)$, the forms on M_1 and M_2 are orthogonal, hence $\sigma(M_1 + M_2) = \sigma(M_1) + \sigma(M_2)$.

(18.7.2) Proposition. *The signature of $\mathbb{C}P^{2n}$ with its natural orientation induced by the complex structure is* 1.

Proof. Since $H^{2n}(\mathbb{C}P^{2n}; \mathbb{Z}))$ is the free abelian group generated by c^n, the claim follows from $\langle c^n, [\mathbb{C}P^n]\rangle = 1$. For $n = 1$ this holds by the definition of the first Chern class (see (19.1.2)). Consider the map $p\colon (\mathbb{C}P^1)^n \to \mathbb{C}P^n$ that sends $([a_1, b_1], \dots, [a_n b_n])$ to $[c_0, \dots, c_n]$ where $\prod_{j=1}^n (a_j x + b_j y) = \sum_{i=0}^n c_j x^j y^{n-j}$. Note that $H^2((\mathbb{C}P^1)^n; \mathbb{Z})$ is the free abelian group with basis $t_1, \dots, t_n$ where t_j is the first Chern class of $\mathrm{pr}_j^*(\eta)$. One verifies that $p^*(c) = t_1 + \dots + t_n$. This implies $p^* c^n = n! \cdot t_1 t_2 \dots t_n$. The map p has degree $n!$. These facts imply

$$n! = \langle p^* c^n, [\mathbb{C}P^1]^n\rangle = \langle c^n, p_*[\mathbb{C}P^1]^n\rangle = \langle c^n, n! \cdot [\mathbb{C}P^n]\rangle = n! \cdot \langle c^n, [\mathbb{C}P^n]\rangle,$$

hence $\langle c^n, [\mathbb{C}P^n]\rangle = 1$. □

(18.7.3) Proposition. *Let M and N be closed oriented manifolds. Give $M \times N$ the product orientation. Then $\sigma(M \times N) = \sigma(M)\sigma(N)$.*

Proof. Let $m = \dim M$ and $n = \dim M$. If $m + n \not\equiv \mathrm{mod} 4$ then $\sigma(M \times N) = 0 = \sigma(M)\sigma(N)$ by definition. In the case that $m + n = 4p$ we use the Künneth isomorphism and consider the decomposition

$$\begin{aligned} H^{2p}(M \times N) = {} & H^{m/2}(M) \otimes H^{n/2}(N) \\ & \textstyle\bigoplus_{2i<m} \left(H^i(M) \otimes H^{2p-i}(N) \oplus H^{m-i}(M) \otimes H^{n-2p+i}(N)\right). \end{aligned}$$

The first summand on the right-hand side is zero if m or n is odd. The form on $H^{2p}(M \times N)$ is transformed via the Künneth isomorphism by the formula

$(x \otimes y) \odot (x' \otimes y') = (-1)^{|y||x'|}(x \odot x')(y \odot y')$. Products of elements in different summands never contribute to the top dimension $m + n$. Therefore the signature to be computed is the sum of the signatures of the forms on the summands.

Consider the first summand. If $m/2$ and $n/2$ are odd, then the form is zero. In the other case let A be a basis of $H^{m/2}(M)$ such that the form has a diagonal matrix with respect to this basis and let B be a basis of $H^{n/2}(N)$ with a similar property. Then $(a \otimes b \mid a \in A, b \in B)$ is a basis of $H^{m/2}(M) \otimes H^{n/2}(N)$ for which the form has a diagonal matrix. Then

$$\begin{aligned}\sigma(M)\sigma(N) &= \left(\textstyle\sum_{a\in A} a \odot a\right)\left(\textstyle\sum_{b\in B} b \odot b\right) \\ &= \textstyle\sum_{(a,b)\in A\times B} (a \otimes b) \odot (a \otimes b) = \sigma(M \times N).\end{aligned}$$

Now consider the summand for $2i < m$. Choose bases A of $H^i(M)$ and B of $H^{2p-i}(N)$ and let A^*, B^* be the dual bases of $H^{m-i}(M)$, $H^{n-2p+i}(N)$ respectively. Then

$$(a \otimes b + a^* \otimes b^*, a \otimes b - a^* \otimes b^* \mid a \in A, b \in B)$$

is a basis of the summand under consideration. The product of different basis elements is zero, and $(a \otimes b + a^* \otimes b^*)^2 = -(a \otimes b - a^* \otimes b^*)^2$ shows that the number of positive squares equals the number of negative squares. Hence these summands do not contribute to the signature. □

There exists a version of the intersection form for cohomology with integral coefficients. We begin again with the bilinear form

$$s\colon H^k(M) \to H^{n-k}(M) \to \mathbb{Z}, \quad (x, y) \mapsto (x \cup y)[M].$$

We denote by $A_\diamond$ the quotient of the abelian group A by the subgroup of elements of finite order. We obtain an induced bilinear form

$$s_\diamond\colon H^k(M)_\diamond \times H^{n-k}(M)_\diamond \to \mathbb{Z}.$$

(18.7.4) Proposition. *The form $s_\diamond$ is regular, i.e., the adjoint homomorphism $H^k(M)_\diamond \to \mathrm{Hom}(H^{n-k}(M)_\diamond, \mathbb{Z})$ is an isomorphism (and not just injective).*

Proof. We use the fact that the evaluation $H^k(M;\mathbb{Z})_\diamond \times H_k(M;\mathbb{Z})_\diamond \to \mathbb{Z}$ is a regular bilinear form over $\mathbb{Z}$. By the universal coefficient formula, the kernel of $H^k(M;\mathbb{Z}) \to \mathrm{Hom}(H_k(M;\mathbb{Z}), \mathbb{Z})$ is a finite abelian group; hence we have isomorphisms

$$H^k(M;\mathbb{Z})_\diamond \cong \mathrm{Hom}(H_k(M;\mathbb{Z}), \mathbb{Z}) \cong \mathrm{Hom}(H_k(M;\mathbb{Z})_\diamond, \mathbb{Z}).$$

Now the proof is finished as before. □

We return to field coefficients. Let M be the oriented boundary of the compact oriented manifold B. We set $A^k = \mathrm{Im}(i^*\colon H^k(B) \to H^k(M))$ with the inclusion $i\colon M \subset B$.

(18.7.5) Proposition. *The kernel of*

$$H^k(M) \xrightarrow{\cong} \mathrm{Hom}(H^{n-k}(M), \mathbb{K}) \to \mathrm{Hom}(A^{n-k}, \mathbb{K})$$

is A^k. The isomorphism is $x \mapsto (y \mapsto \langle y \cup x, [M]\rangle)$; the second map is the restriction to A^{n-k}. In particular $\dim H^k(M) = \dim A^k + \dim A^{n-k}$, and in the case $n = 2t$ we have $\dim H^t(M) = 2\dim A^k$ and $\dim H_t(M) = 2\dim \mathrm{Ker}(i_*\colon H_t(M) \to H_t(B))$.

Proof. Consider the diagram

$$\begin{array}{ccccc} H^k(B) & \xrightarrow{i^*} & H^k(M) & \xrightarrow{\delta} & H^{k+1}(B, M) \\ & & \cong \Big\downarrow \cap[M] & & \cong \Big\downarrow \cap[B] \\ & & H_{n-k}(M) & \xrightarrow{i_*} & H_{n-k}(M). \end{array}$$

By stability of the cap product, the square commutes up to the sign $(-1)^k$. By commutativity and duality

$$x \in A^k \;\Leftrightarrow\; \delta(x) = 0 \;\Leftrightarrow\; \delta(x) \cap [B] = 0 \;\Leftrightarrow\; i_*(x \cap [M]) = 0.$$

The regularity of the pairing $H^j(M) \times H_j(M) \to \mathbb{K}$, $(x, y) \mapsto \langle x, y\rangle$ says that $i_*(x \cap [M]) = 0$ is equivalent to $\langle H^{n-k}(B), i_*(x \cap [M])\rangle = 0$. Properties of pairings yield

$$\begin{aligned} \langle H^{n-k}(B), i_*(x \cap [M])\rangle &= \langle i^* H^{n-k}(B), x \cap [M]\rangle \\ &= \langle A^{n-k}, x \cap [M]\rangle \\ &= \langle A^{n-k} \cup x, [M]\rangle \end{aligned}$$

and we see that $x \in A^k$ is equivalent to $\langle A^{n-k} \cup x, [M]\rangle = 0$, and the latter describes the kernel of the map in the proposition. □

(18.7.6) Example. If $n = 2t$ and $\dim H^t(M)$ is odd, then M is not a boundary of a $\mathbb{K}$-orientable compact manifold. This can be applied to $\mathbb{R}P^{2n}$ (for $\mathbb{K} = \mathbb{Z}/2$) and to $\mathbb{C}P^{2n}$ (for $\mathbb{K} = \mathbb{R}$). ◇

(18.7.7) Proposition. *Let M be the boundary of a compact oriented $(4k+1)$-manifold B. Then the signature of M is zero.*

Proof. It follows from Proposition (18.7.5) that the orthogonal complement of A^{2k} with respect to the intersection form on $H^{2k}(M;\mathbb{R})$ is A^{2k}, and $2\dim A^{2k} = \dim H^{2k}(M;\mathbb{R})$. Linear algebra tells us that a symmetric bilinear form with these properties has signature 0. □

We generalize the preceding by taking advantage of the general duality isomorphism (coefficients in $\mathbb{K}$). Let M be an n-manifold and $K \supset L$ a compact pair in M. Assume that M is $\mathbb{K}$-oriented along K. We define a bilinear form

$$(*) \qquad \check{H}^i(K,L) \times H^j(M \smallsetminus L, M \smallsetminus K) \xrightarrow{\cup} H^{i+j}(M, M \smallsetminus K)$$

as follows: Let (V,W) be a neighbourhood of (K,L). We fix an element $y \in H^j(M \smallsetminus L, M \smallsetminus K)$ and restrict it to $H^j(V \smallsetminus L, V \smallsetminus K)$. Then we have

$$\begin{aligned}&H^i(V,W) \to H^i(V \smallsetminus L, W \smallsetminus L) \xrightarrow{\cup y} H^{i+j}(V \smallsetminus L, W \smallsetminus K \cup V \smallsetminus K) \cong\\ &H^{i+j}(M, W \cup (M \smallsetminus K)) \to H^{i+j}(M, M \smallsetminus K).\end{aligned}$$

The colimit over the neighbourhoods (V,W) yields $\cup y$ in $(*)$.

(18.7.8) Proposition. *Let M be an n-manifold and $K \supset L$ compact ENR in M. Then*

$$H^i(K,L) \times H^{n-i}(M \smallsetminus L, M \smallsetminus K) \xrightarrow{\ \cup\ } H^n(M, M \smallsetminus K) \xrightarrow{\langle -, o_K\rangle} \mathbb{K}$$

is a regular bilinear form. □

(18.7.9) Example. Let M be a compact oriented n-manifold for $n \equiv 2 \bmod (4)$. Then the Euler characteristic $\chi(M)$ is even.

The intersection form $H^{n/2}(M) \times H^{n/2}(M) \to \mathbb{Q}$ with coefficients in $\mathbb{Q}$ is skew-symmetric and regular, since $n/2$ is odd. By linear algebra, a form of this type only exists on even-dimensional vector spaces.

$$\begin{aligned}\chi(M) &= \textstyle\sum_{i=0}^n (-1)^i \dim H_i(M)\\ &= -\dim H^{n/2}(M) + 2\textstyle\sum_{2i<n} (-1)^i \dim H^i(M);\end{aligned}$$

we have used $\dim H^i(M) = \dim H^{n-i}(M)$, and this holds because of $H^i(M) \cong \operatorname{Hom}(H_i(M),\mathbb{Q})$ and hence $\dim H^i(M) = \dim H_i(M) = \dim H^{n-i}(M)$. ◇

18.8 The Euler Number

Let $\xi\colon E(\xi) \to M$ be an n-dimensional real vector bundle over the closed connected orientable manifold M. The manifold is oriented by a fundamental class $[M] \in H_n(M;\mathbb{Z})$ and the bundle by a Thom class $t(\xi) \in H^n(E,E^0;\mathbb{Z})$. Let $s\colon M \to$

$E(\xi)$ be a section of ξ and assume that the zero set $N(s)$ is contained in the disjoint sum $D = D_1 \cup \dots \cup D_r$ of disks D_j. The aim of this section is to determine the Euler number $e(\xi) = \langle s^* t(\xi), [M] \rangle$ by local data. We assume given positive charts $\varphi_j : \mathbb{R}^n \to U_j$ with disjoint images of M such that $\varphi_j(D^n) = D_j$. The bundle is trivial over U_j. Let

$$\begin{array}{ccc} D^n \times \mathbb{R}^n & \xrightarrow{\Phi_j} & E(\xi|D_j) \\ \downarrow{\scriptstyle \mathrm{pr}} & & \downarrow \\ D^n & \xrightarrow{\varphi_j} & D_j \end{array}$$

be a trivialization. We assume that Φ_j is positive with respect to the given orientation of ξ. These data yield a commutative diagram

$$\begin{array}{ccc} H^n(E(\xi), E^0(\xi)) & \xrightarrow{s^*} & H^n(M) \\ \downarrow & & \uparrow{\scriptstyle \alpha^*} \\ H^n(E(\xi|D) \cup E^0(\xi|M \smallsetminus D^\circ, E^0(\xi|D)) & \xrightarrow{s^*} & H^n(, M \smallsetminus D^\circ) \\ \downarrow & & {\scriptstyle \cong}\downarrow{\scriptstyle \beta^*} \\ H^n(E(\xi|D), E^0(\xi|D)) & \xrightarrow{s^*} & H^n(D, S) \\ \downarrow{\scriptstyle \cong} & & \downarrow{\scriptstyle \cong} \\ \bigoplus_j H^n(E(\xi|D_j), E^0(\xi|D_j)) & \xrightarrow{s_j^*} & \bigoplus_j H^n(D_j, S_j). \end{array}$$

Here S_j is the boundary of D_j and $S = \bigcup_j S_j$. The restriction of s to D_j is s_j. The image of $t(\xi)$ in $H^n(E(\xi|D_j), E^0(\xi|D_j))$ is the Thom class $t(\xi|D_j)$. The vertical maps have their counterpart in homology

$$H_n(M) \xrightarrow{\alpha_*} H_n(M, M \smallsetminus D^\circ) \underset{\cong}{\xleftarrow{\beta_*}} H_n(D, S) \underset{\cong}{\longleftarrow} \bigoplus_j H_n(D_j, S_j),$$

and $[M]$ is mapped to $([D_j])$. By commutativity and naturality

$$\langle s^* t(\xi), [M] \rangle = \textstyle\sum_j \langle s_j^* t(\xi|D_j), [D_j] \rangle.$$

The bundle isomorphism (Φ_j, φ_j) transports s_j into a section

$$t_j : D^n \to D^n \times \mathbb{R}^n, \quad x \mapsto (x, u_j(x))$$

of pr. Note that $u_j(S^{n-1}) \subset \mathbb{R}^n \smallsetminus 0$. We have another commutative diagram

$$\begin{array}{ccc}
H^n(E(\xi|D_j), E^0(\xi|D_j)) & \xrightarrow{s_j^*} & H^n(D_j, S_j) \\
\downarrow{\scriptstyle \Phi_j^*} & & \downarrow{\scriptstyle \varphi_j^*} \\
H^n(D^n \times \mathbb{R}^n, D^n \times \mathbb{R}^n_0) & \xrightarrow{t_j^*} & H^n(D^n, S^{n-1}) \\
\uparrow{\scriptstyle \mathrm{pr}_*} & \nearrow{\scriptstyle u_j^*} & \\
H^n(\mathbb{R}^n, \mathbb{R}^n_0). & &
\end{array}$$

The evaluation $\langle s_j^* t(\xi|D_j), [D_j] \rangle$ is the degree of $u_j : S^{n-1} \to \mathbb{R}^n_0$ if we choose the correct orientations. We explain this now and use the following computation:

$$\begin{aligned}
\langle s_j^* t(\xi|D_j), [D_j] \rangle &= \langle s_j^* t(\xi|D_j), \varphi_{j*} \rangle = \langle \varphi_j^* s_j^* t(\xi|D_j), e_n \rangle \\
&= \langle t_j^* \Phi_j^* t(\xi|D_j), e_n \rangle = \langle u_j^* e^n, e_n \rangle \\
&= d(u_j) \langle e^n, e_n \rangle.
\end{aligned}$$

The cohomological degree of u_j is $u_j^* e^n = d(u_j) e^n$. With these definitions we obtain:

(18.8.1) Proposition. $e(\xi) = \left(\sum_{j=1}^r d(u_j)\right) \langle e^n, e_n \rangle$. □

If s has in D_j an isolated zero, then $d(u_j)$ is called the ***index*** of this zero.

Problems

1. There always exists a section with a single zero.
2. The index can be computed for transverse zeros of a smooth section $s \colon M \to E$ of a smooth bundle. Consider the differential

$$T_x s \colon T_x M \to T_x E = T_x M \oplus E_x.$$

Transversality means that the composition with the projection $\mathrm{pr} \circ T_x s \colon T_x M \to E_x$ is an isomorphism between oriented vector spaces. This isomorphism has a sign $\varepsilon(x) \in \{\pm 1\}$, $+1$ if the orientation is preserved. Show that $\varepsilon(x)$ is the local index.
3. The section

$$s \colon S^n \to TS^n \subset S^n \times \mathbb{R}^{n+1}, \quad x = (x_0, \dots, x_n) \mapsto (x, (x_0^2 - 1, x_0 x_1, \dots, x_0 x_n))$$

has the transverse zeros $(1, 0, \dots, 0)$ with index 1 and $(-1, 0, \dots, 0)$ with index $(-1)^n$.
4. Find a vector field on S^{2n} with a single zero (of index 2).
5. There exists a section without zeros if and only if the Euler number is zero.

We know already that the Euler number is zero, if there exists a non-vanishing section. For the converse one has to use two facts: (1) There always exist sections with isolated zeros.

(2) There exists a cell D which contains every zero. Hence one has to consider a single local index. This index is zero, and the corresponding map $u\colon S \to \mathbb{R}^n_0$ is null homotopic. Thus there exists an extension $u\colon D \to \mathbb{R}^n_0$. We use this extension to extend the section s over the interior of D without zeros.

6. Consider the bundle $\xi(k)\colon H(k) \to \mathbb{C}P^1$. Let $P(z_0,z_1) = \sum_{j=0}^k \alpha_k z_0^j z_1^{k-j}$ be a homogeneous polynomial of degree k. Then

$$\sigma\colon \mathbb{C}P^1 \to H(k), \quad [z_0,z_1] \mapsto (z_0,z_1;P(z_0,z_1))$$

is a section of $\xi(k)$. If $P(z_0,z_1) = \prod_j (a_j z_1 - b_j z_0)$ is the factorization into linear factors, then the $[a_j,b_j] \in \mathbb{C}P^1$ are the zeros of σ, with multiplicities.

7. Consider the bundle $\xi\colon S^n \times_{\mathbb{Z}/2} \mathbb{R}^n \to \mathbb{R}P^n$, $(x,z) \mapsto [x]$. Then $\sigma\colon [x_0,\dots,x_n] \mapsto ((x_0,\dots,x_n),(x_1,\dots,x_n))$ is a section with a single zero. The sections correspond to maps $f\colon S^n \to \mathbb{R}^n$ such that $f(-x) = -f(x)$. One form of the theorem of Borsuk–Ulam says that maps of this type always have a zero. We would reprove this result, if we show that the Euler class mod (2) is non-zero. The tautological bundle η over $\mathbb{R}P^n$ has as Euler class the non-zero element w of $H^1(\mathbb{R}P^n;\mathbb{Z}/2)$. The Euler classes are multiplicative and $\xi = n\eta$. Hence $e(\xi) = w^n \neq 0$.

18.9 Euler Class and Euler Characteristic

Let M be a closed connected n-manifold, oriented by a fundamental class $[M] \in H_n(M)$. For $M \times M$ we use the product orientation $[M \times M] = [M] \times [M]$. Let $d\colon M \to M \times M$ be the diagonal map. By naturality of the cap product and Poincaré duality there exists a unique element $\tau(M) \in H^n(M \times M)$ such that $\tau \cap [M \times M] = d_*[M]$. We call $d^*\tau(M) = e(M) \in H^n(M)$ the ***Euler class*** of the topological manifold M and $\langle e(M),[M]\rangle \in \mathbb{Z}$ its ***Euler number***. We now pass to coefficients in a field $\mathbb{K}$. The image of the fundamental class $[M] \in H_n(M;\mathbb{Z})$ in $H_n(M;\mathbb{K})$ is still denoted $[M]$, and similarly for $[M \times M]$.

Let $B = \{\alpha\}$ be a basis of $H^*(M)$ and $\{\alpha^0\}$ the dual basis in $H^*(M)$ with respect to the intersection form $\langle \alpha^0 \cup \beta, [M]\rangle = \delta_{\alpha\beta}$, $|\alpha^0| = n - |\alpha|$.

(18.9.1) Proposition. *The element $\tau(M) \in H^n(M \times M)$ is given by*

$$\tau(M) = \textstyle\sum_{\alpha\in B} (-1)^{|\alpha|} \alpha^0 \times \alpha \in H^n(M \times M).$$

A consequence is

$$e(M) = d^*\tau(M) = \textstyle\sum_{\alpha\in B} (-1)^{|\alpha|} \alpha^0 \cup \alpha,$$

$$\langle e(M),[M]\rangle = \textstyle\sum_\alpha (-1)^{|\alpha|} \langle \alpha^0 \cup \alpha, [M]\rangle = \sum_\alpha (-1)^{|\alpha|} = \chi(M).$$

Proof. The Künneth-isomorphism

$$H^*(M) \otimes H^*(M) \cong H^*(M \times M), \quad u \otimes v \mapsto u \times v$$

tells us that there exists a relation of the form $\tau = \sum_{\gamma,\delta \in B} A(\gamma,\delta)\gamma^0 \times \delta$. The following computations determine the coefficient $A(\gamma,\delta)$. Let α and β be basis elements of degree p. Then

$$\begin{aligned}\langle(\alpha \times \beta^0) \cup \tau, [M \times M]\rangle &= \langle \alpha \times \beta^0, \tau \cap [M \times M]\rangle \\ &= \langle \alpha \times \beta^0, d_*[M]\rangle = \langle d^*(\alpha \times \beta^0), [M]\rangle = \langle \alpha \cup \beta^0, [M]\rangle \\ &= (-1)^{p(n-p)}\langle \beta^0 \cup \alpha, [M]\rangle = (-1)^{p(n-p)}\delta_{\alpha\beta}.\end{aligned}$$

A second computation gives

$$\begin{aligned}\langle(\alpha \times \beta^0) \cup \tau, [M \times M]\rangle &= \langle(\alpha \times \beta^0) \cup \textstyle\sum A(\gamma,\delta)\gamma^0 \times \delta, [M \times M]\rangle \\ &= \textstyle\sum A(\gamma,\delta)(-1)^{|\beta^0||\gamma^0|}\langle(\alpha \cup \gamma^0) \times (\beta^0 \cup \delta), [M] \times [M]\rangle \\ &= \textstyle\sum A(\gamma,\delta)(-1)^{|\beta^0||\gamma^0|}(-1)^{n(|\beta^0|+|\delta|)}\langle \alpha \cup \gamma^0, [M]\rangle\langle \beta^0 \cup \delta, [M]\rangle.\end{aligned}$$

Only summands with $\gamma = \alpha$ and $\delta = \beta$ are non-zero. Thus this evaluation has the value $A(\alpha,\beta) \cdot (-1)^{pn}$ (collect the signs and compute modulo 2). We compare the two results and obtain $A(\alpha,\beta) = (-1)^p \delta_{\alpha\beta}$. □

The last equality in (18.9.1) holds by its proof in each field and therefore also in $\mathbb{Z}$.

We now relate the result to the previously defined Euler class and Euler number for a smooth manifold. Let $\nu \colon E(\nu) \to M$ be the normal bundle of the diagonal $d \colon M \to M \times M$ with disk- and sphere bundle $D(\nu)$ and $S(\nu)$ and tubular map $j \colon D(\nu) \to M \times M$. The fundamental class $[M \times M]$ induces a fundamental class $[D(\nu)] \in H_{2n}(D(\nu), S(\nu))$ via

$$H_{2n}(M \times M) \xrightarrow{\iota_*} H_{2n}(M \times M, M \times M \smallsetminus D) \xleftarrow[\cong]{j_*} H_{2n}(D(\nu), S(\nu)),$$

$\iota_*[M \times M] = z = j_*[D(\nu)]$. The diagram

$$\begin{array}{ccccc} H^n(M \times M) & \xleftarrow{\iota^*} & H^n(M \times M, M \times M \smallsetminus D) & \xrightarrow[\cong]{j^*} & H^n(D(\nu), S(\nu)) \\ \downarrow{\scriptstyle \cap[M\times M]} & & \downarrow{\scriptstyle \cap z} & & \downarrow{\scriptstyle \cap[D(\nu)]} \\ H_n(M \times M) & \xrightarrow{=} & H_n(M \times M) & \xleftarrow{j_*} & H_n(D(\nu)) \end{array}$$

commutes (naturality of the cap-product). From the isomorphisms with the zero section $i \colon M \to D(\nu)$

$$H_n(M) \xrightarrow[\cong]{i_*} H_n(D(\nu)) \xleftarrow[\cong]{\cap[D(\nu)]} H^n(D(\nu), S(\nu))$$

we obtain an element $t(\nu)$ that satisfies $i_*[M] = t(\nu) \cap [D(\nu)]$. It is a generator and therefore a Thom class. Let $j^*t(M) = t(\nu)$. From the definitions we see that $d^*\tau$ is the image of $t(\nu)$ under $H^n(D(\nu), S(\nu)) \to H^n(D(\nu)) \xrightarrow{i^*} H^n(M)$, hence the Euler class $e(\nu)$ of ν. Now we use the fact that the normal bundle ν is isomorphic to the tangent bundle of M. Therefore $e(\nu)$ is an Euler class of M in this sense.

Problems

1. The projection $p\colon (M \times M, M \times M \smallsetminus D) \to M$ onto the first coordinate is locally trivial. The element $t(M) \in H^n(M \times M, M \times M \smallsetminus D)$ restricts to a generator of the group $H^n(x \times M, x \times (M \smallsetminus x))$ for each $x \in M$, hence is a Thom class in this sense.
2. There exists a tubular map $TM \to M \times M$ for the diagonal which is a map over M. For its construction one can use the so-called exponential map, see e.g. [30, §12].

Chapter 19
Characteristic Classes

Characteristic classes are cohomological invariants of bundles which are compatible with bundle maps. Let $h^*(-)$ be a cohomology theory. An h^k-valued characteristic class for numerable n-dimensional complex vector bundles, say, assigns to each such bundle $\xi\colon E(\xi) \to B$ an element $c(\xi) \in h^k(B)$ such that for a bundle map $\xi \to \eta$ over $f\colon B \to C$ the naturality property $f^*c(\eta) = c(\xi)$ holds.

An assignment which has these properties is determined by its value $c(\gamma_n) \in h^k(B\mathrm{U}(n))$ on the universal bundle γ_n, and this value can be prescribed in an arbitrary manner (Yoneda lemma). In other words, the elements of $h^k(B\mathrm{U}(n))$ correspond to this type of characteristic classes.

It turns out that in important cases characteristic classes are generated by a few of them with distinguished properties, essentially a set of generators of the cohomology of classifying spaces.

We work with a multiplicative and additive cohomology theory h^* and bundles are assumed to be numerable. A $\mathbb{C}$-orientation of the theory assigns to each n-dimensional complex vector bundle (numerable, over a CW-complex,...) $\xi\colon E(\xi) \to B$ a Thom class $t(\xi) \in h^{2n}(E(\xi), E^0(\xi))$ such that for a bundle map $f\colon E(\xi) \to E(\eta)$ the naturality $f^*t(\eta) = t(\xi)$ holds and the Thom classes are multiplicative $t(\xi) \times t(\eta) = t(\xi \times \eta)$. If an assignment of this type is given, then the theory is called ***$\mathbb{C}$-oriented***. In a similar manner we call a theory ***$\mathbb{R}$-oriented***, if for each n-dimensional real vector bundle $\xi\colon E(\xi) \to B$ a Thom class $t(\xi) \in h^n(E(\xi), E^0(\xi))$ is given which is natural and multiplicative. It is a remarkable fact that structures of this type are determined by 1-dimensional bundles.

(19.0.1) Theorem. *A $\mathbb{C}$-orientation is determined by its value*

$$t(\gamma_1) \in h^2(E(\gamma_1), E^0(\gamma_1))$$

on the universal 1-dimensional bundle γ_1 over $\mathbb{C}P^\infty$. Each Thom class t of γ_1 determines a $\mathbb{C}$-orientation. A similar bijection exists between Thom classes of the universal 1-dimensional real vector bundle over $\mathbb{R}P^\infty$ and $\mathbb{R}$-orientations.

An example of a $\mathbb{C}$-oriented theory is $H^*(-;\mathbb{Z})$; a complex vector bundle has a canonical Thom class and these Thom classes are natural and multiplicative. One can use an arbitrary commutative ring as coefficient ring.

An example of an $\mathbb{R}$-oriented theory is $H^*(-;\mathbb{Z}/2)$; a real vector bundle has a unique Thom class in this theory. One can use any commutative ring of characteristic 2 as coefficient ring.

Suppose the theory is $\mathbb{C}$-oriented. Then an n-dimensional complex vector bundle ξ over B has an Euler class $e(\xi) \in h^{2n}(B)$ associated to $t(\xi)$. Euler classes are natural, $f^*e(\eta) = e(f^*\eta)$, and multiplicative, $e(\xi \oplus \eta) = e(\xi) \cup e(\eta)$.

The Thom classes have associated Thom homomorphisms. They are defined as before by cup product with the Thom class

$$\Phi(\xi)\colon h^k(B,A) \to h^{k+2n}(E(\xi), E(\xi)^0 \cup E(\xi|A)), \quad x \mapsto \xi^*(x) \cup t(\xi).$$

These Thom homomorphisms are natural and multiplicative as we have explained earlier.

For an $\mathbb{R}$-oriented theory we have natural and multiplicative Euler classes for real vector bundles.

A proof of (19.0.1) is based on a determination of characteristic classes. We present a construction of characteristic classes based on the cohomology of projective bundles. For this purpose, classifying spaces are not used. But they will of course appear and they are necessary for a more global view-point.

19.1 Projective Spaces

Let $\eta_n\colon E_n \to \mathbb{C}P^{n-1}$ be the canonical bundle with total space

$$E_n = \mathbb{C}^n \smallsetminus 0 \times_{\mathbb{C}^*} \mathbb{C}, \quad (z,u) \sim (\lambda z, \lambda u).$$

We have the embedding $\nu_n\colon E_n \to \mathbb{C}P^n$, $(z,u) \mapsto [z,u]$. The image is the complement of the point $* = [0,\dots,0,1]$. Let $t(\eta_n) \in h^2(E_n, E_n^0)$ be a Thom class. The Thom class yields the element $t_n \in h^2(\mathbb{C}P^n)$ as the image under

$$h^2(E_n, E_n^0) \xleftarrow{\cong} h^2(\mathbb{C}P^n, \mathbb{C}P^n \smallsetminus \mathbb{C}P^{n-1}) \cong h^2(\mathbb{C}P^n, *) \to h^2(\mathbb{C}P^n).$$

The first isomorphism is induced by ν_n. Note that ν_n sends the zero section to $\mathbb{C}P^{n-1}$, the image under the embedding $\iota\colon [x_1,\dots,x_n] \mapsto [x_1,\dots,x_n,0]$. The total space E_n of η_n was denoted $H(1)$ in (14.2.6). The bundle η_n is the (complex) normal bundle of the embedding $\iota\colon \mathbb{C}P^{n-1} \to \mathbb{C}P^n$. The embedding ν_n is a tubular map; it also shows that $\mathbb{C}P^n$ is the one-point compactification of E_n (see the definition of a Thom space in the final chapter). The complement $\mathbb{C}P^n \smallsetminus \mathbb{C}P^{n-1}$ is the affine subset $U_{n+1} = \{[x_1,\dots,x_{n+1}] \mid x_{n+1} \neq 0\}$. We obtain a homomorphism $h^*[t_n] \to h^*(\mathbb{C}P^n)$ of graded h^*-algebras; it sends t_n^{n+1} to zero (see (17.2.5)) and induces a homomorphism of the quotient by the principal ideal (t_n^{n+1}).

(19.1.1) Lemma. *Let $t(\eta_{n-1}) \in h^2(E_{n-1}, E_{n-1}^0)$ be the Thom class obtained from $t(\eta_n)$ by restriction along ι. Let $t_{n-1} \in h^2(\mathbb{C}P^{n-1})$ be obtained from $t(\eta_{n-1})$ as explained above. Then:*

(1) $\iota^* t_n = t_{n-1}$.

(2) t_{n-1} *is the Euler class associated to* $t(\eta_n)$.

Proof. (1) The embedding ι is homotopic to the embedding $\iota_1\colon [x_1,\dots,x_n] \mapsto [0,x_1,\dots,x_n]$. Thus it suffices to show $\iota_1^* t_n = t_{n-1}$. We have a bundle map $\iota_2\colon E_{n-1} \to E_n$ which is compatible with the embeddings, i.e., $\nu_n \iota_2 = \iota_1 \nu_{n-1}$. We apply cohomology to this commutativity and obtain the desired result.

(2) With the zero section s the diagram

$$\begin{array}{ccc} h^2(\mathbb{C}P^n, \mathbb{C}P^n \smallsetminus \mathbb{C}P^{n-1}) & \longrightarrow & h^2(\mathbb{C}P^n) \\ \cong \big\downarrow \nu_n^* & & \big\downarrow \iota^* \\ h^2(E_n, E_n^0) & \xrightarrow{\ s^*\ } & h^2(\mathbb{C}P^{n-1}) \end{array}$$

commutes. The definition of the Euler class and (1) now yield the result. □

(19.1.2) Lemma. *For singular homology and cohomology the Kronecker pairing relation* $\langle t_1, [\mathbb{C}P^1]\rangle = 1$ *holds. The element* t_1 *is by* (19.1.1) *also the Euler class of* η_2 *and this is, by definition, the first Chern class.*

Proof. η_1 is a bundle over a point. We have the isomorphism $\varphi\colon \mathbb{C} \to E_1$, $z \mapsto (1,z)$. By definition of the canonical Thom class of a complex vector bundle the Thom class $t(\eta_1) \in H^2(E_1, E_1^0)$ is mapped to the generator $e^{(2)}$ under φ^*, where $e^{(2)}$ is defined by the relation $\langle e^{(2)}, e_2\rangle = 1$ (Kronecker pairing). The element t_1 is the image of $e^{(2)}$ under $H^2(\mathbb{C}, \mathbb{C}\smallsetminus 0) \to H^2(\mathbb{C}P^1, \mathbb{C}P^1 \smallsetminus \mathbb{C}P^0) \to H^2(\mathbb{C}P^1)$ and the fundamental class $[\mathbb{C}P^1]$ is mapped to e_2 under $H_2(\mathbb{C}P^1) \to H_2(\mathbb{C}P^1, \mathbb{C}P^1 \smallsetminus \mathbb{C}P^0) \to H_2(\mathbb{C}, \mathbb{C}\smallsetminus 0)$. Naturality of the Kronecker pairing now gives the desired result. □

(19.1.3) Theorem. *The homomorphism just constructed is an isomorphism* $h^*(\mathbb{C}P^n) \cong h^*[t_n]/(t_n^{n+1})$ *of graded* h^**-algebras. In particular* $h^*(\mathbb{C}P^n)$ *is a free* h^**-module with basis* $1, t_n, t_n^2, \dots, t_n^n$.

Proof. Induction on n. We have the Thom isomorphism

$$h^k(\mathbb{C}P^{n-1}) \to h^{k+2}(E_n, E_n^0), \quad x \mapsto \eta_n^* x \cup t(\eta_n) = x \cdot t(\eta_n).$$

By induction, $h^*(\mathbb{C}P^{n-1})$ is a free h^*-module with basis $1, t_{n-1}, \dots, t_{n-1}^{n-1}$ and therefore $h^*(E_n, E_n^0)$ is a free module with basis

$$1 \cdot t(\eta_n),\ t_{n-1} \cdot t(\eta_n),\ \dots,\ t_{n-1}^{n-1} \cdot t(\eta_n).$$

We apply the isomorphism $h^*(E_n, E_n^0) \cong h^*(\mathbb{C}P^n, *)$ constructed above and claim that it sends $t_{n-1}^k \cdot t(\eta_n)$ to t_n^{k+1}. Unraveling the definitions one shows that this claim is a consequence of the naturality of the cup product and the fact (19.1.1) that t_{n-1} is the Euler class of $t(\eta_n)$. □

We denote by $h^*[[T]]$ the ring of homogeneous formal power series in T over the graded ring h^*. If T has the degree 2, then the power series in $h^*[[T]]$ of degree k consist of the series $\sum_j a_j T^j$ with $a_j \in h^{k-2j}$. If h^* is concentrated in degree zero, then this coincides with the polynomial ring $h^0[T]$.

Let $t(\eta_\infty) \in h^2(E_\infty, E_\infty^0)$ be a Thom class and $t(\eta_n)$ its restriction. Let $t_\infty \in h^2(\mathbb{C}P^\infty)$ and $t_n \in h^2(\mathbb{C}P^n)$ be the corresponding elements. Since t_n is the restriction of t_∞ let us write just t for all these elements. We have a surjective restriction homomorphism $h^*(\mathbb{C}P^{n+1}) \to h^*(\mathbb{C}P^n)$. Thus the restrictions induce an isomorphism (see (17.1.6)and (17.1.7))

$$h^*(\mathbb{C}P^\infty) \cong \lim_n h^*(\mathbb{C}P^n) \cong \lim_n h^*[t]/(t^{n+1}).$$

The algebraic limit is $h^*[[t]]$. This shows:

(19.1.4) Theorem. $h^*(\mathbb{C}P^\infty) \cong h^*[[t]]$. □

We extend the previous results by a formal trick to products $X \times \mathbb{C}P^n$. Let $p\colon X \times \mathbb{C}P^n \to \mathbb{C}P^n$ be the projection. We set $u = u_n = p^*(t_n)$.

(19.1.5) Proposition. *Consider $h^*(X \times \mathbb{C}P^n)$ as a graded $h^*(X)$-algebra. Then $h^*(X \times \mathbb{C}P^n) \cong h^*(X)[u]/(u^{n+1})$ and $h^*(X \times \mathbb{C}P^\infty) \cong h^*(X)[[u]]$.*

Proof. The cohomology theory $k^*(-) = h^*(X \times -)$ is additive and multiplicative, and the coefficient algebra is $h^*(X)$. The multiplicative structure in $k^*(-)$ is induced by the $\times$-product of $h^*(-)$ and the diagonal of X. The element u_∞ now plays the role of t_∞. □

Let $p_i\colon (\mathbb{C}P^\infty)^n \to \mathbb{C}P^\infty$ be the projection onto the i-th factor, and set $T_i = p_i^*(t_\infty)$. Then (19.1.5) implies:

(19.1.6) Proposition. $h^*((\mathbb{C}P^\infty)^\infty) \cong h^*[[T_1, \dots, T_n]]$. □

This statement uses algebraic identities of the type $h^*[[x, y]] \cong (h^*[[x]])[[y]]$ for graded formal power series rings.

Problems

1. An element $\sum_{i=0}^n a_j t_n^j \in h^0(\mathbb{C}P^n)$ is a unit if and only if $a_0 \in h^0$ is a unit.
2. An element $u = \sum_{i=1}^n b_i t_n^i \in h^2(\mathbb{C}P^n)$ is a Thom class of η_{n+1} if and only if b_1 is a unit, and this holds if and only if $u = \varepsilon t_n$ for a unit $\varepsilon \in h^0(\mathbb{C}P^n)$.
3. Let t_{n-1} be a Thom class for η_{n-1} and u_n a Thom class for η_n. Then there exists a Thom class t_n for η_n such that its restriction is t_{n-1}.

19.2 Projective Bundles

Let $\xi\colon E(\xi) \to B$ be an n-dimensional complex vector bundle. (For the moment we work with bundles over spaces of the homotopy type of a CW-complex.) The group $\mathbb{C}^*$ acts fibrewise on $E^0(\xi)$ by scalar multiplication. Let $P(\xi)$ be the orbit space. The projection ξ induces a projection $p_\xi\colon P(\xi) \to B$. The fibre $p_\xi^{-1}(b)$ is the projective space $P(\xi_b)$ of the vector space $\xi^{-1}(b) = \xi_b$. We call p_ξ the ***projective bundle*** associated to ξ.

There exists a canonical line bundle $Q(\xi) \to P(\xi)$ over $P(\xi)$. Its total space is defined as $E^0(\xi) \times_{\mathbb{C}^*} \mathbb{C}$ with respect to the relation $(x, u) \sim (x\lambda, \lambda u)$. Thus over each fibre $P(\xi_b)$ we have a bundle canonically isomorphic to η_n.

The construction of the projective bundle is compatible with bundle maps. Let $\eta\colon E(\eta) \to C$ be a further bundle and $\varphi\colon \xi \to \eta$ a bundle map over $f\colon B \to C$. These data yield an induced bundle map

$$\begin{array}{ccc} Q(\xi) & \xrightarrow{Q(\varphi)} & Q(\eta) \\ \downarrow & & \downarrow \\ P(\xi) & \xrightarrow{P(\varphi)} & P(\eta). \end{array}$$

We now assume that we are given a Thom class $t_\infty \in h^2(\mathbb{C}P^\infty)$ of the universal line bundle η_∞ over $\mathbb{C}P^\infty$. A classifying map $k_\xi\colon P(\xi) \to \mathbb{C}P^\infty$ of the line bundle $Q(\xi) \to P(\xi)$ provides us with the element

$$t_\xi = k_\xi^*(t_\infty) \in h^2(P(\xi)).$$

We consider $h^*(P(\xi))$ in the standard manner as left $h^*(B)$-module, $x \cdot y = p_\xi^*(x) \cup y$.

(19.2.1) Example. Let $\xi = \eta_{n+1}\colon E_{n+1} \to \mathbb{C}P^n$. Then $Q(\xi) \to P(\xi)$ is canonically isomorphic to η_{n+1} and $t_\xi = t_n$. ◇

(19.2.2) Theorem. *The $h^*(B)$-module $h^*(P(\xi))$ is free with basis*

$$1,\ t_\xi,\ t_\xi^2,\ \dots,\ t_\xi^{n-1}.$$

In particular $p_\xi^*\colon h^*(B) \to h^*(P(\xi))$ is injective.

Proof. This is a consequence of the Leray–Hirsch theorem (17.8.4) and the computation (19.1.3). □

(19.2.3) Corollary. *There exist uniquely determined elements $c_j(\xi) \in h^{2j}(B)$ such that*

$$\textstyle\sum_{j=0}^n (-1)^j c_j(\xi) t_\xi^{n-j} = 0,$$

since t_ξ^n is a linear combination of the basis ($c_0(\xi) = 1$). □

(19.2.4) Remark. Here is a justification for the choice of the signs. Let $\xi = \eta_{n+1}$. Then $t_\xi - t_n = 0$ and hence $c_1(\eta_{n+1}) = t_n$. ◇

(19.2.5) Proposition. *Let $\varphi\colon \xi \to \eta$ be a bundle map over $f\colon B \to C$. Then the naturality relation $f^*(c_j(\eta)) = c_j(\xi)$ holds.*

Proof. The homotopy relation $k_\eta \circ P(\varphi) \simeq k_\xi$ implies $P(\varphi)^* t_\eta = t_\xi$. This yields

$$0 = P(\varphi)^*\big(\textstyle\sum_j (-1)^j c_j(\eta) t_\eta^{n-j}\big) = \sum_j (-1)^j f^*(c_j(\eta)) t_\xi^{n-j}.$$

Comparing coefficients gives the claim. We have used the rule $P(\varphi)^*(a \cdot x) = f^*(a) \cdot P(\varphi)^*(x)$, $a \in h^*(C)$, $x \in h^*(P(\eta))$ for the module structure; it is a consequence of the naturality of the cup product. □

(19.2.6) Proposition. *Let ξ and η be bundles over B. Then the sum formula*

$$c_r(\xi \oplus \eta) = \textstyle\sum_{i+j=r} c_i(\xi) c_j(\eta).$$

holds. We set $c_i(\xi) = 0$, if $i > \dim \xi$.

Proof. Consider the subspaces $P(\xi) \subset P(\xi \oplus \eta) \supset P(\eta)$ and their open complements $U = P(\xi \oplus \eta) \smallsetminus P(\eta)$ and $V = P(\xi \oplus \eta) \smallsetminus P(\xi)$. The inclusions $P(\xi) \subset U$ and $P(\eta) \subset V$ are deformation retracts. Let $s\colon P(\xi) \to P(\xi \oplus \eta) \smallsetminus P(\eta)$, $[x] \mapsto [x, 0]$ and $\pi\colon P(\xi \oplus \eta) \smallsetminus P(\eta) \to P(\xi)$, $[x, y] \mapsto [x]$. Then $\pi s = \mathrm{id}$ and $s\pi \simeq \mathrm{id}$ by the homotopy $([x, y], \lambda) \mapsto [x, \lambda y]$.

Let $t = t_{\xi \oplus \eta}$. Consider the elements ($k = \dim \xi, l = \dim \eta$)

$$x = \textstyle\sum_{i=0}^k (-1)^i c_i(\xi) t^{k-i}, \quad y = \sum_{j=0}^l (-1)^j c_j(\eta) t^{l-j}.$$

Under the restriction $h^*(P(\xi \oplus \eta)) \to h^*(U) \cong h^*(P(\xi))$ the element x is sent to zero; this is a consequence of the definition of the $c_i(\xi)$, the deformation retraction and the naturality $t|P(\xi) = t_\xi$. Hence x comes from an $x' \in h^*(P(\xi \oplus \eta), U)$. Similarly y comes from an element $y' \in h^*(P(\xi \oplus \eta), V)$. Since U, V is an open covering of $P(\xi \oplus \eta)$, we see that $x'y' = 0$ and therefore $xy = 0$. We use the definition of the $c_r(\xi \oplus \eta)$ in the relation

$$xy = \textstyle\sum_{r=0}^{k+l} (-1)^r \big(\sum_{i+j=r} c_i(\xi) c_j(\eta)\big) t^r$$

and arrive at the desired sum formula by comparing coefficients. □

19.3 Chern Classes

Let $h^*(-)$ be a cohomology theory with universal element $t = t_\infty \in h^2(\mathbb{C}P^\infty)$. Our first aim is the computation of $h^*(B\mathrm{U}(n))$. Recall that $B\mathrm{U}(1) = \mathbb{C}P^\infty$. The space $B\mathrm{U}(n)$ is the basis of the universal n-dimensional complex vector bundle γ_n

and $\gamma_1 = \eta_\infty$. We use that $h^*(BU(1)^n) \cong h^*[[T_1, \dots, T_n]]$, see (19.1.6). Let us recall the ring of formal graded power series $h^*[[c_1, \dots, c_n]]$. The indeterminate c_j has degree $2j$. The degree of a monomial in the c_j is the sum of the degrees of the factors

$$\text{degree}(c_1^{k(1)} c_2^{k(2)} \dots) = 2k(1) + 4k(2) + \cdots .$$

A homogeneous power series of degree k is the formal sum of terms of the form $\lambda_j M_j$ where M_j is a monomial of degree m and $\lambda_j \in h^{k-m}$. Thus we assign the degree k to the elements in the coefficient group h^k.

(19.3.1) Lemma. *A classifying map $\beta\colon \mathrm{BU}(n-1) \times BU(1) \to BU(n)$ of the product $\gamma_{n-1} \times \gamma_1$ is the projective bundle of γ_n.*

Proof. Let $\mathrm{U}(n-1) \times \mathrm{U}(1) \subset \mathrm{U}(n)$ be the subgroup of block diagonal matrices. We obtain a map

$$\begin{aligned} \alpha\colon B(\mathrm{U}(n-1) \times \mathrm{U}(1)) &= E\mathrm{U}(n)/(\mathrm{U}(n-1) \times \mathrm{U}(1)) \\ &= E\mathrm{U}(n) \times_{\mathrm{U}(n)} (\mathrm{U}(n)/\mathrm{U}(n-1) \times \mathrm{U}(1)) \\ &\to B\mathrm{U}(n). \end{aligned}$$

A model for the universal vector bundle is $\gamma_n\colon E\mathrm{U}(n) \times_{\mathrm{U}(n)} \mathbb{C}^n \to B\mathrm{U}(n)$. The $\mathrm{U}(n)$-matrix multiplication on $\mathbb{C}^n$ induces a $\mathrm{U}(n)$-action on the corresponding projective space $P(\mathbb{C}^n)$. The projective bundle associated to the universal bundle γ_n is $E\mathrm{U}(n) \times_{\mathrm{U}(n)} P(\mathbb{C}^n) \to B\mathrm{U}(n)$. We now use the $\mathrm{U}(n)$-isomorphism $P(\mathbb{C}^n) \cong \mathrm{U}(n)/\mathrm{U}(n-1) \times \mathrm{U}(1)$. Hence α is the projective bundle of γ_n. We compose with a canonical h-equivalence $j\colon B\mathrm{U}(n-1) \times B\mathrm{U}(1) \to B(\mathrm{U}(n-1) \times \mathrm{U}(1))$.

It remains to show that $\beta = \alpha \circ j$ is a classifying map for $\gamma_{n-1} \times \gamma_1$.

Let $E\mathrm{U}(n-1) \times E\mathrm{U}(1) \to E\mathrm{U}(n)$ be a $\mathrm{U}(n-1) \times \mathrm{U}(1)$-map. From it we obtain a bundle map

$$\begin{aligned} E(\gamma_{n-1}) \times E(\gamma_1) &= (E\mathrm{U}(n-1) \times E\mathrm{U}(1)) \times_{\mathrm{U}(n-1)\times\mathrm{U}(1)} (\mathbb{C}^{n-1} \times \mathbb{C}^1) \\ &\to E\mathrm{U}(n) \times_{\mathrm{U}(n-1)\times\mathrm{U}(1)} \mathbb{C}^n \\ &\to E\mathrm{U}(n) \times_{\mathrm{U}(n)} \mathbb{C}^n = E(\gamma_n). \end{aligned}$$

It is a bundle map over $\alpha \circ j$. □

(19.3.2) Theorem. *Let $\kappa\colon \mathrm{BU}(1)^n \to B\mathrm{U}(n)$ be a classifying map of the n-fold Cartesian product of the universal line bundle. Then the following holds: The induced map $\kappa^*\colon h^*(B\mathrm{U}(n)) \to h^*(B\mathrm{U}(1)^n)$ is injective. The image consists of the power series which are symmetric in the variables $T_1, \dots, T_n$. Let $c_i \in h^{2i}(B\mathrm{U}(n))$ be the element such that $\kappa^*(c_i)$ is the i-th elementary symmetric polynomial in $T_1, \dots, T_n$. Then*

$$h^*(B\mathrm{U}(n)) \cong h^*[[c_1, \dots, c_n]].$$

The elements $c_1, \dots, c_n$ are those which were obtained from the projective bundle associated to γ_n by the methods of the previous section.

Proof. Let $\sigma \in S_n$ be a permutation and also the corresponding permutation of the factors of $BU(1)^n$. Then σ is covered by a bundle automorphism of the n-fold product $\gamma_1^n = \gamma_1 \times \cdots \times \gamma_1$. Hence $\kappa \circ \sigma$ is another classifying map of γ_1^n and therefore homotopic to κ. The permutation σ induces on $h^*(BU(1)^n) = h^*[[T_1, \dots, T_n]]$ the corresponding permutation of the T_j. Hence the image of κ^* is contained in the symmetric subring, since $\kappa \circ \sigma \simeq \kappa$. Let $\mathrm{pr}_j : BU(1)^n \to BU(1)$ be the projection onto the j-th factor. We write $\gamma(j) = \mathrm{pr}_j^*(\gamma_1)$ so that $\gamma_1^n = \gamma(1) \oplus \cdots \oplus \gamma(n)$ and $T_j = c_1(\gamma(j))$. We have the relation (naturality)

$$\kappa^* c_i(\gamma_n) = c_i(\kappa^* \gamma_n) = c_i(\gamma_1^n) = c_i(\gamma(1) \oplus \cdots \oplus \gamma(n)).$$

By the sum formula (19.2.6) this equals

$$c_1(\gamma(1)) c_{i-1}(\gamma(2) \oplus \cdots \oplus \gamma(n)) + c_i(\gamma(2) \oplus \cdots \oplus \gamma(n)).$$

This is used to show by induction that this element is the i-th elementary symmetric polynomial σ_i in the variables T_j. We now use the algebraic fact that the symmetric part of $h^*[[T_1, \dots, T_n]]$ equals the ring of graded power series $h^*[[\sigma_1, \dots, \sigma_n]]$ in the elementary symmetric polynomials σ_i. This shows that the image of κ^* is as claimed.

It remains to show that κ^* is injective. From (19.3.1), (19.2.2) and (19.1.5) we obtain an injective map

$$\beta^* \colon h^*(BU(n)) \to h^*(BU(n-1) \times BU(1)) \cong h^*(BU(n-1))[[\gamma(n)]].$$

This fact yields, by induction on n, the claimed injectivity. □

Elements of $h^*(BU(n))$ are called universal $h^*(-)$-valued ***characteristic classes*** for n-dimensional complex vector bundles. Given $c \in h^*(BU(n))$ and a classifying map $f \colon B \to BU(n)$ of the bundle ξ over B we set $c(\xi) = f^*(c)$ and call $c(\xi)$ a characteristic class. With this definition, the naturality $\varphi^* c(\eta) = c(\xi)$ holds for each bundle map $\varphi \colon \xi \to \eta$. Theorem (19.3.2) shows that it suffices to work with c_i. The corresponding characteristic class $c_i(\xi)$ is called the ***i-th Chern class*** of ξ with respect to the chosen Thom class t_∞. It is sometimes useful to consider the ***total Chern class*** $c(\xi) = 1 + c_1(\xi) + c_2(\xi) + \cdots \in h^*(B)$ of a bundle over B; the sum formula then reads $c(\xi \oplus \eta) = c(\xi) \cup c(\eta)$.

The preceding results can in particular be applied to integral singular cohomology. Complex vector bundles have a canonical orientation and a canonical Thom class. There are two choices for the element t_∞, they differ by a sign. We use the element that satisfies $\langle t_1, [\mathbb{C}P^1] \rangle = 1$ (Kronecker pairing), where $[\mathbb{C}P^1] \in H_2(\mathbb{C}P^1; \mathbb{Z})$ denotes the canonical fundamental class determined by the complex structure.

Chern classes are ***stable characteristic classes***, i.e., $c_j(\xi) = c_j(\xi \oplus \varepsilon)$ if ε denotes the trivial 1-dimensional bundle; this follows from the sum formula (19.2.6) and $c_i(\varepsilon) = 0$ for $i > 0$. This fact suggests that we pass to the limit $n \to \infty$.

(19.3.3) Example. The complex tangent bundle $T\mathbb{C}P^n$ of the complex manifold $\mathbb{C}P^n$ satisfies $T\mathbb{C}P^n \oplus \varepsilon \cong (n+1)\eta_{n+1}$, see (15.6.6). Therefore the total Chern class of this bundle is $(1 + c_1(\eta_{n+1}))^{n+1}$. ◇

Let $\omega\colon B\mathrm{U}(n) \to B\mathrm{U}(n+1)$ be a classifying map for $\gamma_n \oplus \varepsilon$. Then $\omega^* c_i = c_i$ for $i \le n$ and $\omega^* c_{n+1} = 0$. Let $\mathrm{U} = \mathrm{colim}_n\, \mathrm{U}(n)$, with respect to the inclusions

$$\mathrm{U}(n) \to \mathrm{U}(n+1), \quad A \mapsto \begin{pmatrix} A & 0 \\ 0 & 1 \end{pmatrix},$$

be the ***stable unitary group***. The classifying space $B\mathrm{U}$ is called the classifying space for stable complex vector bundles. We think of this space as a homotopy colimit (telescope) over the maps $B\mathrm{U}(n) \to B\mathrm{U}(n+1)$. By passage to the limit we obtain (since the $\lim^1$-term vanishes by (17.1.7)):

(19.3.4) Theorem. $h^*(B\mathrm{U}) \cong \lim h^* B\mathrm{U}(n) \cong h^*[[c_1, c_2, \dots]]$. □

(19.3.5) Example. Let $\kappa_{m,n}\colon B\mathrm{U}(m) \times B\mathrm{U}(n) \to B\mathrm{U}(m+n)$ be a classifying map for $\gamma_m \times \gamma_n = \mathrm{pr}_1^* \gamma_m \oplus \mathrm{pr}_2^* \gamma_n$. We use the elements $c'_j = c_j(\mathrm{pr}_1^* \gamma_m)$ and $c''_j = c_j(\mathrm{pr}_2^* \gamma_n)$ and obtain

(1) $$h^*(B\mathrm{U}(m) \times B\mathrm{U}(n)) \cong h^*[[c'_1, \dots, c'_m, c''_1, \dots, c''_n]].$$

Moreover, by the sum formula,

(2) $$\kappa^*_{m,n} c_k = \textstyle\sum_{i+j=k} c'_i c''_j.$$

The map $\kappa^*_{m,n}$ is continuous in the sense that the effect on a formal power series in the variables $c_1, \dots, c_{m+n}$ is obtained by inserting for c_k the value (2).

For the proof of (1) one can use the theory $h^*(B\mathrm{U}(m) \times -)$ and proceed as for (19.3.2). In the case of integral singular cohomology one has the Künneth isomorphism and $\kappa^*_{m,n}$ becomes the homomorphism of algebras

(3) $$\mathbb{Z}[c_1, c_2, \dots, c_{m+n}] \to \mathbb{Z}[c_1, \dots, c_m] \otimes \mathbb{Z}[c_1, \dots, c_n]$$

determined by $c_k \mapsto \sum_{i+j=k} c_i \otimes c_j$. Since formal power series are not compatible with tensor products, one has to use a suitably completed tensor product for general theories if one wants a similar statement.

The maps $\kappa_{m,n}$ combine in the colimit to a map $\kappa\colon B\mathrm{U} \times B\mathrm{U} \to B\mathrm{U}$. It is associative and commutative up to homotopy and there exists a unit element. A classifying map for the "inverse bundle" yields a homotopy inverse for κ. With these structures $B\mathrm{U}$ becomes a group object in h-TOP. The precise definition and the detailed verification of these topological results are not entirely trivial. We are content with the analogous algebraic result that we have a homomorphism

$h^*(BU) \to h^*(BU \times BU)$ determined by the sum formula and continuity. The inverse $\iota^*\colon h^*(BU) \to h^*(BU)$ is determined by the formal relation

$$(1 + c_1 + c_2 + \cdots)(1 + \iota^*(c_1) + \iota^*(c_2) + \cdots) = 1.$$

It allows for an inductive computation $\iota^* c_1 = -c_1, \iota^* c_2 = c_1^2 - c_2$ etc. We also have the homomorphism of algebras (3) for BU. It will be important for the discussion of the Hopf algebra structure. ◇

(19.3.6) Example. We know already $K(\mathbb{Z}, 2) = \mathbb{C}P^\infty = BS^1 = BU(1) \simeq BGL_1(\mathbb{C})$. A numerable complex line bundle ξ over X is determined by its classifying map in

$$H^2(X; \mathbb{Z}) = [X, \mathbb{C}P^\infty] = [X, BU(1)].$$

The corresponding element in $c_1(\xi) \in H^2(X; \mathbb{Z})$ is the first Chern class of ξ. ◇

(19.3.7) Proposition. *The relation* $c_1(\xi \otimes \eta) = c_1(\xi) + c_1(\eta)$ *holds for line bundles* ξ *and* η.

Proof. We begin with the universal situation. We know that $H^2(\mathbb{C}P^\infty; \mathbb{Z}) \cong \mathbb{Z}$ is generated by the first Chern class c of the universal bundle $\gamma = \gamma_1$. Let $k\colon \mathbb{C}P^\infty \times \mathbb{C}P^\infty \to \mathbb{C}P^\infty$ be the classifying map of $\gamma \hat{\otimes} \gamma$. Let $\mathrm{pr}_j\colon \mathbb{C}P^\infty \times \mathbb{C}P^\infty \to \mathbb{C}P^\infty$ be the projection onto the j-th factor. Then $H^2(\mathbb{C}P^\infty \times \mathbb{C}P^\infty; \mathbb{Z})$ has the $\mathbb{Z}$-basis T_1, T_2 with $T_j = \mathrm{pr}_j^*(c)$. There exists a relation $k^* c_1(\gamma) = a_1 e_1 + a_2 e_2$ with certain $a_i \in \mathbb{Z}$. Let $i_1\colon \mathbb{C}P^\infty \to \mathbb{C}P^\infty \times \mathbb{C}P^\infty$, $x \mapsto (x, x_0)$ for fixed x_0. Then $i_1^* e_1 = c_1(\gamma)$, since $\mathrm{pr}_1 i_1 = \mathrm{id}$, and $i_1^* e_2 = 0$ holds, since $\mathrm{pr}_2 i_1$ is constant. We compute

$$a_1 c_1(\gamma) = i_1^* k^* c_1(\gamma) = c_1(i_1^* k^* \gamma) = c_1(i_1^*(\mathrm{pr}_1^* \gamma \otimes \mathrm{pr}_2^* \gamma)) = c_1(\gamma),$$

since $i_1^* \mathrm{pr}_1^* \gamma = \gamma$ and $i_1^* \mathrm{pr}_2^* \gamma$ is the trivial bundle. Hence $a_1 = 1$, and similarly we see $a_2 = 1$.

We continue with the proof. Let $k_\xi, k_\eta\colon B \to \mathbb{C}P^\infty$ be classifying maps of ξ and η. Then $c_1(\xi) = k_\xi^* c_1(\gamma)$ and similarly for η. With the diagonal d the equalities $\xi \otimes \eta = d^*(\xi \hat{\otimes} \eta) = d^*(k_\xi \times k_\eta)^*(\gamma \hat{\otimes} \gamma)$ hold. This yields

$$\begin{aligned} c_1(\xi \otimes \eta) &= c_1(d^*(k_\xi \times k_\eta)^*(\gamma \hat{\otimes} \gamma)) \\ &= d^*(k_\xi \times k_\eta)^*(e_1 + e_2) \\ &= d^*(k_\xi \times k_\eta)^* \mathrm{pr}_1^* c_1(\gamma) + d^*(k_\xi \times k_\eta)^* \mathrm{pr}_2^* c_1(\gamma) \\ &= k_\xi^* c_1(\gamma) + k_\eta^* c_1(\gamma) \\ &= c_1(\xi) + c_2(\eta), \end{aligned}$$

since $k_\xi = \mathrm{pr}_1(k_\xi \times k_\eta) d$ holds. □

(19.3.8) Proposition. *Let $\xi\colon E(\xi) \to B$ be an n-dimensional vector bundle and $p_\xi\colon P(\xi) \to B$ the associated projective bundle. The induced bundle splits $p_\xi^*(\xi) = \xi_1 \oplus \xi'$ into the canonical line bundle ξ_1 over the projective bundle and another $(n-1)$-dimensional bundle ξ'.*

Proof. Think of ξ as associated bundle $E \times_{\mathrm{U}(n)} \mathbb{C}^n \to B$. Let H be the subgroup $\mathrm{U}(n-1) \times \mathrm{U}(1)$ of $\mathrm{U}(n)$. We obtain the pullback

$$\begin{array}{ccc} E \times_H \mathbb{C}^{n-1} \times \mathbb{C} \cong E \times_{\mathrm{U}(n)} (\mathrm{U}(n) \times_H \mathbb{C}^n) & \longrightarrow & E \times_{\mathrm{U}(n)} \mathbb{C}^n \\ \downarrow & & \downarrow \\ E/H \cong E \times_{\mathrm{U}(n)} (\mathrm{U}(n)/H) & \longrightarrow & B \end{array}$$

and this implies the assertion. □

We now iterate this process: We consider over $P(\xi)$ the projective bundle $P(\xi')$, et cetera. Finally we arrive at a map $f(\xi)\colon F(\xi) \to B$ with the properties:

(1) $f(\xi)^*\xi$ splits into a sum of line bundles.

(2) The induced map $f(\xi)^*\colon h^*(B) \to h^*(F(\xi))$ is injective.

Assertion (2) is a consequence of (19.2.2).

A model for $f(\xi)$ is the flag bundle. The ***flag space*** $F(V)$ of the n-dimensional vector space V consists of the sequences (= flags)

$$\{0\} = V_0 \subset V_1 \subset \cdots \subset V_n = V$$

of subspaces V_i of dimension i. Let V carry a Hermitian form. Each flag has an orthonormal basis $b_1, \dots, b_n$ such that V_i is spanned by $b_1, \dots, b_i$. The basis vectors b_i are determined by the flag up to scalars of norm 1. The group $\mathrm{U}(n)$ acts transitively on the set of flags. The isotropy group of the standard flag is the maximal torus $\mathrm{T}(n)$ of diagonal matrices. Hence we can view $F(V)$ as $\mathrm{U}(n)/\mathrm{T}(n)$. The ***flag bundle*** associated to $E \times_{\mathrm{U}(n)} \mathbb{C}^n$ is then

$$f(\xi)\colon F(\xi) = E \times_{\mathrm{U}(n)} \mathrm{U}(n)/\mathrm{T}(n) \cong E/\mathrm{T}(n) \to B.$$

We can apply this construction to a finite number of bundles.

(19.3.9) Theorem (Splitting principle). *Let $\xi_1, \dots, \xi_k$ be complex vector bundles over B. Then there exists a map $f\colon X \to B$ such that $f^*\colon h^*(B) \to h^*(X)$ is injective and $f^*(\xi_j)$ is for each ξ_j a sum of line bundles.* □

We now prove (19.0.1). Consider the exact cohomology sequence of the pair $(E(\gamma_n), E^0(\gamma_n))$. We can use $E^0(\gamma_n)$ as a model for $B\mathrm{U}(n-1)$. The projection $E(\gamma_n) \to B\mathrm{U}(n)$ is an h-equivalence. Our computation of $h^*(B\mathrm{U}(n))$ shows that we have a short exact sequence

$$\begin{array}{ccccccccc} 0 \longrightarrow h^{2n}(E(\gamma_n), E^0(\gamma_n)) \longrightarrow & h^{2n}(E(\gamma_n)) & \xrightarrow{i^*} & h^{2n}(E^0(\gamma_n) & \longrightarrow 0 \\ & \downarrow \cong & & \downarrow \cong & \\ & h^{2n}(B\mathrm{U}(n)) & \longrightarrow & h^{2n}(B\mathrm{U}(n-1)). & \end{array}$$

The element c_n lies in the kernel of i^*. It therefore has a unique pre-image $t(\gamma_n) \in h^{2n}(E(\gamma_n), E^0(\gamma_n))$. For an n-dimensional numerable bundle $\xi\colon E(\xi) \to B$ we define $t(\xi) \in h^{2n}(E(\xi)E^0(\xi))$ to be the element $\kappa^* t(\gamma_n)$ with a classifying map $\kappa\colon \xi \to \gamma_n$. Then the elements $t(\xi)$ are natural with respect to bundle maps. From the relation $\kappa^*_{m,n}(c_{m+n}) = c'_m c''_n$ we conclude (by naturality) $t(\gamma_m) \times t(\gamma_n) = t(\gamma_m \times \gamma_n)$ and then $t(\xi) \times t(\eta) = t(\xi \times \eta)$ for arbitrary numerable bundles. The element $t(\gamma_1) \in h^2(E(\gamma_1), E^0(\gamma_1))$ corresponds to the chosen element $t_\infty \in h^2(\mathbb{C}P^\infty)$. The restriction of $t(\gamma_n)$ to $\gamma_1 \times \cdots \times \gamma_1$ yields $t(\gamma_1) \times \cdots \times t(\gamma_1)$. This is a Thom class, since products of Thom classes are Thom classes. This shows that $t(\gamma_n)$ is a Thom class.

19.4 Stiefel–Whitney Classes

The theory of Chern classes has a parallel theory for real vector bundles. Suppose given an element $t_\infty \in h^1(\mathbb{R}P^\infty, *) \subset h^1(\mathbb{R}P^\infty)$ such that its restriction to $t_1 \in h^1(\mathbb{R}P^1, *)$ is a generator of this h^0-module. Then there exists an isomorphism

$$h^*[T]/(T^{n+1}) \cong h^*(\mathbb{R}P^n)$$

which sends T to the restriction t_n of t_∞. This is then used to derive isomorphisms

$$\begin{aligned} h^*(X \times \mathbb{R}P^n) &\cong h^*(X)[u]/(u^{n+1}), \\ h^*(X \times \mathbb{R}P^\infty) &\cong h^*(X)[[u]], \\ h^*((\mathbb{R}P^\infty)^n) &\cong h^*[[T_1, \dots, T_n]]. \end{aligned}$$

The projective bundle $P(\xi)$ of a real vector bundle ξ over B yields a free $h^*(B)$-module $h^*(P(\xi))$ with basis $1, t_\xi, \dots, t_\xi^{n-1}$, and there exists a relation

$$\textstyle\sum_{j=0}^{\infty} (-1)^j w_j(\xi) t_\xi^{n-j} = 0$$

with elements $w_j(\xi) \in h^j(B)$ which satisfy the sum formula

$$w_r(\xi \oplus \eta) = \textstyle\sum_{i+j=r} w_i(\xi) w_j(\eta)$$

where $w_0(\xi) = 1$ and $w_j(\xi) = 0$ for $j > \dim \xi$. These elements are natural with respect to bundle maps, hence characteristic classes. We obtain an injective map $\kappa^*: h^*(BO(n)) \to h^*(BO(1)^n)$. The classes $w_1, \dots, w_n$ which belong to the universal n-dimensional bundle over $BO(n)$ yield

$$h^*(BO(n)) \cong h^*[[w_1, \dots, w_n]].$$

The image of w_j under κ^* is the j-th elementary symmetric polynomial in the $T_1, \dots, T_n$. We pass to the limit $n \to \infty$ and obtain h^*BO as a ring of graded power series in $w_1, w_2, \dots$ with w_j of degree j. The w_j are the ***universal Stiefel–Whitney classes***. The Stiefel–Whitney classes are natural, stable, and satisfy the sum formula. There holds a splitting principle for real bundles. We write $w(\xi) = 1 + w_1(\xi) + w_2(\xi) + \cdots$ for the total Stiefel–Whitney class of ξ; then the sum formula reads $w(\xi \oplus \eta) = w(\xi) \cup w(\eta)$.

The existence of the universal element has the consequence that the unit element $1 \in h^0$ is of order 2, so that the cohomology groups consist of $\mathbb{Z}/2$-vector spaces and signs can be ignored. This result is due to the fact that multiplication by -1 in the fibres of vector bundles is a bundle map. If we apply this to the universal one-dimensional bundle, then we see that this bundle map preserves the Thom class t_∞. On the other hand, if we restrict to $t_1 \in h^1(\mathbb{R}P^1, *)$, this bundle map is of degree -1 and changes the sign of t_1. Since t_1 corresponds under suspension to a unit of h^0, we conclude that $1 = -1 \in h^0$.

One shows as in the complex case that the theory is $\mathbb{R}$-oriented. One can apply these results to singular cohomology with coefficients in $\mathbb{Z}/2$. There is a unique choice for the universal element $t_\infty \in H^1(\mathbb{R}P^\infty; \mathbb{Z}/2)$. The resulting characteristic classes are the classical ***Stiefel–Whitney classes***.

(19.4.1) Example. The tangent bundle τ of $\mathbb{R}P^n$ satisfies $\tau \oplus \varepsilon \cong (n+1)\eta$ with the canonical line bundle η. The total Stiefel–Whitney class of τ is therefore $(1+w)^{n+1} \in H^*(\mathbb{R}P^n) \cong \mathbb{Z}/2[w]/(w)^{n+1}$. Suppose $\mathbb{R}P^n$ has an immersion into $\mathbb{R}^{n+k}$. Then τ has an inverse bundle of dimension k, the normal bundle of this immersion. Suppose $n = 2^r$. Properties of binomial numbers modulo 2 show that $w(\tau) = 1 + w + w^n$. If ν is inverse to τ, then $w(\tau)w(\nu) = 1$, and this implies in our case $w(\nu) = 1 + w + w^2 + \cdots + w^{n-1}$. This shows that an inverse bundle must have dimension at least $n-1$. Therefore $\mathbb{R}P^n$ has for n of the form 2^k no immersion into $\mathbb{R}^{2n-2}$. ◇

19.5 Pontrjagin Classes

We now discuss characteristic classes for oriented bundles. Suppose $\xi: E(\xi) \to B$ is an oriented n-dimensional real vector bundle. The orientation determines a Thom class $t(\xi) \in H^n(E(\xi), E^0(\xi))$ by the requirement that a positive isomorphism $i_b: \mathbb{R}^n \to E(\xi_b)$ sends $t(\xi)$ to the generator $e^{(n)} \in H^n(\mathbb{R}^n, \mathbb{R}^n \smallsetminus 0)$ which is dual to the generator $e_n = e_1 \times \cdots \times e_1 \in H_n(\mathbb{R}^n, \mathbb{R}^n \smallsetminus 0)$, i.e., $\langle e^{(n)}, e_n \rangle = 1$. This definition implies $e^{(m)} \times e^{(n)} = (-1)^{mn} e^{(m+n)}$ and has the following consequences:

(19.5.1) Lemma. *Let ξ and η be oriented bundles and give $\xi \times \eta$ the sum orientation. Then $t(\xi \times \eta) = (-1)^{|\xi||\eta|} t(\xi) \times t(\eta)$ and $e(\xi \oplus \eta) = (-1)^{|\xi||\eta|} e(\xi)e(\eta)$. Here $|\xi| = \dim \xi$.* □

Let $\xi \colon E \to B$ be a real vector bundle. It has a complexification $\xi_{\mathbb{C}} = \xi \otimes_{\mathbb{R}} \mathbb{C}$. We take $\xi \oplus \xi$ with complex structure $J(x, y) = (-y, x)$ on each fibres as a model for $\xi_{\mathbb{C}}$.

Let ζ be a complex n-dimensional bundle and $\zeta_{\mathbb{R}}$ the underlying $2n$-dimensional real bundle. If $v_1, \dots, v_n$ is a basis of a fibre, then $v_1, iv_1, \dots, v_n, iv_n$ is a basis of the fibre of $\zeta_{\mathbb{R}}$, and it defines the canonical orientation.

(19.5.2) Lemma. *Let ξ be an oriented n-dimensional real bundle. Then $(\xi_{\mathbb{C}})_{\mathbb{R}}$ in our model for $\xi_{\mathbb{C}}$ above is isomorphic to $(-1)^{n(n-1)/2} \xi \oplus \xi$ as an oriented bundle. The factor indicates the change of orientation, and $\xi \oplus \xi$ carries the sum orientation.* □

(19.5.3) Proposition. *Let $\xi \colon E \to B$ be a complex n-dimensional bundle. Consider it as a real bundle with orientation and canonical Thom class induced by the complex structure. Then $c_n(\xi) = e(\xi) \in H^{2n}(B; \mathbb{Z})$.*

Proof. This holds for 1-dimensional bundles by definition of c_1. The general case follows by an application of the splitting principle and the sum formula. □

We set

$$p_i(\xi) = (-1)^i c_{2i}(\xi_{\mathbb{C}}) \in H^{4i}(B; \mathbb{Z})$$

and call this characteristic class the ***i*-th Pontrjagin class** of ξ. The bundle $\xi_{\mathbb{C}}$ is isomorphic to the conjugate bundle $\overline{\xi}_{\mathbb{C}}$. The relation $c_i(\zeta) = (-1)^i c_i(\overline{\zeta})$ holds in general for conjugate bundles. Hence the odd Chern classes of $\xi_{\mathbb{C}}$ are elements of order 2. This is a reason why we ignore them for the moment. The Pontrjagin classes are by definition compatible with bundle maps (naturality) and they do not change by the addition of a trivial bundle (stability). The next proposition justifies the choice of signs in the definition of the p_j.

(19.5.4) Proposition. *Let ξ be an oriented $2k$-dimensional real bundle. Then $p_k(\xi) = e(\xi)^2$.*

Proof. We compute

$$\begin{aligned} p_k(\xi) &= (-1)^k c_{2k}(\xi_{\mathbb{C}}) = (-1)^k e_{2k}((\xi_{\mathbb{C}})_{\mathbb{R}}) = (-1)^{k+2k(2k-1)/2} e(\xi \oplus \xi) \\ &= e(\xi \oplus \xi) = (-1)^{2k \cdot 2k} e(\xi)^2 = e(\xi)^2. \end{aligned}$$

We have used (19.5.1), (19.5.2), and (19.5.3). □

One can remove elements of order 2 if one uses the coefficient ring $R = \mathbb{Z}[\frac{1}{2}]$ of rational numbers with 2-power denominator (or, more generally, assumes that $\frac{1}{2} \in R$). The next theorem shows the universal nature of the Pontrjagin classes.

(19.5.5) Theorem. *Let p_j denote the Pontrjagin classes of the universal bundle and e its Euler class. Then*

$$H^*(BSO(2n+1); R) \cong R[p_1, \dots p_n], \quad H^*(BSO(2n); R) \cong R[p_1, \dots, p_{n-1}, e].$$

Proof. Induction over n. Let $\xi_n \colon ESO(n) \times_{SO(n)} \mathbb{R}^n \to BSO(n)$ be the universal oriented n-bundle and $p \colon BSO(n-1) \to BSO(n)$ the classifying map of $\xi_{n-1} \oplus \varepsilon$. As model for p we take the sphere bundle of ξ_n. Then we have a Gysin sequence at our disposal. Write $B_n = BSO(n)$ for short.

Suppose n is even. Then, by induction, $H^*(B_{n-1})$ is generated by the Pontrjagin classes, and p^* is surjective since the classes are stable. Hence the Gysin sequence decomposes into short exact sequences. Let H_n^* denote the algebra which is claimed to be isomorphic to $H^*(B_n)$. And let $\mu_n \colon H_n^* \to H^*(B_n)$ be the homomorphism which sends the formal elements p_j, e onto the cohomology classes with the same name. We obtain a commutative diagram

$$\begin{array}{ccccccccc}
0 & \longrightarrow & H^i(B_n) & \xrightarrow{\;e\;} & H^{i+n}(B_n) & \xrightarrow{\;p^*\;} & H^{i+n}(B_{n-1}) & \longrightarrow & 0 \\
 & & \uparrow{\scriptstyle \mu_n} & & \uparrow{\scriptstyle \mu_n} & & \uparrow{\scriptstyle \mu_{n-1}} & & \\
0 & \longrightarrow & H_n^i & \xrightarrow{\;e\;} & H_n^{i+n} & \longrightarrow & H_{n-1}^{i+n} & \longrightarrow & 0.
\end{array}$$

By induction, μ_{n-1} is an isomorphism. By a second induction over i the left arrow is an isomorphism. Now we apply the Five Lemma. In order to start the induction, we note that by the Gysin sequence $\mu_n \colon H_n^i \to H^i(B_n)$ is an isomorphism for $i < n$.

Suppose $n = 2m + 1$. The Euler class is zero, since we use the coefficient ring R. Hence the Gysin sequence yields a short exact sequence

$$0 \to H^j(B_n) \xrightarrow{p^*} H^j(B_{n-1}) \to H^{j-2m}(B_n) \to 0.$$

Therefore $H^*(B_n)$ is a subring of $H^*(B_{n-1})$ via p^*. The image of p^* contains the subring P^* generated by $p_1, \dots, p_m$. We use $p_m = e^2$. The induction hypothesis implies

$$\operatorname{rank} H^j(B_{n-1}) = \operatorname{rank} P^j + \operatorname{rank} P^{j-2m}.$$

The Gysin sequence yields

$$\operatorname{rank} H^j(B_{n-1}) = \operatorname{rank} H^j(B_n) + \operatorname{rank} H^{j-2m}(B_n).$$

The equality $\operatorname{rank} P^j = \operatorname{rank} H^j(B_n)$ is a consequence. If $p^* H^j(B_n) \neq P^j$ then the image would contain elements of the form $x + ey$, $x \in P^j$, $y \in P^{j-2m}$. Such an element would be linearly independent of the basis elements of P^*. This contradicts the equality of ranks. □

(19.5.6) Example. Let ζ be a complex bundle. Then $(\zeta_{\mathbb{R}})_{\mathbb{C}}$ is isomorphic to $\zeta \oplus \bar{\zeta}$. An isomorphism from $\zeta_{\mathbb{R}} \oplus \zeta_{\mathbb{R}}$ with the complex structure $(x, y) \mapsto (-y, x)$ is given by

$$(x, y) \mapsto \left(\frac{x + iy}{\sqrt{2}}, \frac{ix + y}{\sqrt{2}} \right).$$

Hence the $p_i(\zeta_{\mathbb{R}}) = (-1)^i c_{2i}(\zeta \oplus \bar{\zeta}) = (-1)^i \sum_{a+b=2i} (-1)^b c_a(\zeta) c_b(\zeta)$. Since $\mathrm{SO}(2) = \mathrm{U}(1)$, an oriented plane bundle ξ has a unique complex structure ζ such that $\zeta_{\mathbb{R}} = \xi$. The total Pontrjagin class of ξ is therefore $1 + c_1(\zeta)^2$. ◇

(19.5.7) Example. Let $\tau = T\mathbb{C}P^n$ denote the complex tangent bundle of $\mathbb{C}P^n$. Then $\tau_{\mathbb{R}}$ is the real tangent bundle. In order to determine the Pontrjagin classes we use $(\tau_{\mathbb{R}})_{\mathbb{C}} = \tau \oplus \bar{\tau}$. The total Chern class of this bundle is $(1+c)^{n+1}(1-c)^{n+1} = (1-c^2)^{n+1}$ if we write $H^*(\mathbb{C}P^n) = \mathbb{Z}[c]/(c^{n+1})$ with $c = c_1(\eta_{n+1})$, see (19.3.3). Hence the total Pontrjagin class of $\mathbb{C}P^n$, i.e., of its tangent bundle with the canonical orientation, is $(1 + c^2)^{n+1}$. ◇

Problems

1. The Pontrjagin classes are stable. Under the hypothesis of (19.5.5) we obtain in the limit $H^*(B\mathrm{SO}; R) \cong R[p_1, p_2, \dots]$. The sum formula $p_k(\xi \oplus \eta) = \sum_{i+j=k} p_i(\xi) p_j(\eta)$ holds ($p_0 = 1$).

19.6 Hopf Algebras

We fix a commutative ring R and work in the category R-MOD of left R-modules. The tensor product of R-modules M and N is denoted by $M \otimes N$. The natural isomorphism $\tau \colon M \otimes N \to N \otimes M$, $m \otimes n \mapsto n \otimes m$ expresses the commutativity of the tensor product. We have canonical isomorphisms $l \colon R \otimes M \to M$, $\lambda \otimes m \mapsto \lambda m$ and $r \colon M \otimes R \to M$, $m \otimes \lambda \mapsto \lambda m$. Co-homology will have coefficients in R, if nothing else is specified.

An ***algebra*** (A, m, e) in R-MOD consists of an R-module A and linear maps $m \colon A \otimes A \to A$ (multiplication), $e \colon R \to A$ (unit) such that $m(e \otimes 1) = l$, $m(1 \otimes e) = r$. If $m(m \otimes 1) = m(1 \otimes m)$ holds, then the algebra is associative, and if $m\tau = m$ holds, the algebra is commutative. Usually we write $m(a \otimes b) = a \cdot b = ab$. We use similar definitions in the category of $\mathbb{Z}$-graded R-modules (with its tensor product and interchange map).

(19.6.1) Example. Let X be a topological space. Then the graded R-module $H^*(X)$ becomes a (graded) associative and commutative algebra with multiplication

$$m \colon H^*(X) \otimes H^*(X) \to H^*(X \times X) \to H^*(X),$$

where the first map is the $\times$-product and the second map is induced by the diagonal $d: X \to X \times X$. The unit is the map induced by the projection $X \to P$ onto a point P. ◇

A ***coalgebra*** (C, μ, ε) in R-MOD consists of an R-module C and linear maps $\mu: C \to C \otimes C$ (comultiplication), $\varepsilon: C \to R$ (counit) such that $(\varepsilon \otimes 1)\mu = l^{-1}$, $(1 \otimes \varepsilon)\mu = r^{-1}$. If $(\mu \otimes 1)\mu = (1 \otimes \mu)\mu$ holds, the coalgebra is coassociative, and if $\tau\mu = \mu$ holds, the coalgebra is cocommutative.

(19.6.2) Example. Let X be a topological space. Suppose $H_*(X)$ is a free R-module. The graded R-module $H_*(X)$ becomes a (graded) coassociative and cocommutative coalgebra with comultiplication

$$\mu: H_*(X) \to H_*(X \times X) \cong H_*(X) \otimes H_*(X)$$

where the first map is induced by the diagonal d and the isomorphism is the Künneth isomorphism. The counit is induced by $X \to P$. ◇

A homomorphism of algebras $\varphi: (A, m, e) \to (A', m', e')$ is a linear map $\varphi: A \to A'$ such that $\varphi m = m(\varphi \otimes \varphi)$ and $e' = \varphi e$. A homomorphism of coalgebras $\psi: (C, \mu, \varepsilon) \to (C', \mu', \varepsilon')$ is a linear map $\psi: C \to C'$ such that $(\psi \otimes \psi)\mu = \mu'\psi$ and $\varepsilon'\psi = \varepsilon$. A continuous map $f: X \to Y$ induces a homomorphism $f^*: H^*(Y) \to H^*(X)$ of the algebras (19.6.1) and a homomorphism $f_*: H_*(X) \to H_*(Y)$ of the coalgebras (19.6.2).

The tensor product of algebras (A_i, m_i, e_i) is the algebra (A, m, e) with $A = A_1 \otimes A_2$ and $m = (m_1 \otimes m_2)(1 \otimes \tau \otimes 1)$ and $e = e_1 \otimes e_2: R \cong R \otimes R \to A_1 \otimes A_2$. The multiplication m is determined by $(a_1 \otimes a_2)(b_1 \otimes b_2) = a_1 b_1 \otimes a_2 b_2$ (with the appropriate signs in the case of graded algebras). The tensor product of coalgebras $(C_i, \mu_i, \varepsilon_i)$ is the coalgebra (C, μ, ε) with $C = C_1 \otimes C_2$, comultiplication $\mu = (1 \otimes \tau \otimes 1)(\mu_1 \otimes \mu_2)$ and counit $\varepsilon = \varepsilon_1 \varepsilon_2: C_1 \otimes C_2 \to R \otimes R \cong R$.

Let (C, μ, ε) be a coalgebra. Let $C^* = \operatorname{Hom}(C, R)$ denote the dual module. The data

$$m: C^* \otimes C^* \to (C \otimes C)^* \xrightarrow{\mu^*} C^*$$

and $e: R \cong R^* \xrightarrow{\varepsilon^*} C^*$ define the ***dual algebra*** (C^*, m, e) of the coalgebra. (The first map is the tautological homomorphism. It is an isomorphism if C is a finitely generated, projective R-module.)

Let (A, m, e) be an algebra with A a finitely generated, projective R-module. The data

$$\mu: A^* \xrightarrow{m^*} (A \otimes A)^* \cong A^* \otimes A^*$$

and $\varepsilon: A^* \xrightarrow{e^*} R^* \cong R$ define the ***dual coalgebra*** (A^*, μ, ε) of the algebra.

In the case of graded modules we take the graded dual; if $A = (A_n \mid n \in \mathbb{N}_0)$, then the dual is $(A^n = \operatorname{Hom}(A_n, R) \mid n \in \mathbb{N}_0)$.

(19.6.3) Example. Let $\kappa\colon H^*(X) \to \operatorname{Hom}(H_*(X), R)$ be the map in the universal coefficient sequence. Then κ is an isomorphism of the algebra (19.6.1) onto the dual algebra of the coalgebra (19.6.2). ◇

(19.6.4) Proposition. *Let C be a coalgebra and A an algebra. Then* $\operatorname{Hom}(C, A)$ *carries the structure of an algebra with product $\alpha * \beta = m(\alpha \otimes \beta)\mu$, for $\alpha, \beta \in$* $\operatorname{Hom}(C, A)$*, and unit $e\varepsilon$. The product $*$ is called* ***convolution****.*

Proof. The map $(\alpha, \beta) \mapsto \alpha \otimes \beta$ is bilinear by construction. The (co-)associativity of m and μ is used to verify that $*$ is associative. The unit and counit axioms yield

$$\alpha * (e\varepsilon) = m(\alpha \otimes e\varepsilon)\mu = m(1 \otimes e)(\alpha \otimes 1)(1 \otimes \varepsilon)\mu = \alpha.$$

Hence $e\varepsilon$ is a right unit. □

A ***bialgebra*** $(H, m, e, \mu, \varepsilon)$ is an algebra (H, m, e) and a coalgebra (H, μ, ε) such that μ and ε are homomorphisms of algebras. (Here $H \otimes H$ carries the tensor product structure of algebras.) The equality $\mu m = (m \otimes m)(1 \otimes \tau \otimes 1)(\mu \otimes \mu)$ expresses the fact that μ is compatible with multiplication. The same equality says that m is compatible with comultiplication. This and a similar interpretation of the identities $\mathrm{id} = \varepsilon e$, $\mu e = (e \otimes e)\mu$, $m(\varepsilon \otimes \varepsilon) = \varepsilon m$ is used to show that a bialgebra can, equivalently, be defined by requiring that m and e are homomorphisms of coalgebras. A homomorphism of bialgebras is an R-linear map which is at the same time a homomorphism of the underlying algebras and coalgebras.

An ***antipode*** for a bialgebra H is an $s \in \operatorname{Hom}(H, H)$ such that s is a two-sided inverse of $\mathrm{id}(H) \in \operatorname{Hom}(H, H)$ in the convolution algebra. A bialgebra with antipode is called ***Hopf algebra***.

(19.6.5) Example. Let X be an H-space with multiplication $\rho\colon X \times X \to X$ and neutral element x. Then

$$m\colon H_*(X) \otimes H_*(X) \to H_*(X \times X) \xrightarrow{\rho_*} H_*(X)$$

is an algebra structure on $H_*(X)$ with unit induced by $\{x\} \subset X$. Suppose $H_*(X)$ is a free R-module. Then the algebra structure m and the coalgebra structure (19.6.2) define on $H_*(X)$ the structure of a bialgebra. An inverse for the multiplication ρ induces an antipode.

Suppose $H^*(X)$ is finitely generated and free in each dimension. Then

$$\mu\colon H^*(X) \xrightarrow{\rho^*} H^*(X \times X) \cong H^*(X) \otimes H^*(X)$$

is a coalgebra structure and together with the algebra structure (19.6.1) we obtain a bialgebra. Again an inverse for ρ induces an antipode. The duality isomorphism $H^*(X) \to \operatorname{Hom}(H_*(X), R)$ is an isomorphism of the bialgebra onto the dual bialgebra of $H_*(X)$.

This situation was studied by Heinz Hopf [91]. The letter Δ for the comultiplication (and even the term "diagonal") has its origin in this topological context. For background on Hopf algebras see [1], [182], [142], [138]. $\Diamond$

(19.6.6) Example. The space $\mathbb{C}P^\infty$ is an H-space with multiplication the classifying map of the tensor product of the universal line bundle. The algebra structure is $H^*(\mathbb{C}P^\infty;\mathbb{Z}) = \mathbb{Z}[c]$ with c the universal Chern class c_1. Let $[\mathbb{C}P^i] \in H_{2i}(\mathbb{C}P^\infty;\mathbb{Z})$ denote the image of the fundamental class of $\mathbb{C}P^i$ under the homomorphism induced by the embedding $\mathbb{C}P^i \to \mathbb{C}P^\infty$. The coalgebra structure is determined by $\mu(c) = c \otimes 1 + 1 \otimes c$, see (19.3.7). Since $\langle c^n, [\mathbb{C}P^n]\rangle = 1$ (see the proof of (18.7.2)), the dual Hopf algebra $H_*(\mathbb{C}P^\infty;\mathbb{Z})$ has an additive basis $x_i = [\mathbb{C}P^i], i \in \mathbb{N}_0$; by dualization of the cohomological coalgebra structure we obtain the multiplicative structure $x_i \cdot x_j = (i,j)x_{i+j}$ with $(i,j) = (i+j)!/(i!j!)$. Geometrically this means that the map $\mathbb{C}P^i \times \mathbb{C}P^j \to \mathbb{C}P^{i+j}$, $([x_0,\dots,x_i],[y_0,\dots,y_j]) \mapsto [z_0,\dots,z_{i+j}]$ with $z_k = \sum_{a+b=k} x_a y_b$ has degree (i,j). The comultiplication in $H_*(\mathbb{C}P^\infty)$ is $\mu(x_n) = \sum_{i+j=n} x_i \otimes x_j$. $\Diamond$

We generalize the Hom-duality of Hopf algebras and define pairings. Let A and B be Hopf algebras. A ***pairing of Hopf algebras*** is a bilinear map $A \times B \to R$, $(a,b) \mapsto \langle a,b\rangle$ with the properties: For $x,y \in A$ and $u,v \in B$

$$\langle xy,u\rangle = \langle x\otimes y, \mu(u)\rangle, \quad \langle x,uv\rangle = \langle \mu(x), u\otimes v\rangle,$$
$$\langle 1,u\rangle = \varepsilon(u), \qquad \langle x,1\rangle = \varepsilon(x).$$

The bilinear form $\langle -,-\rangle$ on $A\times B$ induces a bilinear form on $A\otimes A \times B\otimes B$ by $\langle x\otimes y, u\otimes v\rangle = (-1)^{|y||u|}\langle x,u\rangle\langle y,v\rangle$. This is used in the first two axioms. A pairing is called a ***duality*** between A, B, if $\langle x,u\rangle = 0$ for all $u \in B$ implies $x = 0$, and $\langle x,u\rangle = 0$ for all $x \in A$ implies $u = 0$. An example of a pairing is the Kronecker pairing $H^*(X) \times H_*(X) \to R$ in the case of an H-space X.

An element x of a bialgebra H is called ***primitive***, if $\mu(x) = x\otimes 1 + 1 \otimes x$. Let $P(H) \subset H$ be the R-module of the primitive elements of H. The bracket $(x,y) \mapsto [x,y] = xy - yx$ defines the structure of a Lie algebra on $P(H)$. The inclusion $P(H) \subset H$ yields, by the universal property of the universal enveloping algebra, a homomorphism $\iota\colon U(P(H)) \to H$. For cocommutative Hopf algebras over a field of characteristic zero with an additional technical condition, ι is an isomorphism [1, p. 110].

(19.6.7) Example. A coalgebra structure on the algebra of formal powers series $R[[x]]$ is, by definition, a (continuous) homomorphism $\mu\colon R[[x]] \to R[[x_1,x_2]]$ with $(\mu\otimes 1)\mu = (1\otimes\mu)\mu$ and $\varepsilon(x) = 0$. Here $R[[x_1,x_2]]$ is interpreted as a completed tensor product $R[[x_1]]\hat{\otimes} R[[x_2]]$. Then μ is given by the power series $\mu(x) = F(x_1,x_2)$ with the properties

$$F(x,0) = 0 = F(0,x), \quad F(F(x,y),z) = F(x,F(y,z)).$$

Such power series F are called ***formal group laws***. $\Diamond$

Problems

1. The Group algebra. Let G be a group and RG the group algebra. The R-module RG is the free R-module on the set G, and the multiplication $RG \otimes RG \cong R(G \times G) \to RG$ is the linear extension of the group multiplication. This algebra becomes a Hopf algebra, if we define the comultiplication by $\mu(g) = g \otimes g$ for $g \in G$, the counit by $\varepsilon(g) = 1$, and the antipode by $s(g) = g^{-1}$.

Let G be a finite group and $\mathcal{O}(G)$ the R-algebra of all maps $G \to R$ with pointwise addition and multiplication. Identify $\mathcal{O}(G \times G)$ with $\mathcal{O}(G) \otimes \mathcal{O}(G)$. Show that the group multiplication m induces a comultiplication $\mu = m^*\colon \mathcal{O}(G) \to \mathcal{O}(G \times G)$. The data $\varepsilon(f) = f(1)$ and $s(f)(g) = f(g^{-1})$ complete $\mathcal{O}(G)$ to a Hopf algebra. Evaluation at $g \in G$ defines an algebra homomorphism $\mathcal{O}(G) \to R$. Show that G is canonically isomorphic to the group $A\operatorname{Hom}(\mathcal{O}(G), R)$ of Problem 2.

An element g in a Hopf algebra H is called ***group-like*** if $\mu(g) = g \otimes g$ and $\varepsilon(g) = 1$. The set of group-like elements in H is a group under multiplication. The inverse of g is $s(g)$.

2. Let D be a Hopf algebra and A a commutative algebra. The convolution product induces on the set $A\operatorname{Hom}(D, A)$ of algebra homomorphisms $D \to A$ the structure of a group.

3. Let H be a Hopf algebra with antipode s. Then s is an anti-homomorphism of algebras and coalgebras, i.e., $s(xy) = s(y)s(x)$, $se = e$, $\varepsilon s = s$, $\tau(s \otimes s)\mu = \mu s$. If H is commutative or cocommutative, then $s^2 = \mathrm{id}$.

4. Let H_1 and H_2 be Hopf algebras and $\alpha\colon H_1 \to H_2$ a homomorphism of bialgebras. Then α commutes with the antipodes.

5. Let R be a field of characteristic $p > 0$. Let $A = R[x]/(x^p)$. The following data define a Hopf algebra structure on A: $\mu(x) = x \otimes 1 + 1 \otimes x$, $\varepsilon(x) = 0$, $s(x) = -x$.

19.7 Hopf Algebras and Classifying Spaces

The homology and cohomology of classifying spaces BU, BO, BSO lead to a Hopf algebra which we will study from the algebraic view-point in this section. The polynomial algebra $R[a] = R[a_1, a_2, \dots]$ becomes a Hopf algebra with coassociative and cocommutative comultiplication determined by $\Delta(a_n) = \sum_{p+q=n} a_p \otimes a_q$ and $a_0 = 1$. We consider the algebra as a graded algebra with a_i of degree i. (In the following we disregard the signs which appear in graded situations. Another device would be to assume that the a_i have even degree, say degree $2i$, or that R has characteristic 2.) Let $\rho = (\rho_1, \dots, \rho_r) \in \mathbb{N}_0^r$ be a multi-index with r components. We use the notation $a^\rho = a_1^{\rho_1} \dots a_r^{\rho_r}$. The monomials of type a^ρ (for arbitrary r) form an R-basis of $R[a]$. The homogeneous component $R[a]^n$ of degree n is spanned by the monomials a^ρ with $\|\rho\| = \rho_1 + 2\rho_2 + \dots + r\rho_r = n$.

We have an embedding $R[a_1, \dots, a_n] \xrightarrow{\subset} R[\alpha_1, \dots, \alpha_n]$ where a_j is the j-th elementary symmetric polynomial in the $\alpha_1, \dots, \alpha_n$. The embedding respects the grading if we give α_j the degree 1. The image is the subalgebra of symmetric functions. The a^ρ with $\rho \in \mathbb{N}_0^n$ form an R-basis of the symmetric polynomials.

Another, more obvious, R-basis is obtained by starting with a monomial

$\alpha^I = \alpha_1^{i_1} \dots \alpha_n^{i_n}$ and sum over the S_n-orbit of $I = (i_1, \dots, i_n)$. Let us write $I \sim J$ if $(j_1, \dots, j_n)$ is a permutation of $(i_1, \dots, i_n)$. The polynomials

$$\Sigma_I(\alpha_1, \dots, \alpha_n) = \sum\nolimits_{J \sim I} \alpha^J$$

form an R-basis of the symmetric polynomials in $R[\alpha_1, \dots, \alpha_n]$. The family $I = (i_1, \dots, i_n)$ is called a partition of $|I| = i_1 + i_2 + \dots + i_n$; in the case that $I \sim J$, we say that I and J yield the same unordered partition. We can write Σ_I as a polynomial in the $a_1, \dots, a_n$ and denote it by $\sigma_I(a_1, \dots, a_n)$. The monomials a^ρ which are summands of σ_I have degree $\|\rho\| = |I|$. Thus $\sigma_I(a_1, \dots, a_n) = \sigma_I(a_1, \dots, a_{n-1}, 0) = \sigma_I(a_1, \dots, a_{n-1})$ for $n > |I|$.

(19.7.1) Lemma. *If I is a partition of k and $n \geq k$, then $\sigma_I(a_1, \dots, a_k)$ is independent of n. We consider it as a polynomial in $R[a]$. In this way we obtain another R-basis of $R[a]$ which consists of the polynomials σ_I. The homogeneous component of degree n is spanned by the σ_I with I an (unordered) partition of n.* □

Consider the formal power series

$$U_n = \prod\nolimits_{j=1}^n (1 + a_1\beta_j + a_2\beta_j^2 + \cdots) \in R[a][[\beta_1, \dots, \beta_n]].$$

The series has the form $\sum_\rho a^\rho B_\rho^{(n)}(\beta_1, \dots, \beta_n)$ where the sum is taken over the multi-indices $\rho = (\rho_1, \dots, \rho_r)$ with $|\rho| = \sum_j \rho_j \leq n$. The polynomial $B_\rho^{(n)}$ is symmetric in the $\beta_1, \dots, \beta_n$. Hence we can write it as polynomial $b_\rho^{(n)}(b_1, \dots, b_n)$ where b_k is the k-th elementary symmetric polynomial in the variables $\beta_1, \dots, \beta_n$.

For $\rho = (\rho_1, \dots, \rho_r)$ let $I(\rho)$ denote the multi-index $(i_1, \dots, i_m)$ with $i_\nu = j$ for $\rho_1 + \dots + \rho_{j-1} < \nu \leq \rho_1 + \dots + \rho_j$, i.e., we begin with ρ_1 entries 1, then ρ_2 entries 2 and so on; hence $m = \rho_1 + \dots + \rho_r = |\rho|$ and $\sum_{k=1}^m i_k = |I(\rho)| = \|\rho\|$, i.e., $I(\rho)$ is a partition of $\|\rho\|$ with (weakly) increasing components. In the notation introduced above

$$B_\rho^{(n)}(\beta_1, \dots, \beta_n) = \Sigma_{I(\rho)}(\beta_1, \dots, \beta_n), \quad b_\rho^{(n)}(b_1, \dots, b_n) = \sigma_{I(\rho)}(b_1, \dots, b_n).$$

The polynomial $b_\rho^{(n)}$ only involves the variables $b_1, \dots, b_{|I(\rho)|}$ and is independent of n for $n \geq |I(\rho)|$. We denote this stable version by b_ρ. The b_ρ form an R-basis of the symmetric polynomials in $R[\beta]$. In this sense we can write formally

$$\prod\nolimits_{i=1}^\infty (1 + a_1\beta_j + a_2\beta_j^2 + \cdots) = \sum\nolimits_\rho a^\rho b_\rho = \sum\nolimits_{n \geq 0} U[n]$$

where $U[n]$ is the finite partial sum over the ρ with $\|\rho\| = n$ (although the infinite product itself is not defined). The reader may verify

$$U[1] = a_1 b_1$$
$$U[2] = a_1^2 b_2 + a_2 b_1^2 - 2a_2 b_2$$
$$U[3] = a_1^3 b_3 + a_3 b_1^3 + a_1 b_1 a_2 b_2 + 3a_3 b_3 - 3a_1 a_2 b_3 - 3a_3 b_1 b_2.$$

The polynomials $U[n]$ are symmetric in the a's and the b's

$$U[n](a_1,\dots,a_n;b_1,\dots,b_n) = U[n](b_1,\dots,b_n;a_1,\dots,a_n).$$

In order to see this note that $U = \lim_{m,n} U_{m,n}$ with

$$\begin{aligned} U_{m,n} &= \textstyle\prod_{i=1}^m \prod_{j=1}^n (1+\alpha_i\beta_j) \\ &= \textstyle\prod_{j=1}^m (1 + a_1\beta_j + \dots + a_m\beta_j^m) = \prod_{i=1}^n (1 + b_1\alpha_i + \dots + b_n\alpha_i^n). \end{aligned}$$

(19.7.2) Lemma. *Let us write* $\Delta(a^\rho) = \sum_{\sigma\tau} a^\rho_{\sigma\tau} a^\sigma \otimes a^\tau$ *and* $b_\sigma \cdot b_\tau = \sum_\rho b^\rho_{\sigma\tau} b_\rho$. *Then* $a^\rho_{\sigma\tau} = b^\rho_{\sigma\tau}$.

Proof. The definition of the $b^\rho_{\sigma\tau}$ implies the relation $b^{(n)}_\sigma \cdot b^{(n)}_\tau = \sum_\rho b^\rho_{\sigma\tau} b^{(n)}_\rho$. We compute

$$\begin{aligned} \textstyle\sum_\rho \big(\sum_{\sigma,\tau} a^\rho_{\sigma\tau} a^\sigma \otimes a^\tau\big) b^{(n)}_\rho &= \textstyle\sum_\rho \Delta(a^\rho) b^{(n)}_\rho \\ &= \textstyle\prod (1 + \Delta(a_1)\beta_i + \Delta(a_2)\beta_i^2 + \cdots) \\ &= \textstyle\prod (1 + (a_1 \otimes 1)\beta_i + (a_2 \otimes 1)\beta_i^2 + \cdots) \\ &\quad \cdot \textstyle\prod (1 + (1 \otimes a_1)\beta_i + (1 \otimes a_2)\beta_i^2 + \cdots) \\ &= \textstyle\big(\sum_\sigma (a^\sigma \otimes 1) b^{(n)}_\sigma\big)\big(\sum_\tau (1 \otimes a^\tau) b^{(n)}_\tau\big) = \sum_{\sigma\tau} (a^\sigma \otimes a^\tau) b^{(n)}_\sigma b^{(n)}_\tau \\ &= \textstyle\sum_{\sigma\tau} (a^\sigma \otimes b^\tau) \sum_\rho b^\rho_{\sigma\tau} b^{(n)}_\rho. \end{aligned}$$

Now we compare coefficients and obtain $\sum_\rho a^\rho_{\sigma\tau} b^{(n)}_\rho = \sum_\rho b^\rho_{\sigma\tau} b^{(n)}_\rho$. The sum is finite in each degree. We pass to the stable values b_ρ and compare again coefficients. □

Let $\mathrm{Hom}(R[a], R)$ be the graded dual of $R[a]$. We can view this as the module of R-linear maps $R[a] \to R$ which have non-zero value only at a finite number of monomials. Let a^*_ρ be the dual of a^ρ, i.e., $a^*_\rho(a^\sigma) = \delta^\sigma_\rho$. The Hopf algebra structure of $R[a]$ induces a Hopf algebra structure on $\mathrm{Hom}(R[a], R)$. The basic algebraic result of this section is that the dual Hopf algebra is isomorphic to the original Hopf algebra.

(19.7.3) Theorem. *The homomorphism*

$$\alpha\colon \mathrm{Hom}(R[a], R) \to R[b], \quad f \mapsto \textstyle\sum_\rho f(a^\rho) b_\rho$$

is an isomorphism of Hopf algebras. The generator b_j *is dual to* a_1^j, *that is,* $\alpha((a_1^i)*) = b_i$.

Proof. The dual basis element of a^ρ is mapped to b_ρ. Therefore α is an R-linear isomorphism. It remains to show that α is compatible with the multiplication and the comultiplication.

We verify that α is a homomorphism of algebras.

$$\alpha(f)\alpha(g) = (\sum_\sigma f(a^\sigma)b_\sigma)(\sum_\tau g(a^\tau)b_\tau) = \sum_{\sigma,\tau} f(a^\sigma)g(a^\tau)(\sum_\rho b^\rho_{\sigma,\tau}b_\rho).$$

The coefficient of b_ρ in $\alpha(f \cdot g)$ is $(f \cdot g)(a^\rho)$ and

$$(f\cdot g)(a^\rho) = (f\otimes g)(\Delta a^\rho) = (f\otimes g)(\sum_{\sigma,\tau} a^\rho_{\sigma\tau}a^\sigma \otimes a^\tau) = \sum_{\sigma,\tau} a^\rho_{\sigma\tau} f(a^\sigma)g(a^\tau).$$

Now we use the equality (19.7.2).

The definition of the comultiplication in $\operatorname{Hom}(R[a], R)$ gives for the element a^*_ρ which is dual to a^ρ the relation

$$\Delta a^*_\rho(a^\sigma \otimes a^\tau) = a^*_\rho(a^{\sigma+\tau}) = \begin{cases} 1, & \sigma+\tau=\rho, \\ 0, & \text{otherwise.} \end{cases}$$

This means that $\Delta(a^*_\rho) = \sum_{\sigma+\tau=\rho} a^*_\sigma \otimes a^*_\tau$. Since $\alpha((a^i_1)^*) = b_i$, the generators of the algebras $\operatorname{Hom}(R[a], R)$ and $R[b]$ have the same coproduct. Since we know already that α is a homomorphism of algebras, we conclude that α preserves the comultiplication. In particular we also have for the b_ρ the formula $\Delta(b_\rho) = \sum_{\sigma+\tau=\rho} b_\sigma \otimes b_\tau$. □

The Hopf algebras which we have discussed have other interesting applications, e.g., to the representation theory of symmetric groups, see [113].

(19.7.4) Remark. If we define α in (19.7.3) on the R-module of all R-linear maps, then the image is the algebra $R[[b]]$ of formal power series. The homogeneous components of degree n in $R[b]$ and $R[[b]]$ coincide. ◇

(19.7.5) Remark. The duality isomorphism (19.7.3) can be converted into a symmetric pairing $\tilde{\alpha}\colon R[b]\otimes R[a] \to R$. The pairing is defined by $\tilde{\alpha}(\alpha^*\varphi\otimes y) = \varphi(y)$ and satisfies $\tilde{\alpha}(b_\rho \otimes a^\sigma) = \delta^\sigma_\rho$. ◇

Let $\varphi\colon R[a] \to R$ be a homomorphism of R-algebras. We restrict φ to the component of degree n and obtain $\varphi_n\colon R[a]^n \to R$. We identify φ with the family (φ_n). The duality theorem sets up an isomorphism $\alpha\colon \operatorname{Hom}(R[a]^n, R) \cong R[b]_n$ with the homogeneous part $R[b]_n$ of $R[b]$.

A graded group-like element K of $R[b]$ is defined to be a sequence of polynomials $(K_n(b_1,\dots,b_n) \mid n \in \mathbb{N}_0)$ with $K_0 = 1$ and $K_n \in R[b]_n$ of degree n such that

$$(1) \qquad \Delta K_n = \sum_{i+j=n} K_i \otimes K_j.$$

Since Δ is a homomorphism of algebras, the relation

$$\Delta K_n(b_1,\dots,b_n) = K_n(\Delta b_1,\dots,\Delta b_n)$$

holds. The comultiplication has the form $\Delta b_n = \sum_{i+j=n} b_i \otimes b_j$ (with $b_0 = 1$). If we use two independent sets (b'_i) and (b''_i) of formal variables, we can write the condition (1) in the form

$$K_n(b'_1 + b''_1, b'_2 + b'_1 b''_1 + b''_2, \ldots, \textstyle\sum_{i+j=n} b'_i b''_j)$$

$$= \textstyle\sum_{i+j=n} K_i(b'_1, \ldots b'_i) K_j(b''_1, \ldots, b''_j.)$$

The simplest example is $K_n = b_n$.

(19.7.6) Proposition. *The sequence (φ_n) is an R-algebra homomorphism if and only if the sequence (K_n) with $K_n = \alpha(\varphi_n)$ is a graded group-like element.*

Proof. We use the duality pairing (19.7.5), now with the notation $\tilde{\alpha}(x \otimes y) = \langle x, y \rangle$. Let (K_n) be group-like and define a linear map $\varphi_n \colon R[a]^n \to R$ by $\varphi_n(y) = \langle K_n, y \rangle$. Then for $x \in R[a]^i$ and $y \in R[a]^j$ with $i + j = n$

$$\varphi_n(xy) = \langle K_n, xy \rangle = \langle \Delta K_n, x \otimes y \rangle = \langle K_i, x \rangle \langle K_j, y \rangle = \varphi_i(x)\varphi_j(y).$$

Hence (φ_n) is an algebra homomorphism.

Conversely, let $\varphi \colon R[a] \to R$ be an algebra homomorphism with restriction $\varphi_n \colon R[a]^n \to R$ in degree n. We set $K_n = \alpha(\varphi_n)$. A similar computation as above shows that (K_n) is a group-like element. □

(19.7.7) Remark. The algebra homomorphisms $\varphi \colon R[a] \to R$ correspond to families of elements $(\lambda_i \in R \mid i \in \mathbb{N})$ via $\varphi \mapsto (\varphi(a_i) = \lambda_i)$. Given a family (λ_i) the corresponding group-like element is obtained as follows. From

$$\textstyle\prod_i (1 + \lambda_1 t_i + \lambda_2 t_i^2 + \cdots) = \sum_\rho \lambda^\rho b_\rho = \sum_\rho \varphi(a^\rho) b_\rho = \alpha^*(\varphi)$$

we see that $K_n(b_1, \ldots, b_n)$ is the component of degree n in $\sum_\rho \lambda^\rho b_\rho$. ◇

We now return to classifying spaces and apply the duality theorem (19.7.3). We have the Kronecker pairing $\kappa \colon H^*(BO; \mathbb{F}_2) \otimes H_*(BO; \mathbb{F}_2) \to \mathbb{F}_2, x \otimes y \mapsto \langle x, y \rangle$ and the duality pairing (19.7.5) $\tilde{\alpha} \colon \mathbb{F}_2[w] \otimes \mathbb{F}_2[u] \to \mathbb{F}_2$, now with variables w, u in place of a, b. We also have the isomorphism $\zeta^* \colon \mathbb{F}_2[w] \cong H^*(BO; \mathbb{F}_2)$ from the determination of the Stiefel–Whitney classes. We obtain an isomorphism of Hopf algebras $\zeta_* \colon \mathbb{F}_2[u] \to H_*(BO; \mathbb{F}_2)$ determined via algebraic duality by the compatibility relation $\langle \zeta^* x, \zeta_* y \rangle = \tilde{\alpha}(x \otimes y)$. The generators of a polynomial algebra are not uniquely determined. Our algebraic considerations produce from the universal Stiefel–Whitney classes as canonical generators of $H^*(BO; \mathbb{F}_2)$ canonical generators of $H_*(BO; \mathbb{F}_2)$ via ζ_*.

In a similar manner we obtain isomorphisms $H_*(BU; \mathbb{Z}) \cong \mathbb{Z}[d_1, d_2, \ldots]$ (variables c, d) and $H_*(BSO; R) \cong R[q_1, q_2, \ldots]$ (variables p, q).

Problems

1. Verify the following polynomials b_ρ for $\|\rho\| = 4 = |I(\rho)|$:

$$\begin{aligned} b_{(0,0,0,1)} &= b_1^4 - 4b_1^2 b_2 + 2b_2^2 + 4b_1 b_3 - 4b_4, & I(\rho) &= (4); \\ b_{(1,0,1)} &= b_1^2 b_2 - 2b_2^2 - b_1 b_3 + 4b_4, & I(\rho) &= (1,3); \\ b_{(0,2)} &= b_2^2 - 2b_1 b_3 + 2b_4, & I(\rho) &= (2,2); \\ b_{(2,1)} &= b_1 b_3 - 4b_4, & I(\rho) &= (1,1,2); \\ b_{(4)} &= b_4, & I(\rho) &= (1,1,1,1). \end{aligned}$$

These b_ρ are the coefficients of a^ρ in $U[4] = \sum_\rho a^\rho b_\rho$. Check that $U[4]$ is symmetric in the a's and b's.

2. The assignment $R[a] \otimes R[b] \to R$, $a^\sigma \otimes b_\rho \mapsto (a^\sigma)^*(b_\rho) = \delta^\sigma_\rho$ is a symmetric pairing. (The formal element $U = \sum_\rho a^\rho b_\rho$ could be called a symmetric ***copairing***.)

19.8 Characteristic Numbers

Let $\kappa_\xi\colon X \to BO(n)$ be a classifying map of an n-dimensional bundle. It induces a ring homomorphism $H^*(BO(n); \mathbb{F}_2) \to H^*(X; \mathbb{F}_2)$. We can also pass to the stable classifying map $X \to BO$ and obtain $\kappa_\xi^*\colon H^*(BO; \mathbb{F}_2) \to H^*(X; \mathbb{F}_2)$. This homomorphism codifies the information which is obtainable from the Stiefel–Whitney classes. We use the isomorphism $\mathbb{F}_2[w] \cong H^*(BO; \mathbb{F}_2)$ and the duality theorem (19.7.3). We use a slightly more general form. Let S^* be a graded R-algebra; the grading should correspond to the grading of $R[a]$, there are no signs. We obtain a graded algebra $\mathrm{Hom}_R(R[a], S^*)$ where the component of degree k consists of the homomorphisms of degree k. The product in this algebra is defined by convolution. Then we have:

(19.8.1) Theorem. *There exists a canonical isomorphism*

$$\alpha\colon \mathrm{Hom}_R(R[a], S^*) \cong S^*[[b]]$$

of graded R-algebras. Here $S^*[[b]] = S^*[[b_1, b_2, \dots]]$ *is the algebra of graded formal power series in the* b_i *of degree* $-i$. *The isomorphism* α *sends the R-homomorphism* $\varphi\colon R[a] \to S^*$ *to the series* $\sum_\rho \varphi(a^\rho) b_\rho$. □

In our example we obtain from $\kappa_\xi^*\colon H^*(BO; \mathbb{F}_2) \to H^*(X; \mathbb{F}_2)$ a series $v(\xi) \in H^*(X; \mathbb{F}_2)[[u]]$ of degree zero. The constant term is 1, the multiplicativity relation $v(\xi \oplus \eta) = v(\xi)v(\eta)$ and the naturality $v(f^*\eta) = f^* v(\eta)$ hold. For a line bundle η we have $v(\eta) = 1 + w_1(\eta)u_1 + w_1(\eta)^2 u_2 + \cdots$. These properties characterize the assignment $\xi \mapsto v(\xi)$.

We can apply a similar process to oriented or complex bundles. In the case of a complex oriented theory $h^*(-)$ we obtain series $v(\xi) \in h^*(X)[[d]]$ which are natural, multiplicative and assign to a complex line bundle η the series $v(\eta) = 1 + c_1(\eta)d_1 + c_1(\eta)^2 d_2 + \cdots$.

Interesting applications arise if we apply the process to the tangent bundle of a manifold. Let us consider oriented closed n-manifolds M with classifying map $\kappa_M \colon M \to B\mathrm{SO}$ of the stable oriented tangent bundle. We evaluate the homomorphism κ_M^* on the fundamental class $[M]$

$$H^n(B\mathrm{SO}; R) \to H^n(M; R) \to R, \quad x \mapsto \kappa_M^*(x)[M].$$

By the Kronecker pairing duality $H_n(B\mathrm{SO}; R) \cong \mathrm{Hom}_R(H^n(B\mathrm{SO}; R), R)$ this homomorphism corresponds to an element in $H_n(B\mathrm{SO}; R)$, and this element is $\kappa_{M*}[M]$, the image of the fundamental class $[M] \in H_n(M; R)$ under $(\kappa_M)_*$, by the naturality $\langle \kappa_M^*(p), [M] \rangle = \langle p, (\kappa_M)_*[M] \rangle$ of the pairing. Under the isomorphism $\zeta_* \colon R[q_1, q_2, \ldots] \cong H_*(B\mathrm{SO}; R)$ the element $\kappa_{M*}[M]$ corresponds to an element that we denote $\chi_{\mathrm{SO}}(M) \in R[q_1, q_2, \ldots]$. From the definitions we obtain:

(19.8.2) Proposition. *Let τ_M denote the oriented tangent bundle of M. Then*

$$\chi_{\mathrm{SO}}(M) = \langle v(\tau_M), [M] \rangle,$$

the evaluation of the series $v(\tau_M)$ on the fundamental class. □

If $p \in H^n(M)$ is a polynomial of degree n in the Pontrjagin classes, then the element (number) $p[M]$ is called the corresponding ***Pontrjagin number***. In a similar manner one defines a ***Stiefel–Whitney number*** by evaluating a polynomial in the Stiefel–Whitney classes on the fundamental class. A closed n-manifold M has an associated element $\chi_{\mathrm{O}}(M) \in \mathbb{F}_2[u_1, u_2, \ldots] \cong H_n(B\mathrm{O}; \mathbb{F}_2)$, again the image of the fundamental class under the map induced by the stable classifying map $\kappa_M \colon M \to B\mathrm{O}$ of the tangent bundle.

(19.8.3) Example. Let us consider $M = \mathbb{C}P^{2k}$. The stable tangent bundle is η^{2k+1} where is η is the canonical complex line bundle, now considered as oriented bundle; see (15.6.6). By the multiplicativity of the v-classes, we have for the tangent bundle τ_{2k} of $\mathbb{C}P^{2k}$ the relation

$$\begin{aligned} v(\tau_{2k}) = v(\eta)^{2k+1} &= (1 + p_1(\eta)q_1 + p_1(\eta)^2 q_2 + \cdots)^{2k+1} \\ &= (1 + c^2 q_1 + c^4 q_2 + \cdots)^{2k+1} \end{aligned}$$

where as usual $H^*(\mathbb{C}P^{2k}; R) \cong R[c]/(c^{2k+1})$. Note that $p_1(\eta) = c^2$, by (19.5.6). The evaluation on the fundamental class yields the coefficient of c^{2k} in this series, since $\langle c^{2k}, [\mathbb{C}P^{2k}] \rangle = 1$ (see (18.7.2)). Modulo decomposable elements in the indeterminates q_j, i.e., modulo polynomials in the q_j with $j < k$, this value is $(2k+1)q_k$. ◇

When we pass to rational coefficients $R = \mathbb{Q}$ we can divide by $2k + 1$ and obtain:

(19.8.4) Proposition. *The elements $\chi_{\mathrm{SO}}(\mathbb{C}P^{2k}), k \in \mathbb{N}$ are polynomial generators of $\mathbb{Q}[q]$.* □

In bordism theory it will be shown that the signature of an oriented $4k$-manifold only depends on its image in $H_*(B\mathrm{SO};\mathbb{Q}) \cong \mathbb{Q}[q]$. And from the multiplicativity of the signature it then follows that there exists an algebra homomorphism $s\colon H_*(B\mathrm{SO};\mathbb{Q}) \to \mathbb{Q}$ such that the signature $\sigma(M)$ is obtained as the image of this homomorphism $\sigma(M) = s(\kappa_M)_*[M])$. We know that the generators $\mathbb{C}P^{2k}$ have signature 1; see 18.7.2. The ring homomorphism s is determined by the values $\lambda_i = s(q_i) \in \mathbb{Q}$. Via the duality $\mathbb{Q}[p] \cong \mathrm{Hom}(\mathbb{Q}[q],\mathbb{Q})$ the homomorphism s corresponds to a group-like element $(L_n(p_1,\dots,p_n) \mid n \in \mathbb{N})$ where L_n is a polynomial in the Pontrjagin classes of degree $4n$ such that the evaluation on the fundamental class is the signature, $\langle L_n,[M^{4k}]\rangle = \sigma(M^{4n})$. If we expand the formal product $\prod_i(1+\lambda_1 t_i + \lambda_2 t_i^2 + \cdots)$ and assume that p_k is the k-th elementary symmetric polynomial in the t_i (of degree 4), then L_n is the component of degree $4n$. The t_i are obtained if we split the total Pontrjagin class (formally) into linear factors, $1 + p_1 x + p_2 x^2 + \cdots = \prod_i(1+t_i x)$. Fortunately, nature has already split for us the stable tangent bundle of $\mathbb{C}P^{2n}$, the total Pontrjagin class is $(1+c^2)^{2n+1}$; i.e., we can take $t_i = c^2$ in order to evaluate L_n on $\mathbb{C}P^{2n}$. This allows us to determine the coefficients λ_i: The power series $H(c) = 1 + \lambda_1 c^2 + \lambda_2 c^4 + \cdots$ has the property that the coefficient of c^{2n} in $H(c)^{2n+1}$ is 1. Hirzebruch [81, p. 14] has found this power series

$$H(c) = \frac{c}{\tanh c} = \frac{2c}{e^{2c}-1} + c = 1 + \frac{B_1}{2!}(2c)^2 - \frac{B_2}{4!}(2c)^4 + \frac{B_3}{6!}(2c)^6 - \cdots$$

where the B_j are the so-called ***Bernoulli numbers***. The first four values are

$$B_1 = \frac{1}{6}, \quad B_2 = \frac{1}{30}, \quad B_3 = \frac{1}{42}, \quad B_4 = \frac{1}{30}.$$

The corresponding coefficients in the power series are

$$\lambda_1 = \frac{1}{3}, \quad \lambda_2 = -\frac{1}{3^2\cdot 5}, \quad \lambda_3 = \frac{2}{3^3\cdot 5\cdot 7}, \quad \lambda_4 = -\frac{1}{3^3\cdot 5^2\cdot 7}.$$

From these data we obtain the polynomials L_n if we insert in the universal polynomials $U[n](p_1,\dots,p_n;q_1,\dots,q_n)$ for q_j the value λ_j. We have already listed the polynomials $U[1], U[2], U[3], U[4]$. The result is

$$\begin{aligned}
L_1 &= \frac{1}{3}p_1,\\
L_2 &= \frac{1}{45}(7p_2 - p_1^2),\\
L_3 &= \frac{1}{945}(62p_3 - 13p_2p_1 + 2p_1^3),\\
L_4 &= \frac{1}{14175}(381p_4 - 71p_3p_1 - 19p_2^2 + 22p_2p_1^2 - 3p_1^4).
\end{aligned}$$

The polynomials L_n are called the ***Hirzebruch L-polynomials***.

Problems

1. Show that $\chi_{SO}(M \times N) = \chi_{SO}(M)\chi_{SO}(N)$.
2. Let M be the oriented boundary of a compact manifold. Then $\chi_{SO}(M) = 0$. (See the bordism invariance of the degree.)
3. Show that $\chi_O(\mathbb{R}P^{2n}) = u_{2n}$ modulo decomposable elements. Therefore these elements can serve as polynomial generators of $H_*(BO; \mathbb{F}_2) \cong \mathbb{F}_2[u]$ in even dimensions.
4. The convolution product of the homomorphisms defined at the beginning of the section satisfies $\kappa_\xi^* * \kappa_\eta^* = \kappa_{\xi\oplus\eta}^*$.
5. Determine $\chi_{SO}(\mathbb{C}P^2)$ and $\chi_{SO}(\mathbb{C}P^4)$.

Chapter 20
Homology and Homotopy

We begin this chapter with the theorem of Hurewicz which says in its simplest form that for a simply connected space the first non-zero homotopy group is isomorphic to the first non-zero integral homology group. In the case of the sphere S^n this is essentially the Hopf degree theorem. In our proof we use this theorem and other consequences of the homotopy excision theorem. We indicate an independent proof which only uses methods from homology theory and the Eilenberg subcomplexes introduced earlier. The theorem of Hurewicz has the important consequence that a map between simply connected CW-complexes is a homotopy equivalence if it induces an isomorphism of the integral homology groups (theorem of Whitehead). Another application is to the geometric realization of algebraic chain complexes as cellular chain complexes. We will see that under suitable hypotheses we do not need more cells in a homotopy type than the homology groups predict.

Since homotopy groups are difficult to compute it is desirable to have at least some qualitative information about them. One of the striking results is the famous theorem of Serre that the homotopy groups of spheres are finite groups, except in the few cases already known to Hopf; in particular the stable homotopy groups of spheres are finite (except $\pi_n(S^n)$). Since for a finite abelian group A the tensor product $A \otimes \mathbb{Q} = 0$ and since homology theories are objects of stable homotopy, this theorem has the remarkable consequence that rationalized homology theories $h_*(-) \otimes \mathbb{Q}$ can be reduced to ordinary rational homology.

Along the way we obtain qualitative results in general. They concern, for instance, statements about finiteness or finite generation and are based on qualitative generalizations of the theorem of Hurewicz. For the expert we point out that we do not use the theory of spectral sequences for the proofs. Only elementary methods like induction over skeleta enter. A basic technical theorem relates in a qualitative manner the homology of the total space, fibre and base of a fibration. On the algebraic side we use so-called Serre classes of abelian groups: Properties like "finite generation" are formalized. (In the long run this leads to localization of spaces and categories.)

20.1 The Theorem of Hurewicz

The theorem of Hurewicz relates the homotopy and the homology groups of a space. In this section H_* denotes integral singular homology. Let $(X, A, *)$ be a pointed pair of spaces.

We define natural homomorphisms, called ***Hurewicz homomorphisms***,

$$h_{(X,A,*)} = h\colon \pi_n(X,A,*) \to H_n(X,A), \quad n \ge 2,$$
$$h_{(X,*)} = h\colon \pi_n(X,*) \to H_n(X), \quad n \ge 1,$$

such that the diagrams

$$\begin{array}{ccccc} \pi_n(X,*) & \longrightarrow & \pi_n(X,A,*) & \xrightarrow{\partial} & \pi_{n-1}(A,*) \\ \downarrow h & & \downarrow h & & \downarrow h \\ H_n(X) & \longrightarrow & H_n(X,A) & \xrightarrow{\partial} & H_{n-1}(A) \end{array}$$

commute (compatibility with exact sequences). For this purpose we use the definition $\pi_n(X,*) = [S(n),X]^0$ and $\pi_n(X,A,*) = [(D(n),S(n-1)),(X,A)]^0$ of the homotopy groups (see (6.1.4)). We choose generators $z_n \in H_n(S(n))$ and $\tilde{z}_n \in H_n(D(n),S(n-1))$ such that $\partial\tilde{z}_n = z_{n-1}$ and $q_*(\tilde{z}_n) = z_n$, where $q\colon D(n) \to D(n)/S(n-1) = S(n)$ is the quotient map. If we fix z_1, then the other generators are determined inductively by these conditions. We define $h\colon \pi_n(X,A,*) \to H_n(X,A)$ by $[f] \mapsto f_*(\tilde{z}_n)$ and $h\colon \pi_n(X,*) \to H_n(X)$ by $[f] \mapsto f_*(z_n)$. With our choice of generators the diagram above is then commutative. From (10.4.4) and the analogous result for the relative homotopy groups we see that the maps h are homomorphisms. The singular simplex $\Delta^1 \to I/\partial I = S(1)$, $(t_0,t_1) \mapsto t_1$ represents a generator $z_1 \in H_1(S(1))$. If we use this generator, then $h\colon \pi_1(X,*) \to H_1(X)$ becomes the homomorphism which was shown in (9.2.1) to induce an isomorphism $\pi_1(X,*)^{ab} \cong H_1(X)$ for 0-connected X.

Recall that we have a right action of the fundamental group

$$\pi_n(X,A,*) \times \pi_1(A,*) \to \pi_n(X,A,*), \quad (x,\alpha) \mapsto x \cdot \alpha$$

via transport. We denote by $\pi_n^\#(X,A,*)$ the quotient of $\pi_n(X,A,*)$ by the normal subgroup generated by all elements of the form $x - x \cdot \alpha$ (additive notation in π_n). Recall from (6.2.6) that $\pi_2^\#$ is abelian. Representative elements in π_n which differ by transport are freely homotopic, i.e., homotopic disregarding the base point. Therefore the Hurewicz homomorphism induces a homomorphism

$$h^\#\colon \pi_n^\#(X,A,*) \to H_n(X,A).$$

The transport homomorphism $\pi_n(X,A,a_1) \to \pi_n(X,A,a_2)$ along a path from a_1 to a_2 induces an isomorphism of the $\pi_n^\#$-groups, and this isomorphism is independent of the choice of the path. We can use this remark: An unpointed map $(D(n),S(n-1)) \to (X,A)$ yields in each of the groups $\pi_n^\#(X,A,a)$ a well-defined element (A path connected). Thus, if $\pi_1(A)$ is trivial, we can regard $\pi_n^\#(X,A)$ as the homotopy set $[(D(n),S(n-1)),(X,A)]$. The group $\pi_1^\#(X,*)$ is defined to be the abelianized group $\pi_1(X,*)^{ab}$, i.e., the quotient by the commutator subgroup. We set $\pi_n^\#(X,*) = \pi_n(X,*)$ for $n \ge 2$ and again we have the Hurewicz homomorphism $h^\#\colon \pi_n^\#(X,*) \to H_n(X)$.

(20.1.1) Theorem (Hurewicz). *Let the space X be $(n-1)$-connected $(n \geq 1)$. Then $h^{\#}\colon \pi_n^{\#}(X,*) \to H_n(X)$ is an isomorphism.*

Proof. We have already proved the theorem in the case that $n = 1$. So let $n \geq 2$ and then $\pi_n^{\#} = \pi_n$. Since weak homotopy equivalences induce isomorphisms in homotopy and homology, we only need to prove the theorem for CW-complexes X. We can assume that X has a single 0-cell and no i-cells for $1 \leq i \leq n-1$, see (8.6.2). The inclusion $X^{n+1} \subset X$ induces isomorphisms $\pi_n^{\#}(X) \cong \pi_n^{\#}(X^{n+1})$ and $H_n(X) \cong H_n(X^{n+1})$. Since the Hurewicz homomorphisms h form a natural transformation of functors, it suffices to prove the theorem for $(n+1)$-dimensional complexes. In this case X is h-equivalent to the mapping cone of a map of the form $\varphi\colon A = \bigvee S_j^n \to B = \bigvee S_k^n$.

For $X = S^n$ the theorem holds by (10.5.1). By naturality and additivity it then holds for pointed sums $\bigvee S_j^n$. We have a commutative diagram

$$\begin{array}{ccccccc} \pi_n(A) & \longrightarrow & \pi_n(B) & \longrightarrow & \pi_n(X) & \longrightarrow & 0 \\ \downarrow & & \downarrow & & \downarrow & & \\ H_n(A) & \longrightarrow & H_n(B) & \longrightarrow & H_n(X) & \longrightarrow & 0 \end{array}$$

with exact rows. The exactness of the top row is a consequence of the homotopy excision theorem. □

(20.1.2) Corollary. *Let X be simply connected and suppose that $\tilde{H}_i(X) = 0$ for $i < n$. Then $\pi_i(X,*) = 0$ for $i < n$ and $h\colon \pi_n(X,*) \cong H_n(X)$.*

Proof. (20.1.1) says, in different wording, that $h\colon \pi_j(X) \cong H_j(X)$ for the smallest j such that $\pi_k(X) = 0$ for $1 \leq k < j$. □

(20.1.3) Theorem. *Let (X,A) be a pair of simply connected CW-complexes. Suppose $H_i(X,A) = 0$ for $i < n$, $n \geq 2$. Then $\pi_i(X,A) = 0$ for $i < n$ and $h\colon \pi_n(X,A) \to H_n(X,A)$ is an isomorphism.*

Proof. Induction over $n \geq 2$. We use a consequence of the homotopy excision theorem: Let A be simply connected and $\pi_i(X,A,*) = 0$ for $0 < i < n$. Then $\pi_n(X,A,*) \to \pi_n(X/A,*)$ is an isomorphism. The theorem of Seifert and van Kampen shows $\pi_1(X/A) = \{e\}$. From $H_i(X,A) = \tilde{H}_i(X/A)$ and (20.1.2) we conclude $\pi_i(X/A) = 0$ for $i < n$.

Let $n = 2$. Since X and A are simply connected, $\pi_1(X,A,*) = 0$ and the diagram

$$\begin{array}{ccc} \pi_2(X,A,*) & \xrightarrow{\cong} & \pi_2(X/A,*) \\ \downarrow h & & \downarrow \cong \\ H_2(X,A) & \xrightarrow{\cong} & H_2(X/A) \end{array}$$

shows that h is an isomorphism.

By induction we know that $\pi_j(X, A, *) = 0$ for $0 < j < n$. Then a similar diagram shows that $h\colon \pi_n(X, A, *) \to H_n(X, A)$ is an isomorphism. □

(20.1.4) Theorem. *Let $f\colon X \to Y$ be a map between simply connected spaces. Suppose $f_*\colon H_i(X) \to H_i(Y)$ is bijective for $i < n$ and surjective for $i = n$ ($n \geq 2$). Then $f_*\colon \pi_i(X) \to \pi_i(Y)$ is bijective for $i < n$ and surjective for $i = n$.*

Proof. We pass to the mapping cylinder and assume that f is an inclusion. The hypothesis is then equivalent to $H_i(Y, X) = 0$ and the claim equivalent to $\pi_i(Y, X) = 0$ for $i < n$. Now we use (20.1.3). □

Recall: A map $f\colon X \to Y$ between CW-complexes is an h-equivalence if and only if $f_*\colon \pi_i(X) \cong \pi_i(Y)$ for each i. Together with (20.1.4) we obtain a homological version of this result:

(20.1.5) Theorem (Whitehead). *Let $f\colon X \to Y$ be a map between simply connected CW-complexes, which induces isomorphisms of homology groups. Then f is a homotopy equivalence.* □

In (20.1.5) one cannot dispense with the hypothesis that the spaces are simply connected. There exist, e.g., so-called ***acyclic*** complexes X with reduced homology groups vanishing but with non-trivial fundamental group. Moreover it is important that the isomorphism is induced by a map.

(20.1.6) Proposition. *Let X be a simply connected CW-complex with integral homology of a sphere, $H_*(X) \cong H_*(S^n)$, $n \geq 2$. Then X is h-equivalent to S^n.*

Proof. By (20.1.2), $\pi_n(X) \cong H_n(X)$, and this group is assumed to be isomorphic to $\mathbb{Z}$. Let $f\colon S^n \to X$ represent a generator. Then $f_*\colon \pi_n(S^n) \to \pi_n(X)$ is an isomorphism and also $f_*\colon H_n(S^n) \to H_n(X)$. Now we use (20.1.5). □

The preceding proposition has interesting applications. It is known that a closed connected n-manifold of the homotopy type of the n-sphere is actually homeomorphic to the n-sphere. Therefore these spheres are characterized by invariants of algebraic topology.

(20.1.7) Example. The spaces $S^n \vee S^n \vee S^{2n}$ and $S^n \times S^n$ are for $n \geq 2$ simply connected and have isomorphic homology groups. But they are not h-equivalent, since their cohomology rings are different. ◇

The homological theorem of Whitehead (20.1.5) no longer holds for spaces which are not simply connected, even in the case when the map induces an isomorphism of the fundamental groups. But it suffices to consider the universal covering, as the next theorem shows.

(20.1.8) Theorem. *Let $f: X \to Y$ be a map between connected CW-complexes which induces an isomorphism $f_*: \pi_1(X) \to \pi_1(Y)$. Let $p: \tilde{X} \to X$ and $q: \tilde{Y} \to Y$ be the universal coverings. There exists a lifting $F: \tilde{X} \to \tilde{Y}$ of f, i.e., $qF = fp$. Suppose F induces an isomorphism of the homology groups. Then f is a homotopy equivalence.*

Proof. The maps p and q induce isomorphisms of π_n for $n \geq 2$. The spaces $\tilde{X}$ and $\tilde{Y}$ are simply connected CW-spaces. By (20.1.5) the map F is an h-equivalence. Hence f induces isomorphisms of π_n for $n \geq 2$ and, by hypothesis, also for $n = 1$. Therefore f is an h-equivalence. □

(20.1.9) Theorem. *Let $F: X \to Y$ be a G-map between the total spaces of numerable G-principal bundles. If F is an ordinary homotopy equivalence, then F is a G-equivariant homotopy equivalence.*

Proof. Consider $\mathrm{id} \times_G F: EG \times_G X \to EG \times_G Y$ as a morphism over BG. From (13.3.4) we see that $\mathrm{id} \times_G F$ is a fibre homotopy equivalence. Now we consider the bundles $EG \times_G X \to X/G$ and $EG \times_G Y \to Y/G$ with contractible fibre EG. Using (13.3.4) again, we see that these maps are shrinkable, hence $F/G: X/G \to Y/G$ is a homotopy equivalence. From (14.3.4) we conclude that F is a G-homotopy equivalence. □

Let G be a discrete group which acts on the pair (Y, B). The induced maps of the left translations by group elements yield a left action of G on $H_n(Y, B)$ via homomorphisms, i.e., $H_n(Y, B)$ becomes a module over the integral group ring $\mathbb{Z}G$ of G.

Suppose X is obtained from A by attaching n-cells ($n \geq 3$). Let $p: Y \to X$ be a universal covering and $B = p^{-1}(A)$. Then Y is obtained from B by attaching n-cells. The group $\pi = \pi_1(X)$ of deck transformations acts freely on the set of n-cells in $Y \smallsetminus B$. Hence $H_n(Y, B)$ is a free $\mathbb{Z}\pi$-module, the basis elements correspond bijectively to the n-cells of $X \smallsetminus A$. Theorem (20.1.3) now yields:

(20.1.10) Theorem. *Let X be a connected CW-complex and let $n \geq 3$. Then $\pi_n(X^n, X^{n-1})$ is a free $\mathbb{Z}\pi_1(X^{n-1})$-module. A basis of this module consists of the characteristic maps of the n-cells. The map $h^\#: \pi_n(X^n, X^{n-1}) \to H_n(X^n, X^{n-1})$ is an isomorphism.* □

The exact sequence

$$\pi_{n+1}(X^{n+1}, X^n) \to \pi_n(X^n, X^{n-1}) \to \pi_n(X^{n+1}, X^{n-1}) \to 0$$

is for $n \geq 3$ a sequence of $\mathbb{Z}\pi_1(X^{n-1}) \cong \mathbb{Z}\pi_1(X^n)$-modules. Because of the isomorphism $\pi_n(X, X^{n-1}) \cong \pi_n(X^{n+1}, X^{n-1})$ the sequence is a presentation of the $\mathbb{Z}\pi_1(X^{n-1})$-module $\pi_n(X, X^{n-1})$. The induced sequence of the $\pi^\#$-groups is also exact. This gives the following theorem for $n \geq 3$.

(20.1.11) Theorem (Hurewicz). *Let (X, A) be a CW-pair with connected X and A. Let (X, A) be $(n-1)$-connected $(n \geq 2)$. Then $h^{\#}\colon \pi_n^{\#}(X, A, *) \cong H_n(X, A)$.*

Proof. We now give a purely homological proof of the Hurewicz theorems which also covers the relative case $n = 2$ for spaces which are not simply connected. The proof is by induction. The induction starts with (9.2.1). We assume the absolute theorem for $1 \leq i \leq n-1$ and prove the relative theorem for n. We consider the standard simplex $\Delta[k] = [e_0, \dots, e_k]$ as the usual simplicial complex and denote its l-skeleton by $\Delta[k]^l$. Let $S_k^{n-1}(X, A, *)$ be the chain group spanned by simplices $\sigma\colon \Delta[k] \to X$ such that $\sigma(\Delta[k]^{n-1}) \subset A$ and $\sigma(\Delta[k]^0) = \{*\}$, modulo $S_k(A)$. Let $H_n^{(n-1)}(X, A, *)$ be the n-th homology group of the resulting chain complex. The inclusion of chain complexes induces an isomorphism $H_n^{(n-1)}(X, A, *) \cong H_n(X, A)$ for an $(n-1)$-connected pair (X, A) with path connected A, see (9.5.4).

We have to adapt the homotopy groups to the simplicial setup. We consider elements of $\pi_n(X, A, *)$ as homotopy classes of maps $f\colon (\Delta[n], \partial\Delta[n], e_0) \to (X, A, *)$. For this purpose we fix a homeomorphism $\alpha\colon (D(n), S(n-1), *) \to (\Delta[n], \partial\Delta[n], e_0)$ which sends the generator $\tilde{z}_n$ defined at the beginning to the standard generator $[\mathrm{id}(\Delta[n])] \in H_n(\Delta[n], \partial\Delta[n])$.

The Hurewicz homomorphism then sends the homotopy class of f to $f_*[\mathrm{id}]$, and this class is an element in $H_n^{(n-1)}(X, A, *)$. We now construct an inverse $\psi\colon H_n^{(n-1)}(X, A, *) \to \pi_n^{\#}(X, A, *)$ of h. We assign to a singular simplex

$$\sigma\colon (\Delta[n], \partial\Delta[n], \Delta[n]^0) \to (X, A, *)$$

the element in $\pi_n^{\#}(X, A, *)$ represented by σ. If $\sigma(\Delta[n]) \subset A$, then the corresponding homotopy class is zero. Since the $\pi_n^{\#}$-group is abelian, we obtain a well-defined homomorphism $\psi\colon S_n^{n-1}(X, A, *) \to \pi_n^{\#}(X, A, *)$. The simplices σ are cycles (since $S_{n-1}^{n-1}(X, A, *) = 0$), and thus it remains to show that the composite $\psi \circ \partial\colon S_{n+1}^{n-1}(X, A, *) \to \pi_n^{\#}(X, A, *)$ is trivial, in order to obtain ψ. We reduce the problem to a universal situation. For this purpose we define elements $b_n \in \pi_n(\partial\Delta[n+1], \Delta[n+1]^{n-1}, e_0)$,

$$b_2 = ([d_0^3] \cdot [e_1 e_0])[d_2^3][d_1^3]^{-1}[d_3^3]^{-1},$$

$$b_n = [d_0^{n+1}] \cdot [e_1 e_0] + \sum_{i=1}^{n+1} (-1)^i [d_i^{n+1}], \quad n \geq 3,$$

where $[vw]$ denotes the affine path class in $\Delta[n+1]$ from v to w, and we use the transport along this path. (Multiplicative notation in π_2, additive notation for $n \geq 3$. This definition corresponds to the homological boundary operator, but we have to transport the face maps to the base point e_0. For b_2 we have to pay attention to the order of the factors, since the group is non-abelian.)

Let us write $K = \Delta[n+1]$ and let $\tau\colon (K, K^{n-1}, K^0) \to (X, A, *)$ be a basis element of $S_{n+1}^{n-1}(X, A, *)$. Then

$$\psi\partial[\tau] = \textstyle\sum_i (-1)^i [\tau d_i^{n+1}] = \tau_*^{\#}[b_n]^{\#} = \tau_*^{\#} j_*^{\#}[b_n]^{\#}$$

where j denotes the inclusion $\partial\Delta[n+1] \subset \Delta[n+1]$. Thus it remains to show that $j_*[b_n] = 0$.

The skeleton K^{n-1} is $(n-2)$-connected (use e.g., the induction hypothesis). Hence, by the inductive assumption, $\pi_{n-1}^{\#}(K^{n-1}, e_0) \to H_{n-1}(K^{n-1})$ is an isomorphism. The commutativity $\partial h = h\partial$ now shows that $\partial[b_n] = 0$, since $\partial h[b_n] = 0$ by the fundamental boundary relation for singular homology. We have the factorization

$$\partial\colon \pi_n(K^n, K^{n-1}, e_0) \xrightarrow{j_*} \pi_n(K, K^{n-1}, e_0) \xrightarrow{\partial'} \pi_{n-1}(K^{n-1}, e_0),$$

and ∂' is an isomorphism, since K is contractible. This finishes the inductive step for the relative Hurewicz theorem. For $n \geq 2$ the absolute theorem is a special case of the relative theorem. □

An interesting consequence of the homological proof of the Hurewicz theorem is a new proof of the Brouwer–Hopf degree theorem $\pi_n(S^n) \cong \mathbb{Z}$.

20.2 Realization of Chain Complexes

The computation of homology groups from the cellular chain complex shows that one needs enough cells to realize the homology groups algebraically as the homology groups of a chain complex. It is interesting to know that in certain cases a converse holds. We work with integral homology.

(20.2.1) Theorem (Cell theorem). *Let Y be a 1-connected CW-complex. Suppose $H_j(Y)$ is finitely generated for $j \leq n$. Then Y is homotopy-equivalent to a CW-complex Z with finitely many j-cells for $j \leq n$.*

The proof of this theorem is based on a theorem which says that under suitable hypotheses an algebraic chain complex can be realized as a cellular chain complex. We describe the inductive construction of a realization. We start with the following:

20.2.2 Data and notation.

(1) Y is a CW-complex with i-skeleton Y_i.
(2) Z_r is an r-dimensional CW-complex.
(3) $f\colon Z_r \to Y$ is a cellular map.
(4) $C_i(Z) = H_i(Z_i, Z_{i-1}; \mathbb{Z})$ is the i-th cellular chain group.
(5) f induces a chain map $\varphi_\bullet\colon C_\bullet(Z) \to C_\bullet(Y)$.

(6) We attach $(r+1)$-cells to Z_r such that f can be extended to F:

$$\begin{array}{ccccc} \amalg S_j^r & \longrightarrow & Z_r & \xrightarrow{f} & Y_r \\ \downarrow & & \downarrow\cap & & \downarrow\cap \\ \amalg D_j^{r+1} & \longrightarrow & Z_{r+1} & \xrightarrow{F} & Y_{r+1}. \end{array}$$

(7) From this diagram we obtain a resulting diagram of chain groups

$$\begin{array}{ccccc} A_{r+1} & \xrightarrow{\delta} & C_r(Z) & \xrightarrow{d} & C_{r-1}(Z) \\ \downarrow\psi & & \downarrow\varphi_r & & \downarrow\varphi_{r-1} \\ C_{r+1}(Y) & \xrightarrow{d'_{r+1}} & C_r(Y) & \xrightarrow{d'_r} & C_{r-1}(Y) \end{array}$$

with $A_{r+1} = H_{r+1}(Z_{r+1}, Z_r)$ a free abelian group with a basis given by the $(r+1)$-cells, and ψ induced by (F, f). ◇

We now start from a diagram in which A_{r+1} is a free abelian group with basis $(a_j \mid j \in J)$. The horizontal parts should be chain complexes, i.e., $d\delta = 0$. Can this diagram be realized geometrically?

(20.2.3) Proposition. *A realization exists, if the following holds:*

(1) *$f_*\colon H_i(Z_r) \to H_i(Y)$ is bijective for $i \le r-1$ and surjective for $i = r$;*

(2) *$r \ge 2$;*

(3) *Z_r and Y are 1-connected.*

Proof. Suppose we are given for each $j \in J$ a diagram

$$\begin{array}{ccc} S^r & \xrightarrow{b_j} & Z_r \\ \downarrow & & \downarrow f \\ D^{r+1} & \xrightarrow{B_j} & Y_{r+1}. \end{array}$$

We attach $(r+1)$-cells to Z_r with attaching maps b_j to obtain Z_{r+1} and use the B_j to extend f to $f_{r+1}\colon Z_{r+1} \to Y_{r+1}$. Then $A_{r+1} \cong H_{r+1}(Z_{r+1}, Z_r)$ canonically and basis preserving.

We consider f as an inclusion. The assumption (1) is then equivalent to $H_i(Y, Z_r) = 0$ for $i \le r$. Since $r \ge 2$ and $\pi_1(Y) = 0$ we also have $\pi_1(Y_{r+1}) = 0$. Therefore we have the relative Hurewicz isomorphism $h\colon \pi_{r+1}(Y_{r+1}, Z_r) \cong H_{r+1}(Y_{r+1}, Z_r)$, since $H_i(Y_{r+1}, Z_r) \cong H_i(Y, Z_r) = 0$ for $i \le r$. The diagram represents an element in $\pi_{r+1}(Y_{r+1}, Z_r)$. Let $x_j = h([B_j, b_j]) \in H_{r+1}(Y_{r+1}, Z_r)$ be its image under the relative Hurewicz homomorphism. This element can be determined by homological conditions. The correct maps δ and ψ are obtained, if x_j has the following properties:

(1) The image of x_j under $\partial\colon H_{r+1}(Y_{r+1}, Z_r) \to H_r(Z_r) \to H_r(Z_r, Z_{r-1})$ is $\delta(a_j)$.

(2) The image of x_j under $\gamma\colon H_{r+1}(Y_{r+1}, Z_r) \to H_{r+1}(Y_{r+1}, Y_r)$ is $\psi(a_j)$.

We show that there exists a unique element x_j with these properties. We know $H_{r+1}(Y, r, Z_r) = 0$ and $H_j(Y_{r-1}Z_{r-1}) = 0$, $j \geq r$ for reasons of dimension. The exact sequence of the triple (Y_{r+1}, Y_r, Z_r) shows that γ is injective. Hence there exists at most one x_j with the desired properties. The existence follows if we show that $\mathrm{Im}(\partial, \gamma) = \mathrm{Ker}(\varphi_r - d'_{r+1})$. This follows by diagram chasing in the next diagram with exact rows

$$\begin{array}{ccccccc}
H_{r+1}(Y_{r+1}, Z_{r-1}) & \longrightarrow & H_{r+1}(Y_{r+1}, Z_r) & \xrightarrow{\partial} & H_r(Z_r, Z_{r-1}) & \longrightarrow & H_r(Y_{r+1}, Z_{r-1}) \\
\downarrow{\scriptstyle\alpha} & & \downarrow{\scriptstyle\gamma} & & \downarrow{\scriptstyle\varphi_r} & & \downarrow{\scriptstyle\beta} \\
H_{r+1}(Y_{r+1}, Y_{r-1}) & \longrightarrow & H_{r+1}(Y_{r+1}, Y_r) & \xrightarrow{d'_{r+1}} & H_r(Y_r, Y_{r-1}) & \longrightarrow & H_r(Y_{r+1}, Y_{r-1}).
\end{array}$$

One uses that α is surjective and β injective. □

Proof. Since Y is simply connected, we can assume that Y has a single 0-cell and no 1-cells. We construct Z inductively with $Z_0 = \{*\}$ and $Z_1 = \{*\}$. We choose a finite number of generators for $\pi_2(Y) \cong H_2(Y)$ and representing maps $S^2 \to Y_2$. They yield a cellular map $f_2\colon Z_2 = \bigvee S^2 \to Y$, and the induced map is surjective in H_2 and bijective in H_j, $j < 2$. This starts the inductive construction.

Suppose $f_r\colon Z_r \to Y$ is given such that $f_{r*}\colon H_i(Z_r) \to H_i(Y)$ is bijective for $i \leq r-1$ and surjective for $i = r$. We construct a diagram of type (7) in 20.2.2 as follows. We have $H_r(Z_r) = \mathrm{Ker}(d)$ and $H_r(Y) = \mathrm{Ker}(d'_r)/\mathrm{Im}(d'_{r+1})$ and the map $(f_r)_*\colon H_r(Z_r) = \mathrm{Ker}(d) \to H_r(Y)$ is surjective. Let A_{r+1} be the kernel of $(f_r)_*$ and $\delta\colon H_r(Z_r) \subset C_r(Z_r)$. As a subgroup of $C_r(Z_r)$ it is free abelian. Since Z_r has, by induction, a finite number of r-cells, the group A_{r+1} is finitely generated. By definition of A_{r+1}, the image of $\varphi_r\delta$ is contained in the image of d'_{r+1}. Hence there exists ψ making the diagram commutative. We now apply (20.2.3) in order to attach an $(r+1)$-cell for each basis element of A_{r+1} and to extend f_r to $f'_{r+1}\colon Z'_{r+1} \to Y$. By construction, $(f'_{r+1})_*$ is now bijective on H_r. If this map is not yet surjective on H_{r+1} we can achieve this by attaching more $(r+1)$-cells with trivial attaching maps; if $H_{r+1}(Y)$ is finitely generated, we only need a finite number of cells for this purpose. We continue in this manner as long as $H_*(Y)$ is finitely generated. After that point we do not care about finite generation. The final map $f\colon Z \to Y$ is a homotopy equivalence by (20.1.5). □

20.3 Serre Classes

Typical qualitative results in algebraic topology are statements of the type that the homotopy or homology groups of a space are (in a certain range) finite or finitely generated or that induced maps have finite or finitely generated kernel and cokernel. A famous result of Serre [170] says that the homotopy groups of spheres are finite, except in the cases already known to Hopf.

Here are three basic ideas of Serre's approach:

(1) Properties like 'finite' or 'finitely generated' or 'rational isomorphism' have a formal structure. Only this structure matters – and it is axiomatized in the notion of a Serre class of abelian groups or modules.
(2) One has to relate homotopy groups and homology groups, since qualitative results about homology groups are more accessible. The connection is based on the Hurewicz homomorphism.
(3) For inductive proofs one has to relate the homology groups of the basis, fibre, and total space of a (Serre-)fibration. This is the point where Serre uses the method of spectral sequences.

A non-empty class $\mathfrak{C}$ of modules over a commutative ring R is a ***Serre class*** if the following holds: Let $0 \to A \to B \to C \to 0$ be an exact sequence of R-modules. Then $B \in \mathfrak{C}$ if and only if $A, C \in \mathfrak{C}$. We call $\mathfrak{C}$ ***saturated*** if $A \in \mathfrak{C}$ implies that arbitrary direct sums $\bigoplus_j A$ of copies of A are contained in $\mathfrak{C}$. The class consisting of the trivial module alone is saturated. A morphism $f\colon M \to N$ between R-modules is a $\mathfrak{C}$-epimorphism ($\mathfrak{C}$-monomorphism) if the cokernel (kernel) of f is in $\mathfrak{C}$, and a $\mathfrak{C}$-isomorphism if it is a $\mathfrak{C}$-epi- and -monomorphism. We use certain facts about these notions, especially the $\mathfrak{C}$-Five Lemma. The idea is to neglect modules in $\mathfrak{C}$, or, as one says, to work modulo $\mathfrak{C}$; so, instead of $\mathfrak{C}$-isomorphism, we say isomorphism modulo $\mathfrak{C}$. Here are some examples of Serre classes.

(1) The class containing only the trivial group.
(2) The class $\mathcal{F}$ of finite abelian groups.
(3) The class $\mathcal{G}$ of finitely generated abelian groups.
(4) Let R be a principal ideal domain. The class $\mathfrak{C}$ consists of the (finitely generated) R-modules. If R is a field, then we are considering the class of (finite-dimensional) vector spaces.
(5) Let R be a principal ideal domain. The class $\mathfrak{C}$ consists of the (finitely generated) R-torsion modules. A module M is a torsion module, if for each $x \in M$ there exists $0 \neq \lambda \in R$ such that $\lambda x = 0$.
(6) Let $P \subset \mathbb{N}$ be a set of prime numbers. Let $\mathbb{Z}_P \subset \mathbb{Q}$ denote the subring of rational numbers with denominators not divisible by an element of P. If $P = \emptyset$, then $\mathbb{Z} = \mathbb{Q}$. If $P = \{p\}$, then $\mathbb{Z}_P = \mathbb{Z}_{(p)}$ is the localization of $\mathbb{Z}$ at p. If P contains the primes except p, then $\mathbb{Z}_P = \mathbb{Z}[p^{-1}]$ is the ring of rational numbers with denominators only p-powers. The rings $\mathbb{Z}_P$ are principal ideal domains. Let P' be the complementary set of primes. An abelian group A is a P'-torsion group if and only if $A \otimes_{\mathbb{Z}} \mathbb{Z}_P = 0$.
Let $\mathfrak{C}_P$ be the class of P'-torsion groups. Then a homomorphism $\varphi\colon A \to B$

is a $\mathcal{C}_P$-isomorphism if and only if $\varphi \otimes_{\mathbb{Z}} \mathbb{Z}_P$ is an ordinary isomorphism. Similarly for epi- and monomorphism. This remark reduces the $\mathcal{C}_P$ Five Lemma to the ordinary Five Lemma after tensoring with $\mathbb{Z}_P$. This simplifies working with this class[1].

(20.3.1) Proposition (Five Lemma mod $\mathcal{C}$). *In the next proposition we use the same notation as in* (11.2.7). *The considerations of that section then yield directly a proof of the following assertions.*

(1) *b $\mathcal{C}$-epimorphism $\Rightarrow \tilde{b}$ $\mathcal{C}$-epimorphism.*

(2) *d $\mathcal{C}$-epimorphism, e $\mathcal{C}$-monomorphism $\Rightarrow \tilde{d}$ $\mathcal{C}$-epimorphism.*

(3) *The hypotheses of* (1) *and* (2) *imply: c $\mathcal{C}$-epimorphism.*

(4) *d $\mathcal{C}$-monomorphism $\Rightarrow \tilde{d}$ $\mathcal{C}$-monomorphism.*

(5) *a $\mathcal{C}$-epimorphism, b $\mathcal{C}$-monomorphism $\Rightarrow \tilde{b}$ $\mathcal{C}$-monomorphism.*

(6) *The hypotheses of* (4) *and* (5) *imply: c $\mathcal{C}$-monomorphism.* □

The kernel-cokernel-sequence shows other properties of $\mathcal{C}$-notions.

(20.3.2) Proposition. *Given homomorphisms $f: A \to B$ and $g: B \to C$ between R-modules.*

(1) *If f and g are $\mathcal{C}$-monomorphisms (-epimorphisms), then gf is a $\mathcal{C}$-monomorphism (-epimorphism).*

(2) *If two of the morphisms f, g, and gf are $\mathcal{C}$-isomorphisms so is the third.* □

20.4 Qualitative Homology of Fibrations

In this section we work with singular homology with coefficients in the R-module M.

(20.4.1) Theorem (Fibration theorem). *Let $p: E \to B$ be a (Serre-)fibration with 0-connected fibres. Let (B, A) be a relative CW-complex with t-skeleton B^t. We assume that $A = B^{-1} = B^0 = \cdots = B^{s-1}$ for an $s \geq 0$ and that there are only a finite number of t-cells for $t \leq e$. Finally we assume that $H_i(F; M) \in \mathcal{C}$ for $0 < i < r$ and all fibres F of p. We write $E^t = p^{-1}(B^t)$. Then the following holds:*

(1) *Let $\mathcal{C}$ be saturated. Then $p_*: H_i(E, E^{-1}; M) \to H_i(B, B^{-1}; M)$ is a $\mathcal{C}$-isomorphism for $i \leq r + s - 1 = \alpha$ and a $\mathcal{C}$-epimorphism for $i = \alpha + 1$.*

(2) *Let $\mathcal{C}$ be arbitrary. Then p_* is a $\mathcal{C}$-isomorphism for $i \leq \alpha$ and a $\mathcal{C}$-epimorphism for $i = \alpha + 1$ where now $\alpha = \min(e + 1, r + s - 1)$.*

---

[1] In a more abstract setting one can construct localizations of categories so that $\mathcal{C}$-isomorphisms become isomorphisms in the localized category, etc.

We remark that $r \geq 1$. For $r = 1$ we make no further assumptions about F.

Since a weak homotopy equivalence induces isomorphisms in singular homology, we can assume without essential restriction that B is a CW-complex (pull back the fibration along a CW-approximation). We reduce the proof of the theorem by a Five Lemma argument to the attaching of t-cells. We consider the following situation. Let

$$(\Phi, \varphi)\colon \amalg_a (D^t_a, S^{t-1}_a) \to (B, B')$$

be an attaching of t-cells. Let $p\colon E \to B$ be a fibration and set $E' = p^{-1}(B')$. We assume that the fibres are 0-connected and homotopy equivalent. We pull back the fibration along Φ and obtain two pullback diagrams.

$$\begin{array}{ccc} \amalg E_a & \xrightarrow{\Psi} & E \\ \downarrow{\scriptstyle \amalg p_a} & & \downarrow{\scriptstyle p} \\ \amalg D^t_a & \xrightarrow{\Phi} & B \end{array} \qquad \begin{array}{ccc} \amalg E'_a & \xrightarrow{\Psi} & E' \\ \downarrow{\scriptstyle \amalg p'_a} & & \downarrow{\scriptstyle p'} \\ \amalg S^{t-1}_a & \xrightarrow{\Phi} & B' \end{array}$$

We apply homology (always with coefficients in M) and obtain the diagram

$$\begin{array}{ccc} H_i(E, E') & \xrightarrow{p_*} & H_i(B, B') \\ \uparrow{\scriptstyle \Psi_*} & & \uparrow{\scriptstyle \Phi_*} \\ \bigoplus_a H_i(E_a, E'_a) & \xrightarrow{\oplus_a p_{a*}} & \bigoplus_a H_i(D^t_a, S^{t-1}_a). \end{array}$$

We already know that Φ_* is an isomorphism. In order to show that Ψ_* is an isomorphism, we attach a single t-cell, to simplify the notation. Let B_0 be obtained from B by deleting the center $\Phi(0)$ of the cell.

(20.4.2) Lemma. *Let $p\colon X \to B$ be a fibration with restrictions $p'\colon X' \to B'$ and $p_0\colon X_0 \to B_0$. Let $q\colon Y \to D^t$ be the pullback of p, and similarly $q'\colon Y' \to S^{t-1}$ and $q_0\colon Y_0 \to D^t \smallsetminus 0$. Then*

$$\Psi_*\colon h_i(Y, Y') \to h_i(X, X')$$

is an isomorphism for each homology theory.

Proof. We have a commutative diagram

$$\begin{array}{ccc} h_i(Y, Y') & \xrightarrow{\Psi_*} & h_i(X, X') \\ \downarrow{\scriptstyle (1)} & & \downarrow{\scriptstyle (2)} \\ h_i(Y, Y_0) & \xrightarrow{\Psi_*} & h_i(X, X_0) \\ \uparrow{\scriptstyle (3)} & & \uparrow{\scriptstyle (4)} \\ h_i(Y \smallsetminus Y', Y_0 \smallsetminus Y') & \underset{(5)}{\xrightarrow{\Psi_*}} & h_i(X \smallsetminus X', X_0 \smallsetminus X'). \end{array}$$

Since B' is a deformation retract and p a fibration, also X' is a deformation retract of X_0. Therefore (2) is an isomorphism by homotopy invariance. The map (4) is an isomorphism by excision. For similar reasons (1) and (3) are isomorphisms. Finally (5) is induced by a homeomorphism. □

By the homotopy theorem for fibrations, a fibration over D^t is fibre-homotopy equivalent to a product projection $D^t \times F \to D^t$. We use such equivalences and a suspension isomorphism $H_i(D^t \times F, S^{t-1} \times F) \cong H_{i-t}(F)$ and obtain altogether a commutative diagram (P_a a point and F_a the fibre over P_a):

$$\begin{array}{ccc} H_i(E,E') & \xrightarrow{p_*} & H_i(B,B') \\ \cong\uparrow & & \cong\uparrow \\ \bigoplus_a H_{i-t}(F_a) & \longrightarrow & \bigoplus_a H_{i-t}(P_a). \end{array}$$

(20.4.3) Note. *The considerations so far show that the bottom map has the following properties:*

(1) *Isomorphism for $i \le t$ (since fibres are 0-connected).*

(2) *Epimorphism always.*

(3) *Suppose $H_i(F_a) \in \mathfrak{C}$ for $0 < i < r$. Then each particular map $H_{i-t}(F_a) \to H_{i-t}(P_a)$ is a $\mathfrak{C}$-isomorphism for $0 < i - t < r$. Thus the total map is a $\mathfrak{C}$-isomorphism if either $\mathfrak{C}$ is saturated or if we attach a finite number of cells.* ◇

We apply these considerations to a fibration $p\colon E \to B$ over a relative CW-complex (B, A) as in the statement of the theorem. In this situation the previous considerations yield:

Let $\mathfrak{C}$ be saturated. Then $p_*\colon H_i(E^t, E^{t-1}) \to H_i(B^t, B^{t-1})$ is a $\mathfrak{C}$-isomorphism for each $t \ge 0$, if $i < r + s$. We only have to consider $t \ge s$. By (20.4.3), we have a $\mathfrak{C}$-isomorphism in the cases $i \le t$ and $i > t > i - r$. These conditions hold for each t, if $s > i - r$. If, in addition, there are only finitely many t-cells for $t \le e$, then we have a $\mathfrak{C}$-isomorphism for arbitrary $\mathfrak{C}$ and each t, if $i \le \min(e + 1, r + s - 1)$, by the same argument.

We finish the proof of theorem (20.4.1) with:

(20.4.4) Lemma. *Let (B, A) be a relative CW-complex with t-skeleton B^t. Let $p\colon E \to B$ be a fibration and $E^t = p^{-1}(B^t)$. Suppose $p_*\colon H_i(E^t, E^{t-1}; M) \to H_i(B^t, B^{t-1}; M)$ is a $\mathfrak{C}$-isomorphism for each $t \ge 0$ and each $0 \le i \le \alpha$, then $p_*\colon H_i(E, E^{-1}; M) \to H_i(B, B^{-1}; M)$ is a $\mathfrak{C}$-isomorphism for $0 \le i \le \alpha$ and a $\mathfrak{C}$-epimorphism for $i = \alpha + 1$.*

Proof. We show by induction on $k \ge 0$ that $p_*\colon H_i(E^k, E^{-1}) \to H_i(B^k, B^{-1})$ is a $\mathfrak{C}$-isomorphism ($k \ge 0$). For the induction step one uses the exact homology sequence for the triple (E^k, E^{k-1}, E^{-1}), and similarly for B, and applies the $\mathcal{C}$-Five Lemma to the resulting diagram induced by the various morphisms p_*. For the epimorphism statement we also use part (2) in (20.4.3). The colimit $k \to \infty$ causes no problem for singular homology. □

The statement of (20.4.1) is adapted to the method of proof. The hypotheses can be weakened as follows.

(20.4.5) Remark. Let (X, A) be an $(s-1)$-connected pair of spaces. Then there exists a weak relative homotopy equivalence $(B, A) \to (X, A)$ from a relative CW-complex (B, A) with $A = B^{s-1}$. We pull back a fibration over $q\colon E \to X$ along this equivalence and use the fact that weak equivalences induce isomorphisms in singular homology. Then part (1) of (20.4.1) yields that $q_*\colon H_j(E, q^{-1}(A); M) \to H_j(X, A; M)$ is a $\mathcal{C}$- isomorphism for $j \le r + s - 1$ and a $\mathcal{C}$-epimorphism for $i = r + s$. ◇

(20.4.6) Remark. Let X be a 1-connected space such that $H_j(X; \mathbb{Z})$ is finitely generated for $i \le e$. Then there exists a weak equivalence $B \to X$ such that B has only a finite number of t-cells for $t \le e$. We use this result in the next section. ◇

Problems

1. Suppose that $H_j(F; M) = 0$ for $0 < j < r$. Let B be $(s-1)$-connected. Then $p_*\colon H_j(E, F; M) \to H_j(B, *; M)$ is an isomorphism for $j \le r + s - 1$. We can now insert this isomorphism into the exact homology sequence of the pair (E, F) and obtain an exact sequence

$$\begin{aligned} H_{r+s-1}(F; M) \to H_{r+1-1}(E; M) \to H_{r+s-1}(B; M) \to \cdots \\ \to H_1(F; M) \to H_1(E; M) \to H_1(B; M) \to 0 \end{aligned}$$

which is analogous to the exact sequence of homotopy groups. Compare these sequences via the Hurewicz homomorphism $(M = \mathbb{Z})$.

20.5 Consequences of the Fibration Theorem

We use exact sequences and the fibration theorem to derive a number of results. We consider a fibration $p\colon E \to B$ and assume that B and $F = p^{-1}(*)$ are 0-connected; then E is 0-connected too. We use the notation

$$Z \in \mathcal{C}(r, M) \Leftrightarrow H_j(Z; M) \in \mathcal{C} \text{ for } 0 < j < r \le \infty.$$

(M an R-module. In the case that $M = \mathbb{Z}$ we write $\mathcal{C}(r)$. For $r = 1$ there is no condition.) Let $\mathcal{F}$, $\mathcal{G}$ denote the class of finite, finitely generated abelian groups, respectively. We use homology with coefficients in the R-module M if nothing else is specified.

(20.5.1) Remark. The homomorphism $p_*\colon H_j(E,F;M) \to H_j(B,*;M)$ is always an isomorphism for $j \le 1$ and an epimorphism for $j = 2$ (see (20.4.1)). The homomorphism $p_*\colon H_j(E;M) \to H_j(B;M)$ is an isomorphism for $j = 0$ and an epimorphism for $j = 1$. □

(20.5.2) Theorem. *Suppose $F \in \mathcal{C}(r,M)$. Let $\mathcal{C}$ be saturated. Suppose B is s-connected. Then $p_*\colon H_i(E,F) \to H_i(B,*)$ is a $\mathcal{C}$-isomorphism for $i \le r+s$ and a $\mathcal{C}$-epimorphism for $i = r+s+1$. Moreover $p_*\colon H_i(E) \to H_i(B)$ is a $\mathcal{C}$-isomorphism for $i < r$ and a $\mathcal{C}$-epimorphism for $i = r$.*

Proof. The first statement is (20.4.1). For the second statement we use in addition the exact homology sequence of the pair (E,F). □

(20.5.3) Theorem. *Let $\mathcal{C}$ be saturated.*

(1) $F, B \in \mathcal{C}(r,M) \Rightarrow E \in \mathcal{C}(r,M)$.

(2) $F \in \mathcal{C}(r,M),\ E \in \mathcal{C}(r+1,M) \Rightarrow B \in \mathcal{C}(r+1,M)$.

(3) $B \in \mathcal{C}(r+1,M),\ E \in \mathcal{C}(r,M),\ B$ *1-connected* $\Rightarrow F \in \mathcal{C}(r,M)$.

Proof. (1) and (2) are consequences of the fibration theorem and the exact homology sequence of the pair (E,F). (3) is proved by induction on r. For $r = 1$ there is nothing to prove. For the induction step consider

$$H_r(B,*;M) \xleftarrow{(1)} H_r(E,F;M) \longrightarrow H_{r-1}(F;M) \longrightarrow H_{r-1}(E;M)$$

(1) is a $\mathcal{C}$ isomorphism for $r-1+s-1 = r$, since $s = 2$ and $F \in \mathcal{C}(r-1,M)$ by induction. From the hypotheses $H_r(B;*;M), H_{r-1}(E;M) \in \mathcal{C}$ we conclude $H_{r-1}(F;M) \in \mathcal{C}$. □

(20.5.4) Theorem. *Let $\mathcal{C}$ be arbitrary and assume that B is 1-connected.*

(1) $F, B \in \mathcal{C}(r,M),\ B \in \mathcal{G}(r-1) \Rightarrow E \in \mathcal{C}(r,M)$.

(2) $F \in \mathcal{C}(r,M),\ E \in \mathcal{C}(r+1,M),\ B \in \mathcal{G}(r) \Rightarrow B \in \mathcal{C}(r+1,M)$.

(3) $B \in \mathcal{C}(r+1,M),\ E \in \mathcal{C}(r,M),\ B \in \mathcal{G}(r) \Rightarrow F \in \mathcal{C}(r,M)$.

Proof. As for (20.5.3). The 1-connectedness of B is needed in order to apply the cell theorem (see (20.4.6)). □

(20.5.5) Corollary. *Suppose E is contractible (path fibration over B). Then $F \simeq \Omega B$, the loop space of B. Let B be simply connected. Let $\mathcal{C}$ be saturated. Then $B \in \mathcal{C}(r+1,M)$ if and only if $\Omega B \in \mathcal{C}(r,M)$. Moreover $B \in \mathcal{G}(r+1)$ if and only if $\Omega B \in \mathcal{G}(r)$. Similarly for $\mathcal{F}$ instead of $\mathcal{G}$.* □

(20.5.6) Proposition. *Let A be a finitely generated abelian group. Then the Eilenberg–Mac Lane spaces $K(A,n)$ are contained in $\mathcal{G}(\infty)$. If A is finite, then $K(A,n) \in \mathcal{F}(\infty)$. Moreover $K(A,1) \in \mathcal{G}(\infty)$, $K(A,1) \in \mathcal{C}(\infty,M)$ implies $K(A,n) \in \mathcal{C}(\infty,M)$. If $\mathcal{C}$ is saturated, then $K(A,1) \in \mathcal{C}(\infty,M)$ implies $K(A,n) \in \mathcal{C}(\infty,M)$.*

Proof. We use the path fibration $K(A,n-1) \to P \to K(A,n)$ with contractible P and induction with (20.5.3) and (20.5.4). In the case that $n = 1$, standard constructions yield models for $K(A,1)$ with a finite number of cells in each dimension. One uses $K(A_1,1) \times K(A_2,1) = K(A_1 \times A_2, 1)$, $K(\mathbb{Z},1) = S^1$, and $K(\mathbb{Z}/m,1) = S^\infty/(\mathbb{Z}/m)$. □

Let X be a $(k-1)$-connected space ($k \geq 2$). We attach cells of dimension $j \geq k+2$ to X in order to kill the homotopy groups $\pi_j(X)$ for $j \geq k+1$. The resulting space in an Eilenberg–Mac Lane space $K(\pi,k)$, $\pi = \pi_k(X)$, and the inclusion $\iota\colon X \to K(\pi,k)$ induces an isomorphism $\pi_k(\iota)$. We pull back the path fibration over $K(\pi,k)$ and obtain a fibration

$$K(\pi,k-1) \to Y \to X$$

with k-connected Y, and $q\colon Y \to X$ induces isomorphisms $\pi_j(q)$, $j > k$. This follows from the exact homotopy sequence. If $\pi \in \mathcal{C}$ and $\mathcal{C}$ is saturated, then $q_*\colon H_j(Y;M) \to H_j(X;M)$ is a $\mathcal{C}$-isomorphism for $j > 0$, by the fibration theorem, since $K(\pi,k-1) \in \mathcal{C}(\infty)$. Similarly for arbitrary $\mathcal{C}$ when X is of finite type.

20.6 Hurewicz and Whitehead Theorems modulo Serre classes

Let $\mathcal{C}$ be a Serre class of abelian groups with the additional property: The groups $H_k(K(A,1)) \in \mathcal{C}$ whenever $A \in \mathcal{C}$ and $k > 0$. In this section we work with integral singular homology.

(20.6.1) Theorem (Hurewicz theorem mod $\mathcal{C}$). *Suppose X is 1-connected and $n \geq 2$. Assume that either $\mathcal{C}$ is saturated or $H_i(X)$ is finitely generated for $i < n$ and $\mathcal{C}$ is arbitrary. Then the following assertions are equivalent:*

(1) $\Pi(n)$: $\pi_i(X,*) \in \mathcal{C}$ *for* $1 < i < n$.

(2) $\mathrm{H}(n)$: $H_i(X) \in \mathcal{C}$ *for* $1 < i < n$.

If $\Pi(n)$ or $\mathrm{H}(n)$ holds, the Hurewicz homomorphism $h_n\colon \pi_n(X,*) \to H_n(X)$ is a $\mathcal{C}$-isomorphism.

Proof. The proof is by induction on n.

(1) Let $\Omega X \to PX \xrightarrow{f} X$ be the path fibration with contractible PX. It provides us with a commutative diagram

$$\begin{array}{ccccc}
\pi_n(X,*) & \xleftarrow{\pi_n(f)} & \pi_n(PX,\Omega X,*) & \xrightarrow{\partial} & \pi_{n-1}(\Omega X,*) \\
\downarrow h & & \downarrow h & & \downarrow h \\
H_n(X,*) & \xleftarrow{H_n(f)} & H_n(PX,\Omega X) & \xrightarrow{\partial} & H_{n-1}(\Omega X).
\end{array}$$

The boundary maps ∂ are isomorphisms, since PX is contractible. By a basic property of fibrations $\pi_n(f)$ is an isomorphism.

By the ordinary Hurewicz theorem $h_2\colon \pi_2(X,*) \to H_2(X)$ is an isomorphism. (The method of the following proof can also be used to prove the classical Hurewicz theorem.)

(2) Let $n > 2$. Assume that the theorem holds for $n-1$ and that $\Pi(n)$ holds. We want to show that $H(n)$ holds and that $h_n\colon \pi(X,*) \to H_n(X)$ is a $\mathcal{C}$-isomorphism. We first consider the special (3) case that $\pi_2(X) = 0$ and then reduce the general case (4) to this special.

(3) Thus let $\pi_2(X) = 0$. Then $\pi_i(\Omega X) \cong \pi_{i+1}(X) \in \mathcal{C}$ for $1 < i < n-1$, and ΩX is 1-connected. If $\mathcal{C}$ is saturated, then $h_{n-1}\colon \pi_{n-1}(\Omega X) \to H_{n-1}(\Omega X)$ is a $\mathcal{C}$-isomorphism by induction. If $H_i(X)$ is finitely generated for $i < n$, then by (20.5.3) $H_i(\Omega X)$ is finitely generated for $i < n-1$ so that by induction h_{n-1} is also an isomorphism in this case. The fibration theorem shows that $H_n(f)$ in the diagram is a $\mathcal{C}$-isomorphism. From the diagram we now see that h_n is a $\mathcal{C}$-isomorphism. Also $H_{n-1}(X) \in \mathcal{C}$ by induction.

(4) Let now $\pi_2(X) = \pi_2$ be arbitrary. By assumption, this group is contained in $\mathcal{C}$. There exists a map $\gamma\colon X \to K(\pi_2, 2)$ which induces an isomorphism π_2. We pull back the path fibration along γ

$$\begin{array}{ccc} X_2 & \longrightarrow & PK(\pi_2,2) \\ \downarrow{\scriptstyle\varphi} & & \downarrow{\scriptstyle f} \\ X & \xrightarrow{\gamma} & K(\pi_2,2). \end{array}$$

Since $\pi_2 \in \mathcal{C}$, we have $H_i(K(\pi_2,1)) \in \mathcal{C}$ for $i > 0$, by the general assumption in this section. Note that $K = K(\pi_2,1)$ is the fibre of φ and f. The exact homotopy sequence of γ is used to show that $\varphi_*\colon \pi_i(X_2) \cong \pi_2(X)$ for $i > 2$ and that $\pi_1(X_2) \cong 0 \cong \pi_2(X_2)$. We can therefore apply the special case (3) to X_2. Consider the diagram

$$\begin{array}{ccc} \pi_n(X_2) & \xrightarrow{\cong} & \pi_n(X) \\ \downarrow{\scriptstyle h_n(X_2)} & & \downarrow{\scriptstyle h_n(X)} \\ H_n(X_2) & \xrightarrow{\mathcal{C}\cong} & H_n(X). \end{array}$$

In order to show that $h_n(X)$ is a $\mathcal{C}$-isomorphism we show two things:

(i) $h_n(X_2)$ is a $\mathcal{C}$-isomorphism.

(ii) $\varphi_*\colon H_n(X_2) \to H_n(X)$ is a $\mathcal{C}$-isomorphism.

Part (i) follows from case (3) if we know that $H_i(X_2)$ is finitely generated for $i < n$. This follows from (20.5.3) applied to the fibration $K \to X_2 \to X$, since X is 1-connected and since $\pi_1(K) \cong \pi_2 \cong H_2(X)$ is finitely generated.

For the proof of (ii) we first observe that the canonical map $\beta\colon H_i(X_2) \to H_i(X_2, K)$ is a $\mathcal{C}$-isomorphism for $i > 0$, since $H_i(K) \in \mathcal{C}$ for $i > 0$ by the general assumption of this section. The fibration theorem and (20.4.6) show that $\varphi_*\colon H_i(X_2, K) \to H_i(X, *)$ is a $\mathfrak{C}$-isomorphism for $0 < i \le n$. We now compose with β and obtain (ii).

(5) Now assume that H(k) holds for $2 \le k < n$. Since $H_i(X) \in \mathfrak{C}$ for $2 \le i < k+1$ the Hurewicz map $\pi_i(X) \to H_i(X)$ is a $\mathfrak{C}$-isomorphism for $i \le k$ by H(k), hence $\pi_i(X) \in \mathfrak{C}$ for $i \le k$. By the first part of the proof, h_{k+1} is a $\mathfrak{C}$-isomorphism, hence $\Pi(k+1)$ holds. □

We list some consequences of the Hurewicz theorem. Note that the general assumption of this section holds for the classes $\mathcal{G}$ and $\mathcal{F}$.

(20.6.2) Theorem. *Let X be a 1-connected space.*

(1) *$\pi_i(X)$ is finitely generated for $i < n$ if and only if $H_i(X;\mathbb{Z})$ is finitely generated for $i < n$.*
(2) *$\pi_i(X)$ is finite for $i < n$ if and only if $\tilde{H}_i(X;\mathbb{Z})$ is finite for $i < n$.*
(3) *If X is a finite CW-complex, then its homotopy groups are finitely generated.*

□

(20.6.3) Theorem. *Let $\mathfrak{C}$ be a saturated Serre class. Let $f\colon X \to Y$ be a map between 1-connected spaces with 1-connected homotopy fibre F. Then the following are equivalent:*

(1) *$\pi_k(f)\colon \pi_k(X) \to \pi_k(Y)$ is a $\mathfrak{C}$-isomorphism for $k < n$ and a $\mathfrak{C}$-epimorphism for $k = n$.*
(2) *$H_k(f)\colon H_k(X) \to H_k(Y)$ is a $\mathfrak{C}$-isomorphism for $k < n$ and a $\mathfrak{C}$-epimorphism for $k = n$.*

Proof. The statement (1) is equivalent to $\pi_k(F) \in \mathfrak{C}$ for $k < n$ (exact homotopy sequence). Suppose this holds. Then $H_k(X, F) \to H_k(Y)$ is a $\mathfrak{C}$ isomorphism for $k \le n$, by the fibration theorem (20.4.1). From the exact homology sequence of the pair (X, F) we now conclude that (2) holds, since $H_j(F) \in \mathfrak{C}$ by the Hurewicz theorem. Here we use that $\pi_1(F) = 0$.

Suppose (2) holds. We show by induction that $H_k(F) \in \mathfrak{C}$ for $k < n$. Then we apply again the Hurewicz theorem. The induction starts with $n = 3$. Since Y and F are simply connected, the fibration theorem shows that $H_j(X, F) \to H_j(Y, *)$ is a $\mathfrak{C}$-isomorphism for $j \le 3$. The assumption (2) and the homology sequence of the pair then show $H_2(F) \in \mathfrak{C}$. The general induction step is of the same type. □

Problems

1. Let $T \subset \mathbb{Q}$ be a subring. Let X be a simply connected space such that $H_*(X;T) \cong H_*(S^n;T)$. Then there exists a map $S^n \to X$ which induces an isomorphism in T-homology.

2. The finite CW-complex $X = S^1 \vee S^2$ has $\pi_1(X) \cong \mathbb{Z}$. The group $\pi_2(X)$ is free abelian with a countably infinite number of generators. (Study the universal covering of X.)
3. Use the fibration theorem and deduce for each 0-connected space a natural exact sequence $\pi_2(X) \to H_2(X) \to H_2(K(\pi_1(X),1);\mathbb{Z}) \to 0$.

20.7 Cohomology of Eilenberg–Mac Lane Spaces

We compute the cohomology ring $H^*(K(\mathbb{Z},n);\mathbb{Q})$. Let $f\colon S^n \to K(\mathbb{Z},n) = K(n)$ induce an isomorphism $\pi_n(f)$ of the n-th homotopy groups. Then also $f^*\colon H^n(K(\mathbb{Z},n);\mathbb{Q}) \to H^n(S^n;\mathbb{Q})$ is an isomorphism and ι_n is defined such that $f^*(\iota_n) \in H^n(S^n;\mathbb{Q})$ is a generator.

(20.7.1) Theorem. *If $n \geq 2$ is even, then $H^*(K(\mathbb{Z},n);\mathbb{Q}) \cong \mathbb{Q}[\iota_n]$ (polynomial ring). If n is odd, then $f^*\colon H^*(K(\mathbb{Z},n);\mathbb{Q}) \cong H^*(S^n;\mathbb{Q})$.*

Proof. We work with rational cohomology. Since $K(\mathbb{Z},1) \simeq S^1$ and $K(\mathbb{Z},2) \simeq \mathbb{C}P^\infty$ we know already the cohomology ring for these spaces with coefficients in $\mathbb{Z}$ and this implies (20.7.1) in these cases. We prove the theorem by induction on n, and for this purpose we analyze the path-fibration $K(\mathbb{Z},n-1) \to P \to K(\mathbb{Z},n)$ with contractible P. There are two cases for the induction step, depending on the parity of n.

$2k-1 \Rightarrow 2k$. We have a relative fibration $p\colon (E,E') \to K(n)$ with $E = K(n)^I$, $p(w) = w(1)$, $P = E' = p^{-1}(*)$. The map $p\colon E \to K(n)$ is a homotopy equivalence and E' is contractible. Therefore

$$H^n(E,E') \cong H^n(E) \cong H^n(K(n)), \tag{1}$$

the latter induced by p. The fibres (F,F') of p have a contractible F and $F' = \Omega K(n) = K(n-1)$ is by induction a rational cohomology $(n-1)$-sphere (i.e., has the rational cohomology of S^{n-1}). Hence $H^k(F,F') \cong H^{k-1}(F') \cong \mathbb{Q}$ for $k = n$ and $\cong 0$ for $k \neq n$. Since $K(n)$ is simply connected, we have a Thom class $t_n \in H^n(E,E')$. We can assume that t_n is mapped under (1) to ι_n, hence ι_n is the Euler class e associated to t_n. The Gysin sequence has the form ($n = 2k$)

$$\cdots \to H^j(K(n)) \xrightarrow{e} H^{j+n}(K(n)) \to H^{j+n}(P) \to \cdots.$$

Since P is contractible and $H^j(K(n)) = 0$ for $0 < j < n$, we see inductively that the cup product with the Euler class e is an isomorphism $H^j(K(n)) \to H^{j+2k}(K(n))$. Hence $H^*(K(n)) \cong \mathbb{Q}[\iota_n]$.

$2k \Rightarrow 2k+1$. We reduce the problem to a Wang sequence. Let $n = 2k+1$. We consider a pullback

$$\begin{array}{ccc} Y & \longrightarrow & P \\ \downarrow{\scriptstyle q} & & \downarrow{\scriptstyle p} \\ S^n & \xrightarrow{f} & K(n), \end{array}$$

where f induces an isomorphism $\pi_n(f)$. The Wang sequence for q has the form

$$\cdots \to H^j(Y) \xrightarrow{i^*} H^j(K(n-1)) \xrightarrow{\theta} H^{j+1-n}(K(n-1)) \to \cdots.$$

We use $H^*(K(n-1)) \cong \mathbb{Q}[\iota_{n-1}]$ and the fact that θ is a derivation. From the definition of Y and the exact sequence of homotopy groups we see

$$\pi_j(Y) = 0, \quad j \le n, \quad \pi_j(Y) \cong \pi_j(S^n), \quad j > n.$$

From the Hurewicz theorem and the universal coefficient theorem we conclude that $H^j(Y) = 0$ for $j \le n$. Hence $\theta\colon H^{2k}(K(n-1)) \to H^0(K(n-1))$ is an isomorphism. Using the derivation property of θ we see inductively that $\theta\colon H^{2kr}(K(n-1)) \to H^{2k(r-1)}(K(n-1))$ is an isomorphism, and the Wang sequence then shows us that $\tilde{H}^*(Y) = 0$; and this implies $\tilde{H}_*(Y) = 0$. Since Y is the homotopy fibre of f, we conclude that $f_*\colon H_*(S^n) \to H_*(K(n))$ is an isomorphism (by (20.4.1) say), and similarly for cohomology. This finishes the induction. □

20.8 Homotopy Groups of Spheres

Let $n > 1$ be an odd integer. Let $f: S^n \to K(\mathbb{Z}, n)$ induce an isomorphism $\pi_n(f)$ and denote by Y the homotopy fibre of f. In the previous section we have shown that $H_j(Y;\mathbb{Q}) = 0$ for $j > 0$. The Hurewicz theorem modulo the class of torsion groups (= the rational Hurewicz theorem) shows us that the groups $\pi_j(Y) \otimes \mathbb{Q}$ are zero for $j \in \mathbb{N}$. From $\pi_j(S^n) \cong \pi_j(Y)$ for $j > n$ we see that also $\pi_j(S^n) \otimes \mathbb{Q} = 0$ for $j > n$ and odd n. Since we already know that the homotopy groups of spheres are finitely generated we see:

(20.8.1) Theorem. *Let n be an odd integer. Then the groups $\pi_j(S^n)$ are finite for $j > n$.* □

We now investigate the homotopy groups of S^{2n}. Let $V = V_2(\mathbb{R}^{2n+1})$ denote the Stiefel manifold of orthonormal pairs (x, y) in $\mathbb{R}^{2n+1}$. We have a fibre bundle $S^{2n-1} \to V \to S^{2n}$, and V is the unit-sphere bundle of the tangent bundle of S^{2n}. Recall from 14.2.4 the integral homology of V

$$H_q(V) \cong \begin{cases} \mathbb{Z}, & q = 0, 4n-1, \\ \mathbb{Z}/2, & q = 2n-1, \\ 0, & \text{otherwise.} \end{cases}$$

Let $g\colon V \to S^{4n-1}$ be a map of degree 1. Then g induces an isomorphism in rational homology. Let F be the homotopy fibre of g; it is simply connected. From (20.6.3) we see that $g_* \otimes \mathbb{Q}\colon \pi_j(V) \otimes \mathbb{Q} \to \pi_j(S^{4n-1}) \otimes \mathbb{Q}$ is always an isomorphism. We use (20.8.1) and see that the homotopy groups of V are finite, except $\pi_{4n-1}(V) \cong \mathbb{Z} \oplus E$, E finite. Now we go back to the fibration $V \to S^{2n}$ and its homotopy sequence. It shows:

(20.8.2) Theorem. *$\pi_j(S^{2n})$ is finite for $j \neq 2n, 4n-1$ and $\pi_{4n-1}(S^{2n}) \cong \mathbb{Z} \oplus E$, E finite.* □

The results about the homotopy groups of spheres enable us to prove a refined rational Hurewicz theorem.

(20.8.3) Theorem. *Let X be 1-connected. Suppose $H_i(X;\mathbb{Z})$ is finite for $i < k$ and finitely generated for $i \leq 2k-2$ ($k \geq 2$). Then the Hurewicz homomorphism $h\colon \pi_r(X) \to H_r(X;\mathbb{Z})$ has finite kernel (cokernel) for $r < 2k-1$ ($r \leq 2k-1$).*

Proof. The case $k = 2$ causes no particular problem, since the Hurewicz homomorphism $h\colon \pi_{m+1}(X) \to H_{m+1}(X;\mathbb{Z})$ is surjective for each $(m-1)$-connected space ($m \geq 2$). So let $k \geq 3$. Since X is 1-connected, the Hurewicz theorem shows that $\pi_r(X)$ is finite for $r < k$ and finitely generated for $r \leq 2k-2$. We write $\pi_r(X) = F_r \oplus T_r$, F_r free, and T_r finite. We choose basis elements for F_r and representing maps. These representing elements provide us with a map of the form

$$f\colon S = S^{r(1)} \vee \cdots \vee S^{r(t)} \to X$$

with $k \leq r(j) \leq 2k-2$. The canonical map

$$\textstyle\bigoplus_j \pi_r(S^{r(j)}) \to \pi_r(\bigvee_j S^{n(j)})$$

is an isomorphism for $r \leq 2k-2$ and an epimorphism for $r = 2k-1$. We can now conclude that $f_*\colon \pi_r(S) \to \pi_r(X)$ has finite kernel and cokernel for $r \leq 2k-2$ and finite cokernel for $r = 2k-1$, since the homotopy groups of spheres are finite in the relevant range. The homotopy fibre F of f has finite homotopy groups $\pi_j(F)$ for $j \leq 2k-2$. If, moreover, F is 1-connected, then $H_j(F;\mathbb{Z})$ is finite in the same range, by the Hurewicz theorem. The fibration theorem then yields that $f_*\colon H_r(S) \to H_r(X)$ is an $\mathcal{F}$-isomorphism ($\mathcal{F}$-epimorphism) for $r \leq 2k-2$ ($r = 2k-1$). From our knowledge of the homotopy groups of spheres we see directly that the theorem holds for S. The naturality of the Hurewicz theorem applied to f is now used to see that the desired result also holds for X.

We have used that F is 1-connected. This holds if $\pi_2(X) = 0$. Since $\pi_2(X)$ is finite by assumption ($k \geq 3$), we can pass to the 2-connected cover $q\colon X\langle 2\rangle \to X$ of X. The map q induces $\mathcal{F}$-isomorphisms in homology and homotopy. Therefore it suffices to prove the theorem for $X\langle 2\rangle$ to which the reasoning above applies. □

If one is not interested in finite generation one obtains by a similar reasoning (see also [103]):

(20.8.4) Theorem. *Let X be a 1-connected space. Suppose $H_r(X;\mathbb{Q}) = 0$ for $r < k$. Then the Hurewicz map $\pi_j(X) \otimes \mathbb{Q} \to H_j(X;\mathbb{Q})$ is an isomorphism for $j < 2k-2$ and an epimorphism for $j = 2k-1$.* □

This theorem indicates that homotopy theory becomes simpler "over the rationals". In the so-called rational homotopy theory one constructs algebraic models for the rationalized homotopy theory. For an exposition see [65].

We discuss an example. Consider the path fibration

$$K(\mathbb{Z},2) \simeq \Omega K(\mathbb{Z},3) \to X \to K(\mathbb{Z},3) = K_3$$

with contractible X. Let $f\colon S^3 \to K_3$ induce an isomorphism in π_3 and let $K_2 \xrightarrow{i} Y \xrightarrow{p} S^3$ be the induced fibration. The homotopy groups $\pi_k(Y)$ are zero for $k \le 3$ and p_* induces an isomorphism $\pi_k(Y) \cong \pi_k(S^3)$ for $k \ge 4$.

(20.8.5) Proposition. *$\pi_3(S^2) \cong \mathbb{Z}$ and $\pi_{n+1}(S^n) \cong \mathbb{Z}/2$ for $n \ge 3$.*

Proof. We know already that $\pi_3(S^2) \cong \mathbb{Z}$, generated by the Hopf map $S^3 \to \mathbb{C}P^1 \cong S^2$. From the Freudenthal suspension theorem we know that the suspension $\Sigma_*\colon \pi_{n+1}(S^n) \to \pi_{n+2}(S^{n+1})$ is surjective for $n = 2$ and bijective for $n \ge 3$. Therefore it suffices to determine $\pi_4(S^3)$. By the Hurewicz theorem, $\pi_4(S^3) \cong \pi_4(Y) \cong H_4(Y)$. Thus it remains to compute $H_4(Y)$. We first determine the cohomology.

(20.8.6) Proposition. *The cohomology groups $H^k(Y)$ of Y are: $\mathbb{Z}$ for $k = 0$, 0 for $k \equiv 0 \bmod 2$, and $\mathbb{Z}/n$ for $k = 2n+1$.*

Proof. We use $K_2 = K(\mathbb{Z},2) = \mathbb{C}P^\infty$ and $H^*(K_2) \cong \mathbb{Z}[c]$ with $c \in H^2$. The Wang sequence of $Y \to S^3$ shows that

$$H^{2n}(Y) \cong \operatorname{Ker}\Theta, \quad H^{2n+1}(Y) \cong \operatorname{Coker}\Theta,$$

since $H^*(K_2) = 0$ for odd $*$. The group $H^{2n}(K_2) \cong \mathbb{Z}$ is generated by c^n. By the universal coefficient formula $H^j(Y) = 0$ for $j = 1,2,3$. Hence $\Theta\colon H^2(K_2) \to H^0(K_2)$ is an isomorphism. We can choose c such that $\Theta(c) = 1$. The derivation property of Θ yields then inductively $\Theta(c^n) = nc^{n-1}$. □

The Wang sequence in homology yields in a similar manner

$$H_{2n}(Y) \cong \operatorname{Coker}\Theta_*, \quad H_{2n+1}(Y) \cong \operatorname{Ker}\Theta_*.$$

From the universal coefficient sequence

$$0 \to \operatorname{Ext}(H_{2n}(Y),\mathbb{Z}) \to H^{2n+1}(Y) \to \operatorname{Hom}(H_{2n+1}(Y),\mathbb{Z}) \to 0$$

and the fact that $H_{2n}(Y)$ is a quotient of $\mathbb{Z}$ we obtain

$$H_{2n}(Y) \cong \mathbb{Z}/n, \quad H_{2n+1}(Y) = 0.$$

The special case $H_4(Y) \cong \mathbb{Z}/2$ now proves $\pi_4(S^3) \cong \mathbb{Z}/2$. □

We apply the Hurewicz theorem modulo the class of abelian p-torsion groups (p prime) and see that the p-primary component of $\pi_i(S^3)$ is zero for $3 < i < 2p$ and isomorphic to $\mathbb{Z}/p$ for $i = 2p$. In particular an infinite number of homotopy groups $\pi_n(S^3)$ is non-zero.

The determination of the homotopy groups of spheres is a difficult problem. You can get an impression by looking into [163]. Individual computations are no longer so interesting; general structural insight is still missing. Since the groups $\pi_{n+k}(S^n)$ do not change after suspension for $k \geq n-2$, by the Freudenthal theorem, they are called the ***stable homotopy groups*** π_k^S.

We copy a table from [185]; a denotes a cyclic group of order a, and $a \times b$ is the product of cyclic groups of order a and b, and a^j the j-fold product of cyclic groups of order a.

k	0	1	2	3	4	5	6	7	8	9
π_k^S	∞	2	2	24	0	0	2	240	2^2	2^3
k	10	11	12	13	14	15	16	17	18	19
π_k^S	6	504	0	3	2^2	480×2	2^2	2^4	8×2	264×2

20.8.7 The Hopf invariant. The exceptional case $\pi_{4n-1}(S^{2n})$ is interesting in many respects. Already Hopf constructed a homomorphism $h\colon \pi_{4n-1}(S^{2n}) \to \mathbb{Z}$, now called the ***Hopf invariant***, and gave a geometric interpretation in the simplicial setting [88]. Let $f\colon S^{4n-1} \to S^{2n}$ be a smooth map; and let a, b be two regular values. The pre-images $M_a = f^{-1}(a)$ and $M_b = f^{-1}(b)$ are closed orientable $(2n-1)$-manifolds, they have a linking number, and this number is the Hopf invariant of f. It is easy to define $h(f)$, using cohomology. Let $f\colon S^{2k-1} \to S^k$ be given ($k \geq 2$). Attach a $2k$-cell to S^k by f and call the result $X = X(f)$. The inclusion $i\colon S^k \to X$ induces an isomorphism $H^k(X) \cong H^k(S^k)$, and we also have an isomorphism $H^{2k}(D^{2k}, S^{2k-1}) \cong H^{2k}(X, S^k) \to H^{2k}(X)$. The integral cohomology groups $H^j(X)$ for $j \neq 0, k, 2k$ are zero. Choose generators $x \in H^k(X)$ and $y \in H^{2k}(X)$. Then there holds a relation $x \cup x = h(f)y$ in the cohomology ring. The graded commutativity of the cup product is used to show that $h(f) = 0$ for odd k. In the case $k = 2n$ the integer $h(f)$ is the Hopf invariant. Since $X(f)$ depends up to h-equivalence only on the homotopy class of f, the integer $h(f)$ is a homotopy invariant. One shows the elementary properties of this invariant:

(1) h is a homomorphism.

(2) If $g\colon S^{4n-1} \to S^{4n-1}$ has degree d, then $h(fg) = dh(f)$.

(3) If $k\colon S^{2n} \to S^{2n}$ has degree d, then $h(kf) = d^2h(f)$.

Note that a map of degree d does not induce the multiplication by d, as opposed to the general situation for cohomology theories.

It is an important problem to determine the image of h. Already Hopf showed by an explicit construction that $2\mathbb{Z}$ is always contained in the image. Here is the Hopf construction. Start with a map $u\colon S^k \times S^k \to S^k$. From it we obtain a map $f\colon S^{2k+1} \to S^{k+1}$ via the diagram

$$\begin{array}{ccc} S^k \times S^k \times I & \xrightarrow{u\times \mathrm{id}} & S^k \times I \\ \downarrow{\scriptstyle p} & & \downarrow{\scriptstyle p} \\ S^k \star S^k & \xrightarrow{f} & \Sigma S^k, \end{array}$$

where q is the projection onto the suspension and p the projection onto the join. The map u has a bi-degree (a,b). Hopf shows that (with suitable orientations chosen) $h(f) = ab$. The map $S^{2n-1} \times S^{2n-1} \to S^{2n-1}$, $(x,\xi) \mapsto \xi - \langle x,\xi\rangle x$ has bi-degree $(2,-1)$. If $\alpha\colon \mathbb{R}^n \times \mathbb{R}^n \to \mathbb{R}^n$ is a bilinear map without zero divisors (i.e., $\alpha(x,y) = 0$ implies that x or y is zero), then $(\mathbb{R}^n,\alpha)$ is called an n-dimensional real ***division algebra***. The induced map $\beta\colon S^{n-1} \times S^{n-1} \to S^{n-1}$, $(x,y) \mapsto \alpha(x,y)/\|\alpha(x,y)\|$ satisfies $\beta(\varepsilon x,\eta y) = \varepsilon\eta\beta(x,y)$ for $\varepsilon,\eta \in \{\pm 1\}$. Hence β has a bi-degree (d_1,d_2) with odd d_j. If there exist maps with odd Hopf invariant, then there also exist maps with invariant 1, since $2\mathbb{Z}$ is contained in the image of h. It is a famous result of Adams [2] that maps $f\colon S^{4n-1} \to S^{2n}$ of Hopf invariant 1 only exist for $n = 1,2,4$. Hence there exist n-dimensional real division algebras only for $n = 1,2,4,8$. See [55] for this topic and the classical algebra related to it. Once complex K-theory is established as a cohomology theory, it is fairly easy to solve the Hopf invariant 1 problem [5]. ◇

20.9 Rational Homology Theories

Recall the n-th stable homotopy group $\omega_n(X) = \operatorname{colim}_k \pi_{n+k}(X \wedge S(k))$ of the pointed space X. The Hurewicz homomorphisms are compatible with suspension, i.e., the diagram

$$\begin{array}{ccc} \pi_n(Y) & \xrightarrow{h} & \tilde{H}_n(Y) \\ \downarrow{\scriptstyle \Sigma_*} & & \downarrow{\scriptstyle \Sigma_*} \\ \pi_{n+1}(\Sigma Y) & \xrightarrow{n} & \tilde{H}_{n+1}(\Sigma Y) \end{array}$$

is commutative for each well-pointed space Y. This follows from the inductive definition of the Hurewicz homomorphisms; one has to use the same definition of Σ_* in homotopy and homology via the boundary operator of the pair (CY, Y) and the quotient map $CY \to CY/Y = \Sigma Y$. We pass to the colimit and obtain the stable Hurewicz homomorphism

$$h_n \colon \omega_n(X) \to \tilde{H}_n(X).$$

The h_n are natural transformations of homology theories (on well-pointed spaces). In order to see this one has to verify that the diagram

$$\begin{array}{ccc} \omega_n(X) & \xrightarrow{\quad h_n \quad} & \tilde{H}_n(X) \\ \downarrow \sigma & & \downarrow \Sigma \\ \omega_{n+1}(X \wedge S(1)) & \xrightarrow{h_{n+1}} & \tilde{H}_{n+1}(X \wedge S(1)) \end{array}$$

is commutative. The commutativity of this diagram is a reason for introducing the sign in the definition $\sigma = (-1)^n \tau_* \sigma_l$ of the suspension isomorphism for spectral homology.

The coefficients of the theory $\omega_*(-)$ are the stable homotopy groups of spheres $\omega_n(S^0) = \operatorname{colim}_k \pi_{n+k}(S^k)$. These groups are finite for $n > 0$. Finite abelian groups become zero when tensored with the rational numbers. We thus obtain a natural transformation of homology theories

$$h_* \colon \omega_*(X) \otimes_{\mathbb{Z}} \mathbb{Q} \to \tilde{H}_*(X) \otimes_{\mathbb{Z}} \mathbb{Q} \cong \tilde{H}_*(X; \mathbb{Q})$$

which are isomorphisms on the coefficients and therefore in general for pointed CW-complexes.

This basic rational isomorphism is now used to show that any homology theory $\tilde{h}_*$ with values in $\mathbb{Q}$- MOD can be reduced to ordinary homology. We define natural maps

$$\omega_p(X) \otimes_{\mathbb{Z}} \tilde{h}_q(S^0) \to \tilde{h}_{p+q}(X).$$

Let $x \in \omega_p(X)$ be represented by $f \colon S(p+k) \to X \wedge S(k)$. The image of $y \in \tilde{h}_q(S^0)$ under

$$\tilde{h}_q(S^0) \cong \tilde{h}_{q+p+k}(S(p+k)) \xrightarrow{f_*} \tilde{h}_{q+p+k}(X \wedge S(k)) \cong \tilde{h}_{p+q}(X)$$

is independent of the chosen representative f of x. We combine these homomorphisms

$$\bigoplus_{p+q=n} \omega_p(X) \otimes_{\mathbb{Z}} \tilde{h}_q(S^0) \to \tilde{h}_n(X)$$

and obtain a natural transformation of homology theories. Now assume that the coefficients $\tilde{h}_q(S^0)$ are $\mathbb{Q}$-vector spaces. Then for $X = S^0$ only the groups $\omega_0(S^0) \otimes_{\mathbb{Z}} \tilde{h}_n(S^0)$ are non-zero; and the induced map to $\tilde{h}_n(S^0)$ is an isomorphism, since $\omega_0(S^0) \cong \mathbb{Z}$ by the degree and a map of degree k induces on $\tilde{h}_n(S^0)$ the multiplication by k. We thus have shown:

(20.9.1) Theorem. *Let $\tilde{h}_*$ be an additive homology theory for pointed CW-complexes with values in $\mathbb{Q}$-vector spaces. Then we have an isomorphism*

$$\begin{array}{ccc} \bigoplus_{p+q=n} \tilde{H}_p(X;\tilde{h}_q(S^0)) & \xleftarrow{\cong} & \bigoplus_{p+q=n} \tilde{H}_p(X) \otimes \tilde{h}_q(S^0) \\ \downarrow\cong & & \uparrow\cong \\ \tilde{h}_n(X) & \xleftarrow{\cong} & \bigoplus_{p+q=n} \omega_p(X) \otimes \tilde{h}_q(S^0) \end{array}$$

of homology theories. □

If $\tilde{k}_*(-)$ is an arbitrary additive homology theory we can apply the foregoing to $\tilde{h}_*(-) = \tilde{k}_*(-) \otimes_{\mathbb{Z}} \mathbb{Q}$.

Problems

1. The Eilenberg–Mac Lane spectrum $(K(\mathbb{Z},n) \mid n \in \mathbb{N}_0)$ yields a homology theory which is isomorphic to singular homology with integral coefficients.
2. One can define the stable Hurewicz transformation from a morphism of the sphere spectrum into the Eilenberg–Mac Lane spectrum which is obtained from maps $S^n \to K(\mathbb{Z},n)$ that induce isomorphisms of π_n.
3. Define a stable Hurewicz homomorphism $h^n\colon \omega^n(X) \to \tilde{H}^n(X;\mathbb{Z})$ either from a map of spectra or by an application of cohomology to representing maps of elements in $\omega^n(X)$. Construct a natural commutative diagram

$$\begin{array}{ccc} \omega^n(X) & \xrightarrow{h^n} & \tilde{H}^n(X) \\ \downarrow & & \downarrow \\ \mathrm{Hom}(\omega_n(X),\mathbb{Z}) & \xrightarrow{(h_n)_*} & \mathrm{Hom}(\tilde{H}_n(X),\mathbb{Z}). \end{array}$$

4. Give a proof of the (absolute) Hurewicz theorem by using the $K(\mathbb{Z},n)$-definition of homology. The proof uses: Let $\pi_j(X) = 0$ for $j < n \geq 2$; then $\pi_j(X \wedge K(\mathbb{Z},k)) = 0$ for $j \leq n+k-1$ and $\pi_{n+k}(X \wedge K(\mathbb{Z},k)) \cong \pi_n(X)$.
5. Derive an isomorphism $\tilde{h}^n(X) \cong \prod_{p+q=n} \tilde{H}^p(X;h^q(S^0))$ for cohomology theories with values in $\mathbb{Q}$-vector spaces (X a finite pointed CW-complex).
6. Use a fibration $K(\mathbb{Z},n) \to K(\mathbb{Z},n) \to K(\mathbb{Z}/k,n)$ and derive a universal coefficient sequence for homology with $\mathbb{Z}/k$-coefficients.
7. An interesting example of a rational cohomology isomorphism is given by the ***Chern character***. It is a natural isomorphism of $\mathbb{Q}$-algebras

$$\mathrm{ch}\colon K(X) \otimes \mathbb{Q} \to \textstyle\prod_n H^{2n}(X;\mathbb{Q}),$$

for finite complexes X, say, and which sends a complex line bundle η over X to the power series

$$e^{c_1(\eta)} = \textstyle\sum_{i=0}^{\infty} \frac{1}{i!} c_1(\eta)^i \in H^{2*}(X;\mathbb{Q}).$$

Chapter 21
Bordism

We begin with the definition of bordism homology. The geometric idea of homology is perhaps best understood from the view-point of bordism and manifolds. A "singular" cycle is a map from a closed manifold to a space, and the boundary relation is induced by manifolds with boundary. Several of our earlier applications of homology and homotopy can easily be obtained just from the existence of bordism homology, e.g., the Brouwer fixed point theorem, the generalized Jordan separation theorem and the component theorem, and the theorem of Borsuk–Ulam.

Bordism theory began with the fundamental work of Thom [184]. He determined the bordism ring of unoriented manifolds (the coefficient ring of the associated bordism homology theory). This computation was based on a fundamental relation between bordism and homotopy theory, the theorem of Pontrjagin–Thom. In the chapter on smooth manifolds we developed the material which we need for the present proof of this theorem. One application of this theorem is the isomorphism between the geometric bordism theory and a spectral homology theory via the Thom spectrum. From this reduction to homotopy we compute the rational oriented bordism. Hirzebruch used this computation in the proof of his signature theorem. This proof uses almost everything that we developed in this text.

21.1 Bordism Homology

We define the bordism relation and construct the bordism homology theory. Manifolds are smooth.

Let X be a topological space. An n-dimensional ***singular manifold*** in X is a pair (B, F) which consists of a compact n-dimensional manifold B and a continuous map $F\colon B \to X$. The singular manifold $\partial(B, F) = (\partial B, F|\partial B)$ is the ***boundary*** of (B, F). If $\partial B = \emptyset$, then (B, F) is ***closed***.

A ***null bordism*** of the closed singular manifold (M, f) in X is a triple (B, F, φ) which consists of a singular manifold (B, F) in X and a diffeomorphism $\varphi\colon M \to \partial B$ such that $(F|\partial B) \circ \varphi = f$. If a null bordism exists, then (M, f) is ***null bordant***.

Let (M_1, f_1) and (M_2, f_2) be singular manifolds in X of the same dimension. We denote by $(M_1, f_1) + (M_2, f_2)$ the singular manifold $\langle f_1, f_2\rangle\colon M_1 + M_2 \to X$. We say (M_1, f_1) and (M_2, f_2) are ***bordant***, if $(M_1, f_1) + (M_2, f_2)$ is null bordant. A null bordism of $(M_1, f_1) + (M_2, f_2)$ is called ***bordism*** between (M_1, f_1) and (M_2, f_2). The boundary ∂B of a bordism (B, F, φ) between (M_1, f_1) and (M_2, f_2) thus consists of a disjoint sum $\partial_1 B + \partial_2 B$, and φ decomposes into two diffeomorphisms $\varphi_i\colon M_i \to \partial_i B$.

(21.1.1) Proposition. *"Bordant" is an equivalence relation.*

Proof. Let (M, f) be given. Set $B = M \times I$ and $F = f \circ \mathrm{pr}\colon M \times I \to M \to X$. Then $\partial B = M \times 0 + M \times 1$ is canonically diffeomorphic to $M + M$ and (B, F) is a bordism between (M, f) and (M, f). The symmetry of the relation is a direct consequence of the definition. Let $(B, F, \varphi_i\colon M_i \to \partial_i B)$ be a bordism between (M_1, f_1) and (M_2, f_2) and $(C, G, \psi_i\colon M_i \to \partial_i C)$ a bordism between (M_2, f_2) and (M_3, f_3). We identify in $B + C$ the subset $\partial_2 B$ with $\partial_2 C$ via $x \sim \psi_2\varphi_2^{-1}(x)$ for $x \in \partial_2 B$. The result D carries a smooth structure, and the canonical maps $B \to D \leftarrow C$ are smooth embeddings (15.10.1). We can factor $\langle F, G\rangle\colon B + C \to X$ over the quotient map $B + C \to D$ and get $H\colon D \to X$, and $(D, H, \langle \varphi_1, \psi_3\rangle)$ is a bordism between (M_1, f_1) and (M_3, f_3). □

We denote by $[M, f]$ the bordism class of (M, f) and by $N_n(X)$ the set of bordism classes of n-dimensional closed singular manifolds in X. The set $N_n(X)$ carries an associative and commutative composition law $[M, f]+[N, g] = [M + N, \langle f, g\rangle]$. The reader may check that this is well-defined.

(21.1.2) Proposition. *$(N_n(X), +)$ is an abelian group. Each element has order at most* 2.

Proof. The class of a null bordant manifold serves as neutral element, for example the constant map $S^n \to X$. (For the purpose at hand it is convenient to think of the empty set as an n-dimensional manifold.) For each (M, f) the sum $(M + M, \langle f, f\rangle)$ is null bordant, hence $[M, f] + [M, f] = 0$. □

A continuous map $f\colon X \to Y$ induces a homomorphism

$$N_n(f) = f_*\colon N_n(X) \to N_n(Y), \quad [M, g] \mapsto [M, fg].$$

In this way $N_n(\ )$ becomes a functor from TOP to ABEL. Homotopic maps induce the same homomorphism: If $F\colon X \times I \to Y$, $f \simeq g$ is a homotopy, then $(M \times I, F \circ (h \times \mathrm{id}))$ is a bordism between (M, fh) and (M, gh). If X is empty, we consider $N_n(X)$ as the trivial group.

(21.1.3) Example. A 0-dimensional compact manifold M is a finite discrete set. Hence (M, f) can be viewed as a family $(x_1, \dots, x_r)$ of points in X. Points $x, y \in X$ are bordant if and only if they are contained in the same path component. (Here you have to know 1-dimensional compact manifolds.) One concludes that $N_0(X)$ is isomorphic to the $\mathbb{Z}/2$-vector space over $\pi_0(X)$. ◇

(21.1.4) Proposition. *Let $h\colon K \to L$ be a diffeomorphism. Then $[L, g] = [K, gh]$.*

Proof. Consider the bordism $g \circ \mathrm{pr}\colon L \times I \to X$; on the boundary piece $L \times 1$ we use the canonical diffeomorphism to L, on the boundary piece we $L \times 0$ we compose the canonical diffeomorphism to L with h. □

We now make the functors $N_n(-)$ part of a homology theory. But this time, for variety, we do not begin with the definition of relative homology groups. The exact homology sequence and the excision axiom are now replaced by a Mayer–Vietoris sequence.

Suppose X is the union of open sets X_0 and X_1. We construct the boundary operator

$$\partial\colon N_n(X) \to N_{n-1}(X_0 \cap X_1)$$

of the Mayer–Vietoris sequence. Let $[M, f] \in N_n(X)$ be given. The sets $M_i = f^{-1}(X \smallsetminus X_i)$ are disjoint closed subsets of M.

(21.1.5) Lemma. *There exists a smooth function* $\alpha\colon M \to [0,1]$ *such that:*

(1) $M_i \subset \alpha^{-1}(i)$ *for* $i \in \{0,1\}$.

(2) $\frac{1}{2}$ *is a regular value of* α. □

We call α in (21.1.5) a ***separating function***. If α is a separating function, then $M_\alpha = \alpha^{-1}(\frac{1}{2})$ is a closed submanifold of M of dimension $n-1$ (or empty), and f induces by restriction $f_\alpha\colon M_\alpha \to X_0 \cap X_1$.

If $t \neq 0, 1$ is another regular value of α, then $\alpha^{-1}(t)$ and $\alpha^{-1}(\frac{1}{2})$ are bordant via $\alpha^{-1}[\frac{1}{2}, t]$. The choice of $\frac{1}{2}$ is therefore immaterial. We think of $[M_\alpha, f_\alpha]$ as being given by any choice of a regular value $t \in]0,1[$ of α with $M_\alpha = \alpha^{-1}(t)$.

(21.1.6) Lemma. *Let* $[K, f] = [L, g] \in N_n(X)$ *and let* α, β *be separating functions for* $(K, f), (L, g)$. *Then* $[K_\alpha, f_\alpha] = [L_\beta, g_\beta]$.

Proof. We take advantage of (21.1.4). Let (B, F) be a bordism between (K, f) and (L, g) with $\partial B = K + L$. There exists a smooth function $\gamma\colon B \to [0,1]$ such that

$$\gamma|K = \alpha, \quad \gamma|L = \beta, \quad F^{-1}(X \smallsetminus X_j) \subset \gamma^{-1}(j).$$

We choose a regular value t for γ and $\gamma|\partial B$ and obtain a bordism $\gamma^{-1}(t)$ between some K_α and some L_β. □

From (21.1.6) we obtain a well-defined boundary homomorphism

$$\partial\colon N_n(X) \to N_{n-1}(X_0 \cap X_1), \quad [M, f] \mapsto [M_\alpha, f_\alpha].$$

(21.1.7) Proposition. *Let* X *be the union of open subspaces* X_0 *and* X_1. *Then the sequence*

$$\cdots \xrightarrow{\partial} N_n(X_0 \cap X_1) \xrightarrow{j} N_n(X_0) \oplus N_n(X_1) \xrightarrow{k} N_n(X) \xrightarrow{\partial} \cdots$$

is exact. Here $j(x) = (j_*^0(x), j_*^1(x))$ *and* $k(y,z) = k_*^0 y - k_*^1 z$ *with the inclusions* $j^\nu\colon X_0 \cap X_1 \to X_\nu$ *and* $k^\nu\colon X_\nu \to X$. *The sequence ends with* $\xrightarrow{k} N_0(X) \to 0$.

Proof. (1) Exactness at $N_{n-1}(X_0 \cap X_1)$. Suppose $[M, f] \in N_n(X)$ is given. Then M is decomposed by M_α into the parts $B_0 = \alpha^{-1}[0, \frac{1}{2}]$ and $B_1 = \alpha^{-1}[\frac{1}{2}, 1]$ with common boundary M_α. Since $f(B_0) \subset X_1$, we see via B_0 that $j_*^1 \partial[M, f]$ is in $N_{n-1}(X_1)$ the zero element. This shows $j \circ \partial = 0$.

Suppose, conversely, that $j[K, f] = 0$. Then there exist singular manifolds (B_i, F_i) in X_i such that $\partial B_0 = K = \partial B_1$ and $F_0|K = f = F_1|K$. We identify B_0 and B_1 along K and obtain M; the maps F_0 and F_1 can be combined to $F\colon M \to X$. There exists a separating function α on M such that $M_\alpha = K$: With collars of K in B_0 and B_1 one obtains an embedding $K \times [0, 1] \subset M$ which is the identity on $K \times \{\frac{1}{2}\}$; then one chooses α such that $\alpha(k, t) = t$ for $k \in K$, $\frac{1}{4} \le t \le \frac{3}{4}$. By construction, $\partial[M, F] = [K, f]$.

(2) Exactness at $N_n(X_0) \oplus N_n(X_1)$. By definition, $k \circ j = 0$. Suppose $x_i = [M_i, f_i] \in N_n(X_i)$ are given. If $k(x_0, x_1) = 0$ there exists a bordism (B, F) in X between $(M_0, k^0 f_0)$ and $(M_1, k^1 f_1)$. Choose a smooth function $\psi\colon B \to [0, 1]$ such that:

(1) $F^{-1}(X \smallsetminus X_{1-i}) \cup M_i \subset \psi^{-1}(i)$ for $i = 0, 1$.

(2) ψ has regular value $\frac{1}{2}$.

Let $(N, f) = (\psi^{-1}(\frac{1}{2}), F|\psi^{-1}(\frac{1}{2}))$. Then $(\psi^{-1}[0, \frac{1}{2}], F|\psi^{-1}[0, \frac{1}{2}])$ is a bordism between (N, f) and (M_0, f_0) in X_0; similarly for (M_1, f_1). This shows $j[N, f] = (x_0, x_1)$.

(3) Exactness at $N_n(X)$. The relation $\partial \circ k = 0$ holds, since we can choose on $(M_0, k^0 f_0) + (M_1, k^1 f_1)$ a separating function $\alpha\colon M_0 + M_1 \to [0, 1]$ such that $\alpha^{-1}(\frac{1}{2})$ is empty.

Conversely, let α be a separating function for (M, f) in X and (B, F) a null bordism of (M_α, f_α) in $X_0 \cap X_1$. We decompose M along M_α into the manifolds $B_1 = \alpha^{-1}[0, \frac{1}{2}]$ and $B_0 = \alpha^{-1}[\frac{1}{2}, 1]$ with $\partial B_1 = M_\alpha = \partial B_0$. Then we identify B and B_0 along $M_\alpha = K$ and obtain a singular manifold $(M_0, f_0) = (B_0 \cup_K B, (f|B_0) \cup_K F)$ in X_0, and similarly (M_1, f_1) in X_1. Once we have shown that in $N_n(X)$ the equality $[M_0, f_0] + [M_1, f_1] = [M, f]$ holds, we have verified the exactness. We identify in $M_0 \times I + M_1 \times I$ the parts $B \times 1$ in $M_0 \times 1$ and $M_1 \times 1$. The resulting manifold $L = (M_0 \times I) \cup_{B\times 1} (M_1 \times I)$ has the boundary $(M_0 + M_1) + M$. A suitable map $F\colon L \to X$ is induced by $(f_0, f_1) \circ \mathrm{pr}_1\colon (M_0 + M_1) \times I \to X$. For the smooth structure on L see 15.10.3. □

We now define relative bordism groups $N_n(X, A)$ for pairs (X, A). Elements of $N_n(X, A)$ are represented by maps $f\colon (M, \partial M) \to (X, A)$ from a compact n-manifold M. Again we call $(M, f) = (M, \partial M; f)$ a singular manifold in (X, A). The bordism relation is a little more complicated. A ***bordism*** between (M_0, f_0) and (M_1, f_1) is a pair (B, F) with the following properties:

(1) B is a compact $(n + 1)$-manifold with boundary.

(2) ∂B is the union of three submanifolds with boundary M_0, M_1 and M', where $\partial M' = \partial M_0 + \partial M_1$, $M_i \cap M' = \partial M_i$.

(3) $F|M_i = f_i$.

(4) $F(M') \subset A$.

We call (M_0, f_0) and (M_1, f_1) ***bordant***, if there exists a bordism between them. Again "bordant" is an equivalence relation. For the proof one uses 15.10.3. The sum in $N_n(X, A)$ is again induced by disjoint union. Each element in $N_n(X, A)$ has order at most 2. A continuous map $f: (X, A) \to (Y, B)$ induces a homomorphism $N_n(f) = f_*: N_n(X, A) \to N_n(Y, B)$ by composition with f. If f_0 and f_1 are homotopic as maps between pairs, then $N_n(f_0) = N_n(f_1)$. The assignment $[M, f] \mapsto [\partial M, f|\partial M]$ induces a homomorphism (boundary operator) $\partial: N_n(X, A) \to N_{n-1}(A)$. For $A = \emptyset$ the equality $N_n(X, \emptyset) = N_n(X)$ holds.

(21.1.8) Lemma. *Let M be a closed n-manifold and $V \subset M$ an n-dimensional submanifold with boundary. If $f: M \to X$ is a map which sends $M \smallsetminus V$ into A, then $[M, f] = [V, f|V]$ in $N_n(X, A)$.*

Proof. Consider $F: M \times I \to X, (x,t) \mapsto f(x)$. Then $\partial(M \times I) = M \times \partial I$ and $V \times 1 \cup M \times 0$ is a submanifold of $\partial(M \times I)$ whose complement is mapped under F into A. The definition of the bordism relation now yields the claim. □

(21.1.9) Proposition. *Let $i: A \subset X$ and $j: X = (X, \emptyset) \subset (X, A)$. Then the sequence*

$$\cdots \xrightarrow{\partial} N_n(A) \xrightarrow{i_*} N_n(X) \xrightarrow{j_*} N_n(X, A) \xrightarrow{\partial} \cdots$$

is exact. The sequence ends with $\xrightarrow{j_*} N_0(X, A) \to 0$.

Proof. (1) Exactness at $N_n(A)$. The relation $i_* \circ \partial = 0$ is a direct consequence of the definitions. Let (B, F) be a null bordism of $f: M \to A$ in X. Then $\partial[B, F] = [M, f]$.

(2) Exactness at $N_n(X)$. Let $[M, f] \in N_n(A)$ be given. Choose $V = \emptyset$ in (21.1.8). Then $[M, f] = 0$ in $N_n(X, A)$, and this shows $j_* i_* = 0$.

Let $j_*[M, f] = 0$. A null bordism of $[M, f]$ in (X, A) is a bordism of (M, f) in X to (K, g) such that $g(K) \subset A$. A bordism of this type shows $i_*[K, g] = [M, f]$.

(3) Exactness at $N_n(X, A)$. The relation $\partial \circ i_* = 0$ is a direct consequence of the definitions. Let $\partial[M, f] = 0$. Choose a null bordism $[B, F]$ of $(\partial M, f|\partial M)$. We identify (M, f) and (B, f) along ∂M and obtain (C, g). Lemma (21.1.8) shows $j_*[C, g] = [M, f]$. □

A basic property of the relative groups is the ***excision property***. It is possible to give a proof with singular manifolds.

(21.1.10) Proposition. *The inclusion $i: (X \smallsetminus U, A \smallsetminus U) \to (X, A)$ induces an isomorphism $i_*: N_n(X \smallsetminus U, A \smallsetminus U) \cong N_n(X, A)$, provided $\bar{U} \subset A^\circ$.* □

The bordism notion can be adapted to manifolds with additional structure. Interesting are oriented manifolds. Let M_0 and M_1 be closed oriented n-manifolds. An ***oriented bordism*** between M_0, M_1 is a smooth compact oriented $(n+1)$-manifold B with oriented boundary ∂B together with an orientation preserving diffeomorphism $\varphi\colon M_1 - M_0 \to \partial B$. Here we have to use the convention about the boundary orientation, and $M_1 - M_0$ denotes the disjoint sum of the manifolds M_1 and M_0 where M_1 carries the given and $-M_0$ the opposite orientation. Again this notion of bordism is an equivalence relation. Singular manifolds are defined as before, and we have bordism groups $\Omega_n(X)$ of oriented bordism classes of singular n-manifolds in X. But now elements in the bordism group no longer have order at most 2. For a point P we have $\Omega_0(P) = \mathbb{Z}$, $\Omega_i(P) = 0$ for $1 \le i \le 3$. The assertion about Ω_1 follows from the fact that S^1 is an oriented boundary; the known classification of orientable surfaces as a sphere with handles shows that these surfaces are oriented boundaries. It is a remarkable result that $\Omega_3(P) = 0$: Each oriented closed 3-manifold is an oriented boundary; for a proof of this theorem of Rohlin see [77].

The exact sequences (21.1.7) and (21.1.9) as well as (21.1.10) still hold for the Ω-groups. The definition of the boundary operator $\partial\colon \Omega_n(X,A) \to \Omega_{n-1}(A)$ uses the boundary orientation. In order to define the boundary operator of the MV-sequence we have to orient M_α. There exists an open neighbourhood U of M_α in M and a diffeomorphism $\varphi\colon V =]1/2-\varepsilon, 1/2+\varepsilon[\times M_\alpha \to U$ such that $(\alpha\varphi)(t,x) = t$. If M is oriented, we have the induced orientation of U, and we orient V such that φ preserves the orientation. We orient M_α such that V carries the product orientation.

The idea of bordism can be used to acquire an intuitive understanding of homology. A compact n-manifold M has a fundamental class $z_M \in H_n(M,\partial M;\mathbb{F}_2)$ and $\partial z_M \in H_{n-1}(\partial M;\mathbb{F}_2)$ is again a fundamental class. Let $f\colon (M,\partial M) \to (X,A)$ be a singular n-manifold. We set $\mu(f) = f_* z_M \in H_n(X,A;\mathbb{F}_2)$. In this manner we obtain a well-defined homomorphism

$$\mu\colon N_n(X,A) \to H_n(X,A;\mathbb{F}_2).$$

The morphisms μ constitute a natural transformation of homology theories. One of the basic results of bordism theory says that μ is always surjective. This allows us to view homology classes as being represented by singular manifolds. If, in particular, f is an embedding of manifolds, then we view the image of f as a cycle or a homology class. In bordism theory, the fundamental class of M is M itself, i.e., the identity of M considered as a singular manifold.

The transformation μ can be improved if we take tangent bundle information into account. Let M be a compact n-manifold and denote by $\kappa_M\colon M \to BO$ the classifying map of the stable tangent bundle of M. For a singular manifold $f\colon (M,\partial M) \to (X,A)$ we then have $(f,\kappa_M)\colon (M,\partial M) \to (X,A) \times BO$. Again we take the image of the fundamental class

$$\mu[f] = (f,\kappa_M)_* z_M \in H_n((X,A) \times BO;\mathbb{F}_2).$$

We now obtain a natural transformation of homology theories

$$\mu = \mu(X, A)\colon N_*(X, A) \to H_*((X, A) \times BO; \mathbb{F}_2),$$

and in particular for the coefficient ring, the Thom bordism ring N_* of unoriented manifolds,

$$\mu\colon N_* \to H_*(BO; \mathbb{F}_2).$$

A fundamental result says that $\mu(X, A)$ is always injective [28, p. 185].

This transformation is also compatible with the multiplicative structures. The algebra structure of $H_*(X \times BO; \mathbb{F}_2)$ is induced by the homology product and the H-space structure $m\colon BO \times BO \to BO$ which comes from the Whitney sum of bundles. We obtain a natural homomorphism of graded algebras,

$$\begin{aligned} H_*(X \times BO; \mathbb{F}_2) \otimes H_*(Y \times BO; \mathbb{F}_2) &\to H_*(X \times BO \times Y \times BO; \mathbb{F}_2) \\ &\to H_*(X \times Y \times BO; \mathbb{F}_2); \end{aligned}$$

the first map is the homology product and the second is induced by the permutation of factors and m.

Thom [184] determined the structure of the ring N_*: It is a graded algebra $\mathbb{F}_2[u_2, u_4, u_5, \dots]$ with a generator u_k in each dimension k which does not have the form $k = 2^t - 1$. One can take $u_{2n} = [\mathbb{R}P^{2n}]$ as generators in even dimensions. The ring $H_*(BO; \mathbb{F}_2)$ is isomorphic to $\mathbb{F}_2[a_1, a_2, a_3, \dots]$ with a generator a_i in dimension i.

Another basic result says that there exists a natural isomorphism $N_*(X) \cong N_* \otimes_{\mathbb{F}_2} H_*(X; \mathbb{F}_2)$ of multiplicative homology theories [160], [28, p. 185]. Thus the homology theory $N_*(-)$ can be reduced to the determination of the coefficient ring N_* and singular homology with $\mathbb{F}_2$-coefficients.

For oriented manifolds the situation is more complicated. One can still define a multiplicative natural transformation of homology theories

$$\mu^{\Omega}\colon \Omega_*(X, A) \to H_*((X, A) \times BSO; \mathbb{Z})$$

from the fundamental classes of oriented manifolds as above. But this time the transformation is no longer injective and $\Omega_*(X, A) \to H_*(X, A; \mathbb{Z})$ in general not surjective. Also the theory $\Omega_*(-)$ cannot be reduced to ordinary homology. But the transformation still carries a lot of information. It induces a natural isomorphism

$$\Omega_*(X, A) \otimes \mathbb{Q} \cong H_*((X, A) \times BSO; \mathbb{Q}),$$

and, in particular,

$$\Omega_* \otimes \mathbb{Q} \cong H_*(BSO; \mathbb{Q}) \tag{1}$$

by the stable classifying map of the tangent bundle (see (21.4.2)). The ring $\Omega_* \otimes \mathbb{Q}$ is isomorphic to $\mathbb{Q}[x_4, x_8, \dots]$ with a generator x_n for each $n \equiv 0(4)$. One can take the $x_{4n} = [\mathbb{C}P^{2n}]$ as polynomial generators (see (21.4.4)).

We have seen that the signature of oriented smooth manifolds defines a homomorphism $\sigma\colon \Omega_* \otimes \mathbb{Q} \to \mathbb{Q}$ of $\mathbb{Q}$-algebras with $\sigma[\mathbb{C}P^{2n}] = 1$. The isomorphism (1) tells us that this homomorphism can be determined from the stable tangent bundle. A famous formula of Hirzebruch [81], the so-called Hirzebruch L-genus, gives a polynomial $L_n(p_1, p_2, \dots, p_n)$ in the Pontrjagin classes $p_1, \dots, p_n$ such that the evaluation on the fundamental class of an oriented $4n$-manifold is the signature

$$\langle L_n(p_1(M), \dots, p_n), [M] \rangle = \sigma(M).$$

It is a remarkable fact that the polynomials L_n have rational coefficients with large denominators, but nevertheless the evaluation on the fundamental class is an integer. Such integrality theorems have found a conceptual interpretation in the index theory of Atiyah and Singer [16].

The homomorphisms $N_n \to H_n(BO; \mathbb{F}_2)$ and $\Omega_n \to H_n(BSO; \mathbb{Z})$ have an interpretation in terms of characteristic numbers and can be determined by evaluating polynomials in the Stiefel–Whitney classes or Pontrjagin classes on the fundamental class.

In order to prove some of the results above one starts from the fundamental results of Thom [184], the reduction of bordism to homotopy theory via the Pontrjagin–Thom construction.

Problems

1. $\Omega_0(X)$ is naturally isomorphic to the free abelian group over $\pi_0(X)$.
2. Give a proof of (21.1.10) with singular manifolds.
3. We had defined formally a boundary operator for the MV-sequence from the axioms of a homology theory. Show that the boundary operator defined with separating functions coincides with this formal boundary operator. Pay attention to signs in the oriented case.
4. The homology theories Ω_* and N_* have a product structure. Products of singular manifolds induce a bilinear map

$$\Omega_m(X, A) \times \Omega_n(Y, B) \to \Omega_{m+n}(X \times Y, X \times B \cup A \times Y).$$

Verify the formal properties of a multiplicative structure, in particular the stability axiom.
5. Let M_1 and M_2 be oriented ∂-manifolds which are glued together along a component $N_i \subset \partial M_i$ with a diffeomorphism $\varphi\colon N_1 \to N_2$. Let M be the result. There exists an orientation of M such that the canonical embeddings $M_i \to M$ are orientation preserving, provided φ reverses the boundary orientations of the N_i.

A collar $\kappa\colon \mathbb{R}_- \times \partial M \to M$ of an oriented manifold M is orientation preserving, if $\mathbb{R}_- \subset \mathbb{R}$ carries the standard orientation, ∂M the boundary orientation and $\mathbb{R}_- \times \partial M$ the product orientation.

In order to verify the transitivity of the oriented bordism relation one has to define the orientation of the bordism which is obtained by gluing the given bordisms in such a way that the given bordisms are oriented submanifolds of the glued bordism.
6. The construction of the bordism MV-sequence suggests another set of axioms for a homology theory. A ***one-space homology theory*** consists of a family $h_n\colon \mathrm{TOP} \to R\text{-}\mathrm{MOD}$ of

covariant homotopy invariant functors and a family of boundary operators $\partial\colon h_n(X_0 \cup X_1) \to h_{n-1}(X_0 \cap X_1)$ for each triad $(X; X_0, X_1)$ in which the X_j are open in X. The boundary operators are assumed to be natural with respect to maps of triads and the usual exact MV-sequence should hold for each such triad.

Given these data, one defines relative groups for a pair (X, A) by

$$h_n(X, A) = \operatorname{Coker}(h_n(CA) \to h_n(X \cup CA))$$

where as usual CA is the cone on A. The relative groups are homotopy-invariant functors on TOP(2). Define a boundary operator $\partial\colon h_n(X, A) \to h_{n-1}(A)$ and show that the usual sequence of a pair is exact. In order to derive this sequence, consider the MV-sequence for the triad $(X \cup CA; X \cup CA \smallsetminus X, X \cup CA \smallsetminus *)$. The excision isomorphism $h_n(X \smallsetminus U, A \smallsetminus U) \cong h_n(X, A)$ holds, provided there exists a function $\tau\colon X \to [0, 1]$ with $U \subset \tau^{-1}(0)$ and $\tau^{-1}[0, 1[\subset A$, since under this assumption the canonical map $(X \smallsetminus U) \cup C(A \smallsetminus U) \to X \cup CA$ is a pointed h-equivalence.

21.2 The Theorem of Pontrjagin and Thom

The theorem of Pontrjagin and Thom relates homotopy theory and manifold theory. It describes sets of bordism classes as homotopy sets. We begin by defining the ingredients of the theorem.

Let Q be a smooth manifold without boundary. We denote by $Q^c = Q \cup \{\infty\}$ its one-point compactification. Let $\xi\colon E(\xi) \to B$ be a smooth real vector bundle over a closed manifold B. The one-point compactification $M(\xi) = E(\xi) \cup \{\infty\}$ is called the ***Thom space*** of ξ. The points at infinity serve as base points. The pointed homotopy set $[Q^c, M(\xi)]^0$ will be described as a bordism set.

A ξ-submanifold of Q is a closed submanifold M together with a smooth bundle map $F\colon E(\nu) \to E(\xi)$ from its normal bundle

$$\nu = \nu_M = \nu(M, Q)\colon E(\nu) \to M$$

into the given bundle ξ. A bordism between two ξ submanifolds (M_0, F_0) and (M_1, F_1) is a compact submanifold W of $Q \times I$ (of type I) such that

$$W \cap (Q \times [0, 1/3[) = M_0 \times [0, 1/3[, \quad W \cap (Q \times]2/3, 1]) = M_1 \times]2/3, 1]$$

and $\partial W = M_0 \times 0 \cup M_1 \times 1$ together with a bundle map $F\colon \nu(W, Q \times I) \to \xi$ which extends F_0 and F_1. Note that $\nu(W, Q \times I)|M \times 0$ can be identified with $\nu(M_0, Q)$. The relation of ξ-bordism is an equivalence relation. We denote by $L(Q, \xi)$ the set of ξ-bordism classes.

We define a map $P\colon L(Q, \xi) \to [Q^c, M(\xi)]^0$, and the main theorem then asserts that P is a bijection. We choose an embedding of Q into some Euclidean space. Then Q inherits a Riemannian metric. The normal bundle $\nu(M, Q)\colon E(\nu) \to M$ is the orthogonal complement of $TM \subset TQ|M$. From these data we had constructed a tubular map $\tau\colon E(\nu) \to Q$. The bundle ν inherits a Riemannian metric, and $D_\varepsilon(\nu)$, $E_\varepsilon(\xi)$, $S_\varepsilon(\nu)$ are the subsets of vectors v of norm $\|v\| \le \varepsilon$, $\|v\| < \varepsilon$, $\|v\| = \varepsilon$, respectively. There exists an $\varepsilon > 0$ such that t embeds $E_{2\varepsilon}(\xi)$ onto an open neighbourhood $U_{2\varepsilon}$ of M in Q. We have a fibrewise diffeomorphism $h_\varepsilon\colon E_\varepsilon(\xi) \to E(\xi)$ defined on each fibre by

$$h_\varepsilon(x) = \frac{\varepsilon x}{\sqrt{\varepsilon^2 - \|x\|^2}}.$$

From these data we construct from a ξ-submanifold (M, F) a map

$$g_\varepsilon\colon Q^c \to Q^c/(Q^c \smallsetminus U_\varepsilon) \to D_\varepsilon(\nu)/S_\varepsilon(\nu) \to M(\nu) \to M(\xi).$$

The first map is the quotient map, the second a homeomorphism induced by the inclusion $D_\varepsilon(\nu) \subset Q$ (note that $D_\varepsilon/S_\varepsilon$ is a one-point compactification of E_ε), the third induced by h_ε, and the fourth induced by the bundle map F (a proper map). The pointed homotopy class of g_ε is independent of the chosen (sufficiently small) ε, by a linear homotopy in the fibre. Let us set $P(M, F) = [g_\varepsilon] \in [Q^c, M(\xi)]^0$. We say that a map g_ε is obtained by a ***Pontrjagin–Thom construction***.

(21.2.1) Lemma. *For ξ-bordant manifolds (M_j, F_j) the classes $P(M_j, F_j)$ are equal. Therefore we obtain a well-defined map $P\colon L(Q,\xi) \to [Q^c, M(\xi)]^0$.*

Proof. We apply a Pontrjagin–Thom construction to a ξ-bordism. For this purpose we use for $Q \times I$ the product embedding. Let (W, F) be a ξ-bordism. We obtain a tubular map $\tau\colon E(\nu_W) \to Q \times I$ which is over $M_0 \times [0, 1/3[$ the product of the tubular map for M_0 with $[0, 1/3[$, and similarly for the other end. For sufficiently small ε, again τ embeds $E_{2\varepsilon}(\nu_W)$ onto a neighbourhood of W and we define as above a Pontrjagin–Thom map

$$Q^c \times I \to Q^c \times I/(Q^c \times I \smallsetminus U_\varepsilon) \to D_\varepsilon(\nu_W)/S_\varepsilon(\nu_W) \to M(\nu_W) \to M(\xi),$$

and this map is a homotopy between the Pontrjagin–Thom maps for (M_0, F_0) and (M_1, F_1). □

(21.2.2) Theorem (Pontrjagin, Thom). *The Pontrjagin–Thom map*

$$P\colon L(Q,\xi) \to [Q^c, M(\xi)]^0$$

is a bijection.

Proof. We construct a map in the other direction. Let us observe that the maps $f\colon Q^c \to M(\xi)$ obtained by the Pontrjagin–Thom construction are of a very special type. They have the following properties:

(1) The map $f\colon f^{-1}(E(\xi)) \to E(\xi)$ is proper, smooth and transverse to the zero section $B \subset E(\xi)$.

(2) There exists a tubular neighbourhood U_ε of $M = f^{-1}(B)$ in Q such that $f(x) = \infty \Leftrightarrow x \notin U_\varepsilon$.

(3) The map $f \circ \tau \circ h_\varepsilon^{-1}\colon E(\nu_M) \to E_\varepsilon(\nu_M) \to U_\varepsilon \to E(\xi)$ is a smooth bundle map.

So we have to deform an arbitrary map into one having these three properties.

(i) Let $g\colon Q^c \to M(\xi)$ be given and set $A = g^{-1}(E(\xi))$. Then A is open in Q and $g\colon A \to E(\xi)$ is a proper map. By the approximation theorem (15.8.4) there exists a proper homotopy of g to a smooth map $g_1\colon A \to E(\xi)$. Restrict g_1 to a compact neighbourhood V of $g_1^{-1}(B)$ in A such that V is a manifold and $g_1(\partial V) \subset E(\xi) \smallsetminus B$. By the transversality theorem (15.9.8) we find a smooth homotopy of $g_1|V$ to a map which is transverse to B and such that the homotopy is constant in a neighbourhood of ∂V. We can therefore extend this homotopy to a smooth proper homotopy of g_1 by a constant homotopy in the complement of V. Since both homotopies are proper, they can be extended continuously to Q^c by mapping the complement of A to the base point. The result is a smooth map $g_2\colon Q^c \to M(\xi)$ which has property (1) above.

(ii) Let $M = g_2^{-1}(B)$. Let now ε be small enough such that the tubular neighbourhood $U_{2\varepsilon}$ of M is contained in A. Let $\beta\colon Q^c \to [0,1]$ be a continuous function which is smooth on Q and such that $\beta^{-1}(0) = D_{\varepsilon/2}$ and $\beta^{-1}[0,1[= U_\varepsilon$. We define a homotopy of g_2 by

$$H_t(x) = \begin{cases} (1 - t\beta(x))^{-1} \cdot g_2(x), & x \in A,\ t < 1;\ x \in U_\varepsilon,\ t = 1, \\ \infty, & \text{otherwise.} \end{cases}$$

The map $g_3 = H_1$ has properties (1) and (2) above with $U = U_\varepsilon$.

(iii) Let $f = g_3$ be a map obtained in step (ii). Consider the composition

$$f\tau h_\varepsilon^{-1} = h\colon E(\nu_M) \to E_\varepsilon(\nu_M) \to U_\varepsilon \to E(\xi).$$

This map is proper, smooth, and transverse to $B \subset E(\xi)$. The homotopy $H_t(x) = t^{-1}h(tx)$, defined for $t > 0$ can be extended to $t = 0$ by a bundle map $\Phi\colon \nu_M \to \xi$ such that the resulting homotopy is smooth and proper. The map Φ is the derivative in the direction of the fibres. In order to see what happens in the limit $t \to 0$, we express h in local coordinates. Then h has the form

$$X \times \mathbb{R}^n \to Y \times \mathbb{R}^n, \quad (x, v) \mapsto (a(x, v), b(x, v))$$

with open sets $X \subset M$, $Y \subset B$ and $a(x, 0) = f(x), b(x, 0) = 0$. Then $H_t(x, v) = (a(x, tv), t^{-1}b(x, tv))$. The map $v \mapsto \lim_{t\to 0} t^{-1}b(x, tv)$ is the differential of $b_x\colon v \mapsto B(x, v)$ at $v = 0$. It is a bijective linear map, because f is transverse to the zero section.

(iv) We now construct a map Q which is inverse to P. Suppose $g\colon Q^c \to M(\xi)$ is a map such that $g\colon g^{-1}E(\xi) \to E(\xi)$ is proper, smooth, and transverse to the zero section. Let $M = g^{-1}(B)$. Then the differential of g induces a smooth bundle map $F\colon \nu_M \to \xi$. The map Q sends $[g]$ to the bordism class of the ξ-submanifold (M, F). By (i) we know that each homotopy class has a representative g with the properties just used. If g_0 and g_1 are two such representatives, we choose a homotopy between them which is constant on $[0, 1/3[$ and on $]2/3, 1]$ and apply the method of (i) in order to obtain a homotopy $h\colon Q^c \times I \to M(\xi)$, $g_0 \simeq g_1$ such that $h\colon h^{-1}E(\xi) \to E(\xi)$ is proper, smooth, transverse to the zero section, and constant on $[0, \eta[$ and $]1 - \eta, 1]$. The pre-image of the zero section and the differential of h yield a ξ-bordism. This argument shows that Q is well-defined. By construction, QP is the identity. The arguments of (i)–(iii) show that P is surjective. □

(21.2.3) Example. Let $Q = M$ be a closed connected n-manifold and ξ the n-dimensional bundle over a point. Then $E(\xi) = \mathbb{R}^n$ and $M(\xi) = S^n = \mathbb{R}^n \cup \{\infty\}$. A ξ-submanifold of Q is a finite subset X together with an isomorphism $F_x\colon T_x M \to \mathbb{R}^n$ for each $x \in X$. In the present situation $[Q^c, M(\xi)^0] = [M, S^n]$. We are therefore in the situation of the Hopf degree theorem.

Let M be oriented. If F_x is orientation preserving, we set $\varepsilon(x) = 1$, and $\varepsilon(x) = -1$ otherwise. Let $\varepsilon(X, F) = \sum_{x \in X} \varepsilon(x) \in \mathbb{Z}$. The integer $\varepsilon(X, F)$ characterizes the ξ-bordism class. If we represent a homotopy class in $[M, S^n]$ by a smooth map $f\colon M \to S^n$ with regular value $0 \in \mathbb{R}^n$, then $X = f^{-1}(0)$ and $F_x = T_x f$ and $\varepsilon(X, F)$ is the degree of f, as we have explained earlier. If we show that $\varepsilon(X, F)$ characterizes the bordism class, then the Pontrjagin–Thom theorem gives a proof of the Hopf degree theorem for smooth manifolds.

If M is non-orientable, we have a similar situation, but this time we have to consider $\varepsilon(X, F)$ modulo 2. ◇

(21.2.4) Example. Let $Q = \mathbb{R}^{n+k}$ and $E(\xi) = \mathbb{R}^n$. Then $[Q^c, M(\xi)]^0 = \pi_{n+k}(S^n)$, but we disregard the group structure for the moment. A ξ-submanifold is in this case a closed k-manifold $M \subset \mathbb{R}^{k+n}$ together with a trivialization of the normal bundle. A trivialization of a vector bundle is also called a ***framing*** of the bundle. Since the normal bundle ν is inverse to the tangent bundle τ we see that the tangent bundle is stably trivial and a framing $\nu \to n\varepsilon$ of the normal bundle induces $TM \oplus n\varepsilon \to TM \oplus \nu \to (n+k)\varepsilon$, a stable framing of M.

We denote by $\omega_n(k)$ the bordism set of a closed n-manifold with framing $TM \oplus k\varepsilon \to (n+k)\varepsilon$. The bordism relation is defined as follows. Let W be a bordism between M_0 and M_1 and let $\Phi\colon TW \oplus (k-1)\varepsilon \to (n+k)\varepsilon$ be a framing. Let $\iota_0\colon TW|M_o \cong TM_0 \oplus \varepsilon$ where the positive part of ε corresponds to an inwards pointing vector. Similarly $\iota_1\colon TW|M_1 \cong TM_1 \oplus \varepsilon$ where now the positive part corresponds to an outwards pointing vector. We obtain a framing

$$\varphi_i\colon TM_i \oplus \varepsilon \oplus (k-1)\varepsilon \to TW|M_i \oplus (k-1)\varepsilon \xrightarrow{\Phi} (n+k)\varepsilon$$

of M_i. We say in this case: (W, Φ) is a framed bordism between (M_0, φ_0) and (M_1, φ_1).

The assignment $[M, \varphi_\nu] \to [M, \varphi_\tau]$ is a well-defined map $L(\mathbb{R}^{n+k}, k\varepsilon) \to \omega_n(k)$. It is bijective for $k > n + 1$. $\diamond$

In the Pontrjagin–Thom theorem it is not necessary to assume that ξ is a smooth bundle over a closed manifold. In fact, ξ can be an arbitrary bundle. In this case we have to define the Thom space in a different manner. Let ξ have a Riemannian metric; then we have the unit disk bundle $D(\xi)$ and the unit sphere bundle $S(\xi)$. We define the Thom space now as the quotient space $M(\xi) = D(\xi)/S(\xi)$. A definition that does not use the Riemannian metric runs as follows. The multiplicative group $\mathbb{R}^*_+$ of positive real numbers acts on the subset $E^0(\xi)$ of non-zero vectors fibre-wise by scalar multiplication. Let $S(\xi)$ be the orbit space with induced projection $s_\xi\colon S(\xi) \to B$. The mapping cylinder of s_ξ is a space $d_\xi\colon D(\xi) \to B$ over B and $M(\xi)$ is defined to be the (unpointed) mapping cone of s_ξ. From this definition we see that a bundle map $f : \xi \to \eta$ induces a pointed map $M(f)\colon M(\xi) \to M(\eta)$. In the category of compactly generated spaces we have a canonical homeomorphism $M(\xi \times \eta) \cong M(\xi) \wedge M(\eta)$. If η is the trivial one-dimensional bundle over a point, this homeomorphism amounts to $M(\xi \oplus \varepsilon) \cong M(\xi) \wedge S^{(1)}$ with $S^{(1)} = \mathbb{R} \cup \{\infty\}$.

We can now define as before ξ-submanifolds of Q and ξ-bordisms. Also the Pontrjagin–Thom construction can be applied in this situation, and we obtain a well-defined map

$$P = P_\xi\colon L(Q, \xi) \to [Q^c, M(\xi)]^0.$$

These maps constitute a natural transformation between functors from the category of n-dimensional bundles and bundle maps.

(21.2.5) Theorem (Pontrjagin, Thom). *The Pontrjagin–Thom map P_ξ is for each bundle ξ a bijection.*

Proof. The proof is essentially a formal consequence of the special case (21.2.2), based on the general techniques developed so far.

(i) In the proof of (21.2.2) we used smooth bundles ξ. A bundle over a closed manifold B is induced from a tautological bundle over some Grassmannian $G_n(\mathbb{R}^N)$ by some map f. The map f is homotopic to a smooth map g and the bundle induced by g is therefore smooth and isomorphic to ξ. This fact allows us to work with arbitrary bundles in (21.2.2). The Pontrjagin–Thom construction itself does not use a smooth bundle map.

(ii) Suppose $i : X \to Y$ and $r : Y \to X$ are maps with $ri = \mathrm{id}$ (a retraction). If P is bijective for bundles over Y and ξ is a bundle over X, we pull back this bundle to $\eta = r^*\xi$. From the naturality of P and the fact that P_η is bijective, we conclude that P_ξ is bijective.

(iii) Let C be a compact smooth manifold with boundary. Let B denote the double $D(C)$ of C. Then C is a retract of B. Hence P is bijective for bundles over C.

(iv) Let X be a finite CW-complex. A finite CW-complex is a retract of some open set $U \subset \mathbb{R}^m$. Choose a proper smooth function $t\colon U \to \mathbb{R}_+$ such that $t(X) = \{0\}$. Let $\zeta > 0$ be a regular value of t. Then X is a retract of the compact smooth manifold $t^{-1}[0, \zeta]$ with boundary. Therefore the theorem holds for bundles over finite CW-complexes.

(v) Suppose ξ is a bundle over a CW-complex. By compactness of Q^c we see that a map $Q^c \to M(\xi)$ has an image in the Thom space of the restriction of the bundle to a finite subcomplex. This shows, using (iv), that P_ξ is surjective. An analogous argument shows the injectivity.

(vi) If B is an arbitrary space we choose a CW-approximation $f : C \to B$ and let $\eta = f^*\xi$. One shows that the bundle map $\eta \to \xi$ induces a bijection $L(C, \xi) \to L(B, \eta)$. This is due to the fact that a bundle map $\nu \to \xi$ is up to homotopy the composition of a bundle map $\nu \to \eta$ with the bundle map $\eta \to \xi$ (manifolds have the homotopy type of a CW-complex), and homotopic bundle maps yield, via the Pontrjagin–Thom construction, homotopic maps. A similar argument shows that a bijective map $[Q^c, M(\eta)]^0 \to [Q^c, M(\xi)]^0$ is induced. □

Problems

1. Work out the classification of the ξ-bordism classes in (21.2.3), thus completing the sketched proof of the Hopf degree theorem.

2. Give a proof of (21.2.4). The source of the map in question uses embedded manifolds, the range abstract manifolds. Thus one has to use the Whitney embedding theorem.

3. We use $\omega_1(k)$ to interpret the isomorphism $\pi_{k+1}(S^k) \cong \mathbb{Z}/2$ for $k \geq 3$. Let (A, φ) represent an element of $\omega_1(k)$. The manifold A is a disjoint sum of manifolds A_j diffeomorphic to S^1. Let φ_j be the framing of A_j induced by φ. We assign to (A_j, φ) an element $d(A_j, \varphi_j) \in \mathbb{Z}/2$. Let $h\colon A \to S^1$ be a diffeomorphism. We think of S^1 as boundary of D^2 and give $D^2 \subset \mathbb{R}^2$ the standard framing. This induces a standard framing σ of $TS^1 \oplus \varepsilon$ in which $1 \in \varepsilon$ points outwards. This provides us with a framing

$$\gamma\colon TA \oplus k\varepsilon \xrightarrow{Th \oplus \mathrm{id}} TS^1 \oplus \varepsilon \xrightarrow{\sigma \oplus \mathrm{id}} (k+1)\varepsilon.$$

If A is framed by φ, then $\varphi\colon TA \oplus k\varepsilon \to (k+1)\varepsilon$ orients the bundle TA. We choose the diffeomorphism h such that the composition with γ is orientation preserving. The homotopy class of γ is independent of the chosen h. The framing φ differs from the standard framing by a map

$$A \to \mathrm{GL}^+_{k+1}(\mathbb{R}) \simeq \mathrm{SO}(k+1).$$

Composed with h^{-1} we obtain a well-defined element in

$$[S^1, \mathrm{SO}(k+1)] \cong \pi_1 \mathrm{SO}(k+1) \cong \mathbb{Z}/2, \quad k \geq 2,$$

denoted by $d(A, \varphi) \in \mathbb{Z}/2$. If A consists of the components $(A_j \mid j \in J)$ we set $d(A, \varphi) = \sum_j d(A_j, \varphi_j)$.

Show that $(A, \varphi) \mapsto d(A, \varphi)$ induces for $k \geq 2$ a well-defined isomorphism $d : \omega_1(k) \to \mathbb{Z}/2$.

4. Give a similar interpretation of $\pi_3(S^2) \cong \mathbb{Z}$. The difference to the case of the previous problem is due to $\pi_1 \mathrm{SO}(2) \cong \mathbb{Z}$.

21.3 Bordism and Thom Spectra

The theorem of Pontrjagin–Thom allows us to describe the bordism group $N_n(X)$ as a homotopy group.

Let $\gamma_n : E(\gamma_n) \to BO(n)$ be a universal n-dimensional real vector bundle and $MO(n) = M(\gamma_n)$ its Thom space. A classifying map $\gamma_n \oplus \varepsilon \to \gamma_{n+1}$ induces a pointed map $e_n : \Sigma M(\gamma_n) \cong M(\gamma_n \oplus \varepsilon) \to M(\gamma_{n+1})$. The ***Thom spectrum*** MO consists of the family $(MO(n), e_n)$. The associated homology and cohomology groups of a pointed space Y are denoted $MO_n(Y)$ and $MO^n(Y)$.

(21.3.1) Theorem. *There exists a natural isomorphism*

$$T(X) : N_n(X) \cong MO_n(X^+).$$

We will see that the isomorphism $T(X)$ is obtained by a stable version of the Pontrjagin–Thom construction.

For each space X we denote by $\xi_m(X)$ the product bundle $\mathrm{id}_X \times \gamma_m$. We define a map

$$\Pi_k(X) : L(\mathbb{R}^{n+k}, \xi_k(X)) \to N_n(X).$$

Let $[M, F]$ be an n-dimensional $\xi_k(X)$-submanifold of $\mathbb{R}^{n+k}$. The first component F_1 of the $\xi_k(X)$-structure $F = (F_1, F_2) : E(\nu_M) \to X \times E(\gamma)$ gives us the element $\Pi_k(X)[M, F] = [M, F_1] \in N_n(X)$. It is obvious that we obtain a well-defined map Π_k. There is a kind of suspension map

$$\sigma : L(\mathbb{R}^{n+k}, \xi_k(X)) \to L(\mathbb{R}^{n+k+1}, \xi_{k+1}(X)).$$

For $[M, F]$ consider $M' = M \times \{0\} \subset \mathbb{R}^{n+k} \times \{0\} \subset \mathbb{R}^{n+k+1}$. The normal bundle of M' is $E(\nu_M) \oplus \varepsilon$. We compose $E(\nu_M) \oplus \varepsilon \to E(\gamma_k) \oplus \varepsilon$ with the classifying map $E(\gamma_k) \oplus \varepsilon \to E(\gamma_{k+1})$. From F we thus obtain a new structure $F' = (F_1, F_2') : E(\nu_{M'}) \to X \times E(\gamma_{k+1})$. We set $\sigma[M, F] = [M', F']$. The commutativity $\Pi_{k+1}\sigma = \Pi_k$ holds. Let $L_n(X)$ denote the colimit over the maps $\sigma : L(\mathbb{R}^{n+k}, \xi_k(X)) \to L(\mathbb{R}^{n+k+1}, \xi_{k+1}(X))$. Altogether we obtain $\Pi(X) : L_n(X) \to N_n(X)$.

(21.3.2) Proposition. *The map* $\Pi(X) : L_n(X) \to N_n(X)$ *is bijective.*

Proof. Surjective. Let $[M, f] \in N_n(X)$ be given. We can assume $M \subset \mathbb{R}^{n+k}$ for some k, by the Whitney embedding theorem. Let $\kappa_M : \nu_M \to \gamma_k$ be a classifying map of the normal bundle. Then we have the $\xi_k(X)$-structure $F = (f \circ \nu_M, \kappa_M)$, and $\Pi_k(X)[M, F] = [M, f]$ holds by construction.

Injective. Suppose $[M_0, F_0]$ and $[M_1, F_1]$ have the same image under $\Pi(X)$. We can assume that $M_j \subset \mathbb{R}^{n+k}$ for a suitable k. There exists a bordism B with $\partial B = M_0 + M_1$ and an extension $f: B \to X$ of $\langle f_0, f_1 \rangle$ where f_j is the first component of F_j. There exists an embedding $B \subset \mathbb{R}^{n+k+t} \times [0,1]$ such that

$$B \cap (\mathbb{R}^{n+k} \times \mathbb{R}^t \times [0, 1/3[) = M_0 \times 0 \times [0, 1/3[,$$
$$B \cap (\mathbb{R}^{n+k} \times \mathbb{R}^t \times]2/3, 1]) = M_1 \times 0 \times]2/3, 1].$$

By use of collars we can find a bordism B such that

$$\varphi\colon C = M_0 \times [0, 1/2[+]1/2, 1] \subset B$$

and $\partial B = M_0 \times 0 + M_1 \times 1$. We embed $C \to \mathbb{R}^{n+k} \times \mathbb{R}^t \times [0,1]$, $(m,s) \mapsto (m,0,s)$. Then we choose a continuous function $\Phi\colon B \to \mathbb{R}^{n+k+t} \times [0,1]$ such that Φ extends φ on $D = M_0 \times [0,\eta] + M_1 \times [1-\eta, 1]$ for some $1/3 < \eta < 1/2$ and such that $\Phi(B \smallsetminus D)$ is contained in $\mathbb{R}^{n+k+t} \times]1/3, 2/3[$ (Tietze extension theorem). Suppose $k + t > n + 1$. We now approximate Φ by an embedding $J\colon B \to \mathbb{R}^{n+k+t} \times [0,1]$ such that $J(B \smallsetminus D) \subset \mathbb{R}^{n+k+t} \times]1/3, 2/3[$ and such that J equals φ on $M_0 \times [0, 1/3[+ M_1 \times]2/3, 1]$.

The bundle maps $\nu_{M_j} \to \gamma_k$ yield bundle maps $\nu_{M_j} \oplus t\varepsilon \to \gamma_{k+t}$. Since these classifying maps are unique up to homotopy and since $\partial B \subset B$ is a cofibration, we can extend these maps to a bundle map $\nu_B \to \gamma_{k+t}$. We thus see that the $[M_j, F_j] \in L(\mathbb{R}^{n+k+t}, \xi_k(X))$ have the same image in $L(\mathbb{R}^{n+k+t}, \xi_{k+t}(X))$. □

The Pontrjagin–Thom maps

$$P\colon L(\mathbb{R}^{n+k}, \xi_k(X)) \to [S^{n+k}, M(\xi_k(X))]^0 \cong \pi_{n+k}(X^+ \wedge MO(k))$$

are compatible with suspension $P \circ \sigma = \sigma \circ P$. We obtain a bijection of the colimits $P\colon L_n(X) \cong MO_n(X^+)$ and hence a natural bijection

$$T = P \circ \Pi^{-1}\colon N_n(X) \cong MO_n(X^+).$$

It remains to verify that T is a homomorphism. Let $[M_i, f_i] \in N_n(X)$ be given. Choose embeddings

$$M_1 \subset A_1 = \{x \mid x_{n+k} > 0\} \subset \mathbb{R}^{n+k}, \quad M_2 \subset A_2 = \{x \mid x_{n+k} < 0\} \subset \mathbb{R}^{n+k}$$

and choose tubular neighbourhoods $U_i \subset A_i$ of M_i. The Pontrjagin–Thom construction applied to $M_1 + M_2$ yields a map which factors over the comultiplication $S^{n+k} \to S^{n+k} \vee S^{n+k}$. The restrictions to the summands are representatives of $P[M_i, f_i]$.

For more details on bordism homology and cohomology theories see [37], [181], [160], [28].

Problems

1. Given $f_0, f_1\colon Q^c \to M(\xi)$. Let $U \subset M(\xi)$ be an open neighbourhood of B. Suppose $V = f_0^{-1}(U) = f_1^{-1}(U)$ and $f_0|V = f_1|V$. Then f_0 and f_1 are pointed homotopic.

21.4 Oriented Bordism

The oriented bordism homology theory is isomorphic to the spectral homology theory of the Thom spectrum $M\mathrm{SO} = (M\mathrm{SO}(n), e_n)$ where $M\mathrm{SO}(n)$ is the Thom space of the universal n-dimensional orientable vector bundle over $B\mathrm{SO}(n)$. The isomorphism

$$\Omega_n(X) \cong M\mathrm{SO}_n(X^+)$$

is established as in the case of unoriented bordism. The Pontrjagin–Thom construction uses in this case an orientation of the normal bundle. An embedding $M^n \subset \mathbb{R}^{n+k}$ induces a canonical isomorphism $\tau(M) \oplus \nu(M) \cong (n+k)\varepsilon$ in which the normal bundle $\nu(M)$ is the orthogonal complement of the tangent bundle. An orientation of $\tau(M)$ induces an orientation of $\nu(M)$ such that fibrewise $\tau_x(M) \oplus \nu_x(M) \cong \mathbb{R}^{n+k}$ is orientation preserving.

(21.4.1) Lemma. *$M\mathrm{SO}(k)$ is $(k-1)$-connected.*

Proof. The canonical map $s\colon B\mathrm{SO}(k-1) \to B\mathrm{SO}(k)$ can be taken as the sphere bundle of the universal oriented k-dimensional bundle. From the homotopy sequence of this fibration we see that s is $(k-1)$-connected. The homotopy sequence of s is isomorphic to the sequence of the pair (D_k, S_k) of the universal (disk,sphere)-bundle over $B\mathrm{SO}(k)$. Hence (D_k, S_k) is $(k-1)$-connected. Since $B\mathrm{SO}(k)$ is simply connected, we can apply (6.10.2) and see that $\pi_j(D_k, S_k) \to \pi_j(D_k/S_k) = \pi_j(M\mathrm{SO}(k))$ is an isomorphism for $j \le k-1$. □

The suspension isomorphism $\pi_{n+k}(M\mathrm{SO}(k)) \to \pi_{n+k+1}(\Sigma M\mathrm{SO}(k))$ is an isomorphism for $k \ge n+2$, since $M\mathrm{SO}(k)$ is $(k-1)$-connected (see (6.10.4)). The spectral map $\pi_j(\Sigma M\mathrm{SO}(k)) \to \pi_j(M\mathrm{SO}(k+1))$ is an isomorphism for $j \le 2k-1$. In order to see this, we use the Whitehead theorem (20.1.4): The spaces in question are simply connected. Thus it suffices to see that we have a homology isomorphism in the same range.

$$\begin{array}{ccc} H_j(\Sigma M\mathrm{SO}(k)) & \longrightarrow & H_j(M\mathrm{SO}(k+1)) \\ \downarrow{\scriptstyle \text{Thom}} & & \downarrow{\scriptstyle \text{Thom}} \\ H_{j-k}(B\mathrm{SO}(k)) & \longrightarrow & H_{j-k}(B\mathrm{SO}(k+1)) \end{array}$$

The map $B\mathrm{SO}(k) \to B\mathrm{SO}(k+1)$ is $(k-1)$-connected, hence the vertical maps are isomorphisms for $j-k \le k-1$. These arguments show that we need not pass to the colimit, $\operatorname{colim}_k \pi_{n+k} M\mathrm{SO}(k)$; we already have an isomorphism $\Omega_n \cong \pi_{n+k} M\mathrm{SO}(k)$ for $k \geq n+2$. (The geometric reason for this stability result is the strong form of the Whitney embedding theorem which we had not used in (21.3.1).)

The assignment $[M] \mapsto (\kappa_M)_*[M] \in H_n(B\mathrm{SO}; \mathbb{Q})$ of Section 19.8 induces a ring homomorphism $\Omega_* \to H_*(B\mathrm{SO}; \mathbb{Q})$ which we extend to a homomorphism of $\mathbb{Q}$-algebras $\tau_* \colon \Omega_* \otimes \mathbb{Q} \to H_*(B\mathrm{SO}; \mathbb{Q})$. We can define in a similar manner a homomorphism $\nu_* \colon \Omega_* \otimes \mathbb{Q} \to H_*(B\mathrm{SO}; \mathbb{Q})$ if we use classifying maps of stable normal bundles.

(21.4.2) Theorem. *The homomorphisms*

$$\nu_* \colon \Omega_* \otimes \mathbb{Q} \to H_*(B\mathrm{SO}; \mathbb{Q}) \quad \textit{and} \quad \tau_* \colon \Omega_* \otimes \mathbb{Q} \to H_*(B\mathrm{SO}; \mathbb{Q})$$

are isomorphisms of graded algebras.

Proof. Lemma (21.4.1) and (20.8.3) imply that the Hurewicz homomorphism

$$\pi_r(M\mathrm{SO}(k)) \to H_k(M\mathrm{SO}(k))$$

has for $r < 2k-1$ both a finite kernel and a finite cokernel; hence it induces an isomorphism $\pi_r(M\mathrm{SO}(k)) \otimes \mathbb{Q} \to H_r(M\mathrm{SO}(k); \mathbb{Q})$ in this range. We also have the homological Thom isomorphism $H_r(M\mathrm{SO}(k); \mathbb{Q}) \cong H_{r-k}(B\mathrm{SO}(k); \mathbb{Q})$. The previous considerations now show that we have an isomorphism $\Omega_n \otimes \mathbb{Q} \cong H_n(B\mathrm{SO}(k); \mathbb{Q})$. The computation of $H_n(B\mathrm{SO}(k); \mathbb{Q}) \cong H_n(B\mathrm{SO}; \mathbb{Q})$ for $k \geq n+2$ shows that the $\mathbb{Q}$-vector space $\Omega_{4n} \otimes \mathbb{Q}$ has dimension $\pi(n)$, the number of partitions of n. From our computation of $\tau_*(\mathbb{C}P^{2a})$ we see that $\tau_* \colon \Omega_{4n} \otimes \mathbb{Q} \to H_{4n}(B\mathrm{SO}; \mathbb{Q})$ is a surjective map between vector spaces of the same dimension, hence an isomorphism.

The homomorphism ν_* is obtained from τ_* by composition with the antipode ι of the Hopf algebra $H_*(B\mathrm{SO}; \mathbb{Q})$. It is determined by the formal relation

$$(1 + q_1 + q_2 + \cdots)(1 + \iota(q_1) + \iota(q_2) + \cdots) = 1$$

which expresses the relation between the Pontrjagin classes of a bundle and its inverse. □

Together with our previous computations (19.8.4) we obtain:

(21.4.3) Theorem. *The algebra $\Omega_* \otimes \mathbb{Q}$ is a polynomial $\mathbb{Q}$-algebra in the generators $[\mathbb{C}P^{2n}]$, $n \in \mathbb{N}$.* □

We now collect various results and prove the ***Hirzebruch signature theorem***.

(21.4.4) Theorem. *The signature $\sigma(M)$ of an oriented closed $4n$-manifold is obtained by evaluating the Hirzebruch polynomial L_n in the Pontrjagin classes of M on the fundamental class.*

Proof. From (18.7.3) and (18.7.7) we see that $M \mapsto \sigma(M)$ induces a ring homomorphism $\Omega_* \to \mathbb{Z}$. We extend it to a homomorphism of $\mathbb{Q}$-algebras $\sigma\colon \Omega_* \to \mathbb{Q}$. Via the isomorphism τ_* of Theorem (21.4.2) it corresponds to a homomorphism $s\colon H_*(B\mathrm{SO};\mathbb{Q}) \to \mathbb{Q}$ such that $\sigma = s \circ \tau_*$. The homomorphism s was used at the end of Section 19.8 to determine polynomials $L_n \in H^{4n}(B\mathrm{SO};\mathbb{Q})$ in the Pontrjagin classes such that $\langle L_n(p), [M]\rangle = \sigma(M)$. □

Problems

1. It is not necessary to use the computation of $\tau_*(\mathbb{C}P^{2a})$ in the proof of (21.4.2). The reader is asked to check that the diagram

$$\begin{array}{ccccc} \Omega_n & \xrightarrow{P} & \pi_{n+k}(M\mathrm{SO}(k)) & \xrightarrow{h} & H_{n+k}(M\mathrm{SO}(k)) \\ \downarrow{\scriptstyle \nu_*} & & & & \downarrow{\scriptstyle t\cap} \\ H_n(B\mathrm{SO}) & & \xleftarrow{\quad\cong\quad} & & H_n(B\mathrm{SO}(k)) \end{array}$$

commutes (at least up to sign if one does not care about specific orientations). P is the Pontrjagin–Thom map, h is the Hurewicz homomorphism, $t\cap$ is the homological Thom isomorphism, the bottom map is induced by the stabilization.

Bibliography

[1] Abe, E., *Hopf Algebras*. Cambridge Tracts in Math. 74, Cambridge University Press, 1980. 484

[2] Adams, J. F., On the non-existence of elements of Hopf invariant one. *Ann. of Math.* 72 (1960), 20–104. 517

[3] Adams, J. F., Vector fields on spheres. *Ann. of Math.* 75 (1962), 603–632. 136, 339

[4] Adams, J. F., A variant of E. H. Brown's representability theorem. *Topology* 10 (1971), 185–198. 216

[5] Adams, J. F., Atiyah, M. F., K-theory and the Hopf invariant. *Quart. J. Math. Oxford* 17 (1966), 31–38. 517

[6] Aguilar, M., Gitler, S., Prieto, C., *Algebraic topology from a homotopical viewpoint*. Universitext, New York, Springer 2002. 356

[7] Allday, C., Puppe, V., *Cohomological methods in transformation groups*. Cambridge Stud. Adv. Math. 32, Cambridge University Press, 1993. 328

[8] Araki, S., Toda, H., Multiplicative structure in mod q cohomology theories. I. *Osaka J. Math.* 2 (1965), 71–115; II - *Osaka J. Math.* 3 (1966), 81–120. 256

[9] Atiyah, M. F., K-theory and Reality. *Quart. J. Math. Oxford* (2) 17 (1966), 367–386. 356

[10] Atiyah, M. F., Bott periodicity and the index of elliptic operators. *Quart. J. Math. Oxford* (2) 72 (1968), 113–140. 356

[11] Atiyah, M. F., *K-Theory*. New York, W.A. Benjamin 1967. 356

[12] Atiyah, M. F., *Collected Works*, Vol I–V. Oxford Science Publishers 1988. 356

[13] Atiyah, M. F., Bott, R., On the periodicity theorem for complex vector bundles. *Acta Math.* 112 (1964), 229–247. 356

[14] Atiyah, M. F., Bott, R., Shapiro A., Clifford modules. *Topology* 3 (1964), 3–38. 356

[15] Atiyah, M. F., Hirzebruch, F., Vector bundles and homogeneous spaces. *Proc. Symp. in Pure Math., Amer. Math. Soc.* 3 (1961), 7–38. 356

[16] Atiyah, M. F., Singer, I. M., The index of elliptic operators I. *Ann. of Math.* 87 (1968), 484–534. 527

[17] Barrat, M., Whitehead, J. H. C., The first non-vanishing group of an $(n+1)$-ad. *Proc. London Math. Soc.* (3) 6 (1966), 417–439. 267

[18] Baues, H. J., *Algebraic homotopy.* Cambridge Stud. Adv. Math. 15, Cambridge University Press, 1989. 95

[19] Behrens, M. J., A new proof of the Bott periodicity theorem. *Topology Appl.* 119 (2002), 167 -183. 356

[20] Betti, E., Sopra gli spazi di un numero qualunque dimensioni. *Ann. Mat. Pura Appl.* 2 (4) (1871), 140–158. 310

[21] Blakers, A. L., Some relations between homology and homotopy groups. *Ann. of Math.* 49 (1948), 428–461. 235

[22] Blakers, A. L., Massey, W.S., The homotopy groups of a triad II. *Ann. of Math.* 55 (1952), 192–201. 133

[23] Borsuk, K., Drei Sätze über die n-dimensionale euklidische Sphäre. *Fund. Math.* 20 (1933), 177–190.

[24] Bredon, G. E., *Introduction to compact transformation groups*. Pure Appl. Math. 46, New York, Academic Press 1972.

[25] Bredon, G. E., *Topology and Geometry*. Grad. Texts in Math. 139, New York–Berlin–Heidelberg, Springer 1997. 251

[26] Brinkmann, H.-B., Puppe D., *Kategorien und Funktoren*. Lecture Notes in Math. 18, Berlin, Springer 1966.

[27] Bröcker, Th., *Lineare Algebra und Analytische Geometrie*. Grundstud. Math., Basel, Birkhäuser, 2003. 16

[28] Bröcker, Th., tom Dieck, T., *Kobordismentheorie*. Lecture Notes in Math. 178, Berlin, Springer 1970. 526, 535

[29] Bröcker, Th., tom Dieck, T., *Representations of Compact Lie Groups*. Grad. Texts in Math. 98, New York, Springer 1995. 16, 77, 360, 400

[30] Bröcker, Th., Jänich, K., *Einführung in die Differentialtopologie*. Heidelberger Taschenbücher 143, Berlin, Springer 1973. 357, 364, 377, 465

[31] Brown, E. H., Cohomology theories. *Ann. of Math.* 75 (1962), 467–484; *Ann. of Math.* 78 (1963), 201. 216

[32] Brown, M., A proof of the generalized Schoenflies Theorem. *Bull. Amer. Math. Soc.* 66 (1960), 74–76. 251

[33] Brown, M., Locally flat imbeddings of topological manifolds. *Ann. of Math.* 75 (1962), 331–341.

[34] Brown, R., Groupoids and van Kampen's theorem. *Proc. London Math. Soc.* (3) 17 (1967), 385–401. 45

[35] Burde, G., Zieschang, H., *Knots*. de Gruyter Stud. Math. 5, Berlin–New York, W. de Gruyter 2003.

[36] Čadek, M., Crabb, M., G-structures on spheres. *Proc. London Math. Soc.* (3) 93 (2006), 791–816. 339

[37] Conner, P. E., Floyd, E. E., *Differentiable periodic maps*. Ergeb. Math. Grenzgeb. 33, Berlin–Göttingen–Heidelberg, Springer 1964. 535

[38] Crabb, M., James, I., *Fibrewise Homotopy Theory*. Springer Monogr. Math., Berlin–Heidelberg–New York, Springer 1998. 32

[39] Dicks, W., Dunwoody, M. J., *Groups acting on graphs*. Cambridge Stud. Adv. Math. 17, Cambridge University Press, 1989. 52

[40] tom Dieck, T., Klassifikation numerierbarer Bündel. *Arch. Math.* 17 (1966), 395–399.

[41] tom Dieck, T., Partitions of unity in homotopy theory. *Compositio Math.* 23 (1971), 159–167.

[42] tom Dieck, T., *Transformation groups and representation theory*. Lecture Notes in Math. 766, Berlin, Springer 1979.

[43] tom Dieck, T., *Transformation Groups*. de Gruyter Stud. Math. 8, Berlin–New York, de Gruyter 1987. 328

[44] tom Dieck, T., *Topologie*. (2. Auflage). de Gruyter Lehrbuch, Berlin–New York, de Gruyter 2000. 56, 57, 199, 222, 354, 357, 374, 408, 446

[45] tom Dieck, T., General topology (Preliminary version). www.uni-math.gwdg.de/tammo

[46] tom Dieck, T., Kamps, K. H., Puppe. D., *Homotopietheorie*. Lecture Notes in Math. 157, Berlin, Springer 1970. 118, 149

[47] tom Dieck, T., Petrie, T., Contractible affine surfaces of Kodaira dimension one. *Japan. J. Math.* (1990), 147–169. 361

[48] Dold, A., Partitions of unity in the theory of fibrations. *Ann. of Math.* 78 (1963), 223–255.

[49] Dold, A., *Halbexakte Homotopiefunktoren*. Lecture Notes in Math. 12, Berlin, Springer 1966. 95

[50] Dold, A., Die Homotopieerweiterungseigenschaft (= HEP) ist eine lokale Eigenschaft. *Invent. Math.* 6 (1968), 185–189.

[51] Dold, A., *Lectures on Algebraic Topology*. Grundlehren Math. Wiss. 200, Berlin–Heidelberg–New York, Springer 1972. 178, 311

[52] Dold, A., The fixed point index of fibre-preserving maps. *Invent. Math.* 25 (1974), 281–297. 311

[53] Dold, A., A simple proof of the Jordan-Alexander complement theorem. *Amer. Math. Monthly* 100 (1993), 856–857. 249

[54] Dold, A., Puppe, D., Duality, trace, and transfer. In *Proceedings Int. Conf. Geometric Topology*, PWN - Polish Scientific Publishers, Warszawa 1980, 81- 102. 159, 178

[55] Ebbinghaus, H.-D., et al., *Zahlen*. Grundwissen Math. 1, Berlin, Springer 1983. 517

[56] Eilenberg, S., Singular homology theory. *Ann. of Math.* 45 (1944), 407–447. 223, 235

[57] Eilenberg, S., Mac Lane, S., Acyclic models. *Amer. J. Math.* 79 (1953), 189–199. 286

[58] Eilenberg, S., Steenrod, N., Axiomatic approach to homology theory. *Proc. Nat. Acad. Sci. 31* (1945), 117–120. 244

[59] Eilenberg, S., Steenrod, N., *Foundations of Algebraic Topology*. Princeton University Press, 1952. 244, 245

[60] Eilenberg, S., Zilber, J. A., On products of complexes. *Amer. J. Math.* 75 (1953), 200 -204.

[61] Euler, L., Demonstratio non nullarum insignium proprietatum quibus solida hedris planis indusa sunt praedita. Novi commentarii academiae scientiarum Petropolitanae 4 (1752/3), 140–160. *Opera Mathematica* Vd. 26, 94–108 (1758) 309

[62] Euler, L., Elementa doctrinae solidorum. Novi commentarii academiae scientiarum Petropolitanae 4 (1752/3), 109–140. *Opera Mathematica* Vd. 26, 71–93 (1758) 309

[63] Farkas, H. M., Kra, I., *Riemann Surfaces*. Grad. Texts in Math. 71, New York–Heidelberg–Berlin, Springer 1980. 317

[64] Federico, P. J., *Descartes on Polyhedra*. Sources Hist. Math. Phys. Sci. 4, New York–Heidelberg–Berlin, Springer 1982. 309

[65] Félix, Y., Halperin, S., Thomas, J.-C., *Rational homotopy theory*. Grad. Texts in Math. 205, New York–Berlin–Heidelberg, Springer 2001. 515

[66] Freudenthal, H., Über die Klassen von Sphärenabbildungen I. *Compositio Math.* (1937) 5, 299–314. 135

[67] Fritsch, R., Piccinini, R. A., *Cellular Structures in Topology*. Cambridge University Press, 1990. 199, 218

[68] Fulton, W., *Algebraic Topology*. A first course. Grad. Texts in Math. 153, New York, Springer 1995.

[69] Gabriel, P., M. Zisman, Z., *Calculus of Fractions and Homotopy Theory*. Ergeb. Math. Grenzgeb. 35, Berlin–Heidelberg–New York, Springer 1967. 95

[70] Goebel, K., Kirk, W. A., *Topics in metric fixed point theory*. Cambridge Stud. Adv. Math. 28, Cambridge University Press, 1990. 140

[71] Goerss, R. G., Jardine, J. F., *Simplicial homotopy theory*. Progr. Math. 174, Basel, Birkhäuser 1999.

[72] Gray, B., *Homotopy theory*. Pure and Appl. Math. 64, New York, Academic Press 1975.

[73] Goodwillie, Th. G., Calculus II. Analytic functors. *K-Theory* 5 (1991/1992), 295–332. 152

[74] Greene, J. E., A new short proof of Kneser's conjecture. *Amer. Math. Monthly* (2002), 918–920. 261, 264

[75] Greub, W., Halperin, S., Vanstone R., Pure and Appl. Math. 47, *Connections, Curvature, and Cohomology*. New York, Academic Press 1972.

[76] Griffiths, H. B., The fundamental group of two spaces with a common point. *Quart. J. of Math.* 5 (1954), 175–190.

[77] Guillou, L., Marin, A. (Ed.), *A la Recherche de la Topologie Perdue*. Progr. Math. 62, Boston, Birkhäuser 1986. 525

[78] Hewitt, E., Ross, K. A., *Abstract harmonic analysis* I. Grundlehren Math. Wiss. 115, Berlin–Göttingen–Heidelberg, Springer 1963.

[79] Hirsch, M. W., A proof of the nonretractability of a cell onto its boundary. *Proc. Amer. Math. Soc.* 14 (1963), 364–365. 140

[80] Hirsch, M.W., *Differential topology*. New York–Heidelberg–Berlin, Springer 1976. 311

[81] Hirzebruch, F., *Neue topologische Methoden in der algebraischen Geometrie*. 2. Aufl., Ergeb. Math. Grenzgeb. 9, Berlin, Springer 1962. 492, 527

[82] Hirzebruch, F., *Gesammelte Abhandlungen*. Bde I–II. Berlin–Heidelberg–New York, Springer 1987.

[83] Hirzebruch, F., Mayer K.H., *$O(n)$-Mannigfaltigkeiten, exotische Sphären und Singularitäten*. Lecture Notes in Math. 57, Berlin, Springer 1968.

[84] Hochschild, G., *The structure of Lie groups*. San Francisco, Holden-Day 1965. 348, 360

[85] Hocking, J. G., Young G. S., *Topology*. Reading Mass., Addison-Wesley 1961.

[86] Hopf, H., Über die Abbildungen der dreidimensionalen Sphäre auf die Kugelfläche. *Math. Ann.* 104 (1931), 637–665.

[87] Hopf, H., Die Klassen der Abbildungen der n-dimensionalen Polyeder auf die n-dimensionale Sphäre. *Comment. Math. Helv.* 5 (1933), 39–54.

[88] Hopf, H., Über die Abbildungen von Sphären auf Sphären niedrigerer Dimension. *Fund. Math.* 25 (1935), 427–440. 516

[89] Hopf, H., Systeme symmetrischer Bilinearformen und euklidische Modelle der projektiven Räume. *Vierteljahresschrift der Naturforschenden Gesellschaft in Zürich* 85 (1940), 165–177. 365

[90] Hopf, H., Über den Rang geschlossener Liescher Gruppen. *Comment. Math. Helv.* 13 (1940/41), 119–143. 400

[91] Hopf, H., Über die Topologie der Gruppen-Mannigfaltigkeiten und ihrer Verallgemeinerungen. *Ann. of Math.* 42 (1941), 22–52. 90, 484

[92] Hopf, H., *Selecta Heinz Hopf*. Berlin–Göttingen–Heidelberg, Springer 1964.

[93] Huppert, B., *Endliche Gruppen* I. Grundlehren Math. Wiss. 134, Berlin, Springer 1979.

[94] Hurewicz, W., Wallman, H., *Dimension theory*. Princeton Math. Ser. 4, Princeton University Press, 1948. 140

[95] James, I. M., Euclidean models of projective spaces. *Bull. London Math. Soc.* 3 (1971), 257–276. 365

[96] James, I. M., *General topology and homotopy theory*. Berlin, Springer 1984. 32

[97] James, I. M., *Fibrewise topology*. Cambridge Tracts in Math. 91, Cambridge University Press, 1989. 32

[98] James, I. M. (Editor), *Handbook of Algebraic Topology*. Amsterdam, Elsevier 1995.

[99] James, I. M. (Editor), *History of Topology*. Amsterdam, Elsevier 1999.

[100] van Kampen, E. H., On the connection between the fundamental group of some related spaces. *Amer. J. Math.* 55 (1933), 261–267. 45

[101] Kamps, K. H., Porter, T., *Abstract homotopy and simple homotopy theory*. Singapore, World Scientific 1997. 95

[102] Karoubi, M., *K-Theory*. Grundlehren Math. Wiss. 226, Berlin, Springer 1978. 356

[103] Klaus, S., Kreck, M., A quick proof of the rational Hurewicz theorem and a computation of the rationnal homotopy groups of spheres. *Math. Proc. Camb. Philos. Soc.* 136 (2004), 617–623. 514

[104] Klein, F., *Gesammelte Mathematische Abhandlungen.* Bd. I-III. Berlin, Springer 1921–1923.

[105] Kneser, M., Aufgabe 360. *Jahresbericht der Deutschen Mathematiker-Vereinigung* 58 (1955), 27. 264

[106] Kono, A., Tamaki, D., *Generalized cohomology*. Transl. Math. Monogr. 230, Providence, R.I., Amer. Math. Soc. 2006. 356

[107] Kosinski, A., *Differential Manifolds*. Pure and Appl. Math. 138, Academic Press 1992. 357

[108] Kreck, M., Surgery and duality. *Ann. of Math.* 149 (1999), 707–754. 389

[109] Laitinen, E., Lück, W., Equivariant Lefschetz classes. *Osaka J. Math.* 26 (1989), 491–525. 311

[110] Lamotke, K., *Semisimpliziale algebraische Topologie*. Grundlehren Math. Wiss. 147, Berlin–Heidelberg–New York, Springer 1968.

[111] Lewis Jr., L. G., May, J. P., Steinberger, M., McClure, J. E., *Equivariant stable homotopy theory*. Lecture Notes in Math. 1213, Berlin, Springer 1986.

[112] Lillig, J., A union theorem for cofibrations. *Arch. Math.* 24 (1973), 410–415. 114

[113] Liulevicius, A., Arrows, symmetries and representation rings. *J. Pure Appl. Algebra* 19 (1980), 259–273. 488

[114] Lovász, L., Kneser's conjecture, chromatic number and homotopy. *J. Combin. Theory Ser. A* 25 (1978), 319–324. 264

[115] Lück, W., *Transformation groups and algebraic K-theory*. Lecture Notes in Math. 1408, Berlin, Springer 1989.

[116] Lück, W., The universal functorial Lefschetz invariant. *Fund. Math.* 161 (1999), 167–215. 311

[117] Lück, W., Chern character for proper equivariant homology theories and applications to K- and L-theory. *J. Reine Angew. Math.* 543 (2002), 193–234.

[118] Lück, W., *L^2-invariants: Theory and applications to geometry and K-theory*. Ergeb. Math. Grenzgeb. 44, Berlin–Heidelberg–New York, Springer 2002.

[119] Lück, W., *Algebraische Topologie*. Wiesbaden, Vieweg 2005.

[120] MacLane, S., *Homology*. Grundlehren Mathematischen Wiss. 114, Berlin–Göttingen–Heidelberg, Springer 1963. 294

[121] Magnus, W., *Noneuclidean tesselations and their groups*. Pure and Appl. Math. 61, New York, Academic Press 1974. 316

[122] Magnus, W., Karras, A., Solitar, D., *Combinatorial group theory*. New York, Interscience Publishers 1966. 52

[123] Massey, W. S., *Algebraic Topology: An Introduction.* New York, Harcourt, Brace & World 1967. 56, 311

[124] Massey, W. S., *Homology and cohomology theory*. Monogr. Textbooks Pure Appl. Math. 46, New York, Marcel Dekker 1978.

[125] Matoušek, J., *Using the Borsuk-Ulam theorem*. Universitext, Berlin–Heidelberg–New York, Springer 2003.

[126] May, J. P., Classifying spaces and fibrations. *Mem. Amer. Math. Soc.* 1 (1955), no.1, 155.

[127] May, J. P., *A concise course in algebraic topology*. Chicago Lecture Notes in Math., Chicago University Press, 1999.

[128] May, J. P., Sigurdsson, J., *Parametrized homotopy theory*. Math. Surveys Monogr. 132, Providence, R.I., Amer. Math. Soc. 2006. 32, 194

[129] Mayer, K. H., *Algebraische Topologie*. Basel–Boston–Berlin, Birkhäuser 1989.

[130] McCleary, J., *A user's guide to spectral sequences*. 2. ed., Cambridge Stud. Adv. Math. 58, Cambridge University Press, 2001. 303

[131] Milnor, J., Construction of universal bundles. II. *Ann. of Math.* 63 (1956), 430–436. 345

[132] Milnor, J., On spaces having the homotopy type of a CW-complex. *Trans. Amer. Math. Soc.* 90 (1958), 272–280. 218

[133] Milnor, J., On axiomatic homology theory. *Pacific J. Math.* 12 (1962), 337–342. 407

[134] Milnor, J., *Morse Theory*. Ann. of Math. Stud. 51, Princeton University Press, 1963. 389

[135] Milnor, J., Micro bundles. *Topology* 3, Suppl. 1 (1964), 53–80.

[136] Milnor, J., *Topology from the differentiable view point*. Charlottesville University Press, 1965. 364

[137] Milnor, J., *Lectures on the h-cobordism theorem*. Princeton University Press, 1965. 389

[138] Milnor, J., Moore, J. C., On the structure of Hopf algebras. *Ann. of Math.* 81 (1965), 211–264 484

[139] Milnor, J., Stasheff, J. D., *Characteristic classes*. Ann. Math. Stud. 76, Princeton University Press, 1974.

[140] Möbius, A. F., *Gesammelte Werke*. Bd. I -IV, Leipzig, Hirzel KG 1886. 342

[141] Moise, E. E., *Geometric Topology in Dimensions 2 and 3*. Grad. Texts in Math. 47, Berlin–Heidelberg–New York, Springer 1977. 198, 251

[142] Montgomery, S., *Hopf algebras and their actions on rings*. Amer. Math. Soc. Regional Conf. Series Math. 82, Washington, DC, 1993. 484

[143] Munkres, J. R., *Elementary Differential Topology*. Princeton University Press, 1966. 198

[144] Munkres, J. R., *Topology. A first Course*. Englewood Cliffs, Prentice-Hall 1975.

[145] Munkres, J. R., *Elements of Algebraic Topology*. Reading, Mass., Addison-Wesley 1984.

[146] Nagata, J., *Modern general topology*. Groningen, Nordhoff 1968

[147] Nomura, Y., On mapping sequences. *Nagoya Math.* J. 17 (1960), 111–145. 99

[148] Peano, G., Sur une courbe, qui remplit tout une aire plane. *Math. Ann.* 36 (1890), 157–160.

[149] Pommerenke, Ch., *Boundary behaviour of conformal maps*. Grundlehren Math. Wiss. 299, Berlin, Springer 1992. 251

[150] Poincaré, H., Sur la généralisation d'un théorème d'Euler relatif aux polyèdres. *Compt. Rend. Acad. Sci. Paris* 117 (1893), 144–145. 310

[151] Poincaré, H., Analysis situs. *Journal de l'École Polytechnique* 1 (1895), 1–121. 42

[152] Poincaré, H., Complément à l'Analysis Situs. *Rend. Circ. Mat. Palermo* 13 (1899), 285–343. 310

[153] Poincaré, H., Cinquième complément à l'Analysis Situs. *Rend. Circ. Mat.Palermo* 18 (1904), 45–110.

[154] Poincaré, H., *Œuvres de Henri Poincaré* VI. Paris, Gauthier-Villars 1953.

[155] Puppe, D., Homotopiemengen und ihre induzierten Abbildungen. I. *Math. Z.* 69 (1958), 299–344. 33, 81, 95, 194

[156] Puppe, D., Bemerkungen über die Erweiterung von Homotopien. *Arch. Math.* 18 (1967), 81–88.

[157] Puppe, D., Some well known weak homotopy equivalences are genuine homotopy equivalences. *Symposia Mathematica* 5 (1971), 363–374.

[158] Puppe, D., Homotopy cocomplete classes of spaces and the realization of the singular complex. In: *Topological Topics* (I. M. James ed.), London Math. Soc. Lecture Note Ser. 86, Cambridge University Press, 1983, 55–69.

[159] von Querenburg, B., *Mengentheoretische Topologie*. Berlin–Heidelberg–New York, Springer 1979.

[160] Quillen, D., Elementary proofs of some results of cobordism using Steenrod operations. *Adv. Math.* 7 (1971), 29–56. 526, 535

[161] Radó, T., Über den Begriff der Riemannschen Fläche. *Acta Sci. Math.* (*Szeged*) 2 (1924), 101–121. 198

[162] Ranicki, A., *Algebraic and geometric surgery*. Oxford Math. Monographs, Oxford University Press, 2002. 389

[163] Ravenel, D. C., *Complex cobordism and stable homotopy groups of spheres*. New York, Academic Press 1986. 516

[164] Schubert, H., *Topologie*. Stuttgart, Teubner, 4. Auflage 1975.

[165] Segal, G., Classifying spaces and spectral sequences. *Inst. Hautes Études Sci. Publ. Math.* 34 (1968), 105–112.

[166] Seifert, H., Konstruktion dreidimensionaler geschlossener Räume. *Ber. Sächs. Akad. Wiss.* 83 (1931), 26–66. 45

[167] Seifert, H., Threllfall, W., *Lehrbuch der Topologie*. Leipzig, Teubner 1934. 56, 311, 437

[168] Selick, P., *Introduction to homotopy theory*. Fields Institute Monogr. 9, Amer. Math. Soc., Providence, R.I. , 1997.

[169] Serre, J.-P., Homologie singulière des espaces fibrés. *Ann. of Math.* 54 (1951), 425–505.

[170] Serre, J.-P., Groupes d'homotopie et classes de groupes abéliens. *Ann. of Math.* 58 (1953), 258–294. 503

[171] Serre, J.-P., *Trees*. Berlin–Heidelberg–New York, Springer 1980. 52

[172] Spanier, E. H., *Algebraic Topology*. New York, McGraw–Hill 1966.

[173] Sperner, E., Neuer Beweis für die Invarianz der Dimensionszahl und des Gebietes. *Abh. Math. Sem. Univ. Hamburg* 6 (1928), 265–272. 140

[174] Steen, L. A., Seebach, J. A. jr., *Counterexamples in topology*. Holt, Rinehart ans Winston 1970.

[175] Steenrod, N., *The topology of fibre bundles*. Princeton Math. Ser. 14, Princeton University Press, 1951.

[176] Steenrod, N., A convenient category of topological spaces. *Michigan Math. J.* 14 (1967), 13–152.

[177] Sternberg, S., *Lectures on differential geometry*. Englewood Cliffs, Prentice Hall 1964. 364

[178] Stöcker, R., Zieschang, H., *Algebraische Topologie*. Mathematische Leitfäden, Stuttgart, Teubner 1988.

[179] Strøm, A., Note on cofibrations. *Math. Scand.* 19 (1966), 11–14.

[180] Strøm, A., Note on cofibrations II. *Math. Scand.* 22 (1968), 130–142. 103

[181] Stong, R. E., *Notes on cobordism theory*. Princeton University Press, 1968. 535

[182] Sweedler, M. E., *Hopf algebras*. Math. Lecture Note Ser., New York, Benjamin 1980. 484

[183] Switzer, R. M., *Algebraic Topology – Homotopy and Homology*. Grundlehren Math. Wiss. 212, Berlin–Heidelberg–New York, Springer 1975.

[184] Thom, R., Quelques propriétés globales des variétés différentiables. *Comment. Math. Helv.* 28 (1954), 17–86. 520, 526, 527

[185] Toda, H., *Composition methods in homotopy groups of spheres*. Ann. of Math. Stud. 49, Princeton University Press, 1962. 516

[186] Vogt, R., Convenient categories of topological spaces for homotopy theory. *Arch. Math.* 22 (1971), 545–555.

[187] Waldhausen F., Algebraic K-theory of spaces. In: *Proc. Conf., Rutgers New Brunswick* 1983, Lecture Notes in Math. 1126, Springer, 1985, 318–419.

[188] Wall, C. T. C., Finiteness conditions for CW-complexes. *Ann. of Math.* 81 (1965), 56–69; Finiteness conditions for CW-complexes II, *Proc. London Math. Soc.* 1966, 129–139.

[189] Wall, C. T. C., On the exactness of interlocking sequences. *Enseign. Math.* 12 (1966), 95–100. 245

[190] Wall, C. T. C., *A geometric introduction to topology*. Reading, Mass., Addison-Wesley 1972.

[191] Wall, C. T. C., Ranicki, A., *Surgery on compact manifolds*. Math, Surveys Monogr, 69, Providence, R.I., Amer. Math. Soc. 1999. 389

[192] Whitehead, G. W., *Elements of homotopy theory*. Grad. Texts in Math. 61, New York, Springer 1978. 214

[193] Whitehead, J. H. C., On C^1-complexes. *Ann. of Math.* 41 (1940), 809–824. 198

[194] Whitehead, J. H. C., Combinatorial homotopy. *Bull. Amer. Math. Soc.* 55 (1949), 213–245.

[195] Whitehead, J. H. C., *Mathematical Works*. Vol I– IV. Oxford, Pergamon Press 1962.

[196] Whitney, H., On regular closed curves in the plane. *Compositio Math.* 4 (1937), 276–286.

[197] Zieschang, H., Vogt, E., Coldewey, H.-D., *Surfaces and planar discontinuous groups*. Lecture Notes in Math. 835, Berlin–Heidelberg–New York, Springer 1980. 198

Symbols

Numbers

$\mathbb{N}$	natural numbers $\{1, 2, 3, \dots\}$
$\mathbb{N}_0$	$\mathbb{N} \cup \{0\}$
$\mathbb{Z}$	integers $\{0, \pm 1, \pm 2, \dots\}$
$\mathbb{Q}$	rational numbers
$\mathbb{R}$, $\mathbb{R}_+$, $\mathbb{R}_-$	real numbers, non-negative, non-positive
$\mathbb{C}$	complex numbers
$\mathbb{C}^*$	$\mathbb{C} \smallsetminus 0$, non-zero numbers
$\mathbb{H}$	quaternions
$\mathbb{Z}/m = \mathbb{Z}/m\mathbb{Z}$	integers modulo m
$\mathbb{F}_q$	field with q elements

Categories

TOP	category of topological spaces and continuous maps
TOP^0	pointed spaces and pointed maps
TOP^K	spaces under the space K
TOP_B	spaces over the space B
TOP(2)	pairs (X, A) of a space X and a subspace A
G- TOP	spaces with an action of the topological group G
COV_B	covering spaces of the space B
h-TOP	homotopy category of TOP
h-$\mathcal{C}$	homotopy category associated to a category $\mathcal{C}$ with homotopy notion
R- MOD	left modules over the ring R
ABEL	abelian groups
SET	sets
G- SET	sets with left G-action
$\Pi(X)$	fundamental groupoid of X
$\text{Or}(G)$	orbit category of the group G

$[\mathcal{C}, \mathcal{D}]$	category of functors $\mathcal{C} \to \mathcal{D}$ and natural transformations
TRA_B	transport category $[\Pi(B), \mathrm{SET}]$
Ker, Ke	kernel
Coker, Ko	cokernel
Im	image
id, 1	identity
pr	projection onto a factor of a product
(f_j)	map into a product with components f_j
$\langle f_j \rangle$	map from a sum with components f_j

Spaces

$\mathbb{R}^n$	Euclidean n-space
S^n	unit sphere in $\mathbb{R}^{n+1}$
D^n	unit disk in $\mathbb{R}^n$
E^n	$D^n \smallsetminus S^{n-1}$ unit cell
$I = [0, 1]$	unit interval
I^n	n-fold Cartesian product $I \times I \times \cdots \times I$
∂I^n	combinatorial boundary of I^n
$S(n)$	$I^n / \partial I^n$
$S^{(n)}$	$\mathbb{R}^n \cup \{\infty\}$
$\Delta^n = \Delta[n]$	n-dimensional standard simplex
$\partial \Delta^n$	combinatorial boundary of Δ^n
$\mathbb{R}P^n$	n-dimensional real projective space
$\mathbb{C}P^n$	n-dimensional complex projective space
$V_k(\mathbb{F}^n)$	Stiefel manifold of orthonormal k-frames in $\mathbb{F}^n$
$G_k(\mathbb{F}^n)$	Grassmann manifold of k-dimensional subspaces of $\mathbb{F}^n$
EG	universal free G-space
BG	classifying space of the topological group G
$E/G, G\backslash E$	orbit space of a G-action on E
$E \times_G F$	balanced product
X^H	H-fixed point set
$K(\pi, n)$	Eilenberg–Mac Lane space of type (π, n)

X/A	X with $A \subset X$ identified to a point
X^+	X with additional base point

Groups

$\mathrm{GL}_n(\mathbb{F})$	group of (n, n)-matrices with entries in $\mathbb{F} = \mathbb{R}, \mathbb{C}, \mathbb{H}$
$\mathrm{O}(n)$	group of orthogonal matrices in $\mathrm{GL}_n(\mathbb{R})$
$\mathrm{SO}(n)$	matrices of determinant 1 in $\mathrm{O}(n)$
$\mathrm{U}(n)$	group of unitary matrices in $\mathrm{GL}_n(\mathbb{C})$
$\mathrm{SU}(n)$	matrices of determinant 1 in $\mathrm{U}(n)$
S^1	complex numbers of modulus 1
$\mathrm{Spin}(n)$	Spinor group, double covering of $\mathrm{SO}(n)$
$\mathrm{Sp}(n)$	symplectic group
S_n	symmetric group

Relations

$\Box$	end of proof
$\Diamond$	end of numbered item
$\smallsetminus$	difference set
$\cup$	cup product
$\cap$	cap product
$\sim$	equivalent (equivalence relation)
$\simeq$	homotopy-equivalent, homotopic
$\cong$	isomorphic, homeomorphic, diffeomorphic
$\oplus$	direct sum
$\otimes$	tensor product
$\vee$	pointed sum (bouquet)
$\star$	join
$\wedge$	smash product
$*$	induced morphism, general index, base point, free product of groups
$+, \amalg$	topological sum
$\Vert x \Vert$	Euclidean norm of x

$\lvert z\rvert$	absolute value of complex number z
$\lvert X\rvert$	cardinality of X

Modules

lim	limit (inverse limit)
colim	colimit (direct limit)
$\lim^1$	derived functor of lim
Tor	torsion product, left derived of Hom
Ext, $\mathrm{Ext}^n_\Lambda(A, B)$	module of extensions, right derived of Hom

Bundles

$E(\xi), E^0(\xi)$	total space of vector bundle, without zero section
TM	tangent bundle of the manifold M
$K(X) = K_{\mathbb{C}}(X) = K\mathrm{U}(X)$	Grothendieck ring of complex vector bundles
$K\mathrm{O}(X) = K_{\mathbb{R}}(X)$	Grothendieck ring of real vector bundles
$\varepsilon, n\varepsilon$	trivial bundle, of dimension n
$M(\xi)$	Thom space of the bundle ξ
$M\mathrm{O}(n), M\mathrm{SO}(n)$	Thom space of the universal n-dimensional (oriented) bundle
$c_i(\xi), w_i(\xi), p_i(\xi)$	i-th Chern, Stiefel–Whitney, Pontrjagin class

Homotopy

CX	(pointed) cone on X
ΣX	suspension of X
$Z(f)$	(pointed) mapping cylinder of f
$Z(f, g)$	double mapping cylinder
$C(f)$	mapping cone of f
$C(X, A)$	mapping cone of $A \subset X$
H^-	inverse of the homotopy H
$K * L$	product (concatenation) of homotopies K, L
$[X, Y]$	homotopy classes of maps $X \to Y$
$[X, Y]^0$	pointed homotopy classes
$[X, Y]^K, [X, Y]_B$	homotopy classes under K, over B

$\Pi(X)$	fundamental groupoid of X
$\Pi(X,Y)$	homotopy groupoid
$\pi_n(X,A,*)$	n-homotopy group of the pointed pair $(X,A,*)$
$X^Y = F(X,Y)$	space of maps $X \to Y$ with compact-open topology
ΩX	loop space of X
hocolim	homotopy colimit
$N(\mathcal{U})$	nerve of the covering $\mathcal{U}$
$B(\mathcal{U})$	geometric realization of the nerve

Co-Homology

$S_q(X)$	singular q-chains
$S_\bullet(X,A)$	singular chain complex of (X,A)
$H_q(X,A;G)$	ordinary homology with coefficients in G
$H^q(X,A;G)$	ordinary cohomology with coefficients in G
$[v_0,\dots,v_q]$	affine simplex with vertices v_j
h_*, h^*	general homology, cohomology; or its coefficient groups
$\tilde{h}_*, \tilde{h}^*$	reduced homology, cohomology
∂	boundary operator
δ	coboundary operator
$\chi(X)$	Euler characteristic of X
$N_*(X)$	unoriented bordism
$\Omega_*(X)$	oriented bordism
$M\mathrm{O}_k(X), M\mathrm{SO}_k(X)$	unoriented, oriented bordism via Thom spectra

Index